T0310312

Hypervalent Iodine Chemistry

Hypervalent Iodine Chemistry

Preparation, Structure and Synthetic Applications of Polyvalent Iodine Compounds

Viktor V. Zhdankin
Department of Chemistry and Biochemistry
University of Minnesota Duluth, Minnesota, USA

WILEY

This edition first published 2014

Registered office
John Wiley & Sons Ltd, The Atrium, Southern Gate, Chichester, West Sussex, PO19 8SQ, United Kingdom

For details of our global editorial offices, for customer services and for information about how to apply for permission to reuse the copyright material in this book please see our website at www.wiley.com.

Library of Congress Cataloging-in-Publication Data

Zhdankin, Viktor V., 1956–
Hypervalent iodine chemistry : preparation, structure and synthetic applications of polyvalent iodine compounds / Viktor V. Zhdankin.
pages cm
Includes index.
ISBN 978-1-118-34103-2 (cloth)
1. Iodine compounds. 2. Organoiodine compounds. 3. Hypervalence (Theoretical chemistry) I. Title.
QD181.I1Z43 2014
546′.7342–dc23

2013020439

A catalogue record for this book is available from the British Library.

ISBN: 9781118341032

Typeset in 10/12pt Times by Aptara Inc., New Delhi, India
Printed and bound in Malaysia by Vivar Printing Sdn Bhd

1 2013

Contents

Preface

Iodine is the heaviest non-radioactive element in the Periodic Table that is classified as a nonmetal and it is the largest, the least electronegative and the most polarizable of the halogens. It formally belongs to the main group p-block elements; however, the bonding description, structural features and reactivity of iodine compounds differ from the light main-group elements. The electronic structure of polyvalent iodine is best explained by the hypervalent model of bonding and, therefore, in modern literature organic compounds of trivalent and pentavalent iodine are commonly named as hypervalent iodine compounds. The reactivity pattern of hypervalent iodine in many aspects is similar that of transition metals – the reactions of hypervalent iodine reagents are commonly discussed in terms of oxidative addition, ligand exchange, reductive elimination and ligand coupling, which are typical of transition metal chemistry.

Since the beginning of the twenty-first century, the organic chemistry of hypervalent iodine compounds has experienced an unprecedented, explosive development. Hypervalent iodine reagents are now commonly used in organic synthesis as efficient multipurpose reagents whose chemical properties are similar to derivatives of mercury, thallium, lead, osmium, chromium and other metals, but without the toxicity and environmental problems of these heavy metal congeners. One of the most impressive recent achievements in the field of iodine chemistry has been the discovery of hypervalent iodine catalysis.

This book is the first comprehensive monograph covering all main aspects of the chemistry of organic and inorganic polyvalent iodine compounds, including applications in chemical research, medicine and industry. The introductory chapter (Chapter 1) provides a historical background and describes the general classification of iodine compounds, nomenclature, hypervalent bonding, general structural features and general principles of reactivity of polyvalent iodine compounds. Chapter 2 gives a detailed description of the preparative methods and structural features of all known classes of organic and inorganic derivatives of polyvalent iodine. Chapter 3, the central chapter of the book, deals with the applications of hypervalent iodine reagents in organic synthesis. Chapter 4 describes the most recent achievements in hypervalent iodine catalysis. Chapter 5 deals with recyclable polymer-supported and nonpolymeric hypervalent iodine reagents. Chapter 6 covers the "green" reactions of hypervalent iodine reagents, including solvent-free reactions, reactions in water and reactions in ionic liquids. The final chapter (Chapter 7) provides an overview of important practical applications of polyvalent iodine compounds in medicine and in industry.

This book is aimed at all chemists interested in iodine compounds, including academic and industrial researchers in inorganic, organic, physical, medicinal and biological chemistry. It will be particularly useful to synthetic organic and inorganic chemists, including graduate and advanced undergraduate students. The book also covers the green chemistry aspects of hypervalent iodine chemistry, including the use of water as solvent, reactions under solvent-free conditions, recyclable reagents and solvents and catalytic reactions, which makes it especially useful for industrial chemists. The last chapter provides a detailed summary of practical applications of polyvalent iodine compounds, including various industrial applications, biological activity and applications of iodonium salts in PET (positron emission tomography) diagnostics; this chapter should be especially useful for medical and pharmaceutical researchers. Overall, the book is aimed at a broad, multidisciplinary readership and specialists working in different areas of chemistry, pharmaceutical and medical sciences and industry.

1

Introduction and General Overview of Polyvalent Iodine Compounds

1.1 Introduction

Iodine is a very special element. It is the heaviest non-radioactive element in the Periodic Table classified as a non-metal and it is the largest, the least electronegative and the most polarizable of the halogens. It formally belongs to the main group, p-block elements; however, because of the large atom size, the bonding description in iodine compounds differs from the light main group elements. In particular, the interatomic π-bonding, typical of the compounds of light p-block elements with double and triple bonds, is not observed in the compounds of polyvalent iodine. Instead, a different type of bonding occurs due to the overlap of the 5p orbital on the iodine atom with the appropriate orbitals on the two ligands (L) forming a linear L–I–L bond. Such a three-center-four-electron (3c-4e) bond is commonly referred to as a "hypervalent bond" [1]. The hypervalent bond is highly polarized and is longer and weaker than a regular covalent bond and the presence of hypervalent bonding leads to special structural features and reactivity pattern characteristic of polyvalent iodine compounds. In current literature, synthetically useful derivatives of polyvalent iodine are commonly named as hypervalent iodine reagents. The reactivity pattern of hypervalent iodine in many aspects is similar to the reactivity of transition metals and the reactions of hypervalent iodine reagents are commonly discussed in terms of oxidative addition, ligand exchange, reductive elimination and ligand coupling, which are typical of transition metal chemistry.

Iodine was first isolated from the ash of seaweed by the industrial chemist B. Courtois in 1811 and was named by J. L. Gay Lussac in 1813 [2,3]. Its name derives from the Greek word ιώδες (iodes) for violet, reflecting the characteristic lustrous, deep purple color of resublimed crystalline iodine. Various inorganic derivatives of polyvalent iodine in oxidation states of +3, +5 and +7 were prepared as early as the beginning of the nineteenth century. For example, iodine trichloride was first discovered by Gay Lussac as the result of treating warm iodine or iodine monochloride with an excess of chlorine [4]. In the same paper [4], the preparation of potassium iodate by the action of iodine on hot potash lye was described. The inorganic chemistry of polyvalent iodine has been summarized in numerous well-known texts [3, 5–8]. A detailed review on the history of iodine and all aspects of its chemistry and applications commemorating two centuries of iodine research was published in 2011 by Kuepper and coauthors [9].

Hypervalent Iodine Chemistry: Preparation, Structure and Synthetic Applications of Polyvalent Iodine Compounds, First Edition. Viktor V. Zhdankin.

Most of the world's production of iodine comes from the saltpeter deposits in Chile and natural brines in Japan. In Chile, calcium iodate is found in caliche deposits extracted from open pit mines in the Atacama Desert. Applying an alkaline solution to the caliche yields sodium iodate and iodine is obtained from the sodium iodate by reduction with sulfur dioxide. In Japan, iodine is a by-product of the production of natural gas, which is extracted from brine deposits a mile or two below ground. Iodine is recovered from the brines by one of the following two methods. In the blowout process elemental iodine is liberated as a result of the reaction of chlorine with sodium iodide in the brines. Elemental iodine is blown out of the brine with air and then purified in subsequent reaction steps. The second method, ion exchange, involves recovery of dissolved iodine from oxidized brines using anion-exchange resins packed in columns. In 2010, Chile produced 18 000 metric tons of iodine, compared to Japan's output of 9800 metric tons. Chile has reserves of 9 million metric tons, some 60% of the world's total reserves of iodine [10].

Iodine plays an important role in many biological organisms and is an essential trace element for humans. In the human body, iodine is mainly present in the thyroid gland in the form of thyroxine, a metabolism-regulating hormone. In natural organic compounds, iodine occurs exclusively in the monovalent state. The first polyvalent organic iodine compound, (dichloroiodo)benzene, was prepared by the German chemist C. Willgerodt in 1886 [11]. This was rapidly followed by the preparation of many others, including (diacetoxyiodo)benzene [12] and iodosylbenzene [13] in 1892, 2-iodoxybenzoic acid (IBX) in 1893 [14] and the first examples of diaryliodonium salts reported by C. Hartmann and V. Meyer in 1894 [15]. In 1914 Willgerodt published a comprehensive book describing nearly 500 polyvalent organoiodine compounds known at that time [16].

Research activity in the area of polyvalent organoiodine compounds during the period between 1914 and 1970s was relatively low and represented mainly by valuable contributions from the laboratories of I. Masson, R. B. Sandin, F. M. Beringer, K. H. Pausacker, A. N. Nesmeyanov and O. Neilands. Only three significant reviews were published during this period, most notably the reviews by Sandin [17] and Banks [18] published in *Chemical Reviews* in 1943 and 1966, respectively and a comprehensive tabulation of the physical properties of polyvalent iodine compounds published in 1956 by Beringer [19].

Since the early 1980s interest in polyvalent organoiodine compounds has experienced a renaissance. This resurgence of interest in multivalent organic iodine has been caused by the discovery of several new classes of polyvalent organoiodine compounds and, most notably, by the development of useful synthetic applications of some of these compounds, which are now regarded as valuable organic reagents known under the general name of hypervalent iodine reagents. The foundation of modern hypervalent iodine chemistry was laid in the 1980s by the groundbreaking works of G. F. Koser, J. C. Martin, R. M. Moriarty, P. J. Stang, A. Varvoglis and N. S. Zefirov.

Important contributions to the development of hypervalent iodine chemistry in the 1990s were made by the research groups of A. Varvoglis, N. S. Zefirov, L. M. Yagupolskii, A. R. Katritzky, R. A. Moss, J. C. Martin, D. H. R. Barton, R. M. Moriarty, G. F. Koser, P. J. Stang, H.-J. Frohn, T. Umemoto, M. Yokoyama, Y. Kita, M. Ochiai, T. Okuyama, T. Kitamura, H. Togo, E. Dominguez, I. Tellitu, J. D. Protasiewicz, A. Kirschning, K. S. Feldman, T. Wirth, S. Quideau, S. Hara, N. Yoneda, L. Skulski, S. Spyroudis, V. V. Grushin, V. W. Pike, D. A. Widdowson and others. During the 1980s–1990s, hypervalent iodine research was summarized in several reviews and books. Most notable were the two books published in 1992 and 1997 by A. Varvoglis: the comprehensive monograph *The Organic Chemistry of Polycoordinated Iodine* [20] and a book on the application of hypervalent iodine compounds in organic synthesis [21]. Several general reviews [22–28], numerous book chapters [29–34] and specialized reviews on phenyliodine(III) carboxylates [35, 36], [hydroxy(tosyloxy)iodo]benzene [37], the chemistry of iodonium salts [38], electrophilic perfluoroalkylations [39], application of hypervalent iodine in the carbohydrate chemistry [40], hypervalent iodine oxidations [41–43], fluorinations using hypervalent iodine fluorides [44], hypervalent iodine compounds as free radical precursors [45], synthesis of heterocyclic compounds using organohypervalent iodine reagents [46] and the chemistry of benziodoxoles [47] were also published during 1980s and 1990s.

Since the beginning of the twenty-first century, the chemistry of organohypervalent iodine compounds has experienced explosive development. This surge in interest in iodine compounds is mainly due to the very useful oxidizing properties of hypervalent iodine reagents, combined with their benign environmental character and commercial availability. Iodine(III) and iodine(V) derivatives are now routinely used in organic synthesis as reagents for various selective oxidative transformations of complex organic molecules. Numerous reviews and book chapters summarizing various aspects of hypervalent iodine chemistry have been published since 2000 [48–122]. A book edited by T. Wirth on the application of hypervalent iodine in organic synthesis was published in 2003 [123]. Starting in 2001, the International Conference on Hypervalent Iodine Chemistry has regularly been convened in Europe, the Society of Iodine Science (SIS) holds annual meetings in Japan and the American Chemical Society presents the National Award for Creative Research and Applications of Iodine Chemistry sponsored by SQM S.A. biennially in odd-numbered years. The most impressive modern achievements in the field of organoiodine chemistry include the development of numerous new hypervalent iodine reagents and the discovery of catalytic applications of organoiodine compounds. The discovery of similarities between transition metal chemistry and hypervalent iodine chemistry and, in particular, the development of highly efficient and enantioselective catalytic systems based on the iodine redox chemistry have added a new dimension to the field of hypervalent iodine chemistry and initiated a major increase in research activity, which is expected to continue in the future.

1.2 Classification and Nomenclature of Polyvalent Iodine Compounds

Iodine can form chemical compounds in oxidation states of +3, +5 and +7. The six most common structural types of polyvalent iodine species are represented by structures **1**–**7** (Figure 1.1). Species **2**–**7** can be generally classified using the Martin–Arduengo *N-X-L* designation for hypervalent molecules [124, 125], where *N* is the number of valence electrons formally assignable to the valence shell of the central atom, *X*, either as unshared pairs of electrons or as pairs of electrons in the sigma bonds joining a number, *L*, of ligands to the atom *X*. Structure **1**, the iodonium ion, formally does not belong to hypervalent species since it has only eight valence electrons on the iodine atom; however, in the modern literature iodonium salts are commonly treated as ten-electron hypervalent compounds by taking into account the closely associated anionic part of the molecule. The first three species, structures **1**–**3**, are conventionally considered as derivatives of trivalent iodine, while **4** and **5** represent the most typical structural types of pentavalent iodine. Structural types **6** and **7** are typical of heptavalent iodine; only inorganic compounds of iodine(VII), such as iodine(VII) fluoride (IF_7), iodine(VII) oxyfluorides and the derivatives of periodic acid (HIO_4) are known.

In the older literature, derivatives of iodine(III) were known under the general name of iodinanes, while compounds of pentavalent iodine were called periodinanes. According to the 1983 IUPAC recommendations "Treatment of variable valence in organic nomenclature (lambda convention)" [126], these old names were replaced by λ^3-iodanes for iodine(III) and λ^5-iodanes for iodine(V) compounds. In the lambda nomenclature,

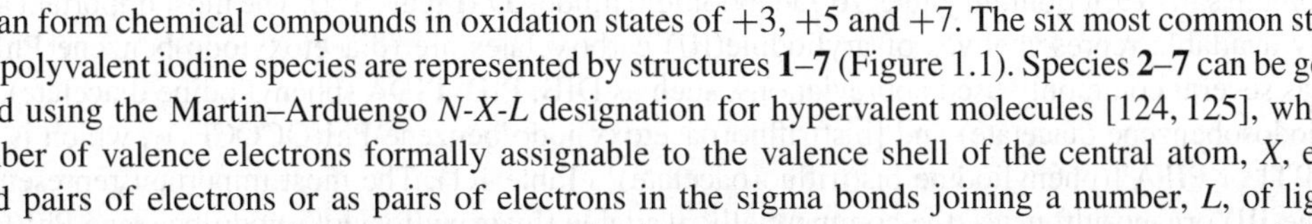

	10-I-3	*12-I-4*	*10-I-4*	*12-I-5*	*14-I-6*	*14-I-7*
1	**2**	**3**	**4**	**5**	**6**	**7**

Figure 1.1 *Typical structural types of polyvalent iodine compounds.*

the symbol λ^n is used to indicate any heteroatom in nonstandard valence states (n) in a formally neutral compound; for iodine the standard valence state is 1. The names λ^3-iodanes and λ^5-iodanes have found broad application in modern literature to indicate the general type of hypervalent iodine compounds and to specify the number of primary bonds at the iodine atom. The λ^3-iodane designation is particularly useful for naming iodonium salts, for example, Ph_2ICl, because it better reflects the actual structure of these compounds with a tricoordinated iodine atom [127].

Notably, however, the lambda nomenclature is not used for naming common hypervalent iodine reagents such as $PhICl_2$, $PhI(OAc)_2$, ArIO, $ArIO_2$ and others. According to the 1979 IUPAC rules [128], "compounds containing the group –$I(OH)_2$ or derivatives of this group are named by adding the prefixes "dihydroxyiodo-", "dichloroiodo-", "diacetoxyiodo-", etc. to the name of the parent compound" (IUPAC Rule C-106.3). Likewise, "compounds containing the group –IO or –IO_2, are named by adding the prefix "iodosyl-" or "iodyl-" (IUPAC Rule C–106.1) [128], which replaces prefixes "iodoso-" and "iodoxy-" used in the older literature. According to IUPAC Rule C-107.1 "cations of the type $R^1R^2I^+$ are given names derived from the iodonium ion H_2I^+ by substitution" [128]. In addition to the IUPAC recommended names, numerous common names and abbreviations are used for polyvalent iodine compounds; for example, about 15 different names have been used in the literature for $PhI(OAc)_2$ [20]. Table 1.1 summarizes commonly used names and abbreviations for several important organic and inorganic polyvalent iodine compounds.

Organoiodine(III) compounds are commonly classified by the type of ligands attached to the iodine atom. The following general classes of iodine(III) compounds have found broad application as reagents in organic synthesis: (difluoroiodo)arenes **8,** (dichloroiodo)arenes **9,** iodosylarenes **10,** [bis(acyloxy)iodo]arenes **11,** aryliodine(III) organosulfonates **12,** five-membered iodine heterocycles (benziodoxoles **13** and benziodazoles **14**), iodonium salts **15,** iodonium ylides **16** and iodonium imides **17** (Figure 1.2). The most important and commercially available representatives of aryliodine(III) carboxylates are (diacetoxyiodo)benzene $PhI(OAc)_2$, which has several commonly used abbreviations, such as DIB, PID, PIDA (phenyliodine diacetate), IBD, or IBDA (iodosobenzene diacetate) and [bis(trifluoroacetoxy)iodo]benzene $PhI(OCOCF_3)_2$, which is abbreviated as BTI or PIFA [(phenyliodine bis(trifluoroacetate)] (Table 1.1). The most important representative of aryliodine(III) organosulfonates, the commercially available [hydroxy(tosyloxy)iodo]benzene PhI(OH)OTs, is abbreviated as HTIB and is also known as Koser's reagent.

Organoiodine(V) compounds are represented by several common classes shown in Figure 1.3; all these compounds have found application as efficient oxidizing reagents. Particularly important in organic synthesis are noncyclic iodylarenes **18,** numerous five-membered heterocyclic benziodoxole derivatives **19** and **20,** including IBX and DMP (Table 1.1), pseudocyclic iodylarenes **21–23** and cyclic or pseudocyclic derivatives of 2-iodylbenzenesulfonic acid, **24–26.**

1.3 Hypervalent Bonding

The definition of "hypervalent" species as ions or molecules of the elements of Groups 15–18 bearing more than eight electrons within a valence shell was established by J. I. Musher in 1969 [129]. General aspects of bonding in hypervalent organic compounds were summarized by K.-y. Akiba in the book *Chemistry of Hypervalent Compounds* [1]. In principle, there are two possible explanations for the ability of main-group elements to hold more than the octet of electrons within a valence shell: (i) by the involvement of the higher-lying d orbitals resulting in dsp^3 or d^2sp^3 hybridization or (ii) by the formation of a new type of highly ionic orbital without involvement of d orbitals. In modern literature, it is generally agreed that the contribution of d orbitals is not essential to form hypervalent compounds and that hypervalent bonding is best explained by a molecular orbital description involving a three-center-four-electron bond.

Table 1.1 *Names and abbreviations of important derivatives of polyvalent iodine.*

Compound	IUPAC names [126, 128]	Common names	Common abbreviations
ICl_3	Iodine trichloride or trichloro-λ^3-iodane	Iodine(III) chloride	None
$PhICl_2$	(Dichloroiodo)benzene	Iodobenzene dichloride Iodosobenzene dichloride Phenyliodo dichloride Phenyliodine(III) dichloride	IBD
$PhI(OAc)_2$	(Diacetoxyiodo)benzene	Iodobenzene diacetate Phenyliodo diacetate Iodosobenzene diacetate Phenyliodine(III) diacetate	DIB, IBD PIDA IBDA
$PhI(OCOCF_3)_2$	[Bis(trifluoroacetoxy) iodo]benzene	Iodobenzene bis(trifluoroacetate) Phenyliodo bis(trifluoroacetate) Phenyliodine(III) bis(trifluoroacetate)	BTI PIFA
PhI(OH)OTs	[Hydroxy(4-methylphenyl sulfonyloxy)iodo]benzene	[Hydroxy(tolsyloxy)iodo] benzene Koser's reagent	HTIB, HTI
PhIO	Iodosylbenzene	Iodosobenzene	IDB
OH, I, O, O	1-Hydroxy-1*H*-1λ^3- benzo[*d*][1,2]iodoxol-3-one	2-Iodosobenzoic acid 2-Iodosylbenzoic acid *o*-Iodosobenzoic acid 1-Hydroxy-1,2-benziodoxol-3-(1*H*)- one	IBA
Ph_2ICl	Diphenyliodonium chloride or chloro(diphenyl)-λ^3-iodane	Chlorodiphenyliodonium	DPI
PhINTs	[*N*-(4-Methylphenylsulfonyl) imino]phenyl-λ^3-iodane	(*N*-Tosylimino)phenyliodinane	None
IF_5	Iodine pentafluoride or pentafluoro-λ^5-iodane	Iodine(V) fluoride Iodic fluoride	None
HIO_3	Iodic acid	Iodic(V) acid	None
$PhIO_2$	Iodylbenzene	Iodoxybenzene	None
O, OH, I, O, O	1-Hydroxy-1-oxo-1*H*-1λ^5- benzo[*d*][1,2]iodoxol-3-one	2-Iodoxybenzoic acid 2-Iodylbenzoic acid *o*-Iodoxybenzoic acid	IBX
OAc, AcO, OAc, I, O, O	1,1,1-Triacetoxy-1*H*-1λ^5- benzo[*d*][1,2]iodoxol-3-one	Dess–Martin periodinane	DMP
IF_7	Iodine heptafluoride or heptafluoro-λ^7-iodane	Iodine(VII) fluoride Heptafluoroiodine	None
HIO_4	Periodic acid	Iodic(VII) acid	None

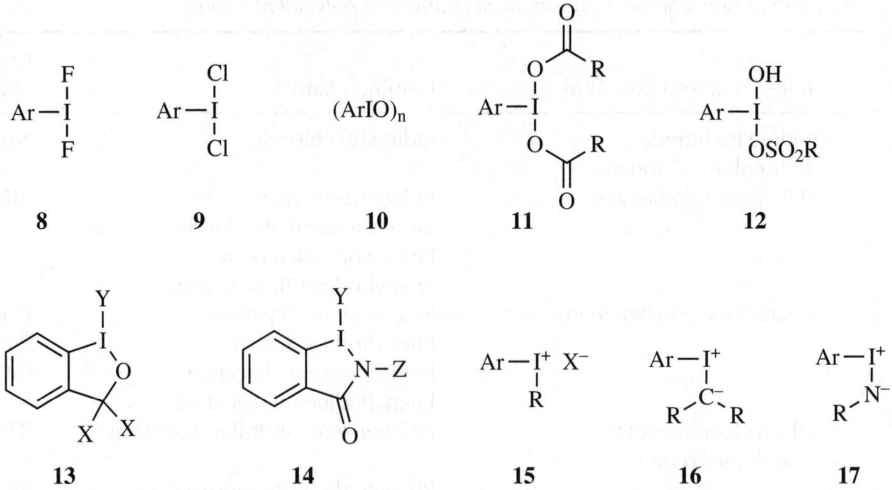

8 9 10 11 12

13 14 15 16 17

X = Me, CF_3 or 2X = O;
Y = OH, OAc, N_3, CN, etc.; Z = H, Ac, etc.

Figure 1.2 *Common classes of organoiodine(III) compounds.*

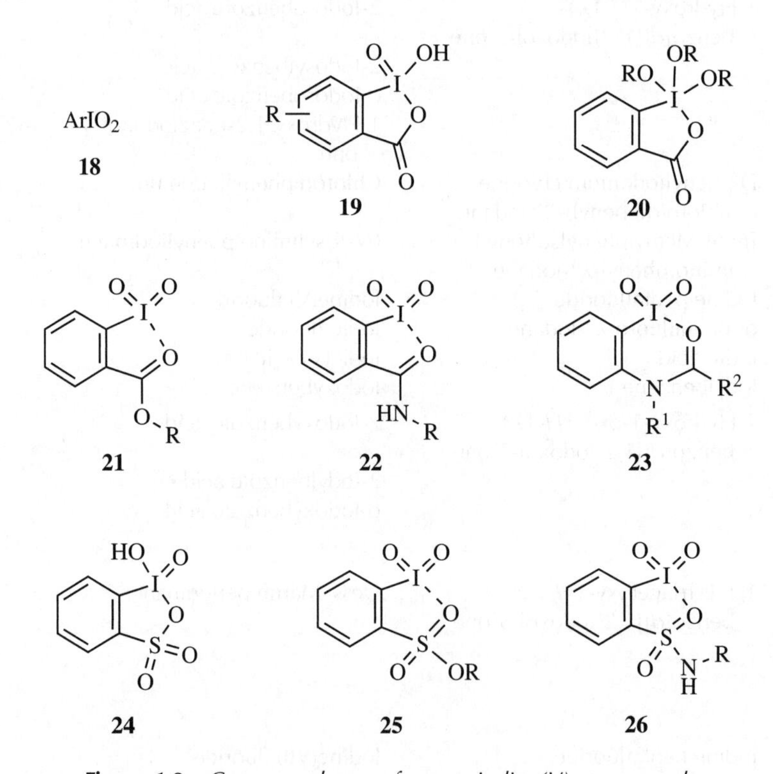

Figure 1.3 *Common classes of organoiodine(V) compounds.*

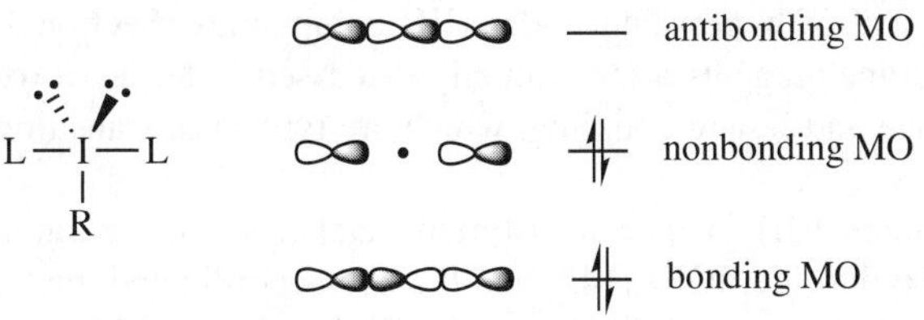

Figure 1.4 *Molecular orbital description of the three-center-four-electron bond in hypervalent iodine(III) molecules RIL_2.*

The idea of a three-center-four-electron (3c-4e) bond was independently proposed by G. C. Pimentel [130] and R. E. Rundle [131] in 1951 on the basis of molecular orbital theory. According to the fundamental description of the 3c-4e bond for L–X–L, one pair of bonding electrons is delocalized to the two ligands L, resulting in the charge distribution of almost –0.5 on each ligand and +1.0 on the central atom X. In iodine(III) molecules RIL_2, the interaction of the filled 5p orbital of the central iodine atom and the half-filled orbitals of the two ligands L *trans* to each other leads to formation of three molecular orbitals: bonding, nonbonding and antibonding (Figure 1.4). Because the highest occupied molecular orbital (HOMO) contains a node at the central iodine, the hypervalent bonds show a highly polarized nature; hence, more electronegative atoms tend to occupy the axial positions formed by the interaction of the orbitals of three collinear atoms. The carbon substituent R is bound by a normal covalent bond and the overall geometry of molecule RIL_2 is a distorted trigonal bipyramid with two heteroatom ligands L occupying the apical positions and the least electronegative carbon ligand R and both electron pairs reside in equatorial positions.

The bonding in iodine(V) compounds, RIL_4, with a square bipyramidal structure may be described in terms of a normal covalent bond between iodine and the organic group R in an apical position and two orthogonal, hypervalent 3c-4e bonds, accommodating four ligands L. The carbon substituent R and unshared electron pair in this case should occupy the apical positions with the electronegative ligands L residing at equatorial positions (Figure 1.5).

Several theoretical computational studies concerning bonding, structure and reactivity of hypervalent iodine compounds were published in the 1990s and 2000s [132–137]. In particular, Reed and Schleyer provided a general theoretical description of chemical bonding in hypervalent molecules in terms of the dominance of ionic bonding and negative hyperconjugation over d-orbital participation [132]. The simple, qualitative bonding concepts for hypervalent molecules developed in this work supersede the inaccurate and misleading dsp^3 and d^2sp^3 models. It has been recognized that there are fundamental similarities in bonding, structure and reactivity of hypervalent λ^3- and λ^5-iodanes with organometallic compounds. In fact, it has been stated in some theoretical studies that, similar to the heavy main group elements, hypervalent bonding commonly occurs in transition metal complexes and the 3c-4e bond is particularly important in the structure of transition metal hydrides [138–142]. The important and well known in transition metal complexes, effect of *trans*

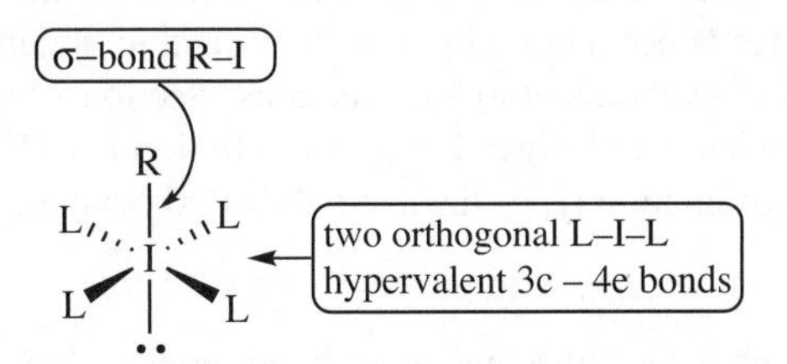

Figure 1.5 *Bonding in hypervalent iodine(V) molecules.*

influence [136] is also typical of hypervalent iodine(III) compounds (Section 1.4.2) [135, 136, 143]. The reactions of hypervalent iodine reagents are commonly discussed in terms of oxidative addition, reductive elimination, ligand exchange and ligand coupling, which are typical of transition metal chemistry (Section 1.5).

Typical structures of iodine(VII) involve a distorted octahedral configuration **6** about iodine in most periodates [144] and the oxyfluoride, IOF_5 [145] and the heptacoordinated, pentagonal bipyramidal species **7** for IF_7 and the IOF_6^- anion (Figure 1.1) [146, 147]. The pentagonal bipyramidal structure **7** has been described as two covalent, collinear, axial bonds between iodine and the ligands in the apical positions and a coplanar, hypervalent 6c-10e bond system for the five equatorial bonds [146].

1.4 General Structural Features

The structural aspects of polyvalent iodine compounds were previously summarized in several books and reviews [20, 30, 32, 127]. In general, the molecular structure of λ^3- and λ^5-iodanes is predetermined by the nature of hypervalent bonding discussed in Section 1.3. The key structural features of the hypervalent organoiodine compounds available from numerous X-ray data may be summarized as follows:

1. λ^3-Iodanes RIX_2 (R = C-ligand, X = heteroatom ligands) have an approximately T-shaped structure with a collinear arrangement of the most electronegative ligands X. Including the nonbonding electron pairs, the geometry about iodine is a distorted trigonal bipyramid with the most electronegative groups occupying the apical positions, while the least electronegative C-ligand R and both electron pairs reside in an equatorial position.
2. The I–C bond lengths in iodonium salts $R_2I^+ X^-$ and λ^3-iodanes RIX_2 are approximately equal to the sum of the covalent radii of iodine and carbon, ranging generally from 2.00 to 2.10 Å.
3. Iodonium salts $R_2I^+ X^-$ generally have a typical distance between iodine and the nearest anion X^- of 2.6–2.8 Å and in principle can be considered as ionic compounds with pseudo-tetrahedral geometry about the central iodine atom. However, with consideration of the anionic part of the molecule, the overall experimentally determined geometry is distorted T-shaped structure similar to the λ^3-iodanes RIX_2.
4. For λ^3-iodanes RI(X)Y with two heteroatom ligands X and Y of the same electronegativity, both I–X and I–Y bonds are longer than the sum of the appropriate covalent radii, but shorter than purely ionic bonds. For example, the I–Cl bond lengths in $PhICl_2$ are 2.45 Å [148] and the I–O bond lengths in $PhI(OAc)_2$ are 2.15–2.16 Å [149], while the sum of the covalent radii of I and O is 1.99 Å. When heteroatom ligands X and Y have different electronegativities, the *trans* influence of ligands has a strong effect on the structure, stability and reactivity of λ^3-iodanes RI(X)Y (Section 1.4.2) [135].
5. Various coordination types have been reported for the organoiodine(V) compounds. Depending on the ligands and taking into account secondary bonding, the overall observed geometry for the structural types **4** and **5** (Figure 1.1) can be pseudo-trigonal–bipyramidal, square bipyramidal and pseudooctahedral.
6. Intramolecular positional isomerization (Berry pseudorotation) resulting in an exchange between the apical and the equatorial ligands occurs rapidly in both λ^3- and λ^5-iodanes. This process is important in explaining the mechanisms of hypervalent iodine reactions (Section 1.5).
7. Only inorganic compounds with O- or F-ligands are known for iodine(VII) structural types **6** and **7** (Figure 1.1). Typical iodine(VII) coordination types involve a distorted octahedral configuration and pentagonal bipyramidal species.

Owing to a highly polarized character of hypervalent bond, noncovalent attractive interactions of a predominantly electrostatic nature are extremely important in the structural chemistry of hypervalent iodine compounds. Such attractive interactions are commonly called secondary bonds. Similarly to hydrogen bonds, secondary bonds involving heavier atoms have strong electrostatic components and show directional preferences [150, 151]. Intermolecular secondary bonding in hypervalent iodine compounds is responsible for crystal packing in the solid state and for the self-assembly of individual molecules into complex supramolecular structures in the solid state and in solution [152, 153]. Intramolecular secondary bonding is commonly observed in the λ^3- and λ^5-aryliodanes, which have a sulfonyl or a carbonyl structural fragment in the *ortho*-position of the phenyl ring [154–159]. The redirection of secondary bonding from intermolecular to intramolecular mode due to the presence of an appropriate *ortho*-substituent leads to a partial disruption of the polymeric network and enhances solubility of a hypervalent iodine compound [154, 155].

1.4.1 Experimental Structural Studies

Numerous X-ray crystal structures have been reported for all main classes of organic polyvalent iodine compounds and the results of these studies are overviewed in the appropriate sections of Chapter 2. Typical coordination patterns in various organic derivatives of iodine(III) in the solid state with consideration of primary and secondary bonding were summarized in 1986 by Sawyer and coworkers [160] and have been updated in several more recent publications [153, 161–165]. Structural features of organic iodine(V) compounds have been discussed in the older papers of Martin and coauthors [166, 167] and in numerous recent publications on IBX and related λ^5-iodanes [155–159, 168–174]. Several general areas of structural research on hypervalent organoiodine compounds have attracted especially active interest. These areas, in particular, include the preparation and structural study of complexes of hypervalent iodine compounds with crown ethers [175–179] or nitrogen ligands [180–182], self-assembly of hypervalent iodine compounds into various supramolecular structures [152, 153, 164, 183, 184] and the intramolecular secondary bonding in *ortho*-substituted aryliodine(V) and aryliodine(III) derivatives [154–159, 168–171, 173, 174, 185–188].

Several important spectroscopic structural studies of polyvalent iodine compounds in solution have been published [108–112, 189]. Reich and Cooperman reported low-temperature NMR study of triaryl-λ^3-iodanes **27** (Scheme 1.1), which demonstrated that these compounds have a nonsymmetrical planar orientation of iodine–carbon bonds and that the barrier to unimolecular degenerate isomerization between **27** and **27′** is greater than 15 kcal mol^{-1}. The exact mechanism of this degenerate isomerization is unknown; both pseudorotation on iodine(III) and intermolecular ligand exchange may account for the isomerization of these compounds [189].

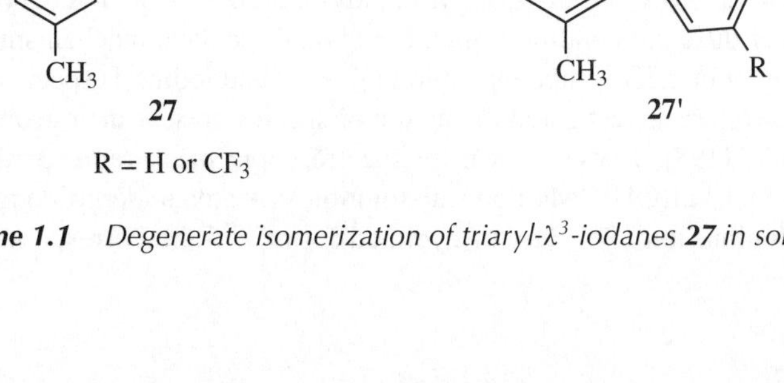

27 **27'**

R = H or CF_3

Scheme 1.1 *Degenerate isomerization of triaryl-λ^3-iodanes **27** in solution.*

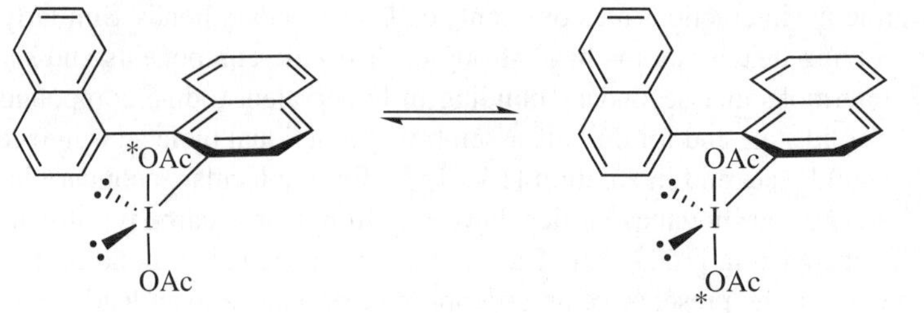

Scheme 1.2 *Degenerate isomerization of (diacetoxyiodo)binaphthyl **28** due to rapid pseudorotation on iodine.*

Ochiai and coworkers observed rapid pseudorotation on iodine(III) for chiral (diacetoxyiodo)binaphthyl **28** (Scheme 1.2) [190]. The two acetoxy groups of compound **28** are anisochronous in $CDCl_3$ at –10 °C and appear as two sharp singlets in ^1H NMR spectra. These two singlets coalesce at 34 °C to one singlet with a free activation energy of 15.1 kcal mol^{-1}. Similar temperature dependence was observed in the ^{13}C NMR spectrum. The authors attributed this degenerate isomerization to rapid pseudorotation on iodine [190].

Amey and Martin have found that cyclic dialkoxy-λ^3-iodanes undergo rapid degenerate ligand exchange on the NMR time scale occurring via an associative mechanism [191]. Cerioni, Mocci and coworkers investigated the structure of bis(acyloxyiodo)arenes and benziodoxolones in chloroform solution by ^{17}O NMR spectroscopy and also by DFT (density functional theory) calculations [192–194]. This investigation provided substantial evidence that the T-shaped structure of iodine(III) compounds observed in the solid state is also adopted in solution. Furthermore, the "free" carboxylic groups of bis(acyloxyiodo)arenes show a dynamic behavior, observable only in the ^{17}O NMR. This behavior is ascribed to a [1,3]-sigmatropic shift of the iodine atom between the two oxygen atoms of the carboxylic groups and the energy involved in this process varies significantly between bis(acyloxyiodo)arenes and benziodoxolones [193].

Hiller and coworkers reported an NMR and LC-MS study on the structure and stability of 1-iodosyl-4-methoxybenzene and 1-iodosyl-4-nitrobenzene in methanol solution [195]. Interestingly, LC-MS analyzes provided evidence that unlike the parent iodosylbenzene, which has a polymeric structure (Section 2.1.4), the 4-substituted iodosylarenes exist in the monomeric form. Both iodosylarenes are soluble in methanol and provide acceptable ^1H and ^{13}C NMR spectra; however, gradual oxidation of the solvent was observed after several hours. Unlike iodosylbenzene, the two compounds did not react with methanol to give the dimethoxy derivative $ArI(OMe)_2$ [195].

Silva and Lopes analyzed solutions of iodobenzene dicarboxylates in acetonitrile, acetic acid, aqueous methanol and anhydrous methanol by electrospray ionization mass spectrometry (ESI-MS) and tandem mass spectrometry (ESI-MS/MS) [196]. The major species found in the solutions of $PhI(OAc)_2$ in acetonitrile, acetic acid and aqueous methanol were $[PhI(OAc)_2Na]^+$, $[PhI(OAc)_2K]^+$, $[PhI]^+$, $[PhIOAc]^+$, $[PhIOH]^+$, $[PhIO_2Ac]^+$, $[PhIO_2H]^+$ and the dimer $[Ph_2I_2O_2Ac]^+$. On the other hand, the anhydrous methanol solutions showed $[PhIOMe]^+$ as the most abundant species. In contrast to the data obtained for $PhI(OAc)_2$, the ESI-MS spectral data of $PhI(O_2CCF_3)_2$ in acetonitrile suggests that the main species in solutions is iodosylbenzene [196]. A similar ESI-MS and ESI-MS/MS study of solutions of [hydroxy(tosyloxy)iodo]benzene has been performed under different conditions and, based on these data, mechanisms were proposed for the disproportionation of the iodine(III) compounds into iodine(V) and iodine(I) species [197].

Richter, Koser and coworkers investigated the nature of species present in aqueous solutions of phenyliodine(III) organosulfonates [198]. It was shown by spectroscopic measurements and potentiometric titrations that PhI(OH)OTs and PhI(OH)OMs upon dissolution in water undergo complete ionization to give the hydroxy(phenyl)iodonium ion (PhI^+OH in hydrated form) and the corresponding sulfonate ions. The hydroxy(phenyl)iodonium ion can combine with [oxo(aquo)iodo]benzene $PhI^+(OH_2)O^-$, a hydrated form of iodosylbenzene that is also observed in the solution, to produce the dimeric μ-oxodiiodine cation $Ph(HO)I–O–I^+(OH_2)Ph$ and dication $Ph(H_2O)I^+–O–I^+(OH_2)Ph$ [198]. Likewise, an ESI-MS study of an aqueous solution of oligomeric iodosylbenzene sulfate, $(PhIO)_3SO_3$, indicated mainly the presence of hydroxy(phenyl)iodonium ion (PhI^+OH) along with dimeric and trimeric protonated iodosylbenzene units [101].

1.4.2 Computational Studies

Relatively few theoretical computational studies concerning the structure and reactivity of hypervalent iodine compounds have appeared. Hoffmann and coworkers analyzed the nature of hypervalent bonding in trihalide anions X_3^- ($X = F, Cl, Br, I$) and related halogen species by applying ideas from qualitative MO theory to computational results from density-functional calculations [133]. This systematic, unified investigation showed that the bonding in all of these systems could be explained in terms of the Rundle–Pimentel scheme for electron-rich three-center bonding (Section 1.3). The same authors reported an analysis of intermolecular interaction between hypervalent molecules, including diaryliodonium halides Ar_2IX, using a combination of density-functional calculations and qualitative arguments [150]. Based on fragment molecular orbital interaction diagrams, the authors concluded that the secondary bonding in these species can be understood using the language of donor–acceptor interactions: mixing between occupied states on one fragment and unoccupied states on the other. There is also a strong electrostatic contribution to the secondary bonding. The calculated strengths of these halogen–halogen secondary interactions are all less than 10 kcal mol^{-1} [150].

The self-assembly of hypervalent iodine compounds to form macrocyclic trimers was studied using MO calculations. The principal driving force for the self-assembly of iodonium units is the formation of secondary bonding interactions between iodonium units as well as a rearrangement of primary and secondary bonding around iodine to place the least electronegative substituent in the equatorial position for every iodine in the trimer [199].

Kiprof has analyzed the iodine–oxygen bonds of hypervalent λ^3-iodanes with T-shaped geometry using the Cambridge Crystallographic Database and *ab initio* MO calculations. Statistical analysis of the I–O bond lengths in $PhI(OR)_2$ revealed an average of 2.14 Å and a strong correlation between the two bond lengths [143]. Further theoretical investigation of the mutual ligand interaction in the hypervalent L–I–L′ system has demonstrated that the ligands' *trans* influences play an important role in the stability of hypervalent molecules [135]. In particular, combinations of ligands with large and small *trans* influences, as in PhI(OH)OTs, or of two moderately *trans* influencing ligands, as in $PhI(OAc)_2$, are favored and lead to higher stability of the molecule. The *trans* influences also seem to explain why iodosylbenzene, $(PhIO)_n$, adopts an oxo-bridged zigzag polymer structure (Section 2.1.4) in contrast to $PhI(OH)_2$, which is monomeric [135].

A theoretical computational study on quantitative measurement of the *trans* influence in hypervalent iodine complexes has been published by Sajith and Suresh [136]. The *trans* influence of various X ligands in hypervalent iodine(III) complexes of the type $CF_3I(X)Cl$ has been quantified using the *trans* I–Cl bond length (d_X), the electron density $\rho(r)$ at the $(3, -1)$ bond critical point of the *trans* I–Cl bond and topological features of the molecular electrostatic potential (MESP). The MESP minimum at the Cl lone pair region (V_{min}) has been found to be a sensitive measure of the *trans* influence. The *trans* influence of X ligands in hypervalent iodine(V) compounds is smaller than that in iodine(III) complexes, while the relative order of this influence is the same in both species. The quantified values of the *trans* influence parameters may find use in assessing the stability of hypervalent iodine compounds as well as in the design of new stable hypervalent complexes [136].

The structure and reactivity of several specific classes of hypervalent iodine compounds have been investigated theoretically. Varvoglis, Tsipis and coauthors have studied the geometry and electronic structure of some hypervalent iodine compounds $PhIX_2$ by means of extended Hückel and CNDO/2 quantum chemical approaches [200]. The bonding was analyzed in terms of both the model of delocalized MOs on the basis of interactions between fragment MOs derived from EHMO–SCCC calculations and that of localized MOs derived by the CNDO/2 method. The ability of these compounds to afford *cis*-addition products with alkenes via a synchronous molecular addition mechanism was found to be theoretically feasible [200].

Okuyama and Yamataka investigated the reactivity of vinyliodonium ions with nucleophiles by *ab initio* MO (MP2) calculations at the double-zeta (DZ) + d level [201]. It was proposed that interaction of methyl(vinyl)iodonium ion with chlorine anion leads to chloro-λ^3-iodane $CH_2{=}CHI(Me)Cl$. Transition states for the S_N2, ligand-coupling substitution and β-elimination were found for reactions at the vinyl group. The barrier to ligand-coupling substitution is usually the lowest in the gas phase, but relative barriers to S_N2 and to β-elimination change with the substituents. Effects of solvent on this reaction were evaluated by a dielectric continuum model and found to be large on S_N2 but small on ligand coupling [201].

Widdowson, Rzepa and coworkers reported *ab initio* and MNDO-d SCF-MO computational studies of the extrusion reactions of diaryliodonium fluorides [202–204]. The results of these studies, in particular, predicted that the intermediates and transition states in these reactions might involve dimeric, trimeric and tetrameric structures. The regioselectivity of nucleophilic substitution in these reactions was investigated theoretically and supported by some experimental observations.

Goddard and Su have investigated theoretically the mechanism of alcohol oxidation with 2-iodoxybenzoic acid (IBX) on the basis of density functional quantum mechanics calculations [134]. It has been found that the rearrangement of hypervalent bonds, so-called hypervalent twisting, is the rate-determining step in this reaction. Based on this mechanism, the authors explain why IBX oxidizes large alcohols faster than small ones and propose a modification to the reagent that is predicted to make it more active [134].

Bakalbassis, Spyroudis and Tsiotra reported a DFT study on the intramolecular thermal phenyl migration in iodonium ylides [205]. The results of this study support a single-step mechanism involving a five-membered ring transition state. The frontier-orbital-controlled migration also confirms the different thermal behavior experimentally observed for two different ylides [205].

Quideau and coworkers presented DFT calculations of spiroheterocylic iodine(III) intermediates to validate their participation in the $PhI(OAc)_2$-mediated spiroketalization of phenolic alcohols [206]. Molecular orbital computational studies of (arylsulfonylimino)iodoarenes ($ArINSO_2Ar'$) [185], benziodazol-3-ones [207] and a series of *ortho*-substituted chiral organoiodine(III) compounds [208] have been reported in the literature. Results of these calculations were found to be in good agreement with X-ray structural data for these compounds.

1.5 General Principles of Reactivity

Hypervalent iodine reagents are used extensively in organic synthesis as efficient and environmentally benign oxidizing reagents whose chemical properties in many aspects are similar to the derivatives of heavy metals. The following general classes of iodine(III) compounds (Figure 1.2) have found broad application in organic synthesis: (difluoroiodo)arenes **8** and (dichloroiodo)arenes **9** are effective fluorinating and chlorinating reagents, respectively, iodosylarenes **10,** aryliodine(III) carboxylates **11** and organosulfonates **12** in general are strong oxidizing agents and have found widespread application as reagents for oxygenation and oxidative functionalization of organic substrates, benziodoxoles **13** and benziodazoles **14** have found synthetic application as efficient group Y transfer reagents, iodonium salts **15** and ylides **16** are used in numerous C–C bond-forming reactions, while iodonium imides **17** are useful reagents for the aziridination of alkenes and the amidation of various organic substrates. Organoiodine(V) compounds (Figure 1.3), especially IBX and DMP, have found application as efficient oxidizing reagents, for example, for the oxidation of alcohols to the respective carbonyl compounds. Inorganic derivatives of iodine(VII), such as periodic acid and periodates, are powerful oxidants useful for glycol cleavage and some other applications.

From the simplified point of view of a synthetic organic chemist, the rich chemistry of hypervalent iodine is explained mainly by its strongly electrophilic character combined with the excellent leaving group ability of the phenyliodonio group. At a more advanced level, the reactions of hypervalent iodine reagents are commonly discussed in terms of ligand exchange, reductive elimination and ligand coupling, which are typical of transition metal chemistry. Homolytic and single-electron transfer (SET) pathways are also frequently observed in the reactions of several classes of hypervalent iodine compounds under appropriate conditions. An excellent, comprehensive survey of hypervalent iodine reactivity patterns has been provided by Ochiai (2003) in *Hypervalent Iodine Chemistry* [127]. The general reactivity features of hypervalent iodine reagents are summarized in the following sections.

1.5.1 Ligand Exchange and Reductive Elimination

Most reactions of λ^3-iodanes $PhIL_2$ involve the initial exchange of ligands on the iodine atom with external nucleophiles (Nu:) followed by reductive elimination of iodobenzene (Scheme 1.3). The second step in this simplified scheme can also proceed as "ligand coupling" [1], if it occurs as a concerted process. A similar general mechanistic description can also be applied to the reactions of λ^5-iodanes.

A detailed mechanism of the process shown in Scheme 1.3 is unknown. Two general mechanistic pathways, dissociative and associative, have been proposed for the ligand exchange reactions of λ^3-iodanes (Scheme 1.4) [26, 127]. The dissociative pathway seems to be less likely to occur, because of the low stability of the dicoordinated iodonium ion $[PhIL]^+$ involved in this mechanism [127]. Such iodonium *8-I-2* species, however, have been frequently observed in the gas phase, for example, in mass spectrometry studies of protonated iodosylbenzene, $[PhIOH]^+$ [101], or in the mass spectra of all known iodonium salts, $[ArIR]^+$. The presence of cationic iodonium species in aqueous solution has been confirmed by spectroscopic measurements and potentiometric titrations of PhI(OH)OTs and PhI(OH)OMs [198]; however, all available experimental data show that the iodonium species in solution are coordinated with solvent molecules or with available counteranions. X-Ray diffraction analysis of the protonated iodosylbenzene aqua complexes $[PhI(H_2O)OH]^+$ isolated from aqueous solutions revealed a T-shaped structure, ligated with one water molecule at the apical site of the iodine(III) atom of hydroxy(phenyl)iodonium ion, with a near-linear O–I–O triad (173.96°), which is in agreement with a regular λ^3-iodane structure [178].

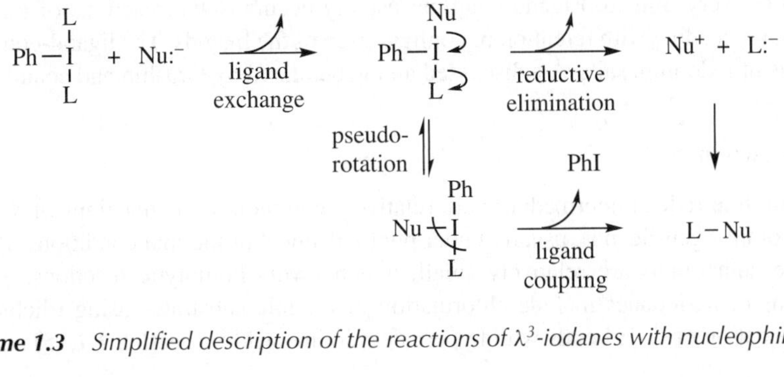

Scheme 1.3 *Simplified description of the reactions of λ^3-iodanes with nucleophiles Nu.*

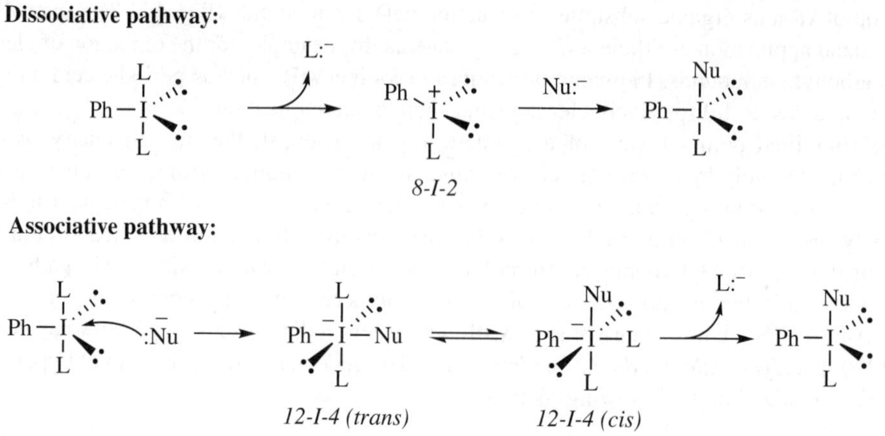

Dissociative pathway:

8-I-2

Associative pathway:

12-I-4 (trans) 12-I-4 (cis)

Scheme 1.4 *Dissociative and associative pathways for the ligand exchange reactions of λ^3-iodanes with nucleophiles Nu.*

The associative pathway of ligand exchange starts from the addition of a nucleophile to the positively charged iodine atom of a λ^3-iodane with the initial formation of a *trans* hypervalent *12-I-4* square-planar species. This intermediate species isomerizes to the *cis 12-I-4* square-planar intermediate and eliminates the ligand L to afford the final product (Scheme 1.4). Such a mechanism has been validated by the isolation and X-ray structural identification of several stable *12-I-4* species. For example, the interaction of ICl_3 with chloride anion affords tetrachloroiodate anion, ICl_4^-, which has a distorted square-planar structure as established by X-ray analysis of the trichlorosulfonium salt, Cl_3S^+ ICl_4^- [209].

The second step of reactions of λ^3-iodanes with nucleophiles (Scheme 1.3) includes elimination of iodobenzene or other reduced iodine species. This is a facile and energetically favorable process. The leaving group in this reaction, PhI, is an excellent leaving group, about million times better than the triflate [210] and Ochiai has suggested calling this group a "hypernucleofuge" [127], which reflects the initial hypervalent character and the exceptional leaving group ability of the phenyliodonio group. Elimination of PhI can occur as reductive elimination or as ligand coupling as shown in Scheme 1.3. Reductive elimination leading to formal umpolung of reactivity of the nucleophile, Nu:$^-$ to Nu$^+$ (Scheme 1.3), is a common process in various reactions of hypervalent iodine reagents; it can result in the formation of products of nucleophilic substitution, α-elimination, β-elimination, rearrangement, or fragmentation.

The ligand coupling pathway requires initial pseudorotation to bring ligands L and Nu to apical and equatorial positions favorable for coupling (Scheme 1.3). Experimental studies on the mechanism of ligand coupling reaction are very limited. Ligand coupling usually occurs in the reactions of iodonium salts as a concerted process, proceeding with retention of configuration of the ligands. The ligand-coupling mechanism for the thermolysis of iodonium salts was discussed and generalized by Grushin and coauthors [48, 211].

1.5.2 Radical Reactions

Processes involving free-radical intermediates are relatively common in the reactions of λ^3-iodanes bearing chloro-, oxygen-, or nitrogen-ligands, usually under photochemical or thermal conditions. Bond dissociation energies in iodine compounds are relatively small, which favors homolytic reactions. Typical examples of radical reactions of λ^3-iodanes include chlorination of organic substrates using (dichloroiodo)benzene (Section 3.1.2), azidation of C–H bonds with hypervalent iodine azides (Section 3.1.15) and various radical

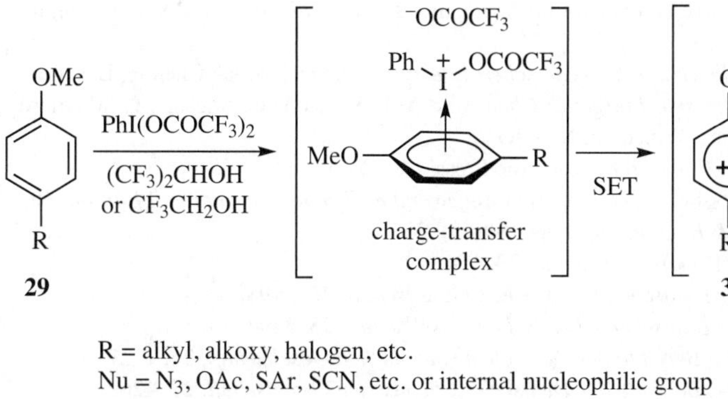

OMe, R, **29** — $PhI(OCOCF_3)_2$, $(CF_3)_2CHOH$ or CF_3CH_2OH → [$^{-}OCOCF_3$, Ph–I^{+}–$OCOCF_3$, MeO–C₆H₄–R, charge-transfer complex] → SET → [OMe, R, **30**] → Nu^{-} → OMe, Nu, R, **31**

R = alkyl, alkoxy, halogen, etc.
Nu = N_3, OAc, SAr, SCN, etc. or internal nucleophilic group

Scheme 1.5 *Simplified single-electron transfer (SET) mechanism for the reaction of* p-*substituted phenol ethers with [bis(trifluoroacetoxy)iodo]benzene.*

fragmentation reactions of alcohols or carboxylic acids with [bis(acyloxy)iodo]arenes in the presence of iodine under photochemical or thermal conditions (Section 3.1.18). These and other homolytic reactions of λ^3-iodanes will be discussed in Chapter 3.

1.5.3 Single-Electron Transfer (SET) Reactions

Processes involving a single-electron transfer (SET) step and cation–radical intermediates can occur in the reactions of λ^3- or λ^5-iodanes with electron-rich organic substrates in polar, non-nucleophilic solvents. Kita and coworkers first found that the reactions of *p*-substituted phenol ethers **29** with [bis(trifluoroacetoxy)iodo]benzene in the presence of some nucleophiles in fluoroalcohol solvents afford products of nucleophilic aromatic substitution **31** via a SET mechanism (Scheme 1.5) [212, 213]. On the basis of detailed UV and ESR spectroscopic measurements, it was confirmed that this process involves the generation of cation-radicals **30** produced by SET oxidation through the charge-transfer complex of phenyl ethers with the hypervalent iodine reagent [213, 214].

A similar SET mechanism involving cation–radical intermediates **30** has also been confirmed for the reactions of phenolic ethers with diaryliodonium salts in hexafluoroisopropanol [215]. The use of fluoroalcohols as solvents in these reactions is explained by their unique ability to stabilize the aromatic cation–radicals [107].

The SET mechanism was also proposed for some oxidations involving λ^5-iodanes. In particular, mechanistic studies involving isotope labeling, kinetic studies, cyclic voltammetry measurements and NMR spectroscopic analysis confirm that SET is a rate-determining step in the IBX-promoted oxidative cyclization of unsaturated anilides in THF–DMSO solutions [216]. The analogous mechanism was proposed for the oxidation of alkylbenzenes at the benzylic position under similar conditions [217].

References

1. Akiba, K.y. (1999) *Chemistry of Hypervalent Compounds*, Wiley-VCH Verlag GmbH, Weinheim.
2. Courtois, B. (1813) *Annali di Chimica*, **88**, 304.
3. Greenwood, N.N. and Earnshaw, A. (1997) *Chemistry of the Elements*, Butterworth-Heinemann, Oxford.
4. Gay Lussac, J.L. (1814) *Annali di Chimica*, **91**, 5.

5. Mellor, J.W. (1922) *A Comprehensive Treatise on Inorganic and Theoretical Chemistry*, vol. **2**, Longmans, Green and Co, London.
6. Meyer, R.J. (1933) *Gmelins Handbuch der Anorganischen Chemie, 8 Auflage*, Verlag Chemie, Berlin.
7. Brasted, R.C. (1954) in *Comprehensive Inorganic Chemistry*, Vol. **3** (eds M.C. Sneed, J.L. Maynard and R.C. Brasted), D. Van Norstrand Co., Inc., Princeton, New Jersey.
8. Gutmann, V. (ed.) (1967) *Halogen Chemistry*, Academic Press, New York.
9. Kuepper, F.C., Feiters, M.C., Olofsson, B., *et al.* (2011) *Angewandte Chemie, International Edition*, **50**, 11598.
10. Tremblay, J.-F. (2011) *Chemical & Engineering News*, **89**(49), 22.
11. Willgerodt, C. (1886) *Journal für Praktische Chemie*, **33**, 154.
12. Willgerodt, C. (1892) *Berichte der Deutschen Chemischen Gesellschaft*, **25**, 3498.
13. Willgerodt, C. (1892) *Berichte der Deutschen Chemischen Gesellschaft*, **25**, 3494.
14. Hartmann, C. and Meyer, V. (1893) *Berichte der Deutschen Chemischen Gesellschaft*, **26**, 1727.
15. Hartmann, C. and Meyer, V. (1894) *Berichte der Deutschen Chemischen Gesellschaft*, **27**, 426.
16. Willgerodt, C. (1914) *Die Organischen Verbindungen mit Mehrwertigen Jod*, Ferdinand Enke Verlag, Stuttgart.
17. Sandin, R.B. (1943) *Chemical Reviews*, **32**, 249.
18. Banks, D.F. (1966) *Chemical Reviews*, **66**, 243.
19. Beringer, F.M. and Gindler, E.M. (1956) *Iodine Abstracts and Reviews*, **3**, 70.
20. Varvoglis, A. (1992) *The Organic Chemistry of Polycoordinated Iodine*, VCH Publishers, Inc., New York.
21. Varvoglis, A. (1997) *Hypervalent Iodine in Organic Synthesis*, Academic Press, London.
22. Varvoglis, A. (1984) *Synthesis*, 709.
23. Varvoglis, A. (1997) *Tetrahedron*, **53**, 1179.
24. Stang, P.J. and Zhdankin, V.V. (1996) *Chemical Reviews*, **96**, 1123.
25. Merkushev, E.B. (1987) *Russian Chemical Reviews*, **56**, 826.
26. Moriarty, R.M. and Prakash, O. (1986) *Accounts of Chemical Research*, **19**, 244.
27. Wirth, T. and Hirt, U.H. (1999) *Synthesis*, 1271.
28. Kitamura, T. and Fujiwara, Y. (1997) *Organic Preparations and Procedures International*, **29**, 409.
29. Nguyen, T.T. and Martin, J.C. (1984) in *Comprehensive Heterocyclic Chemistry*, vol. **1** (eds A.R. Katritzky and C.W. Rees), Pergamon Press, Oxford, p. 563.
30. Koser, G.F. (1983) in *The Chemistry of Functional Groups, Suppl. D: Chem. Halides, Pseudo-Halides, Azides* (eds S. Patai and Z. Rappoport), Wiley-Interscience, Chichester, p. 721.
31. Koser, G.F. (1995) in *Chemistry of Halides, Pseudo-Halides and Azides, Suppl. D2* (eds S. Patai and Z. Rappoport), Wiley-Interscience: Chichester, p. 1173.
32. Koser, G.F. (1983) in *Halides, Pseudo-Halides and Azides*, vol. **2** (eds. S. Patai and Z. Rappoport), Wiley-Interscience, Chichester, p. 1265.
33. Zhdankin, V.V. and Stang, P.J. (1999) in *Chemistry of Hypervalent Compounds* (ed. K.y. Akiba), VCH Publishers, New York, p. 327.
34. Ochiai, M. (1999) in *Chemistry of Hypervalent Compounds* (ed. K.y. Akiba), VCH Publishers, New York, p. 359.
35. Varvoglis, A. (1981) *Chemical Society Reviews*, **10**, 377.
36. Kirschning, A. (1998) *Journal für Praktische Chemie*, **340**, 184.
37. Moriarty, R.M., Vaid, R.K. and Koser, G.F. (1990) *Synlett*, 365.
38. Zhdankin, V.V. and Stang, P.J. (1998) *Tetrahedron*, **54**, 10927.
39. Umemoto, T. (1996) *Chemical Reviews*, **96**, 1757.
40. Kirschning, A. (1998) *European Journal of Organic Chemistry*, 2267.
41. Moriarty, R.M. and Vaid, R.K. (1990) *Synthesis*, 431.
42. Moriarty, R.M. and Prakash, O. (1999) *Organic Reactions*, **54**, 273.
43. Kita, Y., Takada, T. and Tohma, H. (1996) *Pure and Applied Chemistry*, **68**, 627.
44. Hara, S. (2006) *Advances in Organic Synthesis*, **2**, 49.
45. Muraki, T., Togo, H. and Yokoyama, M. (1997) *Reviews on Heteroatom Chemistry*, **17**, 213.
46. Moriarty, R.M. and Prakash, O. (1998) *Advances in Heterocyclic Chemistry*, **69**, 1.
47. Zhdankin, V.V. (1997) *Reviews on Heteroatom Chemistry*, **17**, 133.
48. Grushin, V.V. (2000) *Chemical Society Reviews*, **29**, 315.

49. Pirkuliev, N.S., Brel, V.K. and Zefirov, N.S. (2000) *Russian Chemical Reviews*, **69**, 105.
50. Koser, G.F. (2001) *Aldrichimica Acta*, **34**, 89.
51. Skulski, L. (2000) *Molecules*, **5**, 1331.
52. Pohnert, G. (2000) *Journal für Praktische Chemie*, **342**, 731.
53. Okuyama, T. (2002) *Accounts of Chemical Research*, **35**, 12.
54. Ochiai, M. (2000) *Journal of Organometallic Chemistry*, **611**, 494.
55. Togo, H. and Katohgi, M. (2001) *Synlett*, 565.
56. Zhdankin, V.V. and Stang, P.J. (2002) *Chemical Reviews*, **102**, 2523.
57. Zhdankin, V.V. (2002) *Speciality Chemicals Magazine*, **22**, 38.
58. Togo, H. and Sakuratani, K. (2002) *Synlett*, 1966.
59. Moreno, I., Tellitu, I., Herrero, M.T., *et al.* (2002) *Current Organic Chemistry*, **6**, 1433.
60. Morales-Rojas, H. and Moss, R.A. (2002) *Chemical Reviews*, **102**, 2497.
61. Moore, J.D. and Hanson, P.R. (2002) *Chemtracts*, **15**, 74.
62. Stang, P.J. (2003) *Journal of Organic Chemistry*, **68**, 2997.
63. Dauban, P. and Dodd, R.H. (2003) *Synlett*, 1571.
64. Feldman, K.S. (2003) *ARKIVOC*, (vi), 179.
65. Feldman, K.S. (2004) in *Strategies and Tactics in Organic Synthesis*, vol. **4** (ed. M. Harmata), Elsevier, London, p. 133.
66. Koser, G.F. (2004) *Advances in Heterocyclic Chemistry*, **86**, 225.
67. Tohma, H. and Kita, Y. (2004) *Advanced Synthesis & Catalysis*, **346**, 111.
68. Yoneda, N. (2004) *Journal of Fluorine Chemistry*, **125**, 7.
69. French, A.N., Bissmire, S. and Wirth, T. (2004) *Chemical Society Reviews*, **33**, 354.
70. Quideau, S., Pouysegu, L. and Deffieux, D. (2004) *Current Organic Chemistry*, **8**, 113.
71. Rodriguez, S. and Wipf, P. (2004) *Synthesis*, 2767.
72. Muller, P. (2004) *Accounts of Chemical Research*, **37**, 243.
73. Moriarty, R.M. (2005) *Journal of Organic Chemistry*, **70**, 2893.
74. Wirth, T. (2005) *Angewandte Chemie, International Edition*, **44**, 3656.
75. Zhdankin, V.V. (2005) *Current Organic Synthesis*, **2**, 121.
76. Okuyama, T. and Fujita, M. (2005) *Russian Journal of Organic Chemistry*, **41**, 1245.
77. Kirmse, W. (2005) *European Journal of Organic Chemistry*, 237.
78. Matveeva, E.D., Proskurnina, M.V. and Zefirov, N.S. (2006) *Heteroatom Chemistry*, **17**, 595.
79. Ochiai, M. (2006) *Coordination Chemistry Reviews*, **250**, 2771.
80. Richardson, R.D. and Wirth, T. (2006) *Angewandte Chemie, International Edition*, **45**, 4402.
81. Silva, L.F., Jr. (2006) *Molecules*, **11**, 421.
82. Ladziata, U. and Zhdankin, V.V. (2006) *ARKIVOC*, (ix), 26.
83. Muller, P., Allenbach, Y.F., Chappellet, S. and Ghanem, A. (2006) *Synthesis*, 1689.
84. Zhdankin, V.V. (2007) *Science of Synthesis*, **31a**, 161.
85. Deprez, N.R. and Sanford, M.S. (2007) *Inorganic Chemistry*, **46**, 1924.
86. Ladziata, U. and Zhdankin, V.V. (2007) *Synlett*, 527.
87. Ochiai, M. (2007) *The Chemical Record*, **7**, 12.
88. Ciufolini, M.A., Braun, N.A., Canesi, S., *et al.* (2007) *Synthesis*, 3759.
89. Kita, Y. and Fujioka, H. (2007) *Pure and Applied Chemistry*, **79**, 701.
90. Okuyama, T. and Fujita, M. (2007) *ACS Symposium Series*, **965**, 68.
91. Holsworth, D.D. (2007) in *Name Reactions for Functional Group Transformations* (eds J.J. Li and E.J. Corey), John Wiley & Sons, Inc., Hoboken, N.J., p. 218.
92. Quideau, S., Pouysegu, L. and Deffieux, D. (2008) *Synlett*, 467.
93. Frohn, H.-J., Hirschberg, M.E., Wenda, A. and Bardin, V.V. (2008) *Journal of Fluorine Chemistry*, **129**, 459.
94. Zhdankin, V.V. and Stang, P.J. (2008) *Chemical Reviews*, **108**, 5299.
95. Ochiai, M. and Miyamoto, K. (2008) *European Journal of Organic Chemistry*, 4229.
96. Dohi, T. and Kita, Y. (2009) *Chemical Communications*, 2073.
97. Uyanik, M. and Ishihara, K. (2009) *Chemical Communications*, 2086.

98. Zhdankin, V.V. (2009) *ARKIVOC*, (i), 1.
99. Merritt, E.A. and Olofsson, B. (2009) *Angewandte Chemie, International Edition*, **48**, 9052.
100. Ngatimin, M. and Lupton, D.W. (2010) *Australian Journal of Chemistry*, **63**, 653.
101. Yusubov, M.S., Nemykin, V.N. and Zhdankin, V.V. (2010) *Tetrahedron*, **66**, 5745.
102. Satam, V., Harad, A., Rajule, R. and Pati, H. (2010) *Tetrahedron*, **66**, 7659.
103. Pouysegu, L., Deffieux, D. and Quideau, S. (2010) *Tetrahedron*, **66**, 2235.
104. Dohi, T. (2010) *Chemical & Pharmaceutical Bulletin*, **58**, 135.
105. Kotali, A., Kotali, E., Lafazanis, I.S. and Harris, P.A. (2010) *Current Organic Synthesis*, **7**, 62.
106. Liang, H. and Ciufolini, M.A. (2010) *Tetrahedron*, **66**, 5884.
107. Dohi, T., Yamaoka, N. and Kita, Y. (2010) *Tetrahedron*, **66**, 5775.
108. Quideau, S. and Wirth, T. (2010) *Tetrahedron*, **66**, 5737.
109. Yusubov, M.S. and Zhdankin, V.V. (2010) *Mendeleev Communications*, **20**, 185.
110. Zhdankin, V.V. (2011) *Journal of Organic Chemistry*, **76**, 1185.
111. Merritt, E.A. and Olofsson, B. (2011) *Synthesis*, 517.
112. Brand, J.P., Gonzalez, D.F., Nicolai, S. and Waser, J. (2011) *Chemical Communications*, **47**, 102.
113. Silva, J.L.F. and Olofsson, B. (2011) *Natural Product Reports*, **28**, 1722.
114. Duschek, A. and Kirsch, S.F. (2011) *Angewandte Chemie, International Edition*, **50**, 1524.
115. Turner, C.D. and Ciufolini, M.A. (2011) *ARKIVOC*, (i), 410.
116. Liang, H. and Ciufolini, M.A. (2011) *Angewandte Chemie, International Edition*, **50**, 11849.
117. Yusubov, M.S., Maskaev, A.V. and Zhdankin, V.V. (2011) *ARKIVOC*, (i), 370.
118. Yusubov, M.S. and Zhdankin, V.V. (2012) *Current Organic Synthesis*, **9**, 247.
119. Brand, J.P. and Waser, J. (2012) *Chemical Society Reviews*, **41**, 4165.
120. Fernandez Gonzalez, D., Benfatti, F. and Waser, J. (2012) *ChemCatChem*, **4**, 955.
121. Brown, M., Farid, U. and Wirth, T. (2013) *Synlett*, **24**, 424.
122. Finkbeiner, P. and Nachtsheim, B.J. (2013) *Synthesis*, **45**, 979.
123. Wirth, T. (ed.) (2003) *Hypervalent Iodine Chemistry: Modern Developments in Organic Synthesis*, Topics in Current Chemistry, vol. **224**, Springer Verlag, Berlin.
124. Martin, J.C. (1983) *Science*, **221**, 509.
125. Perkins, C.W., Martin, J.C., Arduengo, A.J., *et al.* (1980) *Journal of the American Chemical Society*, **102**, 7753.
126. Powell, W.H. (1984) *Pure and Applied Chemistry*, **56**, 769.
127. Ochiai, M. (2003) in *Hypervalent Iodine Chemistry: Modern Developments in Organic Synthesis* (ed. T. Wirth), Topics in Current Chemistry, vol. **224**, Springer Verlag, Berlin, p. 5.
128. IUPAC, (1979) *Nomenclature of Organic Chemistry, Sections A, B, C, D, E, F, and H*, Pergamon Press, Oxford.
129. Musher, J.I. (1969) *Angewandte Chemie, International Edition in English*, **8**, 54.
130. Pimentel, G.C. (1951) *Journal of Chemical Physics*, **19**, 446.
131. Hach, R.J. and Rundle, R.E. (1951) *Journal of the American Chemical Society*, **73**, 4321.
132. Reed, A.E. and Schleyer, P.v.R. (1990) *Journal of the American Chemical Society*, **112**, 1434.
133. Landrum, G.A., Goldberg, N. and Hoffmann, R. (1997) *Journal of the Chemical Society, Dalton Transactions*, 3605.
134. Su, J.T. and Goddard, W.A. III (2005) *Journal of the American Chemical Society*, **127**, 14146.
135. Ochiai, M., Sueda, T., Miyamoto, K. *et al.* (2006) *Angewandte Chemie, International Edition*, **45**, 8203.
136. Sajith, P.K. and Suresh, C.H. (2012) *Inorganic Chemistry*, **51**, 967.
137. Minyaev, R.M. and Minkin, V.I. (2000) *Mendeleev Communications*, **10**, 173.
138. Cleveland, T. and Landis, C.R. (1996) *Journal of the American Chemical Society*, **118**, 6020.
139. Firman, T.K. and Landis, C.R. (1998) *Journal of the American Chemical Society*, **120**, 12650.
140. Landis, C.R., Cleveland, T. and Firman, T.K. (1998) *Journal of the American Chemical Society*, **120**, 2641.
141. Landis, C.R., Firman, T.K., Root, D.M. and Cleveland, T. (1998) *Journal of the American Chemical Society*, **120**, 1842.
142. Landis, C.R., Cleveland, T. and Firman, T.K. (1995) *Journal of the American Chemical Society*, **117**, 1859.
143. Kiprof, P. (2005) *ARKIVOC*, (iv), 19.

144. Downs, A.J. and Adams, C.J. (1973) in *Comprehensive Inorganic Chemistry*, vol. **2** (eds J.C. Bailar, H.J. Emeleus, R. Nyholm and A.F. Trotman-Dickenson), Pergamon Press, Oxford, p. 1107.
145. Christe, K.O., Curtis, E.C. and Dixon, D.A. (1993) *Journal of the American Chemical Society*, **115**, 9655.
146. Christe, K.O., Dixon, D.A., Mahjoub, A.R., *et al.* (1993) *Journal of the American Chemical Society*, **115**, 2696.
147. Christe, K.O., Curtis, E.C. and Dixon, D.A. (1993) *Journal of the American Chemical Society*, **115**, 1520.
148. Archer, E.M. and van Schalkwyk, T.G.D. (1953) *Acta Crystallographica*, **6**, 88.
149. Alcock, N.W. and Countryman, R.M. (1979) *Journal of the Chemical Society, Dalton Transactions*, 851.
150. Landrum, G.A., Goldberg, N., Hoffmann, R. and Minyaev, R.M. (1998) *New Journal of Chemistry*, **22**, 883.
151. Landrum, G.A. and Hoffmann, R. (1998) *Angewandte Chemie, International Edition*, **37**, 1887.
152. Zhdankin, V.V., Koposov, A.E., Smart, J.T., *et al.* (2001) *Journal of the American Chemical Society*, **123**, 4095.
153. Nemykin, V.N., Koposov, A.Y., Netzel, B.C., *et al.* (2009) *Inorganic Chemistry*, **48**, 4908.
154. Macikenas, D., Skrzypczak-Jankun, E. and Protasiewicz, J.D. (1999) *Journal of the American Chemical Society*, **121**, 7164.
155. Macikenas, D., Skrzypczak-Jankun, E. and Protasiewicz, J.D. (2000) *Angewandte Chemie, International Edition*, **39**, 2007.
156. Zhdankin, V.V., Koposov, A.Y., Litvinov, D.N., *et al.* (2005) *Journal of Organic Chemistry*, **70**, 6484.
157. Zhdankin, V.V., Koposov, A.Y., Netzel, B.C., *et al.* (2003) *Angewandte Chemie, International Edition*, **42**, 2194.
158. Mailyan, A.K., Geraskin, I.M., Nemykin, V.N. and Zhdankin, V.V. (2009) *Journal of Organic Chemistry*, **74**, 8444.
159. Ladziata, U., Koposov, A.Y., Lo, K.Y., *et al.* (2005) *Angewandte Chemie, International Edition*, **44**, 7127.
160. Batchelor, R.J., Birchall, T. and Sawyer, J.F. (1986) *Inorganic Chemistry*, **25**, 1415.
161. Zhdankin, V.V., Maydanovych, O., Herschbach, J., *et al.* (2002) *Journal of the American Chemical Society*, **124**, 11614.
162. Zhdankin, V.V., Koposov, A.Y., Su, L.S., *et al.* (2003) *Organic Letters*, **5**, 1583.
163. Yusubov, M.S., Funk, T.V., Chi, K.-W., *et al.* (2008) *Journal of Organic Chemistry*, **73**, 295.
164. Nemykin, V.N., Maskaev, A.V., Geraskina, M.R., *et al.* (2011) *Inorganic Chemistry*, **50**, 11263.
165. Koposov, A.Y., Litvinov, D.N., Zhdankin, V.V., *et al.* (2006) *European Journal of Organic Chemistry*, 4791.
166. Dess, D.B., Wilson, S.R. and Martin, J.C. (1993) *Journal of the American Chemical Society*, **115**, 2488.
167. Dess, D.B. and Martin, J.C. (1991) *Journal of the American Chemical Society*, **113**, 7277.
168. Koposov, A.Y., Nemykin, V.N. and Zhdankin, V.V. (2005) *New Journal of Chemistry*, **29**, 998.
169. Zhdankin, V.V., Litvinov, D.N., Koposov, A.Y., *et al.* (2004) *Chemical Communications*, 106.
170. Koposov, A.Y., Karimov, R.R., Geraskin, I.M., *et al.* (2006) *Journal of Organic Chemistry*, **71**, 8452.
171. Nikiforov, V.A., Karavan, V.S., Miltsov, S.A., *et al.* (2003) *ARKIVOC*, (vi), 191.
172. Stevenson, P.J., Treacy, A.B. and Nieuwenhuyzen, M. (1997) *Journal of the Chemical Society, Perkin Transactions 2*, 589.
173. Yoshimura, A., Banek, C.T., Yusubov, M.S., *et al.* (2011) *Journal of Organic Chemistry*, **76**, 3812.
174. Zhdankin, V.V., Nemykin, V.N., Karimov, R.R. and Kazhkenov, Z.-G. (2008) *Chemical Communications*, 6131.
175. Ochiai, M., Miyamoto, K., Shiro, M., *et al.* (2003) *Journal of the American Chemical Society*, **125**, 13006.
176. Ochiai, M., Miyamoto, K., Suefuji, T., *et al.* (2003) *Angewandte Chemie, International Edition*, **42**, 2191.
177. Ochiai, M., Miyamoto, K., Suefuji, T., *et al.* (2003) *Tetrahedron*, **59**, 10153.
178. Ochiai, M., Miyamoto, K., Yokota, Y., *et al.* (2005) *Angewandte Chemie, International Edition*, **44**, 75.
179. Ochiai, M., Suefuji, T., Miyamoto, K., *et al.* (2003) *Journal of the American Chemical Society*, **125**, 769.
180. Suefuji, T., Shiro, M., Yamaguchi, K. and Ochiai, M. (2006) *Heterocycles*, **67**, 391.
181. Ochiai, M., Suefuji, T., Miyamoto, K. and Shiro, M. (2003) *Chemical Communications*, 1438.
182. Zhdankin, V.V., Koposov, A.Y. and Yashin, N.V. (2002) *Tetrahedron Letters*, **43**, 5735.
183. Richter, H.W., Koser, G.F., Incarvito, C.D. and Rheingold, A.L. (2007) *Inorganic Chemistry*, **46**, 5555.
184. Koposov, A.Y., Netzel, B.C., Yusubov, M.S., *et al.* (2007) *European Journal of Organic Chemistry*, 4475.
185. Boucher, M., Macikenas, D., Ren, T. and Protasiewicz, J.D. (1997) *Journal of the American Chemical Society*, **119**, 9366.
186. Meprathu, B.V. and Protasiewicz, J.D. (2003) *ARKIVOC*, (vi), 83.
187. Meprathu, B.V., Justik, M.W. and Protasiewicz, J.D. (2005) *Tetrahedron Letters*, **46**, 5187.
188. Yoshimura, A., Nemykin, V.N. and Zhdankin, V.V. (2011) *Chemistry: A European Journal*, **17**, 10538.

189. Reich, H.J. and Cooperman, C.S. (1973) *Journal of the American Chemical Society*, **95**, 5077.
190. Ochiai, M., Takaoka, Y., Masaki, Y., *et al.* (1990) *Journal of the American Chemical Society*, **112**, 5677.
191. Amey, R.L. and Martin, J.C. (1979) *Journal of Organic Chemistry*, **44**, 1779.
192. Cerioni, G. and Uccheddu, G. (2004) *Tetrahedron Letters*, **45**, 505.
193. Mocci, F., Uccheddu, G., Frongia, A. and Cerioni, G. (2007) *Journal of Organic Chemistry*, **72**, 4163.
194. Fusaro, L., Mocci, F., Luhmer, M. and Cerioni, G. (2012) *Molecules*, **17**, 12718.
195. Hiller, A., Patt, J.T. and Steinbach, J. (2006) *Magnetic Resonance in Chemistry*, **44**, 955.
196. Silva, L.F. and Lopes, N.P. (2005) *Tetrahedron Letters*, **46**, 6023.
197. Vasconcelos, R.S., Silva, L.F., Jr. and Lopes, N.P. (2012) *Quimica Nova*, **35**, 1593.
198. Richter, H.W., Cherry, B.R., Zook, T.D. and Koser, G.F. (1997) *Journal of the American Chemical Society*, **119**, 9614.
199. Kiprof, P. and Zhdankin, V. (2003) *ARKIVOC*, (vi), 170.
200. Mylonas, V.E., Sigalas, M.P., Katsoulos, G.A., *et al.* (1994) *Journal of the Chemical Society, Perkin Transactions 2*, 1691.
201. Okuyama, T. and Yamataka, H. (1999) *Canadian Journal of Chemistry*, **77**, 577.
202. Carroll, M.A., Martin-Santamaria, S., Pike, V.W., *et al.* (1999) *Journal of the Chemical Society, Perkin Transactions 2*, 2707.
203. Martin-Santamaria, S., Carroll, M.A., Carroll, C.M., *et al.* (2000) *Chemical Communications*, 649.
204. Martin-Santamaria, S., Carroll, M.A., Pike, V.W., *et al.* (2000) *Journal of the Chemical Society, Perkin Transactions 2*, 2158.
205. Bakalbassis, E.G., Spyroudis, S. and Tsiotra, E. (2006) *Journal of Organic Chemistry*, **71**, 7060.
206. Pouysegu, L., Chassaing, S., Dejugnac, D., *et al.* (2008) *Angewandte Chemie, International Edition*, **47**, 3552.
207. Zhdankin, V.V., Arbit, R.M., Lynch, B.J., *et al.* (1998) *Journal of Organic Chemistry*, **63**, 6590.
208. Hirt, U.H., Schuster, M.F.H., French, A.N., *et al.* (2001) *European Journal of Organic Chemistry*, 1569.
209. Edwards, A.J. (1978) *Journal of the Chemical Society, Dalton Transactions*, 1723.
210. Okuyama, T., Takino, T., Sueda, T. and Ochiai, M. (1995) *Journal of the American Chemical Society*, **117**, 3360.
211. Grushin, V.V., Demkina, I.I. and Tolstaya, T.P. (1992) *Journal of the Chemical Society, Perkin Transactions 2*, 505.
212. Kita, Y., Tohma, H., Inagaki, M., *et al.* (1991) *Tetrahedron Letters*, **32**, 4321.
213. Kita, Y., Tohma, H., Hatanaka, K., *et al.* (1994) *Journal of the American Chemical Society*, **116**, 3684.
214. Hata, K., Hamamoto, H., Shiozaki, Y., *et al.* (2007) *Tetrahedron*, **63**, 4052.
215. Dohi, T., Ito, M., Yamaoka, N., *et al.* (2010) *Angewandte Chemie, International Edition*, **49**, 3334.
216. Nicolaou, K.C., Baran, P.S., Kranich, R., *et al.* (2001) *Angewandte Chemie, International Edition*, **40**, 202.
217. Nicolaou, K.C., Baran, P.S. and Zhong, Y.-L. (2001) *Journal of the American Chemical Society*, **123**, 3183.

2

Preparation, Structure and Properties of Polyvalent Iodine Compounds

Two general approaches to the synthesis of polyvalent iodine compounds exist: the first is based on the oxidative addition of appropriate ligands to a low-valent iodine species (e.g., I_2 or PhI) and the second is based on ligand exchange in polyvalent iodine compounds. The first approach is generally used to prepare common polyvalent iodine compounds by the oxidation of readily available and cheap precursors with an appropriate oxidant. This approach, in particular, is employed for large-scale preparation of the most important hypervalent iodine reagents, such as (dichloroiodo)benzene, (diacetoxyiodo)benzene and 2-iodoxybenzoic acid (IBX), from the corresponding iodoarenes and appropriate oxidants. Once formed, λ^3- and λ^5-iodanes can readily exchange their ligands by treatment with appropriate nucleophiles. The ligand exchange approach is commonly used for the preparation of a broad variety of λ^3- and λ^5-iodanes, aryliodonium salts and iodonium ylides and imides.

Only several classes of inorganic polyvalent iodine compounds are known: polyvalent iodine fluorides, chlorides, oxides and the derivatives of iodic and periodic acid. Most of the known λ^3- and λ^5-iodanes are organic derivatives with one or two carbon ligands at the iodine, while derivatives of polyvalent iodine with three carbon ligands, R_3I, in general have low thermal stability and cannot be isolated. The overwhelming majority of organic λ^3- and λ^5-iodanes have a benzene ring as a carbon ligand linked to the iodine atom. Derivatives of polyvalent iodine with an alkyl substituent at iodine are highly unstable and generally can exist only as short-lived reactive intermediates in the oxidations of alkyl iodides. However, introduction of an electron-withdrawing substituent into the alkyl moiety leads to significant stabilization of the molecule. Typical representatives of such stabilized compounds with I–Csp^3-bonding are perfluoroalkyl λ^3-iodanes (R_fIL_2), numerous examples of which have been prepared and characterized.

2.1 Iodine(III) Compounds

2.1.1 Inorganic Iodine(III) Derivatives

The known classes of iodine(III) compounds without carbon ligands are represented by the iodine(III) halides and by the derivatives of unstable iodine(III) oxide, I_2O_3, of types OIOR or $I(OR)_3$. Table 2.1 summarizes the known inorganic iodine(III) compounds.

Hypervalent Iodine Chemistry: Preparation, Structure and Synthetic Applications of Polyvalent Iodine Compounds, First Edition. Viktor V. Zhdankin.

Table 2.1 Iodine(III) compounds without carbon ligands.

Compound	Method of synthesis	Properties	Reference
IF_3	$I_2 + F_2$/Ar mixture, –45 °C	Decomposes at –28 °C	[1]
ICl_3	I_2 + liquid Cl_2, –78 °C	Decomposes at 47–62 °C	[2,7]
$KICl_4$	KI/water + Cl_2 gas	Stable golden solid, mp 115 °C	[3]
$(IO)_2SO_4$	$I_2 + I_2O_5$ in conc. H_2SO_4	Stable yellow solid	[4]
$(IO)_2SeO_4$	$I_2 + I_2O_5$ in conc. H_2SeO_4	Stable yellow solid	[4]
$OIOSO_2F$	$I_2 + I_2O_5$ in $HOSO_2F$	Hygroscopic yellow solid	[5]
OIOTf	$I_2 + I_2O_5$ in HOTf, rt	Hygroscopic yellow solid	[5]
$I(OTf)_3$	$I(OCOCF_3)_3$ + HOTf	Colorless solid	[6]
$I(OAc)_3$	I_2 + conc. $HNO_3 + Ac_2O$	Colorless crystals	[6]
$I(OCOCF_3)_3$	$I_2 + CF_3CO_3H$	Colorless solid	[7]

2.1.1.1 Iodine(III) Halides

Iodine(III) halides in general lack stability. Of the four known binary iodine fluorides (IF, IF_3, IF_5 and IF_7) iodine trifluoride is the least stable with a decomposition temperature of –28 °C, as established by differential thermogravimetry [8]. Even at low temperatures IF_3 readily disproportionates to IF_5 and IF or I_2 [9]. However, if iodine is treated with diluted elemental fluorine at low temperatures, iodine trifluoride can be obtained free of IF_5 as an unstable yellow solid insoluble in conventional solvents [10]. Hoyer and Seppelt were able to grow crystals of IF_3 from anhydrous hydrogen fluoride in the presence of traces of water and to perform a single-crystal X-ray structure determination [1]. In crystal form, iodine trifluoride has a polymeric structure **1** (Figure 2.1), assembled from planar distorted T-shaped molecules with one primary I—F_{eq} bond distance of 1.872(4) Å and two I—F_{ax} bonds of 1.983(3) Å each, which have an F_{ax}—I—F_{ax} bond angle of 160.3(2)°. Each iodine atom is linked to a neighboring IF_3 molecule by two intermolecular I···F secondary bonds of 2.769(3) Å, so that the resulting coordination polyhedron around the iodine atom is a planar pentagon [1].

Several computational structural studies of iodine trifluoride have been published [11–16]. According to *ab initio* calculations [16], IF_3 has a distorted T-shaped geometry with the axial I—F bond distance of 1.971 Å, the equatorial I—F bond distance of 1.901 Å and an F_{ax}—I—F_{eq} angle of 81.7°.

The chemical properties of iodine trifluoride are almost unknown. IF_3 forms 1 : 1 complexes with pyrazine or 2,2′-bipyridyl and reacts with CsF in a 1 : 3 molar ratio to give Cs_3IF_6 [8]. The ligand exchange reaction of IF_3 with trifluoroacetic anhydride leading to iodine(III) trifluoroacetates has been reported [17].

Iodine trichloride, ICl_3, is usually prepared by a low-temperature reaction of iodine with liquid chlorine by the method of Booth and Morris [2]. It is obtained in the form of a fluffy orange solid that easily decomposes

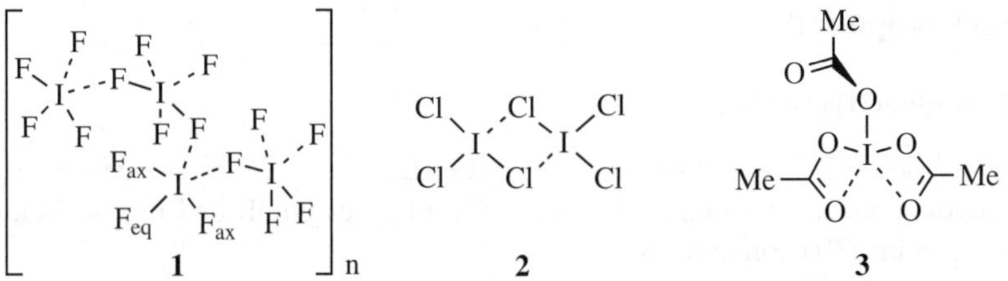

Figure 2.1 Primary and secondary bonding pattern in single-crystal X-ray structures of IF_3 (**1**), I_2Cl_6 (**2**) and $I(OAc)_3$ (**3**).

to ICl and Cl_2 at elevated temperatures. As established by a single-crystal X-ray analysis, iodine trichloride in the solid state exists as a dimer, I_2Cl_6 (**2**, Figure 2.1), with a planar structure containing two bridging I—Cl···I bonds (I—Cl distances 2.68 and 2.72 Å) and four terminal I—Cl bonds (2.38–2.39 Å) [18]. Iodine trichloride forms several addition products, which can be regarded as salts of the acid $HICl_4$. The salts of alkali metals (e.g., $KICl_4$) and ammonia are best prepared by adding chlorine into an aqueous solution of the respective iodide [3].

Iodine tribromide, IBr_3, is unstable and cannot be isolated as an individual compound. A study of solutions of iodine and bromine in hydrobromic acid by electrometric titrations provided experimental evidence for the existence of IBr_3 in the solution [19].

2.1.1.2 Derivatives of Iodine(III) Oxide

The parent iodine(III) oxide, I_2O_3, is unknown; however, several its inorganic derivatives of types OIOR or $I(OR)_3$ have been reported in the literature. Historically, the first of these derivatives was iodosyl sulfate, $(IO)_2SO_4$, which was first isolated as early as 1844 [20]. Iodosyl sulfate and the selenate, $(IO)_2SeO_4$, can be prepared by the interaction of iodine with iodine pentoxide in concentrated sulfuric or selenic acid [4, 21, 22]. A convenient procedure for the preparation of iodosyl sulfate by heating iodine and sodium metaperiodate, $NaIO_4$, in concentrated sulfuric acid was reported by Kraszkiewicz and Skulski in 2008 [23]. X-Ray crystallographic analysis of iodosyl sulfate shows a polymeric structure with infinite $(–O–I^+–O–)_n$ spiral chains linked by SO_4 tetrahedra [24]. Studies of $(IO)_2SO_4$ by IR and Raman spectroscopy in the solid state [22] and by cryoscopic and conductometric measurements of the solution in sulfuric acid [25], were also reported.

Iodosyl fluorosulfate, $OIOSO_2F$ and the triflate, OIOTf, can be prepared as thermally stable, hygroscopic yellow solids by the reaction of iodine with iodine pentoxide or iodic acid in fluorosulfonic or trifluoromethanesulfonic acids, respectively [5]. Raman and infrared spectra of these compounds indicate a polymeric structure analogous to iodosyl sulfate [5]. Iodine tris(fluorosulfate), $I(OSO_2F)_3$ and tris(triflate), $I(OTf)_3$, are also known [6, 26]. $I(OSO_2F)_3$ can be prepared by the reaction of iodine with peroxydisulfuryl difluoride [26]. Salts such as $KI(OSO_2F)_4$ have also been prepared and investigated by Raman spectroscopy [26, 27]. $I(OTf)_3$ was prepared from iodine tris(trifluoroacetate) and trifluoromethanesulfonic acid [6].

Several iodine(III) tris(carboxylate) derivatives, $I(O_2CR)_3$, where R = CH_3, CH_2Cl and CF_3, have been reported in the literature [6]. These compounds are best synthesized by the oxidation of iodine with fuming nitric acid in the presence of the appropriate carboxylic acid and acetic anhydride. Birchall and coworkers reported an X-ray crystal and molecular structure of $I(OAc)_3$ [6]. The geometry about iodine in this compound consists of primary bonds to the three acetate groups (I—O distances 2.159, 2.023 and 2.168 Å) and two strong intramolecular secondary bonds (I···O distances 2.463 and 2.518 Å) to two of the acetate groups, forming a pentagonal-planar arrangement **3** (Figure 2.1). An alternative method for the generation of $I(OAc)_3$ in solution involves the treatment of iodine trichloride with silver acetate in dry acetic acid [28]. $I(O_2CCF_3)_3$ can be prepared similarly from CF_3CO_2Ag and ICl_3 in 90% yield or by the oxidation of iodine with peroxytrifluoroacetic acid in 80% yield [7].

2.1.2 Organoiodine(III) Fluorides

(Difluoroiodo)arenes, $ArIF_2$, can be prepared by two general approaches: (i) oxidative addition of fluorine to iodoarenes using powerful fluorinating reagents and (ii) ligand exchange in iodine(III) compounds, such as ArIO or $ArICl_2$, using HF or SF_4 as a source of fluoride anions. Table 2.2 summarizes the preparation methods for organic iodine(III) difluorides.

Table 2.2 *Preparation of organic iodine(III) difluorides.*

Compound	Method of synthesis	Yield (%)	Reference
$PhIF_2$	PhI, XeF_2, anhyd. HF, CH_2Cl_2, room temp (rt), 1–3 h	95	[29, 37]
$PhIF_2$	PhIO, 46% aq. HF, CH_2Cl_2, rt	86	[30]
$3\text{-}ClC_6H_4IF_2$	ArI, XeF_2, anhyd. HF, CH_2Cl_2, rt, 1–3 h	95	[29]
$3\text{-}NO_2C_6H_4IF_2$	ArI, XeF_2, anhyd. HF, CH_2Cl_2, rt, 1–3 h	95	[29]
$3\text{-}MeOC_6H_4IF_2$	ArI, XeF_2, anhyd. HF, CH_2Cl_2, rt, 1–3 h	95	[29]
$4\text{-}MeOC_6H_4IF_2$	ArI, XeF_2, anhyd. HF, CH_2Cl_2, rt, 1–3 h	95	[29]
$4\text{-}MeC_6H_4IF_2$	ArIO, 46% aq. HF, CH_2Cl_2, rt	86	[30, 31]
$4\text{-}ClC_6H_4IF_2$	ArIO, 46% aq. HF, CH_2Cl_2, rt	79	[30]
$4\text{-}NO_2C_6H_4IF_2$	ArIO, 46% aq. HF, CH_2Cl_2, rt	85	[30]
$2,6\text{-}F_2C_6H_3IF_2$	ArI, F_2/N_2, CCl_3F, –78 °C	Quantitative	[32]
4-*t*-Bu-$2,6\text{-}Me_2C_6H_2IF_2$	ArH, I_2, Selectfluor, MeCN	78	[33]
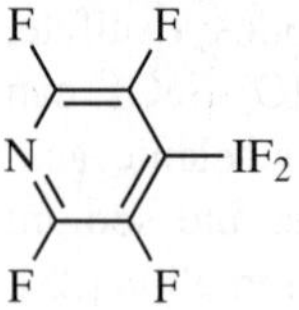 F F N IF_2 F F	RI, XeF_2, CH_2Cl_2, –78 °C to rt, 4 h	84	[34]
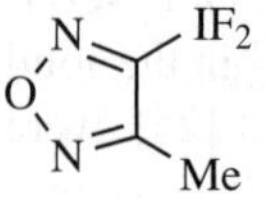 N IF_2 O N Me	RI, XeF_2, MeCN, rt	Quantitative	[35]
CH_3IF_2	CH_3I, XeF_2, no solvent, rt, 20 min	Quantitative	[36]
CF_3IF_2	CF_3I, CF_3OCl, –50 °C, 24 h	Not reported	[37]
 F_2I	RI, XeF_2, CCl_4, rt, 1 h	Quantitative	[38]

2.1.2.1 Preparation by Fluorination of Organic Iodides

Various fluorinating reagents have been used for the fluorination of iodoarenes. A very clean and selective, although relatively expensive, procedure for the preparation of (difluoroiodo)arenes **4** is based on the fluorination of iodoarenes with xenon difluoride in dichloromethane in the presence of anhydrous hydrogen fluoride (Scheme 2.1) [29, 39]. This method works well for the fluorination of iodoarenes with electron-donating or electron-withdrawing substituents; the latter, however, require longer reaction times. (Difluoroiodo)arenes (**4**) are hygroscopic and highly hydrolyzable compounds, which makes their separation and crystallization extremely difficult. Since xenon gas is the only byproduct in this reaction (Scheme 2.1), the

$$\mathrm{ArI} + \mathrm{XeF_2} \xrightarrow[-\,\mathrm{Xe}]{\mathrm{CH_2Cl_2,\ HF\ (anhyd),\ rt,\ 1\text{-}3\ h}} \underset{\mathbf{4}}{\mathrm{ArIF_2}}$$

$$\mathrm{Ar = Ph,\ 3\text{-}ClC_6H_4,\ 3\text{-}NO_2C_6H_4,\ 4\text{-}MeOC_6H_4,\ 3\text{-}MeOC_6H_4}$$

Scheme 2.1

resulting dichloromethane solutions contain essentially pure fluorides **4**, which can be used in the subsequent reactions without additional purification. A similar procedure, but in the absence of anhydrous hydrogen fluoride, has been employed in the synthesis of some heteroaromatic iododifluorides (Table 2.2). 4-(Difluoroiodo)-2,3,5,6-tetrafluoropyridine, 4-(C_5F_4N)IF_2, was prepared in high yield by the reaction of 2,3,5,6-tetrafluoro-4-iodopyridine with xenon difluoride in dichloromethane at room temperature [34]. Likewise, the fluorination of 3-iodo-4-methylfurazan with xenon difluoride in acetonitrile at room temperature was used to prepare 3-(difluoroiodo)-4-methylfurazan (Table 2.2) [35]. A relatively stable 4-(difluoroiodo)tricyclene was prepared in the form of a pale yellow solid by treatment of a solution of 4-iodotricyclene in carbon tetrachloride with an excess of xenon difluoride followed by removal of solvent (Table 2.2) [38].

Various other powerful fluorinating reagents, such as F_2, ClF, CF_3OCl, BrF_5, $C_6F_5BrF_2$, $C_6F_5BrF_4$, XeF_2/BF_3, can be used for the preparation of (difluoroiodo)arenes derived from polyfluoro-substituted iodoarenes [32, 40–42]. Frohn and coworkers investigated the preparation of $C_6F_5IF_2$ and other polyfluorinated (difluoroiodo)arenes by oxidative fluorination of the appropriate iodides using F_2, ClF, CF_3OCl, BrF_5, $C_6F_5BrF_2$, $C_6F_5BrF_4$ and XeF_2 [40, 43, 44]. The highest purity and yield of $C_6F_5IF_2$ was achieved by a low-temperature fluorination with F_2 [40]. The $C_6F_5IF_2$ prepared in this work was fully characterized by multinuclear NMR, IR, Raman spectroscopy and X-ray structural analysis [40]. Another preparation of $C_6F_5IF_2$ in high yield (97%) involved the reaction of IF_3 with $Cd(C_6F_5)_2$ in dichloromethane at –78 °C [44]. Naumann and coworkers prepared 2,6-$F_2C_6H_3IF_2$ in quantitative yield by oxidative fluorination of the corresponding aryl iodide with XeF_2 in acetonitrile or with F_2/N_2 mixtures in CCl_3F [32].

Zefirov, Brel and coworkers developed a procedure for the preparation of [fluoro(trifluoromethylsulfonyloxy)iodo]arenes, ArIF(OTf), by oxidative fluorination of iodoarenes with FXeOTf, which can be generated *in situ* from XeF_2 and triflic acid [45–49]. The analogous mesylate, PhIF(OMs), can be prepared from iodobenzene, XeF_2 and methanesulfonic acid by a similar procedure [50, 51].

Shreeve and coworkers reported a convenient procedure for preparing (difluoroiodo)arenes by direct fluorination of the respective iodoarenes with the commercially available fluorinating reagent Selectfluor® in acetonitrile solution. This procedure has been further improved by using the corresponding arene, elemental iodine and Selectfluor in a one-pot oxidative iodination/fluorination procedure [33].

para-Substituted (difluoroiodo)arenes can be effectively prepared by electrochemical fluorination of the respective iodoarenes [52–54]. In the procedure developed by Fuchigami and Fujita, the electrosynthesis of $ArIF_2$ is accomplished by the anodic oxidation of iodoarenes with $Et_3N{\cdot}3HF$ or $Et_3N{\cdot}5HF$ in anhydrous acetonitrile using a divided cell [52]. This procedure works especially well for the preparation of 4-$NO_2C_6H_4IF_2$, which precipitates from the electrolytic solution in pure form during the electrolysis. Other *para*-substituted (difluoroiodo)arenes, such as $TolIF_2$ and 4-$MeOC_6H_4IF_2$, can be used without isolation as in-cell mediators for subsequent reaction [52–55].

2.1.2.2 Preparation by Ligand Exchange

A classical procedure of Carpenter for the preparation of (difluoroiodo)arenes involves a one-step reaction of (dichloroiodo)arenes with yellow mercuric oxide and 48% aqueous hydrofluoric acid in dichloromethane [56]. The resulting solution of (difluoroiodo)arenes in dichloromethane can be used in subsequent reactions without additional purification. A drawback of this method is the use of a large quantity of harmful HgO to remove the chloride ion from the reaction mixture. A convenient modified procedure without the use of HgO consists of the treatment of iodosylarenes **5** with 40–46% aqueous hydrofluoric acid (Scheme 2.2) followed by crystallization of products **6** from hexane [30, 31]. It is important that the freshly prepared iodosylarenes **5** are used in this procedure.

The methods based on the use of hydrofluoric acid have several general disadvantages. First, (difluoroiodo)arenes are often hygroscopic and highly hydrolyzable compounds, which makes their separation from

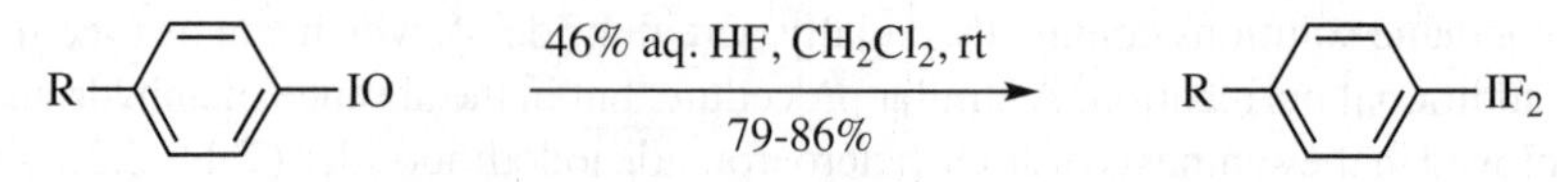

$$R-C_6H_4-IO \xrightarrow[79\text{-}86\%]{46\%\ aq.\ HF,\ CH_2Cl_2,\ rt} R-C_6H_4-IF_2$$

5 R = H, Me, Cl, NO_2 **6**

Scheme 2.2

aqueous hydrofluoric acid and crystallization extremely difficult. Second, the high acidity of the reaction media complicates reactions of base- and acid-sensitive substrates, such as heterocycles. Yagupolskii, Lyalin and coworkers developed a milder procedure based on the reaction of organic iodosyl compounds **7** with SF_4 under neutral conditions [57]. In this method, SF_4 is bubbled at –20 °C through a suspension of the iodosyl compound **7** in dichloromethane (Scheme 2.3). All the byproducts in this reaction are volatile, so evaporation of the solvent under anhydrous conditions affords organic iododifluorides **8** of high purity. Owing to the mild and non-acidic reaction conditions, this method is applicable to the synthesis of the pyridine and perfluoroalkyl derivatives **8** [57].

(Difluoroiodo)arenes are extremely sensitive to moisture and are commonly used as a freshly prepared solution, without isolation. DiMagno and coauthors reported a convenient procedure for almost quantitative generation of $PhIF_2$ in acetonitrile solution by the reaction of $PhI(OAc)_2$ with anhydrous tetrabutylammonium fluoride under absolutely dry conditions [58].

2.1.2.3 Structural Studies

Only several structural studies of organo-iododifluorides, RIF_2, have been reported in the literature. Single-crystal X-ray diffraction studies of trifluoromethyliododifluoride, CF_3IF_2, revealed a distorted T-shaped structure with the two fluorine atoms in the apical positions and the trifluoromethyl group in the equatorial position; I–C bond length 2.174(6) Å, I–F bond distances 1.982(2) Å and the F–I–F angle is 165.4(2)° [37]. The unit cell contains eight CF_3IF_2 molecules and each molecule has contacts to four adjacent molecules via I–F···I bridges resulting in a planar pentagonal coordination around the iodine atom (Figure 2.2). The lengths of all secondary F···I contacts are 2.950 Å [37], while the sum of the van der Waals radii of iodine and fluorine is significantly longer (3.45 Å [59]). Theoretical studies of CF_3IF_2 by *ab initio* and DFT (density functional theory) calculations have also been reported [60].

The X-ray crystal and molecular structures of 4-(difluoroiodo)toluene and 3-(difluoroiodo)nitrobenzene were reported in a PhD dissertation in 1996 [61]. More recently, Shreeve and coworkers have published single-crystal X-ray structures of two (difluoroiodo)arenes, $4\text{-}MeC_6H_4IF_2$ and $4\text{-}Bu^t\text{-}2,6\text{-}Me_2C_6H_2IF_2$ [33]. The single-crystal structure of $4\text{-}MeC_6H_4IF_2$ is similar to CF_3IF_2 and, in particular, has the same secondary bonding pattern resulting in a planar pentagonal coordination around the iodine atom. In contrast, in the crystal structure of $4\text{-}Bu^t\text{-}2,6\text{-}Me_2C_6H_2IF_2$ the iodine atoms are only four coordinated with a distorted square-planar

$$\underset{\mathbf{7}}{RIO} + SF_4 \xrightarrow[82\text{-}100\%]{CH_2Cl_2,\ -20\ to\ -10\ ^\circ C} \underset{\mathbf{8}}{RIF_2}$$

R = Ph, $4\text{-}MeC_6H_4$, $4\text{-}FC_6H_4$, $3\text{-}FC_6H_4$, $2\text{-}NO_2C_6H_4$, 2-pyridyl, C_6F_5, CF_3CF_2

Scheme 2.3

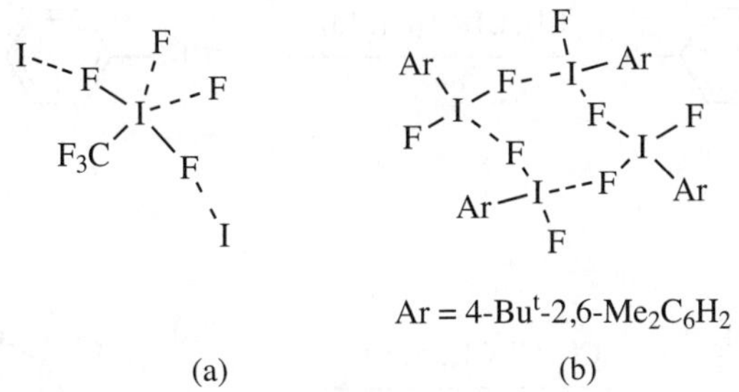

Ar = 4-Bu^t-2,6-$Me_2C_6H_2$

(a) (b)

Figure 2.2 *Primary and secondary bonding pattern in single-crystal X-ray structures of (a) CF_3IF_2 and (b) 4-Bu^t-2,6-$Me_2C_6H_2IF_2$.*

geometry. The intermolecular F···I interactions lead to self-assembly of four molecules of $ArIF_2$ into an unusual eight-membered chair ring formed by repeating F–I···F units (Figure 2.2) [33].

The structure of pentafluorophenyliododifluoride, $C_6F_5IF_2$, has been investigated by single-crystal X-ray crystallography and by multinuclear NMR, IR and Raman spectroscopy [40]. Frohn and coworkers reported the isolation and structural studies of adducts of several (difluoroiodo)arenes with nitrogen bases [62]. In particular, the adducts of Ar_fIF_2, ($Ar_f = C_6F_5, C_6F_2H_3, C_6F_3H_2$) with phenanthroline, 2,2′-bipyridine and quinoline were isolated and characterized by their single-crystal structure, Raman spectra and their multi-NMR spectra in the solution [62].

2.1.3 Organoiodine(III) Chlorides

Organic iodine(III) dichlorides, $RICl_2$, are usually prepared by direct chlorination of organic iodides, or, less commonly, by ligand exchange in other iodine(III) compounds. Table 2.3 summarizes the preparation methods for organic iodine(III) dichlorides.

2.1.3.1 Preparation by Chlorination of Organic Iodides

Historically, (dichloroiodo)benzene, $PhICl_2$, was the first reported organic compound of polyvalent iodine. It was prepared by Willgerodt in 1886 by the reaction of iodobenzene with ICl_3 or, preferably, with chlorine

Table 2.3 *Preparation of organic iodine(III) dichlorides.*

Compound	Method of synthesis	Yield (%)	Reference
$PhICl_2$	PhI, Cl_2, $CHCl_3$, 0 °C, 3 h	94	[63]
$PhICl_2$	PhI, 5.84% aq. NaOCl, conc. HCl, 15 °C	99	[64]
$PhICl_2$	PhI, 30% aq. H_2O_2, conc. HCl, CF_3CH_2OH, rt (room temp.)	89	[65]
4-$MeC_6H_4ICl_2$	ArI, Cl_2, hexane, 3 h, rt	92	[65]
3-$NO_2C_6H_4IF_2$	ArI, Cl_2, hexane, 19 h, rt	76	[65]
3-$HO_2CC_6H_4ICl_2$	ArI, Cl_2, $CHCl_3$, rt, 1 h	95	[66]
2,4,6-$Pr^i_3C_6H_2ICl_2$	ArI, Cl_2, $CHCl_3$, –10 °C, 1 h	86	[67]
$CF_3CH_2ICl_2$	CF_3CH_2I, Cl_2, no solvent, 0 °C, 2 h	85	[68]
$CH_3I(Cl)F$	CF_3I, CF_3OCl, –78 °C	Not reported	[69]
CF_3ICl_2	$CH_3I(Cl)F$, Me_3SiCl, –40 °C	Not reported	[70]
(*E*)-ClCH=CHICl$_2$	ICl_3, HC≡CH, conc. HCl, 0 °C to rt, 2 h	21	[71]

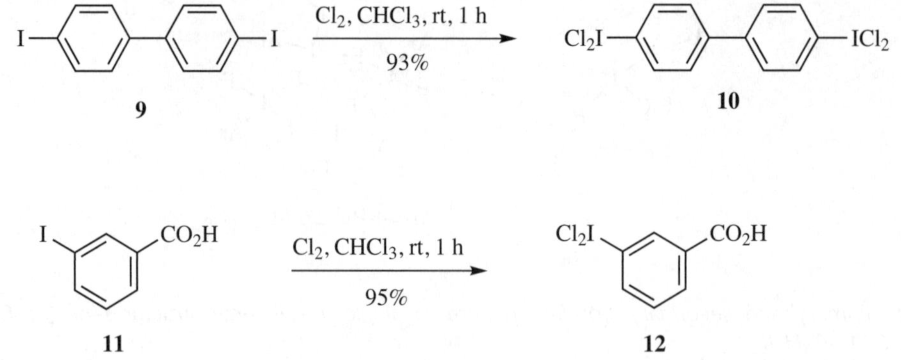

$$\mathrm{I{-}C_6H_4{-}C_6H_4{-}I}\ (\mathbf{9}) \xrightarrow[93\%]{Cl_2,\ CHCl_3,\ rt,\ 1\ h} \mathrm{Cl_2I{-}C_6H_4{-}C_6H_4{-}ICl_2}\ (\mathbf{10})$$

$$\text{3-}\mathrm{I{-}C_6H_4{-}CO_2H}\ (\mathbf{11}) \xrightarrow[95\%]{Cl_2,\ CHCl_3,\ rt,\ 1\ h} \text{3-}\mathrm{Cl_2I{-}C_6H_4{-}CO_2H}\ (\mathbf{12})$$

Scheme 2.4

[72]. Direct chlorination of aryl iodides with chlorine in chloroform or dichloromethane is the most general approach to (dichloroiodo)arenes [63].

This method can be applied to the large-scale (20–25 kg) preparation of $PhICl_2$ by the reaction of iodobenzene with chlorine at –3 to +4 °C in dichloromethane [73]. (Dichloroiodo)arenes are generally isolated as light- and heat-sensitive yellow crystalline solids, which are insufficiently stable for extended storage even at low temperatures.

The direct chlorination of iodoarenes **9** and **11** has been used for the preparation of 4,4′-bis(dichloroiodo)biphenyl (**10**) and 3-(dichloroiodo)benzoic acid (**12**) (Scheme 2.4), which are convenient recyclable hypervalent iodine reagents (Section 5.3) [66].

Gladysz and coworkers reported the synthesis of several fluorous aryl and alkyl iodine(III) dichlorides **14** in 71–98% yields by reactions of chlorine and the corresponding fluorous iodides **13** at room temperature in hexane or chloroform solutions (Scheme 2.5) [74]. A similar chlorination procedure was used to prepare $CF_3CH_2ICl_2$, $CF_3CF_2CH_2ICl_2$, $CF_3CF_2CF_2CH_2ICl_2$ and $H(CF_2)_6CH_2ICl_2$ by Montanari, DesMarteau and coworkers [68,75,76].

Alkyliodine(III) dichlorides **15** and **16**, which are stabilized due to the presence of the electron-withdrawing trialkylammonium or triphenylphosphonium groups, can be prepared as relatively stable, non-hygroscopic, light-yellow microcrystalline solids by the chlorination of corresponding iodomethyl phosphonium and ammonium salts (Scheme 2.6) [77].

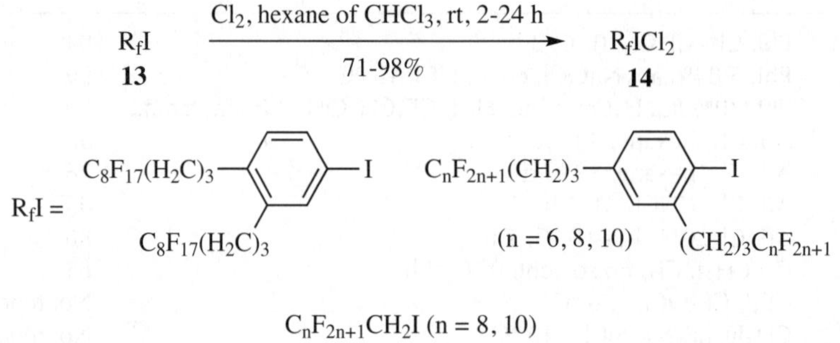

$$\underset{\mathbf{13}}{R_fI} \xrightarrow[71\text{-}98\%]{Cl_2,\ hexane\ of\ CHCl_3,\ rt,\ 2\text{-}24\ h} \underset{\mathbf{14}}{R_fICl_2}$$

R_fI = $C_8F_{17}(H_2C)_3$–C_6H_3(–$(CH_2)_3C_8F_{17}$)–I; $C_nF_{2n+1}(CH_2)_3$–C_6H_3(–$(CH_2)_3C_nF_{2n+1}$)–I (n = 6, 8, 10); $C_nF_{2n+1}CH_2I$ (n = 8, 10)

Scheme 2.5

$$\underset{BF_4^-}{R_3\overset{+}{X}CH_2I} \xrightarrow{Cl_2,\ CH_2Cl_2,\ -10\ ^\circ C} \underset{BF_4^-}{R_3\overset{+}{X}CH_2ICl_2}$$

15, $R_3X = Et_3N$
16, $R_3X = Ph_3P$

Scheme 2.6

Another example of alkyliodine(III) dichlorides stabilized by an electron-withdrawing substituent is represented by (dichloroiodo)methylsulfones, $ArSO_2CH_2ICl_2$, which are prepared similarly by chlorination of the appropriate iodomethylsulfones [78].

The preparation of several thermally unstable trifluoromethyliodine(III) chlorides has been reported. The first spectroscopic indication for the existence of CF_3ICl_2 was reported in 1976 by Naumann and coworkers in a study of the reaction of trifluoromethyl iodide, CF_3I, with chlorine nitrate, $ClONO_2$ [79]. The intermediate in this reaction is a mixed chloride–nitrate $CF_3I(Cl)ONO_2$, which cannot be isolated. More recently, Minkwitz and coworkers developed a better approach to trifluoromethyliodine(III) chlorides. Specifically, the reaction of CF_3I with CF_3OCl at –78 °C affords the mixed dihalide, $CF_3I(Cl)F$, which is unstable even at low temperatures and decomposes to the symmetrical halides, CF_3IF_2 and CF_3ICl_2 [69]. (Trifluoromethyl)iodine dichloride, CF_3ICl_2, can be obtained in high purity and yield by the reaction of $CF_3I(Cl)F$ with chlorotrimethylsilane at –40 °C [70].

To avoid the use of elemental chlorine, iodoarenes can be chlorinated *in situ* in aqueous hydrochloric acid in the presence of an appropriate oxidant, such as $KMnO_4$, activated MnO_2, $KClO_3$, $NaIO_3$, concentrated HNO_3, $NaBO_3$, $Na_2CO_3{\cdot}H_2O_2$, $Na_2S_2O_8$, CrO_3 and the urea–H_2O_2 complex [80–85]. For example, the chlorination of iodoarenes in a biphasic mixture of carbon tetrachloride and concentrated hydrochloric acid in the presence of $Na_2S_2O_8$ affords the corresponding (dichloroiodo)arenes in 60–100% crude yields [81]. A modification of this method consists of the one-pot oxidative iodination/chlorination of arenes with iodine and an appropriate oxidant in hydrochloric acid [81]. Particularly useful is a solvent-free oxidative procedure involving treatment of ArI with the urea–hydrogen peroxide complex followed by addition of aqueous hydrochloric acid [85].

A convenient and mild approach to (dichloroiodo)arenes **17** consists of the chlorination of iodoarenes using concentrated hydrochloric acid and aqueous sodium hypochlorite (Scheme 2.7) [64]. Sodium chlorite, $NaClO_2$, can also be used in this procedure; however, in this case the chlorination takes longer time (3 h at room temperature) and the yields of products **17** are generally lower [64].

Podgorsek and Iskra reported a procedure for the conversion of iodoarenes into (dichloroiodo)arenes in 72–91% preparative yields by the oxidative halogenation approach using 30% aqueous hydrogen peroxide and hydrochloric acid in fluorinated alcohol at room temperature [65]. In this reaction trifluoroethanol is not only the reaction medium but is also an activator of hydrogen peroxide for the oxidation of hydrochloric acid to chlorine.

$$ArI \xrightarrow[94\text{-}100\%]{NaClO\ (5.84\%),\ HCl,\ H_2O,\ 15\ ^\circ C,\ 5\ min} \underset{\mathbf{17}}{ArICl_2}$$

Ar = Ph, 4-MeC_6H_4, 2-FC_6H_4, 2-BrC_6H_4, 3-BrC_6H_4, 4-BrC_6H_4,
4-ClC_6H_4, 3-$NO_2C_6H_4$, 4-$NO_2C_6H_4$, 4-PhC_6H_4, etc.

Scheme 2.7

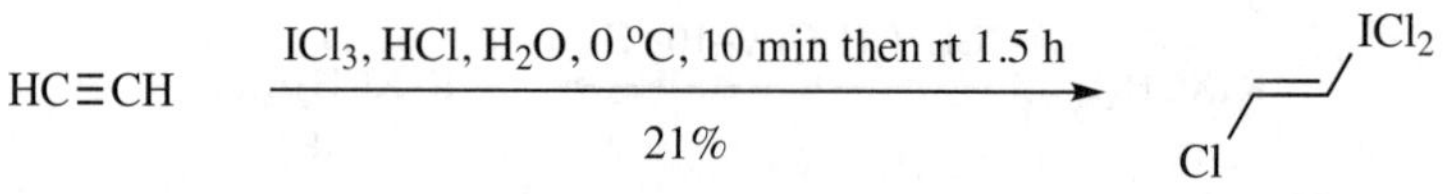

$HC{\equiv}CH \xrightarrow[21\%]{ICl_3,\ HCl,\ H_2O,\ 0\ ^{\circ}C,\ 10\ min\ then\ rt\ 1.5\ h}$ (E)-ClCH=CHICl₂ **18**

Scheme 2.8

2.1.3.2 *Preparation by Ligand Exchange*

The ligand exchange approach is not commonly used to prepare organic iodine(III) dichlorides and only several examples of such reactions are known. (Dichloroiodo)benzene can be conveniently prepared by treatment of iodosylbenzene with chlorotrimethylsilane, Me_3SiCl, in dichloromethane [86]. This is a very clean reaction, which allows us to generate $PhICl_2$ *in situ* and so use it without isolation [87]. A similar approach was employed for the low-temperature preparation of CF_3ICl_2 by the reaction of $CF_3I(Cl)F$ with trimethylchlorosilane at –40 °C [70].

(*E*)-Chlorovinyliodine(III) dichloride (**18**), a useful reagents for the synthesis of aryliodonium salts, is prepared by addition of iodine trichloride to acetylene in concentrated hydrochloric acid (Scheme 2.8) [71,88,89]. **Caution:** product **18** should be handled with a great care; it is extremely unstable and decomposes autocatalytically within seconds. It can, however, be stored for weeks in a freezer at –20 °C or below [71,89].

2.1.3.3 *Structural Studies*

Several X-ray crystallographic studies of organo-iododichlorides, $RICl_2$, have been reported in the literature. The first X-ray crystal structures of $PhICl_2$ [90] and $4\text{-}ClC_6H_4ICl_2$ [91] published in the 1950s were imprecise by modern standards. More recently, Chaloner and coworkers reported a good quality structure of $PhICl_2$ obtained at low temperature [92]. The molecule of $PhICl_2$ has the characteristic T-shape with primary I—Cl bond distances of 2.47 Å and 2.49 Å and Cl—I—C bond angles of 87.8 and 89.2°. In the solid state the molecules form an infinite zigzagged chain, in which one of the chlorine atoms interacts with the iodine of the next unit with an intermolecular I···Cl secondary bond distance of 3.42 Å. The coordination of iodine is distorted square-planar with the lone pairs occupying the *trans*-positions of a pseudo-octahedron [92].

X-Ray structures of two sterically encumbered (dichloroiodo)arenes, $2,4,6\text{-}Pr^i{}_3C_6H_2ICl_2$ [67] and $ArICl_2$ [Ar = 2,6-bis(3,5-dichloro-2,4,6-trimethylphenyl)benzene] [93] have been reported. Both molecules have the expected T-shaped geometry; the latter molecule has Cl—I—C angles of 89.4(3) and 92.1(3)° and I—Cl distances of 2.469(4) and 2.491(4) Å. The secondary I···Cl bond distance in this compound is 3.816 Å, which indicates a significant reduction of intermolecular association as compared to $PhICl_2$ [93]. The X-ray crystal structure of 2-(dichloroiodo)nitrobenzene, $2\text{-}NO_2C_6H_4ICl_2$, does not show any significant intramolecular interaction between the iodine(III) center and the oxygen atom of the nitro group in the *ortho* position (I···O bond distance 3.0 Å) [94].

The X-ray structure of the $PhICl_2$ adduct with tetraphenylphosphonium chloride, $[Ph_4P]^+[PhICl_3]^-$, has been reported [95]. The $[PhICl_3]^-$ anions in this structure have a planar coordination environment at the iodine atom. The I—Cl bond length of the chlorine atom trans to the Ph group is much longer (3.019 Å) than the bond distance to the cis Cl atoms (2.504 Å) [95].

X-Ray crystal structures of two polyfluoroalkyliododichlorides, $CF_3CH_2ICl_2$ and $CHF_2(CF_2)_5CH_2ICl_2$, have been reported [68]. In comparison to $PhICl_2$, which has a simple chain structure, perfluoroalkyliododichlorides have more complicated structures in which weak interactions between chains, coupled with aggregation of perfluoro groups, result in the formation of layers in the solid state.

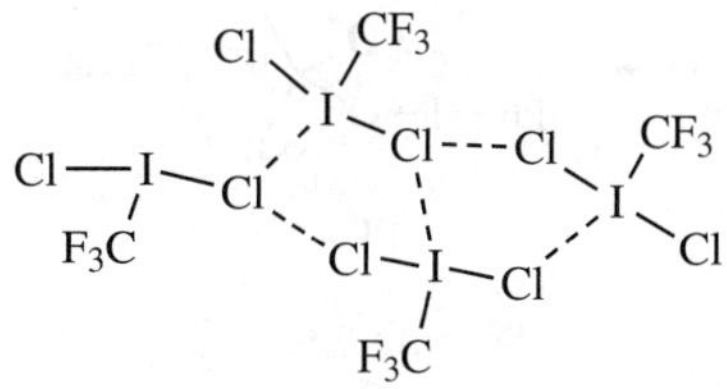

Figure 2.3 *Primary and secondary bonding pattern in a single-crystal X-ray structure of CF_3ICl_2.*

Single-crystal X-ray analysis of (trifluoromethyl)iodine dichloride, CF_3ICl_2, shows a T-shaped molecular structure with the CF_3 group in the equatorial position [70]. The axial I–Cl bonds in CF_3ICl_2 are three-center-four-electron semi-ionic bonds and have lengths of 2.478(2) and 2.457(2) Å and the angle between the iodine and the two apical chlorines is 171.62(9)°. The intermolecular I···Cl contact in CF_3ICl_2 is 3.324 Å long, which is much longer than the bridging bonds in I_2Cl_6 (2.68–2.72 Å) (Section 2.1.1.1), but shorter than in $C_6H_5ICl_2$ (3.40 Å) and 2,4,6-$Pr^i{}_3C_6H_2ICl_2$ (3.49 Å). In addition to the I–Cl contacts, CF_3ICl_2 also contains weak Cl–Cl contacts of 3.361 Å, some 4% under the sum of the van der Waals radii (3.5 Å), which are due to packing effects. All these contacts result in the formation of chains of side-linked five-membered rings (Figure 2.3) [70].

2.1.4 Organo-Iodosyl Compounds

Organic iodosyl compounds usually have a polymeric structure, $(RIO)_n$, with a typical, for λ^3-iodanes, T-shaped geometry at the iodine atom; no structural evidence supporting the existence of an I=O double bond has been reported. Most known iodosyl compounds have low thermal stability and some are explosive upon heating. Iodosyl compounds can be prepared by direct oxidation of organic iodides, or, more commonly, by basic hydrolysis of other iodine(III) compounds. Table 2.4 summarizes the preparation methods for organic iodosyl compounds.

2.1.4.1 Preparation by Oxidation of Organic Iodides

Aryl iodides smoothly react with dimethyldioxirane (DMDO) in acetone to give mainly, depending on the amount of DMDO, iodosyl- or iodylarenes, which precipitate from the solution [97]. Minisci and coworkers

Table 2.4 *Preparation of organo-iodosyl compounds.*

Compound	Method of synthesis	Yield (%)	Reference
PhIO	$PhI(OAc)_2$, NaOH, H_2O, rt (room temp), 1 h	93	[96]
PhIO	PhI, DMDO, acetone, 0 °C to rt, 3 h	26	[97]
PhIO	$PhICl_2$, $NaHCO_3$, NaOH, H_2O, 0 °C	63	[98]
4-MeC_6H_4IO	$ArICl_2$, NaOH, THF, H_2O, rt, 1 min	81	[30]
4-$MeOC_6H_4IO$	$ArI(OAc)_2$, NaOH, THF, H_2O, rt	88	[99]
4-$NO_2C_6H_4IO$	$ArI(OAc)_2$, NaOH, THF, H_2O, rt	90	[99]
2-$Bu^tO_2SC_6H_4IO$	$ArI(OAc)_2$, NaOH, H_2O, rt, 25 min	95	[100]
3-$HO_2CC_6H_4IO$	ArI, AcOOH/AcOH, rt, 12 h	80	[101]
CF_3CF_2IO	$CF_3CF_2I(OCOCF_3)_2$, $NaHCO_3$, ice, 0 °C	95	[102]

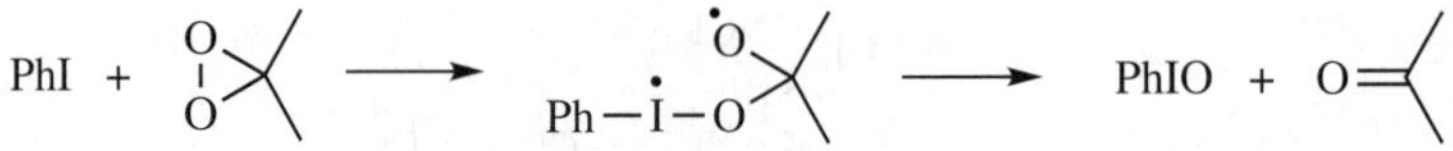

PhI + DMDO ⟶ Ph–I–O (19) ⟶ PhIO + O=C(CH₃)₂

DMDO 19

Scheme 2.9

investigated the mechanism of the reaction of iodobenzene with DMDO and suggested that this reaction occurs as the PhI-induced homolysis of the peroxide bond via the diradical intermediate **19** (Scheme 2.9) [103].

Asensio and coworkers reported a low temperature oxidation of iodomethane with DMDO to afford a pale yellow precipitate of iodosylmethane (**20**, Scheme 2.10) [104]. Upon raising the temperature to –40 °C, in the presence of moisture iodosylmethane decomposes to form the unstable hypoiodous acid, HOI, which can be trapped *in situ* by an alkene to afford iodohydrins. The formation of MeIO has also been detected in the photochemical reaction of iodomethane with ozone in an argon matrix at 17 K [105]. A similar low-temperature reaction of trifluoroiodomethane affords the unstable CF_3IO, which was identified by infrared spectroscopy [106].

The DMDO oxidation of iodocyclohexane affords *trans*-2-iodocyclohexanol as the final product via intermediate formation of iodosylcyclohexane followed by elimination of hypoiodous acid, which then adds to the alkene generated in the elimination step [103]. The oxidative deiodination of iodoalkanes via conversion into iodosylalkanes followed by nucleophilic substitution of the iodosyl group has found some synthetic application, particularly in the synthesis of steroidal products (Section 3.1.19) [107].

Protonated iodosylbenzene species can be generated in solution by the oxidation of iodobenzene with Oxone® ($2KHSO_5{\cdot}KHSO_4{\cdot}K_2SO_4$) in aqueous acetonitrile. The presence of protonated iodosylbenzene $[PhIOH]^+$ in such solutions has been supported by ESI mass spectrometry [108]. This procedure has been utilized in catalytic oxidations using iodobenzene as the pre-catalyst and Oxone as the stoichiometric oxidant (Section 4.1) [109].

3-Iodosylbenzoic acid, $3\text{-}HO_2CC_6H_4IO$, a useful recyclable oxidant, is conveniently prepared by the oxidation of 3-iodobenzoic acid with peracetic acid [101].

2.1.4.2 Preparation from Other Iodine(III) Compounds

Hydrolysis of diacetoxy- or (dichloroiodo)arenes is the most common approach to organic iodosyl compounds. Iodosylbenzene, PhIO, the most important member of the family of iodosyl compounds, is best prepared by hydrolysis of $PhI(OAc)_2$ with aqueous NaOH [96]. The same procedure can be used to prepare various *ortho*-, *meta*- and *para*-substituted iodosylarenes **22** from the respective (diacetoxy)iodoarenes **21** (Scheme 2.11). This procedure, for example, has been used to obtain 4-methoxyiodosylbenzene [99], 4-nitroiodosylbenzene [99] and pseudocyclic iodosylarenes **23–25** bearing *tert*-butylsulfonyl [100, 110], diphenylphosphoryl [111], or nitro [94] groups in the *ortho*-position. The relatively stable iodosylperfluoroalkanes $C_nF_{2n+1}IO$ ($n = 2$, 3, 4, 6) are synthesized similarly by the mild hydrolysis of the respective bis(trifluoroacetates) $C_nF_{2n+1}I(CO_2CF_3)_2$ using $NaHCO_3$ and water with ice [102].

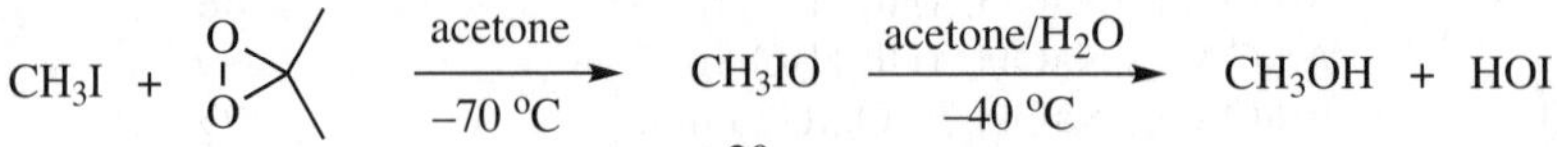

CH_3I + DMDO —(acetone, –70 °C)→ CH_3IO (**20**) —(acetone/H_2O, –40 °C)→ CH_3OH + HOI

Scheme 2.10

$$ArI(OAc)_2 \;(\mathbf{21}) \xrightarrow[65\text{-}95\%]{3N\ NaOH,\ H_2O,\ 0\ ^\circ C\ to\ rt} ArIO \;(\mathbf{22})$$

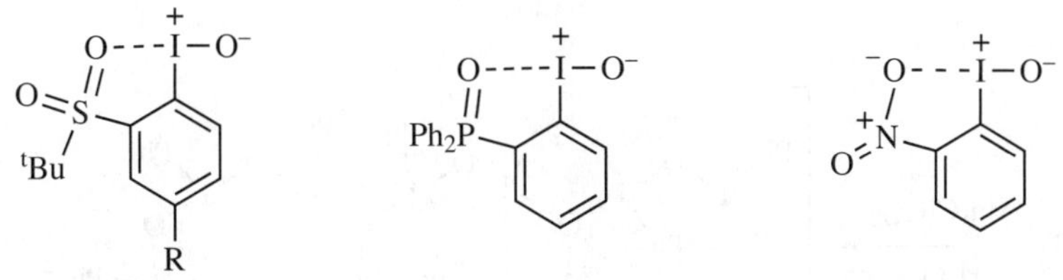

23, R = H or CF_3 **24** **25**

Scheme 2.11

Iodosylbenzene is a yellowish amorphous powder that cannot be recrystallized due to its polymeric nature. It dissolves in methanol with depolymerization to afford $PhI(OMe)_2$ [112]. Heating or extended storage at room temperature results in disproportionation of iodosylbenzene to PhI and colorless, explosive iodylbenzene, $PhIO_2$. Drying iodosylbenzene at elevated temperatures should be avoided because of the possibility of a severe explosion. A violent explosion of iodosylbenzene upon drying at 110 °C in vacuum has been reported [113]. However, handling of even large amounts of iodosylbenzene at room temperature is relatively safe.

An alternative general procedure for the preparation of iodosylarenes is based on the alkaline hydrolysis of (dichloroiodo)arenes under conditions similar to the hydrolysis of (diacetoxyiodo)arenes [98, 114, 115]. A modified procedure employs aqueous tetrahydrofuran as a solvent for the basic hydrolysis of (dichloroiodo)arenes (Scheme 2.12) [30].

Oligomeric iodosylbenzenes **26** and **27** have been prepared by ligand exchange in λ^3-iodanes under moderately acidic conditions. The oligomer **26** was obtained by the treatment of $PhI(OAc)_2$ with aqueous $NaHSO_4$ [116, 117], while product **27** precipitated from dilute aqueous solutions of PhI(OH)OTs and $Mg(ClO_4)_2$ [118] (Scheme 2.13). The formation of both products can be explained by self-assembly of the hydroxy(phenyl)iodonium ions (PhI^+OH in hydrated form) and [oxo(aquo)iodo]benzene $PhI^+(OH_2)O^-$ in aqueous solution under the reaction conditions.

2.1.4.3 Structural Studies

Although iodosylbenzene has been known and widely used as a reagent for over 100 years, structural details are still limited due to its amorphous polymeric nature, insolublility and low stability. On the basis of spectroscopic studies, it was suggested that in the solid state iodosylbenzene exists as a zigzag polymeric,

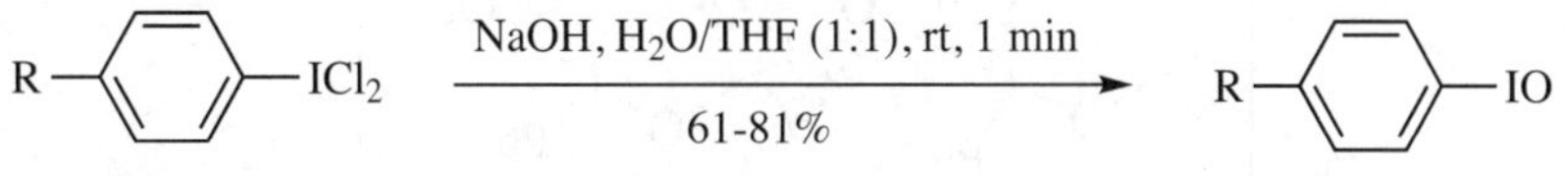

$$R\text{-}C_6H_4\text{-}ICl_2 \xrightarrow[61\text{-}81\%]{NaOH,\ H_2O/THF\ (1:1),\ rt,\ 1\ min} R\text{-}C_6H_4\text{-}IO$$

R = H, Me, Cl, NO_2

Scheme 2.12

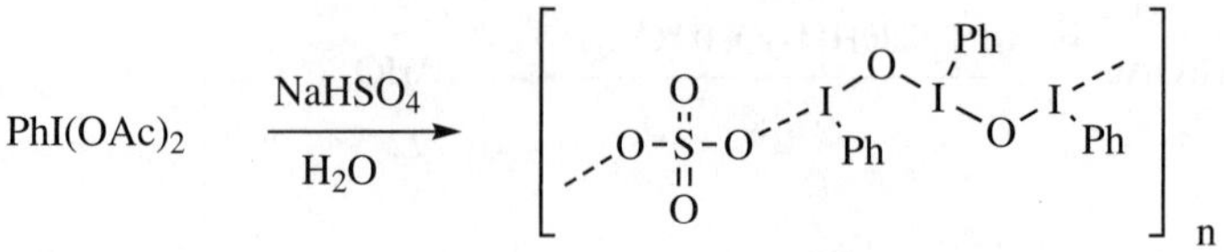

$PhI(OAc)_2$ $\xrightarrow[H_2O]{NaHSO_4}$ **26**

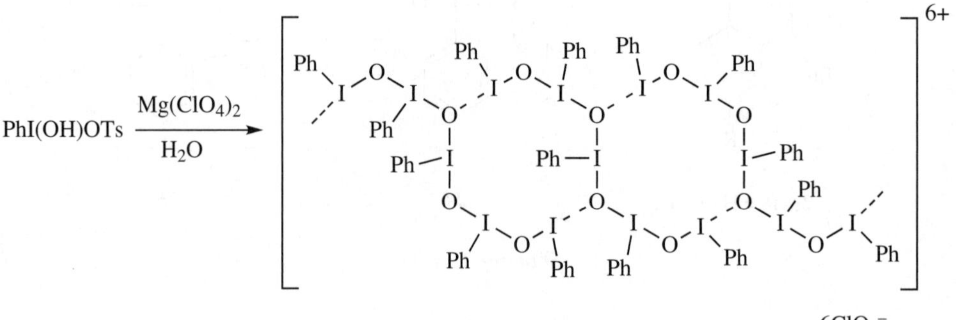

$PhI(OH)OTs$ $\xrightarrow[H_2O]{Mg(ClO_4)_2}$ **27** ($6ClO_4^-$)

Scheme 2.13

asymmetrically bridged structure, in which monomeric units of PhIO are linked by intermolecular I···O secondary bonds [98, 119–121]. Most notable were the studies of $(PhIO)_n$ by EXAFS analysis and by solid-state NMR spectroscopy [98, 122]. In particular, the EXAFS study revealed the T-shaped geometry around iodine centers with the primary I—O single bond of 2.04 Å, the secondary, intermolecular, I···O bond of 2.377(12) Å and an I—O—I angle of 114° (Figure 2.4) [122].

The polymeric structure of iodosylbenzene was also theoretically analyzed by DFT computations at the B3LYP level, establishing in particular the importance of the presence of a terminal hydration water in its zigzag polymeric structure HO–$(PhIO)_n$–H [123]. A solid-state ^{13}C NMR spectroscopy study of $(PhIO)_n$ and several *para*-substituted iodosylarenes ($4\text{-}MeC_6H_4IO$, $4\text{-}MeOC_6H_4IO$, $4\text{-}PrC_6H_4IO$, $4\text{-}Pr^iC_6H_4IO$, $4\text{-}NO_2C_6H_4IO$) confirmed the truly amorphous polymeric structure of iodosylbenzene, but indicated some degree of crystallinity of the *para*-methoxy, propyl and isopropyl derivatives [98].

The zigzag asymmetrically bridged structure of $(PhIO)_n$ has been confirmed by single-crystal X-ray diffraction studies of the oligomeric sulfate **26** and perchlorate **27** derivatives (Scheme 2.13) [116–118]. In particular, iodine atoms in the $(PhIO)_3$ fragment of the oligomeric sulfate **26** exhibit a typical of trivalent iodine T-shaped intramolecular geometry with O—I—O and O—I—C bond angles close to 180° (166.54–177.99°) and 90° (79.18–92.43°), respectively. The I—O bond distances in the $(PhIO)_3$ fragment of sulfate **26** vary in a broad range of 1.95 to 2.42 Å [117]. A single-crystal X-ray crystal study of the oligomeric perchlorate **27**

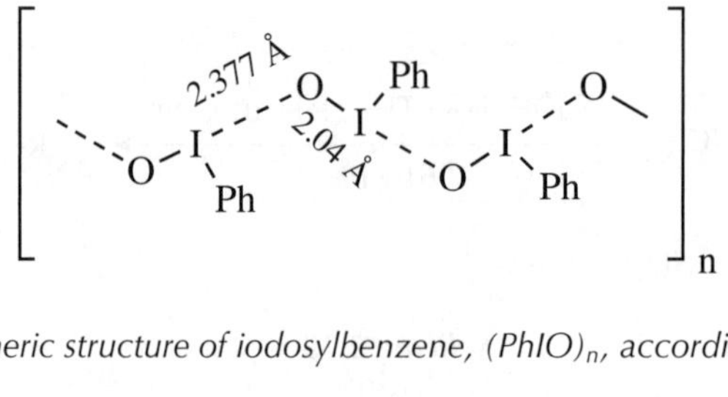

Figure 2.4 *Zigzag polymeric structure of iodosylbenzene, $(PhIO)_n$, according to EXAFS analysis [122].*

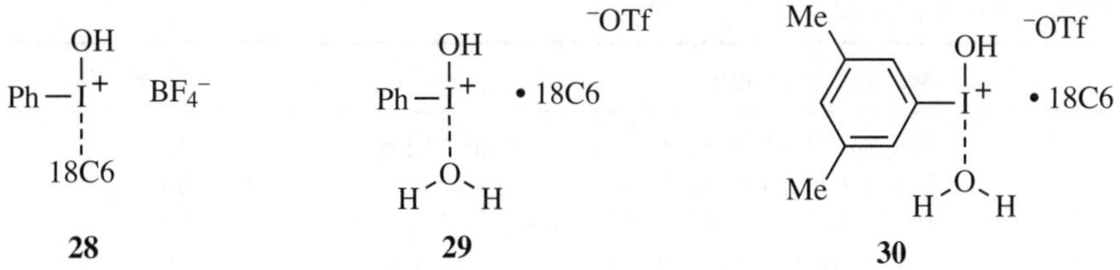

28 **29** **30**

18C6 = 18-Crown-6 ether

Figure 2.5 *Activated iodosylbenzene monomer complexes with 18C6 crown ether.*

revealed a complex structure consisting of pentaiodanyl dicationic units joined by secondary I···O bonds into an infinite linear structure of 12-atom hexagonal rings [118].

Ochiai and coworkers have reported the preparation, X-ray crystal structures and useful oxidizing reactions of activated iodosylbenzene monomer complexes with 18C6 crown ether (Figure 2.5) [124, 125]. Reaction of iodosylbenzene with HBF_4/Me_2O in the presence of equimolar 18C6 in dichloromethane afforded quantitatively the stable, crystalline crown ether complex **28**, which is soluble in acetonitrile, methanol, water and dichloromethane. X-Ray analysis revealed a protonated iodosylbenzene monomer structure **28** stabilized by intramolecular coordination with the crown ether oxygen atoms [125]. The aqua complexes of iodosylarenes **29** and **30** with a water molecule coordinated to iodine(III) were prepared by the reaction of (diacetoxyiodo)benzene with trimethylsilyl triflate in the presence of 18C6 crown ether in dichloromethane. X-Ray analysis of complex **28** confirmed a T-shaped structure, ligated with one water molecule at the apical site of the iodine(III) atom of the hydroxy(phenyl)iodonium ion, with a near–linear O–I–O triad (173.96°). Including a close contact with one of the crown ether oxygens, the complex adopts a distorted square-planar geometry around the iodine [126].

Owing to the polymeric structure, iodosylbenzene is insoluble in all nonreactive solvents; however, soluble derivatives of iodosylbenzene can be realized by placing an appropriate substituent on the *ortho* position of the phenyl ring. Protasiewicz and coworkers have reported the preparation and X-ray structure of the monomeric iodosylarene 2-$Bu^tO_2SC_6H_4IO$, (**23**, R = H, see Scheme 2.11), in which the intramolecular secondary I···O bond replaces the intermolecular interactions that are typical of polymeric iodosylbenzene [100, 127]. Iodosylarene **23** is readily soluble in organic solvents (up to 0.08 M in chloroform) and can be analyzed by NMR in solution [100]. Single-crystal X-ray analysis of **23** showed a structure resembling benziodoxoles with an intramolecular distance of 2.707 Å between one of the sulfone oxygen atoms and the hypervalent iodine center [127]. The I—O bond length in the iodosyl group of **23** is 1.848 Å and the intramolecular O–I–O bond angle is 167.3°. The iodine centers in **23** achieve a pseudo-square-planar geometry by the formation of intermolecular an I···O secondary bond (2.665 Å) to a neighboring iodosyl oxygen atom [127].

2.1.5 Organoiodine(III) Carboxylates

[Bis(acyloxy)iodo]arenes, $ArI(O_2CR)_2$, are the most important, well investigated and practically useful organic derivatives of iodine(III). Two of them, (diacetoxyiodo)benzene and [bis-(trifluoroacetoxy)iodo]benzene, are commercially available and widely used oxidizing reagents. The preferred abbreviations, DIB for (diacetoxyiodo)benzene and BTI for [bis(trifluoroacetoxy)iodo]benzene, were originally suggested by Varvoglis in his book published in 1992 [128]; however, the older abbreviations PIDA (phenyliodine diacetate) and PIFA [phenyliodine bis(trifluoroacetate)] are still frequently used in modern research literature. Two general approaches are used for the preparation of [bis(acyloxy)iodo]arenes: (i) the oxidation of iodoarenes in the presence of a carboxylic acid and (ii) a ligand-exchange reaction of the

Table 2.5 *Preparation of organoiodine(III) carboxylates.*

Compound	Method of synthesis	Yield (%)	Reference
$PhI(OAc)_2$	PhI, AcOOH, AcOH, rt (room temp), 12 h	79	[129]
$PhI(OAc)_2$	PhI, 3-$ClC_6H_4CO_3H$, AcOH, rt, 15 h	83	[130]
$PhI(OAc)_2$	PhI, $NaBO_3$, TfOH, AcOH, 40–45 °C, 3 h	99	[131–139]
4-$MeC_6H_4I(OAc)_2$	ArI, $NaBO_3$, TfOH, AcOH, 40–45 °C, 4 h	96	[131]
3-$MeOC_6H_4I(OAc)_2$	ArI, $NaBO_3$, TfOH, AcOH, 40–45 °C, 4 h	98	[131]
3-$NO_2C_6H_4I(OAc)_2$	ArI, $NaBO_3$, TfOH, AcOH, 40–45 °C, 6 h	94	[131]
4-$FC_6H_4I(OAc)_2$	ArI, $NaBO_3$, TfOH, AcOH, 40–45 °C, 8 h	86	[131]
2,6-$Me_2C_6H_3I(OAc)_2$	ArI, $NaBO_3$, AcOH, 40–45 °C, 4–8 h	75	[132]
N, N, Ts, $I(OAc)_2$	ArI, $NaBO_3$, AcOH, 50–60 °C, 8–24 h	Not reported	[133]
AcO, I, O, I, OAc	ArI, Selectfluor, AcOH/MeCN, rt, 12 h	90	[134]
$C_6F_{13}I(OCOCF_3)_2$	$C_6F_{13}I$, Oxone, CF_3CO_2H, rt, 24 h	88	[135]
$C_6F_5I(OCOCF_3)_2$	C_6F_5I, Oxone, CF_3CO_2H, rt, 2 h	94	[135]
2-$ClC_6H_4I(OCOCF_3)_2$	2-ClC_6H_4I, Oxone, CF_3CO_2H, rt, 1.5 h	97	[135]
4-$CF_3C_6F_4I(OCOCF_3)_2$	ArI, HNO_3, $(CF_3CO)_2O$, −30 °C to rt, 4 days	79	[136]
$CF_3CH_2I(OCOCF_3)_2$	CF_3CH_2I, CF_3CO_3H, CF_3CO_2H, 0 °C to rt, 24 h	89	[137]
$PhSO_2CH_2I(OCOCF_3)_2$	$PhSO_2CH_2I$, CF_3CO_3H, CF_3CO_2H, −30 °C to rt, 4 h	99	[138]
$PhI(OCOCF_3)_2$	$PhI(OAc)_2$, CF_3CO_2H, rt	60	[139]

readily available DIB with an appropriate carboxylic acid. Table 2.5 provides an overview of preparation methods for organoiodine(III) carboxylates.

2.1.5.1 Preparation by Oxidation of Organic Iodides

The most practically important organoiodine(III) carboxylates, DIB, is usually prepared by the oxidation of iodobenzene with peracetic acid in acetic acid according to the procedure of Sharefkin and Saltzman [129]. A similar peracid oxidation of substituted iodobenzenes can be used to furnish numerous other (diacetoxyiodo)arenes, including the polymer-supported analogs of DIB from poly(iodostyrene) or aminomethylated poly(iodostyrene) [140–143] and the ion-supported (diacetoxyiodo)arenes (Chapter 5) [144, 145]. Likewise, various [bis(trifluoroacetoxy)iodo]arenes can be synthesized in high yield by the oxidation of the respective iodoarenes with peroxytrifluoroacetic acid in trifluoroacetic acid [146–148].

A modification of this method consists of the oxidative diacetoxylation of iodoarenes in acetic or trifluoroacetic acid using appropriate oxidants, such as periodates [149–151], chromium(VI) oxide [152], sodium percarbonate [153], *m*-chloroperoxybenzoic acid (*m*CPBA) [130, 154–158], potassium peroxodisulfate [159, 160], H_2O_2–urea [161], Selectfluor [33] and sodium perborate [132]. The oxidation of iodoarenes with sodium perborate in acetic acid at 40 °C is probably the most general procedure that has been used for the small-scale preparation of numerous (diacetoxyiodo)-substituted arenes and hetarenes (Scheme 2.14) [132, 133, 162–166]. For example, the chiral diacetate **31** and several analogous

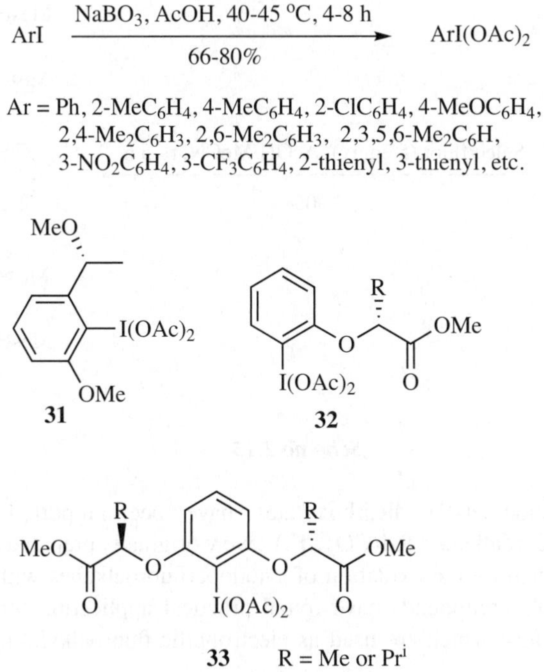

$$\text{ArI} \xrightarrow[\text{66-80\%}]{\text{NaBO}_3\text{, AcOH, 40-45 °C, 4-8 h}} \text{ArI(OAc)}_2$$

Ar = Ph, 2-MeC_6H_4, 4-MeC_6H_4, 2-ClC_6H_4, 4-$MeOC_6H_4$, 2,4-$Me_2C_6H_3$, 2,6-$Me_2C_6H_3$, 2,3,5,6-Me_2C_6H, 3-$NO_2C_6H_4$, 3-$CF_3C_6H_4$, 2-thienyl, 3-thienyl, etc.

MeO, I(OAc)$_2$, OMe **31**

R, OMe, O, I(OAc)$_2$, O **32**

R, R, MeO, OMe, O, O, O, I(OAc)$_2$, O **33** R = Me or Pr^i

Scheme 2.14

products were prepared in 73–87% yield by the perborate oxidation of the appropriate aryl iodides [166]. These acetates serve as important precursors to the chiral [(hydroxy)tosyloxy]iodoarenes (Section 2.1.6). Fujita and coworkers used the perborate oxidation procedure for the synthesis of optically active (diacetoxyiodo)arenes **32** and **33** [165, 167, 168]. Several heteroaromatic derivatives of (diacetoxyiodo)arenes [i.e., *N*-tosyl-4-(diacetoxyiodo)pyrazole, *N*-trifluoromethanesulfonyl-4-(diacetoxyiodo)pyrazole, 2-(diacetoxyiodo)thiophene and 3-(diacetoxyiodo)thiophene] were prepared similarly by using sodium perborate in acetic acid [133]. The perborate oxidation procedure can be further improved by performing the oxidation in the presence of catalytic amounts of trifluoromethanesulfonic acid [131].

A convenient and mild experimental procedure for the preparation of (diacetoxyiodo)arenes using Selectfluor as the oxidant in acetic acid has been developed by Shreeve and coworkers [33]. This method, in particular, has been utilized in the syntheses of a chiral hypervalent iodine(III) reagent having a rigid spirobiindane backbone [134] and the C_2-symmetric chiral (diacetoxyiodo)arene **35** from the respective iodide **34** (Scheme 2.15) [169, 170].

Kitamura and Hossain have found that potassium peroxodisulfate can be used as an efficient oxidant for the preparation of (diacetoxyiodo)arenes and [bis(trifluoroacetoxy)iodo]arenes from iodoarenes [159, 160]. A convenient modification of this approach employs the interaction of arenes with iodine and potassium peroxodisulfate in acetic acid (Scheme 2.16) [171]. The mechanism of this reaction probably includes the oxidative iodination of arenes, followed by oxidative diacetoxylation of ArI *in situ* leading to (diacetoxyiodo)arenes **36**.

Zefirov, Stang and coworkers have developed a general procedure for the preparation of various [bis(trifluoroacetoxy)iodo]arenes by the oxidation of iodoarenes with xenon bis(trifluoroacetate), which can be generated *in situ* from XeF_2 and trifluoroacetic acid (Scheme 2.17) [172].

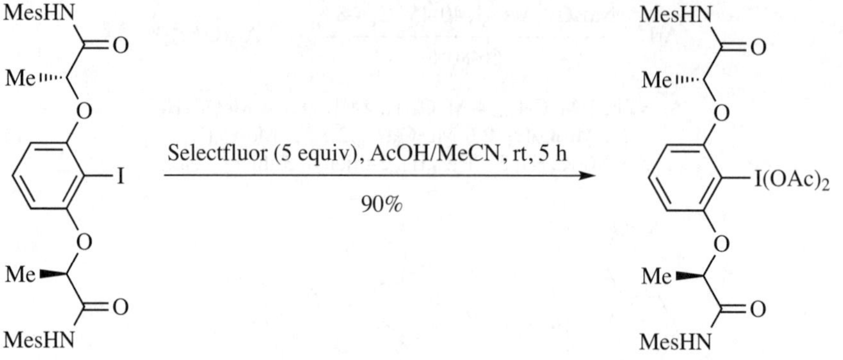

Selectfluor (5 equiv), AcOH/MeCN, rt, 5 h

90%

34 → **35**

Scheme 2.15

Several stabilized alkyliodine(III) dicarboxylates have been reported in the literature. [Bis-(trifluoroacetoxy)iodo]perfluoroalkanes, $R_fI(CO_2CF_3)_2$, were originally prepared by Yagupolskii and coworkers in almost quantitative yield by the oxidation of 1-iodoperfluoroalkanes with peroxytrifluoroacetic acid [173]. These relatively stable compounds have found practical application for the preparation of perfluoroalkyl(phenyl)iodonium salts, which are used as electrophilic fluoroalkylating reagents (FITS reagents) (Section 2.1.9.5) [174].

A convenient newer procedure for preparing various [bis(trifluoroacetoxy)iodo]perfluoroalkanes **37** and also [bis(trifluoroacetoxy)iodo]arenes **38** involves the oxidation of the corresponding aryl and perfluoroalkyl iodides with the commercially available and inexpensive oxidant Oxone ($2KHSO_5{\cdot}3KHSO_4{\cdot}3K_2SO_4$) in trifluoroacetic acid at room temperature (Scheme 2.18) [135].

1-[Bis(trifluoroacetoxy)iodo]-1*H*,1*H*-perfluoroalkanes **39** have been prepared in almost quantitative yield by the oxidation of 1-iodo-1*H*,1*H*-perfluoroalkanes with peroxytrifluoroacetic acid as relatively stable, although moisture sensitive, white microcrystalline solids (Scheme 2.19) [137, 175, 176].

[(Arylsulfonyl)methyl]iodine(III) bis(trifluoroacetates) **41** can be prepared from the respective iodomethyl sulfones **40** using a similar oxidation procedure (Scheme 2.20) [138]. Bis(trifluoroacetates) **41** can be isolated at 0 °C, but are unstable at room temperature and can spontaneously decompose with explosion.

The reaction of 4-iodotricyclene **42** with *m*-chloroperoxybenzoic acid in dichloromethane at room temperature gives a microcrystalline precipitate of 4-[bis(*m*-chlorobenzoyloxy)iodo]tricyclene **43** (Scheme 2.21) [177]. Compound **43** has a melting point of 180–182 °C and is indefinitely stable at room temperature.

2.1.5.2 Preparation by Ligand Exchange

The second common approach to [bis(acyloxy)iodo]arenes is based on the ligand-exchange reaction of a (diacetoxyiodo)arene (usually DIB) with an appropriate carboxylic acid. A typical procedure consists of

$$ArH + I_2 \xrightarrow[69\text{-}73\%]{K_2S_2O_8,\ AcOH,\ H_2SO_4,\ ClCH_2CH_2Cl,\ 40\ ^\circ C,\ 12\text{-}30\ h} ArI(OAc)_2\ \mathbf{36}$$

Ar = Ph, 4-MeC_6H_4, 4-ClC_6H_4, 4-BrC_6H_4, 4-FC_6H_4

Scheme 2.16

$$\mathrm{ArI} \xrightarrow[\text{79-93\%}]{(CF_3CO_2)_2Xe,\ (CF_3CO)_2O,\ CH_2Cl_2,\ -50\ ^\circ C\ \text{to rt},\ 1.5\ h} \mathrm{ArI(OCOCF_3)_2}$$

Ar = Ph, 4-MeC_6H_4, 2-MeC_6H_4, 4-$NO_2C_6H_4$, 3-$NO_2C_6H_4$, 2-Me-4-$NO_2C_6H_3$

Scheme 2.17

$$\mathrm{C_nF_{2n+1}I} \xrightarrow[\text{62-88\%}]{\text{Oxone (0.5-1 equiv)},\ CF_3CO_2H,\ \text{rt},\ 24\text{-}48\ h} \mathrm{C_nF_{2n+1}I(OCOCF_3)_2}\ \mathbf{37}$$

n = 4, 6, 8, 10, 12

$$\mathrm{ArI} \xrightarrow[\text{60-97\%}]{\text{Oxone (1.5 equiv)},\ CF_3CO_2H,\ CHCl_3,\ \text{rt},\ 1.2\text{-}4\ h} \mathrm{ArI(OCOCF_3)_2}\ \mathbf{38}$$

Ar = Ph, 4-FC_6H_4, 4-BrC_6H_4, 4-ClC_6H_4, 3-ClC_6H_4, 2-ClC_6H_4,
4-$NO_2C_6H_4$, 3-$NO_2C_6H_4$, 4-$CF_3C_6H_4$, 3,5-$(CF_3)_2C_6H_3$, etc

Scheme 2.18

$$\mathrm{C_nF_{2n+1}CH_2I} \xrightarrow[\text{89-99\%}]{CF_3CO_3H,\ CF_3CO_2H,\ 0\ \text{to}\ 25\ ^\circ C,\ 24\ h} \mathrm{C_nF_{2n+1}CH_2I(OCOCF_3)_2}\ \mathbf{39}$$

n = 1, 2, 3, 7, etc

Scheme 2.19

$$\underset{\mathbf{40}}{\mathrm{ArSO_2CH_2I}} \xrightarrow[\text{quantitative yield}]{CF_3CO_3H/CF_3CO_2H,\ -30\ ^\circ C\ \text{to rt},\ 4\ h} \underset{\mathbf{41}}{\mathrm{ArSO_2CH_2I(CO_2CF_3)_2}}$$

Ar = Ph or 4-MeC_6H_4

Scheme 2.20

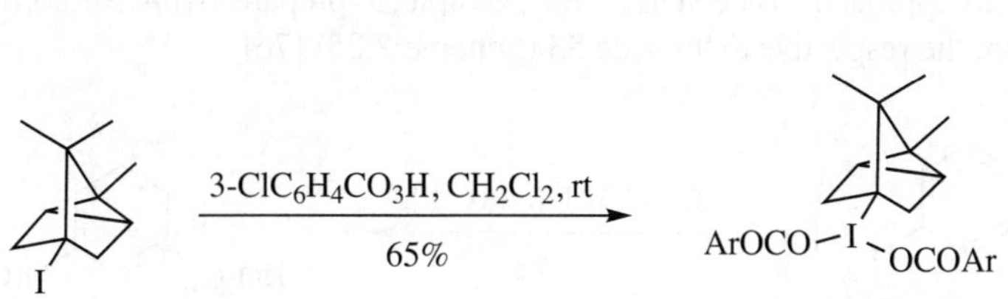

$$\underset{\mathbf{42}}{\text{(bicyclic iodide)}} \xrightarrow[\text{65\%}]{3\text{-}ClC_6H_4CO_3H,\ CH_2Cl_2,\ \text{rt}} \mathrm{ArOCO{-}I{-}OCOAr}$$

42

43, Ar = 3-ClC_6H_4

Scheme 2.21

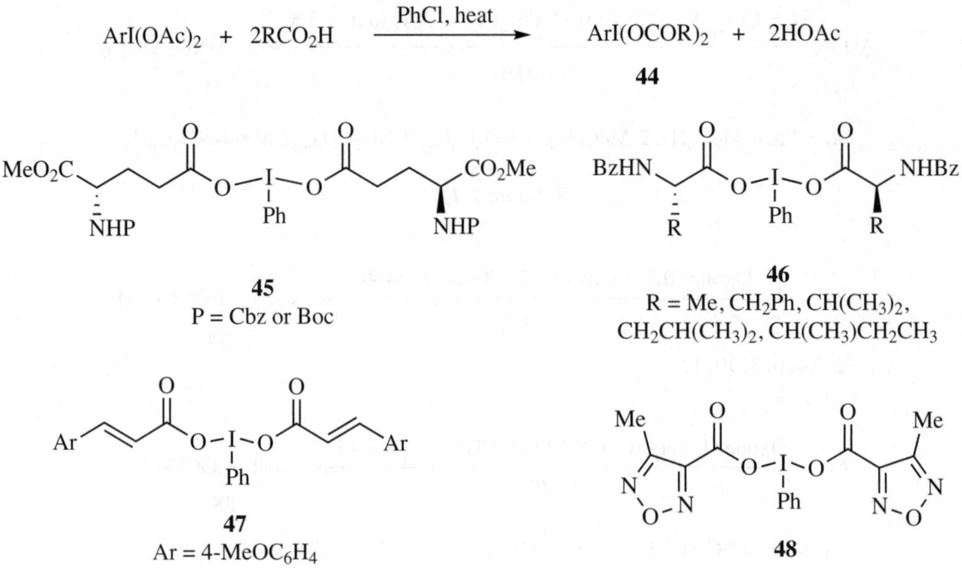

$ArI(OAc)_2 + 2RCO_2H \xrightarrow{PhCl,\ heat} ArI(OCOR)_2 + 2HOAc$

44

45 P = Cbz or Boc

46 R = Me, CH_2Ph, $CH(CH_3)_2$, $CH_2CH(CH_3)_2$, $CH(CH_3)CH_2CH_3$

47 Ar = 4-$MeOC_6H_4$

48

Scheme 2.22

heating DIB with a nonvolatile carboxylic acid RCO_2H in the presence of a high-boiling solvent such as chlorobenzene (Scheme 2.22) [178–183]. The equilibrium in this reversible reaction can be shifted towards the synthesis of the product **44** by distillation under reduced pressure of the relatively volatile acetic acid formed during the reaction. This procedure, for example, has been used to furnish the glutamate-derived (diacyloxyiodo)benzenes **45** [179], protected amino acid derivatives **46** [181], cinnamate derivative **47** [183] and 3-methylfurazan-4-carboxylic acid derivative **48** [184].

Reactions of DIB with stronger carboxylic acids usually proceed under milder conditions at room temperature. A convenient procedure for preparing $PhI(CO_2CF_3)_2$ (BTI) consists of simply dissolving DIB in trifluoroacetic acid and crystallization of the product after evaporating to a small volume [139]. Likewise, 3-[bis(trifluoroacetoxy)iodo]benzoic acid **50**, a useful recyclable oxidizing reagent, has been prepared in high yield by heating 3-iodosylbenzoic acid (which actually exists in the form of a polymeric carboxylate **49**) [185] with trifluoroacetic acid (Scheme 2.23) [186]. A similar approach has been used to prepare a series of $PhI(OCOCO_2R)_2$ by treatment of DIB with oxalyl chloride in the respective alcohol, ROH [187].

Naumann and coworkers prepared [bis(trifluoroacetoxy)iodo]arene **52** in high yield by ligand exchange of the respective (difluoroiodo)arene **51** with trifluoroacetic anhydride (Scheme 2.24) [32].

Another common method involves the reaction of organoiodine(III) dichlorides, $RICl_2$, with silver carboxylates [76,139]. This approach has been used, for example, to prepare 1-[bis(trifluoroacetoxy)iodo]-2,2,2-trifluoroethane **54** from the respective dichloride **53** (Scheme 2.25) [76].

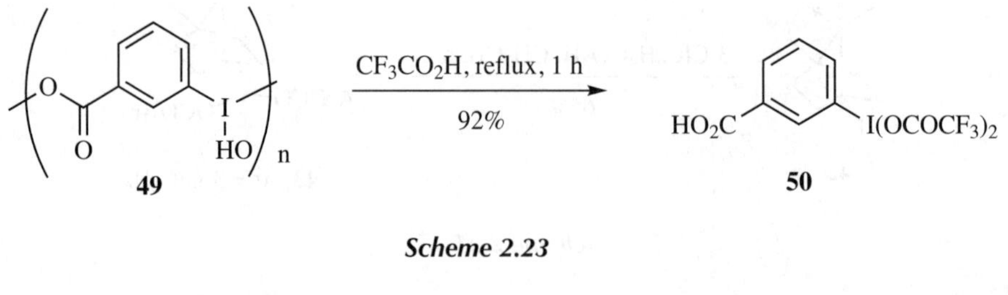

49 $\xrightarrow[92\%]{CF_3CO_2H,\ reflux,\ 1\ h}$ **50** (HO_2C–C_6H_4–$I(OCOCF_3)_2$)

Scheme 2.23

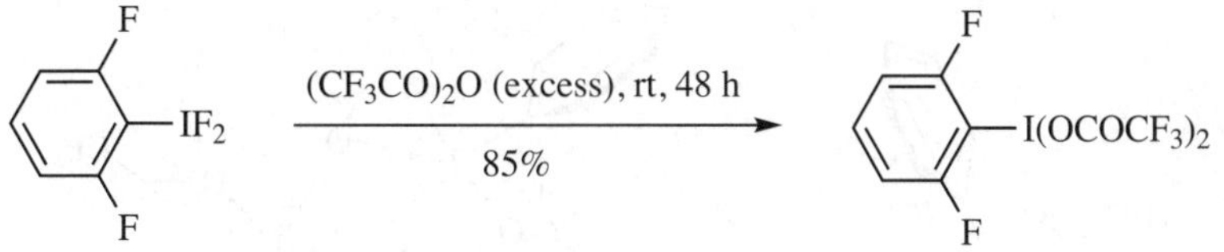

$$\mathbf{51}\ (2,6\text{-}F_2C_6H_3IF_2) \xrightarrow[85\%]{(CF_3CO)_2O\text{ (excess), rt, 48 h}} \mathbf{52}\ (2,6\text{-}F_2C_6H_3I(OCOCF_3)_2)$$

Scheme 2.24

[Bis(acyloxy)iodo]arenes are generally colorless, stable microcrystalline solids, which can be easily recrystallized and stored for extended periods without significant decomposition.

2.1.5.3 *Structural Studies*

Numerous X-ray structural studies of organoiodine(III) carboxylates have been reported in the literature. In general, single-crystal X-ray structural data for [bis(acyloxy)iodo]benzenes indicate a pentagonal planar coordination of iodine within the molecule, combining the primary T-shaped iodine(III) geometry with two secondary intramolecular I···O interactions with the carboxylate oxygens. In 1979 Alcock and coworkers reported single-crystal X-ray structures of (diacetoxyiodo)benzene and [bis(dichloroacetoxy)iodo]benzene, which demonstrated the importance of secondary bonding in the solid state structure of [bis(acyloxy)iodo]benzenes [188]. In the crystal structure of $PhI(OAc)_2$ (**55**, Figure 2.6) the overall geometry of each iodine atom can be described as a pentagonal-planar arrangement of three strong and two weak secondary bonds. The primary I–C distance in **55** of 2.090 Å and the two I–O distances of about 2.156 Å are longer than the sum of the covalent radii of the elements (2.07 Å for I—C and 1.99 Å for I—O) and the O–I–O bond angle is 164°. The secondary intramolecular I···O bonds in the molecule of **55** are 2.817 and 2.850 Å, resulting in the overall pentagonal-planar arrangement [188]. In contrast to $PhI(OAc)_2$ **55**, the molecules of $PhI(OCOCF_3)_2$ (**56**) form a dimer due to the two additional I···O intermolecular secondary bonds (Figure 2.6) [189]. The crystal structure of [bis(trifluoroacetoxy)iodo]pentafluorobenzene, $C_6F_5I(OCOCF_3)_2$, reported in 1989, is characterized by an additional set of I···O intermolecular secondary bonds, bringing the total coordination at the iodine center to seven [190].

Figure 2.7 shows several examples of [bis(acyloxy)iodo]arenes whose X-ray crystal structures have been reported in literature. These structures include the chiral diacetate **31** [191], μ oxodiiodanyl diacetate **57** [117], *N*-tosyl-4-(diacetoxyiodo)pyrazole (**58**) [133], 1,3,5,7-tetrakis[4-(diacetoxyiodo)phenyl]adamantane (**59**) [154], tetrakis[4-(diacetoxyiodo)phenyl]methane (**60**) [155], 3-[bis(trifluoroacetoxy)iodo]benzoic acid (**61**) [186], 1-(diacetoxyiodo)-2-nitrobenzene (**62**) [94] and the chiral (diacetoxyiodo)arene **63** [165] derived from methyl lactate.

In particular, μ-oxo-[bis(acetoxy)iodo]benzene (**57**), prepared by partial hydrolysis of $PhI(OAc)_2$ in the presence of $NaHSO_4$, in the solid state forms an isolated diamond-core dimeric structure with

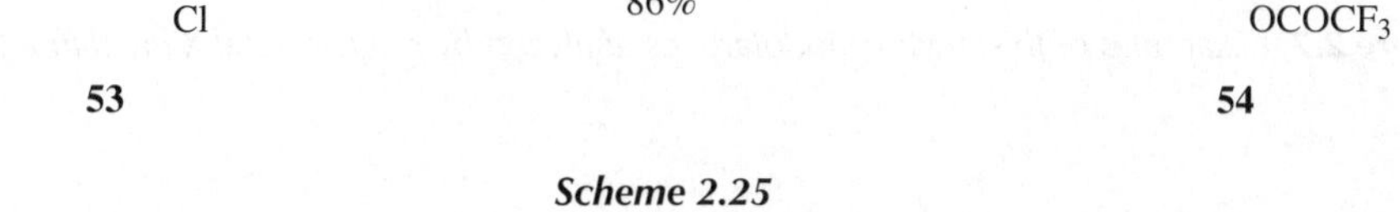

$$\underset{\mathbf{53}}{CF_3CH_2\text{–}ICl_2} \xrightarrow[86\%]{2CF_3CO_2Ag,\ CF_2ClCFCl_2,\ 20\ ^\circ C,\ 3\ h} \underset{\mathbf{54}}{CF_3CH_2\text{–}I(OCOCF_3)_2}$$

Scheme 2.25

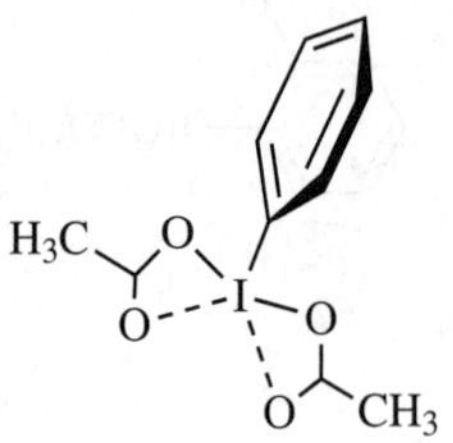

Figure 2.6 Primary and secondary bonding pattern in single-crystal X-ray structures of $PhI(OAc)_2$ (**55**) [188] and $PhI(OCOCF_3)_2$ (**56**) [189].

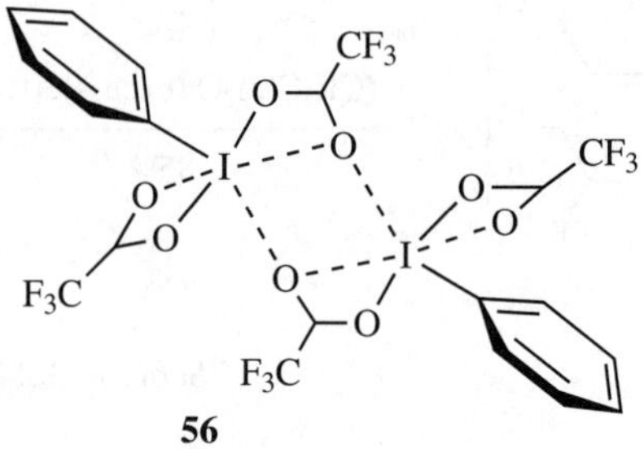

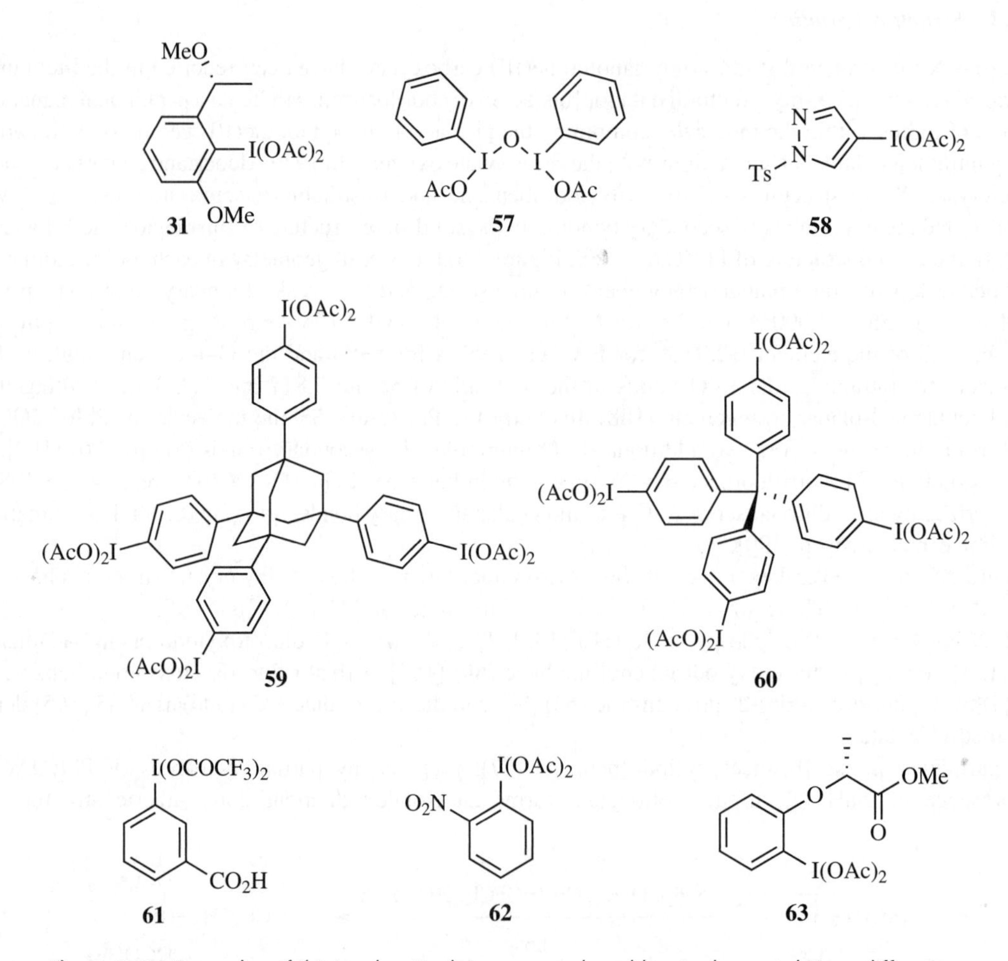

Figure 2.7 Examples of [bis(acyloxy)iodo]arenes analyzed by single-crystal X-ray diffraction.

pentagonal-planar iodine centers [117]. In the structure of **57**, both iodine centers have pentagonal-planar coordination with three short covalent and two long secondary bonds similar to that observed for μ-oxo-[bis(trifluoroacetoxy)iodo]benzene [192]. The I—O—I μ-oxo bridge is bent and consists of two short (2.03 Å, average) I—O bonds, which are close to those observed in μ-oxodiiodanyl di(trifluoroacetate) (2.02 Å) [192].

In the molecule of trifluoroacetate **61**, the C—I bond length is 2.083 Å, the primary I—O bond lengths are 2.149 and 2.186 Å and the intramolecular secondary I···O interactions with the carboxylate oxygens have distances of I(1)···O(5) 3.146 Å and I(1)···O(4) 3.030 Å; these five intramolecular interactions result in the pentagonal-planar coordination of iodine within the molecule [186]. In addition to the five intramolecular interactions, an intermolecular coordination of iodine atom to one the carboxylic oxygens of the neighboring molecule is also present (3.023 Å). Interestingly, the presence of a *meta*-carboxylic group has no noticeable effect on the molecular geometry of **61**, which is very similar to the X-ray crystal structure of [bis(trifluoroacetoxy)iodo]benzene [189]. No intermolecular interaction involving the *meta*-carboxylic group can be found in molecule **61**, which is in sharp contrast with the polymeric structure of 3-iodosylbenzoic acid [185]. The monomeric character of compound **61** leads to its improved solubility and higher reactivity compared to 3-iodosylbenzoic acid [189].

The X-ray crystal structure of 1-(diacetoxyiodo)-2-nitrobenzene (**62**) does not show any significant intramolecular interaction between the iodine(III) center and the oxygen atom of the nitro group in the *ortho* position (I···ONO bond distance 3.11 Å) [94].

Montanari, DesMarteau and Pennington reported the X-ray structural analysis of a fluoroalkyliodine(III) dicarboxylate, 1-[bis(trifluoroacetoxy)iodo]-2,2,2-trifluoroethane (**54**, Scheme 2.25) [68]. Compound **54** has a T-shaped coordination similar to other known dicarboxylates, but forms a previously unknown tetrameric array of molecules due to strong intermolecular I···O contacts.

An ^{17}O NMR study of bis(acyloxy)iodoarenes in chloroform has confirmed that the T-shaped structure of iodine(III) compounds observed in the solid state is also adopted in solution [193, 194]. The carboxylic groups of bis(acyloxy)iodoarenes show a dynamic behavior, which is explained by a [1,3]-sigmatropic shift of the iodine atom between the two oxygen atoms of the carboxylic groups [194].

2.1.6 [Hydroxy(Organosulfonyloxy)Iodo]Arenes

In contrast to carboxylic acids, *p*-toluenesulfonic acid and other organosulfonic acids do not form stable organoiodine(III) derivatives of the type $ArI(OSO_2R)_2$. The existence of several such derivatives has been proposed in the literature; however, none of them was isolated as an individual, stable compound. For example, Koser and Wettach reported in 1977 some evidence for the intermediate formation of the ditosylate $PhI(OTs)_2$ in the reaction of $PhICl_2$ with silver tosylate, but an attempt to isolate this compound failed and the final isolated product was [hydroxy(tosyloxy)iodo]benzene, PhI(OH)OTs [195].

The first preparation of [hydroxy(tosyloxy)iodo]benzene (HTIB) from (diacetoxyiodo)benzene and *p*-toluenesulfonic acid monohydrate was reported in 1970 by Neilands and Karele [196]. A decade later, Koser and coworkers discovered that HTIB is an efficient reagent for the oxytosylation of unsaturated organic substrates and for the synthesis of various iodonium salts [197–201]; consequently, in modern synthetic literature HTIB is often referred for as "Koser's reagent".

[Hydroxy(organosulfonyloxy)iodo]arenes, $ArI(OH)OSO_2R$, represent one the most common, well investigated and practically useful classes of organoiodine(III) compounds. The most important of these compounds, HTIB, is commercially available and commonly used as an oxidizing reagent in organic synthesis. [Hydroxy(organosulfonyloxy)iodo]arenes are usually prepared by a ligand-exchange reaction of (diacetoxyiodo)arenes with an appropriate organosulfonic acid (Table 2.6).

Table 2.6 Preparation of [hydroxy(organosulfonyloxy)iodo]arenes.

Compound	Method of synthesis	Yield (%)	Reference
PhI(OH)OTs	$PhI(OAc)_2$, $TsOH{\cdot}H_2O$, MeCN, rt	93	[195]
PhI(OH)OTs	$PhI(OAc)_2$, $TsOH{\cdot}H_2O$, grinding, rt, 10 min	93	[202]
PhI(OH)OTs	PhI, *m*CPBA, $TsOH{\cdot}H_2O$, $CHCl_3$, rt, 2 h	95	[203]
PhI(OH)OTs	PhH, I_2, *m*CPBA, TfOH, $TsOH{\cdot}H_2O$, CH_2Cl_2, rt	75	[204]
$4\text{-}MeC_6H_4I(OH)OTs$	$ArI(OAc)_2$, $TsOH{\cdot}H_2O$, MeCN, rt	86	[205]
$3\text{-}CF_3C_6H_4I(OH)OTs$	$ArI(OAc)_2$, $TsOH{\cdot}H_2O$, MeCN, rt, 1 h	91	[176, 204, 206]
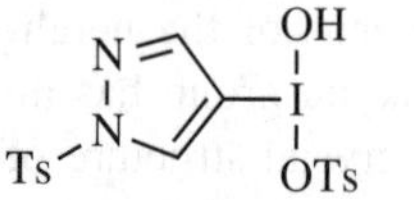 OH, I, OTs, N, N, Ts	$ArI(OAc)_2$, $TsOH{\cdot}H_2O$, MeCN, rt, 1 h	100	[206]
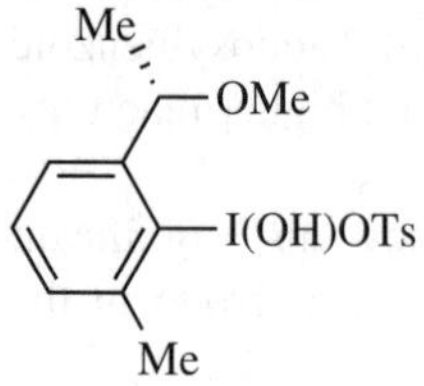 Me, OMe, I(OH)OTs, Me	$ArI(OAc)_2$, $TsOH{\cdot}H_2O$, MeCN, rt, 1 h	90	[166]
$C_4F_9I(OH)OTs$	$C_4F_9I(OCOCF_3)_2$, TsOH, MeCN, –20 °C to rt	91	[207]
$C_6F_{13}I(OH)OTs$	$C_6F_{13}I(OCOCF_3)_2$, TsOH, MeCN, 0 °C to rt, 24 h	100	[135]
$C_6F_5I(OH)OTs$	$C_6F_5I(OCOCF_3)_2$, MeCN, TsOH 0 °C to rt, 24 h	94	[135]
$CF_3CH_2I(OH)OTs$	$CF_3CH_2I(OCOCF_3)_2$, TsOH, MeCN, –30 °C to rt	90	[176]
$CF_3CH_2I(OH)OTs$	CF_3CH_2I, *m*CPBA, TsOH, CH_2Cl_2–CF_3CH_2OH, rt	95	[204]
PhI(OH)OMs	PhH, I_2, *m*CPBA, MsOH, CH_2Cl_2–CF_3CH_2OH, rt	76	[204]
$PhI(OH)OSO_2Ph$	PhH, I_2, *m*CPBA, $PhSO_2OH$, CH_2Cl_2/CF_3CH_2OH, rt	88	[204]
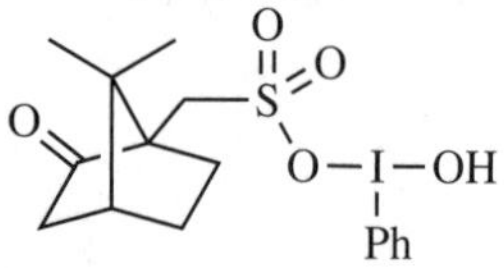 O, O, S, O, O–I–OH, Ph	$PhI(OAc)_2$, RSO_2OH, MeCN, rt, 0.5 h	80	[208]

2.1.6.1 Preparation

Various [hydroxy(tosyloxy)iodo]arenes **64** can be conveniently prepared by a ligand-exchange reaction of (diacetoxyiodo)arenes with *p*-toluenesulfonic acid monohydrate in acetonitrile (Scheme 2.26). This method has been applied to the synthesis of derivatives with various substituted aromatic groups [150, 166, 186, 209, 210], [hydroxy(tosyloxy)iodo]heteroarenes [206] and the recyclable hypervalent iodine reagents **65**–**68** (also see Chapter 5) [154, 155, 158]. Similarly, numerous polyfluoroalkyl derivatives of the types $C_nF_{2n+1}I(OH)OTs$ [135, 176, 211] and $C_nF_{2n+1}CH_2I(OH)OTs$ [176, 212] can be prepared from the respective bis(trifluoroacetates) and *p*-toluenesulfonic acid.

Wirth and coworkers have reported the preparation of a series of *ortho*-substituted chiral [hydroxy(tosyloxy)iodo]arenes **69** starting from the corresponding aryl halides (Scheme 2.27) [166, 191].

A similar procedure using $ArI(OAc)_2$ and 4-nitrobenzenesulfonic acid, methanesulfonic acid, or 10-camphorsulfonic acid leads to the corresponding organosulfonyloxy derivatives [195, 208]. A solvent-free, solid-state version of this reaction can be carried out by simple grinding of $ArI(OAc)_2$ with the appropriate

$$ArI(OAc)_2 + TsOH{\bullet}H_2O \xrightarrow[90\text{-}100\%]{MeCN,\ rt} ArI(OH)OTs$$

64

Ar = Ph, 2-MeC_6H_4, 4-MeC_6H_4, 2-BrC_6H_4, 2-$MeOC_6H_4$, 2,6-$Me_2C_6H_3$, 3-MeC_6H_4, 3-$MeOC_6H_4$, 3-CNC_6H_4, 3-$HOOCC_6H_4$, 3-$NO_2C_6H_4$, 3-$CF_3C_6H_4$, 3-HOC_6H_4, 2-naphthyl, 2-thienyl, 3-thienyl, etc.

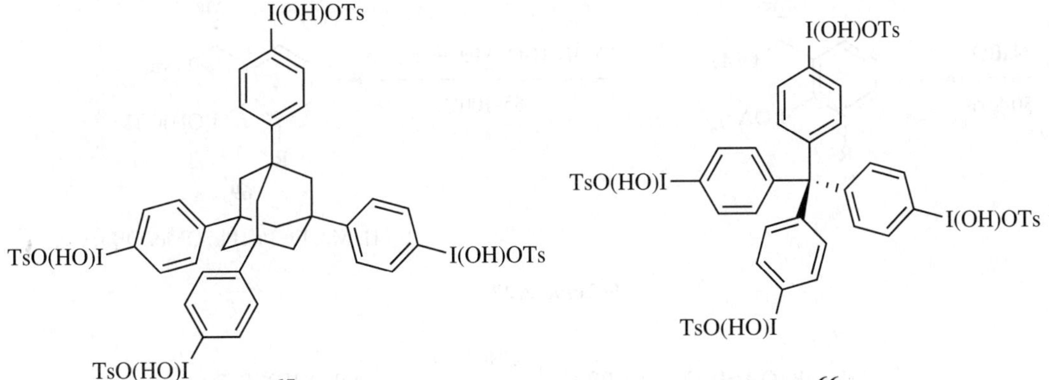

65 **66**

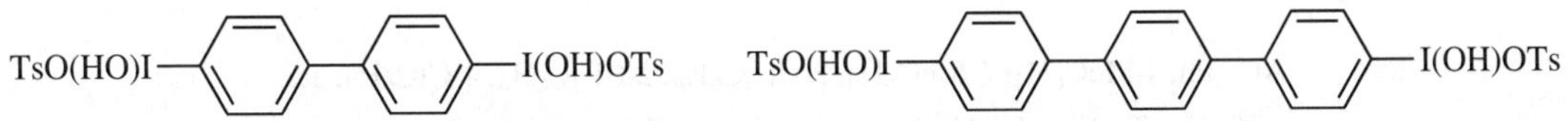

67 **68**

Scheme 2.26

sulfonic acid in an agate mortar followed by washing the solid residue with diethyl ether [202]. This solid-state procedure has been used for the preparation of HTIB and several other [hydroxy(organosulfonyloxy)iodo]arenes in 77–98% yields. A polymer-supported [hydroxy(tosyloxy)iodo]benzene can be prepared similarly by treatment of poly[(diacetoxy)iodo]styrene with *p*-toluenesulfonic acid monohydrate in chloroform at room temperature (Chapter 5) [213, 214].

A convenient modified procedure for the preparation of various [hydroxy(organosulfonyloxy)iodo]arenes **70** consists of the one-pot reaction of iodoarenes and *m*CPBA (*m*-chloroperoxybenzoic acid) in the presence of organosulfonic acids in a small amount of chloroform at room temperature (Scheme 2.28) [203].

Further modification of this procedure, developed by Olofsson and coworkers, involves a one-pot oxidative iodination/oxidation/ligand exchange sequence of reactions leading to the synthesis of [hydroxy(organosulfonyloxy)iodo]arenes **72** from arenes **71**, iodine, *m*CPBA and the respective sulfonic acids (Scheme 2.29) [204].

Several derivatives of HTIB of type PhI(OR)OTs have been reported. In particular, [methoxy(tosyloxy)iodo]benzene, PhI(OMe)OTs, can be prepared by the treatment of HTIB with trimethyl orthoformate [215]. The methoxy ligand in [methoxy(tosyloxy)iodo]benzene (**73**) can be further substituted with a chiral menthyloxy group by treatment with menthol to give product **74** (Scheme 2.30) [216].

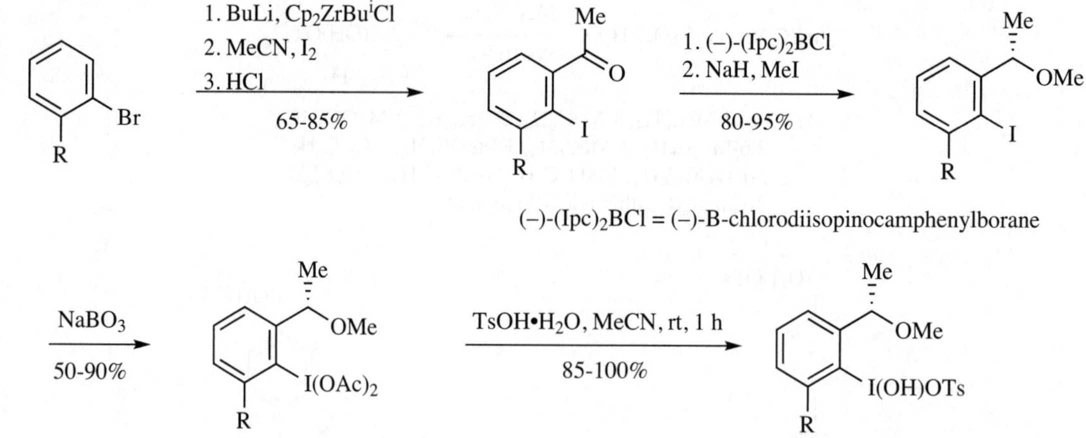

1. BuLi, Cp_2ZrBu^iCl
2. MeCN, I_2
3. HCl
65-85%

1. (–)-$(Ipc)_2BCl$
2. NaH, MeI
80-95%

(–)-$(Ipc)_2BCl$ = (–)-B-chlorodiisopinocamphenylborane

$NaBO_3$
50-90%

$I(OAc)_2$

$TsOH{\bullet}H_2O$, MeCN, rt, 1 h
85-100%

I(OH)OTs

69

R = H, Me, Et, Pr^i, Ph, OMe, OBn, etc.

Scheme 2.27

$$ArI + RSO_3H{\bullet}H_2O + mCPBA \xrightarrow[75\text{-}100\%]{CHCl_3,\ rt} ArI(OH)OSO_2R$$

70

*m*CPBA = *m*-chloroperoxybenzoic acid

Ar = Ph, 4-MeC_6H_4, 2-MeC_6H_4, 3-$CF_3C_6H_4$, 2-$CF_3C_6H_4$, 4-ClC_6H_4, etc.
RSO_3H = TsOH, MsOH, $PhSO_3H$, 4-$ClC_6H_4SO_3H$, 3-$NO_2C_6H_4$,
(*S*)-(+)-camphor-10-sulfonic acid, etc.

Scheme 2.28

$$ArH + I_2 + RSO_3H{\bullet}H_2O + mCPBA \xrightarrow[60\text{-}88\%]{CH_2Cl_2/CF_3CH_2OH,\ rt,\ 30\ min} ArI(OH)OSO_2R$$

71 **72**

ArH = PhH, $PhBu^t$, 1,4-$Me_2C_6H_4$, 1,3,5-$Me_3C_6H_3$, PhOMe
RSO_3H = TsOH, MsOH, $PhSO_3H$, 2-naphthylSO_3H

Scheme 2.29

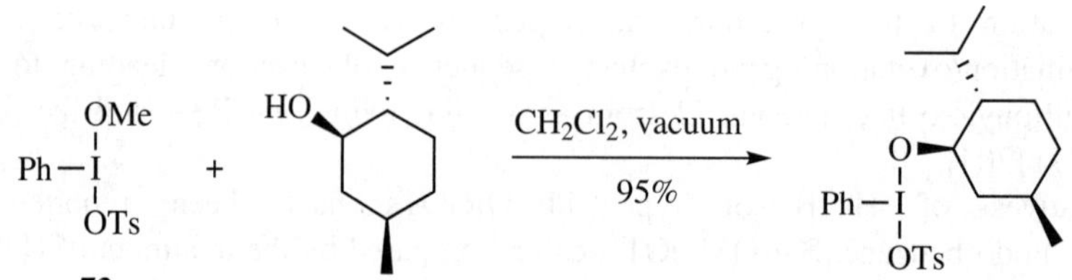

Ph–I(OMe)(OTs) + HO– ... $\xrightarrow[95\%]{CH_2Cl_2,\ vacuum}$ Ph–I(O–)(OTs)

73 **74**

Scheme 2.30

2.1.6.2 *Structural Studies*

Figure 2.8 shows several examples of aryliodine(III) organosulfonates whose X-ray crystal structures have been reported in the literature. Single-crystal X-ray structural data for HTIB (**75**) show the T-shaped geometry around the iodine center with almost collinear O-ligands and two different I—O bonds of 2.47 Å (I—OTs) and 1.94 Å (I—OH) [217]. The presence of a substituent in the phenyl ring has no noticeable effect on the molecular geometry of [hydroxy(tosyloxy)iodo]arenes. In particular, the X-ray single-crystal structure of 3-[hydroxy(tosyloxy)iodo]benzoic acid (**76**) is very similar to that of HTIB. The I—OTs bond in tosylate **76** (2.437 Å) is significantly longer than the I—OH bond (1.954 Å), which is indicative of some ionic character of this compound. In addition to the three intramolecular bonds, a weaker intermolecular coordination of iodine atom to one of the sulfonyl oxygens of the neighboring molecule is found with a distance of 2.931 Å. No intermolecular interaction involving *meta*-carboxylic group is present in molecule **76** [186].

The chiral, *ortho*-substituted compound **77** shows a similar T-shaped structure like HTIB (**75**). However, in contrast to HTIB, the two oxygen atoms nearest to the iodine atom in **77** are the oxygen atom of the hydroxy group (I—OH, 1.94 Å) and the methoxy oxygen atom (I···OMe, 2.47 Å), while the tosylate oxygen atom is further away from the iodine atom (I···OTs, 2.82 Å). Owing to this structural feature, the O–I–O bond angle of 166° in molecule **77** is significantly smaller than the O–I–O bond angle of 179° in HTIB [191].

In a more recent paper, Richter and coauthors reported a single-crystal X-ray structural analysis of [hydroxy(mesyloxy)iodo]benzene (**78**) and the respective oxo-bridged anhydride **79** [218]. Structural parameters of mesylate **78** are very similar to the structure of HTIB (**75**). Analogously to HTIB, compounds **78** and **79** form dimeric structures in the solid state due to the I···O secondary bonding [218].

Richter, Koser and coworkers analyzed the species present in aqueous solutions of [hydroxy(mesyloxy)iodo]benzene and [hydroxy(tosyloxy)iodo]benzene [219]. In particular, it was found that upon solution in water both PhI(OH)OMs and PhI(OH)OTs undergo complete ionization to give the hydroxy(phenyl)iodonium ion (PhI^+OH) and the corresponding sulfonate anion (RSO_3^-) as fully solvated species, that is, free ions, which do not form ion pairs with each other. In addition, the μ-oxo dimer, [Ph(HO)I—O—I^+(OH_2)Ph], is present at significant levels even in relatively dilute solutions [219].

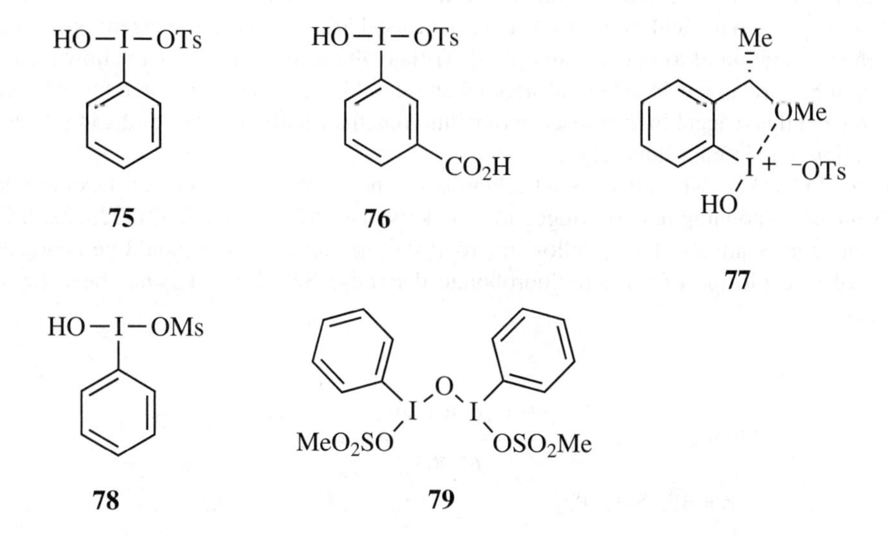

Figure 2.8 *Examples of aryliodine(III) organosulfonates analyzed by single-crystal X-ray diffraction.*

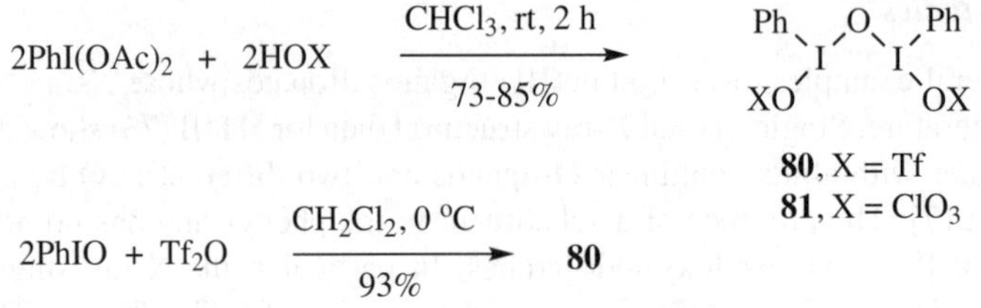

$$2PhI(OAc)_2 + 2HOX \xrightarrow[73\text{-}85\%]{CHCl_3,\ rt,\ 2\ h} Ph(XO)I\text{–}O\text{–}I(OX)Ph$$

80, X = Tf
81, X = ClO_3

$$2PhIO + Tf_2O \xrightarrow[93\%]{CH_2Cl_2,\ 0\ ^\circ C} \mathbf{80}$$

Scheme 2.31

2.1.7 Organoiodine(III) Derivatives of Strong Acids

In contrast to organoiodine(III) dihalides and dicarboxylates, the analogous aryliodine(III) compounds $ArI(OX)_2$ that are derived from strong acids HOX, such as H_2SO_4, HNO_3, $HClO_4$, CF_3SO_3H, $HSbF_6$ and HPF_6, usually lack stability and can only be generated at low temperature, under absolutely dry conditions. The existence of several such derivatives was assumed in the literature; however, none of them was isolated as an individual, stable compound. Specifically, triflate $PhI(OTf)_2$ [86], perchlorate $PhI(OClO_3)_2$ [220], sulfate $PhISO_4$ [221] and nitrate $PhI(ONO_2)_2$ [222] are unstable and can only be generated *in situ*, at low temperature, in the absence of water. Traces of moisture immediately convert these compounds into the μ-oxo-bridged derivatives (e.g., structures **80** and **81** in Scheme 2.31) or more complex polymeric structures, such as the oligomeric sulfate **26** and perchlorate **27** derivatives (Scheme 2.13 in Section 2.1.4.2).

μ-Oxo-bridged iodanes are the most typical organoiodine(III) derivatives of strong inorganic acids, such as sulfuric, nitric, perchloric, trifluoromethanesulfonic and so on; however, compounds of this type are also known for carboxylic acids (e.g., structure **57**, Figure 2.7).

The most general approach to the synthesis of μ-oxo-bridged iodanes is based on the reaction of (diacetoxyiodo)benzene or iodosylbenzene with the respective acids or anhydrides. Triflate **80** and perchlorate **81** are prepared from $PhI(OAc)_2$ and aqueous perchloric or trifluoromethanesulfonic (triflic) acid (Scheme 2.31) [220]. In a convenient modified procedure triflate **80** can be generated *in situ* from iodosylbenzene (two equivalents) and triflic anhydride in dichloromethane at 0 °C. Under these conditions triflate **80** is obtained as a bright yellow suspension in dichloromethane that can be used *in situ* as a useful reagent (known as Zefirov's reagent) for the preparation of iodonium salts [223]. Triflate **80** can be isolated as a yellow microcrystalline solid; however, it has a relatively low thermal stability and should be handled and stored for only brief periods. Extended storage of this reagent in the presence of trifluoromethanesulfonic acid results in self-condensation with the formation of iodonium salts [224].

A reaction of $PhI(OAc)_2$ with aqueous tetrafluoroboric, hexafluoroantimonic and hexafluorophosphoric acids affords the corresponding μ-oxo-bridged iodane derivatives **82** (Scheme 2.32) [225,226]. Compounds **82** can be isolated as relatively stable, yellow microcrystalline solids, which should be handled with care. **Caution**: a violent explosion of the tetrafluoroborate derivative **82** (X = BF_4) has been reported in the literature [227].

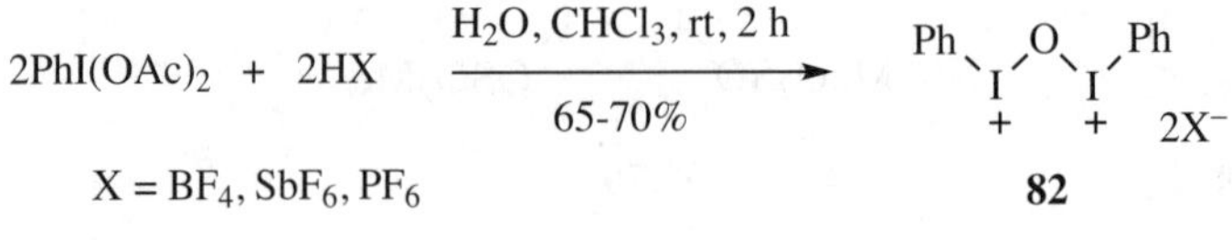

$$2PhI(OAc)_2 + 2HX \xrightarrow[65\text{-}70\%]{H_2O,\ CHCl_3,\ rt,\ 2\ h} Ph\overset{+}{I}\text{–}O\text{–}\overset{+}{I}Ph\ \ 2X^-$$

X = BF_4, SbF_6, PF_6

82

Scheme 2.32

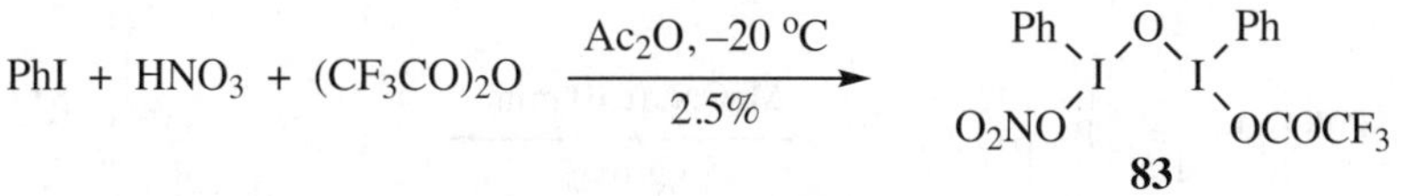

PhI + HNO_3 + $(CF_3CO)_2O$ —(Ac_2O, −20 °C; 2.5%)→ Ph(O₂NO)I–O–I(Ph)(OCOCF₃) **83**

Scheme 2.33

A distinctive feature of all μ-oxo-bridged iodanes is the bright yellow color due to the $^{+}$I—O—I^{+} bridging fragment [119]. The available X-ray structural data for the mixed μ-oxo-bridged iodine(III) nitrato-trifluoroacetate **83** (Scheme 2.33) is also consistent with the partially ionic character of the I—ONO_2 bonds with I—O bond distances of 2.26–2.35 Å as compared to the I—O—I bridge I—O bond distances of 1.99–2.01 Å [228]. Product **83** was isolated in a low yield from the reaction of iodobenzene with nitric acid and trifluoroacetic anhydride in a mixture with the respective symmetrical μ-oxo-bridged iodine(III) bis(nitrate) and bis(trifluoroacetate) [228].

Sulfur trioxide forms two different adducts with iodosylbenzene, PhIO·SO_3 and $(PhIO)_2$·SO_3, depending on the ratio of reactants (Scheme 2.34) [221]. Both sulfates **84** and **85** are hygroscopic solids and are highly reactive toward nucleophilic organic substrates. Taylor and coworkers developed a more convenient procedure for the preparation of phenyliodosulfate (**85**) by the reaction of iodosylbenzene and trimethylsilyl chlorosulfonate followed by removal of the solvent and chlorotrimethylsilane [229]. In addition to phenyliodine(III) sulfates **84** and **85**, the oligomeric iodosylbenzene sulfate **26**, corresponding to a molecular formula of $(PhIO)_3$·SO_3, has been prepared by the reaction of (diacetoxyiodo)benzene with $NaHSO_4$ in water (Scheme 2.13 in Section 2.1.4.2) [116, 117].

Piancatelli and coworkers reported the preparation of an unusual iodosylbenzene derivative of perchloric acid, diperchlorate $PhI(OClO_3)_2$, by the reaction of (diacetoxyiodo)benzene with magnesium perchlorate [230]. The structure of this perchlorate was suggested on the basis of NMR and HPLC-MS data; [230] however, a more recent X-ray single-crystal structural study established a different structure for a product prepared by a similar procedure from PhI(OH)OTs and aqueous magnesium perchlorate (see structure **27** in Scheme 2.13, Section 2.1.4.2) [118].

Several organoiodine(III) derivatives of phosphoric acid are known. In particular, Moriarty and coworkers reported the preparation of [hydroxy(phosphoryloxy)iodo]benzenes **87** by the reaction of iodosylbenzene with the appropriate phosphonic or phosphinic acid **86** (Scheme 2.35) [231].

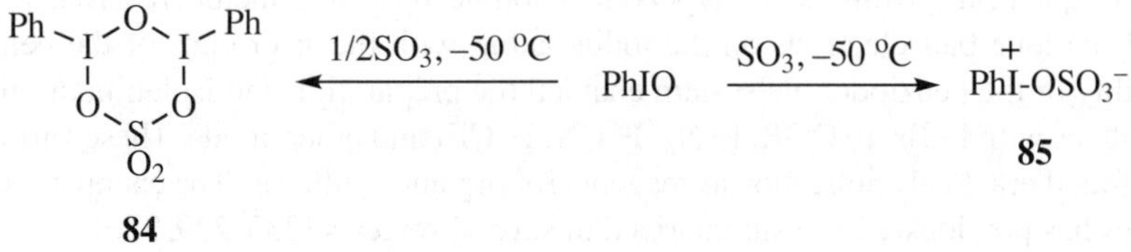

84 (cyclic Ph–I(–O–)I–Ph bridged by O–SO_2–O) ←(1/2SO_3, −50 °C)— PhIO —(SO_3, −50 °C)→ $^{+}$PhI-OSO_3^{-} **85**

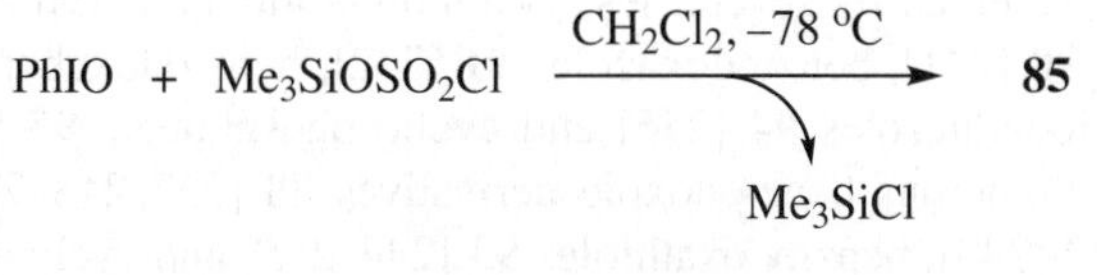

PhIO + Me_3SiOSO_2Cl —(CH_2Cl_2, −78 °C; −Me_3SiCl)→ **85**

Scheme 2.34

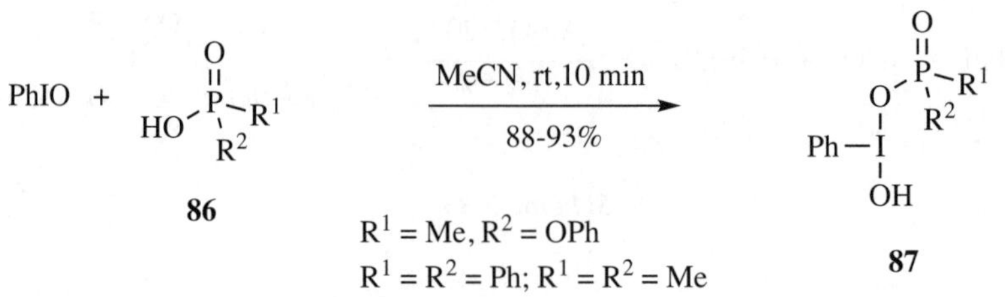

PhIO + HO–P(=O)(R¹)(R²) **86** → (MeCN, rt, 10 min; 88-93%) → Ph–I(OH)–O–P(=O)(R¹)(R²) **87**

R^1 = Me, R^2 = OPh
R^1 = R^2 = Ph; R^1 = R^2 = Me

Scheme 2.35

2.1.8 Iodine(III) Heterocycles

Nearly all of the known heterocyclic λ^3-iodanes are five-membered heterocycles, although several examples of four-membered and six-membered heterocycles with the iodine(III) atom in the ring have also been reported (Section 2.1.8.8). Several types of cyclic iodonium salts are also known (Section 2.1.8.10). Unsaturated heterocyclic systems with a hypervalent iodine atom in the ring generally do not possess any significant aromatic character due to the large iodine atom size precluding p-orbital overlap with much smaller atoms of carbon, oxygen, or nitrogen and also due to the electronic nature and the geometry of hypervalent bonding.

The five-membered iodine(III) heterocycles are represented by various cyclic compounds (**88–95**) incorporating hypervalent iodine and oxygen, nitrogen, or some other elements in the heterocyclic ring (Figure 2.9). The collective name "benziodoxoles" is commonly used for the heterocycles **88** with iodine and oxygen atoms in a five-membered ring and various substituents Y attached to iodine [128]. The first derivatives of benziodoxole, 1-hydroxy-1,2-benziodoxol-3-(1*H*)-one and 1-chloro-1,2-benziodoxol-3-(1*H*)-one, were prepared over 100 years ago by oxidation or chlorination of 2-iodobenzoic acid [232, 233]. In the mid-1980s, 1-hydroxybenziodoxoles attracted considerable interest and research activity mainly due to their excellent catalytic activity in the cleavage of toxic phosphates and reactive esters [234]. More recently, various new benziodoxole derivatives have been synthesized and their usefulness as reagents for organic synthesis and particularly for the atom-transfer reactions [235], was demonstrated. In contrast to benziodoxoles **88**, the analogous five-membered iodine-nitrogen heterocycles, benziodazoles **89**, have received much less attention and, moreover, their structural assignment in some cases was unreliable. The most important and readily available derivative of benziodazole, acetoxybenziodazole (**89**, Y = OAc, Z = H), was first prepared in 1965 by the peracetic oxidation of 2-iodobenzamide [236].

The distinctive feature of five-membered heterocyclic λ^3-iodanes is a considerably higher thermal stability and lower reactivity compared to their acyclic analogs. This stabilization has been explained by the bridging of the apical and equatorial positions on hypervalent iodine by a five-membered ring [237] and also by better overlap of the lone pair electrons on the iodine atom with the π-orbitals of the benzene ring [238]. The greater stability of the benziodoxole system enabled the preparation and isolation of otherwise unstable iodine(III) derivatives with I–Br, I–OOR, I–N_3, I–CN, I–CF_3 and other bonds. These various benziodoxole derivatives have found practical application as reagents for organic synthesis. The chemistry of benziodoxoles and benziodazoles has previously been summarized in several reviews [235, 239, 240].

Besides benziodoxoles and benziodazoles, the other known five-membered heterocyclic systems incorporating hypervalent iodine are represented by several less common compounds (**90–95**, Figure 2.9), which include the fused benziodazoles **90** [241], benziodoxazoles **91** [242], benziodoxaboroles **92** [243], benziodoxathioles **93** [244, 245], benziodathiazoles **94** [246] and cyclic phosphonate **95** [247]. X-Ray molecular structures were reported for numerous benziodoxole derivatives **88** [237, 248–262], benziodazoles **89** [263–266], benziodoxaboroles **92** [243], benziodoxathioles **93** [244, 245] and cyclic phosphonate **95** [247]. In general, the five-membered ring in benziodoxole is highly distorted with an almost linear alignment of the

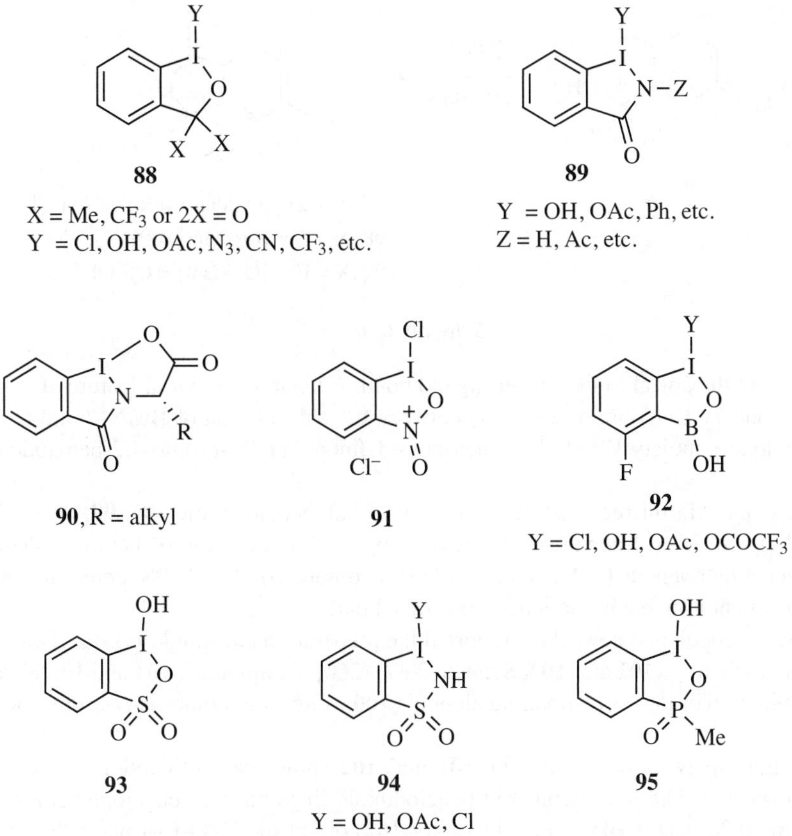

88

X = Me, CF_3 or 2X = O
Y = Cl, OH, OAc, N_3, CN, CF_3, etc.

89

Y = OH, OAc, Ph, etc.
Z = H, Ac, etc.

90, R = alkyl

91

92

Y = Cl, OH, OAc, $OCOCF_3$

93

94

Y = OH, OAc, Cl

95

Figure 2.9 *Known hypervalent iodine(III) five-membered heterocyclic systems.*

two electronegative ligands. The I—O bond length in benziodoxolones (**88**, 2X = O) varies over a wide range from 2.11 Å in carboxylates (**88**, Y = *m*-$ClC_6H_4CO_2$) [255] to 2.48 Å in the phenyl derivative (**88**, Y = Ph) [249], which indicates considerable changes in the ionic character of this bond. The endocyclic C I O bond angle is typically around 80°, which is a significant deviation from the expected angle of 90° for the normal T-shaped geometry of hypervalent iodine. The structural parameters of benziodazoles **89**, Y = OAc or Ph, in general are similar to those of benziodoxoles [263–266].

2.1.8.1 Benziodoxoles

2.1.8.1.1 Halobenziodoxoles 1-Chloro-1,2-benziodoxol-3-(1*H*)-one (**88**, 2X = O, Y = Cl) can be easily prepared by direct chlorination of 2-iodobenzoic acid [233], or by the oxidation of 2-iodobenzoic acid with sodium chlorite ($NaClO_2$) in aqueous hydrochloric acid media [267]. The original X-ray single-crystal analysis of 1-chloro-1,2-benziodoxol-3-(1*H*)-one reported in 1976 was relatively imprecise [268]. More recently, Koser and coworkers reported the single-crystal X-ray structure of a 1 : 1 complex of 1-chloro-1,2-benziodoxol-3-(1*H*)-one and tetra-*n*-butylammonium chloride [262]. The primary bond distances at iodine in this compound are consistent with expectations for a λ^3-iodane. In particular, the I—Cl and I—O bond distances of 2.454 and 2.145 Å, respectively, are greater than the sums of the appropriate covalent radii and reflect the

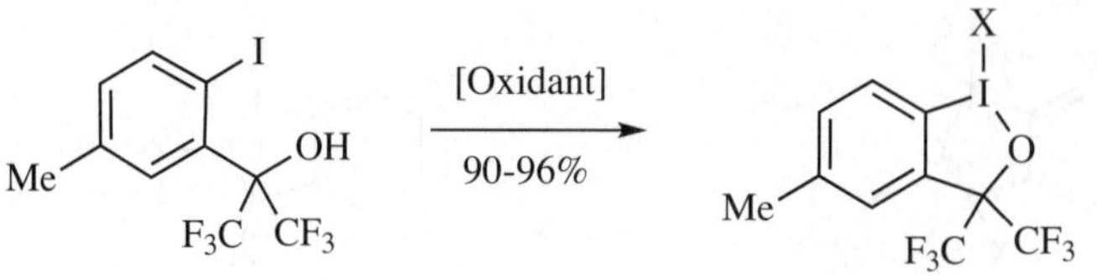

96

97, X = Cl, [Oxidant] = Bu^tOCl or Cl_2
98, X = Br, [Oxidant] = Br_2
99, X = F, [Oxidant] = CF_3OF

Scheme 2.36

hypervalent nature of the apical iodine–heteroligand bond. A notable structural feature of this complex is the existence of a secondary I···Cl bond (2.943 Å) between the chloride ion of Bu_4NCl and the iodine atom of the chlorobenziodoxole moiety [262]. The analogous 1-fluoro- and 1-bromo-1,2-benziodoxol-3-(1*H*)-ones are unknown.

In 1979, Amey and Martin reported the synthesis of halobenziodoxoles **97–99**, derivatives of the 3,3-bis(trifluoromethyl)-3-(1*H*)-1,2-benziodoxole system, by the halogenation of benzylic alcohol **96** with the respective halogenating reagents (Scheme 2.36) [237]. Bromobenziodoxole **98** represents an unusual, stable derivative of polyvalent iodine with an iodine–bromine bond.

More recently, Braddock and coworkers reported the preparation and single-crystal X-ray crystallography of two bromobenziodoxoles (**101** and **102**, Scheme 2.37) [260]. Compounds **101** and **102** were synthesized in one step by bromination of the corresponding alcohols **100** using stoichiometric *N*-bromosuccinimide (NBS) in chloroform.

X-Ray diffraction analyses of crystals of **101** and **102** show that both iodine centers adopt a planar T-geometry, as expected. The five-membered benziodoxole rings have an envelope conformation, with the oxygen atom lying 0.23 Å (for **101**) and 0.44 Å (for **102**) out of the C–C–C–I plane. For both compounds, the closest intermolecular distance to the iodine center is from the oxygen of a centrosymmetrically related counterpart at 3.03 Å (for **101**) and 2.94 Å (for **102**), forming discrete dimer pairs. Most notably, while the I—C bond length is essentially constant [2.120(8) Å in 101 and 2.112(7) Å in 102], the I—O and I—Br bond lengths vary considerably in **101** and **102**, presumably due to the effect of the electron-withdrawing *gem*-trifluoromethyl groups on the three-center-four-electron O—I—Br bond. The I—O bond length is ca. 2.12 Å in **101** and ca. 2.05 Å in **102**. The I—Br bond lengths are ca. 2.59 and ca. 2.69 Å, respectively [260]. The structure of **101** is isomorphous with that of its previously reported chloro analog [269]. The X-ray structure of the fluoro analog of **102**, 1-fluoro-3,3-dimethyl-1,3-dihydro-1λ^3-benzo[*d*][1, 2]iodoxole, is characterized by a very short I–F distance of 2.045 Å, while the I—O and the I—C bonds (2.02 and 2.09 Å, respectively) are comparable with the analogous bonds in **102** [270].

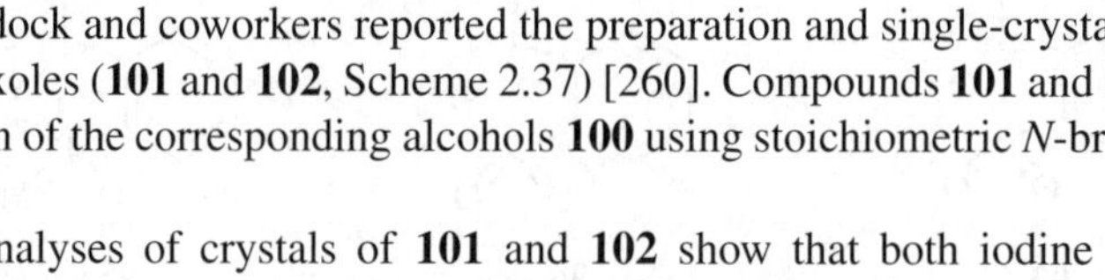

100

101, R = CF_3
102, R = CH_3

Scheme 2.37

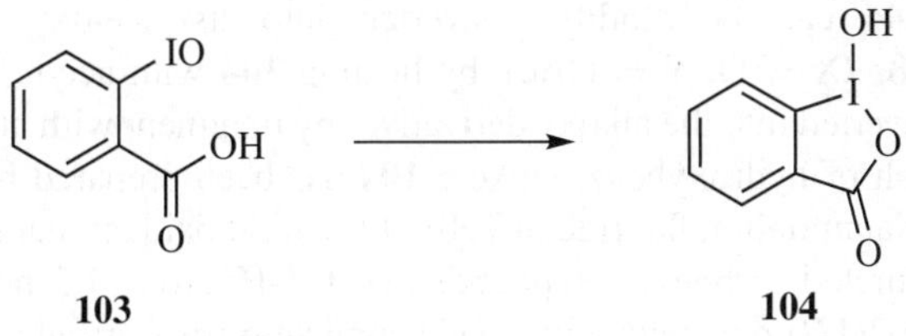

Scheme 2.38

2.1.8.1.2 Hydroxybenziodoxoles The most important and thoroughly investigated benziodoxole derivative is 1-hydroxy-1,2-benziodoxole-3(1*H*)-one (**104**), the cyclic tautomer of 2-iodosylbenzoic acid (**103**) (Scheme 2.38). The tautomeric form **104** represents the actual structure of this compound, as confirmed by its unusually low acidity (pK_a 7.25 against 2.85 for the hypothetical 2-iodosylbenzoic acid) [238] and unambiguously established by a single-crystal X-ray analysis of **104** in the solid state [253, 271]. The I—O bond distance of 2.30 Å in the five-membered ring of **104** is significantly longer than the computed covalent I—O bond length of 1.99 Å, which is indicative of the highly ionic nature of this bond [252]. The early report [272] on the existence of the noncyclic tautomeric form of a ring substituted 2-iodosylbenzoic acid (**103**) was proved to be wrong in more recent work [273].

1-Hydroxy-1,2-benziodoxole-3(1*H*)-one (**104**) is commercially available or can be easily prepared by direct oxidation of 2-iodobenzoic acid or by basic hydrolysis of 2-(dichloroiodo)benzoic acid [232, 233, 274]. A more recent preparative procedure for **104** involves the oxidation of 2-iodobenzoic acid with acetyl nitrate in acetic anhydride at room temperature followed by aqueous work-up [275]. In the 1980s benziodoxole **104** and other hydroxybenziodoxoles attracted considerable research interest due to their excellent catalytic activity in the cleavage of toxic phosphates and reactive esters. This activity is explained by a pronounced O-nucleophilicity of the benziodoxole anion **105** due to the α-effect [234, 272, 276]. Spectroscopic and kinetic mechanistic studies indicate that the highly unstable iodoxole derivatives, such as the phosphate **107**, are reactive intermediates in catalytic cleavage of phosphates, as shown for the catalytic hydrolysis of a typical substrate **106** (Scheme 2.39) [277–279]. This mechanism was proved by the synthesis and reactivity studies of the phosphate intermediate **107**.

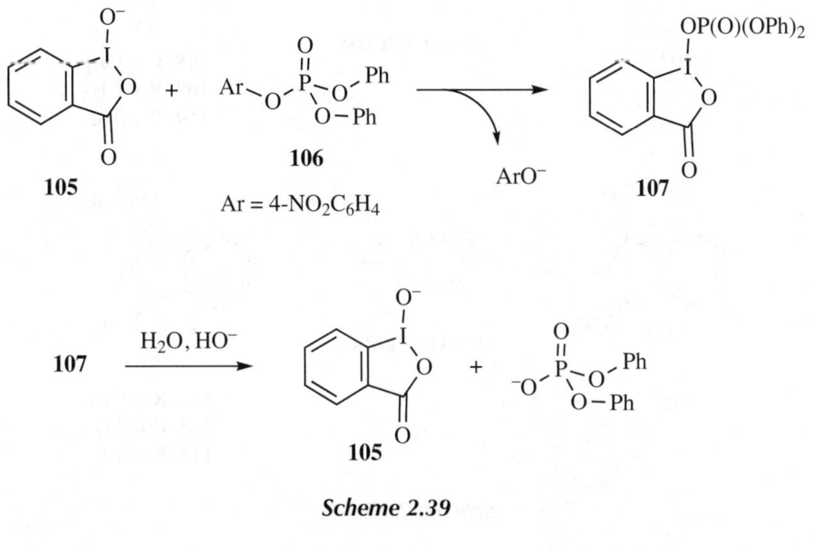

Ar = 4-$NO_2C_6H_4$

Scheme 2.39

Hydroxybenziodoxole **104** can be readily converted into its acetoxy derivative, 1-acetoxy-1,2-benziodoxole-3(1*H*)-one (**88**, 2X = O, Y = OAc), by heating **104** with acetic anhydride and the acetoxy derivative can be further converted into the alkoxy derivatives by treatment with an appropriate alcohol [280]. The tetrabutylammonium salt of hydroxybenziodoxole **104** has been prepared by the reaction of hydroxybenziodoxole with tetrabutylammonium fluoride in THF; it is a mild oxidant that is useful for the preparation of epoxides from α,β-unsaturated carbonyl compounds [281]. 1-Hydroxy-1,2-benziodoxole-3(1*H*)-one and 1-acetoxy-1,2-benziodoxole-3(1*H*)-one have found wide application as starting compounds for the synthesis of various benziodoxole-based hypervalent iodine reagents by ligand exchange on iodine [239].

2.1.8.1.3 Organosulfonate Derivatives of Benziodoxoles 1-Organosulfonyloxy-1,2-benziodoxole-3(1*H*)-ones **108–110** and the analogous sulfonate derivatives **112–114** of the 3,3-bis(trifluoromethyl)-3-(1*H*)-1,2-benziodoxole system can be readily prepared by a simple, one-step procedure starting from **104** or **111** and the corresponding sulfonic acids or trimethylsilyl triflate (Scheme 2.40) [282, 283]. The organosulfonates **108–110** and **112–114** are isolated as moderately hygroscopic, but thermally stable, crystalline solids. The most stable to moisture are the tosylates **110** and **114**. The mesylates and triflates are more hygroscopic and can be isolated only in the form of crystallohydrates; however, for further reactions they can be conveniently used *in situ* [283]. An alternative procedure for preparing benziodoxole triflate **108** in moderate yield consists of the oxidation of 2-iodobenzoic acid with *m*-chloroperoxybenzoic acid in the presence of trifluoromethanesulfonic acid in dichloromethane at room temperature [284].

Like the other iodine(III) derivatives of strong acids, sulfonates **108–110** and **112–114** are highly reactive toward unsaturated organic substrates and other carbon nucleophiles and have found practical application for the preparation of C-substituted benziodoxoles.

2.1.8.1.4 Alkylperoxybenziodoxoles The greater stability of heterocyclic iodanes enables the isolation of otherwise unstable iodine(III) derivatives with I—OOR bonds. Ochiai and coworkers reported the preparation

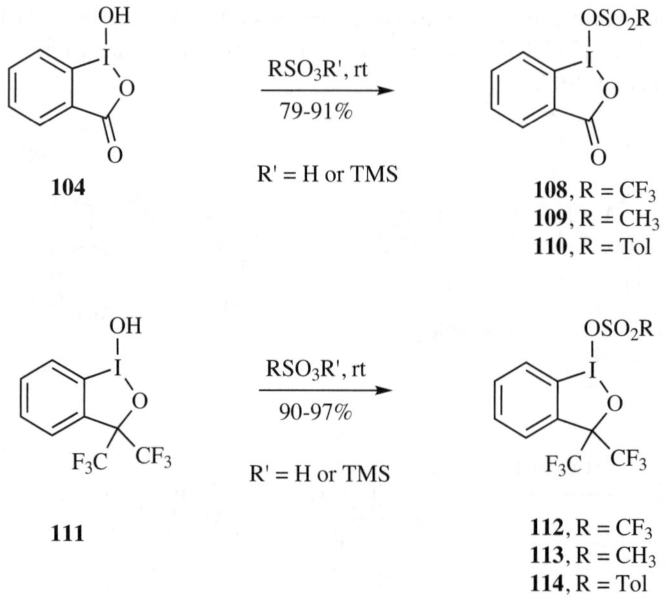

OH
I
O
O
104

RSO_3R', rt
79-91%
R' = H or TMS

OSO_2R
I
O
O
108, R = CF_3
109, R = CH_3
110, R = Tol

OH
I
O
F_3C CF_3
111

RSO_3R', rt
90-97%
R' = H or TMS

OSO_2R
I
O
F_3C CF_3
112, R = CF_3
113, R = CH_3
114, R = Tol

Scheme 2.40

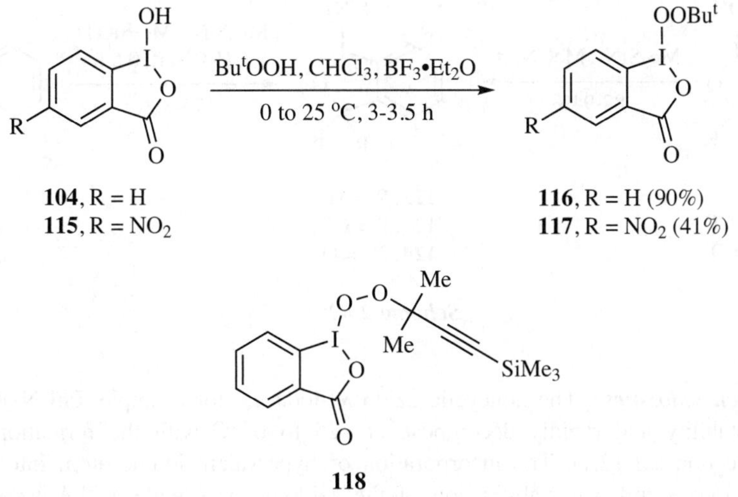
OH
I
O
O
R
Bu^{t}OOH, CHCl$_3$, BF$_3$•Et$_2$O
0 to 25 °C, 3-3.5 h
OOBu^t
I
O
O
R
104, R = H
115, R = NO$_2$
116, R = H (90%)
117, R = NO$_2$ (41%)
Me
O
O
I
Me
O
SiMe$_3$
O
118

Scheme 2.41

of 1-(*tert*-butylperoxy)benziodoxoles **116** and **117** by treatment of the corresponding benziodoxoles **104** and **115** with *tert*-butyl hydroperoxide in the presence of BF_3·etherate (Scheme 2.41) [256, 285]. The structure of 1-(*tert*-butylperoxy)benziodoxole **116** was established by a single-crystal X-ray analysis. In particular, the *tert*-butylperoxy group occupies an apical position of a distorted trigonal-bipyramidal geometry around the iodine. There are two primary hypervalent I—O bonds: a considerably ionic endocyclic bond [2.181 (5) Å] to the ring oxygen and a short exocyclic bond [2.039 (5) Å] to the peroxide oxygen. The peroxide O—O bond in the molecule of **116** has a bond length of 1.459 (7) Å, which is comparable to that in anhydrous hydrogen peroxide [256]. A similar peroxybenziodoxole (**118**) was formed in an unusual reaction of **104** with 1,3-bis(trimethylsilyl)-3-methylbut-1-yne in the presence of $BF_3{\cdot}Et_2O$ in dichloromethane; the structure of **118** was determined by single-crystal X-ray analysis [257]. Peroxides **116–118** are stable, crystalline products, which can be safely stored at room temperature for an indefinite period of time.

Dolenc and Plesnicar reported an alternative procedure for the preparation of 1-(*tert*-butylperoxy)benziodoxoles **120** from the corresponding chlorobenziodoxoles **119** and *tert*-butyl hydroperoxide in the presence of potassium *tert*-butoxide in THF (Scheme 2.42) [286].

Peroxyiodane **116** is a useful reagent acting as a strong oxidant toward various organic substrates, such as ethers, organic sulfides, amides and phenols [285, 287–291].

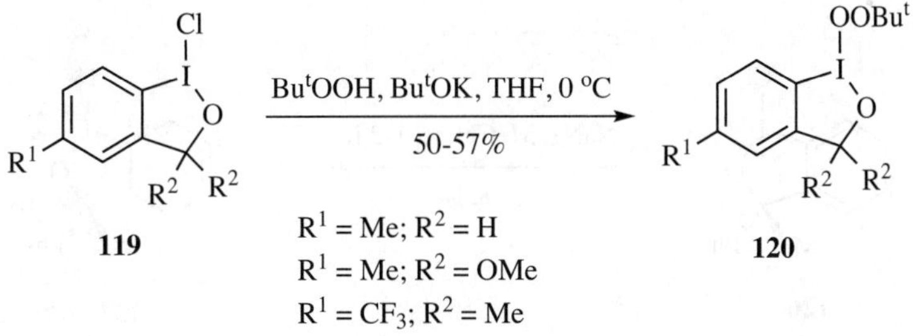
Cl
I
O
R^1
R^2 R^2
119
Bu^{t}OOH, Bu^{t}OK, THF, 0 °C
50-57%
R^1 = Me; R^2 = H
R^1 = Me; R^2 = OMe
R^1 = CF$_3$; R^2 = Me
OOBu^t
I
O
R^1
R^2 R^2
120

Scheme 2.42

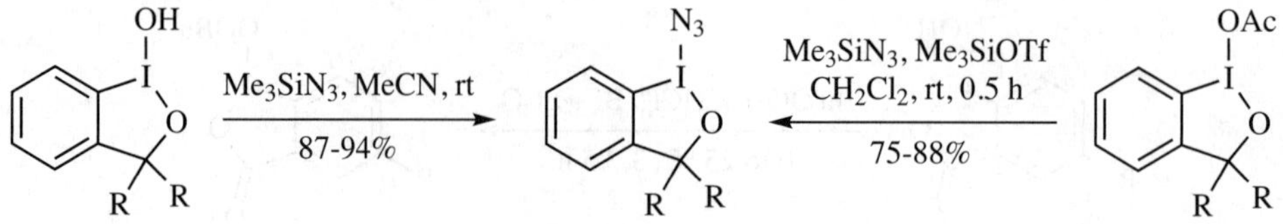

Me_3SiN_3, MeCN, rt
87-94%

Me_3SiN_3, Me_3SiOTf
CH_2Cl_2, rt, 0.5 h
75-88%

121
R = Me, CF_3
or 2R = O

122, R = Me
123, R = CF_3
124, 2R = O

125
R = Me,
or 2R = O

Scheme 2.43

2.1.8.1.5 Azidobenziodoxoles The noncyclic azido λ^3-iodanes, for example, $PhI(N_3)OAc$ or $PhI(N_3)_2$, in general lack stability and rapidly decompose at –25 to 0 °C with the formation of iodobenzene and dinitrogen (Section 2.1.12.1). The incorporation of hypervalent iodine atom into a five-membered heterocycle leads to a significant stabilization of the azidoiodane. Stable azidobenziodoxoles **122–124** can be prepared by the reaction of hydroxybenziodoxoles **121** with trimethylsilyl azide in acetonitrile [251, 292], or by treatment of acetoxybenziodoxoles **125** with trimethylsilyl azide in dichloromethane in the presence of catalytic trimethylsilyl triflate (Scheme 2.43) [259]. All three azides **122–124** were isolated as thermally stable, non-explosive, microcrystalline solids that can be stored indefinitely in a refrigerator.

The structure of azidobenziodoxole **123** was unambiguously established by a single-crystal X-ray analysis [251]. The structural data revealed the distorted T-shaped geometry typical of hypervalent iodine, with an N—I—O bond angle of 169.5°. The lengths of the bonds to the iodine atom, I—N (2.18 Å), I—O (2.13 Å) and I—C (2.11 Å), are within the range of typical single covalent bonds in organic derivatives of polyvalent iodine [251].

Azidobenziodoxoles can be used as efficient radical azidating reagents toward various organic substrates [251,293].

Koser and Rabah have reported the synthesis of optically active 1-azido-1,3-dihydro-3-methyl-3-phenyl-1,2-benziodoxole (**127**) from the corresponding chlorobenziodoxole **126** (Scheme 2.44) [294]. This chiral, non-racemic azidobenziodoxole is a potentially useful reagent for asymmetric azidation reactions.

2.1.8.1.6 Amidobenziodoxoles Amidobenziodoxoles **128–132** are readily synthesized in one step by treatment of the triflate **108** with an appropriate amide, RNH_2 (Scheme 2.45) [295]. All five adducts (**128–132**) were isolated as thermally stable, white, non-hygroscopic, microcrystalline solids. Amidobenziodoxoles can be used as amidating reagents toward polycyclic alkanes under radical conditions [295].

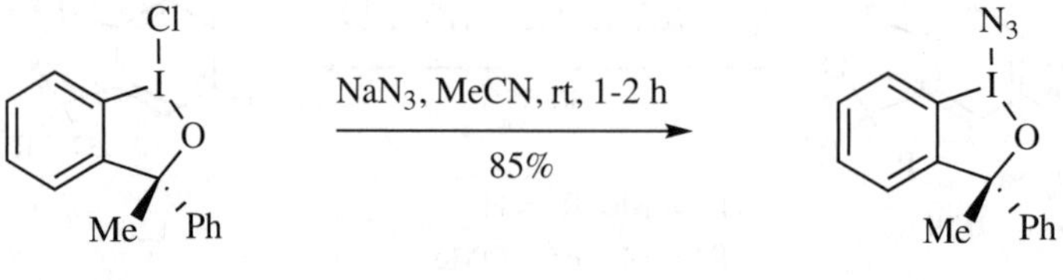

NaN_3, MeCN, rt, 1-2 h
85%

126 **127**

Scheme 2.44

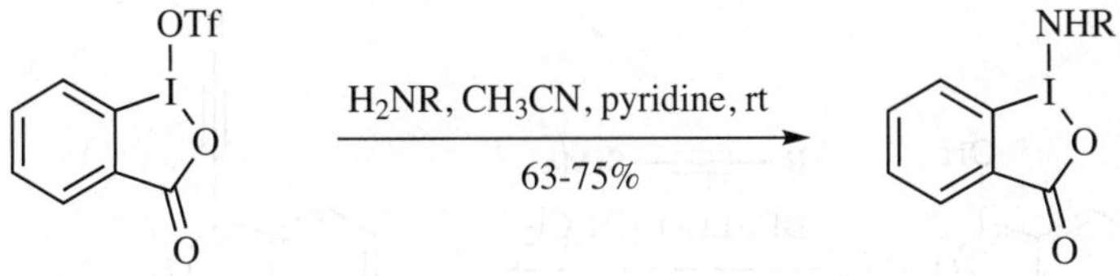

OTf

H_2NR, CH_3CN, pyridine, rt

63-75%

NHR

108

128, R = Ac
129, R = C(O)-4-ClC_6H_4
130, R = C(O)Et
131, R = C(O)NH_2
132, R = Ts

Scheme 2.45

2.1.8.1.7 Cyanobenziodoxoles Cyanobenziodoxoles **133–135** can be synthesized in one step by the reaction of cyanotrimethylsilane with the respective hydroxybenziodoxoles **121** (Scheme 2.46) [258, 296], or from acetoxybenziodoxole and cyanotrimethylsilane [259].

All three products **133–135** were isolated as thermally stable, white, microcrystalline solids. The structures of two of them (**134** and **135**) were unambiguously established by single-crystal X-ray analysis [258, 259]. In particular, X-ray structural data for **134** revealed a distorted T-shaped geometry expected for hypervalent iodine with an endocyclic C—I—O bond angle of 78.2° and a NC—I—O bond angle of 169.5° [258]. The lengths of the bonds to the iodine atom, I—CN (2.167 Å), I—O (2.117 Å) and I—Ar (2.112 Å), are within the range of typical single covalent bond lengths in noncyclic organic derivatives of polyvalent iodine.

The chemical reactivity of cyanobenziodoxoles **133–135** is generally similar to that of azidobenziodoxoles and they can be used as efficient cyanating reagents toward organic substrates [258, 296].

2.1.8.1.8 Alkynylbenziodoxoles The first preparation of 1-alkynylbenziodoxoles **136–138** by a reaction of hydroxybenziodoxole **104** with alkynyltrimethylsilanes and $BF_3{\cdot}Et_2O$ (Scheme 2.47) was reported by Ochiai and coworkers in 1991 [252]. The structure of the cyclohexyl derivative **137** was established by a single-crystal X-ray diffraction analysis [252]. The structural data revealed the usual, distorted T-shaped geometry about the iodine atom. The I—O bond distance of 2.34 Å in **137** is significantly longer than the computed covalent single bond length of 1.99 Å, which is indicative of a highly ionic nature of this bond.

A better procedure for the preparation of various alkynylbenziodoxoles (**136**, **139–146**) in high yields involves the reaction of triflates **108** and **112** with alkynyltrimethylsilanes according to Scheme 2.48 [283]. Further modification of this approach consists of the reaction of 1-acetoxy-1,2-benziodoxol-3(1*H*)-one with 1-alkynyl(diisopropyl)boronates in acetonitrile under reflux conditions [297].

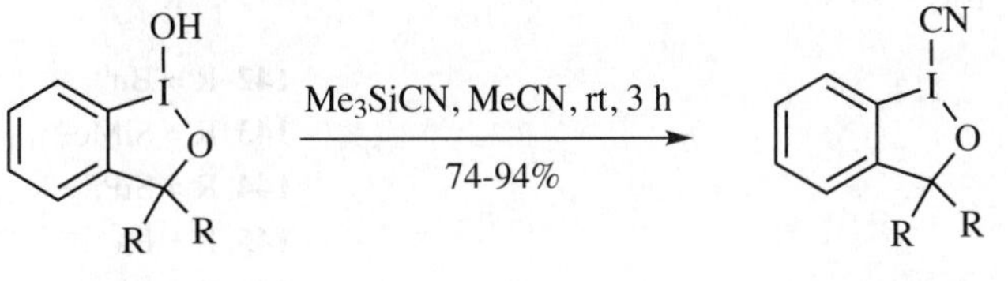

OH

Me_3SiCN, MeCN, rt, 3 h

74-94%

CN

121
R = Me, CF_3
or 2R = O

133, R = Me
134, R = CF_3
135, 2R = O

Scheme 2.46

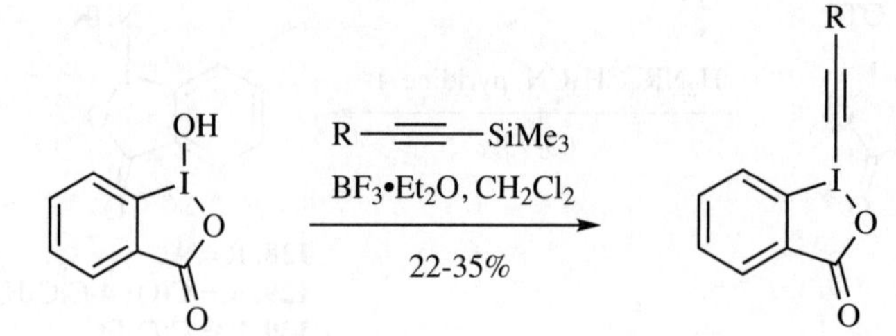

$$\mathbf{104} \xrightarrow[\text{22-35\%}]{\text{R—≡—SiMe}_3,\ \text{BF}_3\text{•Et}_2\text{O, CH}_2\text{Cl}_2} \mathbf{136\text{–}138}$$

136, R = Bu^t
137, R = *cyclo*-C_6H_{11}
138, R = *n*-C_8H_{17}

Scheme 2.47

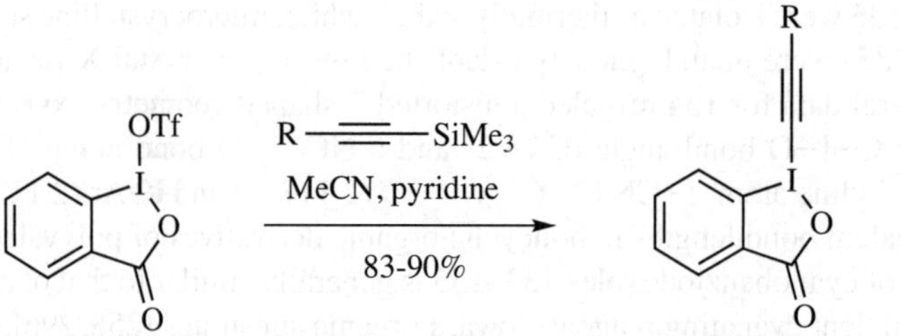

$$\mathbf{108} \xrightarrow[\text{83-90\%}]{\text{R—≡—SiMe}_3,\ \text{MeCN, pyridine}} \mathbf{136, 139\text{–}141}$$

136, R = Bu^t
139, R = $SiMe_3$
140, R = $SiPr^i_3$
141, R = Ph

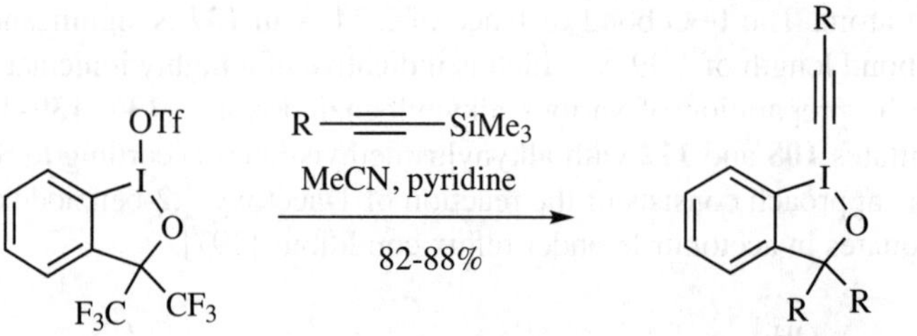

$$\mathbf{112} \xrightarrow[\text{82-88\%}]{\text{R—≡—SiMe}_3,\ \text{MeCN, pyridine}} \mathbf{142\text{–}146}$$

142, R = Bu^t
143, R = $SiMe_3$
144, R = $SiPr^i_3$
145, R = Ph
146, R = 1-cyclohexenyl

Scheme 2.48

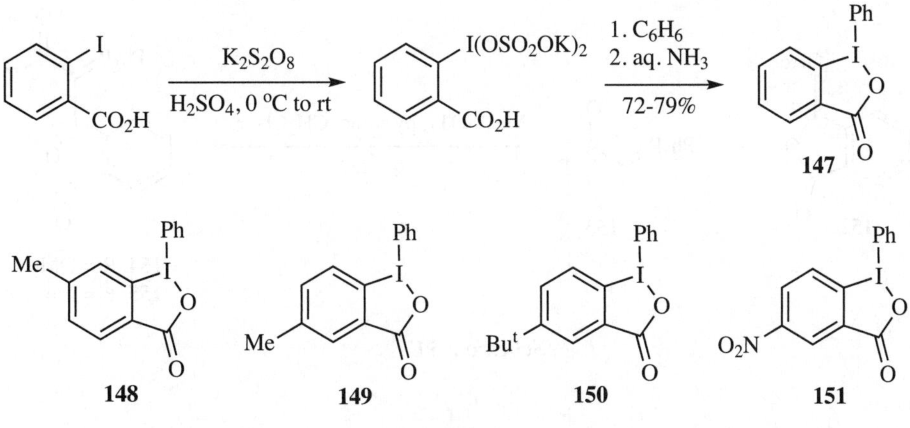

Scheme 2.49

1-[(Triisopropylsilyl)ethynyl]-1,2-benziodoxol-3(1*H*)-one (TIPS-EBX) **140** has found synthetic application as an efficient reagent for ethynylation of various organic substrates [235,298–303].

2.1.8.1.9 Arylbenziodoxoles 1-Phenyl-1,2-benziodoxole-3(1*H*)-one **147**, commonly known under the name of diphenyliodonium-2-carboxylate, is the most important representative of arylbenziodoxoles. Phenylbenziodoxole **147** is commercially available or can be prepared by the oxidation of 2-iodobenzoic acid with potassium persulfate followed by addition of benzene according to the optimized procedure published in *Organic Syntheses* (Scheme 2.49) [304]. The original procedure was reported in 1960 by Beringer and Lillien [305] and it has been used to prepare substituted phenylbenziodoxoles **148** [306], **149** [307] and **150** [308]. A more recent modification of this procedure (Scheme 2.49) consists of the use of Oxone ($2KHSO_5{\cdot}KHSO_4{\cdot}K_2SO_4$) as an oxidant instead of $K_2S_2O_8$ and the use of $NaHCO_3$ as a base instead of NH_3 [309]. This modified method affords various arylbenziodoxoles in high yields and is also applicable for the synthesis of the 7-methylbenziodoxolone ring system using 2-iodo-3-methylbenzoic acid as starting compound [309]. The nitro-substituted phenylbenziodoxole **151** was prepared by the treatment of 5-nitro-2-iodosybenzoic acid with benzene in concentrated sulfuric acid [310].

More recently, Merritt and Olofsson reported a convenient modified procedure for the synthesis of phenylbenziodoxole **147** by a one-pot reaction of 2-iodobenzoic acid with *m*-chloroperoxybenzoic acid, trifluoromethanesulfonic acid (TfOH) and benzene followed by addition of aqueous ammonia (Scheme 2.50) [284].

Batchelor, Birchall and Sawyer in 1986 reported X-ray crystal and molecular structure of phenylbenziodoxole **147** [249]. The geometry about iodine in **147** consists of primary bonds to the two phenyl rings (I—C

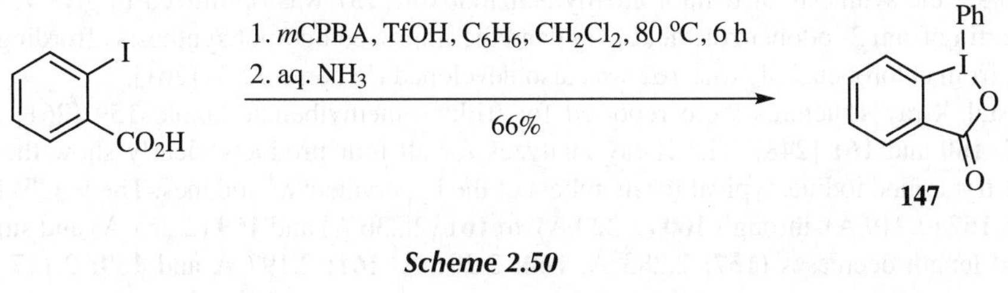

Scheme 2.50

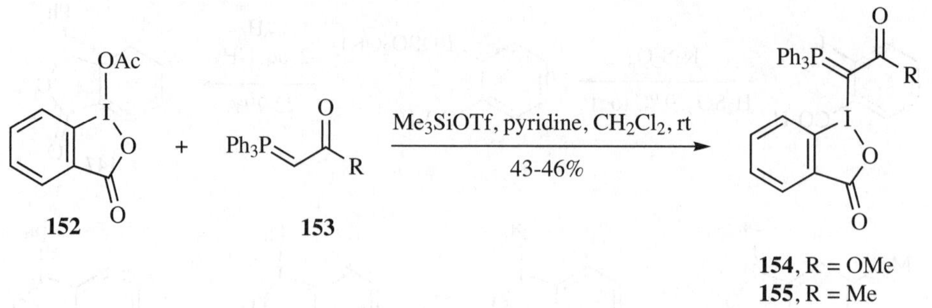

152 + 153 → (Me_3SiOTf, pyridine, CH_2Cl_2, rt; 43-46%) → 154, R = OMe; 155, R = Me

Scheme 2.51

2.105 and 2.119 Å) and a strong intramolecular secondary I···O bond 2.478 Å long to the *ortho*-carboxylate group in an approximate trigonal-bipyramidal arrangement. The C—I—C angle 95.2° is opened to accommodate the I···O interaction and to reduce a C···H interaction of 2.56 Å between the phenyl rings. Unexpectedly, no further I···O secondary bonding occurs and the crystal packing is dictated by hydrogen bonds involving the water molecule in the lattice. The ^{127}I Mössbauer spectral data for the structure of compound **147** is in agreement with that obtained from the crystallographic data [249].

2.1.8.1.10 Phosphoranyl-Derived Benziodoxoles Phosphoranyl-derived benziodoxoles **154** and **155** have been prepared by the reaction of phosphoranes **153** with acetoxybenziodoxole **152** in the presence of trimethylsilyl triflate and pyridine (Scheme 2.51) and isolated in the form of stable, white, microcrystalline solids [250].

Both benziodoxoles **154** and **155** were characterized by a single-crystal X-ray analysis [250]. In particular, X-ray crystallographic analysis of **154** shows that the benziodoxole ring system is essentially planar with the I—O bond length of 2.484 Å, which is very similar to the structure of phenylbenziodoxole **147**. The solid-state packing of **154** demonstrates the significance of secondary bonding interactions, which link individual molecules of **154** into infinite chains via interactions between a carbonyl oxygen of one molecule with the I(III) center of its neighbor. This secondary bonding also affords an approximately square-planar configuration about iodine, with a rather long I···O bond length of 3.270 Å [250].

2.1.8.1.11 Trifluoromethylbenziodoxoles The noncyclic CF_3-substituted λ^3-iodanes in general lack stability and cannot be isolated at room temperature; however, the incorporation of a hypervalent iodine atom into a five-membered heterocycle has a significant stabilization effect. The first synthesis of stable trifluoromethylbenziodoxoles **157** and **159–161** by treatment of corresponding methoxybenziodoxole **156** or acetoxybenziodoxole **158** with trimethyl(trifluoromethyl)silane was reported by Togni and coworkers in 2006 (Scheme 2.52) [248].

In later works, the synthesis of trifluoromethylbenziodoxole **157** was optimized to give 76% yield over three steps starting from 2-iodobenzoic acid [311] and the improved one-pot synthesis affording **159** in 89% yield starting from chlorobenziodoxole **162** was also developed (Scheme 2.53) [261].

Single-crystal X-ray structures were reported for trifluoromethylbenziodoxole **159** [261] and also for products **157**, **160** and **161** [248]. The X-ray analyzes for all four products clearly show the distorted T-shaped geometry around iodine, typical for members of the hypervalent λ^3-iodanes. The I—CF_3 bond lengths increase from **157** (2.219 Å), through **160** (2.229 Å), to **161** (2.236 Å) and **159** (2.267 Å) and simultaneously the I—O bond length decreases (**157:** 2.283 Å, **160:** 2.201 Å, **161:** 2.197 Å and **159:** 2.117 Å), whereas

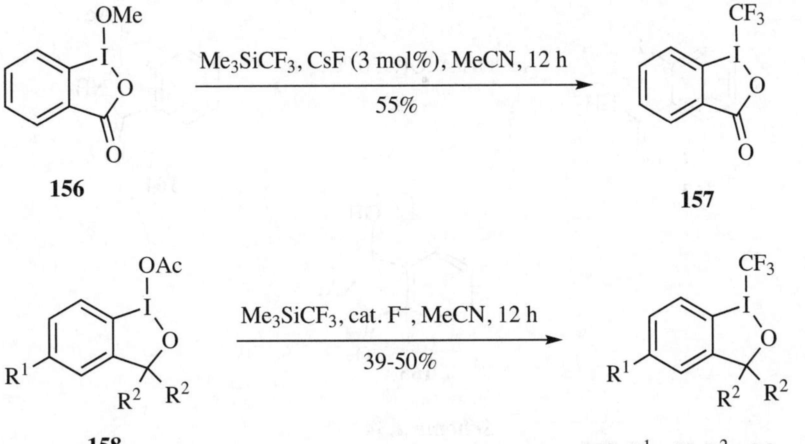

OMe / CF3 — Me_3SiCF_3, CsF (3 mol%), MeCN, 12 h, 55% — **156** → **157**

OAc — Me_3SiCF_3, cat. F^-, MeCN, 12 h, 39-50% — **158** → **159**, R^1 = H, R^2 = Me; **160**, R^1 = H, R^2 = CF_3; **161**, R^1 = Me, R^2 = CF_3

Scheme 2.52

the I—C bond length remains constant within one standard deviation (2.114 Å). The CF_3—I—O angles in trifluoromethylbenziodoxoles are significantly smaller than 180° due to the repulsion of the two lone pairs at iodine as predicted by VSEPR theory and are within the range of those typical of other benziodoxoles (around 170°) [248].

Trifluoromethylbenziodoxoles, especially 1-trifluoromethyl-1,3-dihydro-3,3-dimethyl-1,2-benziodoxole **159**, have found synthetic application as efficient reagents for the trifluoromethylation of various organic substrates [248,261,311–325].

Two other examples of fluoroalkyl-substituted benziodoxoles have been reported in the literature. 1-Pentafluoroethylbenziodoxolone has been prepared similarly to 1-trifluoromethylbenziodoxolone **157** by reacting 1-acetoxybenziodoxolone (**152**) with $C_2F_5SiMe_3$ in the presence of CsF in acetonitrile [326]. Likewise, the analog of reagent **159** bearing a $PhSO_2CF_2$-substituent on the iodine atom was synthesized

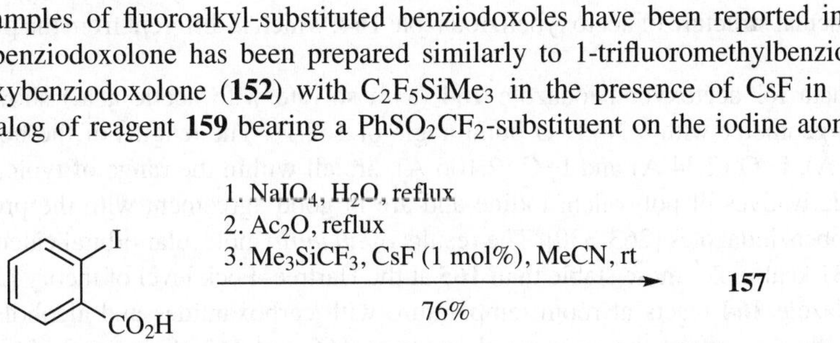

1. $NaIO_4$, H_2O, reflux
2. Ac_2O, reflux
3. Me_3SiCF_3, CsF (1 mol%), MeCN, rt

76% → **157**

KOAc, MeCN then Me_3SiCF_3, cat. F^-, MeCN, rt, 89% — **162** → **159**

Scheme 2.53

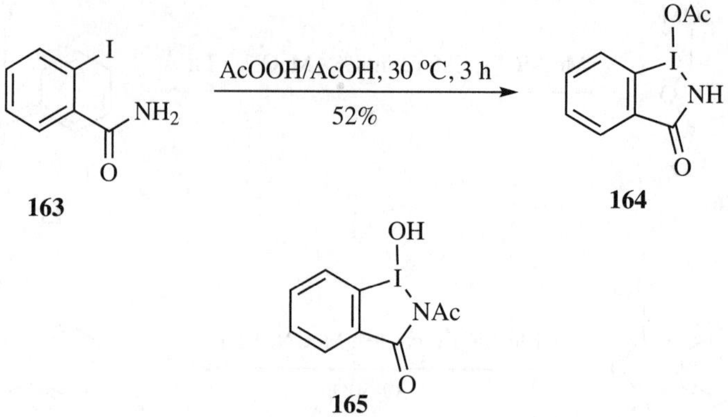

AcOOH/AcOH, 30 °C, 3 h

52%

163 **164**

165

Scheme 2.54

from 1-chloro-1,3-dihydro-3,3-dimethyl-1,2-benziodoxole (**162**) and $PhSO_2CF_2SiMe_3$ using the procedure outlined in Scheme 2.53 [327]. This benziodoxole derivative acts as the electrophilic (phenylsulfonyl) difluoromethylating reagent for various S-nucleophiles under mild reaction conditions.

2.1.8.2 Benziodazoles

The first preparation of a benziodazole heterocyclic system was reported by Wolf and Steinberg in 1965 [236]. The authors of this paper isolated the product of peracetic oxidation of 2-iodobenzamide (**163**) and, based on the IR spectroscopy, incorrectly assigned the structure of *N*-acetyl-1-hydroxy-3-(1*H*)-1,2-benziodazole-3-one **165** to this product. Structure **165** was also adopted in several other studies [259, 328, 329]. More recently, the product of peracetic oxidation of **163** (Scheme 2.54) was investigated by a single-crystal X-ray analysis, which revealed its actual structure of acetoxybenziodazole **164**, which is different from the previously adopted **165** [266].

The structural data for acetoxybenziodazole **164** (as a solvate with acetic acid) showed the expected distorted T-shaped geometry with a N—I—O bond angle of 162.1°. The lengths of the bonds to the iodine atom, I—N (2.101 Å), I—O (2.34 Å) and I—C (2.106 Å), are all within the range of typical single covalent bonds in organic derivatives of polyvalent iodine and are in good agreement with the previously reported structures of chlorobenziodazoles [263, 330]. The results of *ab initio* molecular orbital calculations show that structure **164** is 6.31 kcal mol^{-1} more stable than **165** at the Hartree–Fock level of theory [264].

Acetoxybenziodazole **164** reacts at room temperature with carboxamides and alcohols in the presence of trimethylsilyl triflate to afford the rearranged products **166** and **167** (Scheme 2.55), the structures of which were established by X-ray analysis [264]. A plausible mechanism of this rearrangement most likely includes ring opening and ring closure in the protonated benziodazole. Molecular orbital calculations indicate that the driving force of this rearrangement of benziodazoles to 3-iminiumbenziodoxoles is the greater thermodynamic stability of the N-protonated 3-iminobenziodoxoles **166** and **167** relative to the respective O-protonated benziodazole-3-ones by about 15 kcal mol^{-1} [264]. The reaction of acetoxybenziodazole **164** with azidotrimethylsilane affords the corresponding azide **168** in the form of a yellow, microcrystalline precipitate [266]. Azide **168** has similar reactivity to that of azidobenziodoxoles and can be used as an efficient azidating reagent toward dimethylanilines.

The synthesis and structural studies of several N-functionalized benziodazoles derived from natural amino acids have been reported [265]. Acetoxybenziodazoles **170** and **171** were prepared by the peracetic oxidation

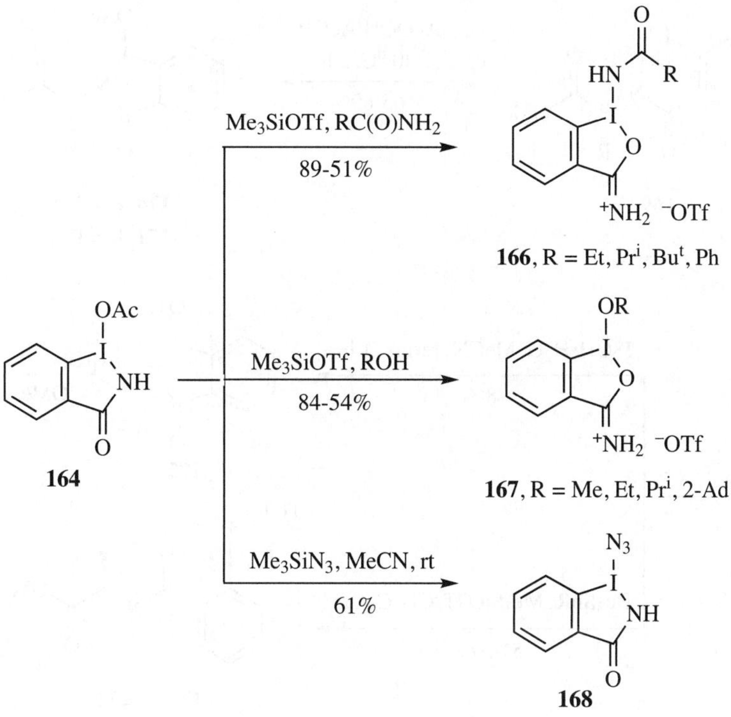

Scheme 2.55

of the readily available 2-iodobenzamides **169** (Scheme 2.56) and isolated as stable, white, microcrystalline solids. Acetoxybenziodazole **170** can be further converted into the tosylate **172** by treatment with *p*-toluenesulfonic acid, or to iodonium salts **173** and **174** by the reaction with tributylphenyltin or tributylphenylethynyltin in the presence of trimethylsilyl triflate.

Treatment of iodonium triflate **173** with aqueous sodium bicarbonate resulted in a restoration of the benziodazole ring with the formation of phenylbenziodazole **175** (Scheme 2.56) [265]. X-Ray crystallographic analysis of **175** shows that the benziodoxole ring system is essentially planar and has a relatively long I—N bond of 2.445 Å. This bond is significantly longer than the analogous I—N bond in acetoxybenziodazole **170**, which is indicative of a substantial ionic character. Overall, the geometry of **175** is similar to that observed for the previously reported structure of phenylbenziodoxole [265].

The oxidation of *N*-(2-iodobenzoyl) amino acids **176** with dimethyldioxirane affords chiral and optically pure hypervalent iodine macrocycles **178–181** as the final isolated products [241]. It is assumed that the initial products in this reaction are the monomeric amino acid derived benziodazoles **177**, subsequent self-assembly of which affords the final products **178–181** (Scheme 2.57). This self-assembly is directed by secondary bonding between hypervalent iodine and oxygen atoms of the amino acid fragment [241].

The structures of macrocycles **180** and **181** were established by X-ray analysis [241]. Molecule **180** consists of a slightly distorted planar macrocyclic system with three oxygens of the amino acid carboxyls inside the ring and all three alkyl groups above the plane. Each iodine atom is covalently bonded to carbon (I—C = 2.092 Å) and nitrogen (I—N = 2.064 Å) and has three longer intramolecular contacts with oxygen atoms (I—O = 2.368, 2.524 and 2.877 Å). With the consideration of primary and secondary bonds, the iodine atoms in **180** have a pentagonal-planar geometry, which is analogous to that found in the solid state for

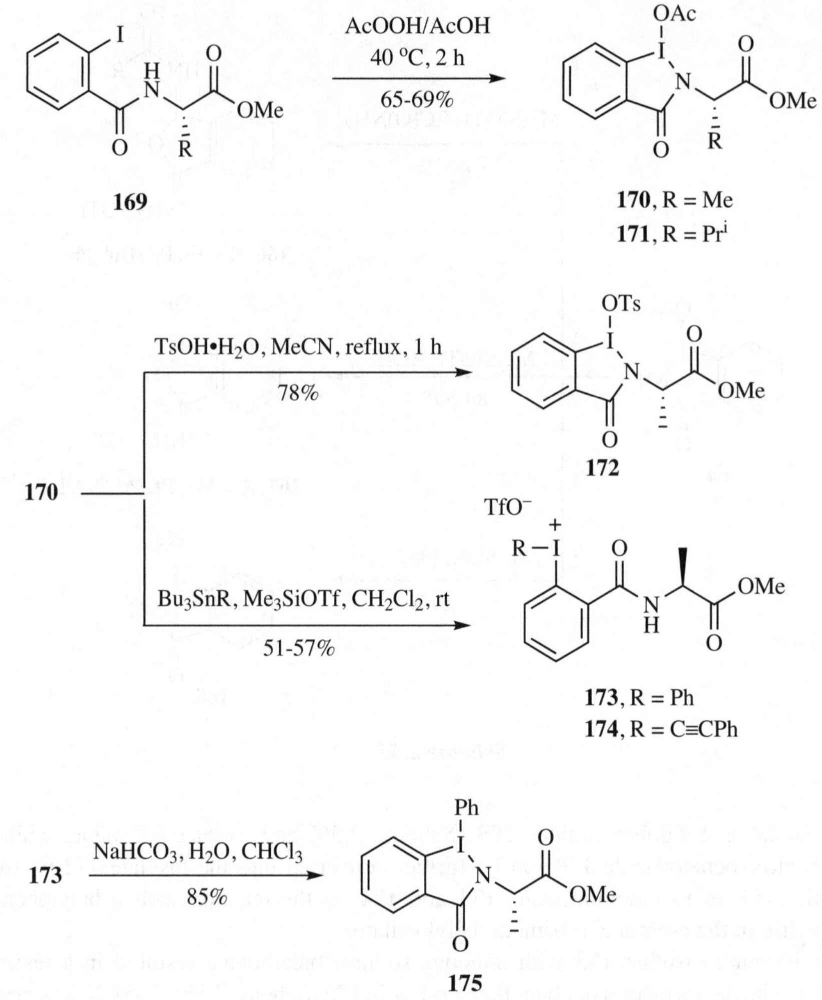

AcOOH/AcOH
40 °C, 2 h
65-69%

169

170, R = Me
171, R = Pr[i]

$TsOH{\bullet}H_2O$, MeCN, reflux, 1 h
78%

172

170

TfO^-

Bu_3SnR, Me_3SiOTf, CH_2Cl_2, rt
51-57%

173, R = Ph
174, R = C≡CPh

$NaHCO_3$, H_2O, $CHCl_3$
85%

173

175

Scheme 2.56

$PhI(OAc)_2$. As a result of the central oxygens, the electron-rich cavity of macrocycle **180** is suitable for complexation of metal cations. Specifically, ESI-MS data indicate that macrocycles **178–181** can selectively form complexes with sodium cations in the presence of K^+, Li^+, Ag^+, or Pb^{2+} [241]. The self-assembly of monomeric benziodazoles **177** into macrocyclic molecules **178–181** was studied using molecular orbital calculations [331]. The driving force for the self-assembly is the formation of secondary bonding interactions between molecules and a rearrangement of primary and secondary bonding around iodine to place the least electronegative substituent in the equatorial position for each iodine in the trimer.

2.1.8.3 Benziodoxaboroles

A series of heterocyclic compounds containing trivalent iodine, oxygen and boron in a five-membered ring has been prepared by oxidative cyclization of commercially available *ortho*-iodophenylboronic acids **182** and **185**

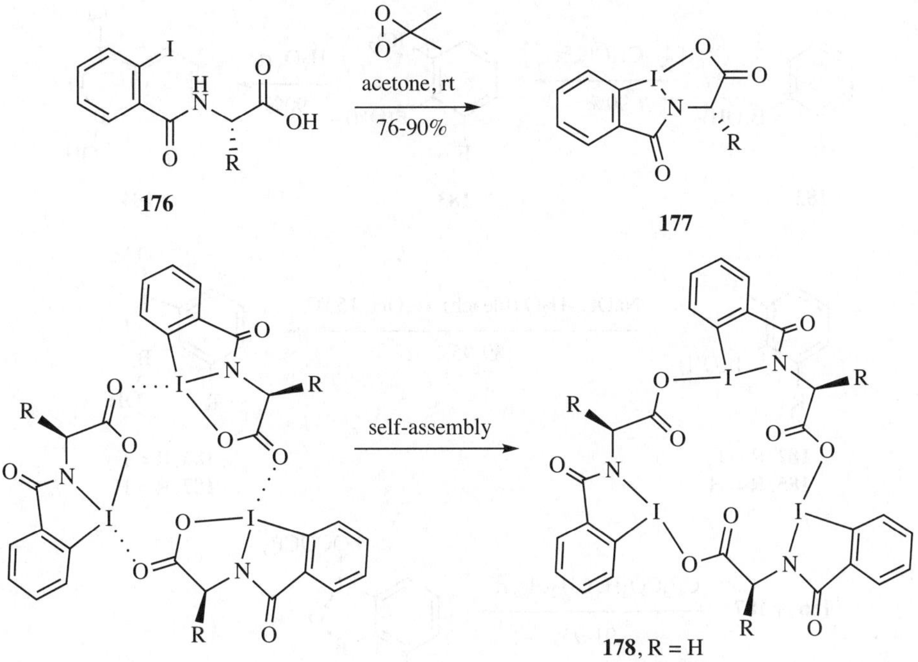

Scheme 2.57

[243]. 1-Chloro-4-fluoro-1*H*-1λ^3-benzo[*d*][1,2,3]iodoxoborol-3-ol (**184**) was synthesized by the chlorination of 2-fluoro-6-iodophenylboronic **182** acid followed by treatment of the intermediate iododichloride **183** with water (Scheme 2.58). The 1-substituted acetoxybenziodoxaboroles **188** and **189** were prepared by hypochlorite oxidation of 2-fluoro-6-iodophenylboronic acid (**182**) or 2-iodophenylboronic acid (**185**) in acetic acid. Acetoxybenziodoxaboroles **188** and **189** can be further converted into the respective trifluoroacetates **188** and **189** by treatment with trifluoroacetic acid and the 1-hydroxy derivative **190** can be prepared by basic hydrolysis of acetoxy- or trifluoroacetoxy-benziodoxaboroles with aqueous $NaHCO_3$. X-Ray structural studies of 1-chloro- and 1-trifluoroacetoxy substituted benziodoxaboroles **184**, **188** and **189** have shown the presence of a planar five-membered heterocyclic ring with an unusually short endocyclic I—O bond distance of 2.04–2.09 Å [243]. Structural parameters of the five-membered iodoxoborole ring, such as the planar geometry and the short B—O and O—I bonds lengths in **184**, **188** and **189** compared to those in known benziodoxoles, are suggestive of a partially aromatic character of this ring. The calculated NIST (0) and NIST (1) indexes for 1-chloro- and 1-trifluoroacetoxy substituted benziodoxaboroles, however, are indicative of a significantly lower aromaticity compared to the classic aromatic systems [243].

Slow crystallization of 1-trifluoroacetoxybenziodoxaborole **188** from methanol afforded the tetrameric macrocyclic structure **192** resulting from self-assembly of the initially formed 4-fluoro-1,3-dimethoxy-1*H*-1λ^3-benzo[*d*][1,2,3]iodoxoborole (**191**) (Scheme 2.59). The structure of macrocycle **192** was established by a single-crystal X-ray analysis [243]. The driving force for formation of the eight-membered cyclic system **192** is the transformation of initial trigonal-planar sp^2 hybrid boron atoms in **191** into tetrahedral sp^3 hybridized atoms. Indeed, each boron atom in tetramer **192** forms one covalent bond with carbon and three covalent

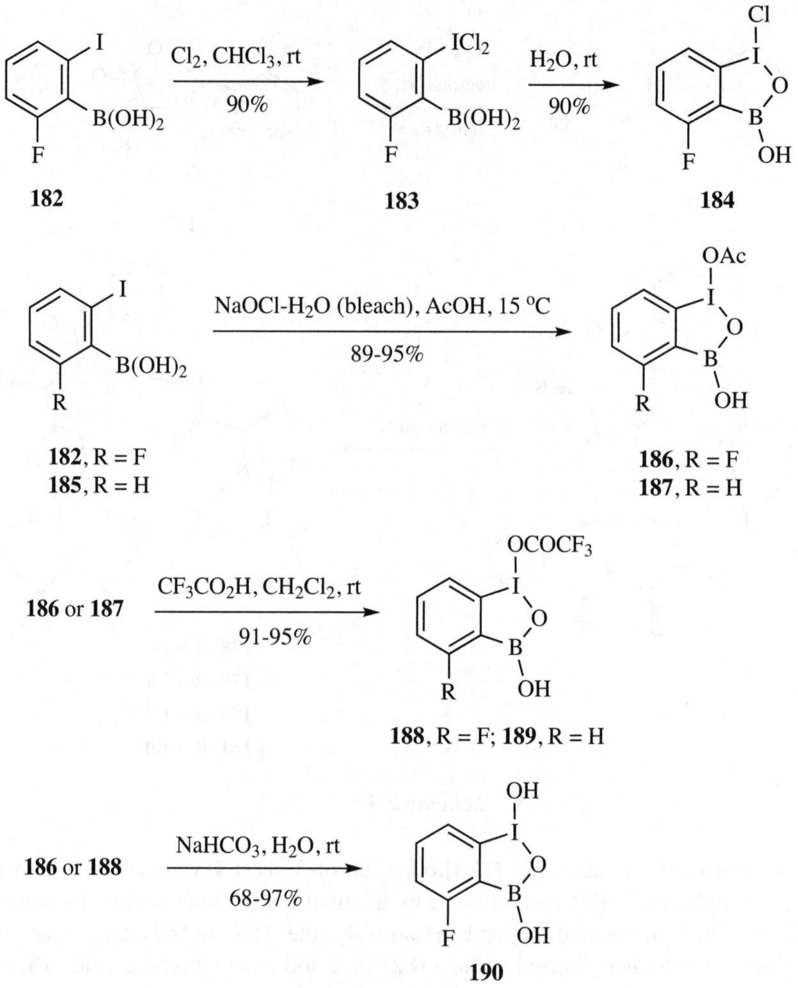

Scheme 2.58

bonds with oxygen atoms. The boron–oxygen bonds, which form the eight-membered cycle (1.48–1.50 Å), are significantly longer than the B—OMe bond distances (1.43–1.44 Å). The eight-membered cycle in **192** is nonplanar with alternating larger B—O—B bond angles (126.2°, average) and smaller O—B—O bond angles (110.7, average), with the latter being close to the expected for the sp^3 hybridized boron atoms. The change in hybridization of the boron atoms in tetramer **192** also leads to a significant elongation of boron–carbon and heterocyclic iodine–oxygen bonds compared to those observed in compounds **184**, **188** and **189**. Similarly to compounds **184**, **188** and **189** the iodine centers in **192** have the usual T-shape geometry for iodine(III) compounds, with O—I—OMe bond angles varying between 167.2 and 170.5° [243].

2.1.8.4 Benziodoxathioles

1*H*-1-Hydroxy-1,2,3-benziodoxathiole 3,3-dioxide (**93** in Figure 2.9) was first described as a sodium salt in 1914 in book by Willgerodt [332]. In 1993 Koser and coworkers reported the preparation of

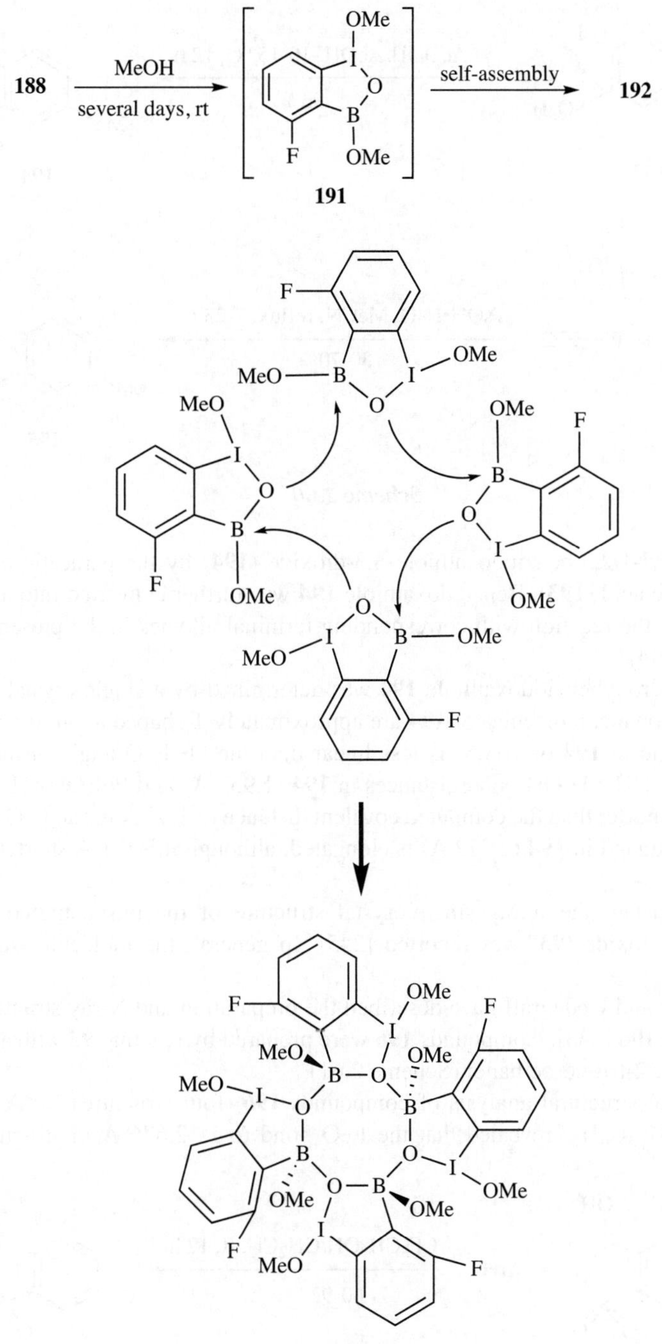

188 —MeOH, several days, rt→ [191] —self-assembly→ 192

Scheme 2.59

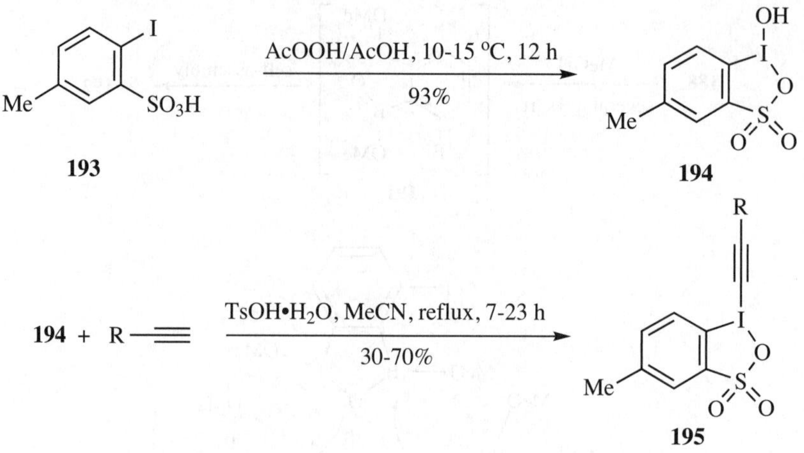

AcOOH/AcOH, 10-15 °C, 12 h
93%

193

194

TsOH•H_2O, MeCN, reflux, 7-23 h
30-70%

194 + R—≡

195

Scheme 2.60

1*H*-1-hydroxy-5-methyl-1,2,3-benziodoxathiole 3,3-dioxide (**194**) by the peracetic oxidation of 2-iodo-5-methylbenzenesulfonic acid (**193**). Benziodoxathiole **194** was further converted into a series of alkynylbenziodoxathioles **195** by the reaction with corresponding terminal alkynes in the presence of toluenesulfonic acid (Scheme 2.60) [244].

The structure of hydroxybenziodoxathiole **194** was determined by a single-crystal X-ray analysis [244]. As expected for a λ^3-iodane, molecules of **194** are approximately T-shaped about the iodine atom, although the O—I—O bond angle in **194** of 168.6° is less linear than the O—I—O angle in the noncyclic sulfonate PhI(OH)OTs (178.80°). The I—OH bond distances in **194** (1.933 Å) and PhI(OH)OTs (1.940 Å) are nearly the same and slightly shorter than the computed covalent distance of 1.99 Å for an I—O single bond. The I—O bond to the sulfonate ligand in **194** (2.372 Å) is elongated, although it is 0.1 Å shorter than in PhI(OH)OTs (2.473 Å).

In a more recent paper, the X-ray single-crystal structure of the unsubstituted 1*H*-1-hydroxy-1,2,3-benziodoxathiole 3,3-dioxide (**93**) was reported [245]. In general, the molecular structure of **93** is very similar to that of **194**.

Justik, Protasiewicz and Updegraff have described the preparation and X-ray structural studies of several arylbenziodoxathioles **196** [333]. Compounds **196** were prepared by reacting **93** with appropriate arenes in a solution containing 2,2,2-trifluoroethanol (Scheme 2.61).

X-Ray single crystal structural analysis of compounds **196** (four structures for Ar = Ph, 4-$MeOC_6H_4$, 4-ClC_6H_4 and 2,4,6-$Me_3C_6H_2$) revealed that the I—O bond (e.g., 2.676 Å in structure **196**, Ar = Ph) is

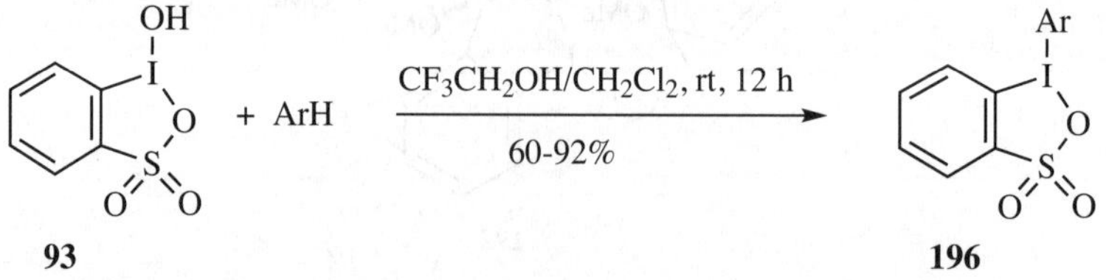

CF_3CH_2OH/CH_2Cl_2, rt, 12 h
60-92%

93 + ArH

196

Ar = Ph, 4-MeC_6H_4, 4-$MeOC_6H_4$, 4-ClC_6H_4, 2,4,6-$Me_3C_6H_2$

Scheme 2.61

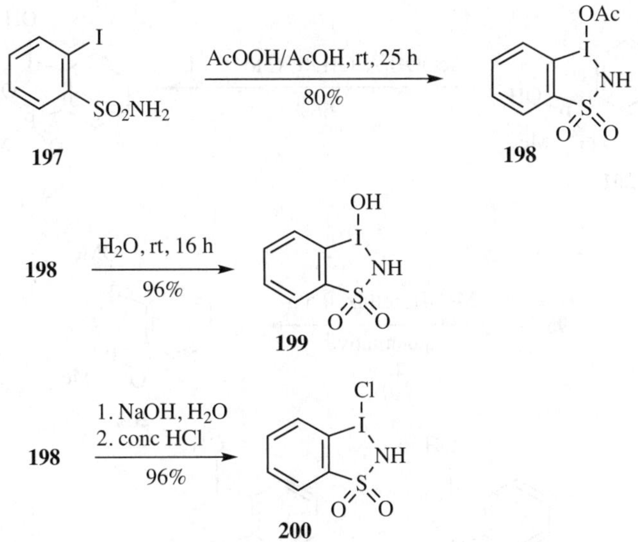

Scheme 2.62

markedly longer than the analogous bond in phenylbenziodoxole **147** (2.478 Å; see Section 2.1.8.1.9), suggesting reduced covalent character and a greater contribution of the betaine form rather than covalent benziodoxathiole [333]. For comparison, in the observed solid-state structure of **93** this bond is markedly shorter (2.381 Å) [245].

2.1.8.5 Benziodathiazoles

Jaffe and Leffler in 1975 achieved the entry into the benziodathiazole system via l-acetoxy-1,2-dihydro-1,3,2-benziodathiazole 3,3-dioxide (**198**, Scheme 2.62) [246]. The acetate **198** was synthesized by the peracetic acid oxidation of 2-iodobenzenesulfonamide (**197**). Hydrolysis of **198** gives l-hydroxy-1,2-dihydro-1,3,2-benziodathiazole 3,3-dioxide (**199**) and l-chloro-1,2-dihydro-1,3,2-benziodathiazole 3,3-dioxide (**200**) can be obtained by treatment of a basic solution of **198** with concentrated hydrochloric acid [246].

2.1.8.6 Benziodoxaphospholes

Balthazor and coworkers reported the synthesis of 1,3-dihydro-1-hydroxy-3-methyl-1,2,3-benziodoxaphosphole 2-oxide (**95**) by the peracetic oxidation of 2-iodophenylmethylphosphinic acid (**201**) (Scheme 2.63) [247]. Benziodoxaphosphole **95** can be converted into the methoxy derivative **202** by treatment with hot methanol. Methoxybenziodoxaphosphole **202** is readily hydrolyzed in moist air to give the initial hydroxybenziodoxaphosphole **201** [247]. Freedman and DeMott reported a similar preparation of benziodoxaphospholes **203** and **204** by a peracid oxidation of 2-iodophenylphosphonic acid or (2-iodophenyl)phenylphosphinic acid, respectively [334].

The molecular structure of hydroxybenziodoxaphosphole **95** was determined by a single-crystal X-ray analysis [247]. Molecules of **95** are T-shaped about the iodine atom with an O—I—O bond angle of 171.3°, I—OH bond distance of 1.952 Å and I—O bond distance to the phosphinate oxygen of 2.286 Å.

Moss and coworkers [335] investigated catalytic activity of hydroxybenziodoxaphospholes in the hydrolysis of phosphate esters and found that the anion of **95** is 44-times less reactive than hydroxybenziodoxole anion **105** (see Scheme 2.39 in Section 2.1.8.1.2).

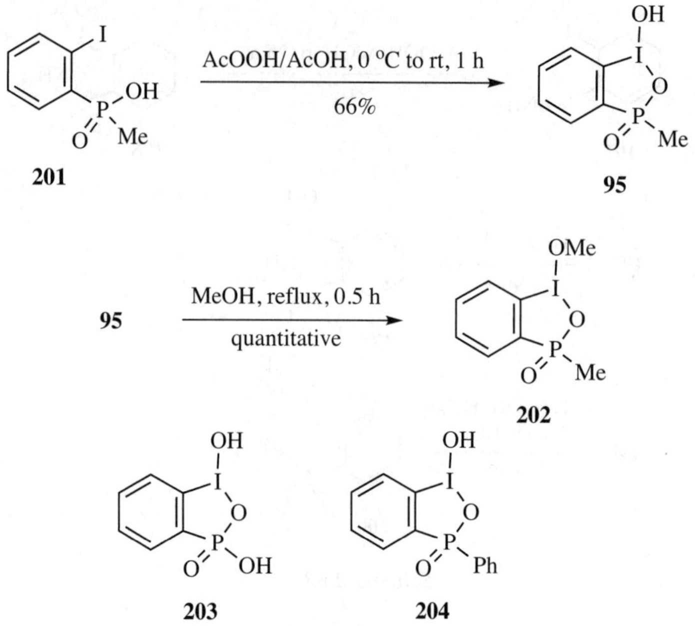

AcOOH/AcOH, 0 °C to rt, 1 h

66%

201 → 95

MeOH, reflux, 0.5 h

quantitative

95 → 202

203 204

Scheme 2.63

2.1.8.7 Iodoxolones

Iodoxolones represent the only known example of hypervalent iodine heterocycles in which the iodine(III) atom is not connected to an aromatic ring. The first representatives of iodoxolones were prepared by Thiele and Peter in the early 1900s by the oxidative cyclization of iodofumaric acid derivatives [336]. More recently, Moss and coworkers have prepared several iodoxolones using modified older procedures. In particular, hydroxyiodoxolone (**207**) was synthesized by the peracetic oxidation of iodofumaric acid (**205**) followed by decarboxylation of the initially formed 5-carboxylic acid derivative **206** (Scheme 2.64) [337]. Another approach to this heterocyclic system (e.g., chloroiodoxolones **208** and **209**) involves the addition of ICl_3 to acetylenedicarboxylic acid followed by a spontaneous cyclization of the initially formed iododichlorides [338].

Several iodoxolones (e.g., **211**, **213**, **215** and **216**) have been prepared by the oxidation of (*Z*)-3-iodo acrylic acid derivatives **210**, **212** and **214** (Scheme 2.65) [339]. Structures of products **211** and **213** were established by X-ray analysis [339]. The angles and distances found in the five-membered iodoxolone system are very similar in structures **211** and **213** and are directly comparable to the X-ray structure of 2-iodosylbenzoic acid (**104**) (Section 2.1.8.1.2) [271] or 3,4,5,6-tetrafluoro-2-iodosylbenzoic acid [340]. Iodoxolones **211**, **213**, **215** and **216** have stabilities and reactivities similar to those of conventional hypervalent iodine(III) reagents [339].

A stable aliphatic iodoxolone **218**, the valence tautomer of *cis*-iodosylcyclopropanecarboxylic acid, has been prepared by two alternative approaches: (i) the peracetic oxidation of *cis*-2-iodocyclopropanecarboxylic acid (**217**) and (ii) by hydrolysis of the corresponding iododichloride **219** (Scheme 2.66) [341].

Compound **218** exists in the hydroxyiodoxolone form, not in the open *cis*-cyclopropanecarboxylic acid form, as indicated by its high pK_a of 7.55 (cf. Section 2.1.8.1.2) and by its ability to cleave phosphate esters in aqueous micellar solution [341].

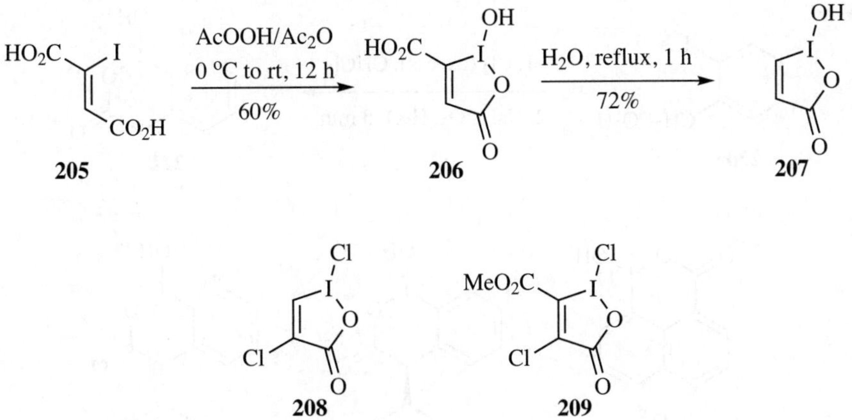

HO_2C I CO_2H **205** → AcOOH/Ac_2O, 0 °C to rt, 12 h, 60% → OH, HO_2C, I, O, O **206** → H_2O, reflux, 1 h, 72% → OH, I, O, O **207**

Cl, I, O, Cl, O **208**

Cl, MeO_2C, I, O, Cl, O **209**

Scheme 2.64

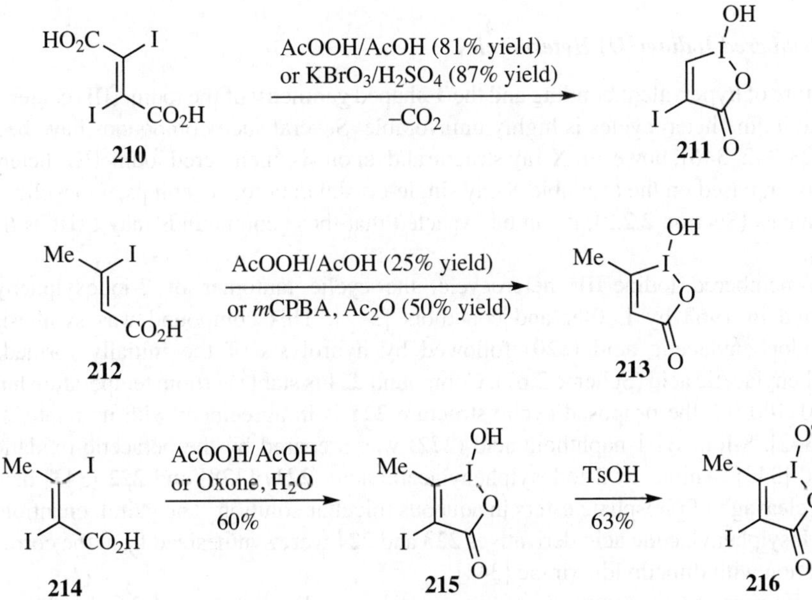

HO_2C I, I CO_2H **210** → AcOOH/AcOH (81% yield) or $KBrO_3/H_2SO_4$ (87% yield), $-CO_2$ → OH, I, O, I **211** O

Me I, CO_2H **212** → AcOOH/AcOH (25% yield) or *m*CPBA, Ac_2O (50% yield) → Me, OH, I, O **213** O

Me I, I CO_2H **214** → AcOOH/AcOH or Oxone, H_2O, 60% → Me, OH, I, O, I, O **215** → TsOH, 63% → Me, OTs, I, O, I, O **216**

Scheme 2.65

I, CO_2H **217** → AcOOH/Ac_2O, 15 °C to rt, 24 h, 59% → OH, I, O **218** O ← $NaHCO_3$, H_2O, 44% ← ICl_2, CO_2H **219**

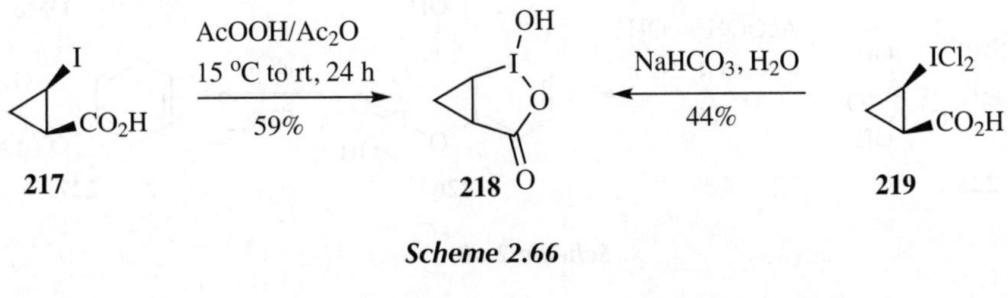

Scheme 2.66

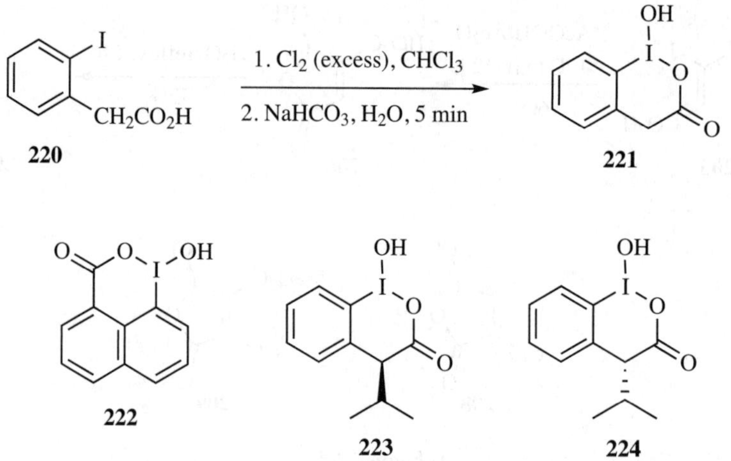

Scheme 2.67

2.1.8.8 Six-Membered Iodine(III) Heterocycles

Owing to the nature of hypervalent bonding and the T-shaped geometry of the iodine(III) center, the formation of six-membered iodine heterocycles is highly unfavorable. Several such compounds have been reported in the literature [328,342–346]; however, X-ray structural data on six-membered iodine(III) heterocycles is not available. Moreover, based on the available X-ray single-crystal data for several pseudocyclic six-membered iodine(V) derivatives (Section 2.2.2), it can be expected that these compounds may exist as their noncyclic tautomers [345].

The first six-membered iodine(III) heterocycle, the cyclic tautomer of 2-iodosylphenylacetic acid, **221**, was reported in 1963 by Leffler and coauthors [342]. This compound was synthesized by chlorination of 2-iodophenylacetic acid (**220**) followed by hydrolysis of the initially formed, unstable 2-(dichloroiodo)phenylacetic acid (Scheme 2.67). Compound **221** is stable at room temperature but decomposes in solution at 80–100 °C; the proposed cyclic structure **221** is in agreement with its relatively low acidity ($pK_a = 7.45$) [342]. 8-Iodosyl-1-naphthoic acid (**222**) was prepared by the peracetic oxidation of 8-iodo-1-naphthoic acid [343]. Anions of 2-iodosylphenylacetic acid (**221**) [328] and **222** [343] have a moderate reactivity in the cleavage of phosphate esters in aqueous micellar solution. The chiral, enantiomerically pure substituted 2-iodosylphenylacetic acid derivatives **223** and **224** were synthesized from the corresponding aryl iodides by oxidation with dimethyldioxirane [344].

Leffler and Jaffe reported the synthesis of a presumably cyclic tautomer of 2-iodosylphenylphosphoric acid (**226**) and its methyl ester **227** starting from 2-iodophenylphosphoric acid (**225**) (Scheme 2.68) [346].

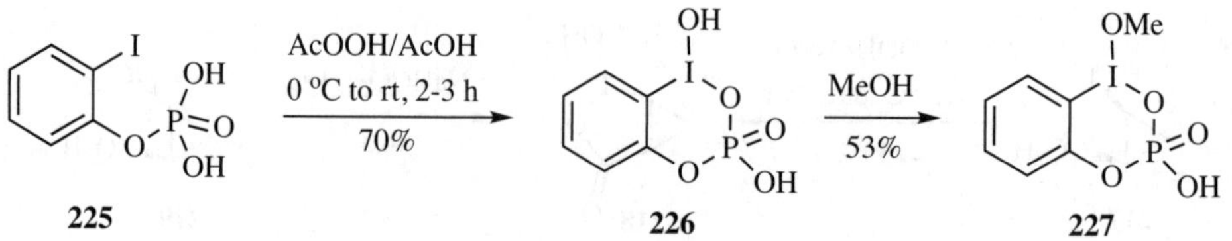

Scheme 2.68

Products **226** and **227** were identified by NMR spectroscopy; an attempted X-ray diffraction analysis of **226** was unsuccessful.

2.1.8.9 *Polycyclic Heterocycles with an Iodine(III) Atom at Ring Junction*

The synthesis of the first representatives of this class of fused hypervalent iodine heterocycles was described by Agosta in 1965 [347]. Compounds **229** and **231** were originally prepared by the peracetic oxidation of corresponding iodides **228** and **230** (Scheme 2.69). A few years later, Martin and Chau reported an alternative way of forming product **229** by the thermal decomposition of the dibutyl ester of 2-iodoisophthalic acid [348]. The benziodoxolone derivative **229** has high thermal stability, with a decomposition point above 260 °C; the formation of this heterocyclic system is particularly favorable due to the T-shaped geometry of hypervalent iodine. The X-ray crystallographic analysis of compound **229** showed a structure of overall planar geometry with the expected T-shape configuration at the iodine(III) center [349]. Several similarly fused iodine(III) heterocycles (compounds **232**) have also been reported [350, 351].

An unusual fused hypervalent iodine heterocycle (**234**) has been prepared by oxidative cyclization of 3-iodo-3-hexene-2,5-diol **233** with neat *tert*-butyl hypochlorite (Scheme 2.70) [352, 353].

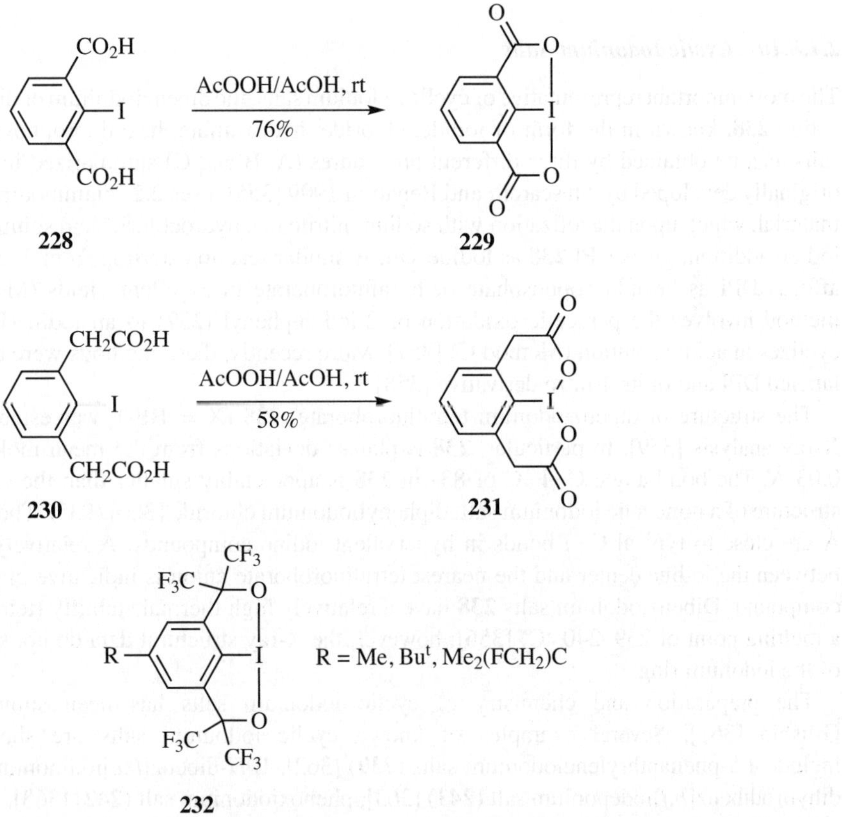

228 → **229**: AcOOH/AcOH, rt, 76%

230 → **231**: AcOOH/AcOH, rt, 58%

232: R = Me, Bu^t, $Me_2(FCH_2)C$

Scheme 2.69

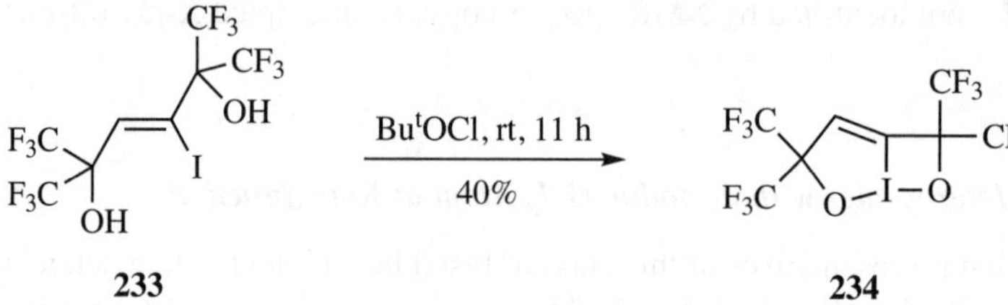

233 → 234 (Bu^tOCl, rt, 11 h; 40%)

Scheme 2.70

The structures of products **232** (R = Bu^t) [354] and **234** [353] were established by single-crystal X-ray analyzes, which showed that both structures have a distorted T-shape geometry around the central iodine atom. Apical bond angles O—I—O for **234** and **232** are 144.5(1)° and 158.2(2)°, respectively, indicating that the deviation (35.5°) of bond angle O—I—O in **234** from 180° is larger than that (21.8°) of **232.** The apical bond length [2.142(3) Å] of the four-membered ring of **234** is longer than that of the five-membered ring [2.070(3) Å]. These results explain the ring strain in the four-membered ring. The four-membered ring is almost planar, judging from the torsion angle I—C—C—O of 0.0(3)° and the sum of bond angles of the four-membered ring of 359.9° [353].

2.1.8.10 Cyclic Iodonium Salts

The most important representative of cyclic iodonium salts, the dibenziodolium or diphenyleneiodonium (DPI) cation **238**, known in the form of iodide, chloride, hydrosulfate, hexafluorophosphate, or tetrafluoroborate salts, can be obtained by three different procedures (A, B and C) summarized in Scheme 2.71. Method A, originally developed by Mascarelli and Benati in 1909 [355], uses 2,2′-diaminodiphenyl (**235**) as the starting material, which upon diazotization with sodium nitrite in a hydrochloric acid solution followed by potassium iodide addition, gives DPI **238** as iodide salt. A similar reaction starting from 2-amino-2′-iododiphenyl **236** affords DPI as hexafluorophosphate or tetrafluoroborate in excellent yields (Method B) [356]. The third method involves the peracetic oxidation of 2-iodobiphenyl (**237**) to an iodine(III) intermediate that then cyclizes in acidic solution (Method C) [357]. More recently, these methods were used to prepare the tritium labeled DPI and of its 4-nitro derivative [358].

The structure of dibenziodolium tetrafluoroborate, **238** (X = BF_4), was established by a single-crystal X-ray analysis [359]. In particular, **238** is planar; deviations from the mean molecular plane are less than 0.03 Å. The bond angle C—I—C of 83° in **238** is appreciably smaller than the corresponding angle in the structure of a noncyclic iodonium salt, diphenyliodonium chloride [360] (93°). The bond lengths C—I of 2.08 Å are close to typical C—I bonds in hypervalent iodine compounds. A relatively long distance of 3.65 Å between the iodine center and the nearest tetrafluoroborate anion is indicative of the ionic character of this compound. Dibenziodolium salts **238** have a relatively high thermal stability (tetrafluoroborate salt **238** has a melting point of 239–240 °C) [356]; however, the X-ray structural data do not support aromatic character of the iodolium ring.

The preparation and chemistry of cyclic iodonium salts has been summarized in a review of Grushin [361]. Several examples of known cyclic iodonium salts are shown in Figure 2.10 and include 4,5-phenanthryleneiodonium salts (**239**) [362], 10*H*-dibenz[*b,e*]iodinonium salt (**240**) [363], 10,11-dihydrodibenz[*b,f*]iodeponium salt (**241**) [363], phenoxiodonium salt (**242**) [363], 10-acetylphenaziodonium salt (**243**) [363], 10-oxidophenothiiodonium salt (**244**) [363], the bicyclic bis-iodonium salt **245** [364], benziodolium chloride **246** [359] and iodolium salt **247** [365].

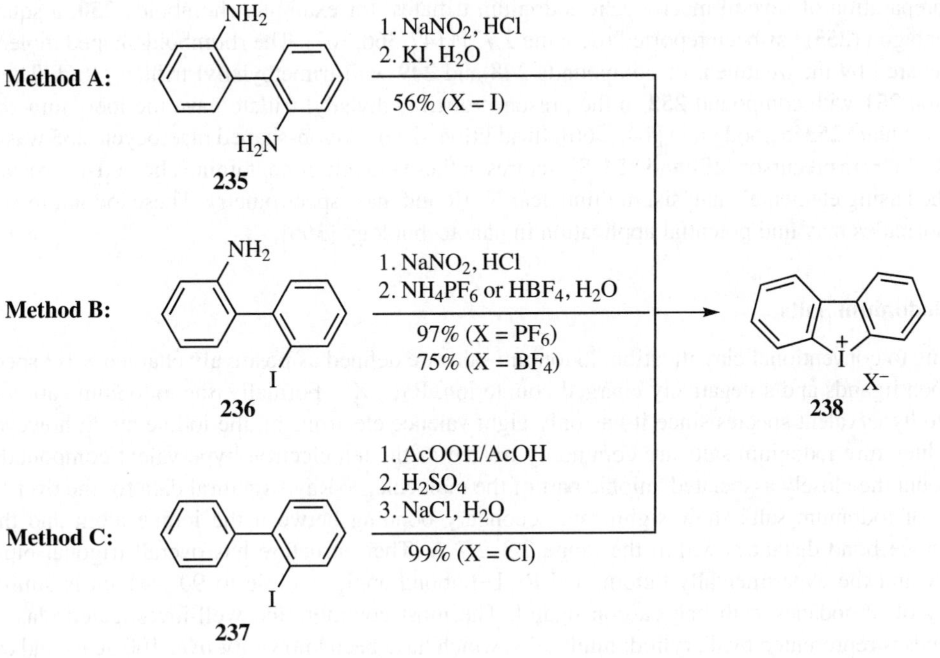

Method A: NH_2, H_2N **235** — 1. $NaNO_2$, HCl; 2. KI, H_2O; 56% (X = I)

Method B: NH_2, I **236** — 1. $NaNO_2$, HCl; 2. NH_4PF_6 or HBF_4, H_2O; 97% ($X = PF_6$); 75% ($X = BF_4$)

Method C: I **237** — 1. AcOOH/AcOH; 2. H_2SO_4; 3. NaCl, H_2O; 99% (X = Cl)

→ I^+ X^- **238**

Scheme 2.71

I^+ Cl^- **239**; I^+ HSO_4^- **240**; I^+ HSO_4^- **241**

O, I^+ Br^- **242**; Ac, N, I^+ Br^- **243**; O, S, I^+ Br^- **244**

I^+, I^+ 2 HSO_4^- **245**; Bu, Ph, I^+ Cl^- **246**; Ph, Ph, Ph, Ph, I^+ $SbCl_6^-$ **247**

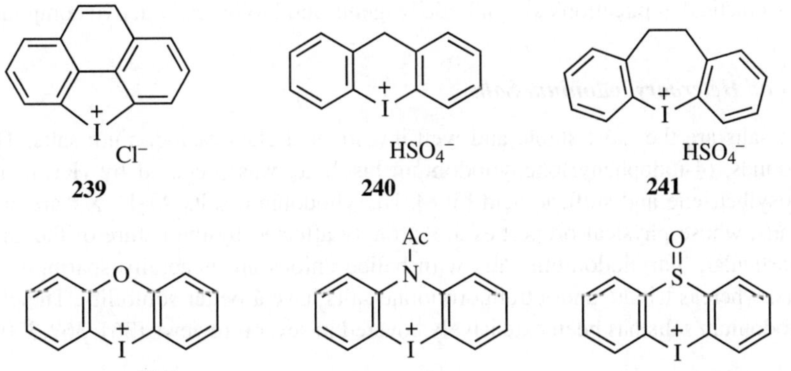

Figure 2.10 *Examples of known cyclic iodonium salts.*

The preparation of several macrocyclic iodonium triflates, for example, rhomboids **250**, a square (**253**) and a pentagon (**255**) has been reported (Scheme 2.72) [147,366,367]. The rhomboid-shaped molecules **250** were prepared by the treatment of compounds **248** and **249** with trimethylsilyl triflate [366]. The reaction of dication **251** with compound **252** in the presence of trimethylsilyl triflate gave the iodonium-containing molecular square **253** in good yield [147,366]. In addition, the pentagon-shaped macrocycle **255** was prepared in 60% yield from precursors **251** and **254**. Structures of these iodonium-containing charged macrocycles were established using elemental analysis, multinuclear NMR and mass spectrometry. These iodonium-containing macromolecules may find potential application in nanotechnology [366].

2.1.9 Iodonium Salts

According to conventional classification, iodonium salts are defined as positively charged 8-I-2 species with two carbon ligands and a negatively charged counterion, $R_2I^+ X^-$. Formally, the iodonium cation does not belong to hypervalent species since it has only eight valence electrons on the iodine atom; however, in the modern literature iodonium salts are commonly treated as the ten-electron hypervalent compounds, taking into account the closely associated anionic part of the molecule. X-Ray structural data for the overwhelming majority of iodonium salts show significant secondary bonding between the iodine atom and the anion, with average bond distances within the range 2.3–2.7 Å. Their structure has overall trigonal-bipyramidal geometry and the experimentally determined R—I—R bond angle is close to 90°, which is similar to the geometry of λ^3-iodanes with one carbon ligand. The most common and well-investigated class of these compounds is represented by diaryliodonium salts, which have been known for over 100 years and especially were extensively studied in the 1950s and 1960s. More recently, significant research activity has focused on aryliodonium salts bearing alkenyl, alkynyl and fluoroalkyl groups as the second ligand. Iodonium salts have found numerous practical applications as synthetic reagents and biologically active compounds.

2.1.9.1 Aryl- and Heteroaryliodonium Salts

Diaryliodonium salts are the most stable and well-investigated class of iodonium salts. The first example of these compounds, (4-iodophenyl)phenyliodonium bisulfate, was prepared by Hartmann and Meyer in 1894 from iodosylbenzene and sulfuric acid [368]. Diaryliodonium salts $Ar_2I^+ X^-$ are air- and moisture-stable compounds, whose physical properties are strongly affected by the nature of the anionic part of the molecule. In particular, diaryliodonium salts with halide anions are generally sparingly soluble in many organic solvents, whereas triflate and tetrafluoroborate salts have a better solubility. The chemistry of aryl- and heteroaryliodonium salts has been extensively covered in several reviews [361,369,370].

2.1.9.1.1 Preparation Synthetic routes to diaryliodonium salts typically involve the initial oxidation of an aryl iodide to a λ^3-iodane and then ligand exchange with an arene or a nucleophilic arylating reagent (e.g., arylborates, arylstannanes, or arylsilanes) to obtain the diaryliodonium salt. In many cases a final anion exchange step is necessary. To shorten the synthetic route to symmetric iodonium salts, preformed inorganic iodine(III) reagents can be employed. A particularly useful approach involves one-pot oxidation and ligand-exchange reactions to obtain the diaryliodonium salts directly from arenes and iodoarenes or molecular iodine. Most of these reactions are performed under acidic conditions, although several neutral or basic methods also exist.

Older methods for the preparation of symmetrical diaryliodonium salts are based on the reaction of arenes with potassium iodate or KIO_3/I_2 in the presence of concentrated sulfuric acid [371, 372]. It is assumed that the mechanisms of these reactions involve initial formation of the electrophilic iodyl or iodosyl species, $IO_2{}^+HSO_4{}^-$ or $IO^+HSO_4{}^-$, which further react with arenes, finally forming diaryliodonium salts,

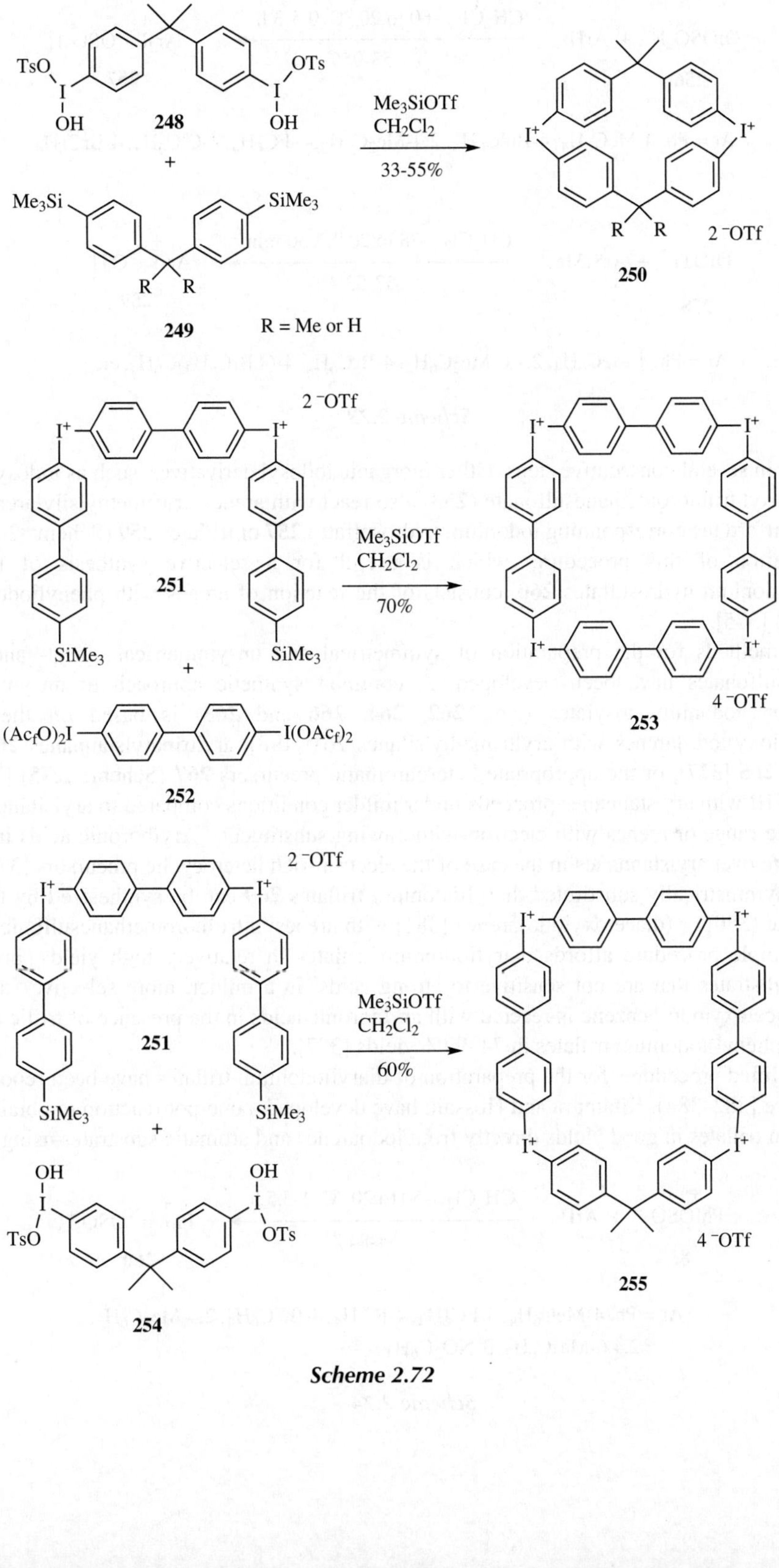

TsO OTs I I OH 248 OH + Me_3Si $SiMe_3$ R R 249 R = Me or H

Me_3SiOTf, CH_2Cl_2, 33-55%

I^+ I^+ R R 2 ^-OTf 250

$2\ ^-OTf$ I^+ I^+ 251 $SiMe_3$ $SiMe_3$ + $(AcfO)_2I$ $I(OAcf)_2$ 252

Me_3SiOTf, CH_2Cl_2, 70%

I^+ I^+ I^+ I^+ 4 ^-OTf 253

$2\ ^-OTf$ I^+ I^+ 251 $SiMe_3$ $SiMe_3$ + OH OH I I TsO OTs 254

Me_3SiOTf, CH_2Cl_2, 60%

I^+ I^+ I^+ I^+ 4 ^-OTf 255

Scheme 2.72

$$\underset{\mathbf{256}}{\mathrm{OIOSO_2F}} + \mathrm{ArH} \xrightarrow[\text{53-95\%}]{\mathrm{CH_2Cl_2}, -60 \text{ to } 20\,^\circ\mathrm{C}, 0.5\text{-}3\text{ h}} \underset{\mathbf{257}}{\mathrm{Ar_2\overset{+}{I}\ {}^{-}OSO_3H}}$$

Ar = Ph, 4-MeC_6H_4, 4-$Bu^tC_6H_4$, 2,4-$Me_2C_6H_3$, 4-FC_6H_4, 4-ClC_6H_4, 4-BrC_6H_4

$$\underset{\mathbf{258}}{\mathrm{OIOTf}} + \mathrm{ArSiMe_3} \xrightarrow[\text{57-93\%}]{\mathrm{CH_2Cl_2}, -78 \text{ to } 20\,^\circ\mathrm{C}, 30\text{ min}} \underset{\mathbf{259}}{\mathrm{Ar_2\overset{+}{I}\ {}^{-}OTf}}$$

Ar = Ph, 4-MeC_6H_4, 2,4,6-$Me_2C_6H_2$, 4-BrC_6H_4, 4-(4'BrC_6H_4)C_6H_4, etc.

Scheme 2.73

$Ar_2I^+HSO_4^-$, in several consecutive steps. Other inorganic iodosyl derivatives, such as iodosyl fluorosulfate (**256**) and iodosyl trifluoromethanesulfonate (**258**) also react with arenes or trimethylsilylarenes under mild conditions to afford the corresponding iodonium hydrosulfates **257** or triflates **259** (Scheme 2.73) [373, 374].

A modification of this procedure, which is useful for a selective synthesis of unsymmetrical aryl(phenyl)iodonium hydrosulfates **260**, consists of the reaction of arenes with phenyliodosyl sulfate **85** (Scheme 2.74) [375].

Numerous methods for the preparation of symmetrical and unsymmetrical diaryl- and hetaryliodonium organosulfonates have been developed. A common synthetic approach to unsymmetric diaryl- and hetaryl(aryl)iodonium tosylates (e.g., **262**, **264**, **266** and **268**) is based on the reactions of [hydroxy(tosyloxy)iodo]arenes with aryltrimethylsilanes **261** [198], aryltributylstannanes **263** [376], arylboronic acids **265** [377], or the appropriate heteroaromatic precursors **267** (Scheme 2.75) [378, 379]. The reaction of HTIB with arylstannanes proceeds under milder conditions compared to arylsilanes and is applicable to a wide range of arenes with electron-withdrawing substituents. Arylboronic acids in general have some advantage over arylstannanes in the case of the electron-rich heterocyclic precursors [377].

Various unsymmetrically substituted diaryliodonium triflates **269** can be synthesized by the reaction of iodosylbenzene [380] or (diacetoxyiodo)arenes [381] with arenes in trifluoromethanesulfonic acid (Scheme 2.76). This simple procedure affords diaryliodonium triflates in relatively high yields, but it is limited to aromatic substrates that are not sensitive to strong acids. In a milder, more selective variation of this procedure (diacetoxyiodo)benzene is reacted with arylboronic acids in the presence of triflic acid at –30 °C to afford aryl(phenyl)iodonium triflates in 74–97% yields [377].

Several modified procedures for the preparation of diaryliodonium triflates have been reported in a more recent literature [382–384]. Kitamura and Hossain have developed a one-pot reaction to obtain symmetrical diaryliodonium triflates in good yields directly from iodoarenes and aromatic substrates using $K_2S_2O_8$ as an

$$\underset{\mathbf{85}}{\mathrm{Ph\overset{+}{I}OSO_3^-}} + \mathrm{ArH} \xrightarrow[\text{54-84\%}]{\mathrm{CH_2Cl_2}, -50 \text{ to } 20\,^\circ\mathrm{C}, 1\text{-}3.5\text{ h}} \underset{\mathbf{260}}{\mathrm{Ph\overset{+}{I}Ar\ {}^{-}OSO_3H}}$$

Ar = Ph, 4-MeC_6H_4, 4-FC_6H_4, 4-IC_6H_4, 4-$Bu^tC_6H_4$, 2,4-$Me_2C_6H_3$,
2,4,6-$Me_3C_6H_2$, 3-$NO_2C_6H_4$

Scheme 2.74

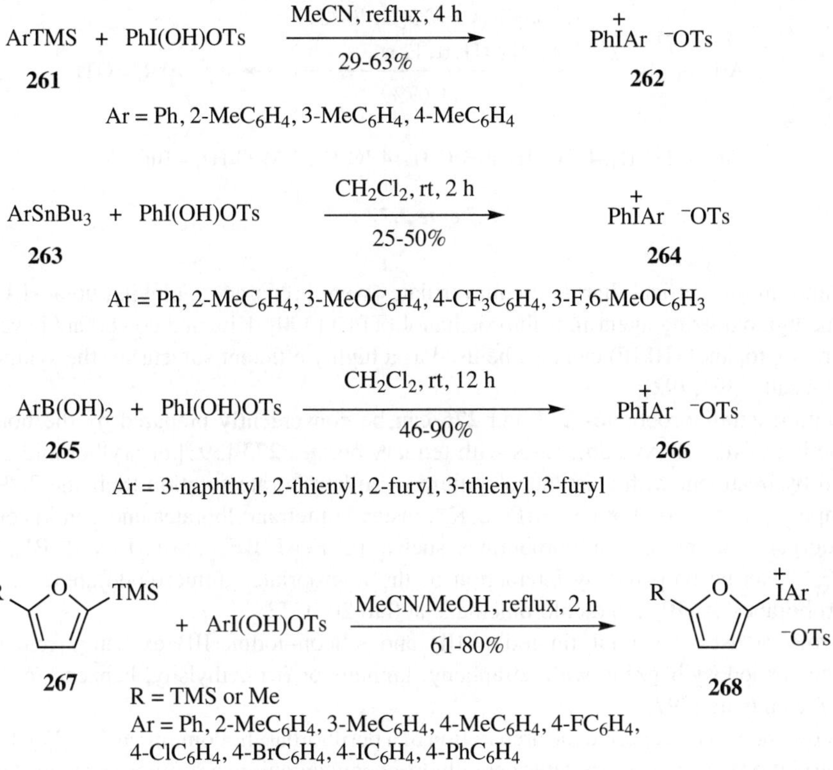

$$\underset{\mathbf{261}}{\text{ArTMS}} + \text{PhI(OH)OTs} \xrightarrow[\text{29-63\%}]{\text{MeCN, reflux, 4 h}} \underset{\mathbf{262}}{\text{Ph}\overset{+}{\text{I}}\text{Ar}\ ^{-}\text{OTs}}$$

Ar = Ph, 2-MeC_6H_4, 3-MeC_6H_4, 4-MeC_6H_4

$$\underset{\mathbf{263}}{\text{ArSnBu}_3} + \text{PhI(OH)OTs} \xrightarrow[\text{25-50\%}]{\text{CH}_2\text{Cl}_2\text{, rt, 2 h}} \underset{\mathbf{264}}{\text{Ph}\overset{+}{\text{I}}\text{Ar}\ ^{-}\text{OTs}}$$

Ar = Ph, 2-MeC_6H_4, 3-$MeOC_6H_4$, 4-$CF_3C_6H_4$, 3-F,6-$MeOC_6H_3$

$$\underset{\mathbf{265}}{\text{ArB(OH)}_2} + \text{PhI(OH)OTs} \xrightarrow[\text{46-90\%}]{\text{CH}_2\text{Cl}_2\text{, rt, 12 h}} \underset{\mathbf{266}}{\text{Ph}\overset{+}{\text{I}}\text{Ar}\ ^{-}\text{OTs}}$$

Ar = 3-naphthyl, 2-thienyl, 2-furyl, 3-thienyl, 3-furyl

$$\mathbf{267} + \text{ArI(OH)OTs} \xrightarrow[\text{61-80\%}]{\text{MeCN/MeOH, reflux, 2 h}} \mathbf{268}$$

(267: 2-R-5-TMS-furan; 268: 2-R-5-($\overset{+}{\text{I}}$Ar)-furan $^{-}$OTs)

R = TMS or Me
Ar = Ph, 2-MeC_6H_4, 3-MeC_6H_4, 4-MeC_6H_4, 4-FC_6H_4,
4-ClC_6H_4, 4-BrC_6H_4, 4-IC_6H_4, 4-PhC_6H_4

Scheme 2.75

oxidant [382]. Additional improvements of this procedure consist of the reaction of arenes with elemental iodine and $K_2S_2O_8$ in trifluoroacetic acid, followed by treatment with sodium triflate (Scheme 2.77) [383,384].

Olofsson and coworkers have developed several efficient one-pot syntheses of diaryliodonium salts [385–390]. A general and universal procedure provides both symmetrical and unsymmetrical diaryliodonium triflates **272** from both electron-deficient and electron-rich arenes **271** and aryl iodides **270** using *m*CPBA as the oxidant and triflic acid (Scheme 2.78) [385–387]. The electron-rich diaryliodonium tosylates are prepared similarly using toluenesulfonic acid instead of triflic acid as the additive [387]. Symmetrical diaryliodonium triflates can be synthesized by a modified one-pot procedure from iodine, arenes, *m*CPBA and triflic acid under similar conditions [374, 375]. A similar procedure based on a one-pot reaction of arylboronic acids, aryl iodides, *m*CPBA and $BF_3{\cdot}Et_2O$ has been used for regioselective synthesis of unsymmetrical diaryliodonium tetrafluoroborates [388, 389]. In a further improvement of this approach, a range of

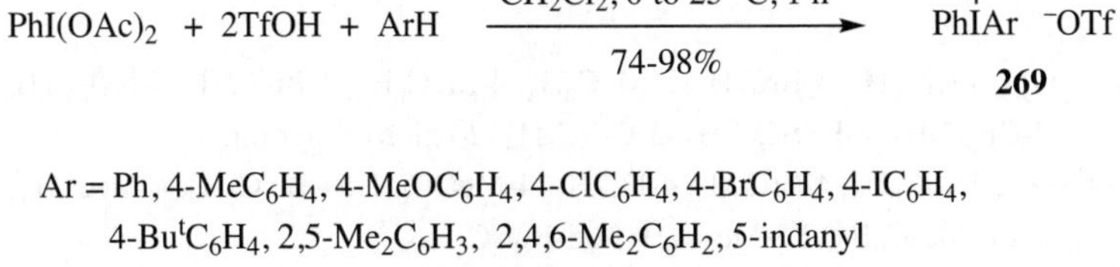

$$\text{PhI(OAc)}_2 + 2\text{TfOH} + \text{ArH} \xrightarrow[\text{74-98\%}]{\text{CH}_2\text{Cl}_2\text{, 0 to 25 °C, 1 h}} \underset{\mathbf{269}}{\text{Ph}\overset{+}{\text{I}}\text{Ar}\ ^{-}\text{OTf}}$$

Ar = Ph, 4-MeC_6H_4, 4-$MeOC_6H_4$, 4-ClC_6H_4, 4-BrC_6H_4, 4-IC_6H_4,
4-$Bu^tC_6H_4$, 2,5-$Me_2C_6H_3$, 2,4,6-$Me_2C_6H_2$, 5-indanyl

Scheme 2.76

$$\mathrm{ArH + I_2 \xrightarrow[\text{11-69\%}]{\text{1. } K_2S_2O_8,\ CF_3CO_2H,\ 40\ ^\circ C,\ 72\ h \quad 2. \text{ aq. NaOTf, rt, 12 h}} Ar_2I^+\ ^-OTf}$$

Ar = 4-FC_6H_4, 4-ClC_6H_4, 4-BrC_6H_4, 4-IC_6H_4, 4-MeC_6H_4, 4-$Bu^tC_6H_4$

Scheme 2.77

symmetrical and unsymmetrical diaryliodonium triflates were prepared employing urea–H_2O_2 as the environmentally benign oxidizing agent in trifluoroethanol (TFE) [390]. Kita and coworkers have demonstrated that hexafluoroisopropanol (HFIP) can also be used as a highly efficient solvent for the synthesis of various diaryliodonium salts [391,392].

Diaryliodonium tetrafluoroborates **274** and **276** can be conveniently prepared by the boron–iodine(III) exchange reaction of (diacetoxyiodo)arenes with tetraarylborates **273** [393] or arylboronic acids **275** [394, 395] followed by treatment with a saturated sodium tetrafluoroborate solution (Scheme 2.79). A modified procedure employs aryltrifluoroborates, $ArBF_3{}^-K^+$, instead of tetraarylborates under mild conditions [396]. Several fluoroorgano-iodonium tetrafluoroborates, such as $(C_6F_5)_2I^+BF_4{}^-$, $(4\text{-}C_5F_4N)_2I^+BF_4{}^-$ and $[C_6F_5(4\text{-}C_5F_4N)I]^+BF_4{}^-$, can be prepared by interaction of the appropriate (difluoroiodo)arenes with fluorinated organodifluoroboranes, Ar_fBF_2, in dichloromethane at 0 to 20 °C [34].

Another protocol uses a similar tin–iodine(III) and silicon–iodine(III) exchange reaction of (diacetoxyiodo)arenes or iodosylbenzene with tetraphenylstannane or (trimethylsilyl)benzene in the presence of boron trifluoride etherate [397].

Frohn and coworkers have reported the preparation of a perfluoroaryliodonium salt, $(C_6F_5)_2I^+AsF_6{}^-$, by the electrophilic arylation of C_6F_5I with a stable pentafluorophenylxenonium hexafluoroarsenate, $C_6F_5Xe^+AsF_6{}^-$ [398].

Skulski and Kraszkiewicz have also reported a new direct method for the preparation of various symmetrical diaryliodonium bromides (15–88% crude yields) from arenes by the reaction of ArH with $NaIO_4$ in sulfuric acid followed by the addition of KBr [23].

A very mild and general method for the preparation of diaryl- and heteroaryliodonium triflates is based on iodonium transfer reactions of iodine(III) cyanides with the respective aryl- or heteroarylstannanes [146,148, 399–401]. Specifically, (dicyano)iodonium triflate (**277**), generated *in situ* from iodosyl triflate and Me_3SiCN, reacts with tributyltin derivatives of aromatic and heteroaromatic compounds to afford the corresponding symmetrical iodonium salts under very mild conditions (Scheme 2.80) [389,390].

$$\underset{\mathbf{270}}{\mathrm{Ar^1I}} + \underset{\mathbf{271}}{\mathrm{Ar^2H}} \xrightarrow[\text{51-91\%}]{m\text{CPBA, TfOH, } CH_2Cl_2,\ \text{rt to } 80\ ^\circ C,\ 1\text{-}21\ h} \underset{\mathbf{272}}{\mathrm{Ar^1I^+Ar^2\ ^-OTf}}$$

Ar^1 = Ph, 4-ClC_6H_4, 4-BrC_6H_4, 2-MeC_6H_4, 4-MeC_6H_4, 4-$Bu^tC_6H_4$, 4-$NO_2C_6H_4$, 4-$CF_3C_6H_4$, 4-$HO_2CC_6H_4$, 3-$CF_3C_6H_4$, 2-chloro-5-pyridinyl

Ar^2 = Ph, 4-ClC_6H_4, 4-BrC_6H_4, 4-FC_6H_4, 4-$MeOC_6H_4$, 4-$Bu^tC_6H_4$, 4-MeC_6H_4, 2,4,6-$Me_3C_6H_2$, 2,5-$Me_2C_6H_3$, 2,5-$Bu^t{}_2C_6H_3$

Scheme 2.78

$$ArI(OAc)_2 + (4\text{-}MeOC_6H_4)_4B^- Na^+ \xrightarrow[49\text{-}63\%]{1.\ AcOH,\ rt;\ 2.\ NaBF_4,\ H_2O} 4\text{-}MeOC_6H_4I^+Ar\ ^-BF_4$$

Ar = Ph, Tol **273** **274**

$$PhI(OAc)_2 + ArB(OH)_2 \xrightarrow[73\text{-}83\%]{1.\ BF_3\bullet Et_2O,\ CH_2Cl_2;\ 2.\ NaBF_4,\ H_2O} PhI^+Ar\ ^-BF_4$$

275 **276**

Ar = Ph, 4-FC_6H_4, 4-ClC_6H_4, 4-$MeOC_6H_4$, Tol

Scheme 2.79

A similar iodonium exchange reaction involves aryl(cyano)iodonium triflates **278** and stannylated aromatic precursors providing many kinds of diaryl or aryl(heteroaryl) iodonium salts [145, 147, 401]. Tykwinski, Hinkle and coworkers have reported an application of such iodonium transfer reaction for obtaining of a series of mono- and bithienyl(aryl)iodonium triflates **279** with increasingly electron-withdrawing substituents on the aryl moiety (Scheme 2.81) [401]. Thienyl and bithienyl iodonium salts prepared by iodonium transfer reaction using PhI(CN)OTf are potentially useful as nonlinear optical materials [402].

A very mild and selective approach to aryl- and hetaryliodonium chlorides **282** is based on the reaction of aryllithium **280** (generated *in situ* from bromoarenes and butyllithium) with (*E*)-chlorovinyliodine(III) dichloride (**18**) (Scheme 2.82) [71, 88, 89, 403, 404]. The iodonium transfer reagent **18** is prepared by the reaction of iodine trichloride with acetylene in concentrated hydrochloric acid (Scheme 2.8 in Section 2.1.3.2) [403]; **caution:** this compound is highly unstable and should be handled and stored with proper safety precautions [71]. However, the iodonium transfer procedure with reagent **18** is particularly useful for the preparation of bis(hetaryl)iodonium chlorides **283** from the appropriate nitrogen heterocycles **282** (Scheme 2.82) [71].

A similar approach to aryl- and heteroaryl(phenyl)iodonium triflates **285** involves the ligand-transfer reaction between vinyliodonium salt **284** with aryllithiums (Scheme 2.83) [405]. Likewise, the reaction of (*E*)-[β-(trifluoromethanesulfonyloxy)ethenyl](aryl)iodonium triflates **286** with aryllithiums or alkynyllithiums can be used for a selective preparation of the respective diaryl- or alkynyl(aryl)iodonium triflates in high yields [406].

Peacock and Pletcher have reported a simple, one-step procedure for the synthesis of diaryliodonium salts by the electrochemical oxidation of aryl iodide at a carbon felt anode in acetic acid in the presence of an arene [407, 408]. This reaction gives good yields of diaryliodonium salts for aryl iodides and arenes with alkyl substituents in the benzene ring.

Several heteroaryl(phenyl)iodonium organosulfonates have been prepared by the [3+2] cycloaddition reactions of alkynyliodonium salts with 1,3-dipolar reagents. The reaction of (arylethynyl)iodonium tosylates

$$(NC)_2I^+\ ^-OTf + 2ArSnBu_3 \xrightarrow[41\text{-}75\%]{CH_2Cl_2,\ -40\ to\ 20\ ^\circ C} Ar_2I^+\ ^-OTf$$

277

Ar = Ph, 3-$MeOC_6H_4$, 4-$MeOC_6H_4$, 2-furyl, 2-thienyl, 4-pyrazolyl, etc.

Scheme 2.80

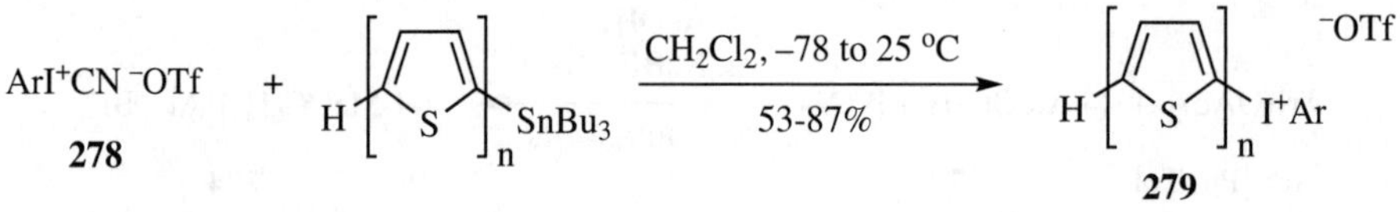

$ArI^{+}CN\ ^{-}OTf$ + H[thienyl]$_n$SnBu$_3$ —(CH_2Cl_2, –78 to 25 °C; 53-87%)→ H[thienyl]$_n$I$^+$Ar $^{-}$OTf

278 **279**

Ar = Ph, n = 1; Ar = Ph, n = 2
Ar = 4-$CF_3C_6H_4$, n = 2; Ar = 3-ClC_6H_4, n = 2

Scheme 2.81

287 with nitrile oxides **288** affords cycloadducts **289** as sole products in good yield (Scheme 2.84) [409]. The nitrone **291** reacts with (phenylethynyl)iodonium tosylate **290** to give cycloaddition product **292** in moderate yield [409]. A similar cycloaddition of alkynyliodonium triflate **293** with diazocarbonyl compounds **294** affords pyrazolyliodonium salts **295** as sole regioisomers [410,411]. Methyl and phenyl azides **297** react with alkynyliodonium triflates **296** upon heating in tetrahydrofuran or acetonitrile to give triazolyliodonium salts **298** as sole regioisomers in low yields (Scheme 2.84) [410].

Single-crystal X-ray structures have been reported for the following aryl- and heteroaryliodonium salts: diphenyliodonium triiodide [412], (2-methylphenyl)(2-methoxyphenyl)iodonium chloride [413], (2-methoxy-5-methylphenyl)(4-methoxy-2-methylphenyl)iodonium trifluoroacetate [414], (2-methoxy-5-methylphenyl)(4-methoxyphenyl)iodonium trifluoroacetate [415], a complex of diphenyliodonium tetrafluoroborate with pyridine [416], a complex of diphenyliodonium tetrafluoroborate with 1,10-phenanthroline [417], a complex of diphenyliodonium tetrafluoroborate with 18-crown-6 [418], 1-naphthylphenyliodonium tetrafluoroborate [419], 3,10-dimethyl-10*H*-dibenzo[*b,e*]iodinium tetrafluoroborate [420], aryl(pentafluorophenyl)iodonium tetrafluoroborates [421], 4,4′-[bis(phenyliodonium)]-diphenylmethane ditriflate [422], [bis(4-methoxyphenyl)](diethylaminocarbodithioato)iodine(III) [423], di(*p*-tolyl)iodonium bromide [424], diphenyliodonium chloride, bromide and iodide [425,426], diphenyliodonium nitrate [427], diphenyliodonium tetrafluoroborate [428], thienyl(phenyl)iodonium salts [401], (*anti*-dimethanoanthracenyl)phenyliodonium tosylate and hexafluorophosphate [429] and 3-mesityl-5-phenylisoxazol-4-yl(phenyl)iodonium *p*-toluenesulfonate [409].

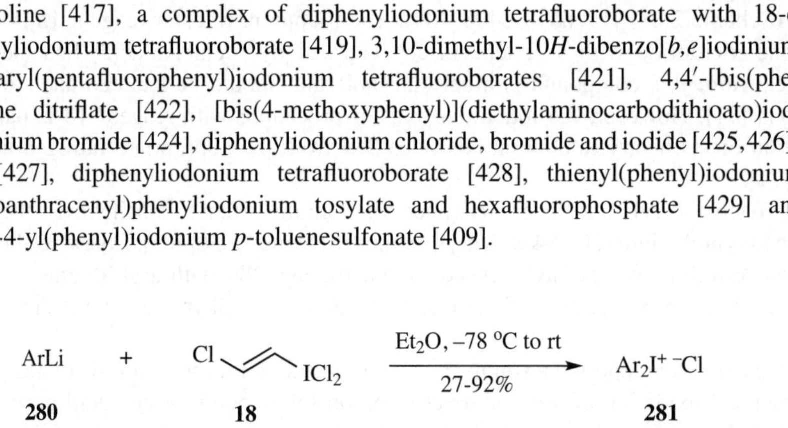

ArLi + Cl–CH=CH–ICl_2 —(Et_2O, –78 °C to rt; 27-92%)→ $Ar_2I^{+}\ ^{-}Cl$

280 **18** **281**

Ar = Ph, Tol, 1-naphthyl, 2-naphthyl, 2-thienyl, 2-furanyl, etc.

R-pyridyl-Br —(1. BuLi, Et_2O, –78 °C, 40 min; 2. **18**, –78 °C to rt, 4 h; 71%)→ (R-pyridyl)$_2I^{+}\ ^{-}Cl$

282 R = H or Cl **283**

Scheme 2.82

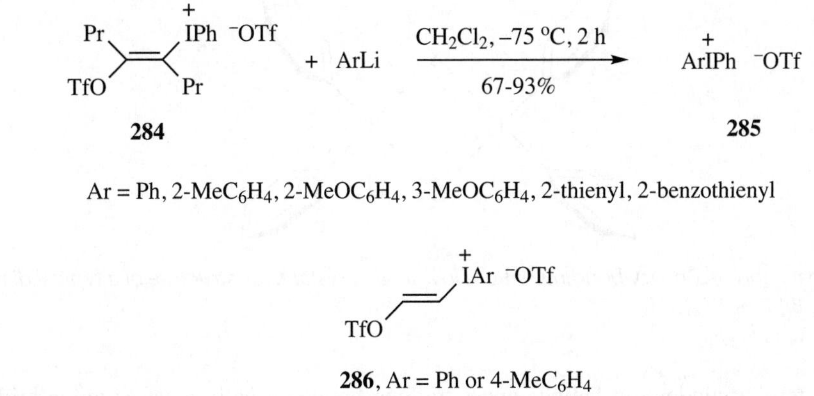

$$\text{(Pr)(TfO)C=C(Pr)}\overset{+}{\text{I}}\text{Ph}\ ^{-}\text{OTf}\ \mathbf{284} + \text{ArLi} \xrightarrow[67\text{-}93\%]{CH_2Cl_2,\ -75\ ^\circ C,\ 2\ h} \text{Ar}\overset{+}{\text{I}}\text{Ph}\ ^{-}\text{OTf}\ \mathbf{285}$$

Ar = Ph, 2-MeC_6H_4, 2-$MeOC_6H_4$, 3-$MeOC_6H_4$, 2-thienyl, 2-benzothienyl

$$\text{TfOCH=CH}\overset{+}{\text{I}}\text{Ar}\ ^{-}\text{OTf}$$

286, Ar = Ph or 4-MeC_6H_4

Scheme 2.83

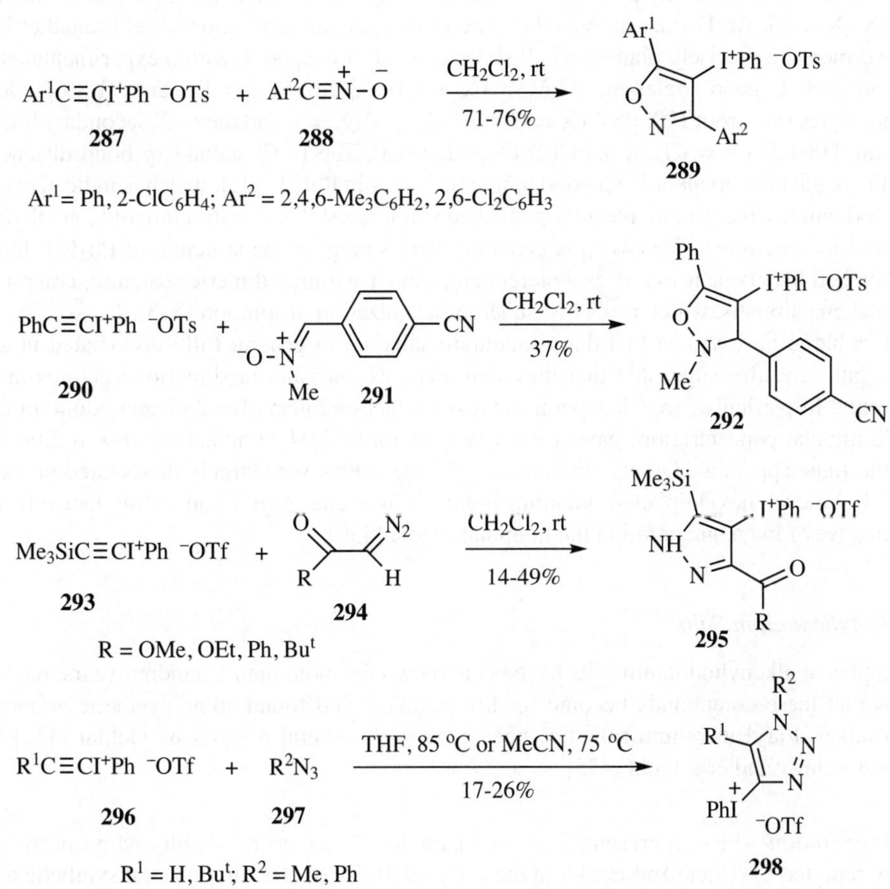

$$\text{Ar}^1\text{C}\equiv\text{CI}^+\text{Ph}\ ^{-}\text{OTs}\ \mathbf{287} + \text{Ar}^2\text{C}\equiv\overset{+}{\text{N}}-\overset{-}{\text{O}}\ \mathbf{288} \xrightarrow[71\text{-}76\%]{CH_2Cl_2,\ rt} \mathbf{289}$$

Ar¹ = Ph, 2-ClC_6H_4; Ar² = 2,4,6-$Me_3C_6H_2$, 2,6-$Cl_2C_6H_3$

$$\text{PhC}\equiv\text{CI}^+\text{Ph}\ ^{-}\text{OTs}\ \mathbf{290} + \mathbf{291} \xrightarrow[37\%]{CH_2Cl_2,\ rt} \mathbf{292}$$

$$\text{Me}_3\text{SiC}\equiv\text{CI}^+\text{Ph}\ ^{-}\text{OTf}\ \mathbf{293} + \text{RC(O)CHN}_2\ \mathbf{294} \xrightarrow[14\text{-}49\%]{CH_2Cl_2,\ rt} \mathbf{295}$$

R = OMe, OEt, Ph, Bu^t

$$\text{R}^1\text{C}\equiv\text{CI}^+\text{Ph}\ ^{-}\text{OTf}\ \mathbf{296} + \text{R}^2\text{N}_3\ \mathbf{297} \xrightarrow[17\text{-}26\%]{THF,\ 85\ ^\circ C\ or\ MeCN,\ 75\ ^\circ C} \mathbf{298}$$

R¹ = H, Bu^t; R² = Me, Ph

Scheme 2.84

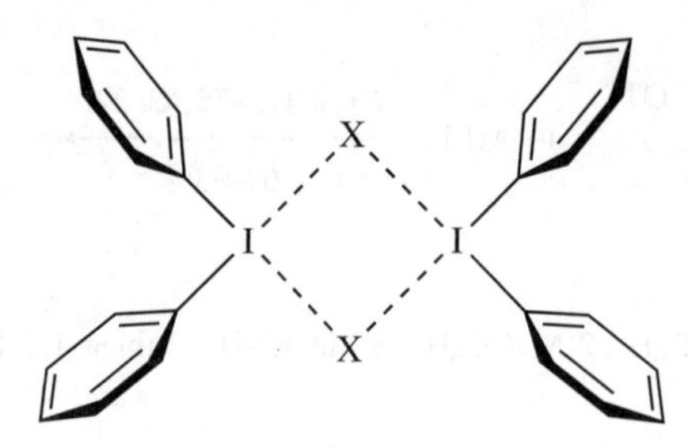

Figure 2.11 Primary and secondary bonding pattern in a single-crystal X-ray structure of a typical diaryliodonium salt Ph_2IX (X = Cl, Br, I, or BF_4).

In the solid state, diaryliodonium halides and tetrafluoroborates usually exist as anion-bridged dimers held together by iodine–halogen bonds (Figure 2.11). The overall iodine coordination is square planar and the C—I—C bond angle is close to 90°. A particularly detailed X-ray structural study of diphenyliodonium chloride, bromide and iodide was published by Alcock and Countryman in 1977 [425]. All three compounds Ph_2IX (X = Cl, Br, I) exist in the solid state as centrosymmetric dimers, held together by halogen bridges. The dimers are precisely planar and all six I—C bonds are equal to within experimental error (mean 2.090 Å). The C—I—C bond angles are 92.6(3)° (X = Cl), 93.2(5)° (X = Br) and 91.8(6)° (X = I), the C—I—X bond angles vary from 179.3(3)° (X = Cl) to 174.8(4)° (X = I) and the I···X secondary bond distance increases from 3.064(3) (X = Cl) to 3.463(2) (X = I) [373]. The I···Cl secondary bond distance is much longer than the regular hypervalent I—Cl bond length of 2.47 Å in $PhICl_2$ [92], which is indicative of the ionic character of iodonium salts. A more recently published single-crystal X-ray structure of (2-methylphenyl)(2-methoxyphenyl)iodonium chloride [413], is generally very similar to the structure of Ph_2ICl. Interestingly, this study unveiled hypervalent iodine as a stereogenic center within a dimeric structure, composed of two conformational enantiomers, which probably undergo racemization in solution [413].

Studies in solution have shown that diaryliodonium salts are in general fully dissociated in aqueous or polar media, but have also suggested that they may exist as ion pairs or dimeric structures in less polar organic solvents. In particular, freezing point depression measurements for diphenyliodonium chloride in water (25–50 mmolar concentration) gave a van't Hoff factor of 2.04, compared to one of 2.06 for sodium chloride in the same apparatus. Diphenyliodonium chloride is thus very largely dissociated in water. Dilute solutions of 4,4′-dicyclohexyldiphenyliodonium iodide in benzene gave a van't Hoff factor of about 0.5, which is indicative of ion aggregation in the nonpolar media [430].

2.1.9.2 *Alkenyliodonium Salts*

Several examples of alkenyliodonium salts have been known for more than a hundred years, but it was only in the 1990s that these compounds become readily available and found some synthetic application. The chemistry of alkenyl(aryl)iodonium salts has been covered in several reviews by Ochiai [431], Okuyama [432–434] and Zefirov and coauthors [435].

2.1.9.2.1 Preparation First representatives of alkenyliodonium salts, dichlorovinyl(phenyl)iodonium species, were reported by Thiele and Haakh in the early 1900s [436]. The first general synthetic approach to alkenyl(phenyl)iodonium salts was developed by Ochiai in the mid-1980s [437, 438]. This method is based on the reaction of silylated alkenes **299** with iodosylbenzene in the presence of Lewis acids, leading to the stereoselective formation of various alkenyliodonium tetrafluoroborates **300** in good yield (Scheme 2.85).

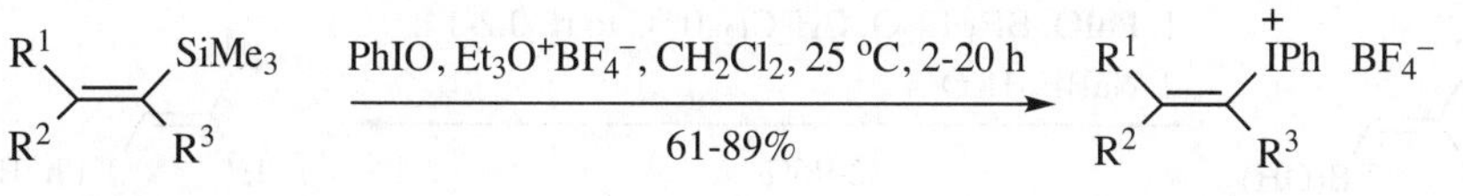

$$\mathrm{R^1(R^2)C{=}C(R^3)SiMe_3}\ \mathbf{299} \xrightarrow[61\text{-}89\%]{PhIO,\ Et_3O^+BF_4^-,\ CH_2Cl_2,\ 25\ ^\circ C,\ 2\text{-}20\ h} \mathrm{R^1(R^2)C{=}C(R^3)\overset{+}{I}Ph\ BF_4^-}\ \mathbf{300}$$

R^1, R^2, R^3 = H, alkyl, Ph

Scheme 2.85

The boron trifluoride-catalyzed silicon–iodine(III) exchange reaction of alkenylsilanes **301** with iodosylarenes followed by treatment with aqueous $NaBF_4$ is the most general method for the synthesis of alkenyl(aryl)iodonium tetrafluoroborates **302** (Scheme 2.86) [394, 439]. This reaction proceeds under mild conditions and in a stereoselective manner with retention of the geometry of organosilanes.

A similar boron–iodine(III) exchange of alkenylboronic acids **303** with iodosylbenzene or (diacetoxyiodo)benzene in the presence of boron trifluoride etherate is an efficient alternative method for a selective preparation of alkenyl(phenyl)iodonium tetrafluoroborates **304** in excellent yields (Scheme 2.87) [440, 441].

Likewise, the reaction of vinylzirconium derivatives **305** with (diacetoxyiodo)benzene followed by anion exchange affords alkenyl(phenyl)iodonium salts **306** stereoselectively with retention of configuration (Scheme 2.88) [442].

A very general and mild procedure for the stereospecific synthesis of alkenyl(aryl)iodonium triflates **309** involves aryl(cyano)iodonium triflates **308** as iodonium transfer reagents in reactions with stannylated alkenes **307** (Scheme 2.89) [367, 443, 444]. This method was also applied to the preparation of the parent vinyliodonium triflate from tributyl(vinyl)tin [445].

Several types of functionalized alkenyl(aryl)iodonium salts have been prepared by the addition of hypervalent iodine reagents to alkynes. Reactions of terminal alkynes with iodosylbenzene and triflic acid proceed as a highly stereoselective *anti*-addition to afford (*E*)-(β-trifluoromethanesulfonyloxyalkenyl)phenyliodonium triflates **310** in high yield (Scheme 2.90) [446–448]. Similar products were obtained from the reactions of internal alkynes and parent acetylene [447].

Similarly functionalized alkenyl(aryl)iodonium triflates **312** were prepared by the addition of (aryl)fluoroiodonium triflates **311** to terminal alkynes (Scheme 2.91) [45, 48].

Likewise, various alkenyliodonium fluorides **314** were prepared by the addition of 4-(difluoroiodo)toluene **313** to alkynes (Scheme 2.92) [449–452].

Alkynyl(aryl)iodonium salts can serve as precursors to some alkenyl(aryl)iodonium salts [453–455]. For example, (*E*)-β-fluoroalkenyl(phenyl)iodonium tetrafluoroborates **316** can be stereoselectively prepared by

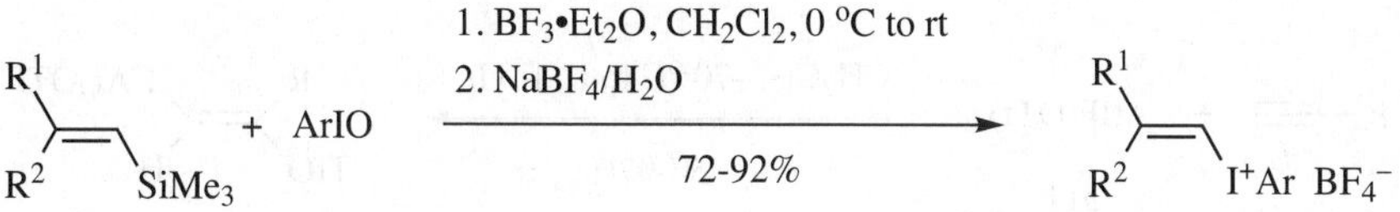

$$\mathrm{R^1(R^2)C{=}CHSiMe_3}\ \mathbf{301} + \mathrm{ArIO} \xrightarrow[72\text{-}92\%]{1.\ BF_3\bullet Et_2O,\ CH_2Cl_2,\ 0\ ^\circ C\ to\ rt;\ 2.\ NaBF_4/H_2O} \mathrm{R^1(R^2)C{=}CHI^+Ar\ BF_4^-}\ \mathbf{302}$$

R^1 = 4-$BrC_6H_4OCH_2$, $PhCH_2CH_2$, 4-$ClC_6H_4OCH_2$, C_8H_{17}, etc.
R^2 = H, Me; Ar = Ph, 2,4,6-$Me_3C_6H_2$, etc.

Scheme 2.86

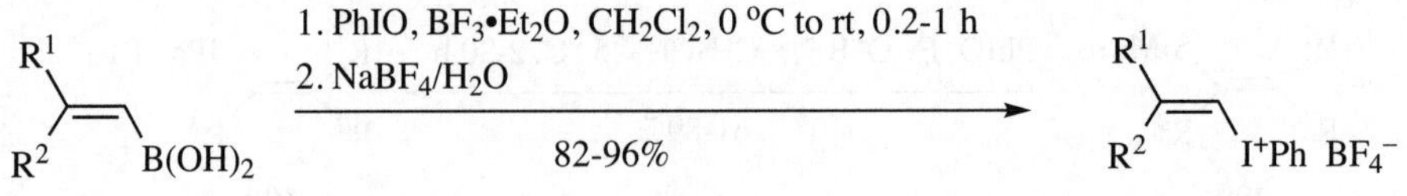

$R^1 = Bu, Bu^t, Ph(CH_2)_3, (CH_3)_2CH(CH_2)_2$, etc.; R^2 = H, Me

1. PhIO, $BF_3{\bullet}Et_2O$, CH_2Cl_2, 0 °C to rt, 0.2-1 h
2. $NaBF_4/H_2O$
82-96%

303 → **304** ($I^+Ph\ BF_4^-$)

Scheme 2.87

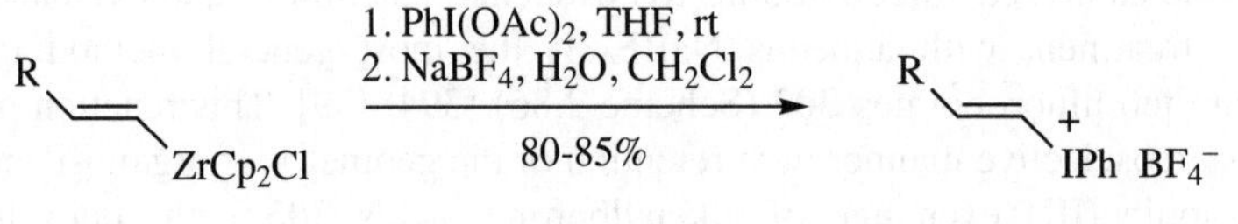

1. $PhI(OAc)_2$, THF, rt
2. $NaBF_4$, H_2O, CH_2Cl_2
80-85%

305 ($ZrCp_2Cl$) → **306** ($IPh\ BF_4^-$)

R = Bu, CH_3OCH_2, Ph

Scheme 2.88

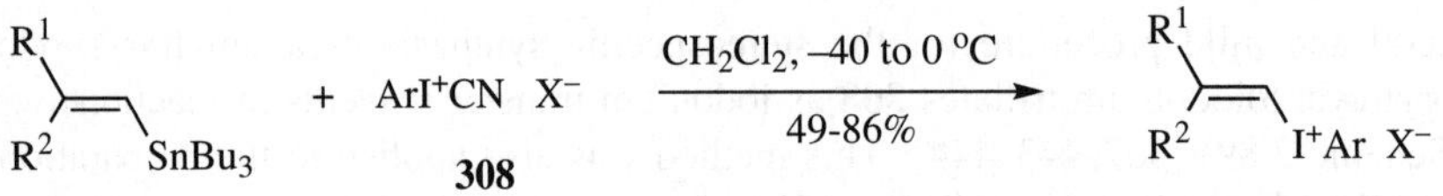

307 ($SnBu_3$) + $ArI^+CN\ X^-$ **308** → **309** ($I^+Ar\ X^-$)

CH_2Cl_2, –40 to 0 °C
49-86%

R^1 = Me, Et, Bu, Ph; R^2 = Me, Et, Bu
Ar = Ph, $4\text{-}CF_3C_6H_4$, $3,5\text{-}(CF_3)_2C_6H_3$; X = OTf or OTs

Scheme 2.89

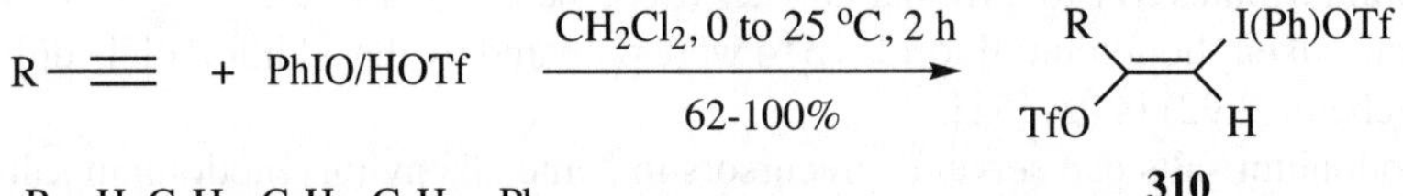

R—≡ + PhIO/HOTf → **310**

CH_2Cl_2, 0 to 25 °C, 2 h
62-100%

R = H, C_3H_7, C_4H_9, C_6H_{13}, Ph

Scheme 2.90

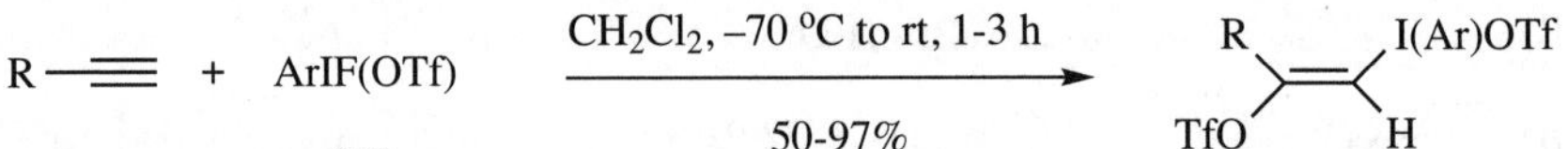

R—≡ + ArIF(OTf) **311** → **312**

CH_2Cl_2, –70 °C to rt, 1-3 h
50-97%

R = H, C_3H_7, C_4H_9, C_8H_{17}, CH_2OCH_3, CH_2OH, CH_2Cl
Ar = Ph, $4\text{-}MeC_6H_4$, $2\text{-}MeC_6H_4$, $4\text{-}NO_2C_6H_4$

Scheme 2.91

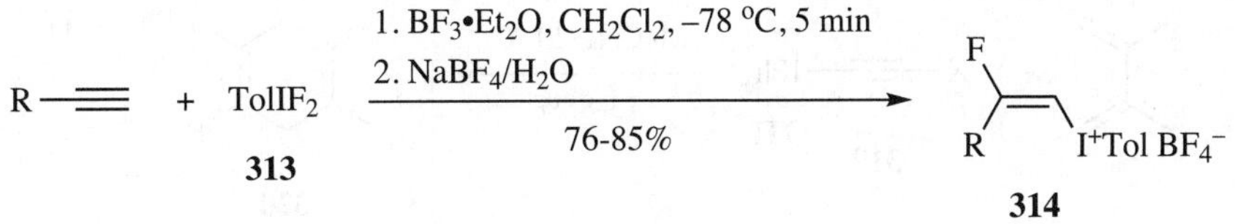

R—≡ + $TolIF_2$ (**313**) —[1. $BF_3 \cdot Et_2O$, CH_2Cl_2, –78 °C, 5 min; 2. $NaBF_4/H_2O$; 76-85%]→ F(R)C=CH–$I^+Tol\ BF_4^-$ (**314**)

R = $AcO(CH_2)_9$, $Cl(CH_2)_9$, $MeOOC(CH_2)_8$, $Bu^tCO(CH_2)_8$, (*cyclo*-C_6H_{11})CH_2

Scheme 2.92

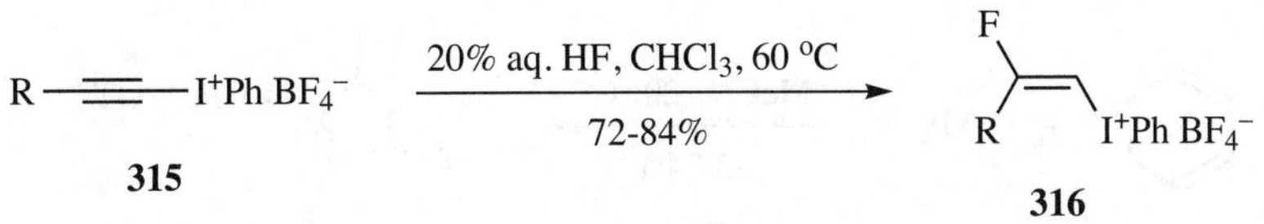

R—≡—$I^+Ph\ BF_4^-$ (**315**) —[20% aq. HF, $CHCl_3$, 60 °C; 72-84%]→ F(R)C=CH–$I^+Ph\ BF_4^-$ (**316**)

R = Bu^t, $C_{10}H_{21}$, (*cyclo*-C_6H_{11})CH_2, $Cl(CH_2)_9$, $Bu^tCO(CH_2)_8$, $Pr^iOCO(CH_2)_8$

Scheme 2.93

the reaction of alkynyl(phenyl)iodonium salts **315** with aqueous HF in good yields (Scheme 2.93) [453,454]. This method is applicable to the synthesis of fluoroalkenyliodonium salts having functional groups such as ketone, ester and chloride.

(*Z*)-β-Substituted vinyliodonium salts **318** can be prepared by a Michael-type nucleophilic addition of halide anions to alkynyliodonium salts **317** (Scheme 2.94) [456,457].

Various cyclohexadienyl(phenyl)iodonium salts have been prepared by the [4+2] Diels–Alder cycloadditions of alkynyliodonium salts **319**, functionalized with electron-withdrawing substituents in the β-position, with a wide range of dienes. Scheme 2.95 shows several examples of these cycloadditions, affording adducts **320–322** as stable microcrystalline solids [458].

The reaction of alkynyliodonium salts **319** with unsymmetrically substituted dienes **323** results in a mixture of two regioisomeric cyclohexadienes **324** and **325** (Scheme 2.96) [459]. In general, this cycloaddition shows low regioselectivity in the case of 2-substituted dienes and has a better degree of regioselectivity in the case of 1-substituted dienes. Moreover, the reaction of 1-methylbutadiene (**326**) with alkynyliodonium salt **327** selectively affords a single regioisomer (**328**), whose structure was established by X-ray analysis (Scheme 2.96) [459].

Several iodonium- and bis(iodonium) norbornadienes and other polycyclic adducts have been synthesized by [2+4] cycloaddition reactions of alkynyliodonium triflates with cyclic 1,3-dienes [458, 460–464]. In particular, the bis-iodonium acetylene **331** undergoes Diels–Alder reactions with cyclopentadiene **329**, furan

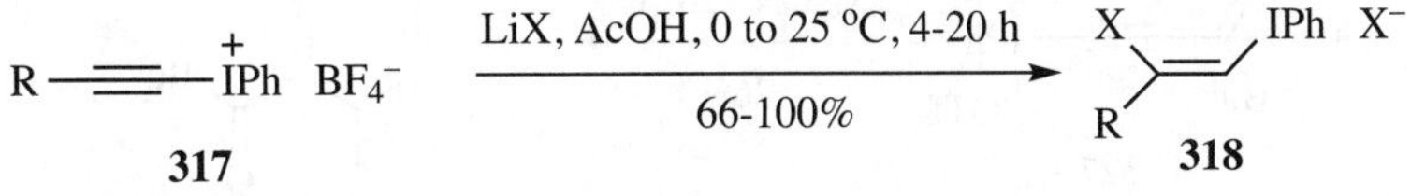

R—≡—$\overset{+}{I}Ph\ BF_4^-$ (**317**) —[LiX, AcOH, 0 to 25 °C, 4-20 h; 66-100%]→ X(R)C=CH–$\overset{+}{I}Ph\ X^-$ (**318**)

R = alkyl; X = F, Cl, Br

Scheme 2.94

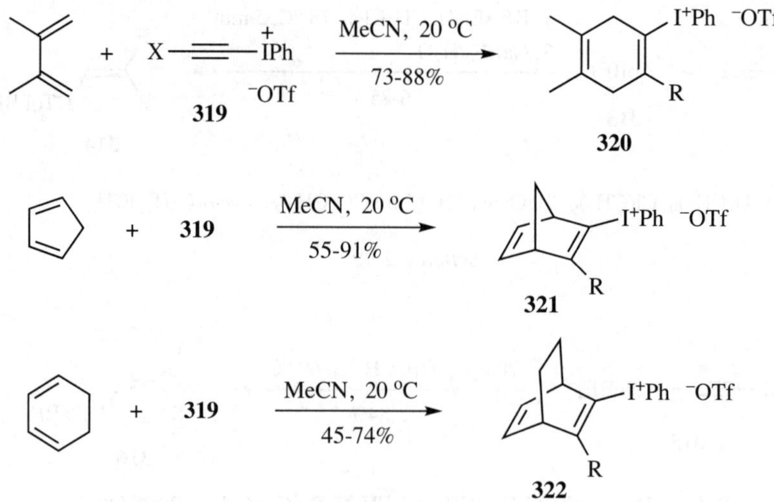

X = CN, Ts, PhC(O), 2-furyl-C(O), 2-thienyl-C(O)

Scheme 2.95

330 and 1,3-diphenylisobenzofuran **334** in acetonitrile under very mild conditions (Scheme 2.97) [460]. All adducts (**332**, **333** and **335**) were isolated in the form of stable microcrystalline solids; the structure of adduct **332** was established by single-crystal X-ray analysis [460].

2.1.9.2.2 Structural Studies Examples of alkenyliodonium salts whose X-ray crystal structures have been reported in literature include the following: (*E*)-(β-trifluoromethanesulfonyloxyalkenyl)phenyliodonium triflate **336** [465], 2,3-bis(iodonium)norbornadiene **332** [460], 4-*tert*-butylcyclohexenyl(phenyl)iodonium

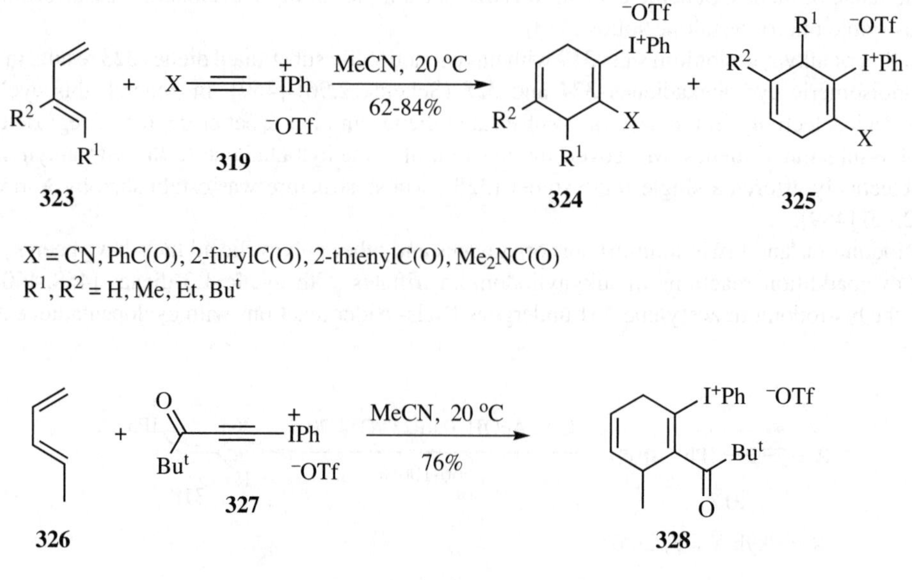

X = CN, PhC(O), 2-furylC(O), 2-thienylC(O), $Me_2NC(O)$
R^1, R^2 = H, Me, Et, Bu^t

Scheme 2.96

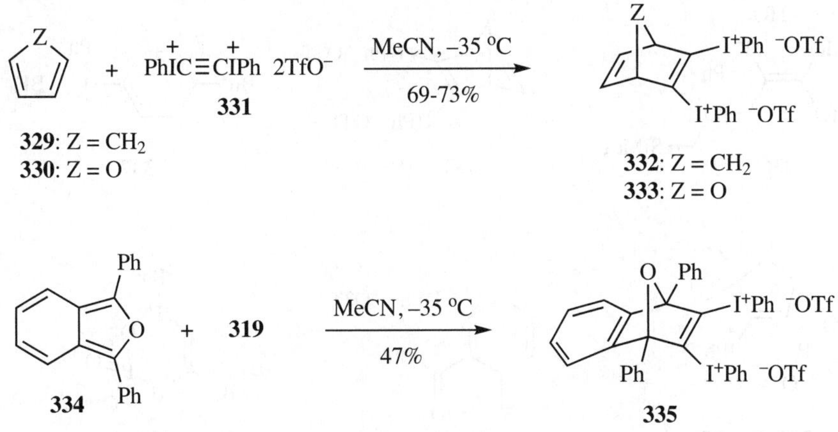

Scheme 2.97

tetrafluoroborate (**337**) [437], phenyl(2,2-dimethyl-4-(diethylphosphono)-2,5-dihydro-3-furyl)iodonium salts **338** and **339** [466], 1-(trimethylacetyl)-2-(phenyl{[(trifluoromethyl)sulfonyl]oxy}iodo)-6-methyl-1,4-cyclohexadiene (**328**) [459], internal alkenyliodonium salt **340** [467], substituted 1-{[(*Z*)-1-phenyl-2-phenyliodonio)vinyl]oxy}pyridinium ditosylate **341** [455], (*E*)-(β-chloroalkenyl)phenyliodonium tetrafluoroborates **342** and **343** [452] and a complex of **337** with 18-crown-6 [468] (Figure 2.12).

The structures of alkenyliodonium salts can generally be considered as ionic with a typical distance between iodine and the nearest anion of 2.8–3.1 Å. However, with consideration of the anionic part of the molecule, the overall experimentally determined geometry is T-shaped, similar to the hypervalent 10-I-3 species. The formation of dimeric structures with bridging anions is typical in the solid state; for example, 4-*tert*-butylcyclohexenyl(phenyl)iodonium tetrafluoroborate **337** has a centrosymmetric dimeric structure analogous to diphenyliodonium salts (cf. Figure 2.11) [437]. The carbon–carbon double bonds in alkenyliodonium salts have normal lengths of 1.31–1.34 Å, the C—I distances can vary from 2.07 to 2.13 Å and the C—I—C angles are in the range 91–99°.

2.1.9.3 *Alkynyliodonium Salts*

The first alkynyliodonium salt, phenylethynyl(phenyl)iodonium chloride, was obtained in the mid-1960s by Beringer and Galton by interaction of $PhICl_2$ with lithium phenylacetylide [469]. However, this product was unstable due to the relatively high nucleophilicity of the chloride counter ion and decomposed in a few hours at room temperature into a 1 : 1 mixture of iodobenzene and phenyl chloroacetylene. In 1979, Merkushev and coworkers reported the formation of the crystalline, relatively stable, albeit hygroscopic, alkynyliodonium trifluoroacetates but characterized them only by IR spectroscopy [470]. The first fully characterized, stable alkynyliodonium salts were alkynyl(phenyl)iodonium tosylates prepared by Koser and coworkers via the interaction of [hydroxy(tosyloxy)iodo]benzene with terminal alkynes [199, 471]. From then on it has been recognized that the stability of alkynyliodonium salts depends on the nature of the counterion. In general, non-nucleophilic anions such as sulfonate and tetrafluoroborate are required to stabilize the iodonium salt. The chemistry of alkynyl(aryl)iodonium salts has been previously covered in several specialized reviews [370, 472–475].

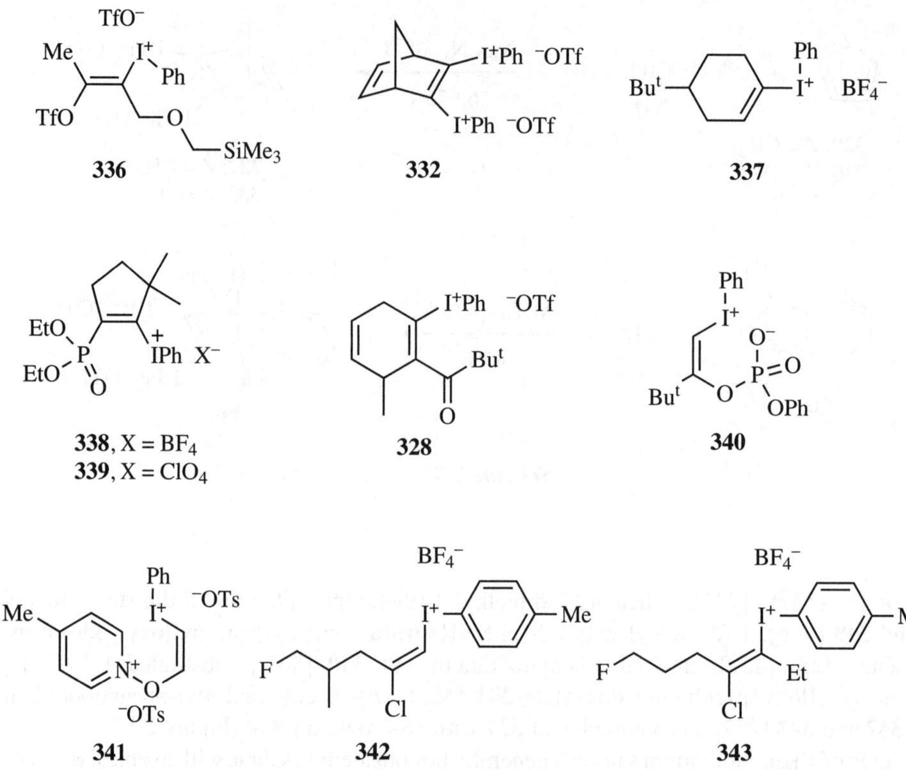

336 332 337

338, X = BF_4
339, X = ClO_4

328 340

341 342 343

Figure 2.12 Examples of alkenyliodonium salts analyzed by single-crystal X-ray diffraction.

2.1.9.3.1 Preparation A common synthetic approach to alkynyliodonium salts involves the reaction of an electrophilic λ^3-iodane with a terminal alkyne or its silylated, stannylated, or lithiated derivative. In the early 1980s, Koser and coworkers found that [hydroxy(tosyloxy)iodo]benzene **75** reacts with terminal alkynes **344** upon gentle heating in chloroform or dichloromethane to form alkynyliodonium tosylates **345** in moderate to low yield (Scheme 2.98) [199,471,476].

This reaction (Scheme 2.98) works well only for alkynes **344** with a bulky alkyl or an aryl group as the substituent R. The addition of a desiccant to the reaction mixture results in broader applicability of this procedure with a greater variety of alkynes [477,478]. This method is also applicable to the preparation of alkynyliodonium mesylates and *p*-nitrobenzenesulfonates by the reaction of the appropriate [hydroxy(organosulfonyloxy)iodo]benzenes with terminal alkynes under similar conditions [477,478]. This procedure has been applied for the preparation of arylethynyl(phenyl)iodonium tosylates **347** bearing long

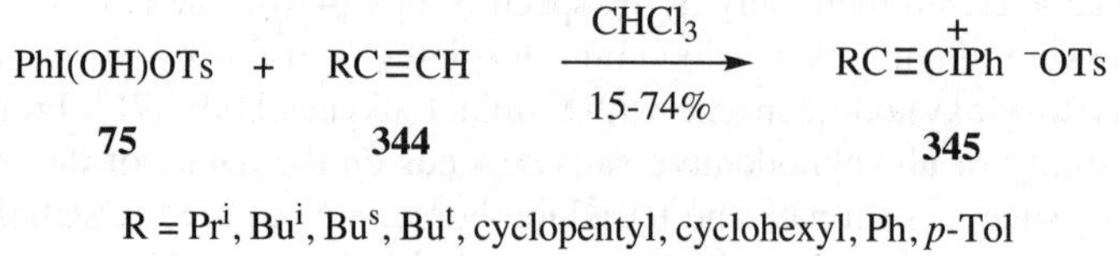

$$\underset{\mathbf{75}}{\mathrm{PhI(OH)OTs}} + \underset{\mathbf{344}}{\mathrm{RC{\equiv}CH}} \xrightarrow[15\text{-}74\%]{\mathrm{CHCl_3}} \underset{\mathbf{345}}{\mathrm{RC{\equiv}C\overset{+}{I}Ph\ {}^{-}OTs}}$$

R = Pr^i, Bu^i, Bu^s, Bu^t, cyclopentyl, cyclohexyl, Ph, *p*-Tol

Scheme 2.98

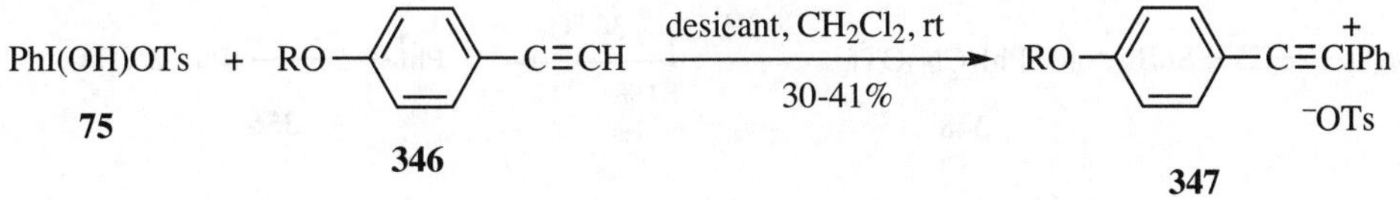

PhI(OH)OTs + RO–C_6H_4–C≡CH (desicant, CH_2Cl_2, rt; 30-41%) → RO–C_6H_4–C≡C$\overset{+}{I}$Ph $^{-}$OTs

75, 346, 347

R = C_8H_{17}, $C_{10}H_{21}$, $C_{12}H_{25}$, $C_{14}H_{29}$, (*S*)-$C_2H_5CH(CH_3)CH_2$

Scheme 2.99

alkoxy chains and chiral alkynyl ligands using Koser' reagent **75** and the appropriate alkynes **346** (Scheme 2.99) [479–481].

The most versatile method for preparing alkynyl(phenyl)iodonium triflates employs the iodonium transfer reaction between cyano(phenyl)iodonium triflate (**348**) and alkynylstannanes. The interaction of a large variety of readily available β-functionalized alkynylstannanes **349** with reagent **348** under very mild conditions provides ready access to diverse β-functionalized alkynyliodonium salts **350** in excellent yields (Scheme 2.100) [458, 482, 483]. This procedure is particularly useful for the preparation of various complex,

PhI(CN)OTf + RC≡CSnBu_3 (CH_2Cl_2, –42 °C; - R_3SnCN) → RC≡C$\overset{+}{I}$Ph $^{-}$OTf

348, 349, 350

R = H, Me, Bu, 1-cyclohexenyl, $MeOCH_2$, $ClCH_2$, $BrCH_2$, CN, Cl, MeC(OH)Ph, Ts, t-BuC(O), PhC(O), MeOC(O), $Me_2NC(O)$, 1-adamantyl-C(O), 2-furyl-C(O), 2-thienyl-C(O), cyclopropyl-C(O), (pyrrolidin-1-yl)N–C(O), (piperidin-1-yl)N–C(O), (azepan-1-yl)N–C(O), (morpholin-4-yl)N–C(O)

351: $CH(OMe)_2$, NHTs, $Ph\overset{+}{I}$ $^{-}$OTf

352: NHCONHTs, $Ph\overset{+}{I}$ $^{-}$OTf

353: O, N-BOC, $^{+}$IPh $^{-}$OTf

354: Bu_3Sn, $SnBu_3$, N, O, $^{+}$IPh $^{-}$OTf

355: OBn, BnO, BnO, O, O, OBn, $^{+}$IPh $^{-}$OTf

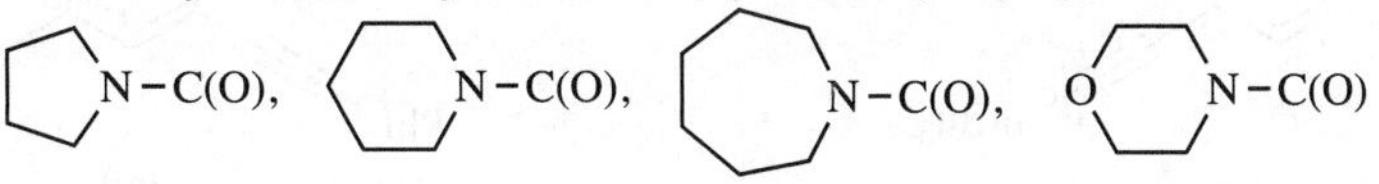

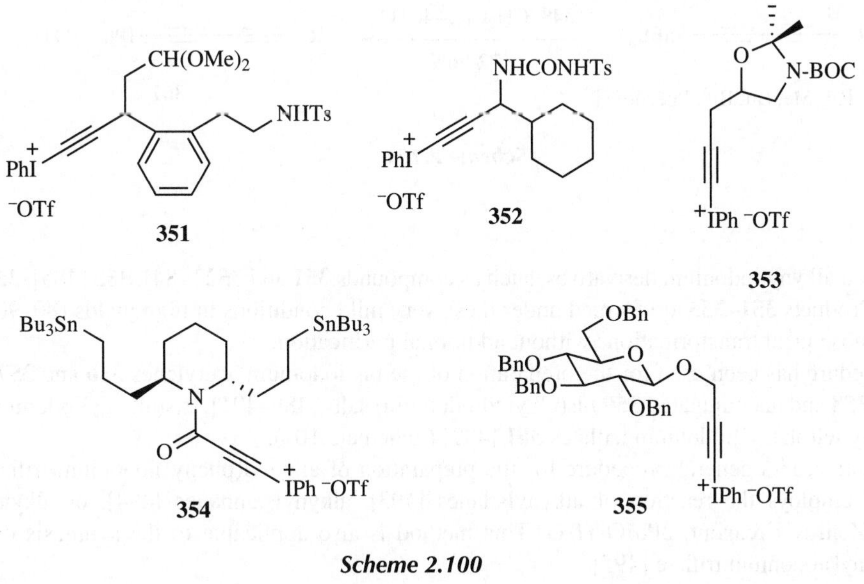

Scheme 2.100

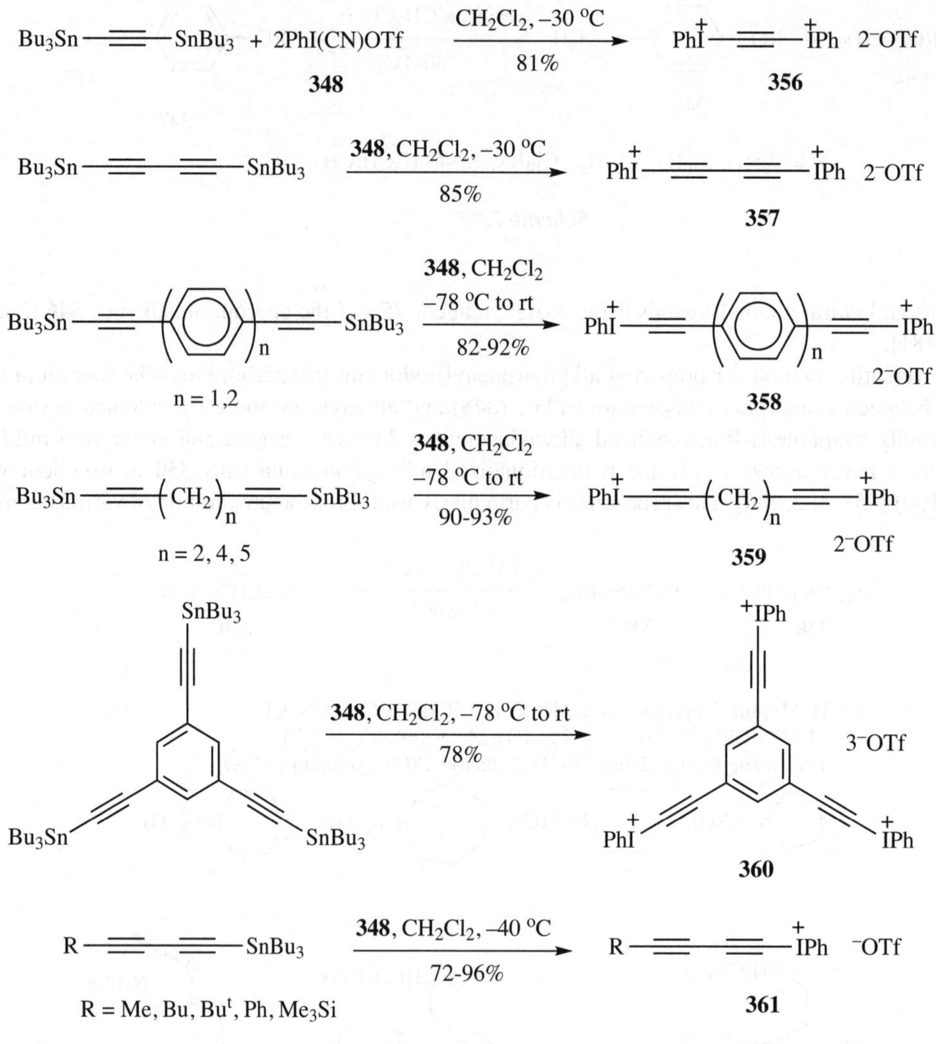

Scheme 2.101

functionalized alkynyliodonium derivatives, such as compounds **351** and **352** [484], **353** [485], **354** [486] and **355** [487]. Products **351–355** are formed under these very mild conditions in high yields (80–90%) and can be used in subsequent transformations without additional purification.

This procedure has been used for the preparation of the bis-iodonium acetylenes **356** and **357** [460, 488], conjugated **358** and unconjugated **359** bis(alkynyliodonium) salts [489–491], tris(alkynyliodonium) salts **360** [491] and diynyl(phenyl)iodonium triflates **361** [492] (Scheme 2.101).

An alternative, less general procedure for the preparation of alkynyl(phenyl)iodonium triflates in moderate yields employs the reaction of alkynylsilanes [493], alkynylstannanes [494], or alkynylboronates [297] with Zefirov's reagent, $2PhIO{\cdot}Tf_2O$. This method is also applicable to the synthesis of the parent ethynyl(phenyl)iodonium triflate [495].

$$\underset{\mathbf{362}}{PhIO{\bullet}Et_3O^+BF_4^-} + RC{\equiv}CSiMe_3 \xrightarrow[56\text{-}75\%]{CH_2Cl_2,\ rt} \underset{\mathbf{363}}{RC{\equiv}C\overset{+}{I}Ph\ BF_4^-}$$

R = Ph, $PhCH_2$, $PhCH_2CH_2$, C_8H_{17}, *cyclo*-C_6H_{11}

$$\underset{\mathbf{364}}{PhIO{\bullet}BF_3} + RC{\equiv}CSiMe_3 \xrightarrow[54\text{-}85\%]{1.\ CH_2Cl_2,\ rt;\ 2.\ NaBF_4,\ H_2O} \mathbf{363}$$

R = Ph, $PhCH_2$, C_8H_{17}, Me_3Si

$$RC{\equiv}CSiMe_3 + \mathbf{364} \xrightarrow[62\text{-}89\%]{1.\ CHCl_3,\ rt;\ 2.\ NaOSO_2Ar,\ H_2O} \underset{\mathbf{365}}{RC{\equiv}C\overset{+}{I}Ph\ ArSO_3^-}$$

R = Me, Et, Pr, Bu, Me_3Si; Ar = 4-MeC_6H_4 or 4-$NO_2C_6H_4$

Scheme 2.102

A common method for preparing alkynyl(phenyl)iodonium tetrafluoroborates involves the reaction of iodosylbenzene in the presence of triethyloxonium tetrafluoroborate or boron trifluoride etherate with alkynylsilanes. For example, the complex of iodosylbenzene with triethyloxonium tetrafluoroborate (**362**) reacts with alkynylsilanes in dichloromethane at room temperature to afford alkynyl(phenyl)iodonium tetrafluoroborates **363** in good yield [496]. A variation of this procedure employs the complex of iodosylbenzene with boron trifluoride etherate (**364**) followed by treatment with aqueous $NaBF_4$ [493,497] or sodium arylsulfonates to furnish, respectively, the appropriate alkynyl(phenyl)iodonium tetrafluoroborates **363** [483,487] or organosulfonates **365** [488,489] (Scheme 2.102) [498,499].

A modified method for the synthesis of alkynyl(phenyl)iodonium tetrafluoroborates **363** consists of the direct reaction of terminal alkynes with iodosylbenzene, a 42% aqueous solution of tetrafluoroboric acid and a catalytic amount of mercury oxide (Scheme 2.103) [500].

Yoshida and coauthors have reported a facile preparation of iodonium salts **367** by the reaction of potassium organotrifluoroborates **366** with (difluoroiodo)arenes under mild conditions (Scheme 2.104) [396]. A similar approach to alkynyliodonium salts by the reaction of alkynyldifluoroboranes with polyfluoroorganyliodine difluorides was developed by Frohn and Bardin [501].

$$RC{\equiv}CH \xrightarrow[54\text{-}86\%]{PhIO,\ HBF_4,\ cat.\ HgO;\ CH_2Cl_2,\ rt,\ 0.5\text{-}1\ h} \underset{\mathbf{363}}{RC{\equiv}C\overset{+}{I}Ph\ BF_4^-}$$

R = Bu, Bu^t, $C_{10}H_{21}$, (*cyclo*-C_6H_{11})CH_2, $Cl(CH_2)_9$,
$MeO_2C(CH_2)_9$, $AcO(CH_2)_9$, $Bu^tCO(CH_2)_9$

Scheme 2.103

$$\underset{\mathbf{366}}{RC{\equiv}CBF_3^- K^+} + ArIF_2 \xrightarrow[62\text{-}95\%]{\text{MeCN, rt, 15 min}} \underset{\mathbf{367}}{RC{\equiv}C\overset{+}{I}Ar\ BF_4^-}$$

$R = C_{10}H_{21}, BnOCH_2$; Ar = Ph, Tol, 4-$ClC_6H_4$

Scheme 2.104

2.1.9.3.2 Properties and Structure Most known alkynyl(aryl)iodonium salts are prepared as white microcrystalline products, which are insoluble in water and nonpolar organic solvents and moderately soluble in acetonitrile and other polar organic solvents. Their thermal stability varies over a broad range depending on the nature of the counterion and the substituent on the acetylenic β-carbon. Alkynyl(aryl)iodonium triflates and tosylates bearing an aryl or unsubstituted alkyl group at the acetylenic β-carbon, as well as the bis-iodonium acetylene **356**, the unconjugated bis(alkynyliodonium) salts **359** and the parent ethynyl(phenyl)iodonium triflate, in general have a decomposition point in the range 120–140 °C and can be stored for several months in a refrigerator. Functionalized iodonium alkynyliodonium triflates (e.g., **351–355**) generally have lower thermal stability, while diynyliodonium triflates **357** and **361** are unstable and should be handled only at low temperature.

Alkynyliodonium salts can be conveniently identified by IR and NMR spectroscopy. In the infrared spectrum, the most characteristic absorption is the triple bond band between 2120 and 2190 cm^{-1}. In the ^{13}C NMR spectrum, the most distinctive signals are the acetylenic α- and β-carbons, with the former generally between 10 and 40 ppm and the latter at 110–120 ppm.

Single-crystal X-ray structural data have been reported for the following alkynyliodonium compounds: the parent ethynyl(phenyl)iodonium triflate (**368**) [495], phenylethynyl(phenyl)iodonium tosylate (**369**) [477], cyanoethynyl(phenyl)iodonium triflate (**370**) [458], propynyl(phenyl)iodonium periodate (**371**) [502], trimethylsilylethynyl(phenyl)iodonium triflate (**372**) [503], 3,3,3-trifluoropropynyl(phenyl)iodonium triflate (**373**) [504], bis(alkynyl)iodonium triflate **374** [505] and complexes of ethynyl(phenyl)iodonium [506] and 1-decynyl(phenyl)iodonium [507] tetrafluoroborates with 18-crown-6 (Figure 2.13).

Single-crystal X-ray structural data are all consistent with the pseudo-trigonal bipyramidal, or T-shaped geometry, of alkynyliodonium species. In all known cases, the aryl group occupies an equatorial position, whereas the alkynyl moiety and the counter ion occupy apical positions. The alkynyl–iodine bond length is about 2.0 Å and the I···O distances to the nearest sulfonate anion vary from 2.56 to 2.70 Å. The C—I—O bond angles vary from 166° to 172° and the C—I—C bond angles are between 90° and 95°.

2.1.9.4 Cyanoiodonium Salts

Two structural types of cyanoiodonium salts are known: (dicyano)iodonium triflate, $(NC)_2IOTf$ [399,400] and aryl(cyano)iodonium derivatives, ArI(CN)X [146,460,508,509]. (Dicyano)iodonium triflate **277** can be prepared by the reaction of iodosyl triflate (**375**) (Section 2.1.1.2) with cyanotrimethylsilane in dichloromethane (Scheme 2.105). In the solid state, compound **277** is thermally unstable and air-sensitive; it completely decomposes at room temperature in 2–5 min forming cyanogen iodine, ICN and explodes when exposed to air. However, it can be stored at –20 °C under nitrogen for several days [400]. Despite its low stability, cyanide **277** can be used *in situ* for the very mild and efficient preparation of various bis(heteroaryl)iodonium salts by an iodonium transfer reaction with the respective stannylated heteroarenes (Section 2.1.9.1.1).

Aryl(cyano)iodonium triflates **348** and **278** are prepared by reactions of iodosylbenzene or [bis(trifluoroacetoxy)iodo]arenes with trimethylsilyl triflate and cyanotrimethylsilane (Scheme 2.106).

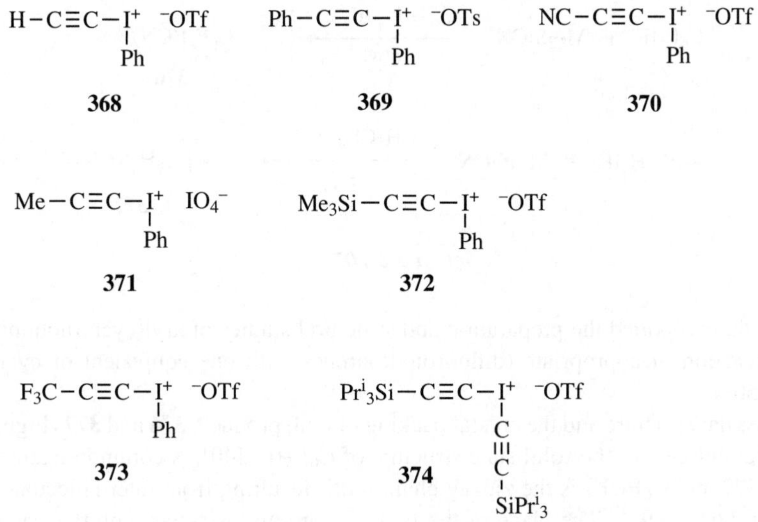

$H{-}C{\equiv}C{-}I^+(Ph)\ {}^-OTf$ **368**

$Ph{-}C{\equiv}C{-}I^+(Ph)\ {}^-OTs$ **369**

$NC{-}C{\equiv}C{-}I^+(Ph)\ {}^-OTf$ **370**

$Me{-}C{\equiv}C{-}I^+(Ph)\ IO_4^-$ **371**

$Me_3Si{-}C{\equiv}C{-}I^+(Ph)\ {}^-OTf$ **372**

$F_3C{-}C{\equiv}C{-}I^+(Ph)\ {}^-OTf$ **373**

$Pr^i_3Si{-}C{\equiv}C{-}I^+(C{\equiv}C{-}SiPr^i_3)\ {}^-OTf$ **374**

Figure 2.13 *Examples of alkynyliodonium salts analyzed by single-crystal X-ray diffraction.*

Cyanides **278** are relatively stable, white microcrystalline solids that decompose over several days at room temperature, but can be stored for extended periods in a refrigerator without change [146, 508]. Phenyl(cyano)iodonium triflate (**348**) has found some synthetic application as the iodonium transfer reagent useful for preparation of various iodonium salts (Sections 2.1.9.1.1, 2.1.9.2.1 and 2.1.9.3.1).

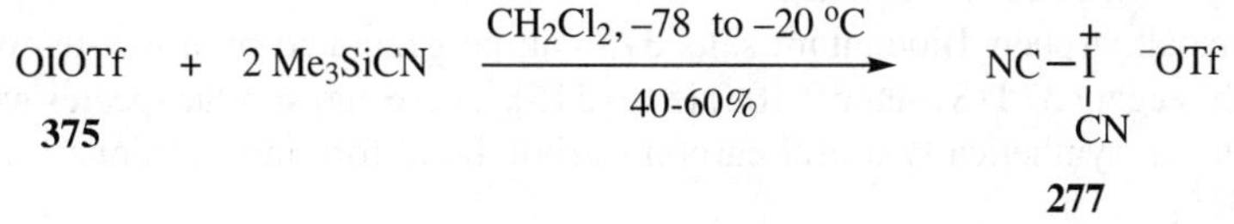

$$\underset{\mathbf{375}}{OIOTf} + 2\,Me_3SiCN \xrightarrow[40\text{-}60\%]{CH_2Cl_2,\,-78\text{ to }-20\,^{\circ}C} \underset{\mathbf{277}}{NC{-}\overset{+}{I}(CN)\ {}^-OTf}$$

Scheme 2.105

$$PhIO + Me_3SiOTf + Me_3SiCN \xrightarrow[89\%]{CH_2Cl_2,\,-20\text{ to }0\,^{\circ}C} \underset{\mathbf{348}}{Ph{-}I^+(CN)\ {}^-OTf}$$

$$ArI(CO_2CF_3)_2 + Me_3SiOTf + Me_3SiCN \xrightarrow[67\text{-}84\%]{CH_2Cl_2,\,25\,^{\circ}C} \underset{\mathbf{278}}{Ar{-}I^+(CN)\ {}^-OTf}$$

$Ar = Ph, 4\text{-}FC_6H_4, 4\text{-}(CF_3)C_6H_4, 3,5\text{-}(CF_3)_2C_6H_3$

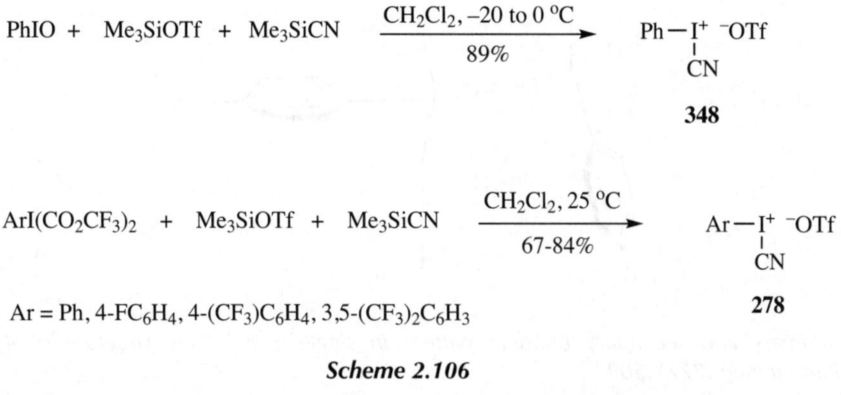

Scheme 2.106

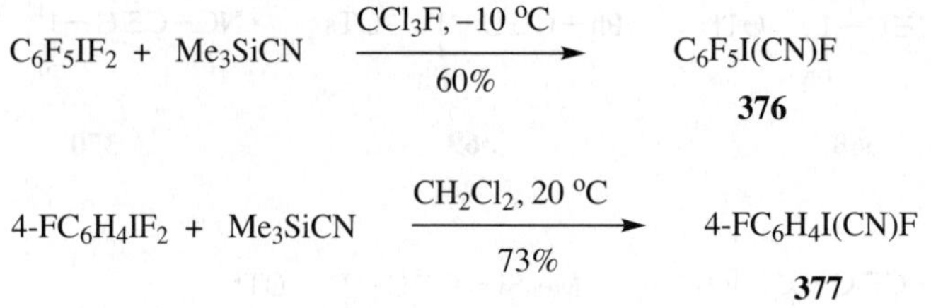

$$C_6F_5IF_2 + Me_3SiCN \xrightarrow[60\%]{CCl_3F,\ -10\ ^\circ C} C_6F_5I(CN)F \quad \mathbf{376}$$

$$4\text{-}FC_6H_4IF_2 + Me_3SiCN \xrightarrow[73\%]{CH_2Cl_2,\ 20\ ^\circ C} 4\text{-}FC_6H_4I(CN)F \quad \mathbf{377}$$

Scheme 2.107

Frohn and coauthors reported the preparation and structural studies of aryl(cyano)iodonium fluorides **376** and **377** by the reaction of appropriate (difluoroiodo)arenes with one equivalent of cyanotrimethylsilane (Scheme 2.107) [509].

The X-ray molecular structure and the crystal packing of both products **376** and **377** (Figure 2.14) [509] are very similar and reminiscent of the solid-state structure of $C_6F_5IF_2$ [40]. A common feature in the structures of products **376**, **377** and $C_6F_5IF_2$ is the zigzag chain motif resulting from intermolecular I···F interactions with a distance of 2.64–2.69 Å. The plane of the F—I—CN group intersects with the aryl plane at angle of approximately 65.5°. The I—CN distance of 2.112 Å is relatively short and the I—F distance (2.110–2.135 Å) is elongated compared to that in $C_6F_5IF_2$ (average 1.99 Å). The Ar—I—CN angle is close to 90° (86.3° in **376** and 89.0° in **377**) [509].

2.1.9.5 Alkyl- and Fluoroalkyliodonium Salts

Iodonium salts with one or two non-substituted aliphatic alkyl groups generally lack stability. However, several examples of these unstable species have been generated and investigated by NMR spectroscopy at low temperatures and some of them (e.g., $Me_2I^+ X^-$, where $X = AsF_6, SbF_6, Sb_2F_{11}$) were even isolated in the form of labile crystalline salts [510–512].

The unstable β-oxoalkyl(phenyl)iodonium salts **379** can be generated by a low-temperature reaction of silyl enol ethers with reagent **378** (Scheme 2.108) [513–515]. These and similar species were proposed as the reactive intermediates in synthetically useful carbon–carbon bond forming reactions reviewed by Moriarty and Vaid in 1990 [515].

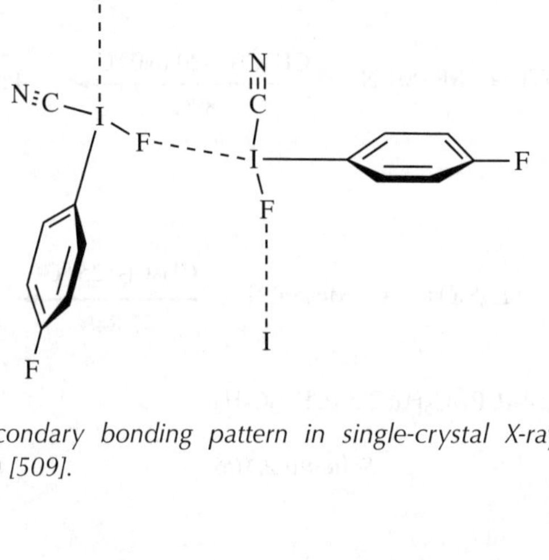

Figure 2.14 *Primary and secondary bonding pattern in single-crystal X-ray structure of 4-fluorophenyl(cyano)iodonium fluoride (**377**) [509].*

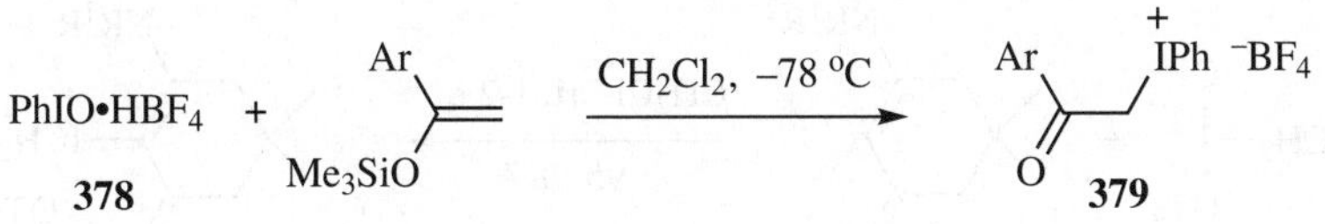

$PhIO{\bullet}HBF_4$ + $Ar(Me_3SiO)C{=}CH_2$ —(CH_2Cl_2, –78 °C)→ $ArC(O)CH_2I^{+}Ph\ ^{-}BF_4$

378 **379**

Ar = Ph, $4\text{-}MeC_6H_4$, $4\text{-}ClC_6H_4$, $4\text{-}NO_2C_6H_4$, $4\text{-}MeOC_6H_4$

Scheme 2.108

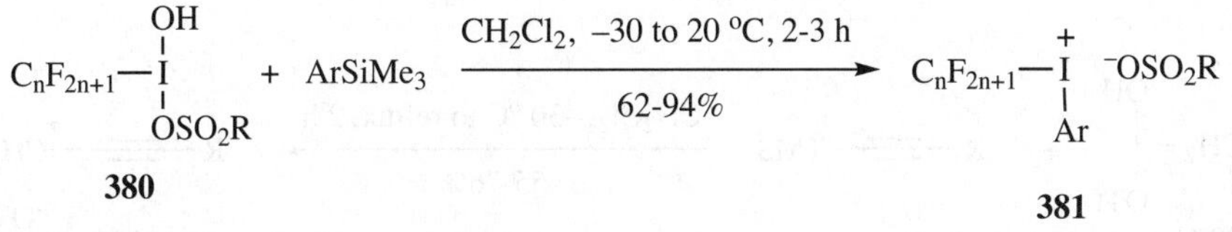

$C_nF_{2n+1}I(OH)OSO_2R$ + $ArSiMe_3$ —(CH_2Cl_2, –30 to 20 °C, 2-3 h; 62-94%)→ $C_nF_{2n+1}I^{+}Ar\ ^{-}OSO_2R$

380 **381**

n = 4 or 6; R = p-Tol, CH_3, CF_3
Ar = Ph, $4\text{-}MeC_6H_4$, $4\text{-}MeOC_6H_4$, $2\text{-}MeC_6H_4$, $4\text{-}Me_3SiC_6H_4$

Scheme 2.109

The presence of electron-withdrawing groups in the alkyl group of iodonium salts usually has a pronounced stabilizing effect. The most stable and important derivatives of this type are polyfluoroalkyl(aryl)iodonium salts [174, 516]. These salts as chlorides were first prepared in 1971 by Yagupolskii and coworkers [517] and later widely applied as electrophilic fluoroalkylating reagents by Umemoto and coworkers [518–526]. Perfluoroalkyl(phenyl)iodonium triflates, $PhIC_nF_{2n+1}OTf$, were originally synthesized by the reaction of [bis(trifluoroacetoxy)iodo]perfluoroalkanes with benzene in the presence of triflic acid [527, 528]. A more recent, general method for preparing various perfluoroalkyl(aryl)iodonium sulfonates **381** involves the reaction of [hydroxy(sulfonyloxy)iodo]perfluoroalkanes **380** with arylsilanes under Lewis acid catalysis (Scheme 2.109) [207, 211, 529].

In a similar manner, 1*H*,1*H*-perfluoroalkyl(aryl)iodonium triflates **383** are prepared by the reaction of triflates **382** with trimethylsilylarenes under mild conditions (Scheme 2.110) [176, 212].

Several fluoroalkyliodonium salts of other types have been reported [77, 530, 531]. The reaction of tosylate **384** with cyclic enaminones **385** affords stable iodonium salts **386** (Scheme 2.111) [530]. Mild thermolysis of salts **386** in boiling acetonitrile cleanly affords 2,2,2-trifluoroethyl tosylate and the respective iodoenaminone [530].

Fluoroalkyl(alkynyl)iodonium triflates **388** can be prepared by the reaction of triflates **387** and (trimethylsilyl)acetylenes (Scheme 2.112) [176].

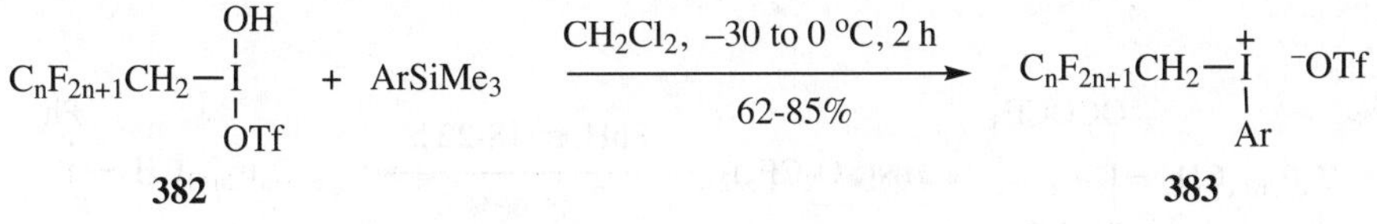

$C_nF_{2n+1}CH_2I(OH)OTf$ + $ArSiMe_3$ —(CH_2Cl_2, –30 to 0 °C, 2 h; 62-85%)→ $C_nF_{2n+1}CH_2I^{+}Ar\ ^{-}OTf$

382 **383**

n = 1 or 2
Ar = Ph, $4\text{-}MeC_6H_4$, $2\text{-}MeC_6H_4$, $4\text{-}Me_3SiC_6H_4$, $4\text{-}MeOC_6H_4$

Scheme 2.110

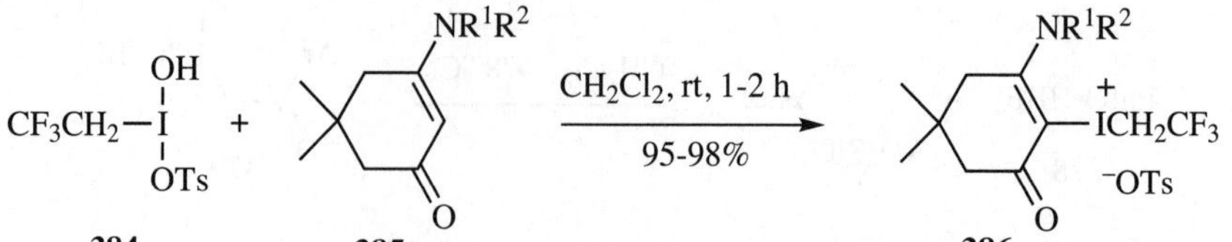

CF_3CH_2–I(OH)(OTs) (384) + 385 (5,5-dimethyl-3-NR^1R^2-cyclohex-2-enone) → [CH_2Cl_2, rt, 1-2 h; 95-98%] → 386 (2-($\overset{+}{I}CH_2CF_3$) $^-$OTs)

R^1 = H or Me; R^2 = H or Bu

Scheme 2.111

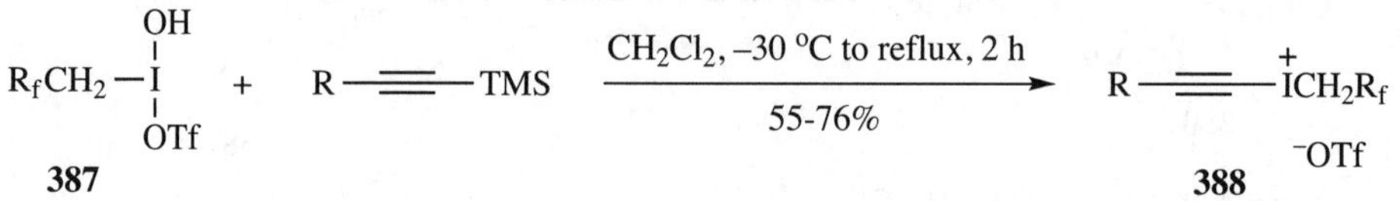

R_fCH_2–I(OH)(OTf) (387) + R–≡–TMS → [CH_2Cl_2, –30 °C to reflux, 2 h; 55-76%] → R–≡–$\overset{+}{I}CH_2R_f$ $^-$OTf (388)

$R_f = CF_3, C_2F_5$; R = Ph, TMS, Bu^t

Scheme 2.112

DesMarteau and coworkers reported the preparation, X-ray crystal structure and chemistry of trifluoroethyliodonium salts **391** by the reaction of fluoroalkyliodo-bis(trifluoroacetates) **389** with benzene and triflimide acid (**390**) (Scheme 2.113) [68,531,532]. The structure of trifluoroethyl(phenyl)iodonium salt **391** ($n = 1$) was established by a single-crystal X-ray analysis [68]. In contrast to fluoroalkyliodonium triflates **383**, compounds **391** are stable to water and can be used as reagents for fluoroalkylation reactions in aqueous media.

(Arylsulfonylmethyl)iodonium salts **394** and **395**, which are stabilized due to the presence of the electron-withdrawing sulfonyl group, can be prepared in two steps starting from the readily available iodomethyl sulfones **392** (Scheme 2.114) [138]. In the first step, starting iodides **392** are oxidized with peroxytrifluoroacetic acid to trifluoroacetates **393** in almost quantitative yield. Subsequent treatment of **393** with benzene and trimethylsilyl (TMS) triflate in dichloromethane affords products **394** and **395** in good yields. Both iodonium salts **394** and **395** are not moisture sensitive, can be purified by crystallization from acetonitrile and can be stored for several months in a refrigerator.

The structure of iodonium triflate **395** was unambiguously established by a single-crystal X-ray analysis [138]. The structural data revealed the expected geometry for iodonium salts with a C–I–C bond angle of 91.53°. The I–C bond distances of 2.131 and 2.209 Å are longer than the typical bond length in diaryliodonium salts (2.0–2.1 Å). The I···O distance between the iodine atom and the nearest oxygen of the triflate anion is 2.797 Å.

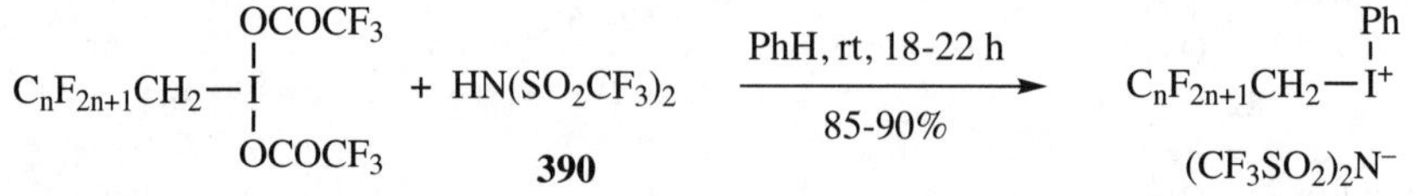

$C_nF_{2n+1}CH_2$–I($OCOCF_3$)$_2$ (389, n = 1, 7, 9) + $HN(SO_2CF_3)_2$ (390) → [PhH, rt, 18-22 h; 85-90%] → $C_nF_{2n+1}CH_2$–I^+(Ph) $(CF_3SO_2)_2N^-$ (391)

Scheme 2.113

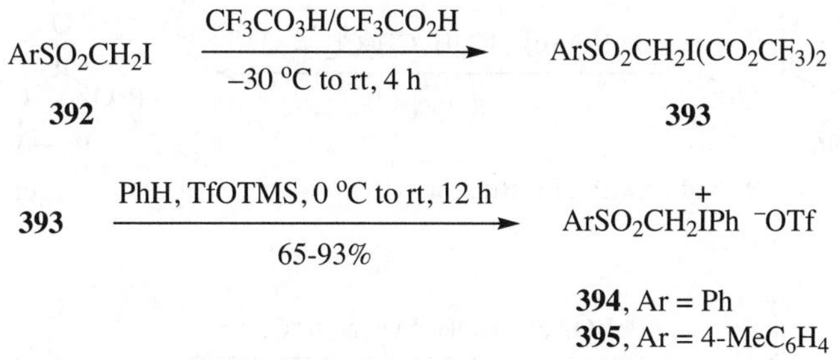

$ArSO_2CH_2I$ **392** —(CF_3CO_3H/CF_3CO_2H, –30 °C to rt, 4 h)→ $ArSO_2CH_2I(CO_2CF_3)_2$ **393**

393 —(PhH, TfOTMS, 0 °C to rt, 12 h; 65-93%)→ $ArSO_2CH_2\overset{+}{I}Ph\ ^{-}OTf$

394, Ar = Ph
395, Ar = 4-MeC_6H_4

Scheme 2.114

2.1.10 Iodonium Ylides

Aryliodonium ylides, $ArICX_2$, where X is an electron-withdrawing substituent (e.g., carbonyl or sulfonyl group), represent the most important and relatively stable class of iodonium ylides. Alkyliodonium ylides such as dimethyliodonium ylide, CH_3ICH_2, are unknown, although they have been a subject of computational studies [533,534]. The first example of a stable iodonium ylide (see structure **405**, R = Me, below in Figure 2.15), prepared by the reaction of dimedone (5,5-dimethyl-1,3-cyclohexanedione) and (difluoroiodo)benzene, was reported by Neiland and coworkers in 1957 [535]. Since then, numerous stable aryliodonium ylides have been prepared and structurally investigated. Single X-ray crystallographic studies demonstrate that the geometry of aryliodonium ylides is similar to that of iodonium salts with a C—I—C angle close to 90°, which is indicative of a zwitterionic nature of the ylidic C—I bond. The chemistry of aryliodonium ylides has been summarized in several reviews mainly devoted to their use as precursors for the generation of singlet carbene or carbenoid species [536–539].

2.1.10.1 Preparation

Most iodonium ylides have low thermal stability and can be handled only at low temperature or generated and used *in situ*. The relatively stable and practically important iodonium ylides, the dicarbonyl derivatives $PhIC(COR)_2$ [535,540–543] and bis(organosulfonyl)(phenyliodonium)methanides, $PhIC(SO_2R)_2$ [544–547], are prepared by the reaction of (diacetoxyiodo)benzene with the appropriate dicarbonyl compound or disulfone under basic conditions. A general procedure for the synthesis of phenyliodonium ylides **397** from malonate esters **396** is based on the treatment of esters **396** with (diacetoxyiodo)benzene in dichloromethane in the presence of potassium hydroxide (Scheme 2.115) [542]. An optimized method for preparing bis(methoxycarbonyl)(phenyliodonium)methanide (**399**) by using reaction of dimethyl malonate

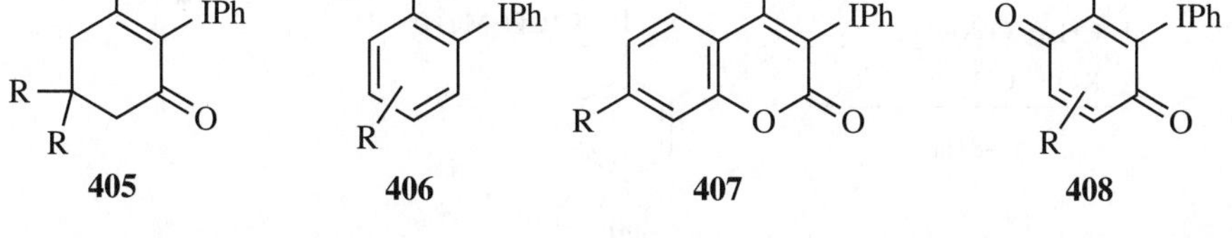

405 **406** **407** **408**

Figure 2.15 *Iodonium ylides derived from cyclic 1,3-dicarbonyl compounds (**405**), phenols (**406**), coumarin derivatives (**407**) and hydroxyquinones (**408**).*

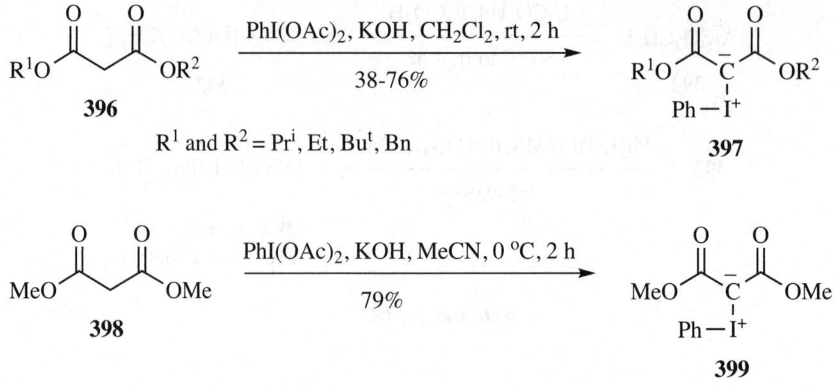

R^1O–C(O)–CH_2–C(O)–OR^2 (**396**) $\xrightarrow[38\text{-}76\%]{PhI(OAc)_2,\ KOH,\ CH_2Cl_2,\ rt,\ 2\ h}$ R^1O–C(O)–$\bar{C}(I^+{-}Ph)$–C(O)–OR^2 (**397**)

R^1 and R^2 = Pr^i, Et, Bu^t, Bn

MeO–C(O)–CH_2–C(O)–OMe (**398**) $\xrightarrow[79\%]{PhI(OAc)_2,\ KOH,\ MeCN,\ 0\ ^\circ C,\ 2\ h}$ MeO–C(O)–$\bar{C}(I^+{-}Ph)$–C(O)–OMe (**399**)

Scheme 2.115

ester (**398**) in acetonitrile solution was published in *Organic Syntheses* in 2010 [548]. Ylides **397** and **399** decompose slowly at room temperature and can be kept for several weeks at –20 °C.

Phenyliodonium ylide **399** has found some synthetic use as an efficient carbene precursor and has been especially useful as a reagent for cyclopropanation of alkenes. Practical application of ylide **399** is, however, limited by its poor solubility [insoluble in most organic solvents except DMSO (dimethyl sulfoxide)] and low stability. 2-Alkoxyphenyliodonium ylides **400** derived from malonate methyl ester and bearing an *ortho* alkoxy substituent on the phenyl ring, can be synthesized from commercially available 2-iodophenol according to the procedure shown in Scheme 2.116. Ylides **400** are relatively stable, have good solubility in dichloromethane, chloroform, or acetone (e.g., the solubility of ylide **400**, R = Pr, in dichloromethane is 0.56 g ml^{-1}) and have higher reactivity than common phenyliodonium ylides in the Rh-catalyzed cyclopropanation, C–H insertion and transylidation reactions under homogeneous conditions [549].

Bis(perfluoroalkanesulfonyl)(phenyliodonium)methanides **401**, synthesized from (diacetoxyiodo)benzene and bis[(perfluoroalkyl)sulfonyl]methane (Scheme 2.117), have unusually high thermal stability; they can be stored without decomposition at room temperature for several months [546]. The non-symmetric, cyano[(perfluoroalkyl)sulfonyl]-substituted ylides **402** (Scheme 2.117) [545] or bis(fluorosulfonyl)-(phenyliodonium)methanide, $PhIC(SO_2F)_2$ [550], were prepared by a similar reaction using appropriate bis(sulfonyl)methanes as starting materials.

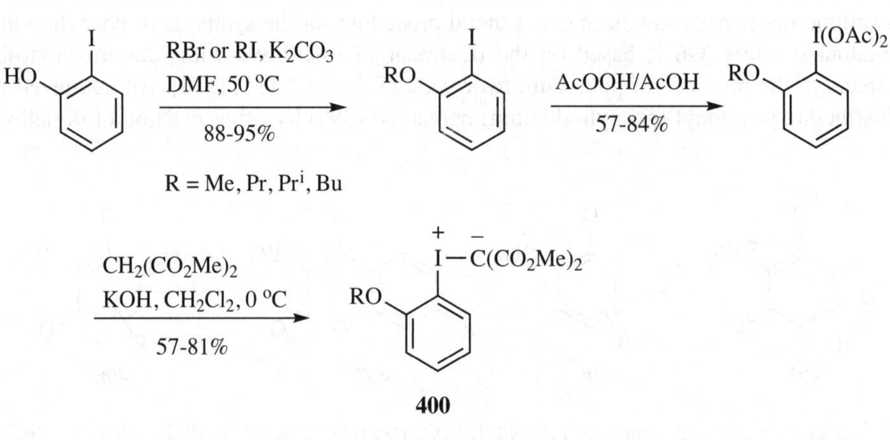

2-HO-C_6H_4-I $\xrightarrow[88\text{-}95\%]{RBr\ or\ RI,\ K_2CO_3,\ DMF,\ 50\ ^\circ C}$ 2-RO-C_6H_4-I $\xrightarrow[57\text{-}84\%]{AcOOH/AcOH}$ 2-RO-C_6H_4-$I(OAc)_2$

R = Me, Pr, Pr^i, Bu

$\xrightarrow[57\text{-}81\%]{CH_2(CO_2Me)_2,\ KOH,\ CH_2Cl_2,\ 0\ ^\circ C}$ 2-RO-C_6H_4-$\overset{+}{I}$–$\overset{-}{C}(CO_2Me)_2$ (**400**)

Scheme 2.116

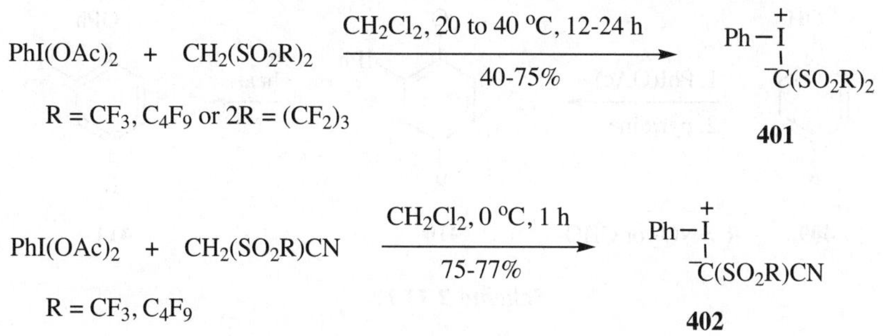

$PhI(OAc)_2$ + $CH_2(SO_2R)_2$ —(CH_2Cl_2, 20 to 40 °C, 12-24 h; 40-75%)→ $Ph-\overset{+}{I}-\overset{-}{C}(SO_2R)_2$ **401**

R = CF_3, C_4F_9 or 2R = $(CF_2)_3$

$PhI(OAc)_2$ + $CH_2(SO_2R)CN$ —(CH_2Cl_2, 0 °C, 1 h; 75-77%)→ $Ph-\overset{+}{I}-\overset{-}{C}(SO_2R)CN$ **402**

R = CF_3, C_4F_9

Scheme 2.117

Cyclic iodonium ylides, in which the iodonium atom is incorporated in a five-membered ring, have much higher thermal stability. The unusually stable cyclic iodonium ylides **404** can be synthesized via intramolecular transylidation of a preformed acyclic ylide **403** (Scheme 2.118) [551].

Iodonium ylides derived from cyclic 1,3-dicarbonyl compounds, phenols, coumarin derivatives and hydroxyquinones represent a particularly stable and synthetically important class of zwitterionic iodonium compounds [539, 552]. Several examples of these compounds are shown in Figure 2.15 in the major resonance form as enolate or phenolate zwitterionic structures **405–408**.

The preparation of iodonium phenolates **410** was first reported in 1977 via a reaction of phenols **409** with (diacetoxyiodo)benzene followed by treatment with pyridine (Scheme 2.119) [553]. The system of an iodonium phenolate is stabilized by the presence of at least one electron-withdrawing substituent on the aromatic ring. Monosubstituted iodonium phenolates **410** are relatively unstable and easily rearrange to iodo ethers **411** under heating. Such a 1,4 aryl migration is a very common phenomenon for iodonium ylides of types **405–408**; according to mechanistic and computational studies it is an intramolecular rearrangement via a concerted mechanism [554, 555].

Ylides of hydroxycoumarins (**413**) were prepared by the reaction of (diacetoxyiodo)arenes with 4-hydroxycoumarins **412** (Scheme 2.120) [556, 557]. X-Ray crystal structure determination [558] and the Mössbauer spectra [559] indicate that ylides **413** exist in predominantly zwitterionic form. These ylides are

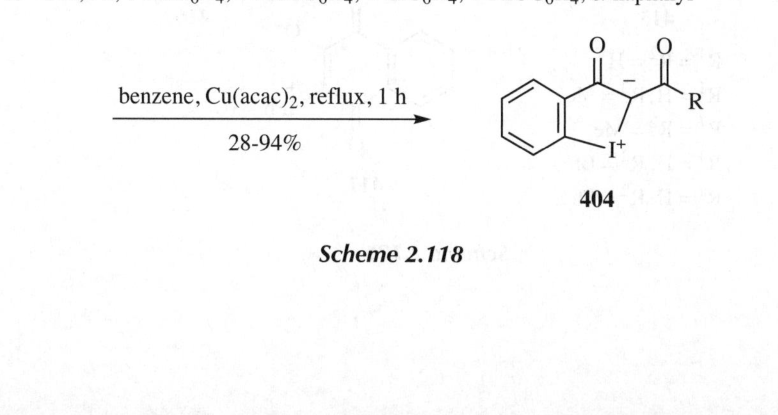

—($PhI(OAc)_2$, KOH, MeOH, 0 °C)→ **403**

R = Me, Ph, 4-MeC_6H_4, 4-$MeOC_6H_4$, 4-ClC_6H_4, 4-$PhOC_6H_4$, α-naphthyl

—(benzene, $Cu(acac)_2$, reflux, 1 h; 28-94%)→ **404**

Scheme 2.118

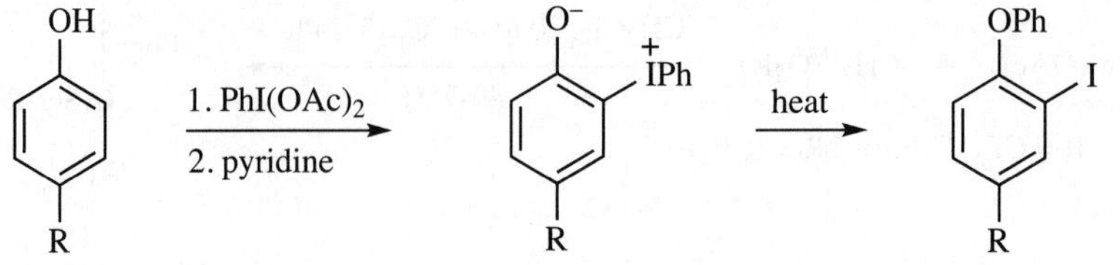

409 R = NO_2 or CHO → 1. $PhI(OAc)_2$ 2. pyridine → 410 → heat → 411

Scheme 2.119

easily transformed into the corresponding iodo ethers **414** through the usual thermal aryl migration from iodine to oxygen [557].

Ylides of hydroxyquinones **416** were synthesized by treatment of 2-hydroxy-1,4-benzoquinones **415** with an equimolar amount of (diacetoxyiodo)benzene (Scheme 2.121) [560]. The most important member of this class, 2-oxido-3-(phenyliodonio)-1,4-naphthoquinone (**417**), was prepared similarly by treatment of 2-hydroxy-1,4-naphthoquinone (lawsone) with an equimolar amount of (diacetoxyiodo)benzene [561].

The aza analogues of **417**, iodine–nitrogen zwitterions **419**, have been prepared by the reaction of 2-amino-1,4-naphthoquinone (**418**) with [hydroxy(tosyloxy)iodo]arenes (Scheme 2.122) [205, 562, 563]. Ylides **419** show interesting reactivity: upon heating, aryl migration from iodine to nitrogen occurs, giving product **420**,

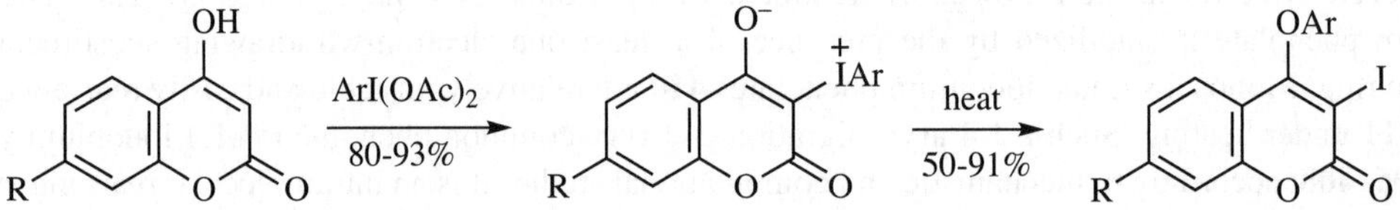

412 → $ArI(OAc)_2$, 80-93% → 413 → heat, 50-91% → 414

R = H, OMe

Ar = Ph, 4-$MeOC_6H_4$, 3-$MeOC_6H_4$

Scheme 2.120

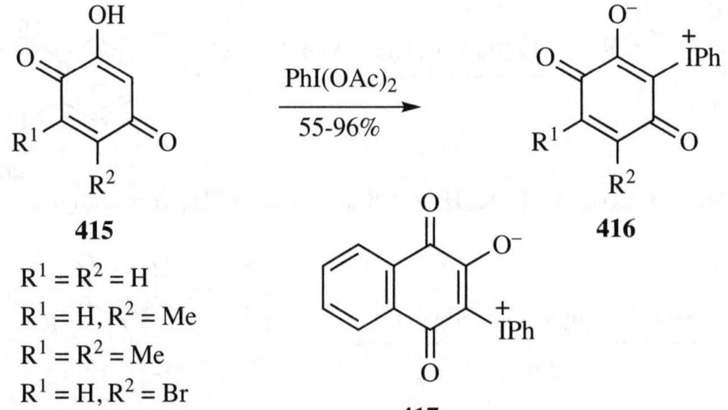

415 → $PhI(OAc)_2$, 55-96% → 416

$R^1 = R^2 = H$

$R^1 = H, R^2 = Me$

$R^1 = R^2 = Me$

$R^1 = H, R^2 = Br$

$R^1 = H, R^2 = Ph$

417

Scheme 2.121

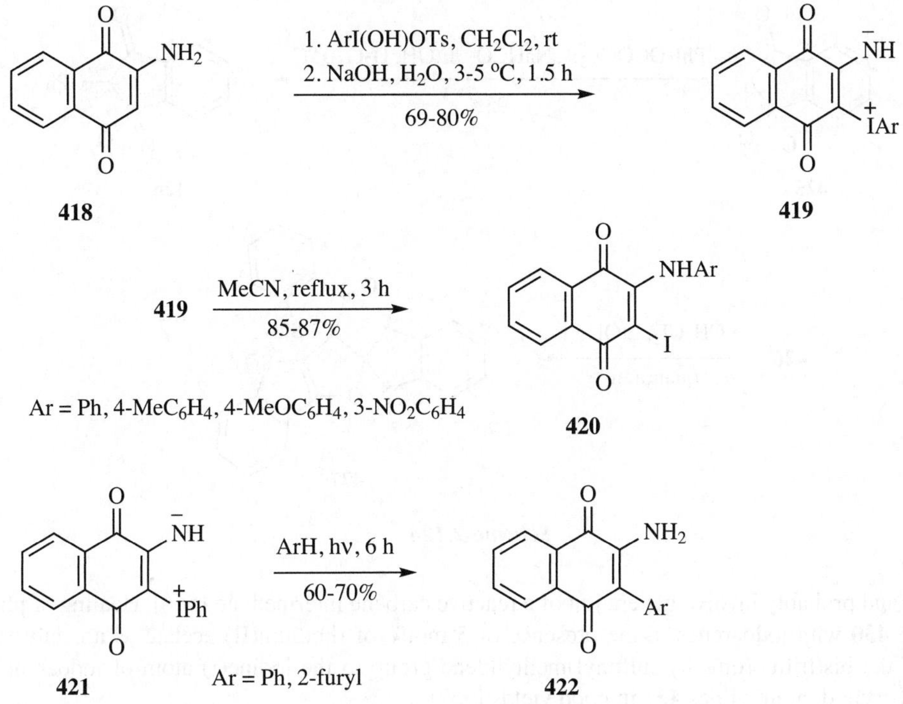

1. ArI(OH)OTs, CH_2Cl_2, rt
2. NaOH, H_2O, 3-5 °C, 1.5 h
69-80%

418 **419**

419 MeCN, reflux, 3 h, 85-87% → **420**

Ar = Ph, 4-MeC_6H_4, 4-$MeOC_6H_4$, 3-$NO_2C_6H_4$

421 ArH, hν, 6 h, 60-70% → **422**

Ar = Ph, 2-furyl

Scheme 2.122

while the photochemical reaction of ylide **421** with aromatic compounds and furan leads to substitution products **422** (Scheme 2.122).

Stable zwitterionic aryliodonium compounds **424** have been prepared by the reaction of 2,4-dihydroxyacetophenones **423** with (diacetoxyiodo)benzene under basic conditions (Scheme 2.123) [564]. Heating of these compounds leads to phenyl migration from iodine to oxygen with formation of the respective *o*-iodophenoxy ethers.

A relatively stable iodonium ylide (**426**) was prepared by the reaction of β-ketosulfone **425** with [bis(trifluoroacetoxy)iodo]benzene (Scheme 2.124) [565]. Ylide **426** decomposes in a solution of dichloromethane/ethanol to form quantitatively, the trimer **427**.

Ochiai and coworkers developed a new synthetic approach to various iodonium ylides **429** by the intermolecular transylidation reactions between halonium ylides under thermal or catalytic conditions (Scheme 2.125) [566, 567]. The transylidations of bromonium **428** to iodonium **429** ylides proceed under thermal

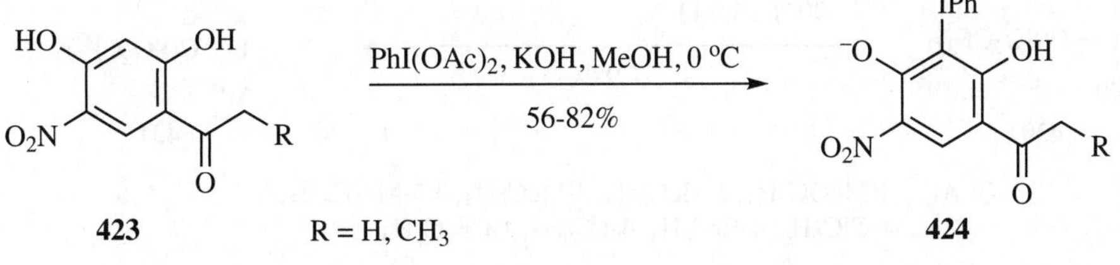

423 $PhI(OAc)_2$, KOH, MeOH, 0 °C, 56-82% → **424**

R = H, CH_3

Scheme 2.123

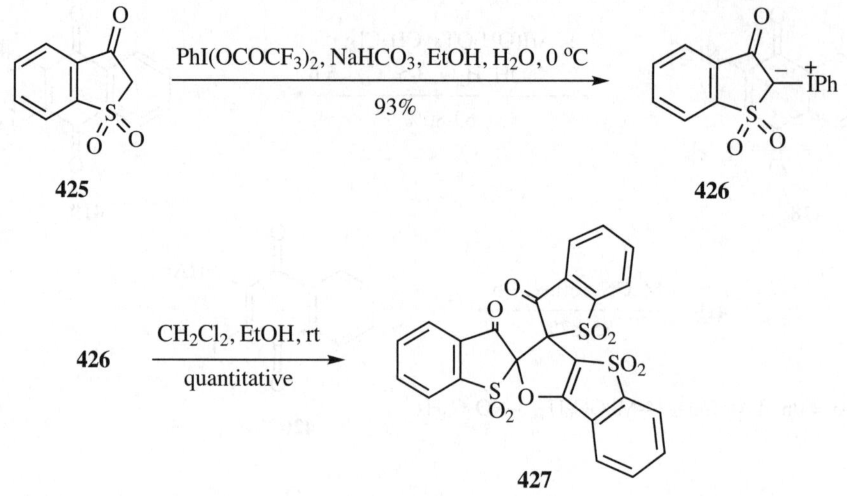

$PhI(OCOCF_3)_2$, $NaHCO_3$, EtOH, H_2O, 0 °C
93%

425 → 426

426 → 427
CH_2Cl_2, EtOH, rt
quantitative

Scheme 2.124

conditions and probably involve generation of a reactive carbene intermediate [566]. Heating of phenyliodonium ylide **430** with iodoarenes in the presence of 5 mol% of rhodium(II) acetate as a catalyst results in transfer of the bis(trifluoromethylsulfonyl)methylidene group to the iodine(I) atom of iodoarene to afford substituted aryliodonium ylides **431** in good yields [567].

Mixed phosphonium–iodonium ylides **432** represent a useful class of reagents that combine in one molecule the synthetic advantages of a phosphonium ylide and an iodonium salt. The preparation of the tetrafluoroborate derivatives **432** by the reaction of phosphonium ylides with (diacetoxyiodo)benzene in the presence of HBF_4

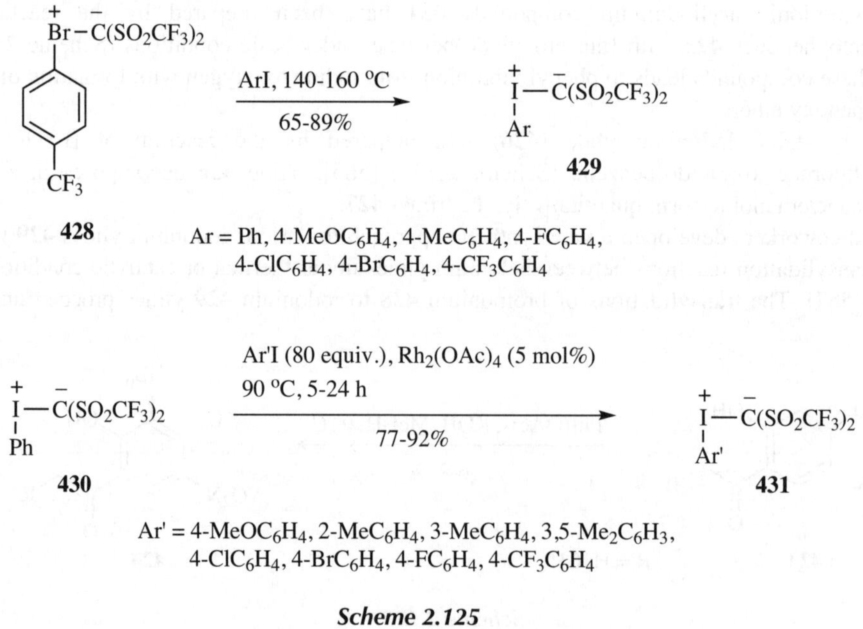

428 → **429**
ArI, 140-160 °C
65-89%

Ar = Ph, 4-$MeOC_6H_4$, 4-MeC_6H_4, 4-FC_6H_4, 4-ClC_6H_4, 4-BrC_6H_4, 4-$CF_3C_6H_4$

430 → **431**
Ar'I (80 equiv.), $Rh_2(OAc)_4$ (5 mol%)
90 °C, 5-24 h
77-92%

Ar' = 4-$MeOC_6H_4$, 2-MeC_6H_4, 3-MeC_6H_4, 3,5-$Me_2C_6H_3$, 4-ClC_6H_4, 4-BrC_6H_4, 4-FC_6H_4, 4-$CF_3C_6H_4$

Scheme 2.125

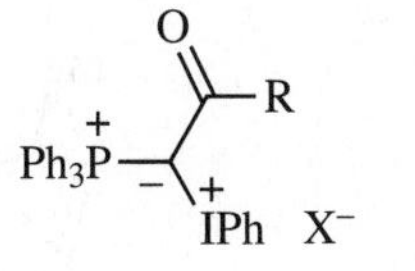

$$Ph_3\overset{+}{P}-\overset{-}{C}(C(O)R)-\overset{+}{I}Ph\ X^-$$

432

R = OEt, Ph
X = BF_4^- or Br^-

$Ph_3P{=}CH(Y)$ **433** → ($PhI(OTf)_2{\bullet}2Py$, CH_2Cl_2, 56-89%) → $Ph_3\overset{+}{P}-\overset{-}{C}(Y)-\overset{+}{I}Ph$ ^-OTf **434**

$Ph_3P{=}CH(Y)$ **433** → ($PhI(OH)OTs$, CH_2Cl_2, 47-51%) → $Ph_3\overset{+}{P}-\overset{-}{C}(Y)-\overset{+}{I}Ph$ ^-OTs **435**

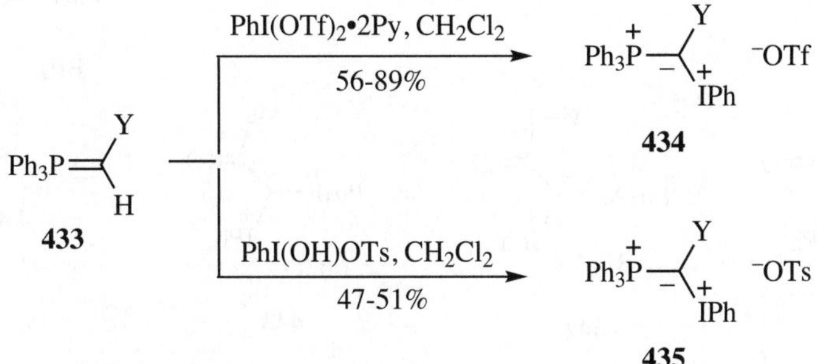

Y = C(O)Me, CO_2Me, CN, CHO

Scheme 2.126

was first reported by Neilands and Vanags in 1964 [568]. Later, in 1984, Moriarty and coworkers reported the preparation of several new tetrafluoroborate derivatives **432** and the X-ray crystal structure for one of the products [569]. The triflate (**434**) and tosylate (**435**) derivatives of phosphonium–iodonium ylides have been prepared in good yields by the reaction of phosphonium ylides **433** with the pyridinium complex of iodobenzene ditriflate, $PhI(OTf)_2{\cdot}2Py$ [570], or with [hydroxy(tosyloxy)iodo]benzene, respectively (Scheme 2.126) [571–574]. The analogous mixed arsonium–iodonium ylides (see structure **442** below in Figure 2.16) were synthesized using a similar procedure [574]. Preparation of hetaryl-substituted phosphonium–iodonium ylides was also reported [575].

The unstable monocarbonyl iodonium ylides **437** can be generated from (*Z*)-(2-acetoxyvinyl)iodonium salts **436** by ester exchange reaction with lithium ethoxide in tetrahydrofuran (THF) at –78 °C (Scheme 2.127). ^{1}H NMR measurements indicate that ylides **437** are stable up to –30 °C and they can be conveniently used in subsequent transformations without isolation [576–580].

The unstable ylides $PhIC(H)NO_2$ [581, 582] and $PhIC(CO_2Me)NO_2$ [583, 584] can be generated *in situ* from nitromethane and methyl nitroacetate, respectively and used in rhodium(II) carbenoid reactions without isolation.

2.1.10.2 Structural Studies

Single-crystal X-ray structures have been reported for the following iodonium ylides: 3-phenyliodonio-1,2,4-trioxo-1,2,3,4-tetrahydronaphthalenide (**438**) [558], 3-phenyliodonio-2,4-dioxo-1,2,3,4-tetrahydro-1-oxanaphthalenide (**439**) [558], mixed phosphonium–iodonium tetrafluoroborates **440** [569] and **441** [572], mixed arsonium–iodonium tetrafluoroborate **442** [585], mixed phosphonium iodonium triflate **443** [571], cyclic iodonium ylide **444** [551], 2-methoxyphenyliodonium bis(methoxycarbonyl)methanide (**445**) [549] and phenyliodonium bis(trifluoromethanesulfonyl)methide **446** [544] (Figure 2.16). In particular, the X-ray structural analysis for phenyliodonium ylide **446** shows a geometry typical for an iodonium ylide with the I–C ylide bond length of about 1.9 Å and an C–I–C bond angle of 98° [544]. X-Ray structural

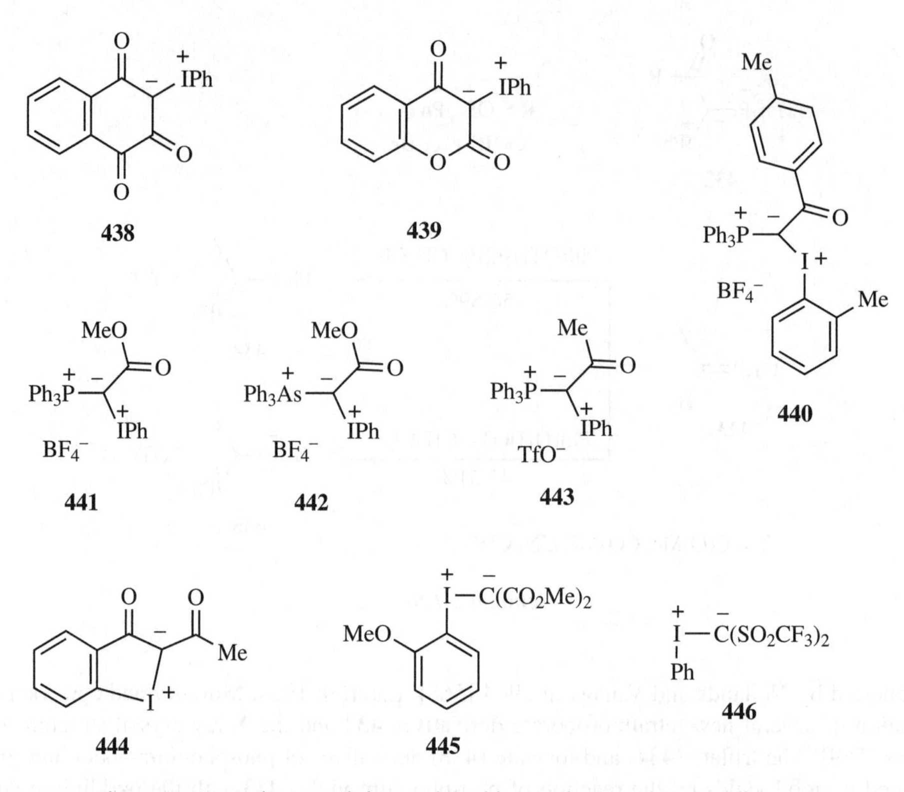

Figure 2.16 Iodonium ylides analyzed by single-crystal X-ray diffraction.

analysis for cyclic ylide **444** reveals a distorted five-membered ring with an ylidic bond length about 2.1 Å and a C—I—C bond angle of 82°, which is smaller than the usual 90° [551]. 2-Methoxyphenyliodonium bis(methoxycarbonyl)methanide **445** in the solid state has a polymeric, asymmetrically bridged structure with a hexacoordinated geometry around the iodine centers formed by two short C—I bonds [2.117 Å for I—C(Ph) and 2.039 Å for I—C(malonate)] and two relatively long iodine–oxygen intramolecular interactions between iodine and the oxygen atom of the *ortho* substituent (2.928 Å) and the carbonyl oxygen atom of the methoxycarbonyl group (3.087 Å). In addition, a relatively weak intermolecular I···O secondary interaction of 2.933 Å between the iodine center and the carbonyl oxygen atom of the neighboring molecule is also present in the solid state structure of **445** [549].

436: R, AcO, I(Ph)X; EtOLi, THF, −78 °C → [intermediate: R, O, I(Ph)X, EtO—C(OLi)(Me)] → EtOAc, LiX; 437: R–C(O)–CH⁻–I⁺Ph

X = BF_4 or Br
R = Me, C_8H_{17}, Bu^t

Scheme 2.127

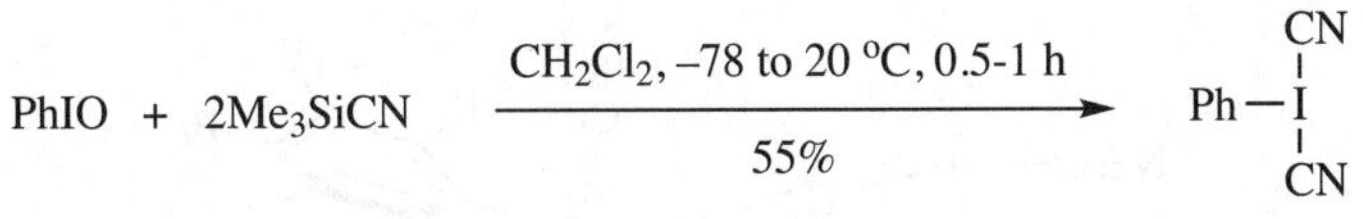

$$PhIO + 2Me_3SiCN \xrightarrow[55\%]{CH_2Cl_2,\,-78 \text{ to } 20\,^{\circ}C,\,0.5\text{-}1\,h} Ph-I(CN)_2$$

447

Scheme 2.128

2.1.11 Iodine(III) Species with Three Carbon Ligands

In general, iodine(III) species with three carbon ligands are unstable at room temperature. Only a few examples of isolable compounds of this type were reported in the earlier literature, namely, triphenyl iodine, Ph_3I [586,587] and 5-aryl-5*H*-dibenziodoles [588–591]. All triaryliodanes are highly unstable and air sensitive compounds.

In 1960 Beringer and coworkers reported that diphenyliodonium chloride interacts with *n*-butyllithium to form an unstable trivalent organoiodine, presumably Ph_2IBu, which decomposes in solution at approximately –40 °C [592]. More recently, in 1992, Barton and coworkers investigated a carbon–carbon bond forming reaction of dichloroiodobenzene or (diacetoxyiodo)benzene with alkyllithiums [593]. One of the possible mechanistic descriptions of this reaction involves ligand coupling in the triorgano-iodane intermediate $PhIR_2$; however, these species cannot be detected even at –80 °C [593].

More recent examples of relatively stable, isolable compounds of this type are represented by (dicyanoiodo)arenes, $ArI(CN)_2$ [509, 580]. (Dicyanoiodo)benzene (**447**) was prepared in good yield by the reaction of iodosylbenzene with trimethylsilyl cyanide in methylene chloride (Scheme 2.128) [594]. Cyanide **447** was isolated as an air-sensitive, crystalline solid that is stable at room temperature under nitrogen for several weeks.

Frohn and coworkers have synthesized and investigated structure and properties of several fluoroaryliodine(III) cyano compounds [509]. In particular, $C_6F_5I(CN)_2$ and *ortho*, *meta* and *para* isomers of $FC_6F_4I(CN)_2$ were isolated from reactions of the corresponding aryliodine difluorides $ArIF_2$ and a stoichiometric excess of Me_3SiCN in Freon® at 0 °C or in dichloromethane at 20 °C. An alternative preparation of these fluorophenyliodine(III) cyanides involves the reaction of $ArI(OAc)_2$, $ArI(OCOCF_3)_2$, or ArIO with trimethylsilyl cyanide in dichloromethane at room temperature. All fluoroaryliodine(III) cyano compounds $ArI(CN)_2$ are relatively unstable and exothermically decompose in the solid state at 72–100 °C [509].

Single-crystal X-ray structural data are available for $C_6F_5I(CN)_2$, 3-$FC_6F_4I(CN)_2$ and 4-$FC_6F_4I(CN)_2$. All three products have the expected T-shape molecular structures with similar parameters. In $C_6F_5I(CN)_2$ the C(aryl)—I distance (2.074 Å) is significantly shorter than both C(CN)—I distances (average 2.25 Å) [509]. Repulsion of the two ICN units by the two lone pairs of electrons reduces the aryliodine–CN angles to 81.72° and 82.69°. The plane defined by the $I(CN)_2$ group and the aryl plane intersects by 70.7°. The nitrogen terminus of one CN group interacts with the hypervalent iodine center of a neighboring molecule with the intermolecular I···N distance of 2.933 Å, which is 17% shorter than the sum of van der Waals radii for iodine and nitrogen (3.53 Å), resulting in a zigzag chain arrangement (Figure 2.17) [509].

Fluoroaryliodine(III) cyano compounds $ArI(CN)_2$ can form relatively stable adducts with nitrogen bases (phenanthroline, 2,2′-bipyridine and quinoline); the adduct of 4-$FC_6F_4I(CN)_2$ with 1,10-phenanthroline has been characterized by single-crystal X-ray diffraction [62].

2.1.12 Iodine(III) Species with I–N Bonds

Iodanes with I–N bonds are generally less common than those with I–O bonds. Many of these compounds lack stability and are sensitive to moisture. In addition to the previously discussed N-substituted benziodoxoles

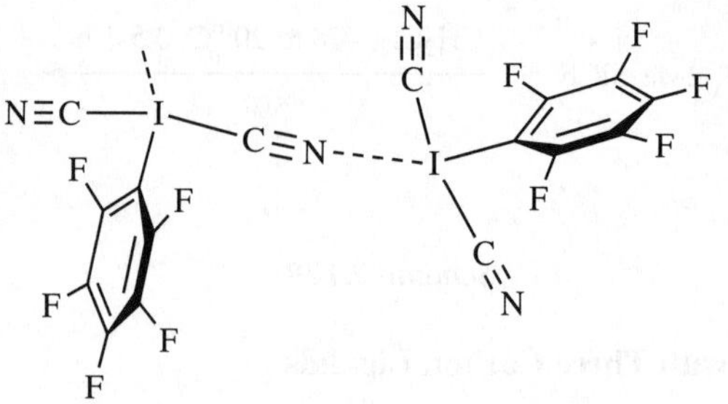

Figure 2.17 Primary and secondary bonding pattern in a single-crystal X-ray structure of $C_6F_5I(CN)_2$.

(Sections 2.1.8.1.5 and 2.1.8.1.6), benziodazoles (Section 2.1.8.2) and benziodathiazoles (Section 2.1.8.5), several other structural types of iodine(III) derivatives with one carbon ligand and one or two I–N bonds are known: azidoiodanes **448** and **449**, amidoiodanes **450–452**, bis(pyridinium) aryliodine(III) salts **453** and imidoiodanes **454** (Figure 2.18).

2.1.12.1 Azidoiodanes

Azidoiodanes **448** and **449** were proposed as reactive intermediates in synthetically useful azidation reactions involving the combination of PhIO, $PhI(OAc)_2$, or $PhICl_2$ with trimethylsilyl azide or NaN_3 [595–610]. Attempts to isolate these intermediates always resulted in rapid decomposition at –25 to 0 °C with the formation of iodobenzene and dinitrogen; however, low-temperature spectroscopy and the subsequent chemical reactions *in situ* provided some experimental evidence for the existence of these species. A systematic study of structure and reactivity of these unstable azidoiodanes was attempted by Zbiral and coworkers in the early 1970s [595–598]. In particular, the IR spectroscopic measurements at –60 to 0 °C indicated the existence of $PhI(N_3)_2$ and $PhI(N_3)OAc$ species in the mixture of $PhI(OAc)_2$ with trimethylsilyl azide [595]. Interestingly, the absorption band of 2040 cm^{-1} for the azido group in $PhI(N_3)OAc$ is very close to the absorption stretch of 2046–2048 cm^{-1} in the stable, heterocyclic azidoiodanes **122–124** (Section 2.1.8.1.5).

448

449

450

451

452

X = OAc, Cl, $OSiMe_3$, OTs, etc.
NY_2 = phthalimidate, succimidate, saccharinate   R = alkyl, aryl

453

454

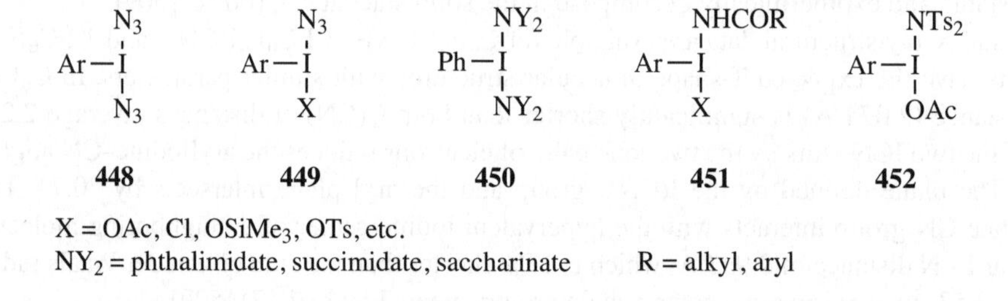

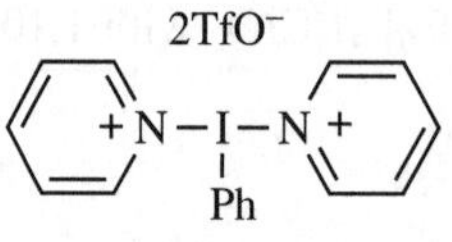

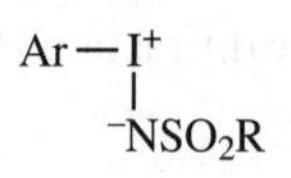

Figure 2.18 Iodine(III) derivatives with I–N bonds.

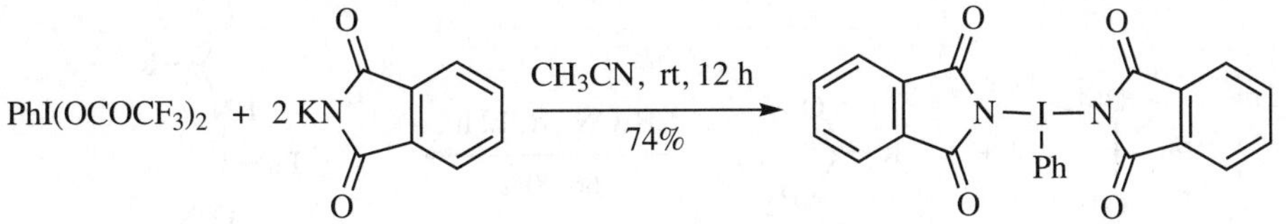

$PhI(OCOCF_3)_2$ + 2 KN(phthalimide) —(CH_3CN, rt, 12 h; 74%)→ PhI(N-phthalimide)$_2$ **455**

Scheme 2.129

The unstable and highly reactive azidoiodanes generated *in situ* from PhIO, $PhI(OAc)_2$, or $PhICl_2$ and trimethylsilyl azide or NaN_3 have found some practical application as efficient reagents for the introduction of the azido function into organic molecules.

2.1.12.2 *Amidoiodanes*

The only known amidoiodanes of type **450** with two nitrogen ligands on iodine are represented by the derivatives of cyclic imides, such as phthalimide, succinimide, glutarimide and saccharine [611–613]. For example, the thermally stable, but water-sensitive, bis(phthalimidate) **455** was prepared by treatment of [bis(trifluoroacetoxy)iodo]benzene with potassium phthalimidate in acetonitrile (Scheme 2.129) [611].

Amidoiodanes **451** bearing one nitrogen ligand at hypervalent iodine center are plausible intermediates in the Hofmann-type degradation of amides with [bis(acyloxy)iodo]arenes or [hydroxy(tosyloxy)iodo]benzene [614]. In most cases, these intermediates are highly unstable and instantaneously rearrange at room temperature with loss of iodobenzene to give isocyanates. However, *N*-phenyliodonio carboxamide tosylates **456** can be isolated as relatively stable, crystalline products in the reaction of [methoxy(tosyloxy)iodo]benzene with amides (Scheme 2.130) [615]. Koser and coworkers have investigated some chemical reactions of amidoiodane **456**. In particular, it decomposes in a $CDCl_3$ solution under anhydrous conditions at room temperature with the formation of benzyl isocyanate or gives the ammonium salt upon decomposition in the presence of water. Concentrated hydrochloric acid converts compound **456** into dichloroiodobenzene and the respective amide [615]. The unstable amidoiodanes **457** can be generated by the reaction of [bis(trifluoroacetoxy)iodo]benzene with *N*-methoxyamides [215].

More recently, Muñiz and coworkers prepared the stable bis(sulfonylimide) derived amidoiodanes **458** and **459** by treatment of (diacetoxyiodo)benzene with bis(tosylimide) or bis(methanesulfonylimide) (Scheme 2.131) [616–618]. The structure of bis(tosylimide) **458** was established by single-crystal X-ray diffractometry [616]. Generated *in situ* amidoiodanes **458** and **459** are useful reagents for metal-free direct allylic amination or diamination of alkenes [616, 617].

2.1.12.3 *Bis(Pyridinium) Aryliodine(III) Salts*

Bis(onio)-substituted aryliodine(III) salts of type **453** were originally reported in 1994 by Weiss and Seubert [570]. Treatment of (diacetoxyiodo)benzene or iodosylbenzene with trimethylsilyl triflate and a nitrogen heterocycle (L) in a molar ratio 1 : 2 at room temperature leads to the precipitation of triflate salts **453** and **460–462** (Scheme 2.132), which are sparingly soluble in dichloromethane and are stable in the absence of moisture [570]. The structures of products **453** and **460** were established by single-crystal X-ray diffraction studies [619]. The alternative procedure for the preparation of the triflate salt **453** involves the reaction of PhI(F)OTf with pyridine [620]. Analogous complexes of hypervalent iodine compounds with nitrogen heterocycles have also been reported for benziodoxoles and iodonium salts [416, 417, 621].

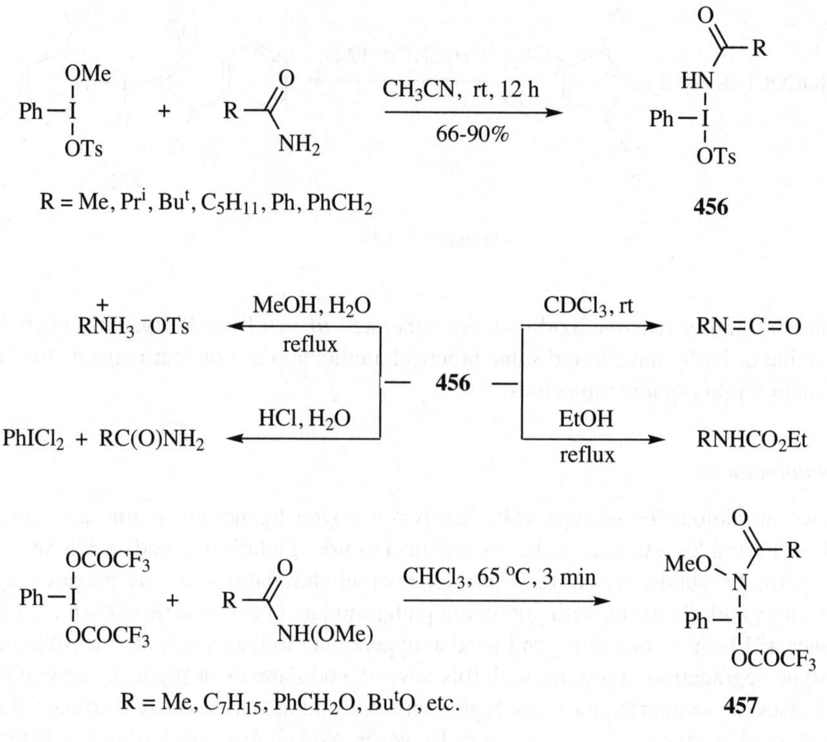

$PhI(OMe)(OTs)$ + $RC(O)NH_2$ → (CH_3CN, rt, 12 h; 66-90%) → $PhI(NHC(O)R)(OTs)$ **456**

R = Me, Pr^i, Bu^t, C_5H_{11}, Ph, $PhCH_2$

456 + MeOH, H_2O, reflux → $RNH_3^+ \, ^-OTs$

456 + HCl, H_2O → $PhICl_2$ + $RC(O)NH_2$

456 + $CDCl_3$, rt → $RN{=}C{=}O$

456 + EtOH, reflux → $RNHCO_2Et$

$PhI(OCOCF_3)_2$ + $RC(O)NH(OMe)$ → ($CHCl_3$, 65 °C, 3 min) → $PhI(N(OMe)C(O)R)(OCOCF_3)$ **457**

R = Me, C_7H_{15}, $PhCH_2O$, Bu^tO, etc.

Scheme 2.130

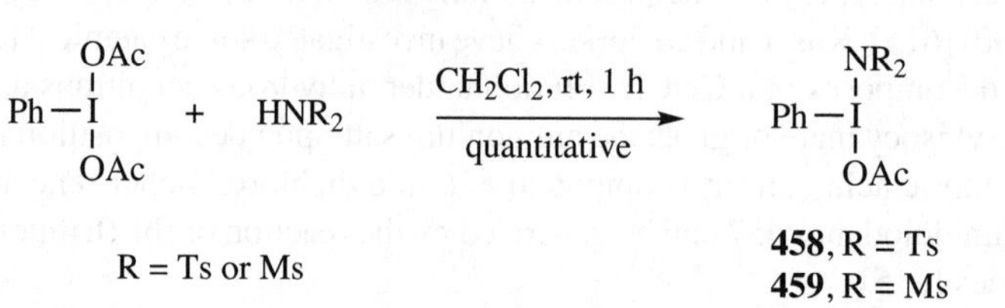

$PhI(OAc)_2$ + HNR_2 → (CH_2Cl_2, rt, 1 h; quantitative) → $PhI(NR_2)(OAc)$

R = Ts or Ms

458, R = Ts
459, R = Ms

Scheme 2.131

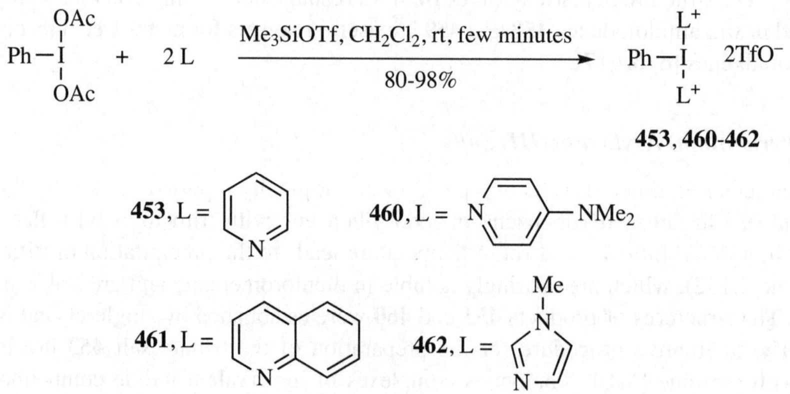

$PhI(OAc)_2$ + 2 L → (Me_3SiOTf, CH_2Cl_2, rt, few minutes; 80-98%) → $PhI(L^+)_2 \; 2TfO^-$

453, 460-462

453, L = pyridine
460, L = 4-(dimethylamino)pyridine ($N{-}C_5H_4{-}NMe_2$)
461, L = quinoline
462, L = N-methylimidazole

Scheme 2.132

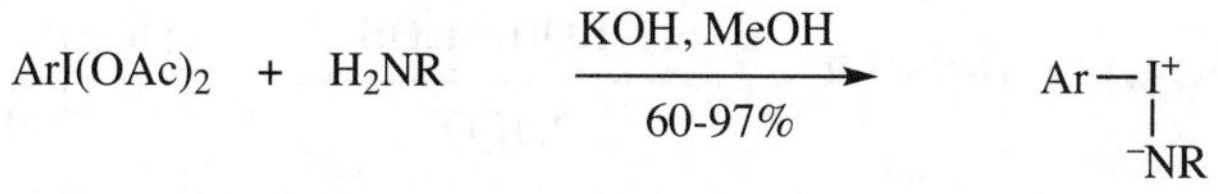

$$ArI(OAc)_2 + H_2NR \xrightarrow[60\text{-}97\%]{KOH,\ MeOH} Ar{-}I^{+}{-}{}^{-}NR$$

463

Ar = Ph, 3-MeC_6H_4, 2,4,6-$Me_3C_6H_2$, etc.
R = Ts, 4-$MeOC_6H_4SO_2$, 4-$CF_3C_6H_4SO_2$,
4-$NO_2C_6H_4SO_2$, 2-$NO_2C_6H_4SO_2$, 4-$FC_6H_4SO_2$,
4-$IC_6H_4SO_2$, $PhSO_2$, $MeSO_2$, CF_3CO, etc.

Scheme 2.133

Bis(onio)-substituted aryliodine(III) salts of type **453** are powerful oxidants. According to polarographic measurements, the $E_{\frac{1}{2}}$ value for the acceptance of the first electron by bis(pyridinium) salt **453** is 1.3 V greater than for $PhI(OAc)_2$, which is indicative of a dramatically increased oxidation power of **453** [570]. In particular, bis(pyridinium) salt **453** can instantaneously oxidize 1,2-bishydrazones to alkynes [570] and it reacts with diazoacetate esters at room temperature to give the corresponding α-aryliodonio diazo compounds [622].

2.1.12.4 Imidoiodanes

Imidoiodanes or iodonium imides, ArINR, are the I–N analogues of iodonium ylides and iodosylbenzene. The best known and widely used iodonium imide is *N*-tosyliminophenyliodane (PhINTs), which has found synthetic application as a useful nitrene precursor under thermal or catalytic conditions in the aziridination of alkenes and the amidation reactions of various organic substrates. The chemistry of iodonium imides was reviewed by Dauban and Dodd in 2003 [623].

2.1.12.4.1 Preparation Aryliodonium imides **463** are usually prepared by the reaction of (diacetoxyiodo)arenes with the respective amides under basic conditions (Scheme 2.133) [623]. Most iodonium imides are relatively unstable at room temperature and their storage under an inert atmosphere at low temperature is recommended. Exothermic decomposition frequently occurs at the melting point of ylides and some of them were even claimed to be explosive [623].

Examples of the known iodonium imides, ArINR, include *N*-tosyl- [67, 624–627], *N*-methanesulfonyl [628] and *N*-trifluoroacetyl derivatives [629]. *N*-(Trifluoroacetyl) and *N*-(methanesulfonyl) iminoiodanes are relatively unstable and explosive, while the tosylate derivatives are stable, crystalline compounds that can be stored for extended periods.

Various arenesulfonyl phenyliodonium imides, $PhINSO_2Ar$ (Ar = Ph, 4-MeC_6H_4, 4-$NO_2C_6H_4$, 4-$MeOC_6H_4$, 4-$CF_3C_6H_4$, 2-$NO_2C_6H_4$, 4-FC_6H_4, 4-BrC_6H_4 and 4-IC_6H_4), have been prepared from (diacetoxyiodo)benzene and the appropriate sulfonylamides under basic conditions [626].

Several phenyliodonium imides **465** derived from heteroarenesulfonylamides have been synthesized from (diacetoxyiodo)benzene and the respective amides **464** (Scheme 2.134) [630]. Imides **465** can be used as sources of the corresponding heterocycle-containing nitrenes in the copper-catalyzed aziridination and sulfimidization reactions.

Imidoiodane **466** [PhINSes, where Ses = (trimethylsilyl)ethanesulfonyl] was prepared by a similar procedure from (diacetoxyiodo)benzene and the respective sulfonamide [631]. This reagent is useful for the copper-catalyzed aziridination of olefins leading to the synthetically versatile Ses-protected aziridines.

Protasiewicz and coworkers have reported the preparation and X-ray crystal structure of a highly soluble nitrene precursor **467**, in which the intramolecular secondary I···O bond replaces intermolecular interactions

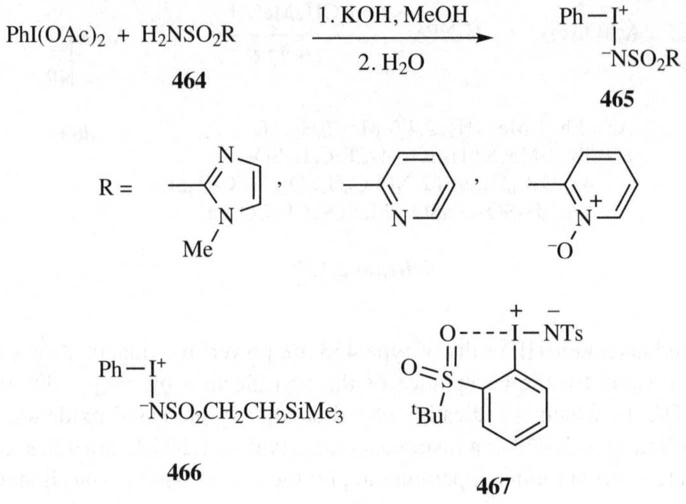

$PhI(OAc)_2$ + H_2NSO_2R (464) → [1. KOH, MeOH; 2. H_2O] → $Ph-I^+$ $^-NSO_2R$ (465)

R = 1-methylimidazol-2-yl, 2-pyridyl, 2-pyridyl N-oxide

$Ph-I^+$ $^-NSO_2CH_2CH_2SiMe_3$ (466)

467

Scheme 2.134

that are typical of iodonium imides [100, 632]. Imide **467** is readily soluble in organic solvents (up to 0.14 M in chloroform, which is a 50-fold increase over PhINTs) and it can be analyzed by NMR in solution [100]. Solubilization of various amides ArINTs in organic solvents can also be achieved by the addition of organic N-oxides, such as Me_3NO [633].

The highly soluble imidoiodanes **473–476**, which are derived from *ortho*-alkoxyiodobenzenes, were synthesized in two simple steps starting from readily available 2-iodophenol ethers **468** (Scheme 2.135) [634]. In the first step, iodides **468** were oxidized by peracetic acid to form diacetoxyiodo derivatives **469–472**; the structures of products **469** and **472** were established by X-ray analysis. In the second step, diacetates **469–472** were converted into imidoiodanes **473–476** by treatment with tosylamide under basic conditions in methanol. Compounds **473–476** are relatively stable at room temperature and can be stored for several weeks in a refrigerator. They also have good solubility in dichloromethane, chloroform and acetonitrile (e.g., the solubility of **472** in dichloromethane is 0.25 g ml^{-1}) [634].

2.1.12.4.2 ***Structural Studies*** Single-crystal X-ray structural data have been reported for the following iodonium imides (Figure 2.19): phenyl(*N*-tosylimino)iodane (**477**) [67, 635], mesityl(*N*-tosylimino)iodane

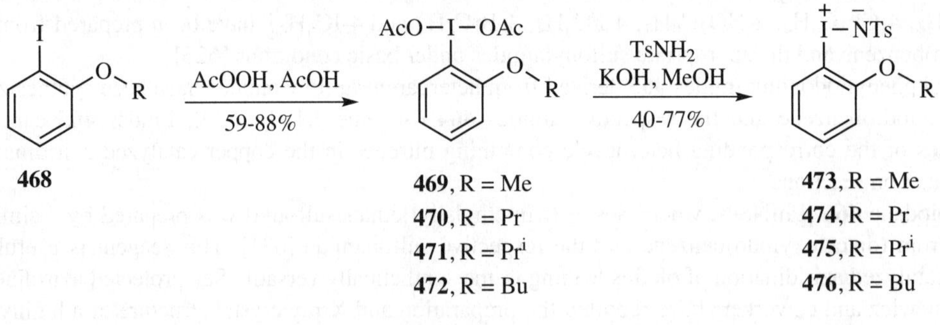

468 → [AcOOH, AcOH; 59-88%] → AcO–I–OAc (**469**, R = Me; **470**, R = Pr; **471**, R = Pr^i; **472**, R = Bu) → [$TsNH_2$, KOH, MeOH; 40-77%] → I–NTs (**473**, R = Me; **474**, R = Pr; **475**, R = Pr^i; **476**, R = Bu)

Scheme 2.135

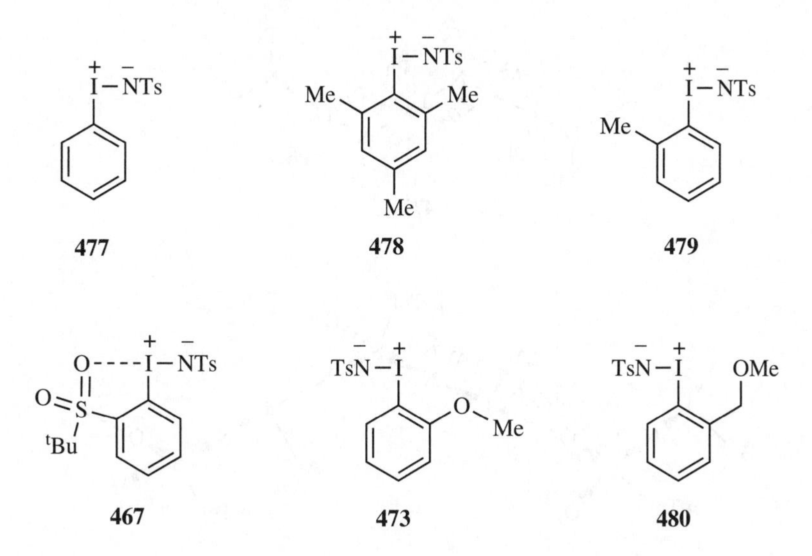

Figure 2.19 *Iodonium imides analyzed by a single-crystal X-ray diffraction.*

(**478**) [67], *o*-tolyl(*N*-tosylimino)iodane (**479**) [627], *ortho*-sulfonyl substituted phenyliodonium imide **467** [100], *ortho*-methoxy substituted phenyliodonium imide **473** [634] and *ortho*-methoxymethyl substituted phenyliodonium imide **480** [636].

Similar to iodosylbenzene (Section 2.1.4.3), aryl(*N*-tosylimino)iodanes have a linear polymeric, asymmetrically bridged structure with the T-shaped geometry around the iodine centers. In the case of PhINTs **477**, the monomeric units are bridged by I–N interactions, while in the more sterically hindered MesINTs **478** the bridging atom is the oxygen of the tosyl group (Figure 2.20) [67]. The structure of 2-TolINTs **479** is intermediate between the structures **477** and **478:** it can form two different polymorphic modifications, one with nitrogen and the second with oxygen as the bridging atom in the polymeric chain [627]. A polymeric, nitrogen-bridged structure was determined for 4-MeC_6H_4INTs by X-ray powder diffraction and EXAFS analyses [122]. The intramolecular I–N distance of 2.01–2.04 Å in *N*-tosyliminoiodanes is consistent with a single bond and the positive charge at iodine and the negative charge delocalized at the nitrogen and oxygen atoms of the tosylimino group.

A single-crystal X-ray analysis of *ortho*-sulfonyl substituted phenyliodonium imide **467** showed a structure of loosely associated centrosymmetric dimers with a long-range intramolecular I–N and I–O distance of more than 3.0 Å, quite unlike the infinite polymeric chains adopted in the solid state for PhINTs [100]. One of the sulfonyl oxygen atoms forms a short intramolecular I···O secondary bond to the iodine atom with a bond length of 2.667 Å. Because of the non-polymeric structure, imide **467** has excellent solubility in common organic solvents.

Similar to the structure of PhINTs (**477**, Figure 2.20), molecules of *ortho*-methoxy substituted phenyliodonium imide **473** [634] have a polymeric, asymmetrically bridged structure with a T-shaped geometry around the iodine centers formed by two iodine–nitrogen bonds and one iodine–carbon bond. However, in contrast to PhINTs, compound **473** has two additional weak intra- and intermolecular I···O contacts between the iodine center and the oxygen atoms of the alkoxy and sulfonyl groups. These weak interactions lead to an elongation of the I···N intermolecular bond in **473** (2.735 Å) compared with that observed in PhINTs (2.482 Å). As a result, the polymeric structure of **473** is weakened and the solubility is significantly increased [634].

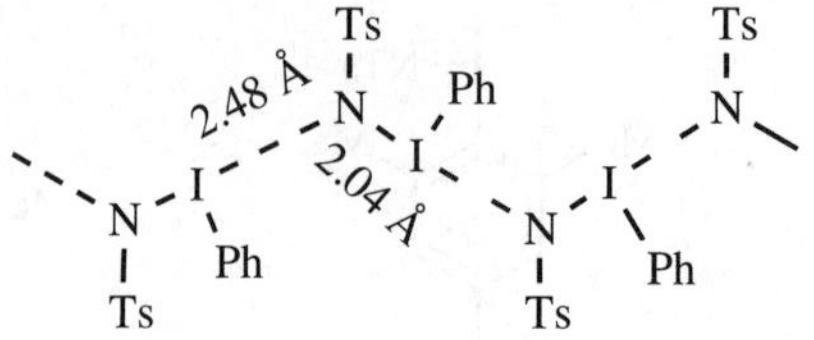

*Figure 2.20 Primary and secondary bonding pattern in single-crystal X-ray structures of iodonium imides **477** and **478**.*

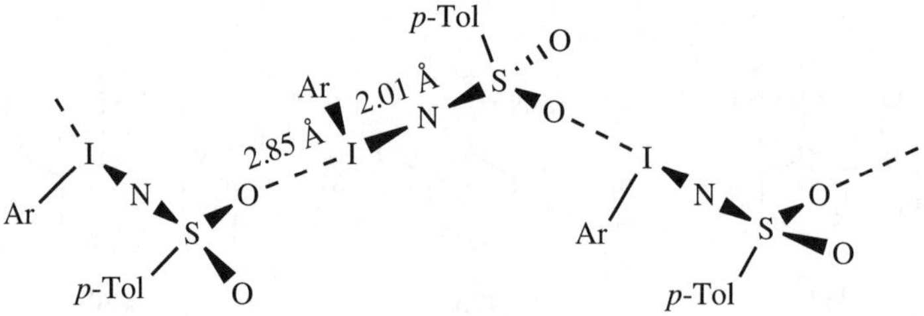

2.2 Iodine(V) Compounds

Inorganic derivatives of iodine(V), such as salts of iodic acid, are the most common, naturally occurring derivatives of hypervalent iodine. In fact, most of the world's production of iodine comes from calcium iodate found in caliche deposits in Chile. Organic iodine(V) derivatives represent a less common, but practically important, class of hypervalent iodine compounds. Particularly useful are the cyclic derivatives of 2-iodoxybenzoic acid (IBX) and its analogues, which have found broad practical application as mild and selective oxidizing reagents in organic synthesis. The chemistry of organic iodine(V) compounds has been summarized in several reviews [637–640].

2.2.1 Inorganic Iodine(V) Derivatives

Iodine pentafluoride, the only known binary interhalogen compound of iodine(V), was first prepared in 1871 by heating iodine with silver fluoride [641–643]. This compound is a highly reactive, colorless liquid with a boiling point of 98 °C and a freezing point of 9.6 °C. According to microwave and electron-diffraction measurements, the molecule of IF_5 has a tetrahedral pyramidal geometry with four equatorial and one axial fluorine atoms with bond distances F_{eq}—I = 1.869 Å and F_{ax}—I = 1.844 Å (structure **481**, Figure 2.21) [644]. The X-ray crystal structure of the molecular addition compound $XeF_2 \cdot IF_5$ with more-or-less discrete IF_5 molecules was also reported [645]. The salts $M^+IF_6^-$ can be prepared by the reaction of IF_5 with the respective fluorides, MF; crystal structures of $(CH_3)_4N^+$ IF_6^- and NO^+ IF_6^- have been reported [646]. Reaction of IF_5 with I_2O_5 gives stable, solid oxyfluorides IOF_3 and IO_2F [641–643].

Various oxygen-bonded iodine(V) derivatives are known [641–643]. Iodine pentoxide, I_2O_5, the most important and thermally stable iodine oxide, is prepared in the form of a hygroscopic, white solid by dehydration of iodic acid, HIO_3, at 200 °C. It readily absorbs water from the atmosphere, giving the hydrate,

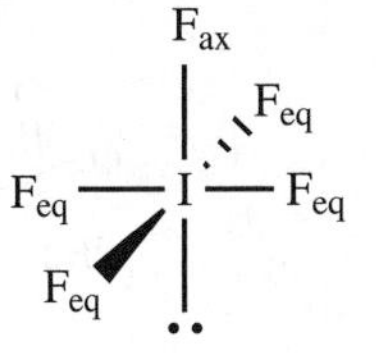

Figure 2.21 Structures of iodine pentafluoride (**481**) and iodine pentoxide (**482**).

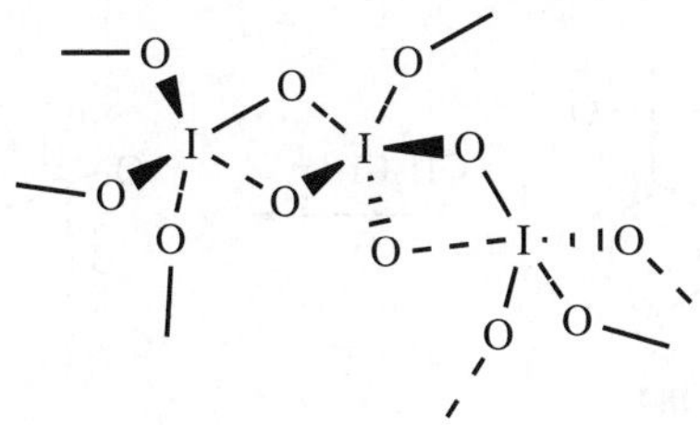

HI_3O_8. On dissolution in water, I_2O_5 regenerates iodic acid [643]. Iodine pentoxide has a polymeric structure (**482**, Figure 2.21) with primary I–O bonds of 1.77–1.95 Å in the distinguishable molecular I_2O_5 units and intermolecular I···O distances as short as 2.23 Å [647]. Several mixed iodine(III/V) oxides, such as I_2O_4 and I_4O_9, are also known [643]; the structure of a mixed iodine(V/VII) oxide, I_4O_{12}, has been reported [648].

Several iodyl derivatives, IO_2X, of strong inorganic acids have been synthesized and characterized. The earliest reports on iodyl derivatives of sulfuric or polysulfuric acids date well back into the twentieth century [641–643]. More recently, a crystal structure of a mixed iodine(III/V) sulfate, $(IO_2)_3HSO_4$, was reported [649]. The fluorosulfate IO_2OSO_2F [650] and the triflate $IO_2OSO_2CF_3$ [5] are also known. Raman and IR spectra of these compounds indicate the presence of discrete IO_2 groups with relatively weak secondary I···O bonding [5].

2.2.2 Noncyclic and Pseudocyclic Iodylarenes

Iodylarenes, $ArIO_2$, which are also known as iodoxyarenes, are commonly prepared by direct oxidation of iodoarenes, ArI, with strong oxidants or by disproportionation of iodosylarenes, ArIO. The initial oxidation of ArI usually leads to ArIO, which then slowly disproportionates to ArI and $ArIO_2$ upon moderate heating, or even at room temperature [111, 149, 651]. Common oxidizing reagents that are used for the preparation of iodylarenes from iodoarenes include sodium hypochlorite, sodium periodate, dimethyldioxirane and Oxone. For example, various iodylarenes have been prepared from the corresponding iodoarenes using sodium periodate as the oxidant in boiling 30% aqueous acetic acid [651].

Noncyclic iodylarenes **483**, in particular iodylarenes bearing strongly electron-withdrawing groups in the aromatic ring, can be efficiently prepared using peracetic acid as an oxidant in the presence of catalytic amounts of ruthenium trichloride (Scheme 2.136) [652, 653]. The mechanism of this reaction consists of the initial oxidation of iodobenzene by peracetic acid followed by $RuCl_3$-catalyzed disproportionation of the intermediate (diacetoxyiodo)benzene to iodylbenzene and iodobenzene.

$$\text{ArI} + \text{AcOOH/AcOH} \xrightarrow[\text{64-94\%}]{\text{RuCl}_3\ (0.08\ \text{mol\%}),\ 40\ ^\circ\text{C},\ 16\ \text{h}} \underset{\mathbf{483}}{\text{ArIO}_2}$$

Ar = Ph, 4-MeC_6H_4, 2-MeC_6H_4, 2-ClC_6H_4, 3-ClC_6H_4, 4-ClC_6H_4, 4-BrC_6H_4, 4-FC_6H_4, 4-$CF_3C_6H_4$, 3,5-$CF_3C_6H_3$, etc.

Scheme 2.136

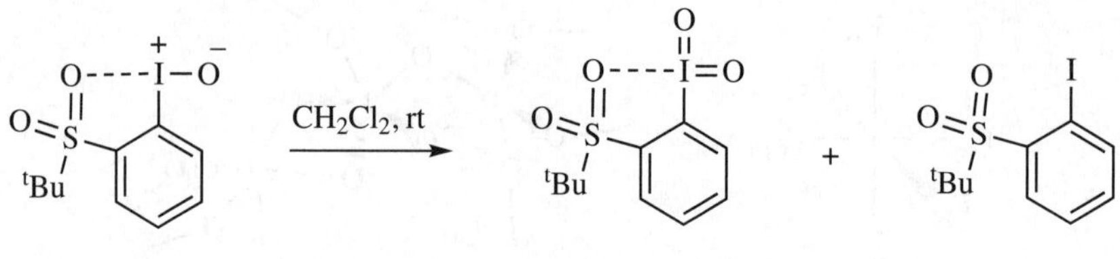

Scheme 2.137

Iodylarenes usually precipitate from the reaction mixture and can be additionally purified by recrystallization from hot water or other solvents. Iodylarenes are potentially hazardous compounds, which may explode upon impact, scratching with a spatula, or heating and therefore should be handled with appropriate precautions. A violent explosion of 2 g of 3-iodyltoluene during crystallization from boiling water with vigorous stirring has been reported [654].

The first noncyclic iodylarene, iodylbenzene $PhIO_2$, was reported by Willgerodt in 1892 [655]. Iodylbenzene is a colorless, microcrystalline solid, which detonates violently upon heating. A single-crystal X-ray structural analysis of $PhIO_2$ revealed a polymeric structure with a distorted octahedral arrangement at the iodine(V) center with three primary I—O bonds in the range of 1.92–2.01 Å and three secondary I···O intermolecular interactions of 2.57–2.73 Å [656]. Because of the polymeric structure, iodylbenzene is insoluble in most organic solvents, with the exception of DMSO. Iodylbenzene in general has found only very limited practical application due to its explosive properties.

Aryliodyl derivatives bearing an appropriate substituent in the *ortho*-position to the iodine are characterized by the presence of a pseudocyclic structural moiety due to a strong intramolecular secondary bonding between the hypervalent iodine center and the oxygen atom in the *ortho*-substituent. Compared to the noncyclic aryliodyl derivatives, pseudocyclic iodine(V) compounds have much better solubility, which is explained by a partial disruption of their polymeric nature due to the redirection of secondary bonding. The first pseudocyclic iodylarene of this type, 1-(*tert*-butylsulfonyl)-2-iodylbenzene **485**, was prepared by Protasiewicz and coworkers in 2000 by the disproportionation of iodosylarene **484** (Scheme 2.137) [127]. A single-crystal X-ray structure of product **485** showed a pseudooctahedral geometry with I—O bond lengths in the iodyl group of 1.796 and 1.822 Å and an intramolecular distance of 2.693 Å between one of the sulfone oxygen atoms and the hypervalent iodine center [127].

A similar, soluble *ortho*-phosphoryl substituted aryliodyl derivative **487** was obtained by the hypochlorite oxidation of the appropriate aryl iodide **486** (Scheme 2.138) [111]. A single-crystal X-ray analysis of compound **487** has shown a close contact of the phosphoryl oxygen atom and the iodine(V) atom with a distance of 2.612 Å, which is significantly shorter than the I···O distance of 3.291 Å determined for the unoxidized aryl iodide **486** [111].

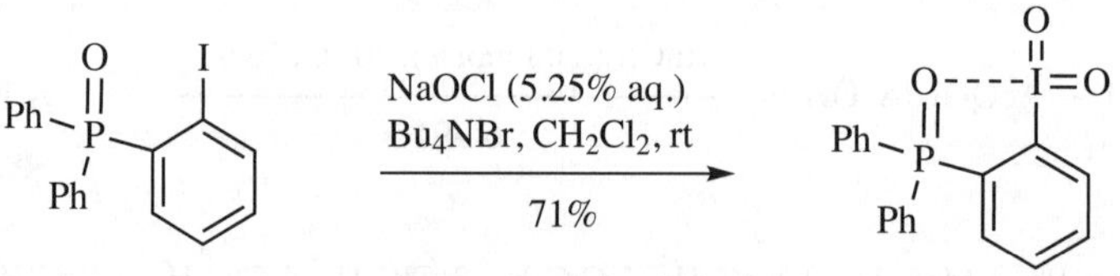

Scheme 2.138

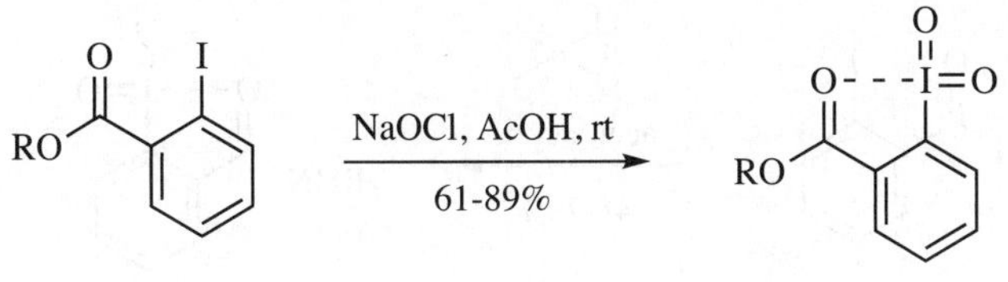

488 —NaOCl, AcOH, rt; 61-89%→ **489**

R = Me, Et, Pr^i, (–)-menthyl, (+)-menthyl, (±)-menthyl, [(1*S*)-endo]-(–)-bornyl, 2-adamantyl, 1-adamantyl, Bu^t

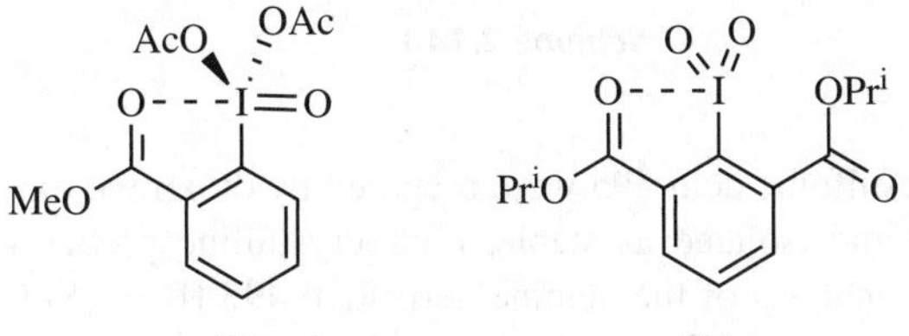

490 **491**

Scheme 2.139

Esters of 2-iodoxybenzoic acid (IBX-esters) **489** have been prepared by the hypochlorite oxidation of the readily available 2-iodobenzoate esters **488** (Scheme 2.139) and isolated in the form of stable microcrystalline solids [657,658]. This procedure has been used for the synthesis of IBX-esters **489** derived from various types of alcohols, such as primary, secondary and tertiary alcohols, adamantanols, optically active menthols and borneol. Single-crystal X-ray data on products **489** revealed a pseudo-benziodoxole structure in which the intramolecular I···O secondary bonds partially replace the intermolecular I···O secondary bonds, disrupting the polymeric structure characteristic of $PhIO_2$ and other previously reported iodylarenes [658]. This structural feature substantially increases the solubility of these compounds in comparison to other iodine(V) reagents and affects their oxidizing reactivity.

Methyl 2-iodoxybenzoate can be further converted into the diacetate **490** or a similar bis(trifluoroacetate) derivative by treatment with acetic anhydride or trifluoroacetic anhydride, respectively [658]. Single-crystal X-ray diffraction of methyl 2-[(diacetoxy)iodosyl]benzoate **490** revealed a pseudo-benziodoxole structure with three relatively weak intramolecular I···O interactions. The esters of 2-iodoxyisophthalic acid (e.g., **491**) have been prepared by oxidation of the respective iodoarenes with dimethyldioxirane. X-Ray structural analysis of diisopropyl 2-iodoxyisophthalate **491** showed intramolecular I···O interaction with the carbonyl oxygen of only one of the two ester groups, while NMR spectra in solution indicated equivalency of both ester groups [658].

Amides of 2-iodoxybenzoic acid (IBX-amides) **493** have been prepared by the dioxirane oxidation of the appropriate 2-iodobenzamides **492** (Scheme 2.140) in the form of stable, microcrystalline solids moderately soluble in dichloromethane and chloroform [659]. This procedure has been used for the synthesis of amides **493** derived from various types of amino compounds, such as esters of α-amino acids, esters of β-amino acids and (*R*)-1-phenylethylamine. A single-crystal X-ray analysis of the phenylalanine derivative [**493**, R = (*S*)-$CH(CH_2Ph)CO_2Me$] revealed a close intramolecular contact of 2.571 Å between the hypervalent iodine center and the oxygen atom of the amido group within each molecule. This enforces a planar geometry of the resulting five-membered ring, a geometry that is analogous to that observed for IBX and other benziodoxoles [659].

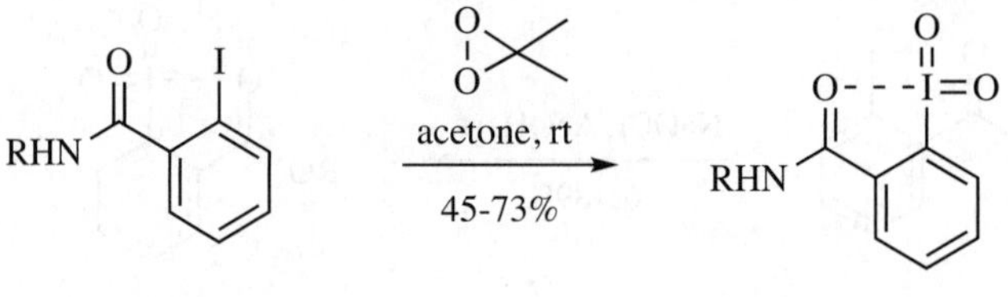

492 → 493 (acetone, rt, 45-73%)

R = (S)-CH(CH$_3$)CO$_2$CH$_3$, (R)-CH(CH$_3$)CO$_2$CH$_3$, (S)-CH(CH$_2$Ph)CO$_2$CH$_3$, (S)-CH(Bu^i)CO$_2$CH$_3$, CH$_2$CH$_2$CO$_2$H, CH(CH$_3$)CH$_2$CO$_2$H, (R)-CH(Ph)CH$_3$

Scheme 2.140

Amides of 2-iodoxybenzenesulfonic acid **495** were prepared by dioxirane oxidation of the corresponding 2-iodobenzenesulfamides **494** and isolated as stable, microcrystalline products (Scheme 2.141) [660]. A single-crystal X-ray structural analysis of the alanine derivative **495** [R = (S)-CH(CH$_3$)CO$_2$Me] showed a combination of intra- and intermolecular I···O interactions leading to a unique heptacoordinated iodine(V) center in this molecule [661]. The analogous esters **497** were prepared similarly from the respective sulfonate esters **496** [662].

The soluble and stable IBX analogues having pseudo-benziodoxazine structure, *N*-(2-iodylphenyl)acylamides **499**, have been prepared in good yields by the oxidation of 2-iodoaniline derivatives **498** with 3,3-dimethyldioxirane under mild conditions (Scheme 2.142) [663]. X-Ray data on compounds **499** revealed a unique pseudo-benziodoxazine structure with an intramolecular secondary I···O

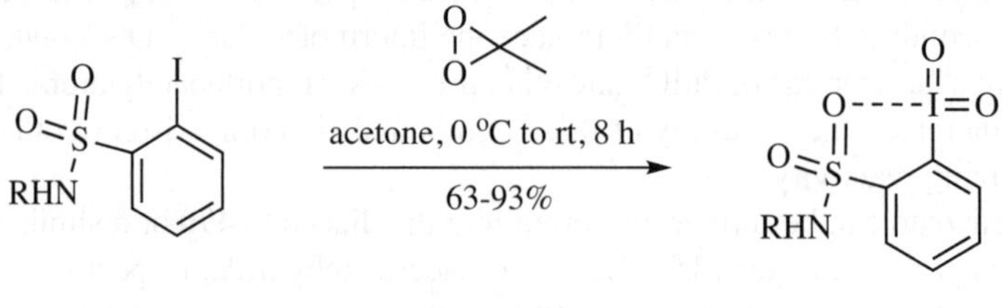

494 → 495 (acetone, 0 °C to rt, 8 h, 63-93%)

R = (S)-CH(CH$_3$)CO$_2$CH$_3$, (S)-CH(CH$_2$Ph)CO$_2$CH$_3$, (S)-CH(Pr^i)CO$_2$CH$_3$, (S)-CH(Bu^i)CO$_2$CH$_3$, (R)-CH(Ph)CH$_3$

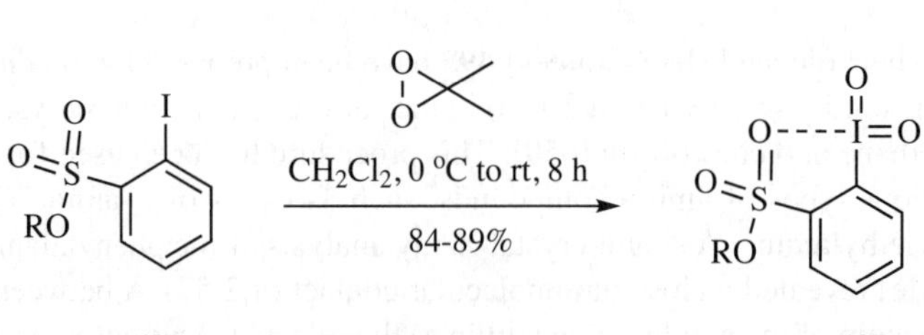

496 → 497 (CH$_2$Cl$_2$, 0 °C to rt, 8 h, 84-89%)

R = Me, Et

Scheme 2.141

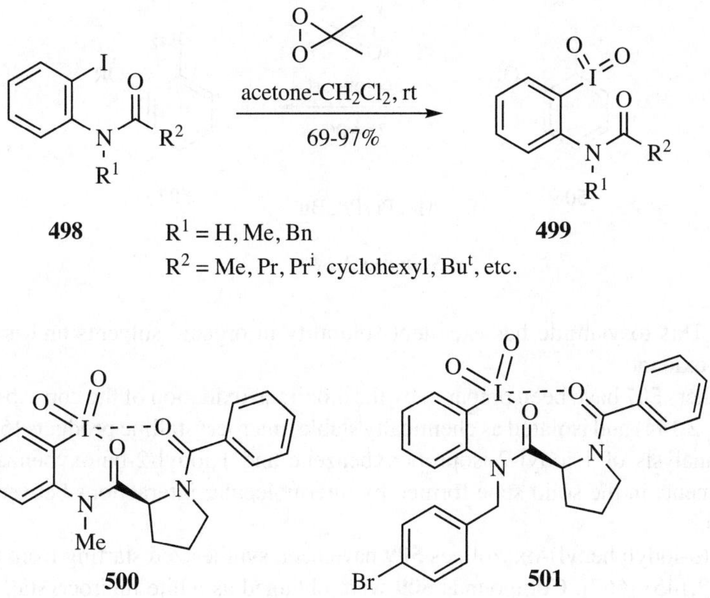

498 $R^1 = H, Me, Bn$ 499
$R^2 = Me, Pr, Pr^i$, cyclohexyl, Bu^t, etc.

acetone-CH_2Cl_2, rt
69-97%

500 501

Scheme 2.142

bond distance of 2.647 Å. The synthesis of chiral pseudo-benziodoxazine reagents **500** and **501** has been achieved based on commercially available and inexpensive (*S*)-proline [664].

Similarly to *N*-(2-iodylphenyl)acylamides, the tosyl derivatives of 2-iodylaniline **503** and 2-iodylphenol **505** were prepared by the dimethyldioxirane oxidation of the corresponding 2-iodophenyltosylamides **502** or 2-iodophenyl tosylate (**504**) (Scheme 2.143) and isolated as stable, microcrystalline products [665]. A single-crystal X-ray diffraction analysis of tosylamide **503** (R = Me) revealed a pseudocyclic structure formed by intramolecular I···O interactions between the hypervalent iodine center and the sulfonyl oxygens in the

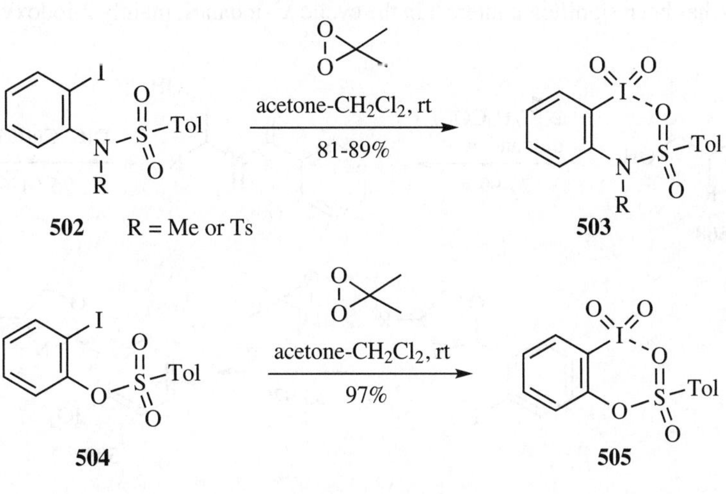

acetone-CH_2Cl_2, rt
81-89%

502 R = Me or Ts 503

acetone-CH_2Cl_2, rt
97%

504 505

Scheme 2.143

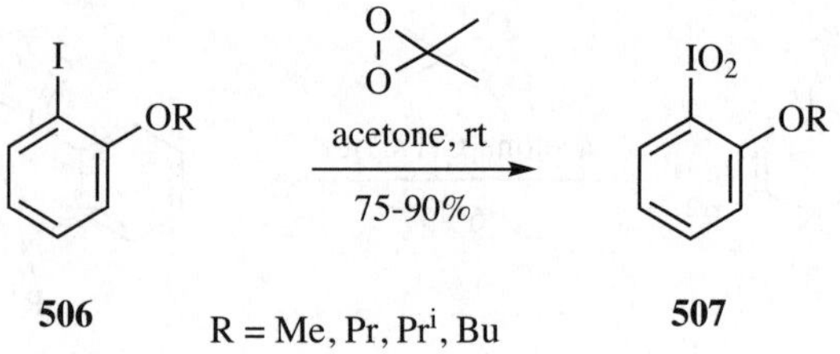

Scheme 2.144

tosyl group [665]. This tosylamide has excellent solubility in organic solvents and is a potentially useful hypervalent iodine oxidant.

2-Iodylphenol ethers **507** have been prepared by the dioxirane oxidation of the corresponding 2-iodophenol ethers **506** (Scheme 2.144) and isolated as chemically stable, microcrystalline products [666]. A single-crystal X-ray diffraction analysis of 1-iodyl-2-isopropoxybenzene and 1-iodyl-2-butoxybenzene showed pseudo-polymeric arrangements in the solid state formed by intermolecular interactions between the IO_2 groups of different molecules.

Several chiral 2-(*o*-iodylphenyl)-oxazolines **509** have been synthesized starting from chiral 2-amino alcohols **508** (Scheme 2.145) [667]. Compounds **509** were obtained as white microcrystalline powders that are soluble in most organic solvents.

Alkyl iodyl derivatives, RIO_2, usually lack stability and cannot be isolated. For example, the matrix isolation and FTIR spectra was reported of the unstable iodylalkanes, generated by the co-deposition and photolysis of ozone with iodoethane, 2-iodopropane, pentafluoroiodoethane, 1,1,1-trifluoroiodoethane, 1,1,2,2-tetrafluoroiodoethane, 1,1,1,2-tetrafluoroiodoethane, or iodine cyanide in an argon matrix at 14–16 K [668–670]. A relatively stable iodyltrifluoromethane, CF_3IO_2, was prepared by the reaction of CF_3IF_4 with silicon dioxide [671].

2.2.3 Iodine(V) Heterocycles

Within the broad field of hypervalent iodine chemistry, five-membered iodine(V) heterocycles occupy a special place. There has been significant interest in the cyclic λ^5-iodanes, mainly 2-iodoxybenzoic acid (IBX)

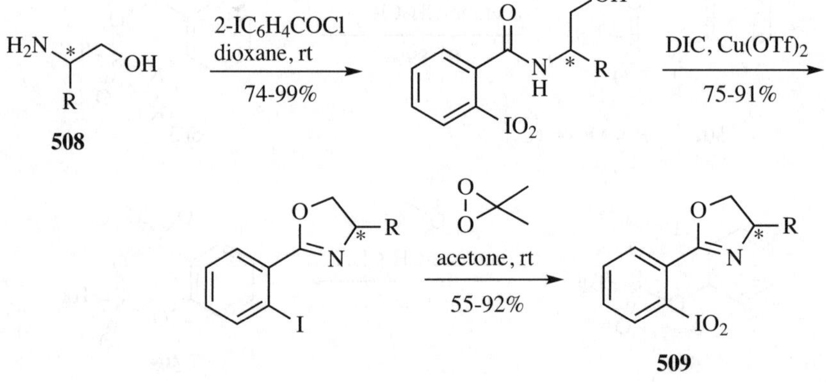

Scheme 2.145

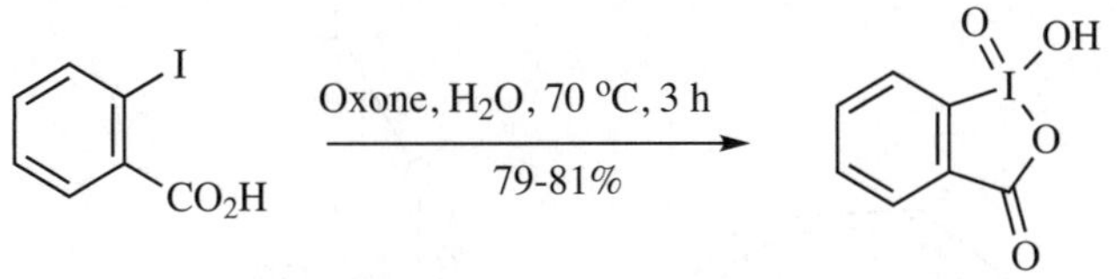

2-iodobenzoic acid → Oxone, H_2O, 70 °C, 3 h; 79-81% → **510**

Scheme 2.146

and Dess–Martin periodinane (DMP), which have found broad practical application as mild and selective reagents for the oxidation of alcohols and some other useful oxidative transformations. Several comprehensive reviews on the chemistry and synthetic applications of IBX and DMP have been published [637–640, 672].

2.2.3.1 2-Iodoxybenzoic Acid (IBX) and Analogues

The most important pentavalent iodine heterocycle, 2-iodoxybenzoic acid (IBX, **510**), was first prepared in 1893 by Hartman and Mayer [673]. IBX has the structure of the cyclic benziodoxole oxide (1-hydroxy-1-oxo-1*H*-1λ^5-benzo[*d*][1, 2]iodoxol-3-one according to the IUPAC nomenclature), as determined by X-ray structural analysis [674–676]. The original preparation of IBX involved the oxidation of 2-iodobenzoic acid with potassium bromate in an aqueous solution of sulfuric acid [673]; more recently, an optimized procedure for the bromate oxidation was published in *Organic Syntheses* [677]. The samples of IBX prepared by this procedure were reported to be explosive under heating or impact [678], possibly due to the presence of bromate impurities [679]. In 1999 Santagostino and coworkers published a convenient and safe procedure for the preparation of IBX (**510**) by the oxidation of 2-iodobenzoic acid using Oxone ($2KHSO_5 \cdot KHSO_4 \cdot K_2SO_4$) in water at 70 °C (Scheme 2.146) [680]. This convenient protocol has become the most commonly used method for the large-scale preparation of IBX.

IBX in the solid state has a three-dimensional polymeric structure due to strong intermolecular secondary I···O contacts and hydrogen bonding. A detailed X-ray diffraction study of IBX samples, prepared by the oxidation of 2-iodobenzoic acid with potassium bromate, revealed the presence of the powder and the macrocrystalline forms of IBX [674]. It was also noticed that the powder form of IBX is more reactive in the reaction with acetic anhydride than the macrocrystalline form and thus is more useful as the Dess–Martin periodinane precursor (Section 2.2.3.2). Treatment of macrocrystalline IBX with aqueous sodium hydroxide and then with hydrochloric acid can be used to convert it into the more reactive powder form [674].

A single-crystal X-ray structural study of the macrocrystalline form of IBX has shown that the iodine(V) atom is chiral and in the solid state IBX is a racemic mixture of two enantiomers. Figure 2.22 shows the molecular geometry of IBX (**510**), with primary and secondary bonds at the iodine(V) center [674]. The crystal structure of racemic IBX consists of chains of molecules linked by I···O secondary bonds (2.808 Å) and also by additional hydrogen bonding. There is also a second iodine–oxygen contact of 2.782 Å linking the chains to form a two-dimensional sheet. Overall, the molecules of IBX are linked together via one O—H···O, four C—H···O hydrogen bonds and two I···O secondary bonds to form a three-dimensional, hydrogen and iodine–oxygen bonded network in which the iodine atoms have distorted octahedral coordination (Figure 2.22). This three-dimensional network is further stabilized by π–π stacking of the phenyl rings with C···C contacts of 3.2 Å and H–ring center contacts of 2.4 Å [674].

IBX is a potentially explosive compound and even the bromate-free samples of IBX are not safe. Santagostino and coworkers reported that as a rule pure IBX explodes at 233 °C [680]. The explosibility tests of analytically pure IBX samples (over 99% purity) confirmed the earlier observations by Plumb and Harper

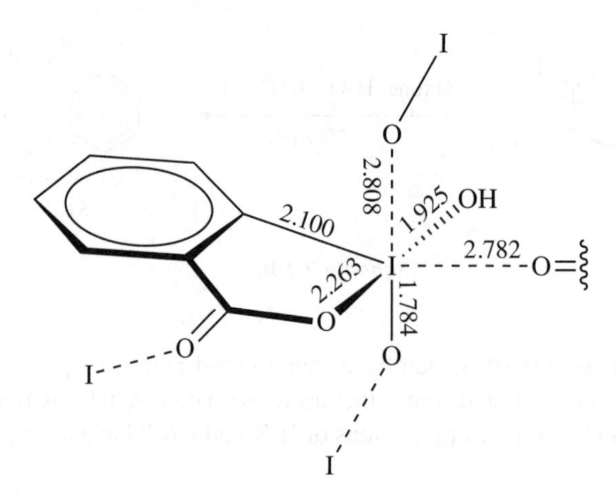

Figure 2.22 *Molecular structure of IBX (**510**) with primary and secondary bonds at the iodine(V) center (bond lengths in Å) [674].*

[678] that IBX is explosive under impact or heating above 200 °C. A non-explosive formulation of IBX (SIBX), consisting of IBX, benzoic acid and isophthalic acid, was introduced by Quideau and coworkers in 2003 [681]. SIBX has similar reactivity to IBX in the reactions of hydroxylative phenol dearomatization [682–684], oxidation of sulfides into sulfoxides [685], oxidative demethylation of phenolic methyl aryl ethers [681] and other useful oxidative transformations [681].

Theoretical and experimental studies of the pK_a value and proton affinity of IBX solutions in aqueous media and DMSO have been published. In particular, the aqueous pK_a value of 2.40 for IBX was obtained by using standard potentiometric titration methods [686]. The relatively high acidity of IBX should be taken into consideration while using this important reagent in the oxidation of complex organic molecules. The gas-phase proton affinities of the anions of IBX (1300 ± 25 kJ mol^{-1}) and 2-iodosylbenzoic acid (1390 ± 10 kJ mol^{-1}) using mass spectrometry-based experiments were reported [687]. The experimental results were supported by theoretical calculations, which yielded proton affinities of 1336 and 1392 kJ mol^{-1} for the anions of IBX and 2-iodosylbenzoic acid, respectively, at the B3LYP/aug-cc-PVDZ level of theory [687].

Numerous analogues of IBX have been reported in the literature (Figure 2.23). In the early 1990s, Martin and coworkers reported the synthesis, structure and properties of several cyclic λ^5-iodanes **511–513** [350,679,688]. Martin and coworkers also introduced bis(trifluoromethyl)benziodoxole oxides **514** and **515**, which are stable and non-explosive oxidizing reagents that are soluble in a wide range of organic solvents [688, 689]. Chiral derivatives of pentavalent iodine, benziodazole oxides **516**, were prepared by the oxidation of the corresponding amino acid derived 2-iodobenzamides with potassium bromate [690]. Vinod and coworkers have developed water-soluble analogs of IBX, *m*-iodoxyphthalic acid (mIBX) **517** [691] and a similar derivative of terephthalic acid (**518**) [692], which are useful oxidizing reagents in aqueous solutions. A zwitterionic, water-soluble IBX analog (**519**) was prepared by oxidation of the corresponding iodide with dimethyldioxirane and was characterized by single-crystal X-ray diffraction [693]. Wirth and coworkers have developed the tetrafluoro-IBX derivative (FIBX, **520**), which is more soluble and has higher reactivity than its non-fluorinated counterpart [340]. Moorthy and coworkers have prepared *o*-methyl-substituted IBX (Me-IBX, **521**) [694] and tetramethyl-IBX (TetMe-IBX, **522**) [695], the modified analogues of IBX that can oxidize alcohols in common organic solvents at room temperature due to the hypervalent twisting-promoted rate enhancement. Fluorous IBX **523** is a recyclable oxidant that can also be used as a catalyst for the

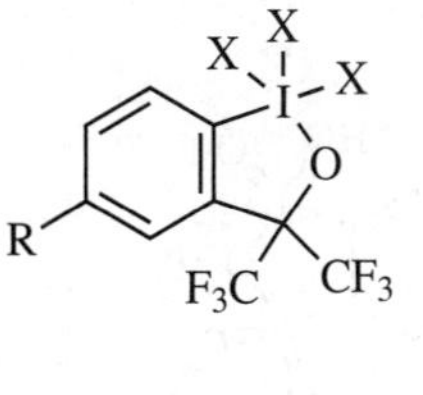

511

R = H, Me
X = F, OAc, $OCOCF_3$

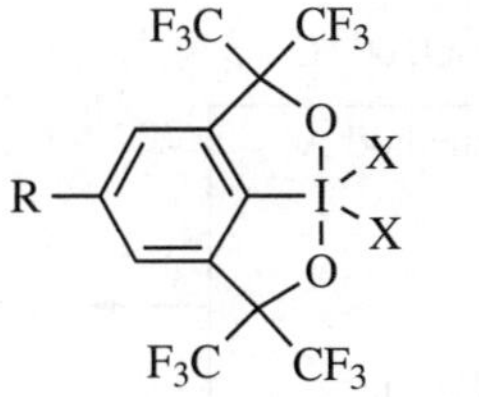

512

R = Me, Bu^t
2X = O or
X = F, OTf, $OCOCF_3$

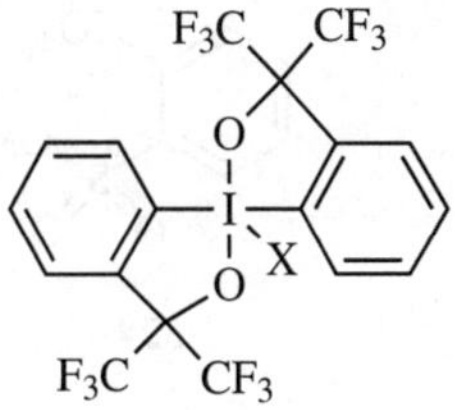

513

X = F, Cl, Br, O^-, $NHBu^t$,
OTf, $OCOCF_3$, etc.

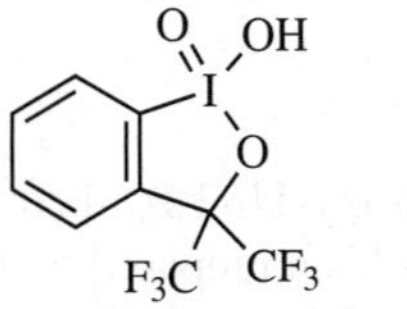

514

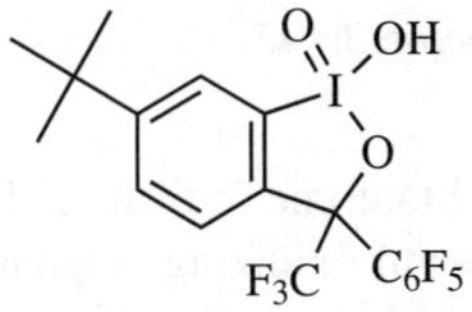

515

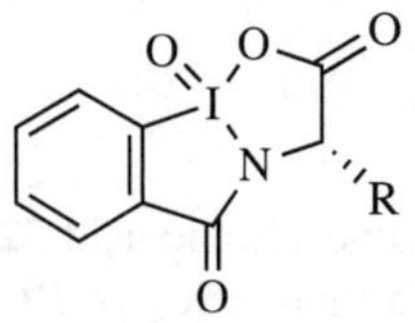

516

R = Me, Pr^i, Bu^i

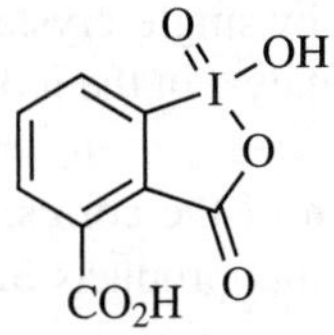

517

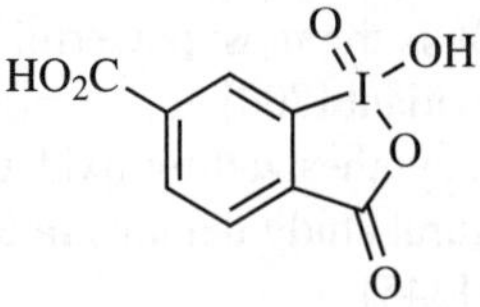

518

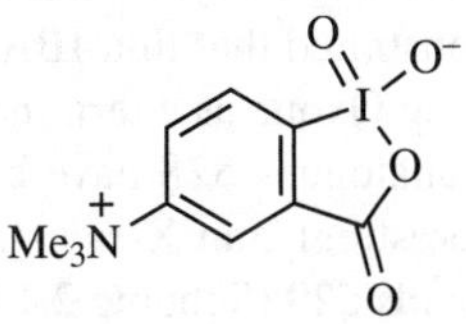

519

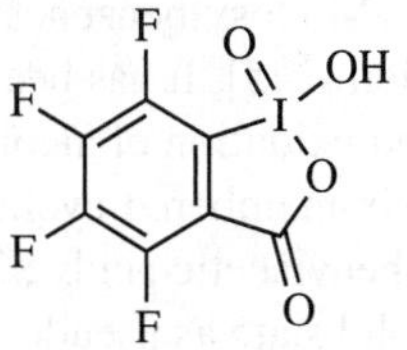

520

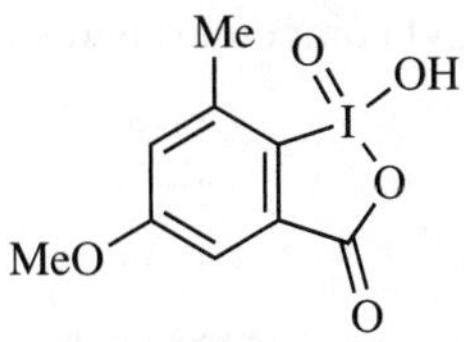

521

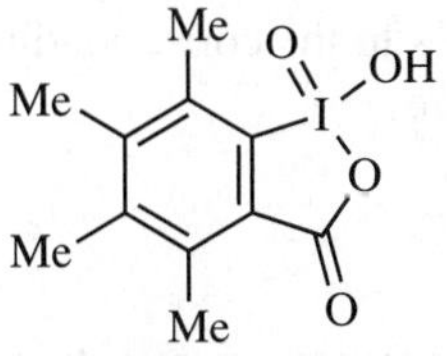

522

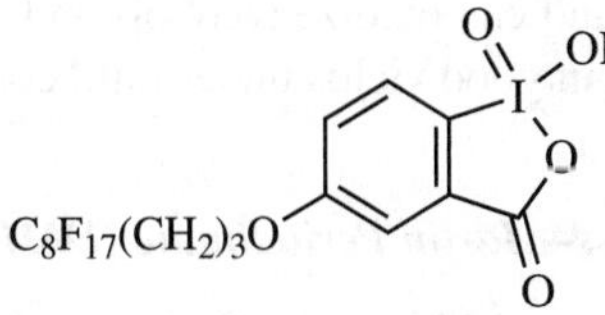

523

Figure 2.23 *IBX analogues.*

oxidation of alcohols to the corresponding carbonyl compounds [696]. Several recyclable polymer-supported IBX derivatives are discussed in Section 5.2.

A pyridinium salt of IBX has been prepared, by treatment of IBX with pyridine, in the form of stable, non-explosive, colorless crystals [697]. Pyridinium 2-iodobenzoate has a similar reactivity pattern to that of IBX and can be used as a safe oxidant that is soluble in organic solvents. IBX can also form relatively stable complexes with *N*-oxides, sulfoxides (DMSO) and tetrahydrofuran [698, 699]. Such ligand complexation allows us to fine-tune the reactivity profile of IBX [698].

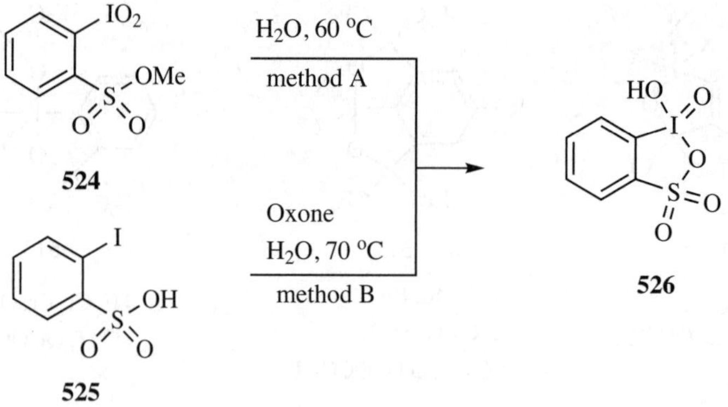

Scheme 2.147

2-Iodoxybenzenesulfonic acid **526** (in a cyclic tautomeric form of 1-hydroxy-1*H*-1,2,3-benziodoxathiole 1,3,3-trioxide), a thia-analog of IBX and a powerful oxidizing reagent, has been prepared by two different pathways: (A) hydrolysis of the methyl ester of 2-iodylbenzenesulfonic acid (**524**) and (B) direct oxidation of 2-iodobenzenesulfonic acid (**525**) (Scheme 2.147) [245]. 2-Iodoxybenzenesulfonic acid **526** was found to be thermally unstable and highly reactive towards organic solvents. The structure of its reductive decomposition product, 2-iodosylbenzenesulfonic acid in its cyclic tautomeric form, was established by single-crystal X-ray diffraction [245]. It has been demonstrated that thia-IBX **526** is the most powerful catalyst in the iodine(V)-catalyzed oxidation of alcohols using Oxone as a terminal oxidant [700].

The six-membered cyclic IBX analogues **528** have been synthesized by oxidation of the corresponding 2-iodophenylacetic acids **527.** Interestingly, an X-ray structural study demonstrated that products **528** exist in the solid state as pseudocyclic acids **529** (Scheme 2.148) [345].

The bicyclic fused iodoxole oxide **531**, the aliphatic analog of IBX, has been prepared by the fluorination of a tricoordinate 1,2-iodoxetane **234** (Section 2.1.8.9) with xenon difluoride followed by hydrolysis of the intermediate, non-isolable, difluoride **530** (Scheme 2.149) [701]. Compound **531** has a similar reactivity to that of IBX and can oxidize alcohols and sulfides to the corresponding carbonyl compounds and sulfoxides, respectively, in good yields under mild conditions.

2.2.3.2 Dess–Martin Periodinane (DMP)

In 1983 Dess and Martin first reported the preparation of triacetoxybenziodoxolone (**532**) by heating IBX with acetic anhydride to 100 °C [702]. In subsequent years the triacetate **532** has emerged as a reagent of choice for the oxidation of alcohols to the respective carbonyl compounds [679] and in the present literature

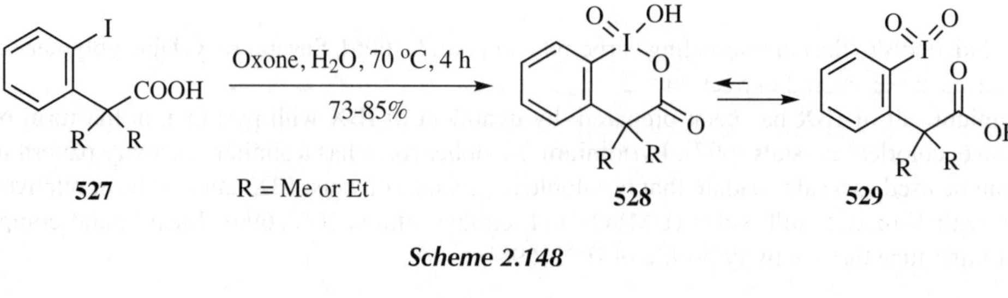

Scheme 2.148

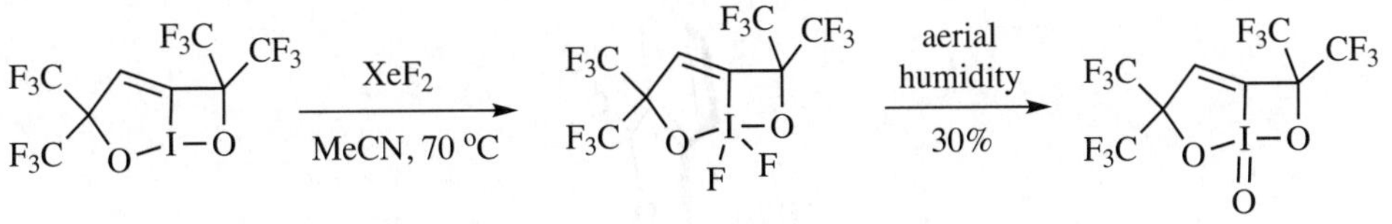

Scheme 2.149

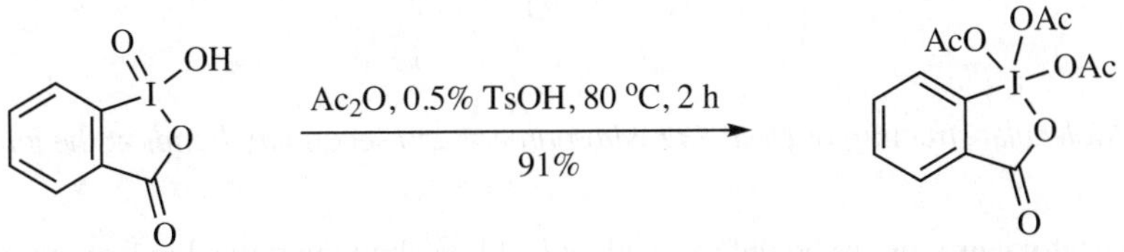

Scheme 2.150

it is commonly referred to as Dess–Martin periodinane (DMP). An improved procedure for the preparation of **532** consists of the reaction of IBX with acetic anhydride in the presence of *p*-toluenesulfonic acid at 80 °C (Scheme 2.150) [703]. DMP is also commercially available from several chemical companies. An optimized procedure for the preparation of DMP has been published in *Organic Syntheses* [677]. The synthetic applications of DMP are highlighted in two overviews [704, 705].

Freshly prepared, crystalline DMP is completely soluble in dichloromethane and chloroform. It should be stored and handled under dry conditions; exposure to atmosphere should be avoided. Careful hydrolysis of DMP (**532**) by slow addition of one equivalent of water in dichloromethane furnishes the monoacetate **533**, which can be isolated in 80% yield (Scheme 2.151) [706]. The freshly prepared or generated *in situ* monoacetate **533** is a stronger oxidant than DMP.

A single-crystal X-ray structural study has shown that DMP (**532**) crystallizes as a dimer; the molecular geometry of the iodine(V) center with primary and secondary bonds is shown in Figure 2.24 [707]. The central iodine atom resides in a distorted octahedral environment that is in accordance with a simple VSEPR model. The equatorial positions are occupied by acetoxy groups, whereas the apical positions are occupied by the phenyl ring and the lone electron pair, respectively. Owing to the steric demands of the electron pair, which is engaged in supramolecular interactions, the acetoxy substituents are pushed toward the phenyl ring. As a consequence, the iodine atom lies 0.315(1) Å below a plane formed by oxygens of the three acetoxy

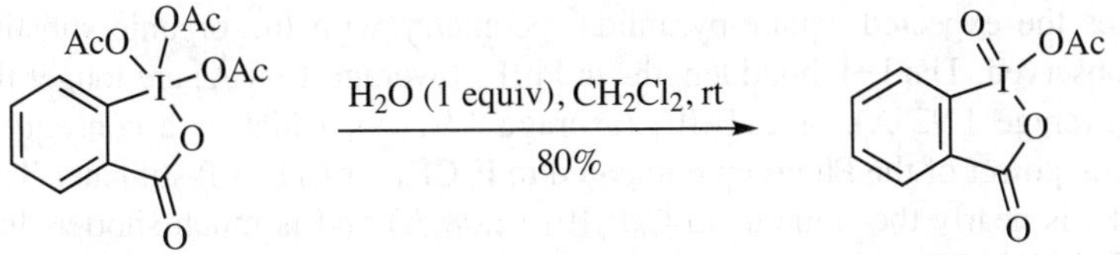

Scheme 2.151

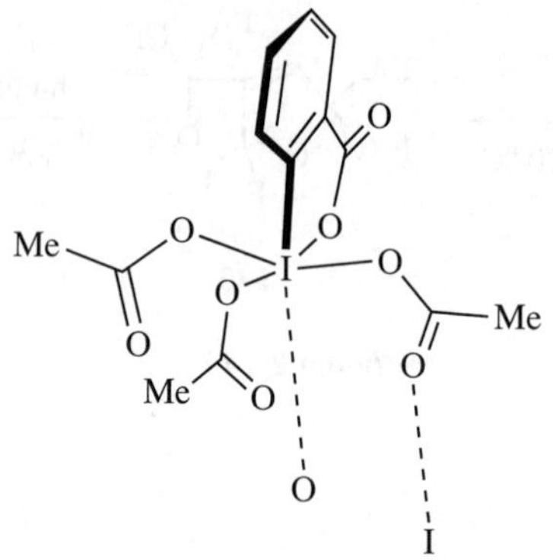

Figure 2.24 Molecular structure of DMP **532** with primary and secondary bonds at the iodine(V) center.

groups. All three acetoxy groups are bound in covalent fashion, showing typical iodine–oxygen bond lengths between 2.06 and 2.11 Å. The length of the iodine–oxygen bond of the iodoxolone ring (2.089 Å) is also well within the range of a covalent iodine–oxygen bond. The iodine–carbon bond is 2.10 Å long and forms a C—I—O angle of 79.6° within the iodoxolone ring. The unit cell is occupied by a centrosymmetric dimer that is held together, in part, by two intermolecular I···O bonds between the iodine atom and a carbonyl group of the adjacent molecule. The intermolecular iodine–oxygen distance of 3.3 Å is well below the sum of the van der Waals radii (3.46 Å). In addition to I···O bonds, the dimer is stabilized by two weak hydrogen bonds of C—H···O type.

2.2.4 Organoiodine(V) Fluorides

Several organoiodine(V) fluorides RIF_4 are known. The trifluoromethyl derivative, CF_3IF_4, has been prepared by the reaction of ClF_3 with CF_3I in perfluorohexane at –78 °C as a white, moisture sensitive solid, which decomposes at 20 °C [708, 709]. It was characterized by NMR spectroscopy [710] and single-crystal X-ray diffractometry [711]. Other perfluoroalkyliodine(V) fluorides R_fIF_4 [R_f = C_2F_5, $(CF_3)_2CF$, or n-C_4F_9] can be prepared similarly as relatively stable products [708, 709].

Aryl- or perfluoroaryl substituted organoiodine(V) fluorides are thermally stable compounds. (Tetrafluoroiodo)benzene, $C_6H_5IF_4$, was first reported in 1968 by Yagupolskii and coworkers [712]. Aryl iodine tetrafluorides can be prepared by three different methods: (i) fluorination of ArI [713], (ii) arylation of IF_5 [714] and (iii) conversion of $ArIO_2$ into $ArIF_4$ with SF_4 [712]. The reaction of $PhIO_2$ with SF_4 is the method of choice for the preparation of $PhIF_4$ [712]. The reaction is nearly quantitative and the gaseous byproducts can be removed easily. $PhIF_4$ is crystalline material, thermally stable to about 300 °C, that is sensitive to hydrolysis.

Single-crystal structures of $C_6F_5IF_4$ [715], CF_3IF_4 [711] and $PhIF_4$ [716] have been reported. In all three molecular structures the expected square-pyramidal geometry with the organic substituent in the apical position has been observed. The I—F bond lengths in $PhIF_4$ (average 1.94 Å) are longer than in IF_5 (average 1.87 Å), CF_3IF_4 (average 1.92 Å), or $C_6F_5IF_4$ (average 1.91 Å), which is a consequence of the weaker electron-withdrawing power of the Ph group compared to F, CF_3, or C_6F_5 substituents. The I—C bond length of 2.077 Å in $PhIF_4$ is nearly the same as in $C_6F_5IF_4$ (2.08 Å) and is much shorter than in CF_3IF_4 (2.22 Å). The average C—I—F angle of 85.36° in $PhIF_4$ is slightly larger than the average C—I—F angles in IF_5 (81.98°), CF_3IF_4 (82.18°) and $C_6F_5IF_4$ (84.38°). Molecules of $PhIF_4$ form a zigzag chain due to the I···F intermolecular interactions [716].

The iodine atom in $PhIF_4$ can act as a fluoride ion acceptor in the reaction with 1,1,3,3,5,5-hexamethylpiperidinium fluoride (pip^+F^-) as a source of nearly naked fluoride ions. $PhIF_4$ reacts with pip^+F^- in acetonitrile at –30 °C to form the colorless, hydrolytically sensitive adduct $pip^+C_6H_5IF_5^-$, which has been characterized by Raman and NMR spectroscopy and its single X-ray crystal structure [716].

2.3 Iodine(VII) Compounds

Only inorganic compounds of iodine(VII) are known. Iodine(VII) fluoride, IF_7, has been known since 1930 [717] and numerous papers dealing with its properties and structure have been published. It can be prepared, by heating IF_5 with fluorine at 150 °C, as a colorless gas with a musty odor [718]. On cooling this gas, colorless crystals are formed, which melt under slight pressure at 6.5 °C. Iodine(VII) fluoride is a unique, neutral, simple binary compound in which heptacoordination is present. Numerous attempts at crystal structure determination of solid IF_7 were inconclusive due to disorder problems. A powder neutron diffraction study demonstrated that there are three solid-state phases of IF_7 at ambient pressure. Although the neutron diffraction study is more sensitive to the fluorine atom distribution than earlier X-ray diffraction studies, the high degree of thermal motion prohibits a definitive assessment of the molecular geometry of IF_7 [719]. However, important structural information was obtained from ^{19}F NMR, microwave, vibrational spectra and gas-phase electron diffraction data [720]. The isolated IF_7 molecule **534** (Figure 2.25) has the form of a pentagonal bipyramid belonging to the symmetry group D_{5h}, in which the bonds of the axial F—I—F unit (1.786 Å) are shorter than those of the more congested equatorial IF_5 unit (1.858 Å). An attempt to prepare an iodine(VII) compound with a carbon ligand (e.g., Ar_fIF_6) by the reaction of IF_7 with $C_6F_5SiX_3$ (X = Me, F), $C_6F_5BF_2$, or 1,4-$C_6F_4(BF_2)_2$ was unsuccessful [721].

Two oxyfluorides, IO_2F_3 [722, 723] and IOF_5 [724], were reported in the literature. IO_2F_3 can be synthesized by the reaction of tetrafluoroorthoperiodic acid, $HOIOF_4$ and SO_3 in the form of a yellow sublimable solid with a melting point of 41 °C [723]. According to ^{19}F NMR and Raman spectroscopy, in the liquid state IO_2F_3 exists as a cyclic trimer with *cis*-oxygen bridges, in the boat conformation [722]. The second oxyfluoride, IOF_5, can be prepared by the interaction of IF_7 with SiO_2 [718]. The high symmetry, C_{4v}, of IOF_5 renders its structure determination very difficult. Nevertheless, on the basis of a combined electron diffraction–microwave study [725] and *ab initio* calculations [724], the structure **535** (Figure 2.25) was determined. In this structure, the axial and the equatorial I—F bonds are of comparable lengths (about 1.826 Å), the I—O bond distance is 1.725 Å and the O—I—F_{eq} bond angle is close to 97.2° [724]. IOF_5 reacts with tetramethylammonium fluoride to give a stable, crystalline salt, $Me_4N^+IOF_6^-$ [726]. X-Ray analysis of this salt showed the expected, pentagonal bipyramidal structure for the IOF_6^- anion with the oxygen atom occupying one of the axial positions (structure **536**, Figure 2.25). The I—O bond in this compound is

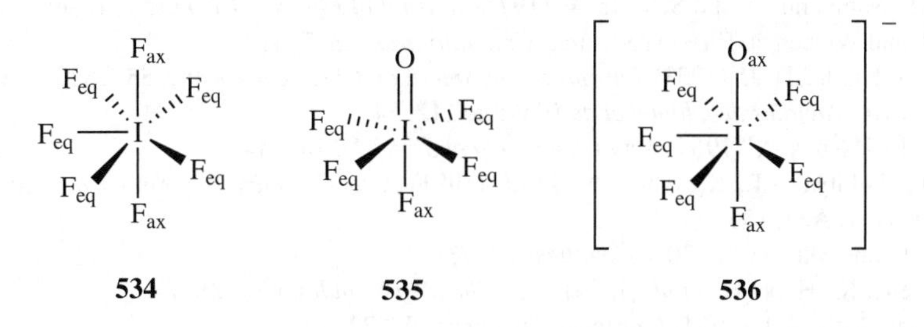

Figure 2.25 Derivatives of iodine(VII).

relatively short (1.75–1.77 Å), which is indicative of a significant double bond character. The axial I—F bonds (1.823 Å) and the equatorial I—F bonds (average 1.88 Å) in IOF_6^- are significantly longer than the corresponding bonds in IF_7, which can be attributed to greater I—F bond polarities due to the formal negative charge on IOF_6^- [726].

Iodine(VII) oxide, I_2O_7, is unknown. A mixed iodine(V/VII) oxide, I_4O_{12}, has been reported [648]. The crystal structure of this oxide includes the basic molecular unit I_4O_{12} with two octahedrally coordinated iodine(VII) atoms and two trigonal pyramidal iodine(V) atoms.

Periodic acid and periodates are the most important, commercially available compounds of iodine(VII). Periodates were first prepared in 1833 by oxidation of $NaIO_3$ with chlorine in alkaline solutions [727]. The industrial preparation of periodates is based on electrolytic oxidation of iodates. Aqueous solutions of periodic acid are best obtained by treating barium paraperiodate with concentrated nitric acid. In solution it exists as a fairly weak orthoperiodic acid, H_5IO_6, which can be dehydrated to HIO_4 by heating to 100 °C in vacuum [643]. Structural investigations indicate an octahedral configuration about iodine in most periodates [643,728].

Periodic acid and periodates have found wide application in organic chemistry as powerful oxidizing reagents [729–731]. The well-known periodate glycol oxidation involves the cyclic periodate esters as key intermediates; however, none of the organic periodate esters have been isolated as stable compounds.

References

1. Hoyer, S. and Seppelt, K. (2000) *Angewandte Chemie, International Edition*, **39**, 1448.
2. Booth, H.S. and Morris, W.C. (1939) *Inorganic Syntheses*, **1**, 167.
3. Zefirov, N.S., Sereda, G.A., Sosonuk, S.E., *et al.* (1995) *Synthesis*, 1359.
4. Ahmed, M.A.K., Fjellvag, H. and Kjekshus, A. (1994) *Acta Chemica Scandinavica*, **48**, 537.
5. Dalziel, J.R., Carter, H.A. and Aubke, F. (1976) *Inorganic Chemistry*, **15**, 1247.
6. Birchall, T., Frampton, C.S. and Kapoor, P. (1989) *Inorganic Chemistry*, **28**, 636.
7. Schmeisser, M., Dahmen, K. and Sartori, P. (1970) *Chemische Berichte*, **103**, 307.
8. Schmeisser, M., Sartori, P. and Naumann, D. (1970) *Chemische Berichte*, **103**, 590.
9. Lehmann, E., Naumann, D. and Schmeisser, M. (1976) *Journal of Fluorine Chemistry*, **7**, 33.
10. Schmeisser, M., Ludovici, W., Naumann, D., *et al.* (1968) *Chemische Berichte*, **101**, 4214.
11. Dixon, D.A., Grant, D.J., Christe, K.O. and Peterson, K.A. (2008) *Inorganic Chemistry*, **47**, 5485.
12. Kim, H., Choi, Y.J. and Lee, Y.S. (2008) *Journal of Physical Chemistry, B*, **112**, 16021.
13. Bae, C., Han, Y.-K. and Lee, Y.S. (2003) *Journal of Physical Chemistry, A*, **107**, 852.
14. Baran, E.J. (2001) *Zeitschrift fur Naturforschung Section A – A Journal of Physical Sciences*, **56**, 333.
15. Bartell, L.S. and Gavezzotti, A. (1983) *THEOCHEM*, **8**, 331.
16. Boldyrev, A.I., Zhdankin, V.V., Simons, J. and Stang, P.J. (1992) *Journal of the American Chemical Society*, **114**, 10569.
17. Schmeisser, M., Naumann, D. and Scheele, R. (1972) *Journal of Fluorine Chemistry*, **1**, 369.
18. Boswijk, K.H. and Wiebenga, E.H. (1954) *Acta Crystallographica*, **7**, 417.
19. Forbes, G.S. and Faull, J.H. Jr. (1933) *Journal of the American Chemical Society*, **55**, 1820.
20. Millon, M.E. (1844) *Annales de Chimie et de Physique*, **12**, 345.
21. Daehlie, G. and Kjekshus, A. (1964) *Acta Chemica Scandinavica*, **18**, 144.
22. Ellestad, O.H., Woldbaek, T., Kjekshus, A., *et al.* (1981) *Acta Chemica Scandinavica – Series A: Physical & Inorganic Chemistry*, **A35**, 155.
23. Kraszkiewicz, L. and Skulski, L. (2008) *Synthesis*, 2373.
24. Furuseth, S., Selte, K., Hope, H., *et al.* (1974) *Acta Chemica Scandinavica*, **28**, 71.
25. Gillespi, R.J. and Senio, J.B. (1964) *Inorganic Chemistry*, **3**, 972.
26. Gillespie, R.J. and Milne, J.B. (1966) *Inorganic Chemistry*, **5**, 1236.

27. Aubke, F., Carter, H.A. and Jones, S.P.L. (1970) *Inorganic Chemistry*, **9**, 2485.
28. Cambie, R.C., Chambers, D., Rutledge, P.S. and Woodgate, P.D. (1977) *Journal of the Chemical Society, Perkin Transactions 1*, 2231.
29. Zupan, M. and Pollak, A. (1976) *Journal of Fluorine Chemistry*, **7**, 445.
30. Sawaguchi, M., Ayuba, S. and Hara, S. (2002) *Synthesis*, 1802.
31. Arrica, M.A. and Wirth, T. (2005) *European Journal of Organic Chemistry*, 395.
32. Padelidakis, V., Tyrra, W. and Naumann, D. (1999) *Journal of Fluorine Chemistry*, **99**, 9.
33. Ye, C., Twamley, B. and Shreeve, J.M. (2005) *Organic Letters*, **7**, 3961.
34. Abo-Amer, A., Frohn, H.-J., Steinberg, C. and Westphal, U. (2006) *Journal of Fluorine Chemistry*, **127**, 1311.
35. Sheremetev, A.B., Dmitriev, D.E. and Konkina, S.M. (2004) *Russian Chemical Bulletin*, **53**, 1130.
36. Gibson, J.A. and Janzen, A.F. (1973) *Journal of the Chemical Society, Chemical Communications*, 739.
37. Minkwitz, R. and Berkei, M. (1998) *Inorganic Chemistry*, **37**, 5247.
38. Bradley, G.W., Holloway, J.H., Koh, H.J., *et al.* (1992) *Journal of the Chemical Society, Perkin Transactions 1*, 3001.
39. Gregorcic, A. and Zupan, M. (1977) *Bulletin of the Chemical Society of Japan*, **50**, 517.
40. Bailly, E., Barthen, P., Breuer, W., *et al.* (2000) *Zeitschrift für Anorganische und Allgemeine Chemie*, **626**, 1406.
41. Frohn, H.J. and Bardin, V.V. (2005) *Journal of Fluorine Chemistry*, **126**, 1036.
42. Naumann, D. and Ruether, G. (1980) *Journal of Fluorine Chemistry*, **15**, 213.
43. Frohn, H.J. and Nielinger, R. (1996) *Journal of Fluorine Chemistry*, **77**, 143.
44. Frohn, H.J. and Schrinner, K. (1997) *Zeitschrift für Anorganische und Allgemeine Chemie*, **623**, 1847.
45. Kasumov, T.M., Pirguliyev, N.S., Brel, V.K., *et al.* (1997) *Tetrahedron*, **53**, 13139.
46. Kasumov, T.M., Brel, V.K., Koz'min, A.S., *et al.* (1997) *New Journal of Chemistry*, **21**, 1347.
47. Pirguliyev, N.S., Brel, V.K., Kasumov, T.M., *et al.* (1999) *Synthesis*, 1297.
48. Pirguliyev, N.S., Brel, V.K., Zefirov, N.S. and Stang, P.J. (1999) *Mendeleev Communications*, **9**, 189.
49. Pirkuliyev, N.S., Brel', V.K., Zhdankin, V.V. and Zefirov, N.S. (2002) *Russian Journal of Organic Chemistry*, **38**, 1224.
50. Pirkuliev, N.S., Brel, V.K., Kasumov, T.M., *et al.* (1999) *Russian Journal of Organic Chemistry*, **35**, 1600.
51. Pirkuliev, N.S., Brel, V.K., Akhmedov, N.G. and Zefirov, N.S. (2000) *Russian Journal of Organic Chemistry*, **36**, 1282.
52. Fuchigami, T. and Fujita, T. (1994) *Journal of Organic Chemistry*, **59**, 7190.
53. Hara, S., Hatakeyama, T., Chen, S.-Q., *et al.* (1998) *Journal of Fluorine Chemistry*, **87**, 189.
54. Fujita, T. and Fuchigami, T. (1996) *Tetrahedron Letters*, **37**, 4725.
55. Hara, S., Sekiguchi, M., Ohmori, A., *et al.* (1996) *Chemical Communications*, 1899.
56. Carpenter, W.R. (1966) *Journal of Organic Chemistry*, **31**, 2688.
57. Lyalin, V.V., Orda, V.V., Alekseeva, L.A. and Yagupolskii, L.M. (1970) *Zhurnal Organicheskoi Khimii*, **6**, 329.
58. Sun, H., Wang, B. and DiMagno, S.G. (2008) *Organic Letters*, **10**, 4413.
59. Bondi, A. (1964) *Journal of Physical Chemistry*, **68**, 441.
60. Choo, J., Kim, S., Joo, H. and Kwon, Y. (2002) *THEOCHEM*, **587**, 1.
61. Porter, C.W. (1996) Structural analyses using X-ray crystallography ((difluoroiodo)toluene, hydroxymethylbenziodoxathiole), PhD Thesis, The University of Akron.
62. Frohn, H.-J., Hirschberg, M.E., Westphal, U., *et al.* (2009) *Zeitschrift für Anorganische und Allgemeine Chemie*, **635**, 2249.
63. Lucas, H.J. and Kennedy, E.R. (1955) *Organic Syntheses Collective Volume III*, 482.
64. Zhao, X.-F. and Zhang, C. (2007) *Synthesis*, 551.
65. Podgorsek, A. and Iskra, J. (2010) *Molecules*, **15**, 2857.
66. Yusubov, M.S., Drygunova, L.A. and Zhdankin, V.V. (2004) *Synthesis*, 2289.
67. Mishra, A.K., Olmstead, M.M., Ellison, J.J. and Power, P.P. (1995) *Inorganic Chemistry*, **34**, 3210.
68. Montanari, V., DesMarteau, D.D. and Pennington, W.T. (2000) *Journal of Molecular Structure*, **550–551**, 337.
69. Minkwitz, R. and Berkei, M. (2001) *Inorganic Chemistry*, **40**, 36.
70. Minkwitz, R. and Berkei, M. (1999) *Inorganic Chemistry*, **38**, 5041.
71. Aggarwal, V.K. and Olofsson, B. (2005) *Angewandte Chemie, International Edition*, **44**, 5516.

72. Willgerodt, C. (1886) *Journal für Praktische Chemie*, **33**, 154.
73. Zanka, A., Takeuchi, H. and Kubota, A. (1998) *Organic Process Research & Development*, **2**, 270.
74. Podgorsek, A., Jurisch, M., Stavber, S., *et al.* (2009) *Journal of Organic Chemistry*, **74**, 3133.
75. Bravo, P., Montanari, V., Resnati, G. and DesMarteau, D.D. (1994) *Journal of Organic Chemistry*, **59**, 6093.
76. Montanari, V. and Resnati, G. (1994) *Tetrahedron Letters*, **35**, 8015.
77. Zhdankin, V.V., Callies, J.A., Hanson, K.J. and Bruno, J. (1999) *Tetrahedron Letters*, **40**, 1839.
78. Cotter, J.L., Andrews, L.J. and Keefer, R.M. (1962) *Journal of the American Chemical Society*, **84**, 4692.
79. Naumann, D., Heinsen, H.H. and Lehmann, E. (1976) *Journal of Fluorine Chemistry*, **8**, 243.
80. Obeid, N. and Skulski, L. (2000) *Polish Journal of Chemistry*, **74**, 1609.
81. Baranowski, A., Plachta, D., Skulski, L. and Klimaszewska, M. (2000) *Journal of Chemical Research. Synopses*, 435.
82. Krassowska-Swiebocka, B., Prokopienko, G. and Skulski, L. (1999) *Synlett*, 1409.
83. Obeid, N. and Skulski, L. (2001) *Molecules*, **6**, 869.
84. Kazmierczak, P., Skulski, L. and Obeid, N. (1999) *Journal of Chemical Research. Synopses*, 64.
85. Zielinska, A. and Skulski, L. (2004) *Tetrahedron Letters*, **45**, 1087.
86. Zefirov, N.S., Safronov, S.O., Kaznacheev, A.A. and Zhdankin, V.V. (1989) *Zhurnal Organicheskoi Khimii*, **25**, 1807.
87. Salamant, W. and Hulme, C. (2006) *Tetrahedron Letters*, **47**, 605.
88. Stang, P.J., Olenyuk, B. and Chn, K. (1995) *Synthesis*, 937.
89. Beringer, F.M. and Nathan, R.A. (1969) *Journal of Organic Chemistry*, **34**, 685.
90. Archer, E.M. and van Schalkwyk, T.G.D. (1953) *Acta Crystallographica*, **6**, 88.
91. Bekoe, D.A. and Hulme, R. (1956) *Nature*, **177**, 1230.
92. Carey, J.V., Chaloner, P.A., Hitchcock, P.B., *et al.* (1996) *Journal of Chemical Research. Synopses*, 358.
93. Protasiewicz, J.D. (1995) *Journal of the Chemical Society, Chemical Communications*, 1115.
94. Nikiforov, V.A., Karavan, V.S., Miltsov, S.A., *et al.* (2003) *ARKIVOC*, (vi), 191.
95. Grebe, J., Geiseler, G., Harms, K. and Dehnicke, K. (1999) *Zeitschrift für Naturforschung B. A Journal of Chemical Sciences*, **54**, 140.
96. Saltzman, H. and Sharefkin, J.G. (1973) Organic Syntheses Collective Volume **5**, 658.
97. Bravo, A., Fontana, F., Fronza, G., *et al.* (1995) *Tetrahedron Letters*, **36**, 6945.
98. Carey, J.V., Chaloner, P.A., Seddon, K.R. and Taylor, M. (1996) *Journal of Chemical Research (S)*, 156.
99. Hiller, A., Patt, J.T. and Steinbach, J. (2006) *Magnetic Resonance in Chemistry*, **44**, 955.
100. Macikenas, D., Skrzypczak-Jankun, E. and Protasiewicz, J.D. (1999) *Journal of the American Chemical Society*, **121**, 7164.
101. Yusubov, M.S., Gilmkhanova, M.P., Zhdankin, V.V. and Kirschning, A. (2007) *Synlett*, 563.
102. Zhdankin, V.V., Kuehl, C.J. and Simonsen, A.J. (1996) *Main Group Chemistry*, **1**, 349.
103. Bravo, A., Fontana, F., Fronza, G., Minisci, F. and Zhao, L. (1998) *Journal of Organic Chemistry*, **63**, 254.
104. Asensio, G., Andreu, C., Boix-Bernardini, C., *et al.* (1999) *Organic Letters*, **1**, 2125.
105. Hawkins, M. and Andrews, L. (1985) *Inorganic Chemistry*, **24**, 3285.
106. Andrews, L., Hawkins, M. and Withnall, R. (1985) *Inorganic Chemistry*, **24**, 4234.
107. Nicoletti, D., Ghini, A.A., Baggio, R.F., *et al.* (2001) *Journal of the Chemical Society, Perkin Transactions 1*, 1511.
108. Yusubov, M.S., Nemykin, V.N. and Zhdankin, V.V. (2010) *Tetrahedron*, **66**, 5745.
109. Yoshimura, A., Neu, H.M., Nemykin, V.N. and Zhdankin, V.V. (2010) *Advanced Synthesis & Catalysis*, **352**, 1455.
110. Meprathu, B.V. and Protasiewicz, J.D. (2003) *ARKIVOC*, (vi), 83.
111. Meprathu, B.V., Justik, M.W. and Protasiewicz, J.D. (2005) *Tetrahedron Letters*, **46**, 5187.
112. Schardt, B.C. and Hill, C.L. (1983) *Inorganic Chemistry*, **22**, 1563.
113. McQuaid, K.M. and Pettus, T.R.R. (2004) *Synlett*, 2403.
114. Lucas, H.J., Kennedy, E.R. and Formo, M.W. (1955) *Organic Syntheses Collective Volume* **3**, 483.
115. Stang, P.J., Zhdankin, V.V. and Zefirov, N.S. (1992) *Mendeleev Communications*, **2**, 159.
116. Koposov, A.Y., Netzel, B.C., Yusubov, M.S., *et al.* (2007) *European Journal of Organic Chemistry*, 4475.
117. Nemykin, V.N., Koposov, A.Y., Netzel, B.C., *et al.* (2009) *Inorganic Chemistry*, **48**, 4908.
118. Richter, H.W., Koser, G.F., Incarvito, C.D. and Rheingold, A.L. (2007) *Inorganic Chemistry*, **46**, 5555.

119. Dasent, W.E. and Waddington, T.C. (1960) *Journal of the Chemical Society*, 3350.
120. Bell, R. and Morgan, K.J. (1960) *Journal of the Chemical Society*, 1209.
121. Siebert, H. and Handrich, M. (1976) *Zeitschrift für Anorganische und Allgemeine Chemie*, **426**, 173.
122. Carmalt, C.J., Crossley, J.G., Knight, J.G., *et al.* (1994) *Journal of the Chemical Society, Chemical Communications*, 2367.
123. Barea, G., Maseras, F. and Lledos, A. (2003) *New Journal of Chemistry*, **27**, 811.
124. Ochiai, M. (2006) *Coordination Chemistry Reviews*, **250**, 2771.
125. Ochiai, M., Miyamoto, K., Shiro, M., *et al.* (2003) *Journal of the American Chemical Society*, **125**, 13006.
126. Ochiai, M., Miyamoto, K., Yokota, Y., *et al.* (2005) *Angewandte Chemie, International Edition*, **44**, 75.
127. Macikenas, D., Skrzypczak-Jankun, E. and Protasiewicz, J.D. (2000) *Angewandte Chemie, International Edition*, **39**, 2007.
128. Varvoglis, A. (1992) *The Organic Chemistry of Polycoordinated Iodine*, VCH Publishers, Inc., New York.
129. Sharefkin, J.G. and Saltzman, H. (1973) *Organic Syntheses Collective Volume* **5**, 660.
130. Iinuma, M., Moriyama, K. and Togo, H. (2012) *Synlett*, **23**, 2663.
131. Hossain, M.D. and Kitamura, T. (2005) *Journal of Organic Chemistry*, **70**, 6984.
132. McKillop, A. and Kemp, D. (1989) *Tetrahedron*, **45**, 3299.
133. Togo, H., Nabana, T. and Yamaguchi, K. (2000) *Journal of Organic Chemistry*, **65**, 8391.
134. Dohi, T., Maruyama, A., Takenage, N., *et al.* (2008) *Angewandte Chemie, International Edition*, **47**, 3787.
135. Zagulyaeva, A.A., Yusubov, M.S. and Zhdankin, V.V. (2010) *Journal of Organic Chemistry*, **75**, 2119.
136. Schaefer, S. and Wirth, T. (2010) *Angewandte Chemie, International Edition*, **49**, 2786.
137. Mironova, A.A., Soloshonok, I.V., Maletina, I.I., *et al.* (1988) *Zhurnal Organicheskoi Khimii*, **24**, 593.
138. Zhdankin, V.V., Erickson, S.A. and Hanson, K.J. (1997) *Journal of the American Chemical Society*, **119**, 4775.
139. Spyroudis, S. and Varvoglis, A. (1975) *Synthesis*, 445.
140. Togo, H. and Sakuratani, K. (2002) *Synlett*, 1966.
141. Ficht, S., Mulbaier, M. and Giannis, A. (2001) *Tetrahedron*, **57**, 4863.
142. Chen, F.-E., Xie, B., Zhang, P., *et al.* (2007) *Synlett*, 619.
143. Shang, Y., But, T.Y.S., Togo, H. and Toy, P.H. (2007) *Synlett*, 67.
144. Qian, W., Jin, E., Bao, W. and Zhang, Y. (2005) *Angewandte Chemie, International Edition*, **44**, 952.
145. Handy, S.T. and Okello, M. (2005) *Journal of Organic Chemistry*, **70**, 2874.
146. Zhdankin, V.V., Scheuller, M.C. and Stang, P.J. (1993) *Tetrahedron Letters*, **34**, 6853.
147. Stang, P.J. and Zhdankin, V.V. (1993) *Journal of the American Chemical Society*, **115**, 9808.
148. Gallop, P.M., Paz, M.A., Fluckiger, R., *et al.* (1993) *Journal of the American Chemical Society*, **115**, 11702.
149. Kazmierczak, P., Skulski, L. and Kraszkiewicz, L. (2001) *Molecules*, **6**, 881.
150. Lee, B.C., Lee, K.C., Lee, H., *et al.* (2007) *Bioconjugate Chemistry*, **18**, 514.
151. Ross, T.L., Ermert, J., Hocke, C. and Coenen, H.H. (2007) *Journal of the American Chemical Society*, **129**, 8018.
152. Kazmierczak, P. and Skulski, L. (1998) *Synthesis*, 1721.
153. Zielinska, A. and Skulski, L. (2002) *Molecules*, **7**, 806.
154. Tohma, H., Maruyama, A., Maeda, A., *et al.* (2004) *Angewandte Chemie, International Edition*, **43**, 3595.
155. Dohi, T., Maruyama, A., Yoshimura, M., *et al.* (2005) *Chemical Communications*, 2205.
156. Dohi, T., Morimoto, K., Takenaga, N., *et al.* (2006) *Chemical & Pharmaceutical Bulletin*, **54**, 1608.
157. Dohi, T., Morimoto, K., Takenaga, N., *et al.* (2007) *Journal of Organic Chemistry*, **72**, 109.
158. Moroda, A. and Togo, H. (2006) *Tetrahedron*, **62**, 12408.
159. Hossain, M.D. and Kitamura, T. (2006) *Bulletin of the Chemical Society of Japan*, **79**, 142.
160. Hossain, D. and Kitamura, T. (2005) *Synthesis*, 1932.
161. Page, T.K. and Wirth, T. (2006) *Synthesis*, 3153.
162. Ochiai, M., Takaoka, Y., Masaki, Y., *et al.* (1990) *Journal of the American Chemical Society*, **112**, 5677.
163. Ochiai, M., Oshima, K., Ito, T., *et al.* (1991) *Tetrahedron Letters*, **32**, 1327.
164. Rocaboy, C. and Gladysz, J.A. (2003) *Chemistry: A European Journal*, **9**, 88.
165. Fujita, M., Okuno, S., Lee, H.J., *et al.* (2007) *Tetrahedron Letters*, **48**, 8691.
166. Hirt, U.H., Schuster, M.F.H., French, A.N., *et al.* (2001) *European Journal of Organic Chemistry*, 1569.
167. Fujita, M., Wakita, M. and Sugimura, T. (2011) *Chemical Communications*, **47**, 3983.

168. Fujita, M., Yoshida, Y., Miyata, K., *et al.* (2010) *Angewandte Chemie, International Edition*, **49**, 7068.
169. Uyanik, M., Yasui, T. and Ishihara, K. (2010) *Angewandte Chemie, International Edition*, **49**, 2175.
170. Uyanik, M., Yasui, T. and Ishihara, K. (2010) *Tetrahedron*, **66**, 5841.
171. Hossain, M.D. and Kitamura, T. (2006) *Tetrahedron Letters*, **47**, 7889.
172. Kasumov, T.M., Brel, V.K., Grishin, Y.K., *et al.* (1997) *Tetrahedron*, **53**, 1145.
173. Yagupolskii, L.M., Maletina, I.I., Kondratenko, N.V. and Orda, V.V. (1978) *Synthesis*, 835.
174. Umemoto, T. (1996) *Chemical Reviews*, **96**, 1757.
175. Umemoto, T. and Goto, Y. (1987) *Bulletin of the Chemical Society of Japan*, **60**, 3307.
176. Zhdankin, V.V., Kuehl, C.J. and Simonsen, A.J. (1996) *Journal of Organic Chemistry*, **61**, 8272.
177. Morris, D.G. and Shepherd, A.G. (1981) *Journal of the Chemical Society, Chemical Communications*, 1250.
178. Stang, P.J., Boehshar, M., Wingert, H. and Kitamura, T. (1988) *Journal of the American Chemical Society*, **110**, 3272.
179. Sutherland, A. and Vederas, J.C. (2002) *Chemical Communications*, 224.
180. Ray, D.G. III and Koser, G.F. (1992) *Journal of Organic Chemistry*, **57**, 1607.
181. Koposov, A.Y., Boyarskikh, V.V. and Zhdankin, V.V. (2004) *Organic Letters*, **6**, 3613.
182. Merkushev, E.B., Novikov, A.N., Makarchenko, S.S., *et al.* (1975) *Journal of Organic Chemistry USSR (English Translation)*, **11**, 1246.
183. Das, J.P., Roy, U.K. and Roy, S. (2005) *Organometallics*, **24**, 6136.
184. Sheremetev, A.B. and Konkina, S.M. (2003) *Mendeleev Communications*, **13**, 277.
185. Katritzky, A.R., Savage, G.P., Gallos, J.K. and Durst, H.D. (1990) *Journal of the Chemical Society, Perkin Transactions 2*, 1515.
186. Yusubov, M.S., Funk, T.V., Chi, K.-W., *et al.* (2008) *Journal of Organic Chemistry*, **73**, 295.
187. Togo, H., Aoki, M. and Yokoyama, M. (1993) *Tetrahedron*, **49**, 8241.
188. Alcock, N.W., Countryman, R.M., Esperas, S. and Sawyer, J.F. (1979) *Journal of the Chemical Society, Dalton Transactions*, 854.
189. Alcock, N.W., Harrison, W.D. and Howes, C. (1984) *Journal of the Chemical Society, Dalton Transactions*, 1709.
190. Bardan, M., Birchall, T., Frampton, C.S. and Kapoor, P. (1989) *Canadian Journal of Chemistry*, **67**, 1878.
191. Hirt, U.H., Spingler, B. and Wirth, T. (1998) *Journal of Organic Chemistry*, **63**, 7674.
192. Gallos, J., Varvoglis, A. and Alcock, N.W. (1985) *Journal of the Chemical Society, Perkin Transactions 1*, 757.
193. Cerioni, G. and Uccheddu, G. (2004) *Tetrahedron Letters*, **45**, 505.
194. Mocci, F., Uccheddu, G., Frongia, A. and Cerioni, G. (2007) *Journal of Organic Chemistry*, **72**, 4163.
195. Koser, G.F. and Wettach, R.H. (1977) *Journal of Organic Chemistry*, **42**, 1476.
196. Neilands, O. and Karele, B. (1970) *Zhurnal Organicheskoi Khimii*, **6**, 885.
197. Koser, G.F. and Wettach, R.H. (1980) *Journal of Organic Chemistry*, **45**, 1542.
198. Koser, G.F., Wettach, R.H. and Smith, C.S. (1980) *Journal of Organic Chemistry*, **45**, 1543.
199. Koser, G.F., Rebrovic, L. and Wettach, R.H. (1981) *Journal of Organic Chemistry*, **46**, 4324.
200. Rebrovic, L. and Koser, G.F. (1984) *Journal of Organic Chemistry*, **49**, 2462.
201. Koser, G.F., Relenyi, A.G., Kalos, A.N., *et al.* (1982) *Journal of Organic Chemistry*, **47**, 2487.
202. Yusubov, M.S. and Wirth, T. (2005) *Organic Letters*, **7**, 519.
203. Yamamoto, Y. and Togo, H. (2005) *Synlett*, 2486.
204. Merritt, E.A., Carneiro, V.M.T., Silva, L.F. Jr. and Olofsson, B. (2010) *Journal of Organic Chemistry*, **75**, 7416.
205. Papoutsis, I., Spyroudis, S., Varvoglis, A. and Raptopoulou, C.P. (1997) *Tetrahedron*, **53**, 6097.
206. Nabana, T. and Togo, H. (2002) *Journal of Organic Chemistry*, **67**, 4362.
207. Kuehl, C.J., Bolz, J.T. and Zhdankin, V.V. (1995) *Synthesis*, 312.
208. Hatzigrigoriou, E., Varvoglis, A. and Bakola-Christianopoulou, M. (1990) *Journal of Organic Chemistry*, **55**, 315.
209. Chun, J.-H., Lu, S. and Pike, V.W. (2011) *European Journal of Organic Chemistry*, 4439.
210. Chun, J.-H., Lu, S., Lee, Y.-S. and Pike, V.W. (2010) *Journal of Organic Chemistry*, **75**, 3332.
211. Zhdankin, V.V. and Kuehl, C. (1994) *Tetrahedron Letters*, **35**, 1809.
212. Zhdankin, V.V., Kuehl, C.J. and Simonsen, A.J. (1995) *Tetrahedron Letters*, **36**, 2203.
213. Abe, S., Sakuratani, K. and Togo, H. (2001) *Synlett*, 22.
214. Abe, S., Sakuratani, K. and Togo, H. (2001) *Journal of Organic Chemistry*, **66**, 6174.

215. Kikugawa, Y. and Kawase, M. (1990) *Chemistry Letters*, 581.
216. Ray, D.G. III and Koser, G.F. (1990) *Journal of the American Chemical Society*, **112**, 5672.
217. Koser, G.F., Wettach, R.H., Troup, J.M. and Frenz, B.A. (1976) *Journal of Organic Chemistry*, **41**, 3609.
218. Richter, H.W., Paul, N.M., Ray, D.G., *et al.* (2010) *Inorganic Chemistry*, **49**, 5413.
219. Richter, H.W., Cherry, B.R., Zook, T.D. and Koser, G.F. (1997) *Journal of the American Chemical Society*, **119**, 9614.
220. Zefirov, N.S., Zhdankin, V.V. and Koz'min, A.S. (1983) *Izvestiya Akademii Nauk SSSR, Seriya Khimicheskaya*, 1682.
221. Zefirov, N.S., Sorokin, V.D., Zhdankin, V.V. and Koz'min, A.S. (1986) *Russian Journal of Organic Chemistry*, **22**, 450.
222. Schmeisser, M., Lehmann, E. and Naumann, D. (1977) *Chemische Berichte*, **110**, 2665.
223. Hembre, R.T., Scott, C.P. and Norton, J.R. (1987) *Journal of Organic Chemistry*, **52**, 3650.
224. Kitamura, T., Inoue, D., Wakimoto, I., *et al.* (2004) *Tetrahedron*, **60**, 8855.
225. Zhdankin, V.V., Tykwinski, R., Caple, R., *et al.* (1988) *Tetrahedron Letters*, **29**, 3717.
226. Zhdankin, V.V., Tykwinski, R., Berglund, B., *et al.* (1989) *Journal of Organic Chemistry*, **54**, 2609.
227. Van Look, G. (1989) *Chemical & Engineering News*, **67**, 2.
228. Bushnell, G.W., Fischer, A. and Ibrahim, P.N. (1988) *Journal of the Chemical Society. Perkin Transactions 2*, 1281.
229. Bassindale, A.R., Katampe, I. and Taylor, P.G. (2000) *Canadian Journal of Chemistry*, **78**, 1479.
230. De Mico, A., Margarita, R., Parlanti, L., *et al.* (1997) *Tetrahedron*, **53**, 16877.
231. Moriarty, R.M., Condeiu, C., Tao, A. and Prakash, O. (1997) *Tetrahedron Letters*, **38**, 2401.
232. Meyer, V. and Wachter, W. (1892) *Chemische Berichte*, **25**, 2632.
233. Willgerodt, C. (1894) *Journal für Praktische Chemie*, **49**, 466.
234. Morales-Rojas, H. and Moss, R.A. (2002) *Chemical Reviews*, **102**, 2497.
235. Brand, J.P., Gonzalez, D.F., Nicolai, S. and Waser, J. (2011) *Chemical Communications*, 102.
236. Wolf, W. and Steinberg, L. (1965) *Journal of the Chemical Society, Chemical Communications*, 449.
237. Amey, R.L. and Martin, J.C. (1979) *Journal of Organic Chemistry*, **44**, 1779.
238. Koser, G.F. (1983) in *The Chemistry of Functional Groups, Suppl. D: Chem. Halides, Pseudo-Halides, Azides* (eds S. Patai and Z. Rappoport), Wiley-Interscience, Chichester, p. 721.
239. Zhdankin, V.V. (2005) *Current Organic Synthesis*, **2**, 121.
240. Zhdankin, V.V. (1997) *Reviews on Heteroatom Chemistry*, **17**, 133.
241. Zhdankin, V.V., Koposov, A.E., Smart, J.T., *et al.* (2001) *Journal of the American Chemical Society*, **123**, 4095.
242. Jeffery, E.A., Andrews, L.J. and Keefer, R.M. (1965) *Journal of Organic Chemistry*, **30**, 617.
243. Nemykin, V.N., Maskaev, A.V., Geraskina, M.R., *et al.* (2011) *Inorganic Chemistry*, **50**, 11263.
244. Koser, G.F., Sun, G., Porter, C.W. and Youngs, W.J. (1993) *Journal of Organic Chemistry*, **58**, 7310.
245. Koposov, A.Y., Litvinov, D.N., Zhdankin, V.V., *et al.* (2006) *European Journal of Organic Chemistry*, 4791.
246. Jaffe, H. and Leffler, J.E. (1975) *Journal of Organic Chemistry*, **40**, 797.
247. Balthazor, T.M., Miles, J.A. and Stults, B.R. (1978) *Journal of Organic Chemistry*, **43**, 4538.
248. Eisenberger, P., Gischig, S. and Togni, A. (2006) *Chemistry: A European Journal*, **12**, 2579.
249. Batchelor, R.J., Birchall, T. and Sawyer, J.F. (1986) *Inorganic Chemistry*, **25**, 1415.
250. Zhdankin, V.V., Maydanovych, O., Herschbach, J., *et al.* (2002) *Journal of the American Chemical Society*, **124**, 11614.
251. Zhdankin, V.V., Krasutsky, A.P., Kuehl, C.J., *et al.* (1996) *Journal of the American Chemical Society*, **118**, 5192.
252. Ochiai, M., Masaki, Y. and Shiro, M. (1991) *Journal of Organic Chemistry*, **56**, 5511.
253. Shefter, E. and Wolf, W. (1965) *Journal of Pharmaceutical Sciences*, **54**, 104.
254. Etter, M.C. (1976) *Journal of the American Chemical Society*, **98**, 5326.
255. Gougoutas, J.Z. and Lessinger, L. (1974) *Journal of Solid State Chemistry*, **9**, 155.
256. Ochiai, M., Ito, T., Masaki, Y. and Shiro, M. (1992) *Journal of the American Chemical Society*, **114**, 6269.
257. Ochiai, M., Ito, T. and Shiro, M. (1993) *Journal of the Chemical Society, Chemical Communications*, 218.
258. Zhdankin, V.V., Kuehl, C.J., Arif, A.M. and Stang, P.J. (1996) *Mendeleev Communications*, **6**, 50.
259. Akai, S., Okuno, T., Takada, T., *et al.* (1996) *Heterocycles*, **42**, 47.
260. Braddock, D.C., Cansell, G., Hermitage, S.A. and White, A.J.P. (2006) *Chemical Communications*, 1442.

261. Kieltsch, I., Eisenberger, P. and Togni, A. (2007) *Angewandte Chemie, International Edition*, **46**, 754.
262. Koser, G.F., McConville, D.B., Rabah, G.A. and Youngs, W.J. (1995) *Journal of Chemical Crystallography*, **25**, 857.
263. Balthazor, T.M., Godar, D.E. and Stults, B.R. (1979) *Journal of Organic Chemistry*, **44**, 1447.
264. Zhdankin, V.V., Arbit, R.M., Lynch, B.J., *et al.* (1998) *Journal of Organic Chemistry*, **63**, 6590.
265. Zhdankin, V.V., Koposov, A.Y., Su, L.S., *et al.* (2003) *Organic Letters*, **5**, 1583.
266. Zhdankin, V.V., Arbit, R.M., McSherry, M., *et al.* (1997) *Journal of the American Chemical Society*, **119**, 7408.
267. Li, X.-Q. and Zhang, C. (2009) *Synthesis*, 1163.
268. Prout, K., Stevens, N.M., Coda, A., *et al.* (1976) *Zeitschrift für Naturforschung B. A Journal of Chemical Sciences*, **31**, 687.
269. Takahashi, M., Nanba, H., Kitazawa, T., *et al.* (1996) *Journal of Coordination Chemistry*, **37**, 371.
270. Legault, C.Y. and Prevost, J. (2012) *Acta Crystallographica, Section E: Structure Reports*, **E68**, o1238.
271. Shefter, E. and Wolf, W. (1964) *Nature*, **203**, 512.
272. Panetta, C.A., Garlick, S.M., Durst, H.D., *et al.* (1990) *Journal of Organic Chemistry*, **55**, 5202.
273. Moss, R.A. and Emge, T.J. (1998) *Chemical Communications*, 1559.
274. Gavina, F., Luis, S.V., Costero, A.M. and Gil, P. (1986) *Tetrahedron*, **42**, 155.
275. Folsom, H.E. and Castrillon, J. (1992) *Synthetic Communications*, **22**, 1799.
276. Moss, R.A., Bracken, K. and Emge, T.J. (1995) *Journal of Organic Chemistry*, **60**, 7739.
277. Moss, R.A., Scrimin, P. and Rosen, R.T. (1987) *Tetrahedron Letters*, **28**, 251.
278. Moss, R.A. and Zhang, H. (1994) *Journal of the American Chemical Society*, **116**, 4471.
279. Moss, R.A., Kim, K.Y. and Swarup, S. (1986) *Journal of the American Chemical Society*, **108**, 788.
280. Baker, G.P., Mann, F.G., Sheppard, N. and Tetlow, A.J. (1965) *Journal of the Chemical Society*, 3721.
281. Ochiai, M., Nakanishi, A. and Suefuji, T. (2000) *Organic Letters*, **2**, 2923.
282. Zhdankin, V.V., Kuehl, C.J., Bolz, J.T., *et al.* (1994) *Tetrahedron Letters*, **35**, 7323.
283. Zhdankin, V.V., Kuehl, C.J., Krasutsky, A.P., *et al.* (1996) *Journal of Organic Chemistry*, **61**, 6547.
284. Merritt, E.A. and Olofsson, B. (2011) *European Journal of Organic Chemistry*, 3690.
285. Ochiai, M., Ito, T., Takahashi, H., *et al.* (1996) *Journal of the American Chemical Society*, **118**, 7716.
286. Dolenc, D. and Plesnicar, B. (1997) *Journal of the American Chemical Society*, **119**, 2628.
287. Sueda, T., Fukuda, S. and Ochiai, M. (2001) *Organic Letters*, **3**, 2387.
288. Ochiai, M., Nakanishi, A. and Ito, T. (1997) *Journal of Organic Chemistry*, **62**, 4253.
289. Ochiai, M., Kajishima, D. and Sueda, T. (1999) *Tetrahedron Letters*, **40**, 5541.
290. Ochiai, M., Kajishima, D. and Sueda, T. (1997) *Heterocycles*, **46**, 71.
291. Ochiai, M., Nakanishi, A. and Yamada, A. (1997) *Tetrahedron Letters*, **38**, 3927.
292. Zhdankin, V.V., Kuehl, C.J., Krasutsky, A.P., *et al.* (1994) *Tetrahedron Letters*, **35**, 9677.
293. Krasutsky, A.P., Kuehl, C.J. and Zhdankin, V.V. (1995) *Synlett*, 1081.
294. Rabah, G.A. and Koser, G.F. (1996) *Tetrahedron Letters*, **37**, 6453.
295. Zhdankin, V.V., McSherry, M., Mismash, B., *et al.* (1997) *Tetrahedron Letters*, **38**, 21.
296. Zhdankin, V.V., Kuehl, C.J., Krasutsky, A.P., *et al.* (1995) *Tetrahedron Letters*, **36**, 7975.
297. Zhdankin, V.V., Persichini, P.J. III, Cui, R. and Jin, Y. (2000) *Synlett*, 719.
298. Brand, J.P., Charpentier, J. and Waser, J. (2009) *Angewandte Chemie, International Edition*, **48**, 9346.
299. Fernandez Gonzalez, D., Brand, J.P. and Waser, J. (2010) *Chemistry: A European Journal*, **16**, 9457.
300. Nicolai, S., Erard, S., Gonzalez, D.F. and Waser, J. (2010) *Organic Letters*, **12**, 384.
301. Brand, J.P., Chevalley, C. and Waser, J. (2011) *Beilstein Journal of Organic Chemistry*, **7**, 565.
302. Nicolai, S., Piemontesi, C. and Waser, J. (2011) *Angewandte Chemie, International Edition*, **50**, 4680.
303. Brand, J.P., Chevalley, C., Scopelliti, R. and Waser, J. (2012) *Chemistry: A European Journal*, **18**, 5655.
304. Fieser, L.F. and Haddadin, M.J. (1966) *Organic Syntheses*, **46**, 107.
305. Beringer, F.M. and Lillien, I. (1960) *Journal of the American Chemical Society*, **82**, 725.
306. Bonilha, J.B.S., Petragnani, N. and Toscano, V.G. (1978) *Chemische Berichte*, **111**, 2510.
307. Del Mazza, D. and Reinecke, M.G. (1988) *Journal of Organic Chemistry*, **53**, 5799.
308. Miller, R.D., Franz, L. and Fickes, G.N. (1985) *Journal of Organic Chemistry*, **50**, 3200.
309. Yusubov, M.S., Yusubova, R.Y., Nemykin, V.N. and Zhdankin, V.V. (2013) *Journal of Organic Chemistry*, **78**, 3767.

310. Morrison, G.F. and Hooz, J. (1970) *Journal of Organic Chemistry*, **35**, 1196.
311. Stanek, K., Koller, R. and Togni, A. (2008) *Journal of Organic Chemistry*, **73**, 7678.
312. Armanino, N., Koller, R. and Togni, A. (2010) *Organometallics*, **29**, 1771.
313. Capone, S., Kieltsch, I., Flogel, O., *et al.* (2008) *Helvetica Chimica Acta*, **91**, 2035.
314. Cvengros, J., Stolz, D. and Togni, A. (2009) *Synthesis*, 2818.
315. Eisenberger, P., Kieltsch, I., Armanino, N. and Togni, A. (2008) *Chemical Communications*, 1575.
316. Fantasia, S., Welch, J.M. and Togni, A. (2010) *Journal of Organic Chemistry*, **75**, 1779.
317. Ibrahim, H., Kleinbeck, F. and Togni, A. (2004) *Helvetica Chimica Acta*, **87**, 605.
318. Kieltsch, I., Eisenberger, P., Stanek, K. and Togni, A. (2008) *Chimia*, **62**, 260.
319. Koller, R., Huchet, Q., Battaglia, P., *et al.* (2009) *Chemical Communications*, 5993.
320. Niedermann, K., Frueh, N., Vinogradova, E., *et al.* (2011) *Angewandte Chemie, International Edition*, **50**, 1059.
321. Niedermann, K., Welch, J.M., Koller, R., *et al.* (2010) *Tetrahedron*, **66**, 5753.
322. Santschi, N. and Togni, A. (2011) *Journal of Organic Chemistry*, **76**, 4189.
323. Wiehn, M.S., Vinogradova, E.V. and Togni, A. (2010) *Journal of Fluorine Chemistry*, **131**, 951.
324. Mejia, E. and Togni, A. (2012) *ACS Catalysis*, **2**, 521.
325. Allen, A.E. and MacMillan, D.W.C. (2010) *Journal of the American Chemical Society*, **132**, 4986.
326. Li, Y. and Studer, A. (2012) *Angewandte Chemie, International Edition*, **51**, 8221.
327. Zhang, W., Zhu, J. and Hu, J. (2008) *Tetrahedron Letters*, **49**, 5006.
328. Moss, R.A., Chatterjee, S. and Wilk, B. (1986) *Journal of Organic Chemistry*, **51**, 4303.
329. Barber, H.J. and Henderson, M.A. (1970) *Journal of the Chemical Society, C*, 862.
330. Naae, D.G. and Gougoutas, J.Z. (1975) *Journal of Organic Chemistry*, **40**, 2129.
331. Kiprof, P. and Zhdankin, V. (2003) *ARKIVOC*, (vi), 170.
332. Willgerodt, C. (1914) *Die Organischen Verbindungen mit Mehrwertigen Jod*, Ferdinand Enke Verlag, Stuttgart.
333. Justik, M.W., Protasiewicz, J.D. and Updegraff, J.B. (2009) *Tetrahedron Letters*, **50**, 6072.
334. Freedman, L.D. and DeMott, R.P. (1974) *Phosphorus and Related Group V Elements*, **3**, 277.
335. Moss, R.A., Bose, S. and Krogh-Jespersen, K. (1997) *Journal of Physical Organic Chemistry*, **10**, 27.
336. Thiele, J. and Peter, W. (1910) *Justus Liebigs Annalen der Chemie*, **369**, 119.
337. Moss, R.A., Wilk, B., Krogh-Jespersen, K., *et al.* (1989) *Journal of the American Chemical Society*, **111**, 250.
338. Brainina, E.M. and Freidlina, R.K. (1950) *Izvestiya Akademii Nauk SSSR, Seriya Khimicheskaya*, 315.
339. Shah, A.-u.-H.A., Khan, Z.A., Choudhary, N., *et al.* (2009) *Organic Letters*, **11**, 3578.
340. Richardson, R.D., Zayed, J.M., Altermann, S., *et al.* (2007) *Angewandte Chemie, International Edition*, **46**, 6529.
341. Moss, R.A., Wilk, B., Krogh-Jespersen, K. and Westbrook, J.D. (1989) *Journal of the American Chemical Society*, **111**, 6729.
342. Leffler, J.E., Dyall, L.K. and Inward, P.W. (1963) *Journal of the American Chemical Society*, **85**, 3443.
343. Moss, R.A., Zhang, H., Chatterjee, S. and Krogh-Jespersen, K. (1993) *Tetrahedron Letters*, **34**, 1729.
344. Quideau, S., Lyvinec, G., Marguerit, M., *et al.* (2009) *Angewandte Chemie, International Edition*, **48**, 4605.
345. Moorthy, J.N., Senapati, K. and Parida, K.N. (2010) *Journal of Organic Chemistry*, **75**, 8416.
346. Leffler, J.E. and Jaffe, H. (1973) *Journal of Organic Chemistry*, **38**, 2719.
347. Agosta, W.C. (1965) *Tetrahedron Letters*, 2681.
348. Martin, J.C. and Chau, M.M. (1974) *Journal of the American Chemical Society*, **96**, 3319.
349. Tian, J., Gao, W.-C., Zhou, D.-M. and Zhang, C. (2012) *Organic Letters*, **14**, 3020.
350. Weclas-Henderson, L., Nguyen, T.T., Hayes, R.A. and Martin, J.C. (1991) *Journal of Organic Chemistry*, **56**, 6565.
351. Nguyen, T.T., Amey, R.L. and Martin, J.C. (1982) *Journal of Organic Chemistry*, **47**, 1024.
352. Kawashima, T., Hoshiba, K. and Kano, N. (2001) *Phosphorus, Sulfur Silicon and the Related Elements*, **168–169**, 141.
353. Kawashima, T., Hoshiba, K. and Kano, N. (2001) *Journal of the American Chemical Society*, **123**, 1507.
354. Nguyen, T.T., Wilson, S.R. and Martin, J.C. (1986) *Journal of the American Chemical Society*, **108**, 3803.
355. Mascarelli, L. and Benati, G. (1909) *Gazzetta Chimica Italiana*, **38**, 619.
356. Heaney, H. and Lees, P. (1968) *Tetrahedron*, **24**, 3717.
357. Collette, J., McGreer, D., Crawford, R., *et al.* (1956) *Journal of the American Chemical Society*, **78**, 3819.
358. Berthon, J.-L., Dias, M., Mornet, R. and Camadro, J.-M. (2000) *Journal of Labelled Compounds and Radiopharmaceuticals*, **43**, 515.

359. Beringer, F.M., Ganis, P., Avitabile, G. and Jaffe, H. (1972) *Journal of Organic Chemistry*, **37**, 879.
360. Khotsyanova, T.L. (1956) *Doklady Akademii Nauk SSSR*, **110**, 71.
361. Grushin, V.V. (2000) *Chemical Society Reviews*, **29**, 315.
362. Beringer, F.M., Chang, L.L., Fenster, A.N. and Rossi, R.R. (1969) *Tetrahedron*, **25**, 4339.
363. Beringer, F.M., Kravetz, L. and Topliss, G.B. (1965) *Journal of Organic Chemistry*, **30**, 1141.
364. Sandin, R.B. (1969) *Journal of Organic Chemistry*, **34**, 456.
365. Sandel, V.R., Buske, G.R., Maroldo, S.G., *et al.* (1981) *Journal of Organic Chemistry*, **46**, 4069.
366. Radhakrishnan, U. and Stang, P.J. (2003) *Journal of Organic Chemistry*, **68**, 9209.
367. Hinkle, R.J., Poulter, G.T. and Stang, P.J. (1993) *Journal of the American Chemical Society*, **115**, 11626.
368. Hartmann, C. and Meyer, V. (1894) *Berichte der Deutschen Chemischen Gesellschaft*, **27**, 426.
369. Merritt, E.A. and Olofsson, B. (2009) *Angewandte Chemie, International Edition*, **48**, 9052.
370. Yusubov, M.S., Maskaev, A.V. and Zhdankin, V.V. (2011) *ARKIVOC*, (i), 370.
371. Beringer, F.M., Falk, R.A., Karniol, M., *et al.* (1959) *Journal of the American Chemical Society*, **81**, 342.
372. Beringer, F.M., Drexler, M., Gindler, E.M. and Lumpkin, C.C. (1953) *Journal of the American Chemical Society*, **75**, 2705.
373. Zefirov, N.S., Kasumov, T.M., Koz'min, A.S., *et al.* (1993) *Synthesis*, 1209.
374. Stang, P.J., Zhdankin, V.V., Tykwinski, R. and Zefirov, N.S. (1991) *Tetrahedron Letters*, **32**, 7497.
375. Kasumov, T.M., Brel, V.K., Koz'min, A.S. and Zefirov, N.S. (1995) *Synthesis*, 775.
376. Pike, V.W., Butt, F., Shah, A. and Widdowson, D.A. (1999) *Journal of the Chemical Society, Perkin Transactions 1*, 245.
377. Carroll, M.A., Pike, V.W. and Widdowson, D.A. (2000) *Tetrahedron Letters*, **41**, 5393.
378. Carman, C.S. and Koser, G.F. (1983) *Journal of Organic Chemistry*, **48**, 2534.
379. Margida, A.J. and Koser, G.F. (1984) *Journal of Organic Chemistry*, **49**, 3643.
380. Kitamura, T., Matsuyuki, J., Nagata, K., *et al.* (1992) *Synthesis*, 945.
381. Shah, A., Pike, V.W. and Widdowson, D.A. (1997) *Journal of the Chemical Society, Perkin Transactions 1*, 2463.
382. Hossain, M.D. and Kitamura, T. (2006) *Tetrahedron*, **62**, 6955.
383. Hossain, M.D., Ikegami, Y. and Kitamura, T. (2006) *Journal of Organic Chemistry*, **71**, 9903.
384. Hossain, M.D. and Kitamura, T. (2007) *Bulletin of the Chemical Society of Japan*, **80**, 2213.
385. Bielawski, M., Zhu, M. and Olofsson, B. (2007) *Advanced Synthesis & Catalysis*, **349**, 2610.
386. Bielawski, M. and Olofsson, B. (2007) *Chemical Communications*, 2521.
387. Zhu, M., Jalalian, N. and Olofsson, B. (2008) *Synlett*, 592.
388. Bielawski, M., Aili, D. and Olofsson, B. (2008) *Journal of Organic Chemistry*, **73**, 4602.
389. Jalalian, N. and Olofsson, B. (2010) *Tetrahedron*, **66**, 5793.
390. Merritt, E.A., Malmgren, J., Klinke, F.J. and Olofsson, B. (2009) *Synlett*, 2277.
391. Dohi, T., Yamaoka, N., Itani, I. and Kita, Y. (2011) *Australian Journal of Chemistry*, **64**, 529.
392. Dohi, T., Yamaoka, N. and Kita, Y. (2010) *Tetrahedron*, **66**, 5775.
393. Ochiai, M., Toyonari, M., Sueda, T. and Kitagawa, Y. (1996) *Tetrahedron Letters*, **37**, 8421.
394. Chen, D.-W. and Ochiai, M. (1999) *Journal of Organic Chemistry*, **64**, 6804.
395. Kalyani, D., Deprez, N.R., Desai, L.V. and Sanford, M.S. (2005) *Journal of the American Chemical Society*, **127**, 7330.
396. Yoshida, M., Osafune, K. and Hara, S. (2007) *Synthesis*, 1542.
397. Ochiai, M., Ito, T., Takaoka, Y. and Masaki, Y. (1991) *Journal of the American Chemical Society*, **113**, 1319.
398. Helber, J., Frohn, H.-J., Klose, A. and Scholten, T. (2003) *ARKIVOC*, (vi), 71.
399. Stang, P.J., Tykwinski, R. and Zhdankin, V. (1992) *Journal of Heterocyclic Chemistry*, **29**, 815.
400. Stang, P.J., Zhdankin, V.V., Tykwinski, R. and Zefirov, N.S. (1992) *Tetrahedron Letters*, **33**, 1419.
401. Bykowski, D., McDonald, R., Hinkle, R.J. and Tykwinski, R.R. (2002) *Journal of Organic Chemistry*, **67**, 2798.
402. Tykwinski, R.R., Kamada, K., Bykowski, D., *et al.* (2000) *Advanced Materials*, **12**, 133.
403. Beringer, F.M. and Nathan, R.A. (1970) *Journal of Organic Chemistry*, **35**, 2095.
404. Stang, P.J. and Chen, K. (1995) *Journal of the American Chemical Society*, **117**, 1667.
405. Kitamura, T., Kotani, M. and Fujiwara, Y. (1996) *Tetrahedron Letters*, **37**, 3721.
406. Pirguliyev, N.S., Brel, V.K., Akhmedov, N.G. and Zefirov, N.S. (2000) *Synthesis*, 81.

407. Peacock, M.J. and Pletcher, D. (2001) *Journal of The Electrochemical Society*, **148**, D37.
408. Peacock, M.J. and Pletcher, D. (2000) *Tetrahedron Letters*, **41**, 8995.
409. Kotali, E., Varvoglis, A. and Bozopoulos, A. (1989) *Journal of the Chemical Society, Perkin Transactions 1*, 827.
410. Maas, G., Regitz, M., Moll, U., *et al.* (1992) *Tetrahedron*, **48**, 3527.
411. Stang, P.J. and Murch, P. (1997) *Tetrahedron Letters*, **38**, 8793.
412. Chernov'yants, M.S., Burykin, I.V., Starikova, Z.A. and Rostovskaya, A.A. (2012) *Russian Journal of Inorganic Chemistry*, **57**, 193.
413. Lee, Y.-S., Hodoscek, M., Chun, J.-H. and Pike, V.W. (2010) *Chemistry: A European Journal*, **16**, 10418.
414. Liu, C.Y., Li, H. and Meng, A.G. (2007) *Acta Crystallographica, Section E: Structure Reports*, **63**, o3647.
415. Li, H. and Jiang, S. (2007) *Acta Crystallographica, Section E: Structure Reports*, **63**, o83.
416. Suefuji, T., Shiro, M., Yamaguchi, K. and Ochiai, M. (2006) *Heterocycles*, **67**, 391.
417. Ochiai, M., Suefuji, T., Miyamoto, K. and Shiro, M. (2003) *Chemical Communications*, 1438.
418. Ochiai, M., Suefuji, T., Miyamoto, K., *et al.* (2003) *Journal of the American Chemical Society*, **125**, 769.
419. Kaafarani, B.R., Gu, H., Pinkerton, A.A. and Neckers, D.C. (2002) *Journal of the Chemical Society, Dalton Transactions*, 2318.
420. Zhukhlistova, N.E., Tishchenko, G.N., Tolstaya, T.P. and Asulyan, L.D. (2001) *Kristallografiya*, **46**, 698.
421. Bailly, F., Barthen, P., Frohn, H.J. and Kockerling, M. (2000) *Zeitschrift für Anorganische und Allgemeine Chemie*, **626**, 2419.
422. Kasumov, T.M., Brel, V.K., Potekhin, K.A., *et al.* (1996) *Doklady Akademii Nauk*, **349**, 634.
423. Bozopoulos, A.P. and Rentzeperis, P.J. (1987) *Acta Crystallographica, Section C Crystal Structure Communications*, **43**, 914.
424. Alcock, N.W. and Countryman, R.M. (1987) *Journal of the Chemical Society, Dalton Transactions*, 193.
425. Alcock, N.W. and Countryman, R.M. (1977) *Journal of the Chemical Society, Dalton Transactions*, 217.
426. Khotsyanova, T.L., Babushkina, T.A., Saatsazov, V.V., *et al.* (1976) *Koordinatsionnaya Khimiya*, **2**, 1567.
427. Wright, W.B. and Meyers, E.A. (1972) *Crystal Structure Communications*, **1**, 95.
428. Struchkov, Y.T. and Khotsyanova, T.L. (1960) *Izvestiya Akademii Nauk SSSR, Seriya Khimicheskaya*, 821.
429. Murray, S.J., Mueller-Bunz, H. and Ibrahim, H. (2012) *Chemical Communications*, **48**, 6268.
430. Beringer, F.M. and Mausner, M. (1958) *Journal of the American Chemical Society*, **80**, 4535.
431. Ochiai, M. (2000) *Journal of Organometallic Chemistry*, **611**, 494.
432. Okuyama, T. (2002) *Accounts of Chemical Research*, **35**, 12.
433. Okuyama, T. and Fujita, M. (2005) *Russian Journal of Organic Chemistry*, **41**, 1245.
434. Okuyama, T. and Fujita, M. (2007) *ACS Symposium Series*, **965**, 68.
435. Pirkuliev, N.S., Brel, V.K. and Zefirov, N.S. (2000) *Russian Chemical Reviews*, **69**, 105.
436. Thiele, J. and Haakh, H. (1910) *Justus Liebigs Annalen der Chemie*, **369**, 131.
437. Ochiai, M., Sumi, K., Takaoka, Y., *et al.* (1988) *Tetrahedron*, **44**, 4095.
438. Ochiai, M., Sumi, K., Nagao, Y. and Fujita, E. (1985) *Tetrahedron Letters*, **26**, 2351.
439. Ochiai, M., Sueda, T., Noda, R. and Shiro, M. (1999) *Journal of Organic Chemistry*, **64**, 8563.
440. Ochiai, M., Toyonari, M., Nagaoka, T., *et al.* (1997) *Tetrahedron Letters*, **38**, 6709.
441. Fujita, M., Lee, H.J. and Okuyama, T. (2006) *Organic Letters*, **8**, 1399.
442. Huang, X. and Xu, X.-H. (1998) *Journal of the Chemical Society, Perkin Transactions 1*, 3321.
443. Hinkle, R.J. and Stang, P.J. (1994) *Synthesis*, 313.
444. McNeil, A.J., Hinkle, R.J., Rouse, E.A., *et al.* (2001) *Journal of Organic Chemistry*, **66**, 5556.
445. Stang, P.J. and Ullmann, J. (1991) *Angewandte Chemie, International Edition*, **30**, 1469.
446. Kitamura, T., Furuki, R., Nagata, K., *et al.* (1992) *Journal of Organic Chemistry*, **57**, 6810.
447. Kitamura, T., Furuki, R., Taniguchi, H. and Stang, P.J. (1992) *Tetrahedron*, **48**, 7149.
448. Kitamura, T., Furuki, R., Taniguchi, H. and Stang, P.J. (1990) *Tetrahedron Letters*, **31**, 703.
449. Hara, S., Yoshida, M., Fukuhara, T. and Yoneda, N. (1998) *Chemical Communications*, 965.
450. Hara, S., Yamamoto, K., Yoshida, M., *et al.* (1999) *Tetrahedron Letters*, **40**, 7815.
451. Yoshida, M., Hara, S., Fukuhara, T. and Yoneda, N. (2000) *Tetrahedron Letters*, **41**, 3887.
452. Ochiai, M., Hirobe, M., Yoshimura, A., *et al.* (2007) *Organic Letters*, **9**, 3335.
453. Yoshida, M. and Hara, S. (2003) *Organic Letters*, **5**, 573.

454. Yoshida, M., Komata, A. and Hara, S. (2006) *Tetrahedron*, **62**, 8636.
455. Justik, M.W., Kristufek, S.L., Protasiewicz, J.D. and Deligonul, N. (2010) *Synthesis*, 2345.
456. Ochiai, M., Uemura, K., Oshima, K., *et al.* (1991) *Tetrahedron Letters*, **32**, 4753.
457. Ochiai, M., Oshima, K. and Masaki, Y. (1991) *Tetrahedron Letters*, **32**, 7711.
458. Williamson, B.L., Stang, P.J. and Arif, A.M. (1993) *Journal of the American Chemical Society*, **115**, 2590.
459. Murch, P., Arif, A.M. and Stang, P.J. (1997) *Journal of Organic Chemistry*, **62**, 5959.
460. Stang, P.J. and Zhdankin, V.V. (1991) *Journal of the American Chemical Society*, **113**, 4571.
461. Kitamura, T., Kotani, M., Yokoyama, T., *et al.* (1999) *Journal of Organic Chemistry*, **64**, 680.
462. Ryan, J.H. and Stang, P.J. (1996) *Journal of Organic Chemistry*, **61**, 6162.
463. Stang, P.J., Schwarz, A., Blume, T. and Zhdankin, V.V. (1992) *Tetrahedron Letters*, **33**, 6759.
464. Stang, P.J., Blume, T. and Zhdankin, V.V. (1993) *Synthesis*, 35.
465. Gately, D.A., Luther, T.A., Norton, J.R., *et al.* (1992) *Journal of Organic Chemistry*, **57**, 6496.
466. Zefirov, N.S., Koz'min, A.S., Kasumov, T., *et al.* (1992) *Journal of Organic Chemistry*, **57**, 2433.
467. Stang, P.J., Wingert, H. and Arif, A.M. (1987) *Journal of the American Chemical Society*, **109**, 7235.
468. Ochiai, M., Suefuji, T., Miyamoto, K. and Shiro, M. (2005) *Organic Letters*, **7**, 2893.
469. Beringer, F.M. and Galton, S.A. (1965) *Journal of Organic Chemistry*, **30**, 1930.
470. Merkushev, E.B., Karpitskaya, L.G. and Novosel'tseva, G.I. (1979) *Doklady Akademii Nauk SSSR*, **245**, 607.
471. Margida, A.J. and Koser, G.F. (1984) *Journal of Organic Chemistry*, **49**, 4703.
472. Zhdankin, V.V. and Stang, P.J. (1998) *Tetrahedron*, **54**, 10927.
473. Feldman, K.S. (2003) *ARKIVOC*, (vi), 179.
474. Brand, J.P. and Waser, J. (2012) *Chemical Society Reviews*, **41**, 4165.
475. Feldman, K.S. (2004) in *Strategies and Tactics in Organic Synthesis* vol. **4** (ed. M. Harmata), Elsevier, London, p. 133.
476. Rebrovic, L. and Koser, G.F. (1984) *Journal of Organic Chemistry*, **49**, 4700.
477. Stang, P.J., Surber, B.W., Chen, Z.C., *et al.* (1987) *Journal of the American Chemical Society*, **109**, 228.
478. Stang, P.J. and Surber, B.W. (1985) *Journal of the American Chemical Society*, **107**, 1452.
479. Kitamura, T., Lee, C.H., Taniguchi, Y., *et al.* (1996) *Molecular Crystals and Liquid Crystals Science and Technology, Section A: Molecular Crystals and Liquid Crystals*, **287**, 93.
480. Kitamura, T., Lee, C.H., Taniguchi, Y., *et al.* (1997) *Journal of the American Chemical Society*, **119**, 619.
481. Kitamura, T., Lee, C.H., Taniguchi, H., *et al.* (1994) *Journal of Organic Chemistry*, **59**, 8053.
482. Stang, P.J., Williamson, B.L. and Zhdankin, V.V. (1991) *Journal of the American Chemical Society*, **113**, 5870.
483. Williamson, B.L., Tykwinski, R.R. and Stang, P.J. (1994) *Journal of the American Chemical Society*, **116**, 93.
484. Feldman, K.S., Bruendl, M.M., Schildknegt, K. and Bohnstedt, A.C. (1996) *Journal of Organic Chemistry*, **61**, 5440.
485. Feldman, K.S., Saunders, J.C. and Wrobleski, M.L. (2002) *Journal of Organic Chemistry*, **67**, 7096.
486. Feldman, K.S., Perkins, A.L. and Masters, K.M. (2004) *Journal of Organic Chemistry*, **69**, 7928.
487. Wardrop, D.J. and Fritz, J. (2006) *Organic Letters*, **8**, 3659.
488. Stang, P.J. and Zhdankin, V.V. (1990) *Journal of the American Chemical Society*, **112**, 6437.
489. Stang, P.J., Tykwinski, R. and Zhdankin, V.V. (1992) *Journal of Organic Chemistry*, **57**, 1861.
490. Stang, P.J. and Tykwinski, R. (1992) *Journal of the American Chemical Society*, **114**, 4411.
491. Tykwinski, R.R. and Stang, P.J. (1994) *Organometallics*, **13**, 3203.
492. Stang, P.J. and Ullmann, J. (1991) *Synthesis*, 1073.
493. Koumbis, A.E., Kyzas, C.M., Savva, A. and Varvoglis, A. (2005) *Molecules*, **10**, 1340.
494. Bachi, M.D., Bar-Ner, N., Crittell, C.M., *et al.* (1991) *Journal of Organic Chemistry*, **56**, 3912.
495. Stang, P.J., Arif, A.M. and Crittell, C.M. (1990) *Angewandte Chemie, International Edition*, **29**, 287.
496. Ochiai, M., Kunishima, M., Sumi, K., *et al.* (1985) *Tetrahedron Letters*, **26**, 4501.
497. Ochiai, M., Ito, T., Takaoka, Y., *et al.* (1990) *Journal of the Chemical Society, Chemical Communications*, 118.
498. Kitamura, T. and Stang, P.J. (1988) *Journal of Organic Chemistry*, **53**, 4105.
499. Stang, P.J. and Kitamura, T. (1992) *Organic Syntheses*, **70**, 215.
500. Yoshida, M., Nishimura, N. and Hara, S. (2002) *Chemical Communications*, 1014.
501. Frohn, H.-J. and Bardin, V.V. (2008) *Zeitschrift für Anorganische und Allgemeine Chemie*, **634**, 82.

502. Ochiai, M., Kunishima, M., Fuji, K., *et al.* (1989) *Chemical & Pharmaceutical Bulletin*, **37**, 1948.
503. Bykowski, D., McDonald, R. and Tykwinski, R.R. (2003) *ARKIVOC*, (vi), 21.
504. Shimizu, M., Takeda, Y. and Hiyama, T. (2008) *Chemistry Letters*, **37**, 1304.
505. Stang, P.J., Zhdankin, V.V. and Arif, A.M. (1991) *Journal of the American Chemical Society*, **113**, 8997.
506. Ochiai, M., Miyamoto, K., Suefuji, T., *et al.* (2003) *Angewandte Chemie, International Edition*, **42**, 2191.
507. Ochiai, M., Miyamoto, K., Suefuji, T., *et al.* (2003) *Tetrahedron*, **59**, 10153.
508. Zhdankin, V.V., Crittell, C.M. and Stang, P.J. (1990) *Tetrahedron Letters*, **31**, 4821.
509. Frohn, H.-J., Hirschberg, M.E., Boese, R., *et al.* (2008) *Zeitschrift für Anorganische und Allgemeine Chemie*, **634**, 2539.
510. Olah, G.A. (1975) *Halonium Ions*, John Wiley & Sons, Inc., New York.
511. Minkwitz, R. and Gerhard, V. (1991) *Zeitschrift für Naturforschung B. A Journal of Chemical Sciences*, **46**, 561.
512. Olah, G.A. and DeMember, J.R. (1969) *Journal of the American Chemical Society*, **91**, 2113.
513. Zhdankin, V.V., Tykwinski, R., Caple, R., *et al.* (1988) *Tetrahedron Letters*, **29**, 3703.
514. Zhdankin, V.V., Tykwinski, R., Mullikin, M., *et al.* (1989) *Journal of Organic Chemistry*, **54**, 2605.
515. Moriarty, R.M. and Vaid, R.K. (1990) *Synthesis*, 431.
516. Maletina, I.I., Mironova, A.A., Orda, V.V. and Yagupolskii, L.M. (1993) *Reviews on Heteroatom Chemistry*, **8**, 232.
517. Lyalin, V.V., Orda, V.V., Alekseeva, L.A. and Yagupolskii, L.M. (1971) *Zhurnal Organicheskoi Khimii*, **7**, 1473.
518. Umemoto, T. (1984) *Tetrahedron Letters*, **25**, 81.
519. Umemoto, T., Kuriu, Y. and Nakayama, S. (1982) *Tetrahedron Letters*, **23**, 4101.
520. Umemoto, T., Kuriu, Y. and Shuyama, H. (1981) *Chemistry Letters*, 1663.
521. Umemoto, T., Kuriu, Y. and Miyano, A. (1982) *Tetrahedron Letters*, **23**, 3579.
522. Umemoto, T. and Kuriu, Y. (1981) *Tetrahedron Letters*, **22**, 5197.
523. Umemoto, T. and Gotoh, Y. (1986) *Bulletin of the Chemical Society of Japan*, **59**, 439.
524. Umemoto, T., Kuriu, Y. and Nakayama, S. (1982) *Tetrahedron Letters*, **23**, 1169.
525. Umemoto, T., Kuriu, Y., Nakayama, S.-i. and Miyano, O. (1982) *Tetrahedron Letters*, **23**, 1471.
526. Umemoto, T. and Gotoh, Y. (1991) *Bulletin of the Chemical Society of Japan*, **64**, 2008.
527. Umemoto, T., Kuriu, Y., Shuyama, H., *et al.* (1982) *Journal of Fluorine Chemistry*, **20**, 695.
528. Umemoto, T., Kuriu, Y., Shuyama, H., *et al.* (1986) *Journal of Fluorine Chemistry*, **31**, 37.
529. Zhdankin, V.V., Kuehl, C.J., Bolz, J.T. and Zefirov, N.S. (1994) *Mendeleev Communications*, **4**, 165.
530. Papoutsis, I., Spyroudis, S., Varvoglis, A., *et al.* (1997) *Tetrahedron Letters*, **38**, 8401.
531. DesMarteau, D.D. and Montanari, V. (1998) *Chemical Communications*, 2241.
532. Montanari, V. and Kumar, K. (2004) *Journal of the American Chemical Society*, **126**, 9528.
533. Jubert, A., Okulik, N., Michelini, M.d.C. and Mota, C.J.A. (2008) *Journal of Physical Chemistry A*, **112**, 11468.
534. Noronha, L.A., Judson, T.J.L., Dias, J.F., *et al.* (2006) *Journal of Organic Chemistry*, **71**, 2625.
535. Gudriniece, E., Neiland, O. and Vanags, G. (1957) *Zhurnal obshchei khimii*, **27**, 2737.
536. Kirmse, W. (2005) *European Journal of Organic Chemistry*, 237.
537. Muller, P., Allenbach, Y.F., Chappellet, S. and Ghanem, A. (2006) *Synthesis*, 1689.
538. Muller, P. (2004) *Accounts of Chemical Research*, **37**, 243.
539. Malamidou-Xenikaki, E. and Spyroudis, S. (2008) *Synlett*, 2725.
540. Hadjiarapoglou, L., Spyroudis, S. and Varvoglis, A. (1985) *Journal of the American Chemical Society*, **107**, 7178.
541. Hatjiarapoglou, L., Varvoglis, A., Alcock, N.W. and Pike, G.A. (1988) *Journal of the Chemical Society, Perkin Transactions 1*, 2839.
542. Goudreau, S.R., Marcoux, D. and Charette, A.B. (2009) *Journal of Organic Chemistry*, **74**, 470.
543. Cardinale, J. and Ermert, J. (2013) *Tetrahedron Letters*, **54**, 2067.
544. Zhu, S.-Z. (1994) *Heteroatom Chemistry*, **5**, 9.
545. Hackenberg, J. and Hanack, M. (1991) *Journal of the Chemical Society, Chemical Communications*, 470.
546. Zhu, S. and Chen, Q. (1990) *Journal of the Chemical Society, Chemical Communications*, 1459.
547. Zhu, S., Chen, Q. and Kuang, W. (1993) *Journal of Fluorine Chemistry*, **60**, 39.
548. Goudreau, S.R., Marcoux, D., Charette, A.B. and Hughes, D. (2010) *Organic Syntheses*, **87**, 115.
549. Zhu, C., Yoshimura, A., Ji, L., *et al.* (2012) *Organic Letters*, **14**, 3170.

550. Maletina, I.I., Orda, V.V. and Yagupolskii, Y.L. (1995) *Journal of Fluorine Chemistry*, **70**, 85.
551. Yang, R., Dai, L. and Chen, C. (1992) *Journal of the Chemical Society, Chemical Communications*, 1487.
552. Neilands, O. (2003) *Chemistry of Heterocyclic Compounds (English Translation)*, **39**, 1555.
553. Kokil, P.B. and Nair, P.M. (1977) *Tetrahedron Letters*, 4113.
554. Moriarty, R.M., Tyagi, S., Ivanov, D. and Constantinescu, M. (2008) *Journal of the American Chemical Society*, **130**, 7564.
555. Bakalbassis, E.G., Spyroudis, S. and Tsiotra, E. (2006) *Journal of Organic Chemistry*, **71**, 7060.
556. Kappe, T., Korbuly, G. and Stadlbauer, W. (1978) *Chemische Berichte*, **111**, 3857.
557. Laschober, R. and Kappe, T. (1990) *Synthesis*, 387.
558. Alcock, N.W., Bozopoulos, A.P., Hatzigrigoriou, E. and Varvoglis, A. (1990) *Acta Crystallographica, Section C: Crystal Structure Communications*, **46**, 1300.
559. Nishimura, T., Iwasaki, H., Takahashi, M. and Takeda, M. (2003) *Journal of Radioanalytical and Nuclear Chemistry*, **255**, 499.
560. Papoutsis, I., Spyroudis, S. and Varvoglis, A. (1994) *Tetrahedron Letters*, **35**, 8449.
561. Hatzigrigoriou, E., Spyroudis, S. and Varvoglis, A. (1989) *Liebigs Annalen der Chemie*, 167.
562. Papoutsis, I., Spyroudis, S. and Varvoglis, A. (1996) *Tetrahedron Letters*, **37**, 913.
563. Papoutsis, I., Spyroudis, S. and Varvoglis, A. (1996) *Journal of Heterocyclic Chemistry*, **33**, 579.
564. Prakash, O., Sharma, V. and Tanwar, M.P. (1999) *Canadian Journal of Chemistry*, **77**, 1191.
565. Hadjiarapoglou, L.P. and Schank, K. (1997) *Tetrahedron*, **53**, 9365.
566. Ochiai, M., Tada, N., Okada, T., *et al.* (2008) *Journal of the American Chemical Society*, **130**, 2118.
567. Ochiai, M., Okada, T., Tada, N. and Yoshimura, A. (2008) *Organic Letters*, **10**, 1425.
568. Neilands, O. and Vanags, G. (1964) *Doklady Akademii Nauk SSSR*, **159**, 373.
569. Moriarty, R.M., Prakash, I., Prakash, O. and Freeman, W.A. (1984) *Journal of the American Chemical Society*, **106**, 6082.
570. Weiss, R. and Seubert, J. (1994) *Angewandte Chemie, International Edition*, **33**, 891.
571. Zhdankin, V.V., Maydanovych, O., Herschbach, J., *et al.* (2002) *Tetrahedron Letters*, **43**, 2359.
572. Matveeva, E.D., Podrugina, T.A., Grishin, Y.K., *et al.* (2003) *Russian Journal of Organic Chemistry*, **39**, 536.
573. Zhdankin, V.V., Maydanovych, O., Herschbach, J., *et al.* (2003) *Journal of Organic Chemistry*, **68**, 1018.
574. Huang, Z.-Z., Yu, X.-C. and Huang, X. (2002) *Tetrahedron Letters*, **43**, 6823.
575. Matveeva, E.D., Podrugina, T.A., Taranova, M.A., *et al.* (2012) *Journal of Organic Chemistry*, **77**, 5770.
576. Ochiai, M. and Kitagawa, Y. (1998) *Tetrahedron Letters*, **39**, 5569.
577. Ochiai, M. and Kitagawa, Y. (1999) *Journal of Organic Chemistry*, **64**, 3181.
578. Ochiai, M., Kitagawa, Y. and Yamamoto, S. (1997) *Journal of the American Chemical Society*, **119**, 11598.
579. Ochiai, M., Tuchimoto, Y. and Higashiura, N. (2004) *Organic Letters*, **6**, 1505.
580. Ochiai, M., Nishitani, J. and Nishi, Y. (2002) *Journal of Organic Chemistry*, **67**, 4407.
581. Bonge, H.T. and Hansen, T. (2007) *Synlett*, 55.
582. Bonge, H.T. and Hansen, T. (2008) *Tetrahedron Letters*, **49**, 57.
583. Moreau, B. and Charette, A.B. (2005) *Journal of the American Chemical Society*, **127**, 18014.
584. Wurz, R.P. and Charette, A.B. (2003) *Organic Letters*, **5**, 2327.
585. Huang, Z., Yu, X. and Huang, X. (2002) *Journal of Organic Chemistry*, **67**, 8261.
586. Wittig, G. and Clauss, K. (1952) *Justus Liebigs Annalen der Chemie*, **578**, 136.
587. Wittig, G. and Rieber, M. (1949) *Justus Liebigs Annalen der Chemie*, **562**, 187.
588. Reich, H.J. and Cooperman, C.S. (1973) *Journal of the American Chemical Society*, **95**, 5077.
589. Beringer, F.M. and Chang, L.L. (1971) *Journal of Organic Chemistry*, **36**, 4055.
590. Clauss, K. (1955) *Chemische Berichte*, **88**, 268.
591. Beringer, F.M. and Chang, L.L. (1972) *Journal of Organic Chemistry*, **37**, 1516.
592. Beringer, F.M., Dehn, J.W. Jr. and Winicov, M. (1960) *Journal of the American Chemical Society*, **82**, 2948.
593. Barton, D.H.R., Jaszberenyi, J.C., Lessmann, K. and Timar, T. (1992) *Tetrahedron*, **48**, 8881.
594. Zhdankin, V.V., Tykwinski, R., Williamson, B.L., *et al.* (1991) *Tetrahedron Letters*, **32**, 733.
595. Cech, F. and Zbiral, E. (1975) *Tetrahedron*, **31**, 605.
596. Zbiral, E. and Ehrenfreund, J. (1971) *Tetrahedron*, **27**, 4125.

597. Ehrenfreund, J. and Zbiral, E. (1973) *Liebigs Annalen der Chemie*, 290.
598. Zbiral, E. and Nestler, G. (1970) *Tetrahedron*, **26**, 2945.
599. Moriarty, R.M. and Khosrowshahi, J.S. (1986) *Tetrahedron Letters*, **27**, 2809.
600. Moriarty, R.M. and Khosrowshahi, J.S. (1987) *Synthetic Communications*, **17**, 89.
601. Kita, Y., Tohma, H., Inagaki, M., *et al.* (1991) *Tetrahedron Letters*, **32**, 4321.
602. Kita, Y., Tohma, H., Hatanaka, K., *et al.* (1994) *Journal of the American Chemical Society*, **116**, 3684.
603. Kita, Y., Tohma, H., Takada, T., *et al.* (1994) *Synlett*, 427.
604. Fontana, F., Minisci, F., Yan, Y.M. and Zhao, L. (1993) *Tetrahedron Letters*, **34**, 2517.
605. Tingoli, M., Tiecco, M., Chianelli, D., *et al.* (1991) *Journal of Organic Chemistry*, **56**, 6809.
606. Czernecki, S. and Randriamandimby, D. (1993) *Tetrahedron Letters*, **34**, 7915.
607. Magnus, P., Evans, A. and Lacour, J. (1992) *Tetrahedron Letters*, **33**, 2933.
608. Magnus, P., Roe, M.B. and Hulme, C. (1995) *Journal of the Chemical Society, Chemical Communications*, 263.
609. Magnus, P. and Hulme, C. (1994) *Tetrahedron Letters*, **35**, 8097.
610. Kirschning, A., Domann, S., Draeger, G. and Rose, L. (1995) *Synlett*, 767.
611. Hadjiarapoglou, L., Spyroudis, S. and Varvoglis, A. (1983) *Synthesis*, 207.
612. Papadopoulou, M. and Varvoglis, A. (1983) *Journal of Chemical Research (S)*, 66.
613. Papadopoulou, M. and Varvoglis, A. (1984) *Journal of Chemical Research (S)*, 166.
614. Lazbin, I.M. and Koser, G.F. (1986) *Journal of Organic Chemistry*, **51**, 2669.
615. Lazbin, I.M. and Koser, G.F. (1987) *Journal of Organic Chemistry*, **52**, 476.
616. Roeben, C., Souto, J.A., Gonzalez, Y., *et al.* (2011) *Angewandte Chemie, International Edition*, **50**, 9478.
617. Souto, J.A., Zian, D. and Muñiz, K. (2012) *Journal of the American Chemical Society*, **134**, 7242.
618. Souto, J.A., Martinez, C., Velilla, I. and Muñiz, K. (2013) *Angewandte Chemie, International Edition*, **52**, 1324.
619. Pell, T.P., Couchman, S.A., Ibrahim, S., *et al.* (2012) *Inorganic Chemistry*, **51**, 13034.
620. Pirkuliyev, N.S., Brel, V.K., Zhdankin, V.V. and Zefirov, N.S. (2002) *Russian Journal of Organic Chemistry*, **38**, 1224.
621. Zhdankin, V.V., Koposov, A.Y. and Yashin, N.V. (2002) *Tetrahedron Letters*, **43**, 5735.
622. Weiss, R., Seubert, J. and Hampel, F. (1994) *Angewandte Chemie, International Edition*, **33**, 1952.
623. Dauban, P. and Dodd, R.H. (2003) *Synlett*, 1571.
624. Yamada, Y., Yamamoto, T. and Okawara, M. (1975) *Chemistry Letters*, 361.
625. Boucher, M., Macikenas, D., Ren, T. and Protasiewicz, J.D. (1997) *Journal of the American Chemical Society*, **119**, 9366.
626. Sodergren, M.J., Alonso, D.A., Bedekar, A.V. and Andersson, P.G. (1997) *Tetrahedron Letters*, **38**, 6897.
627. Cicero, R.L., Zhao, D. and Protasiewicz, J.D. (1996) *Inorganic Chemistry*, **35**, 275.
628. Abramovitch, R.A., Bailey, T.D., Takaya, T. and Uma, V. (1974) *Journal of Organic Chemistry*, **39**, 340.
629. Mansuy, D., Mahy, J.P., Dureault, A., *et al.* (1984) *Journal of the Chemical Society, Chemical Communications*, 1161.
630. Meprathu, B.V., Diltz, S., Walsh, P.J. and Protasiewicz, J.D. (1999) *Tetrahedron Letters*, **40**, 5459.
631. Dauban, P. and Dodd, R.H. (1999) *Journal of Organic Chemistry*, **64**, 5304.
632. Meprathu, B.V. and Protasiewicz, J.D. (2010) *Tetrahedron*, **66**, 5768.
633. Macikenas, D., Meprathu, B.V. and Protasiewicz, J.D. (1998) *Tetrahedron Letters*, **39**, 191.
634. Yoshimura, A., Nemykin, V.N. and Zhdankin, V.V. (2011) *Chemistry–A European Journal*, **17**, 10538.
635. Protasiewicz, J.D. (1996) *Acta Crystallographica, Section C: Crystal Structure Communications*, **52**, 1570.
636. Blake, A.J., Novak, A., Davies, M., *et al.* (2009) *Synthetic Communications*, **39**, 1065.
637. Zhdankin, V.V. (2011) *Journal of Organic Chemistry*, **76**, 1185.
638. Ladziata, U. and Zhdankin, V.V. (2006) *ARKIVOC*, (ix), 26.
639. Satam, V., Harad, A., Rajule, R. and Pati, H. (2010) *Tetrahedron*, **66**, 7659.
640. Duschek, A. and Kirsch, S.F. (2011) *Angewandte Chemie, International Edition*, **50**, 1524.
641. Brasted, R.C. (1954) in *Comprehensive Inorganic Chemistry*, vol. **3** (eds M.C. Sneed, J.L. Maynard and R.C. Brasted), D. Van Norstrand Co., Inc., Princeton, New Jersey.
642. Greenwood, N.N. and Earnshaw, A. (1997) *Chemistry of the Elements*, Butterworth-Heinemann, Oxford.

643. Downs, A.J. and Adams, C.J. (1973) in *Comprehensive Inorganic Chemistry*, vol. **2** (eds J.C. Bailar, H.J. Emeleus, R. Nyholm and A.F. Trotman-Dickenson), Pergamon Press, Oxford, p. 1107.
644. Robiette, A.G., Bradley, R.H. and Brier, P.N. (1971) *Journal of the Chemical Society D, Chemical Communications*, 1567.
645. Jones, G.R., Burbank, R.D. and Bartlett, N. (1970) *Inorganic Chemistry*, **9**, 2264.
646. Mahjoub, A.R. and Seppelt, K. (1991) *Angewandte Chemie, International Edition*, **30**, 323.
647. Selte, K. and Kjekshus, A. (1970) *Acta Chemica Scandinavica*, **24**, 1912.
648. Kraft, T. and Jansen, M. (1995) *Journal of the American Chemical Society*, **117**, 6795.
649. Rehr, A. and Jansen, M. (1992) *Zeitschrift für Anorganische und Allgemeine Chemie*, **608**, 159.
650. Carter, H.A. and Aubke, F. (1971) *Inorganic Chemistry*, **10**, 2296.
651. Kraszkiewicz, L. and Skulski, L. (2003) *ARKIVOC*, (vi), 120.
652. Yusubov, M.S., Chi, K.-W., Park, J.Y., *et al.* (2006) *Tetrahedron Letters*, **47**, 6305.
653. Koposov, A.Y., Karimov, R.R., Pronin, A.A., *et al.* (2006) *Journal of Organic Chemistry*, **71**, 9912.
654. Fletcher, J.C. (1967) *The Biochemical journal*, **102**, 815.
655. Willgerodt, C. (1892) *Chemische Berichte*, **25**, 3494.
656. Alcock, N.W. and Sawyer, J.F. (1980) *Journal of the Chemical Society, Dalton Transactions*, 115.
657. Zhdankin, V.V., Litvinov, D.N., Koposov, A.Y., *et al.* (2004) *Chemical Communications*, 106.
658. Zhdankin, V.V., Koposov, A.Y., Litvinov, D.N., *et al.* (2005) *Journal of Organic Chemistry*, **70**, 6484.
659. Zhdankin, V.V., Koposov, A.Y., Netzel, B.C., *et al.* (2003) *Angewandte Chemie, International Edition*, **42**, 2194.
660. Koposov, A.Y., Litvinov, D.N. and Zhdankin, V.V. (2004) *Tetrahedron Letters*, **45**, 2719.
661. Koposov, A.Y., Nemykin, V.N. and Zhdankin, V.V. (2005) *New Journal of Chemistry*, **29**, 998.
662. Zhdankin, V., Goncharenko, R.N., Litvinov, D.N. and Koposov, A.Y. (2005) *ARKIVOC*, (iv), 8.
663. Ladziata, U., Koposov, A.Y., Lo, K.Y., *et al.* (2005) *Angewandte Chemie, International Edition*, **44**, 7127.
664. Ladziata, U., Carlson, J. and Zhdankin, V.V. (2006) *Tetrahedron Letters*, **47**, 6301.
665. Mailyan, A.K., Geraskin, I.M., Nemykin, V.N. and Zhdankin, V.V. (2009) *Journal of Organic Chemistry*, **74**, 8444.
666. Koposov, A.Y., Karimov, R.R., Geraskin, I.M., *et al.* (2006) *Journal of Organic Chemistry*, **71**, 8452.
667. Boppisetti, J.K. and Birman, V.B. (2009) *Organic Letters*, **11**, 1221.
668. Clark, R.J.H. and Dann, J.R. (1996) *Journal of Physical Chemistry*, **100**, 532.
669. Clark, R.J.H., Dann, J.R. and Foley, L.J. (1997) *Journal of Physical Chemistry A*, **101**, 9260.
670. Clark, R.J.H., Foley, L.J. and Price, S.D. (2000) *Journal of Physical Chemistry A*, **104**, 10675.
671. Naumann, D., Deneken, L. and Renk, E. (1975) *Journal of Fluorine Chemistry*, **5**, 509.
672. Wirth, T. (2001) *Angewandte Chemie, International Edition*, **40**, 2812.
673. Hartman, C. and Mayer, V. (1893) *Chemische Berichte*, **26**, 1727.
674. Stevenson, P.J., Treacy, A.B. and Nieuwenhuyzen, M. (1997) *Journal of the Chemical Society, Perkin Transactions 2*, 589.
675. Gougoutas, J.Z. (1981) *Crystal Structure Communications*, **10**, 489.
676. Katritzky, A.R., Savage, G.P., Palenik, G.J., *et al.* (1990) *Journal of the Chemical Society, Perkin Transactions 2*, 1657.
677. Boeckman, R.K., Shao, P. and Mullins, J.J. (2000) *Organic Syntheses*, **77**, 141.
678. Plumb, J.B. and Harper, D.J. (1990) *Chemical & Engineering News*, **68**, 3.
679. Dess, D.B., Wilson, S.R. and Martin, J.C. (1993) *Journal of the American Chemical Society*, **115**, 2488.
680. Frigerio, M., Santagostino, M. and Sputore, S. (1999) *Journal of Organic Chemistry*, **64**, 4537.
681. Ozanne, A., Pouysegu, L., Depernet, D., *et al.* (2003) *Organic Letters*, **5**, 2903.
682. Quideau, S., Pouysegu, L., Deffieux, D., *et al.* (2003) *ARKIVOC*, (vi), 106.
683. Lebrasseur, N., Gagnepain, J., Ozanne-Beaudenon, A., *et al.* (2007) *Journal of Organic Chemistry*, **72**, 6280.
684. Gagnepain, J., Castet, F. and Quideau, S. (2007) *Angewandte Chemie, International Edition*, **46**, 1533.
685. Ozanne-Beaudenon, A. and Quideau, S. (2006) *Tetrahedron Letters*, **47**, 5869.
686. Gallen, M.J., Goumont, R., Clark, T., *et al.* (2006) *Angewandte Chemie, International Edition*, **45**, 2929.
687. Waters, T., Boulton, J., Clark, T., *et al.* (2008) *Organic & Biomolecular Chemistry*, **6**, 2530.
688. Dess, D.B. and Martin, J.C. (1991) *Journal of the American Chemical Society*, **113**, 7277.
689. Stickley, S.H. and Martin, J.C. (1995) *Tetrahedron Letters*, **36**, 9117.

690. Zhdankin, V.V., Smart, J.T., Zhao, P. and Kiprof, P. (2000) *Tetrahedron Letters*, **41**, 5299.
691. Thottumkara, A.P. and Vinod, T.K. (2002) *Tetrahedron Letters*, **43**, 569.
692. Kommreddy, A., Bowsher, M.S., Gunna, M.R., *et al.* (2008) *Tetrahedron Letters*, **49**, 4378.
693. Cui, L.-Q., Dong, Z.-L., Liu, K. and Zhang, C. (2011) *Organic Letters*, **13**, 6488.
694. Moorthy, J.N., Singhal, N. and Senapati, K. (2008) *Tetrahedron Letters*, **49**, 80.
695. Moorthy, J.N., Senapati, K., Parida, K.N., *et al.* (2011) *Journal of Organic Chemistry*, **76**, 9593.
696. Miura, T., Nakashima, K., Tada, N. and Itoh, A. (2011) *Chemical Communications*, 1875.
697. Kumanyaev, I.M., Lapitskaya, M.A., Vasiljeva, L.L. and Pivnitsky, K.K. (2012) *Mendeleev Communications*, **22**, 129.
698. Nicolaou, K.C., Montagnon, T. and Baran, P.S. (2002) *Angewandte Chemie, International Edition*, **41**, 993.
699. Nicolaou, K.C., Gray, D.L.F., Montagnon, T. and Harrison, S.T. (2002) *Angewandte Chemie, International Edition*, **41**, 996.
700. Uyanik, M., Akakura, M. and Ishihara, K. (2009) *Journal of the American Chemical Society*, **131**, 251.
701. Kano, N., Ohashi, M., Hoshiba, K. and Kawashima, T. (2004) *Tetrahedron Letters*, **45**, 8173.
702. Dess, D.B. and Martin, J.C. (1983) *Journal of Organic Chemistry*, **48**, 4155.
703. Ireland, R.E. and Liu, L. (1993) *Journal of Organic Chemistry*, **58**, 2899.
704. Speicher, A., Bomm, V. and Eicher, T. (1996) *Journal für Praktische Chemie*, **338**, 588.
705. Chaudhari, S.S. (2000) *Synlett*, 278.
706. Meyer, S.D. and Schreiber, S.L. (1994) *Journal of Organic Chemistry*, **59**, 7549.
707. Schroeckeneder, A., Stichnoth, D., Mayer, P. and Trauner, D. (2012) *Beilstein Journal of Organic Chemistry*, **8**, 1523.
708. Chambers, O.R., Oates, G. and Winfield, J.M. (1972) *Journal of the Chemical Society, Chemical Communications*, 839.
709. Oates, G. and Winfield, J.M. (1974) *Journal of the Chemical Society, Dalton Transactions*, 119.
710. Tyrra, W., Miczka, M. and Naumann, D. (1997) *Zeitschrift für Anorganische und Allgemeine Chemie*, **623**, 1857.
711. Minkwitz, R., Broechler, R. and Preut, H. (1995) *Zeitschrift für Anorganische und Allgemeine Chemie*, **621**, 1247.
712. Yagupolskii, L.M., Lyalin, V.V., Orda, V.V. and Alekseeva, L.A. (1968) *Zhurnal obshchei khimii*, **38**, 2813.
713. Maletina, I.I., Orda, V.V., Aleinikov, N.N., *et al.* (1976) *Zhurnal Organicheskoi Khimii*, **12**, 1371.
714. Frohn, H.J. (1984) *Chemiker-Zeitung*, **108**, 146.
715. Frohn, H.J., Goerg, S., Henkel, G. and Laege, M. (1995) *Zeitschrift für Anorganische und Allgemeine Chemie*, **621**, 1251.
716. Hoyer, S. and Seppelt, K. (2004) *Journal of Fluorine Chemistry*, **125**, 989.
717. Ruff, O. and Keim, R. (1930) *Zeitschrift für Anorganische und Allgemeine Chemie*, **193**, 176.
718. Schack, C.J., Pilipovich, D., Cohz, S.N. and Sheehan, D.F. (1968) *Journal of Physical Chemistry*, **72**, 4697.
719. Vogt, T., Fitch, A.N. and Cockcroft, J.K. (1993) *Journal of Solid State Chemistry*, **103**, 275.
720. Christe, K.O., Curtis, E.C. and Dixon, D.A. (1993) *Journal of the American Chemical Society*, **115**, 1520.
721. Frohn, H.-J. and Bardin, V.V. (2010) *Journal of Fluorine Chemistry*, **131**, 1000.
722. Gillespie, R.J. and Krasznai, J.P. (1976) *Inorganic Chemistry*, **15**, 1251.
723. Engelbrecht, A. and Peterfy, P. (1969) *Angewandte Chemie, International Edition in English*, **8**, 768.
724. Christe, K.O., Curtis, E.C. and Dixon, D.A. (1993) *Journal of the American Chemical Society*, **115**, 9655.
725. Bartell, L.S., Clippard, F.B. and Jacob, E.J. (1976) *Inorganic Chemistry*, **15**, 3009.
726. Christe, K.O., Dixon, D.A., Mahjoub, A.R., *et al.* (1993) *Journal of the American Chemical Society*, **115**, 2696.
727. Mellor, J.W. (1922) *A Comprehensive Treatise on Inorganic and Theoretical Chemistry*, Vol. **2**, Longmans, Green and Co, London.
728. Levason, W. (1997) *Coordination Chemistry Reviews*, **161**, 33.
729. Kristiansen, K.A., Potthast, A. and Christensen, B.E. (2010) *Carbohydrate Research*, **345**, 1264.
730. Wee, A.G., Slobodian, J., Fernandez-Rodriguez, M.A. and Aguilar, E. (2001) in *e-EROS Encyclopedia of Reagents for Organic Synthesis* (editor in chief D. Crich), John Wiley & Sons, Inc., Hoboken, doi: 10.1002/047084289X.
731. Stengel, J.H. and McMills, M.C. (2001) in *e-EROS Encyclopedia of Reagents for Organic Synthesis* (editor in chief D. Crich), John Wiley & Sons, Inc., Hoboken, doi: 10.1002/047084289X.

3

Hypervalent Iodine Reagents in Organic Synthesis

Compounds of polyvalent iodine have found broad practical application in organic synthesis due to their diverse reactivity combined with benign environmental character and commercial availability. Particularly useful are the organic derivatives of λ^3- and λ^5-iodanes, commonly known as hypervalent iodine reagents. Several inorganic compounds of trivalent, pentavalent and heptavalent iodine have also found wide application in organic chemistry and can be included in this class of reagents. Numerous aryl-substituted λ^3- and λ^5-iodanes, aryliodonium salts and inorganic iodates and periodates are commercially available and used as common reagents in organic synthesis. Typical synthetic applications of these reagents include the following: oxidative halogenation of organic substrates, oxidative functionalization of unsaturated compounds, oxidative conversions of C—H to C—O bonds, various oxidative rearrangements and numerous reactions resulting in the formation of new C—C, C—N and other C—heteroatom or heteroatom—heteroatom bonds [1–3]. Hypervalent iodine compounds have found a particularly important application as selective reagents in the total synthesis of natural products [4], as they are efficient alternatives to toxic and expensive heavy metals for a large range of synthetic transformations. An overview of general principles of reactivity of hypervalent iodine compounds can be found in Section 1.5 of this book. This chapter summarizes specific synthetic applications of polyvalent iodine compounds.

3.1 Reactions of Iodine(III) Compounds

The general reactivity pattern of λ^3-iodanes, $ArIX_2$, is determined by the hypervalent, loose nature of I—X bonds and by a distinct positive charge on the iodine atom. These structural features are responsible for the enhanced electrophilic properties of $ArIX_2$ and explain such typical pathways as ligand exchange and reductive ligand transfer in their reactions with organic substrates (Section 1.5). Owing to these properties, hypervalent iodine(III) compounds have found broad application in organic synthesis as selective oxidants and electrophilic ligand transfer reagents. The growing interest in synthetic application of these compounds is also explained by their ready availability and absence of toxic properties and associated environmental concerns. Synthetic applications of hypervalent iodine(III) compounds have previously been summarized in several books [1–3] and numerous comprehensive reviews [4–12].

Hypervalent Iodine Chemistry: Preparation, Structure and Synthetic Applications of Polyvalent Iodine Compounds, First Edition. Viktor V. Zhdankin.

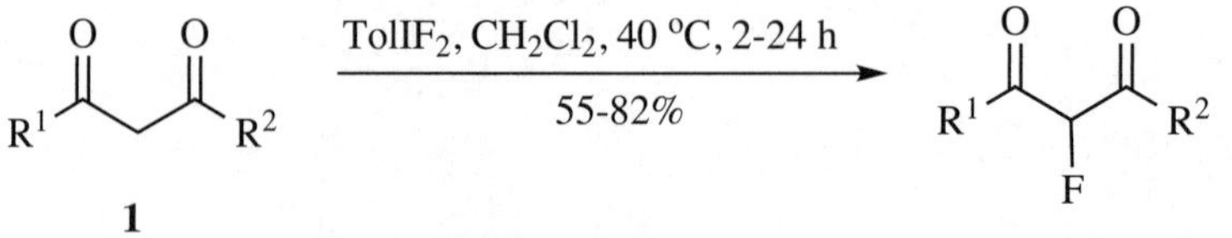

$TolIF_2$, CH_2Cl_2, 40 °C, 2-24 h; 55-82%

1 → 2

R^1 = Me, Pr, Ph, etc.
R^2 = OC_5H_{11}, Ph, Et, Bu^t, NMe_2, NPr^i_2, etc.

Scheme 3.1

3.1.1 Fluorinations

Various fluorinated compounds can be prepared by the fluorination of organic substrates with (difluoroiodo)arenes, which are powerful and selective electrophilic fluorinating reagents. Comparison of several $ArIF_2$ in fluorination reactions revealed that 4-alkyl-substituted (difluoroiodo)benzenes, $4\text{-}MeC_6H_4IF_2$ and $4\text{-}Bu^tC_6H_4IF_2$, are the preferred reagents, owing to their facile preparation and purification by crystallization, relatively high stability and solubility in organic solvents [13–16]. 4-(Difluoroiodo)toluene (also known as difluoroiodotoluene, $TolIF_2$) has become one of the commonly used fluorinating reagents [17].

β-Ketoesters and β-dicarbonyl compounds can be selectively fluorinated at the α-position by difluoroiodotoluene [18–20]. In a specific example, the monofluorinated products **2** can be prepared from β-ketoesters, β-ketoamides, or β-diketones **1** in good yields using difluoroiodotoluene under mild conditions (Scheme 3.1) [20]. A practical and convenient variation of the procedure for fluorination of 1,3-dicarbonyl compounds consists of the generation of $PhIF_2$ *in situ* from aqueous hydrofluoric acid and iodobenzene in dichloromethane [21]. For example, the reaction of ethyl benzoylacetate with the reagent system of aqueous HF and iodobenzene in CH_2Cl_2 affords ethyl 2-fluoro-2-benzoylacetate in 98% yield [21].

A similar fluorination can be performed electrochemically using difluoroiodotoluene as the mediator generated *in situ* from iodotoluene. Thus, the electrolysis of a 1 : 1 mixture of iodotoluene and various β-dicarbonyl compounds **3** in $Et_3N/5HF$ in an undivided cell under constant potential affords the respective α-fluoro-β-dicarbonyl compounds **4** in good yields (Scheme 3.2) [19].

Ketones cannot be directly fluorinated by (difluoroiodo)arenes; however, α-fluoroketones can be prepared by the reaction of silyl enol ethers with difluoroiodotoluene in the presence of $BF_3{\cdot}OEt_2$ and the $Et_3N\text{–}HF$ complex [22]. Some steroidal silyl enol ethers can be converted into the respective α-fluoroketones in a moderate yield [23].

Several synthetically useful fluorinations of various organosulfur compounds with difluoroiodotoluene have been reported [13, 24–26]. A clean and selective preparation of *gem*-difluorides **6** can be achieved by the

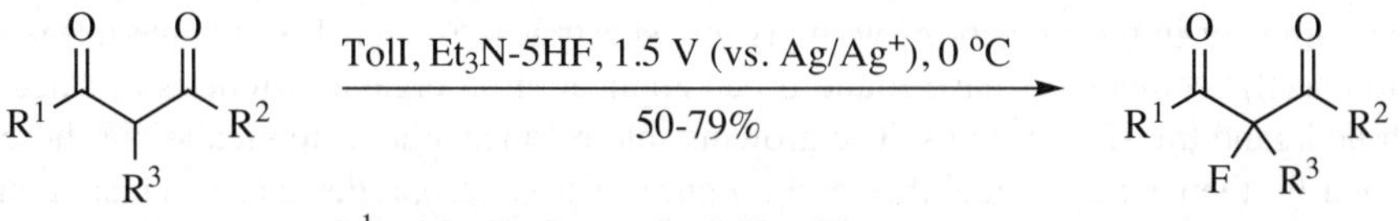

TolI, $Et_3N\text{-}5HF$, 1.5 V (vs. Ag/Ag^+), 0 °C; 50-79%

3 → 4

R^1 = Me, Et, Pr, *cyclo*-C_6H_{11}, Ph
R^2 = OEt, OBu, Pr, Pr^i, Ph
R^3 = H, Me

Scheme 3.2

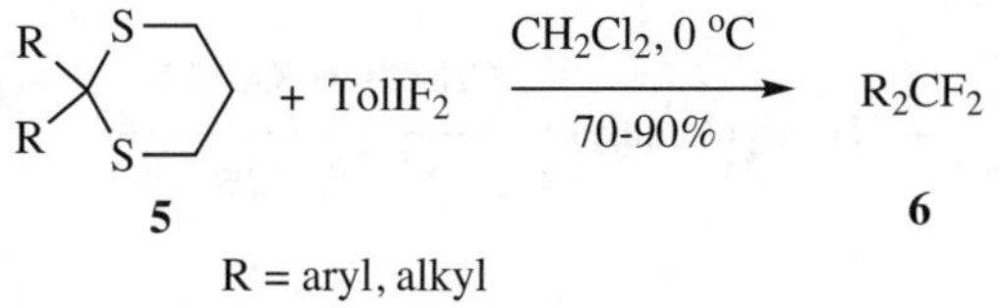

R, R (dithiane) **5** + $TolIF_2$ —(CH_2Cl_2, 0 °C; 70-90%)→ R_2CF_2 **6**

R = aryl, alkyl

Scheme 3.3

reaction of difluoroiodotoluene with dithioketals **5** (Scheme 3.3) [13]. This fluorination can also be performed electrochemically using a difluoroiodoarene as the in-cell mediator generated *in situ* from catalytic amounts (0.05–0.2 equiv) of 4-methoxyiodobenzene [27].

α-(Phenylthio)acetamides **7** are fluorinated in the α-position to give products **8** when treated with difluoroiodotoluene under mild conditions (Scheme 3.4) [24, 28].

Under similar conditions α-phenylthio esters **9** afford fluorides **10** (Scheme 3.5) [25]. The mechanism of this Pummerer-type reaction involves initial nucleophilic addition of the sulfur atom to the electrophilic iodine center to form the iodosulfonium salt **11.** The liberated fluoride anion acts as a base with resultant formation of the classical Pummerer intermediate **12.** Subsequent trapping of cation **12** with fluoride anion yields the final product **10** (Scheme 3.5) [25].

Addition of a second equivalent of difluoroiodotoluene in this reaction (Scheme 3.5) affords α,α-difluoro sulfides and the reaction with three equivalents of $TolIF_2$ leads to α,α-difluoro sulfoxides. This sequential fluorination–oxidation reaction has been exploited in the one-pot synthesis of 3-fluoro-2(5*H*)-furanone starting from (3*R*)-3-fluorodihydro-2(3*H*)-furanone [29].

The reaction of α-phenylthio lactones **13** with two equivalents of difluoroiodotoluene results in fluorination–oxidation to give α-fluoro sulfoxides **14,** which then undergo thermal *syn* elimination to produce vinyl fluorides **15** (Scheme 3.6) [25].

Treatment of the readily available thio- and seleno-glycosides with difluoroiodotoluene gives the corresponding fluoroglycosides in moderate to good yields [15, 26]. In a typical representative example, treatment of the glucose derivative **16** with difluoroiodotoluene under mild conditions affords anomeric fluoroglycosides **17** and **18** in a 3 : 2 ratio (Scheme 3.7) [26].

The reaction of 4-*tert*-butyl(difluoroiodo)benzene with esters of cephalosporin V (**19**) is solvent-dependent: the major product in dichloromethane is fluoroazetidine **20,** while the same reactants in acetonitrile afford oxazoline disulfide **21** (Scheme 3.8) [16].

The monofluorination of a series of α-acceptor-substituted selenides **22** using difluoroiodotoluene (Scheme 3.9) has been reported [30]. Although the yields of products **23** are only moderate, the reactions are usually very clean and, under the reaction conditions used, no further oxidized products are observed.

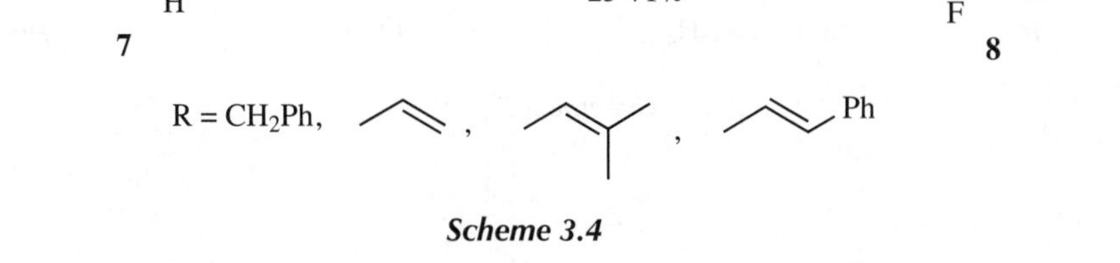

PhS–CH2–C(O)–NH–R **7** + $TolIF_2$ —(CH_2Cl_2, 0 °C or reflux, 12 h; 25-71%)→ PhS–CHF–C(O)–NH–R **8**

R = CH_2Ph, allyl, prenyl, cinnamyl (Ph)

Scheme 3.4

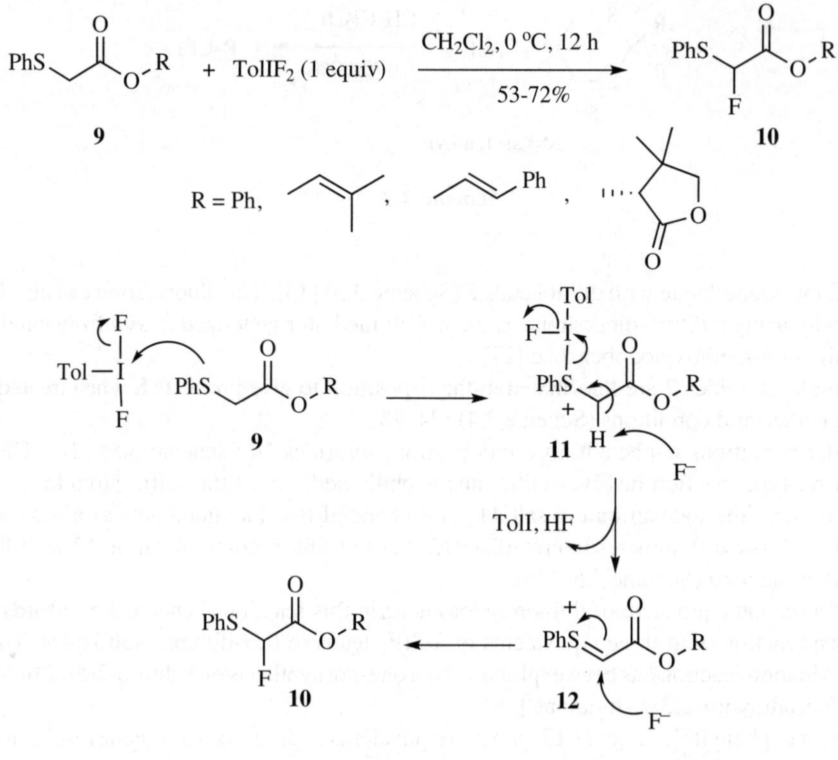

CH$_2$Cl$_2$, 0 °C, 12 h

53-72%

PhS + TolIF$_2$ (1 equiv)

9

10

R = Ph,

Tol

F

TolI, HF

F$^-$

9

11

12

10

Scheme 3.5

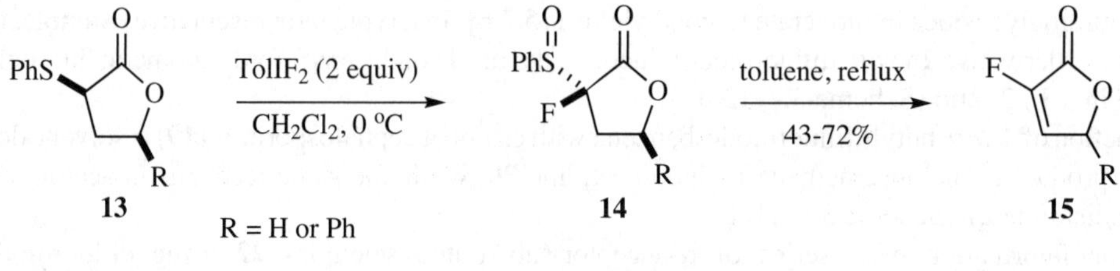

TolIF$_2$ (2 equiv)

CH$_2$Cl$_2$, 0 °C

toluene, reflux

43-72%

13

14

15

R = H or Ph

Scheme 3.6

PhIF$_2$, CH$_2$Cl$_2$

–78 °C to rt, 12 h

78%

Ar = 4-ClC$_6$H$_4$

16

17

18

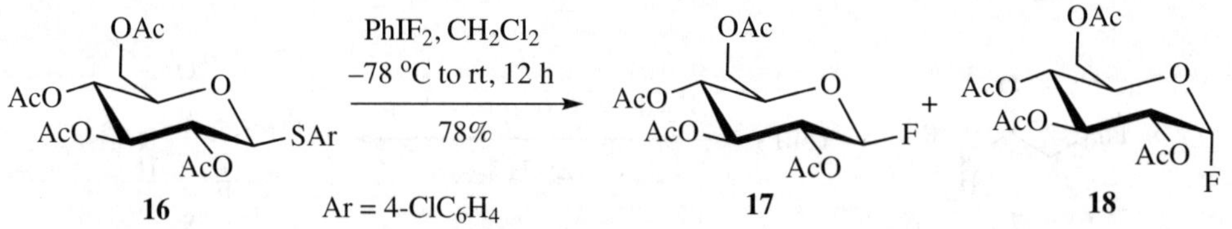

Scheme 3.7

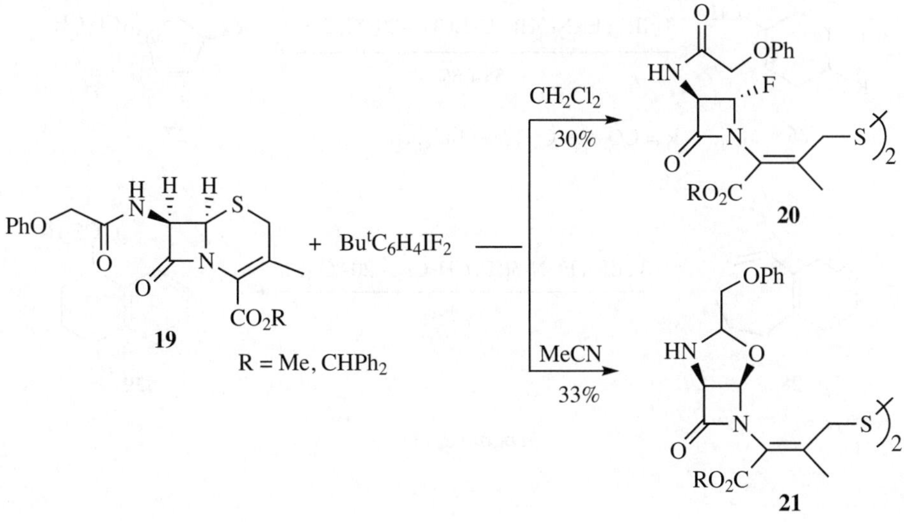

19 R = Me, $CHPh_2$ **20** **21**

Scheme 3.8

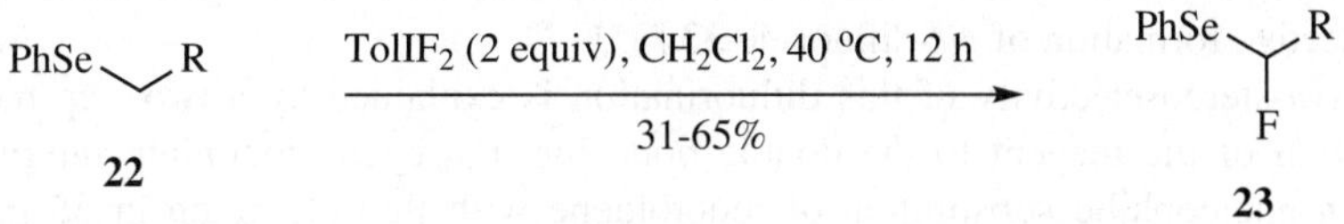

22 **23**

R = CO_2Et, CO_2Ph, $CO_2CH_2CH{=}CH_2$, CONHMe, $CONMe_2$, CN, etc.

Scheme 3.9

Fluorinated five- to seven-membered cyclic ethers have been stereoselectively synthesized from iodoalkyl substituted four- to six-membered cyclic ethers by fluorinative ring-expansion reaction using difluoroiodotoluene [31]. Scheme 3.10 shows a specific example of a fluorinative ring-expansion reaction of oxetanes **24** leading to five-membered cyclic ethers **25** [31].

(Difluoroiodo)arenes react with aryl-substituted alkenes to afford the rearranged, geminal difluorides, owing to the migration of the aryl group [32, 33]. Likewise, the reaction of substituted cyclic alkenes with difluoroiodotoluene and $Et_3N/5HF$ results in a fluorinative ring-contraction with the selective formation of difluoroalkyl substituted cycloalkanes. Thus, the fluorination of 1-methylcyclohexene derivatives **26** affords

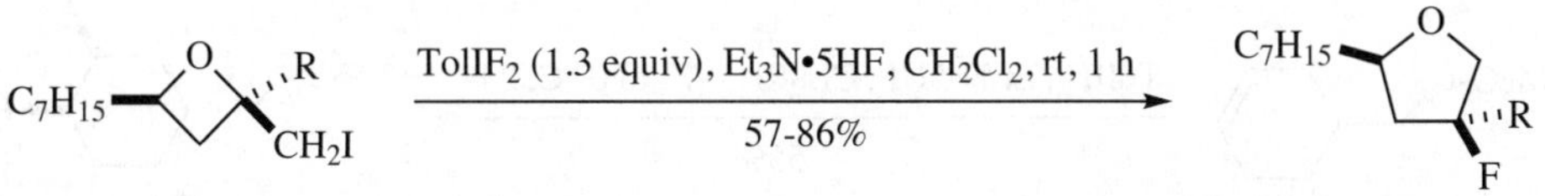

24 R = H or Me **25**

Scheme 3.10

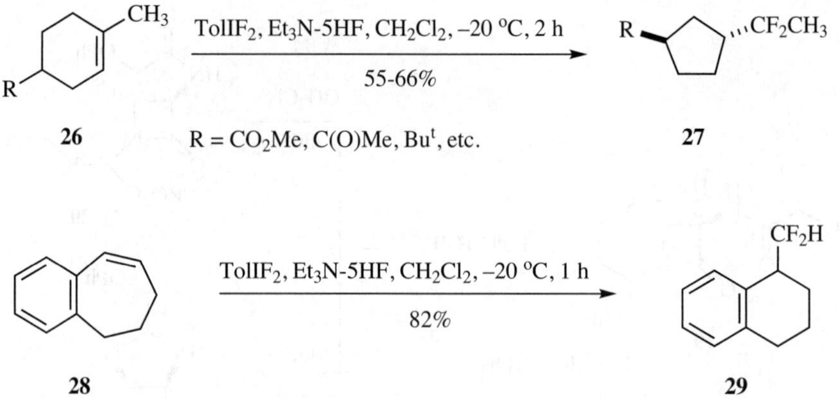

$TolIF_2$, Et_3N-5HF, CH_2Cl_2, –20 °C, 2 h

55-66%

26 R = CO_2Me, C(O)Me, Bu^t, etc. **27**

$TolIF_2$, Et_3N-5HF, CH_2Cl_2, –20 °C, 1 h

82%

28 **29**

Scheme 3.11

(1,1-difluoroethyl)cyclopentanes **27,** while a similar reaction of benzocycloheptane **28** gives the respective difluoromethyl-substituted benzocyclohexane **29** in high yield (Scheme 3.11) [34].

The reaction of difluoroiodotoluene with terminal alkenes **30** furnishes *vic*-difluoroalkanes **31** in moderate yields (Scheme 3.12) [35]. The cyclohexene derivative **32** reacts with this reagent under similar conditions with the stereoselective formation of *cis*-difluoride **33** [35].

The observed *syn*-stereoselectivity of this difluorination is explained by a two-step mechanism involving the *anti*-addition of the reagent to the double bond through cyclic iodonium intermediate **34** in the first step and then nucleophilic substitution of iodotoluene with fluoride anion in **35** in the second step (Scheme 3.13) [35].

Steroidal dienes **36** react with (difluoroiodo)arenes **37** to afford fluorinated product **38** with a high degree of regioselectivity and stereoselectivity (Scheme 3.14) [14]. (Difluoroiodo)arenes react with terminal alkynes with stereo- and regioselective formation of synthetically useful (*E*)-2-fluoro-1-alkenyliodonium salts [36–39]. A convenient procedure for the preparation of various (*E*)-2-fluoroalkenyliodonium fluorides **39** is based on the addition of difluoroiodotoluene to terminal acetylenes (Scheme 3.15) [37–39]. Products **39** can be

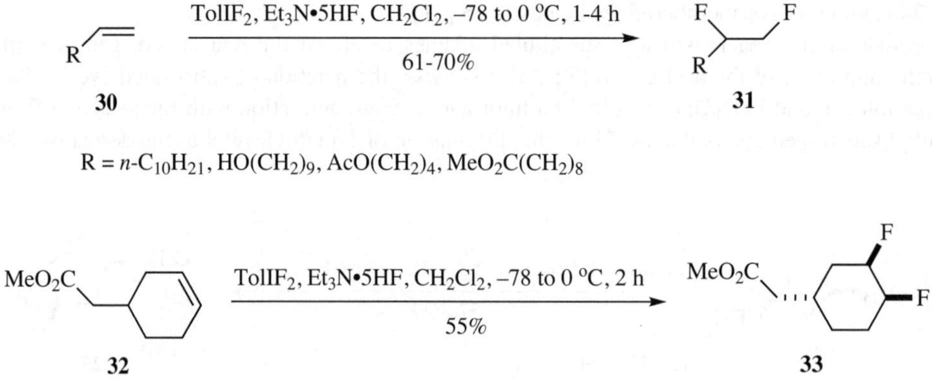

$TolIF_2$, Et_3N•5HF, CH_2Cl_2, –78 to 0 °C, 1-4 h

61-70%

30 **31**

R = n-$C_{10}H_{21}$, $HO(CH_2)_9$, $AcO(CH_2)_4$, $MeO_2C(CH_2)_8$

$TolIF_2$, Et_3N•5HF, CH_2Cl_2, –78 to 0 °C, 2 h

55%

32 **33**

Scheme 3.12

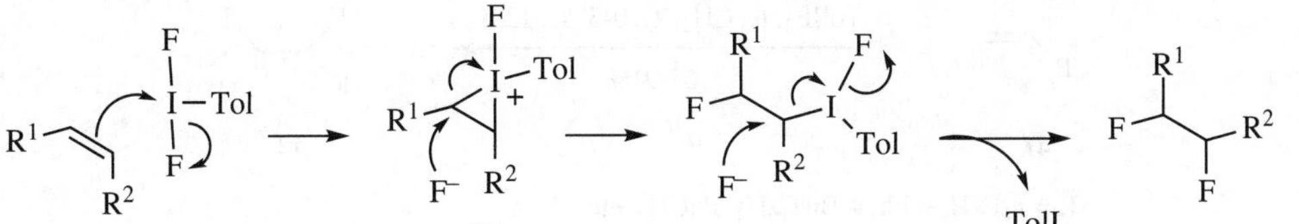

Scheme 3.13

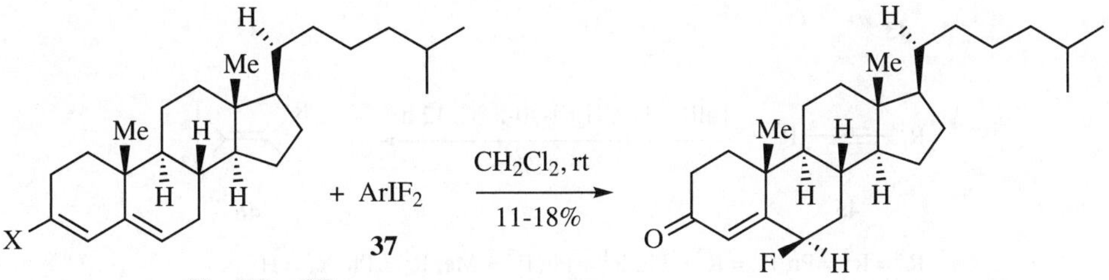

$X = NR_2$, OR

Ar = Ph, 4-$Bu^tC_6H_4$, 4-FC_6H_4, 2,6-$Pr^i{}_2C_6H_3$

Scheme 3.14

further converted into (*E*)-2-fluoro-1-iodo-1-alkenes **40** without isolation by treatment with KI/CuI [37], or can be used as reagents in the Pd-catalyzed coupling reactions [38, 39].

The fluorination of alkenes **41** and **43** and alkynes **45** with difluoroiodotoluene in the presence of iodine affords *vic*-fluoroiodoalkanes **42** and **44** and fluoroiodoalkenes **46** in moderate to good yields (Scheme 3.16) [40]. This reaction proceeds in a Markovnikov fashion and with prevalent *anti*-stereoselectivity via initial addition of the electrophilic iodine species followed by nucleophilic attack of fluorine anion. The analogous reaction of alkenes and alkynes with difluoroiodotoluene in the presence of diphenyl diselenides affords the respective products of phenylselenofluorination in good yields [41].

The reaction of difluoroiodotoluene with a four-, five-, or six-membered carbocycles **47** affords the ring-expanded (*E*)-δ-fluoro-β-halovinyliodonium tetrafluoroborates **48** stereoselectively in high yields (Scheme 3.17) [42]. This reaction proceeds via a sequence of λ^3-iodanation-1,4-halogen shift–ring enlargement–fluorination steps.

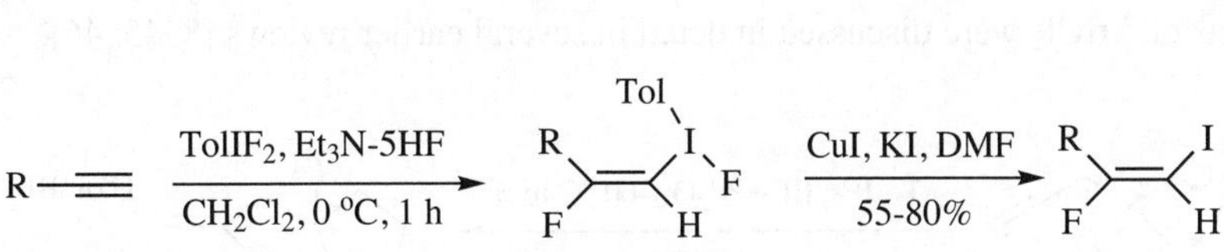

R = MeC_9H_{18}, HOC_9H_{18}, ClC_9H_{18}, $MeO_2CC_8H_{16}$, $Bu^tC(O)C_8H_{16}$, etc.

Scheme 3.15

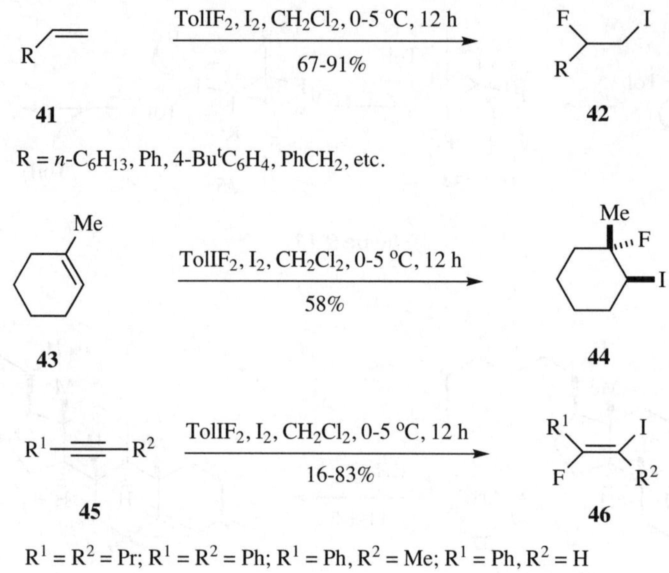

$TolIF_2$, I_2, CH_2Cl_2, 0-5 °C, 12 h; 67-91%

41 → **42**

R = n-C_6H_{13}, Ph, 4-$Bu^tC_6H_4$, $PhCH_2$, etc.

$TolIF_2$, I_2, CH_2Cl_2, 0-5 °C, 12 h; 58%

43 → **44**

$TolIF_2$, I_2, CH_2Cl_2, 0-5 °C, 12 h; 16-83%

45 → **46**

$R^1 = R^2$ = Pr; $R^1 = R^2$ = Ph; R^1 = Ph, R^2 = Me; R^1 = Ph, R^2 = H

Scheme 3.16

N-Protected anilines can be selectively fluorinated at the *para*-position by bis(*tert*-butylcarbonyloxy)-iodobenzene, $PhI(OPiv)_2$ and hydrogen fluoride/pyridine. This convenient procedure provides facile access to various *para*-fluorinated anilides in moderate to good yields [43].

3.1.2 Chlorinations

(Dichloroiodo)arenes are widely used as reagents for chlorination of various organic substrates. Among (dichloroiodo)arenes, (dichloroiodo)benzene, $PhICl_2$, is the most commonly used reagent, which can be conveniently prepared by direct chlorination of iodobenzene (Section 2.1.3). The preparation and reactions of several recyclable, polymer-supported or nonpolymeric iodine(III) chlorides are discussed in Chapter 5.

Typical chlorinations of alkanes or alkenes with (dichloroiodo)benzene proceed via a radical mechanism and generally require photochemical conditions or the presence of radical initiators in solvents of low polarity, such as chloroform or carbon tetrachloride. However, the alternative ionic pathways are also possible due to the electrophilic properties of the iodine atom in $PhICl_2$ or electrophilic addition of Cl_2 generated by the dissociation of the reagent. An alternative synchronous molecular addition mechanism in the reactions of $PhICl_2$ with alkenes has also been discussed and was found to be theoretically feasible [44]. The general reactivity patterns of $ArICl_2$ were discussed in detail in several earlier reviews [8, 45, 46].

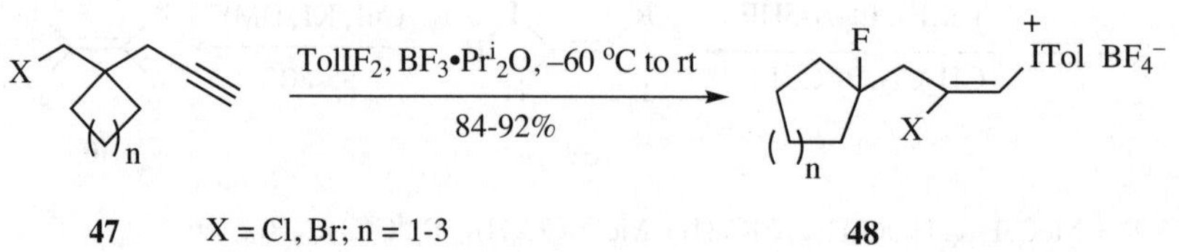

$TolIF_2$, $BF_3 \cdot Pr^i_2O$, –60 °C to rt; 84-92%

47 X = Cl, Br; n = 1-3 → **48** ($\overset{+}{I}Tol$ BF_4^-)

Scheme 3.17

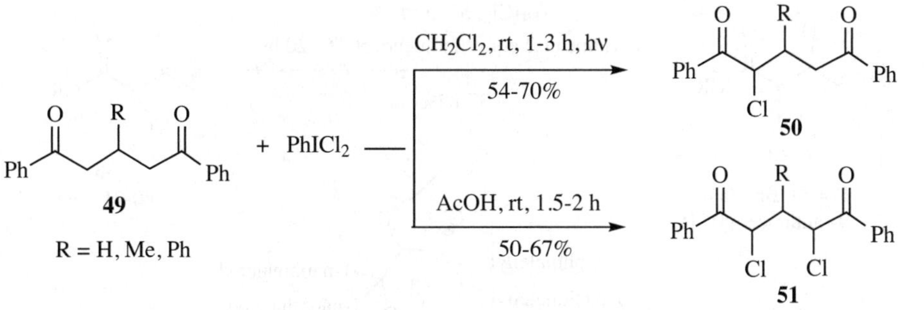

Ph(C=O)CH₂CH(R)CH₂(C=O)Ph **49**, R = H, Me, Ph + $PhICl_2$ → (CH_2Cl_2, rt, 1-3 h, hν; 54-70%) → **50** (monochloride); → (AcOH, rt, 1.5-2 h; 50-67%) → **51** (dichloride)

Scheme 3.18

(Dichloroiodo)benzene has been applied for a substitutive chlorination at sp^3-carbon of various organic substrates, such as alkanes, ethers, esters, thioethers, ketones, sulfoxides and so on [45–51]. Ketones can be chlorinated at the α-position under either radical or ionic conditions. In a typical example, 1,5-diketones **49** react with $PhICl_2$ in dichloromethane under radical conditions (dichloromethane, UV-irradiation) to form, predominately, monochlorides **50,** while the same reagents in acetic acid in the dark (ionic conditions) selectively afford dichlorides **51** (Scheme 3.18) [47].

Various ketones, including aliphatic and aromatic ketones **52,** can be directly converted into their corresponding α-chloroketone acetals **53** in high to excellent yields using $PhICl_2$ in ethylene glycol in the presence of molecular sieves at room temperature (Scheme 3.19) [52]. For comparison, a similar reaction using Cl_2 as chlorinating reagent under similar conditions affords only trace amount of chlorides **53.**

(Dichloroiodo)toluene is a suitable chlorinating reagent in the catalytic asymmetric chlorination of β-keto esters **54,** catalyzed by the titanium complex **55,** leading to the respective α-chlorinated products **56** in moderate to good yields and enantioselectivities (Scheme 3.20) [53]. Interestingly, the enantioselectivity of this reaction shows a strong temperature dependence, with the maximum selectivity reached at 50 °C.

Depending on the conditions, reactions of $ArICl_2$ with alkenes may follow a radical or ionic mechanism [49, 54–56]. Under radical conditions, the reaction gives exclusively the products of 1,2-addition of chlorine, often with high *trans*-stereoselectivity, while under polar conditions chlorination is usually accompanied by skeletal rearrangements. For example, norbornene reacts with $PhICl_2$ in nonpolar solvents to give 1,2-adducts **57** and **58** (Scheme 3.21) as the only detectable products [49]. The same reaction in the presence of trifluoroacetic acid affords exclusively the products of skeletal rearrangement **59–61.** Bicyclic diene **62** reacts with $PhICl_2$ under similar polar conditions to afford the tricyclic, rearranged products **63** and **64** in low yield [50].

Reactions of (dichloroiodo)benzene with various monoterpenes in methanol proceed via the ionic mechanism and afford the respective products of chloromethoxylation of the double bond with high regio- and stereoselectivity [56]. Likewise, the reaction of a recyclable chlorinating reagent, 4,4′-bis(dichloroiodo)biphenyl

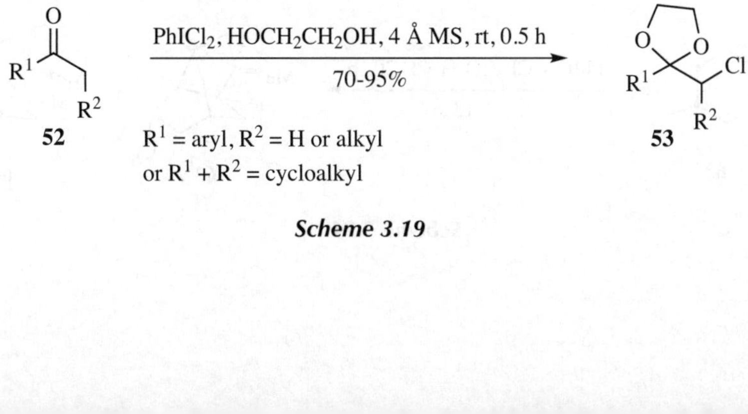

52 → ($PhICl_2$, $HOCH_2CH_2OH$, 4 Å MS, rt, 0.5 h; 70-95%) → **53**

R^1 = aryl, R^2 = H or alkyl
or $R^1 + R^2$ = cycloalkyl

Scheme 3.19

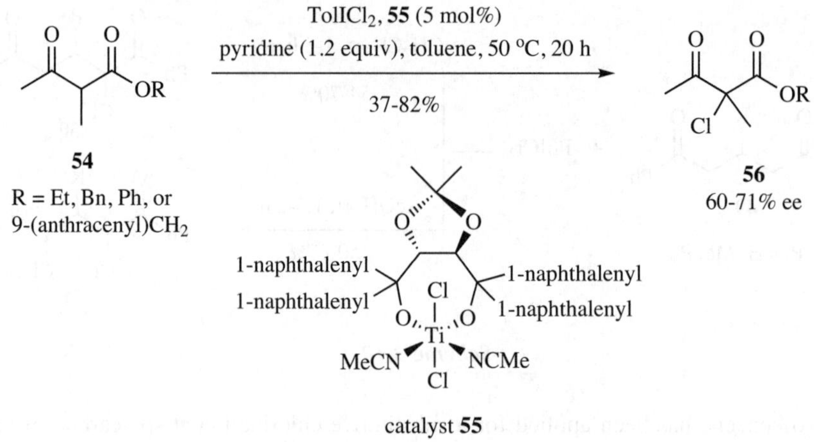

$TolICl_2$, **55** (5 mol%)
pyridine (1.2 equiv), toluene, 50 °C, 20 h
37-82%

54
R = Et, Bn, Ph, or 9-(anthracenyl)CH_2

56
60-71% ee

1-naphthalenyl, 1-naphthalenyl, 1-naphthalenyl, 1-naphthalenyl, Cl, Ti, MeCN, NCMe, Cl

catalyst **55**

Scheme 3.20

66 (Section 5.3.1), with styrene derivatives **65** in methanol affords exclusively the products of electrophilic chloromethoxylation (**67**) (Scheme 3.22) [54].

The electrophilic reactivity of $PhICl_2$ can be increased by adding Lewis acids or $AgBF_4$ to the reaction mixture. Under these conditions, the reaction of $PhICl_2$ with diene **68** gives exclusively rearranged products **69** and **70,** which clearly result from an ionic mechanism of addition (Scheme 3.23) [51].

An enantioselective dichlorination of allylic alcohols using (dichloroiodo)arenes as chlorine sources in the presence of catalytic amounts of a dimeric cinchona alkaloid derivative $(DHQ)_2PHAL$ has been developed [57]. For example, the dichlorination of *trans*-cinnamyl alcohols **71** with 4-Ph(C_6H_4)ICl_2 catalyzed by $(DHQ)_2PHAL$ affords products **72** in good yields and enantioselectivities (Scheme 3.24).

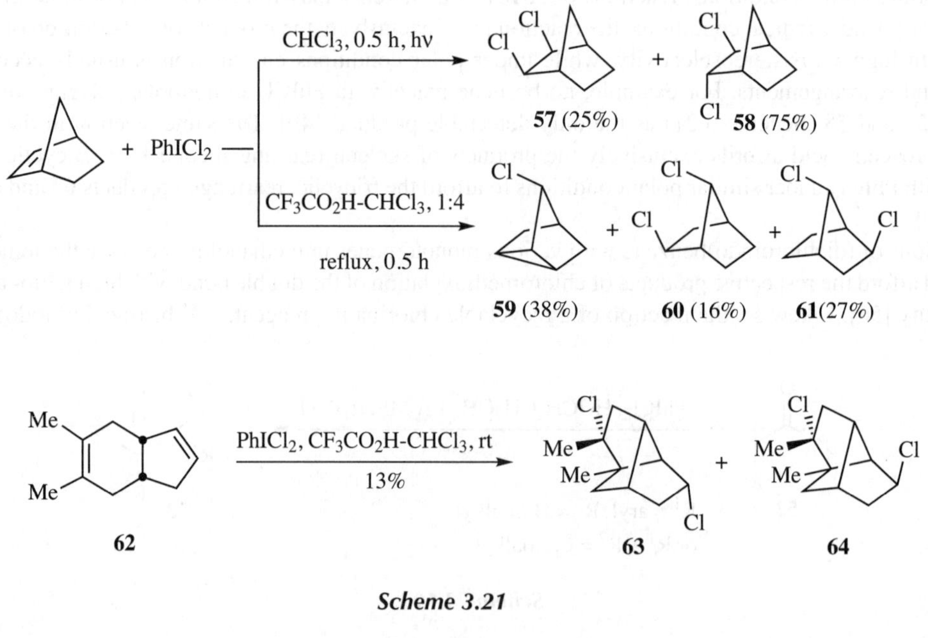

+ $PhICl_2$

$CHCl_3$, 0.5 h, hν

57 (25%) + **58** (75%)

CF_3CO_2H-$CHCl_3$, 1:4
reflux, 0.5 h

59 (38%) + **60** (16%) + **61**(27%)

62

$PhICl_2$, CF_3CO_2H-$CHCl_3$, rt
13%

63 + **64**

Scheme 3.21

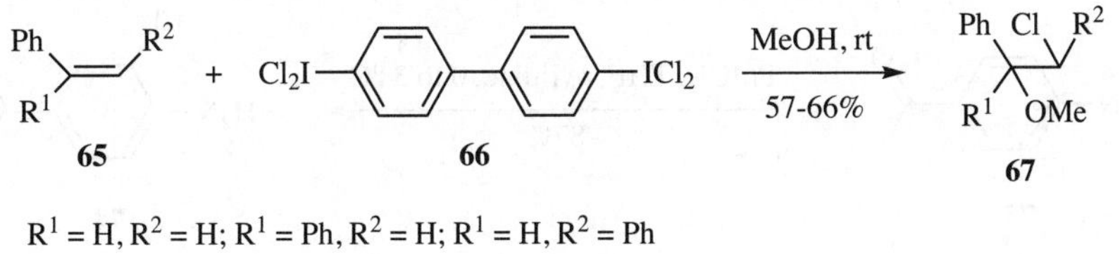

$R^1 = H, R^2 = H; R^1 = Ph, R^2 = H; R^1 = H, R^2 = Ph$

Scheme 3.22

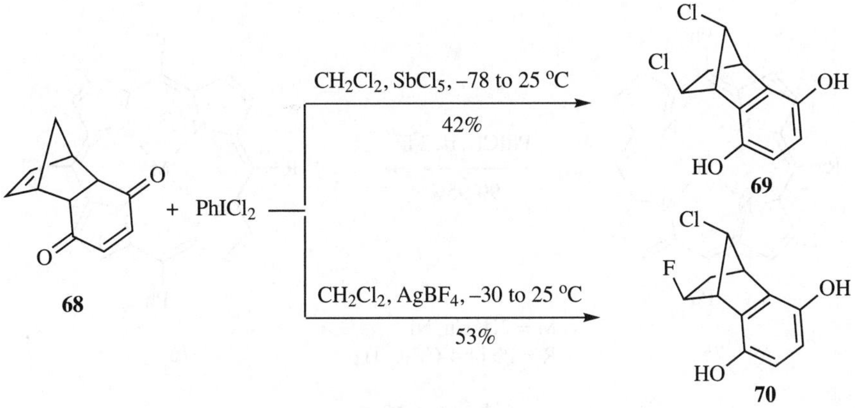

Scheme 3.23

(Dichloroiodo)arenes can also be used for the chlorination of aromatic compounds. Aminoacetophenone **73** is selectively chlorinated with (dichloroiodo)benzene to give product **74** in good yield (Scheme 3.25). This process can be scaled up to afford 24.8 kg of product **74** with 94% purity [58].

Treatment of 5,10,15-trisubstituted porphyrins **75** with (dichloroiodo)benzene affords the corresponding *meso*-chlorinated porphyrins **76** (Scheme 3.26) [59]. The chlorination of 5,10,15,20-tetraarylporphyrins, in

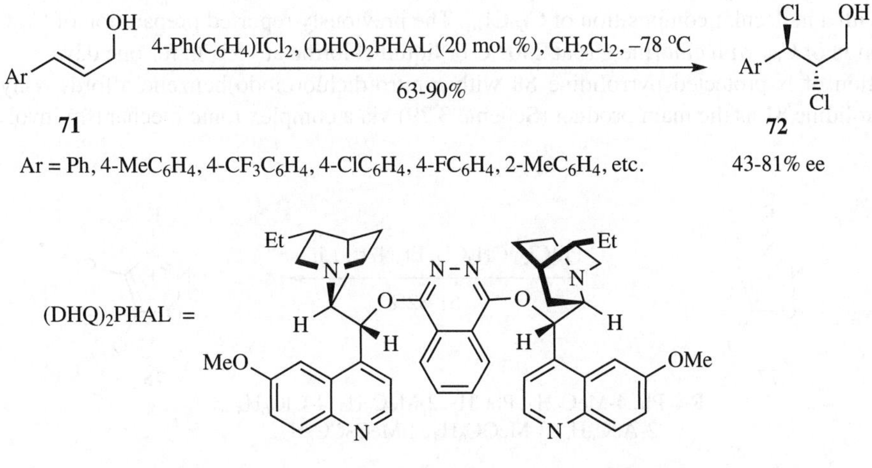

Ar = Ph, 4-MeC_6H_4, 4-$CF_3C_6H_4$, 4-ClC_6H_4, 4-FC_6H_4, 2-MeC_6H_4, etc. 43-81% ee

Scheme 3.24

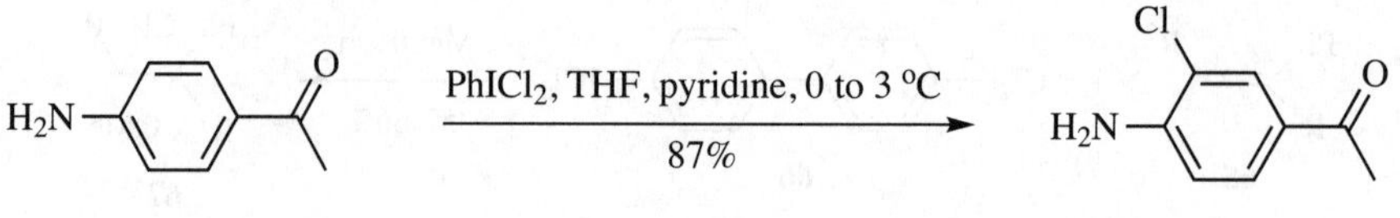

$PhICl_2$, THF, pyridine, 0 to 3 °C
87%

73 74

Scheme 3.25

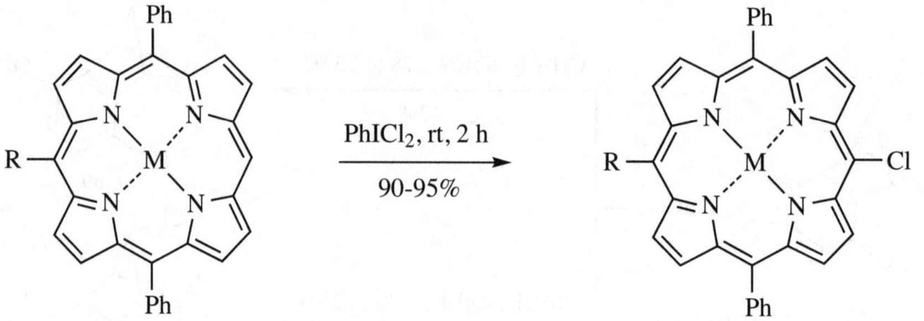

$PhICl_2$, rt, 2 h
90-95%

M = 2H, Cu, Ni
R = Ph or 4-$CF_3C_6H_4$

75 76

Scheme 3.26

which all *meso*-positions are substituted, under similar conditions affords β-monochlorinated products in high yields.

(Dichloroiodo)benzene in the presence of triethylamine has been used for the selective chlorination of 3-substituted sydnones **77** to furnish the 4-chloro substituted products **78** in good yield (Scheme 3.27) [60].

Fullerene (C_{60}) reacts smoothly with (dichloroiodo)benzene in tetrachloroethane at –25 to 25 °C to give polychlorinated fullerenes **79** (Scheme 3.28) [61]. Laser desorption mass-spectrometry analysis of product **79** shows $C_{60}Cl_7$ and $C_{60}Cl_9$ as the most prevalent species, while elemental analysis of the product is consistent with a molecular composition of $C_{60}Cl_{16}$. The previously reported preparation of $C_{60}Cl_n$ involved the chlorination of C_{60} with chlorine gas at 250 °C or liquid chlorine at –35 °C for one day.

The reaction of N-protected pyrrolidine **80** with *p*-nitro(dichloroiodo)benzene affords α-hydroxy-β,β-dichloropyrrolidine **81** as the main product (Scheme 3.29) via a complex ionic mechanism involving a triple

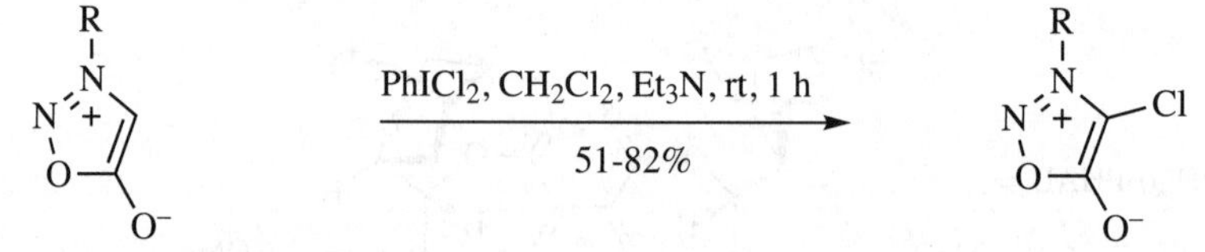

$PhICl_2$, CH_2Cl_2, Et_3N, rt, 1 h
51-82%

77 78

R = Ph, 4-MeC_6H_4, $PhCH_2$, 2-MeC_6H_4, 4-ClC_6H_4
2-AcC_6H_4, 3-$MeOC_6H_4$, 2-$MeO_2CC_6H_4$

Scheme 3.27

C_{60} + $PhICl_2$ (10 equiv) —($C_2H_2Cl_4$, –25 to 25 °C, 120 h)→ $C_{60}Cl_n$ **79** n = 8-16

Scheme 3.28

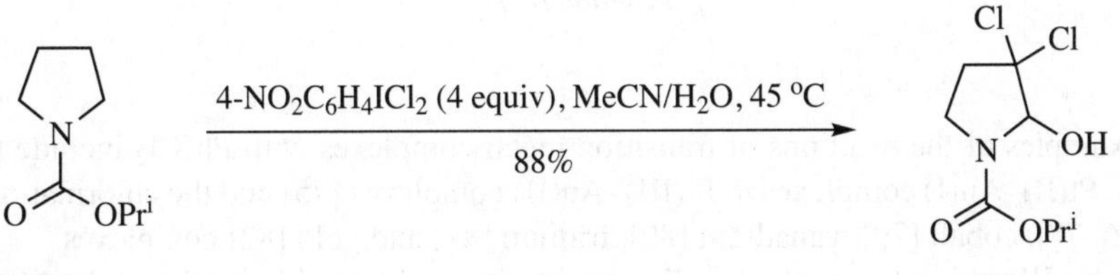

80 —(4-$NO_2C_6H_4ICl_2$ (4 equiv), MeCN/H_2O, 45 °C; 88%)→ **81**

Scheme 3.29

C—H bond activation. This oxidative pathway has been demonstrated to be general for several saturated, urethane-protected nitrogen heterocyclic systems [62].

(Dichloroiodo)benzene can be conveniently generated *in situ* from other hypervalent iodine reagents and used for subsequent chlorination of organic substrates. In a specific example, an efficient chlorination of β-keto esters, 1,3-diketones and alkenes has been performed using iodosylbenzene with concentrated HCl, selectively giving α-chloro-β-keto esters, 2-chloro-1,3-diketones and 1,2-dichloroalkanes, respectively [63]. A stereoselective *anti*-addition was observed in the chlorination of indene under these conditions.

Various organic compounds containing heteroatoms can be effectively chlorinated by (dichloroiodo)arenes. Numerous examples of chlorinations at sulfur, phosphorus, selenium, arsenic, antimony and some metals have been reported in older reviews [45, 46]. More recent examples include the oxidation of sulfides to the corresponding sulfoxides and sulfones by 2-(dichloroiodo)benzoic acid [64], the preparation of *N*-chlorotriphenylphosphinimine, Ph_3PNCl, in good yield by the reaction of $Ph_3PNSiMe_3$ with $PhICl_2$ [65] and the synthesis of trichlorophosphine imides **82** (Scheme 3.30) [66].

(Dichloroiodo)benzene is commonly used as a reagent for the chlorination or oxidation of various transition metal complexes. Numerous examples of oxidative chlorination of various molybdenum and tungsten complexes with $PhICl_2$ have been reported in the literature [67–74]. In a representative example, oxidative decarbonylation of the cyclopentadienyl complexes **83** with one equivalent of $PhICl_2$ selectively yields the respective molybdenum(IV) or tungsten(IV) complexes **84** in excellent yields (Scheme 3.31) [73].

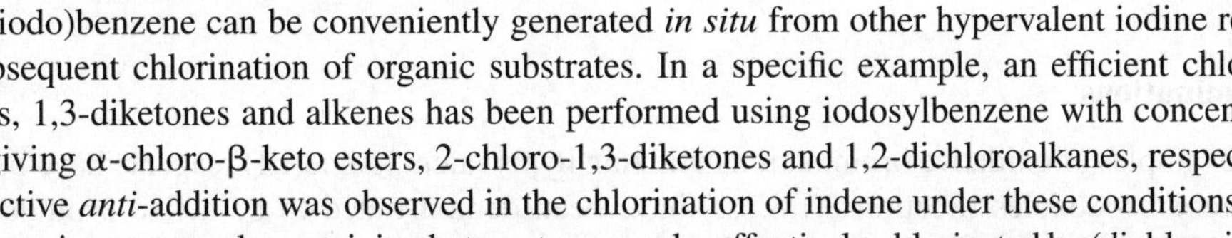

2,4,6-$Bu^t_3C_6H_2$–N=PCl + $PhICl_2$ —(CH_2Cl_2, rt, 30 min; 50%)→ 2,4,6-$Bu^t_3C_6H_2$–N=PCl_3 **82**

Scheme 3.30

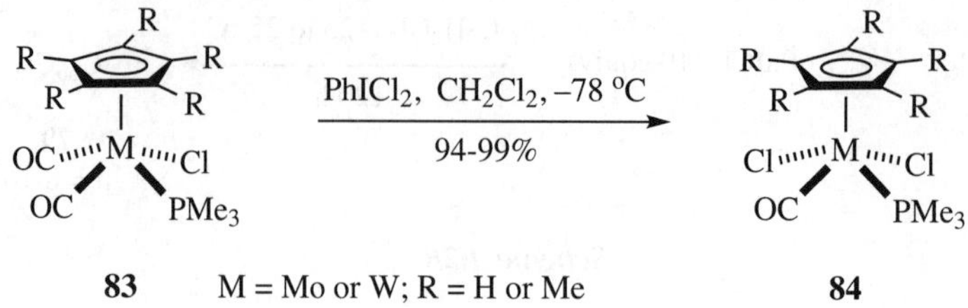

$Cp^{R}M(CO)_2(PMe_3)Cl$ (**83**) $\xrightarrow[\text{94-99\%}]{PhICl_2,\ CH_2Cl_2,\ -78\ ^\circ C}$ $Cp^{R}M(CO)(PMe_3)Cl_2$ (**84**)

83 M = Mo or W; R = H or Me **84**

Scheme 3.31

Additional examples of the reactions of transition metal complexes with $PhICl_2$ include the oxidation of heterobimetallic Pt(II)–Au(I) complexes to Pt(III)–Au(II) complexes [75] and the chlorinations or oxidations of palladium [76–78], cobalt [79], vanadium [80], iridium [81] and gold [82] complexes.

Inorganic iodine(III) chlorides, such as iodine trichloride and tetrachloroiodate salts $M^+ICl_4^-$, can also be used as chlorinating reagents. It has been demonstrated that ICl_3 reacts with alkanes in the presence of light to form chloroalkanes via a radical mechanism [83–85]. Benzyltrimethylammonium tetrachloroiodate, $PhCH_2Me_3N^+ICl_4^-$, is an effective reagent for benzylic chlorination of alkylaromatic compounds under reflux conditions in carbon tetrachloride in the presence of radical initiators [86]. Potassium tetrachloroiodate, $K^+ICl_4^-$, has been used for the chlorination of fullerenes [87, 88].

3.1.3 Brominations

Only a few examples of oxidative brominations utilizing hypervalent iodine(III) reagents have been reported. Stable bromobenziodoxoles (Section 2.1.8.1.1) have been demonstrated to be active brominating agents. In particular, benziodoxole **85** can be used for selective allylic and benzylic brominations under radical conditions [89] and bromobenziodoxole **86** has been utilized for electrophilic bromination of anisole or bromolactonization of 4-pentenoic acid (Scheme 3.32) [90].

Common hypervalent iodine(III) reagents, such as $PhI(OAc)_2$ and PhI(OH)OTs, in the presence of bromide anions can be used for oxidative bromination of organic substrates. The oxidative halogenation of 1,4-dimethoxynaphthalene derivatives with (diacetoxyiodo)benzene and trimethylsilyl bromide or chloride affords the corresponding halogenated or haloacetylated products [91, 92]. For example, the treatment of 1,4-dimethoxynaphthalene (**87**) with $PhI(OAc)_2$ and trimethylsilyl bromide affords 3-bromo-1,4-dimethoxynaphthalene (**88**) in quantitative yield (Scheme 3.33) [92].

Polyalkylbenzenes react with [hydroxy(tosyloxy)iodo]benzene in the presence of ionic halides or *N*-halosuccinimide to afford the products of ring halogenation in good yields [93–95]. For example, the reaction of mesitylene with PhI(OH)OTs in the presence of sodium bromide selectively gives monobrominated product **89** in excellent yield (Scheme 3.34) [95].

A convenient procedure for the aminobromination of electron-deficient olefins **90** using Bromamine-T (TsNBrNa) as nitrogen and bromine source promoted by (diacetoxyiodo)benzene has been developed [96]. This heavy-metal-free protocol is highly efficient and affords the vicinal bromamines **91** with excellent regio- and stereoselectivities (Scheme 3.35).

Carboxylic acids are bromodecarboxylated by reaction with (diacetoxyiodo)benzene and bromine under irradiation with a tungsten lamp, leading to respective alkyl or aryl bromides in 50–79% yield [97]. The reaction works very well with carboxylic acids having a primary, secondary or tertiary α-carbon atom, although diphenylacetic acid gives benzophenone. Benzoic acid derivatives are bromodecarboxylated in

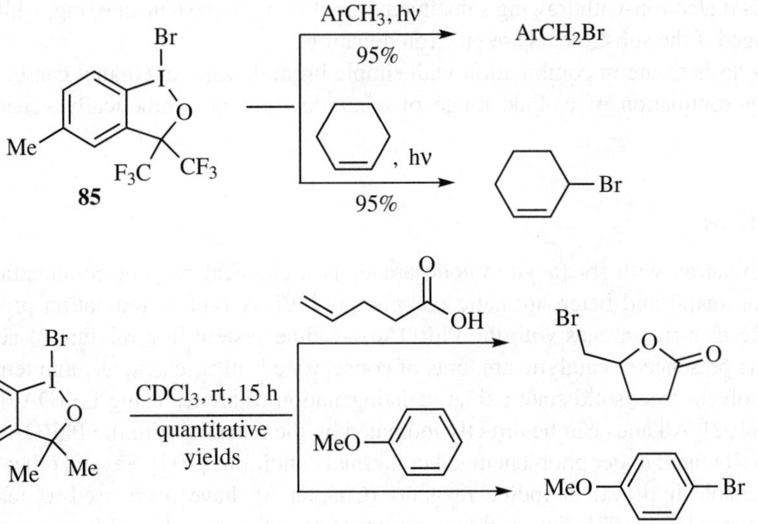

Scheme 3.32

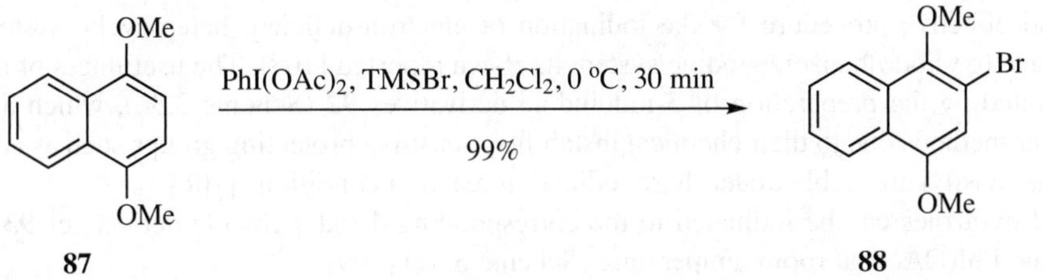

Scheme 3.33

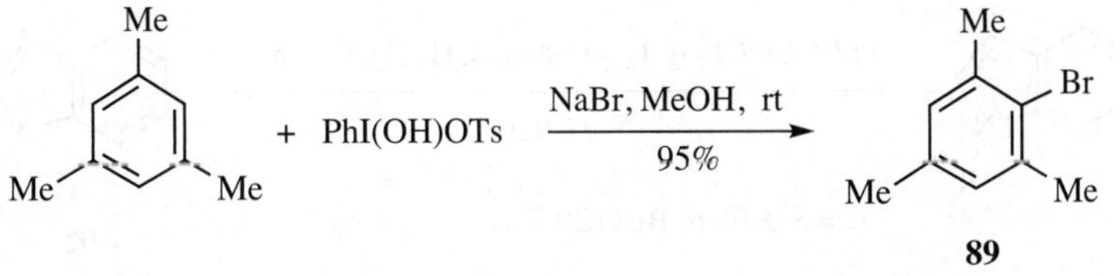

Scheme 3.34

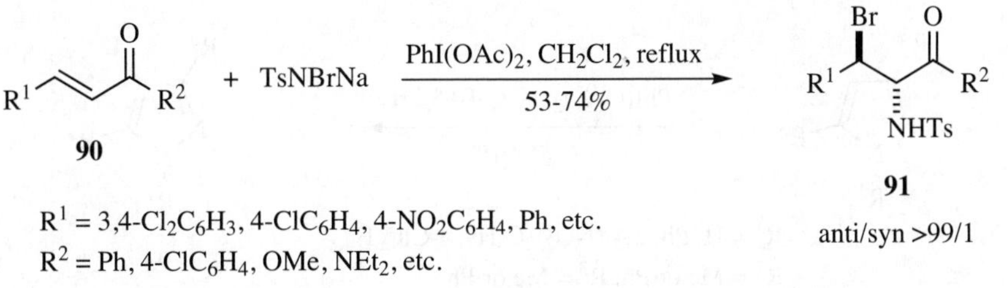

R^1 = 3,4-$Cl_2C_6H_3$, 4-ClC_6H_4, 4-$NO_2C_6H_4$, Ph, etc.
R^2 = Ph, 4-ClC_6H_4, OMe, NEt_2, etc.

anti/syn >99/1

Scheme 3.35

moderate yields if electron-withdrawing substituents are present in the benzene ring, while they are recovered mostly unchanged if the substituents are electron donating.

(Diacetoxyiodo)benzene in combination with simple bromide salts in ethanol can be used for the regioselective ethoxybromination of a wide range of enamides, giving synthetically versatile α-bromo hemiaminals [98].

3.1.4 Iodinations

Iodine in combination with [bis(acyloxy)iodo]arenes is a classical reagent combination for the oxidative iodination of aromatic and heteroaromatic compounds [99]. A typical iodination procedure involves the treatment of electron-rich arenes with the $PhI(OAc)_2$–iodine system in a mixture of acetic acid and acetic anhydride in the presence of catalytic amounts of concentrated sulfuric acid at room temperature for 15 min [100, 101]. A solvent-free, solid state oxidative halogenation of arenes using $PhI(OAc)_2$ as the oxidant has been reported [102]. Alkanes can be directly iodinated by the reaction with the $PhI(OAc)_2$–iodine system in the presence of *t*-butanol under photochemical or thermal conditions [103]. Several other iodine(III) oxidants, including recyclable hypervalent iodine reagents (Chapter 5), have been used as reagents for oxidative iodination of arenes [104–107]. For example, a mixture of iodine and [bis(trifluoroacetoxy)iodo]benzene in acetonitrile or methanol iodinates the aromatic ring of methoxy substituted alkyl aryl ketones to afford the products of electrophilic mono-iodination in 68–86% yield [107].

A mild and effective procedure for the iodination of electron-deficient heterocyclic systems using the [bis(trifluoroacetoxy)iodo]benzene–iodine system has been reported [108]. The usefulness of this procedure can be illustrated by the preparation of 3-iodoindole derivatives **92** (Scheme 3.36), which are difficult to obtain by other methods due to their chemical instability. Sensitive protecting groups such as acetyl, Boc and *tert*-butyldimethylsilyl are stable under these iodination reaction conditions [108].

Substituted pyrazoles can be iodinated to the corresponding 4-iodopyrazole derivatives **93** by treatment with iodine and $PhI(OAc)_2$ at room temperature (Scheme 3.37) [109].

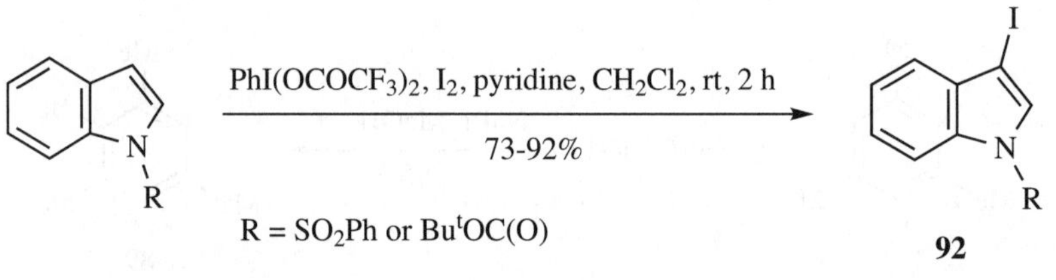

$PhI(OCOCF_3)_2$, I_2, pyridine, CH_2Cl_2, rt, 2 h; 73-92%

R = SO_2Ph or $Bu^tOC(O)$

92

Scheme 3.36

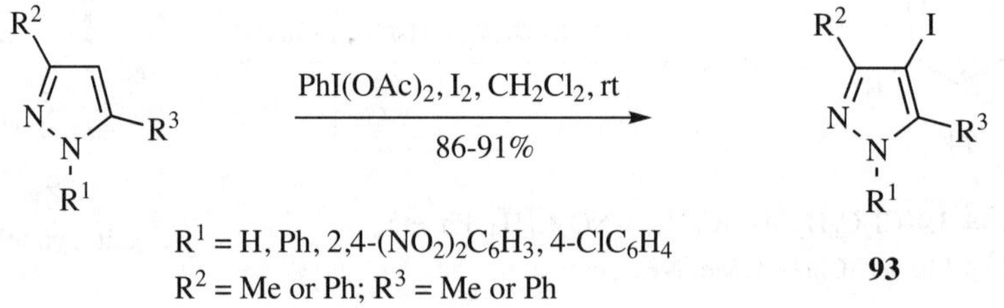

$PhI(OAc)_2$, I_2, CH_2Cl_2, rt; 86-91%

R^1 = H, Ph, 2,4-$(NO_2)_2C_6H_3$, 4-ClC_6H_4
R^2 = Me or Ph; R^3 = Me or Ph

93

Scheme 3.37

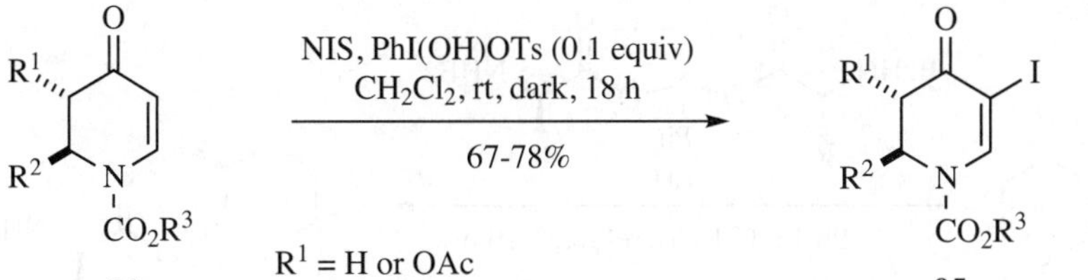

NIS, PhI(OH)OTs (0.1 equiv)
CH_2Cl_2, rt, dark, 18 h
67-78%

94 → **95**

R^1 = H or OAc
R^2 = Ph or CH_2=CH
R^3 = Ph or Bn

Scheme 3.38

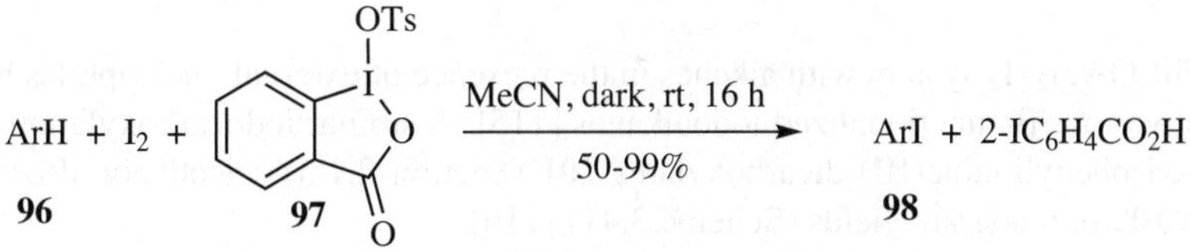

ArH + I_2 + **97** (OTs) → MeCN, dark, rt, 16 h, 50-99% → ArI + 2-$IC_6H_4CO_2H$

96 **97** **98**

ArH = 1,3,5-$(MeO)_3C_6H_3$, 1,3,5-$(i\text{-}Pr)_3C_6H_3$, 1,3,5-$Me_3C_6H_3$,
1-MeO-4-$MeCO_2C_6H_4$, 1-MeO-4-BrC_6H_4, 1,4-$Me_2C_6H_4$, 1,3-$Me_2C_6H_4$,
$MeOC_6H_5$, $Bu^tC_6H_5$, $AcOC_6H_5$, naphthalene, 2,3-benzothiophene, etc.

Scheme 3.39

Various dihydropyridone derivatives **94** can be efficiently iodinated by treatment with *N*-iodosuccinimide (NIS) in the presence of [hydroxy(tosyloxy)iodo]benzene to give products **95** (Scheme 3.38) [110].

It has been demonstrated that tosyloxybenziodoxole **97** (Section 2.1.8.1.3) can be used as an effective reagent for the oxidative iodination of aromatic compounds [111, 112]. Treatment of various aromatic compounds **96** with reagent **97** and I_2 gives the corresponding iodinated compounds **98** in good yields (Scheme 3.39). As compared with other trivalent iodine compounds, the tosylate **97** shows the best reactivity as an oxidant for oxidative halogenation [112].

In addition, the reagent **97**–iodine system can be used for the iodotosyloxylation of alkynes **99** to give the addition products **100** in good yields (Scheme 3.40) [112]. These reactions presumably proceed via the intermediate formation of arenesulfonyl hypoiodites.

Kirschning and coworkers have developed several experimental procedures for the stereoselective bromoacetoxylation or iodoacetoxylation of alkenes based on the interaction of $PhI(OAc)_2$ with iodide or bromide anions [113, 114]. The actual reacting electrophilic species in these reactions are the diacetylhalogen(I) anions, $(AcO)_2I^-$ and $(AcO)_2Br^-$, which can also be prepared as the polymer-supported variant [114].

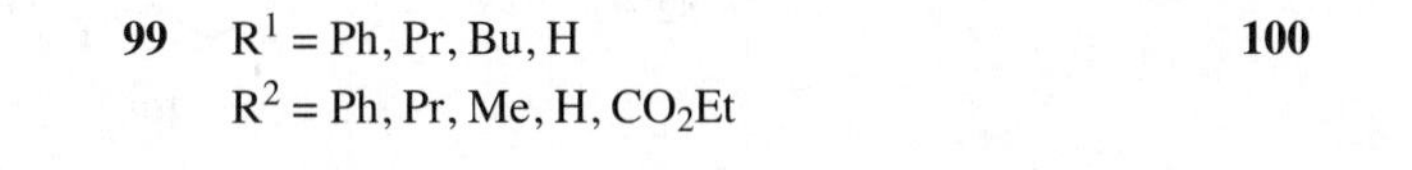

R^1—≡—R^2 → **97**, I_2, $ClCH_2CH_2Cl$, dark, rt, 16 h, 29-93% → **100** (R^1, I, TsO, R^2)

99 R^1 = Ph, Pr, Bu, H
R^2 = Ph, Pr, Me, H, CO_2Et

Scheme 3.40

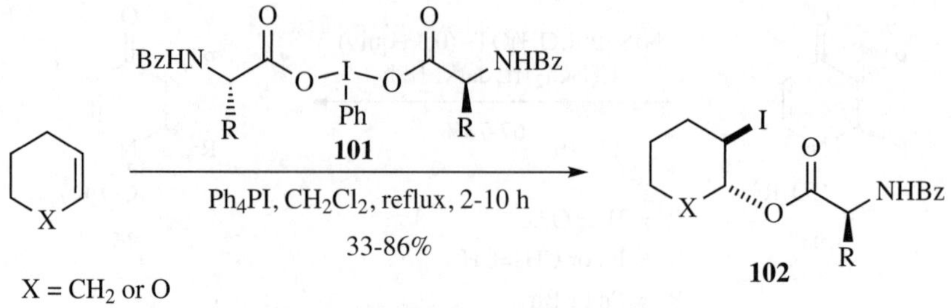

$X = CH_2$ or O
R = Me, CH_2Ph, $CH(CH_3)_2$, $CH_2CH(CH_3)_2$, $CH(CH_3)CH_2CH_3$

Ph_4PI, CH_2Cl_2, reflux, 2-10 h
33-86%

Scheme 3.41

The reaction of $PhI(OAc)_2$–I_2 system with alkenes in the presence of external nucleophiles has been used for the preparation of various β-functionalized iodoalkanes [115]. A similar iodocarboxylation of alkenes using amino acid-derived phenyliodine(III) dicarboxylates **101** (Section 2.1.5) selectively affords the respective amino acid esters **102** in moderate yields (Scheme 3.41) [116].

Similarly, [hydroxy(phosphoryloxy)iodo]benzenes **104** (Section 2.1.7) are useful reagents for iodophosphoryloxylation of alkynes and alkenes [117]. Specifically, alkynes **103** can be converted into the corresponding 1,2-iodophosphoryloxylated compounds **105** in moderate to good yields upon treatment with reagents **104** in the presence of iodine (Scheme 3.42). Under similar conditions, cyclohexene is converted into the corresponding adduct **106** in high yield [117].

The reaction of ketones **107** with [hydroxy(4-nitrobenzenesulfonyloxy)iodo]benzene and subsequent treatment with samarium iodide has been used for a one-pot preparation of α-iodoketones **108** (Scheme 3.43) in high yields [118]. 2-Iodosylbenzoic acid can also be used as a convenient recyclable hypervalent iodine oxidant for the synthesis of α-iodoketones by oxidative iodination of ketones [119].

The $PhI(OAc)_2$–I_2 system has been used for the oxidative decarboxylation/iodination of carboxylic acids [120–123]. This reaction was employed in the efficient syntheses of enantiopure 1-benzoyl-2(*S*)-*tert*-butyl-3-methylperhydropyrimidin-4-one [120] and 2-substituted-5-halo-2,3-dihydro-4(*H*)-pyrimidin-4-ones [123].

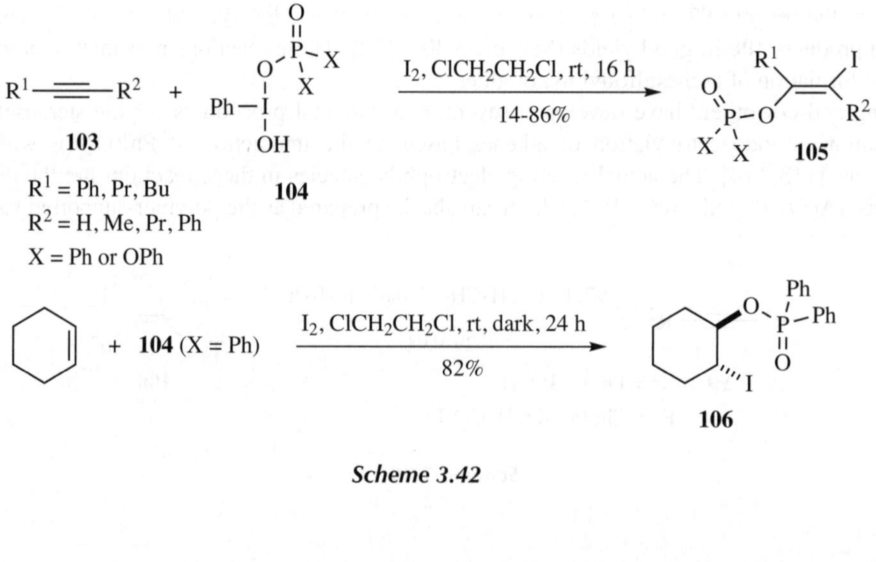

I_2, $ClCH_2CH_2Cl$, rt, 16 h
14-86%

R^1 = Ph, Pr, Bu
R^2 = H, Me, Pr, Ph
X = Ph or OPh

+ **104** (X = Ph)
I_2, $ClCH_2CH_2Cl$, rt, dark, 24 h
82%

Scheme 3.42

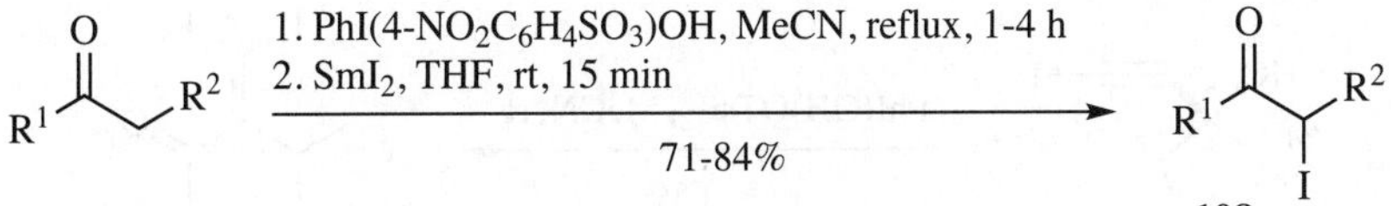

1. $PhI(4\text{-}NO_2C_6H_4SO_3)OH$, MeCN, reflux, 1-4 h
2. SmI_2, THF, rt, 15 min

71-84%

107 → **108**

R^1 = aryl, alkyl, cyclopropyl; R^2 = H, CH_3
or $R^1 + R^2$ = cycloalkyl

Scheme 3.43

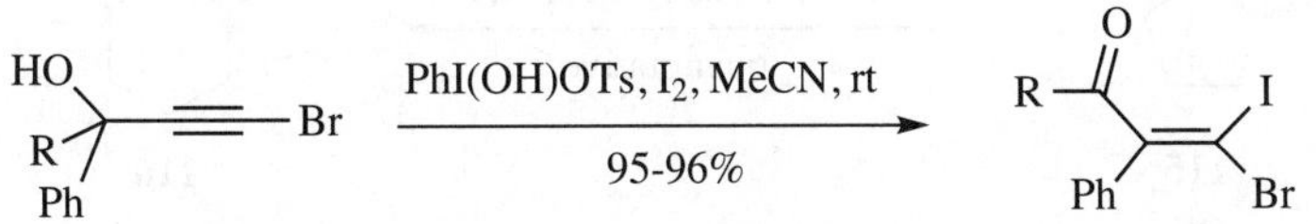

PhI(OH)OTs, I_2, MeCN, rt

95-96%

109 R = Me, H → **110**

Scheme 3.44

1-Iodoalkynes can be prepared in good to excellent yields by the oxidative iodination of terminal alkynes with $PhI(OAc)_2$, potassium iodide and copper(I) iodide [124].

In a series of communications, McNelis and coworkers have reported the scope of reactions of the reagent combination PhI(OH)OTs–halogen or PhI(OH)OTs–*N*-halosuccinimide with alkynols, resulting in the formation of various haloenones [125–133]. All these oxidative rearrangements involve iodonium intermediates and are highly regio- and stereoselective. For example, bromoethynyl alcohols **109** are cleanly converted into β,β-dihaloenones **110** by reaction with the PhI(OH)OTs–I_2 system (Scheme 3.44) [127, 128].

Reactions of this type (Scheme 3.44) are especially useful for the preparation of cyclic β,β-dihaloenones by ring expansions of alkynyl cyclopentanols or alkynyl cyclohexanols. 1-Bromoethynylcyclopentanols **111** react with equimolar amounts of iodine and PhI(OH)OTs under mild conditions to produce cyclohexanone derivatives **112** with a high degree of stereoselectivity (Scheme 3.45) [129].

Likewise, stereoselective ring expansions in 1-iodoethynyl-2-methylcyclopentanols **113** and **115** affords β,β-diiodoenones upon treatment with PhI(OH)OTs and iodine (Scheme 3.46) [134]. Depending on the relative stereochemistry of the methyl and the hydroxyl groups in the starting cyclopentanol, the products are 2-(diiodomethylidene)-3-methylcyclohexanone (**114**), from *cis*-cyclopentanol **113** and 2-(diiodomethylidene)-6-methylcyclohexanone (**116**) from *trans*-cyclopentanol **115** [134].

A similar reaction was used in the synthesis of phenanthrenone **118** from the readily available fluorenone derivative **117** (Scheme 3.47) [130].

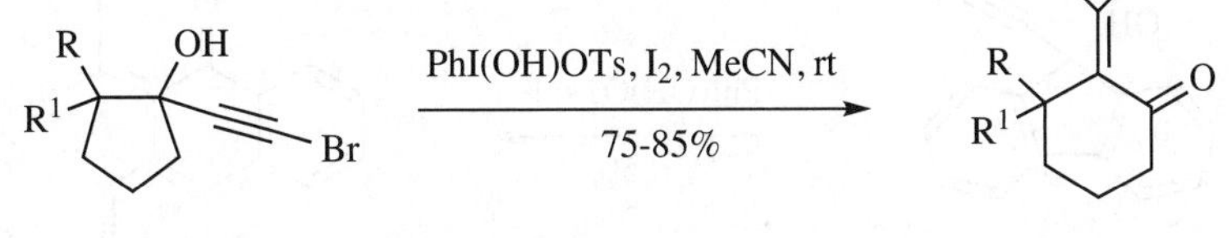

PhI(OH)OTs, I_2, MeCN, rt

75-85%

111 R, R^1 = H, alkyl → **112**

Scheme 3.45

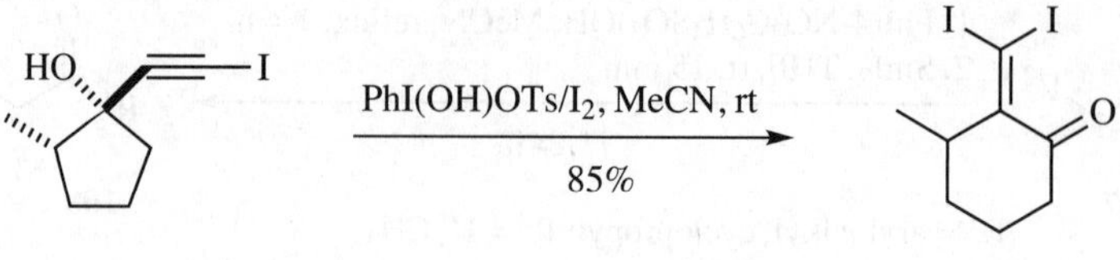

$PhI(OH)OTs/I_2$, MeCN, rt
85%

113 114

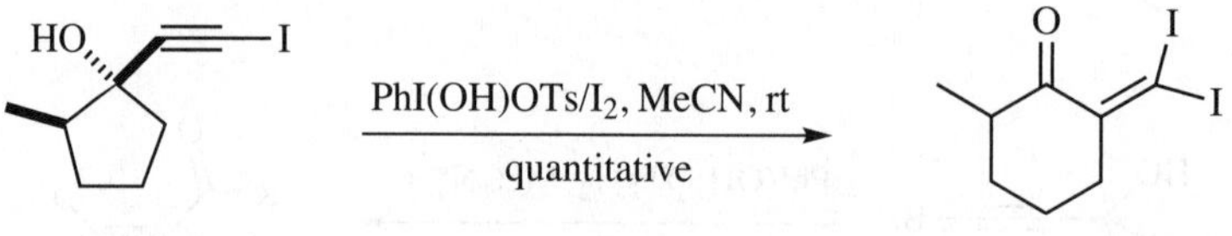

$PhI(OH)OTs/I_2$, MeCN, rt
quantitative

115 116

Scheme 3.46

Ring expansion of camphor (**119**) and adamantanone (**121**) derivatives under the same conditions afforded the respective (*Z*)-bromoiodoenones **120** and **122** in good yields and in high stereoselectivity (Scheme 3.48) [131–133].

The treatment of iodoalkynol derivatives of xylose **123** with PhI(OH)OTs and iodine under similar conditions furnishes β,β-diiodoenol ethers **124** with a furo[3,4-*b*]furan heterocyclic core (Scheme 3.49) [135].

3.1.5 Oxidation of Alcohols

In contrast with λ^5-iodanes (Section 3.2), hypervalent iodine(III) reagents are not effective oxidants of alcohols in the absence of catalysts. Kita and coworkers were the first to find that, in the presence of bromide salts, iodosylbenzene or (diacetoxyiodo)benzene can be used as an efficient reagent for selective oxidation of alcohols [136, 137]. The iodosylbenzene–KBr system is applicable to the oxidation of various primary and secondary alcohols, even in the presence of sensitive functional groups such as ether, ester, sulfonamide and azido groups. Primary alcohols under these conditions afford carboxylic acids (Scheme 3.50), while the oxidation of various secondary alcohols under similar conditions affords the appropriate ketones in almost quantitative yield [136].

[Bis(acyloxy)iodo]arenes in the presence of KBr in water can oxidize primary and secondary alcohols analogously to the iodosylbenzene–KBr system [137]. The oxidation of primary alcohols affords carboxylic acids or esters [136, 138], while the oxidation of secondary alcohols under same conditions

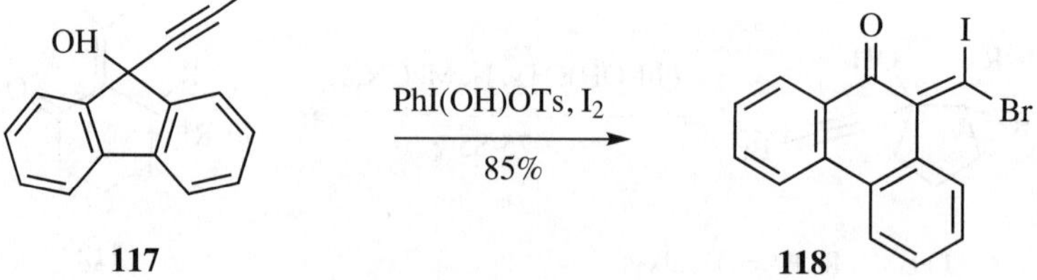

$PhI(OH)OTs, I_2$
85%

117 118

Scheme 3.47

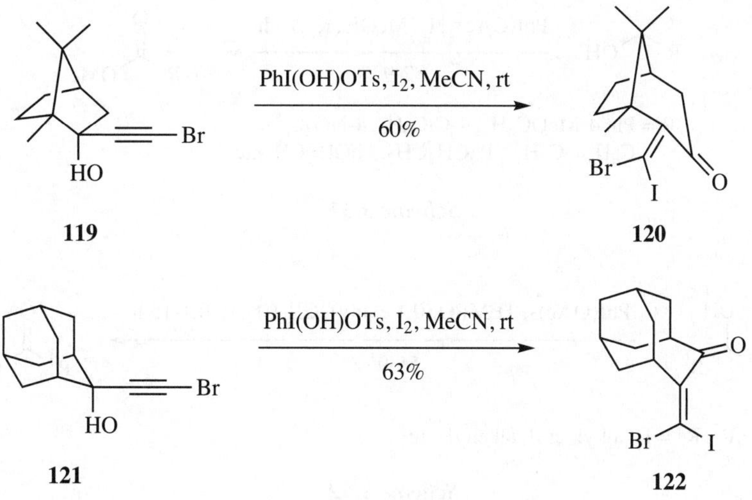

PhI(OH)OTs, I_2, MeCN, rt
60%

119 → **120**

PhI(OH)OTs, I_2, MeCN, rt
63%

121 → **122**

Scheme 3.48

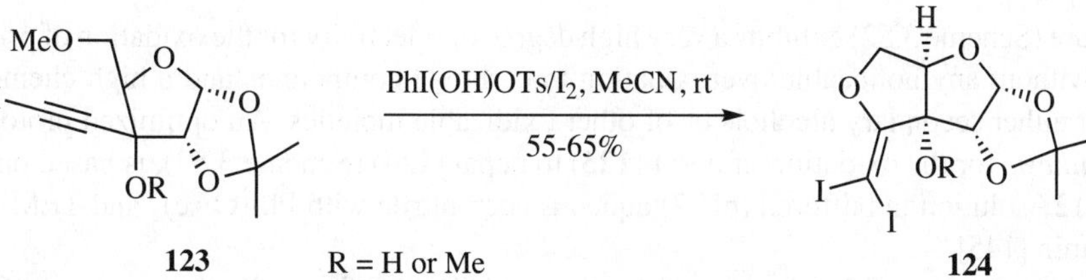

PhI(OH)OTs/I_2, MeCN, rt
55-65%

123 R = H or Me → **124**

Scheme 3.49

leads to the respective ketones in excellent yields [139]. Aldehydes can be converted into methyl esters by a similar procedure using $PhI(OAc)_2$ and NaBr [140]. Molecular iodine can serve as an efficient catalyst in the oxidation of secondary alcohols to ketones and primary alcohols to carboxylic acids using $PhI(OAc)_2$ as an oxidant in acetonitrile solution [141]. The oxidation of primary alcohols or aldehydes with the $PhI(OAc)_2$–I_2 system in methanol solution yields the respective methyl esters in excellent yields (Scheme 3.51) [141, 142].

An efficient procedure for the oxidation alcohols with $PhI(OAc)_2$ in the presence of catalytic amounts of TEMPO (2,2,6,6-tetramethyl-1-piperidinyloxyl), originally developed by Piancatelli, Margarita and coworkers [143], has been frequently used in recent years [144–150]. This procedure works well for the

$$RCH_2OH \xrightarrow[76\text{-}92\%]{(PhIO)_n\ (2.2\ equiv),\ KBr\ (0.2\text{-}1\ equiv),\ H_2O,\ rt,\ 2\ h} RCO_2H$$

R = $Ph(CH_2)_2$, $BnO(CH_2)_3$, $EtO_2C(CH_2)_4$, $N_3(CH_2)_4$

Scheme 3.50

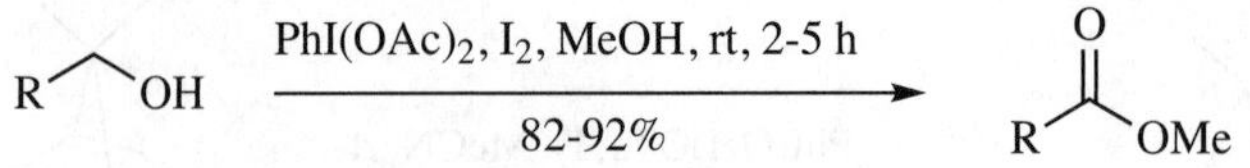

R⌒OH —[$PhI(OAc)_2$, I_2, MeOH, rt, 2-5 h; 82-92%]→ RC(O)OMe

R = Ph, 4-$MeOC_6H_4$, 4-ClC_6H_4, 4-$NO_2C_6H_4$,
C_9H_{19}, C_7H_{15}, $PhCH_2CH_2$, PhCH=CH, etc.

Scheme 3.51

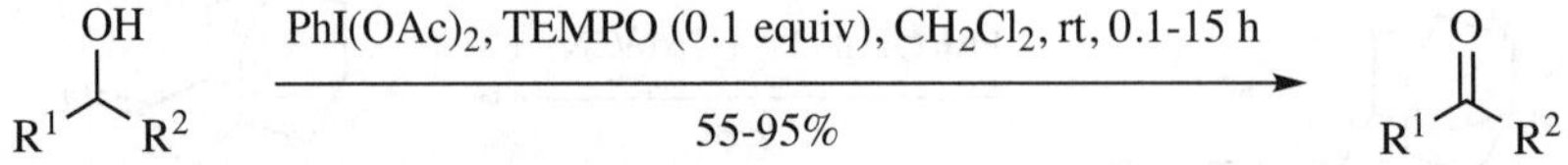

R^1CH(OH)R^2 —[$PhI(OAc)_2$, TEMPO (0.1 equiv), CH_2Cl_2, rt, 0.1-15 h; 55-95%]→ R^1C(O)R^2

R^1, R^2 = H, alkyl, aryl, alkenyl, etc.

Scheme 3.52

conversion of various primary and secondary alcohols into carbonyl compounds in generally high yields (Scheme 3.52) [143].

This procedure (Scheme 3.52) exhibits a very high degree of selectivity for the oxidation of primary alcohols to aldehydes, without any noticeable overoxidation to carboxyl compounds and a high chemoselectivity in the presence of either secondary alcohols or of other oxidizable moieties. An optimized protocol, published in *Organic Synthesis* for the oxidation of nerol (**125**) to nepal (**126**) (Scheme 3.53), is based on the treatment of the alcohol **125** solution in buffered (pH 7) aqueous acetonitrile with $PhI(OAc)_2$ and TEMPO (0.1 equiv) at 0 °C for 20 min [145].

A similar oxidative protocol has been used for the oxidation of (fluoroalkyl)alkanols, $R_F(CH_2)_nCH_2OH$, to the respective aldehydes [146], in the one-pot selective oxidation/olefination of primary alcohols using the $PhI(OAc)_2$–TEMPO system and stabilized phosphorus ylides [147] and in the chemo-enzymatic oxidation–hydrocyanation of γ,δ-unsaturated alcohols [148]. Other [bis(acyloxy)iodo]arenes can be used instead of $PhI(OAc)_2$ in the TEMPO-catalyzed oxidations, in particular the recyclable monomeric and the polymer-supported hypervalent iodine reagents (Chapter 5). Further modifications of this method include the use of polymer-supported TEMPO [151], fluorous-tagged TEMPO [152, 153], ion-supported TEMPO [154] and TEMPO immobilized on silica [148].

Based on the ability of the $PhI(OAc)_2$–TEMPO system to selectively oxidize primary alcohols to the corresponding aldehydes in the presence of secondary alcohols, Forsyth and coworkers have developed the selective oxidative conversion of various highly functionalized 1°,2°-1,5-diols into the corresponding δ-lactones [155]. A representative example, showing the conversion of substrate **127** into the δ-lactone

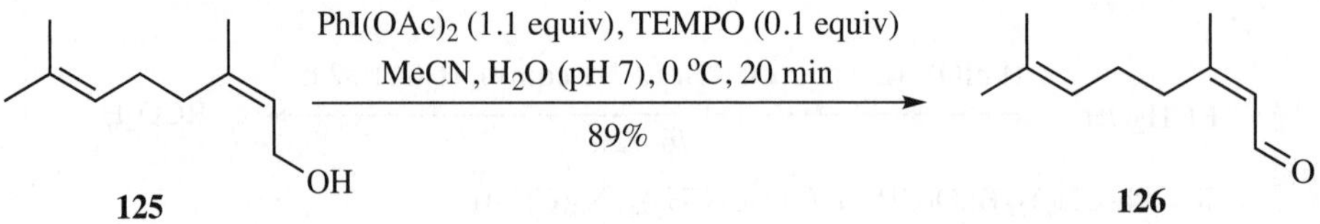

125 (OH) —[$PhI(OAc)_2$ (1.1 equiv), TEMPO (0.1 equiv); MeCN, H_2O (pH 7), 0 °C, 20 min; 89%]→ **126** (O)

Scheme 3.53

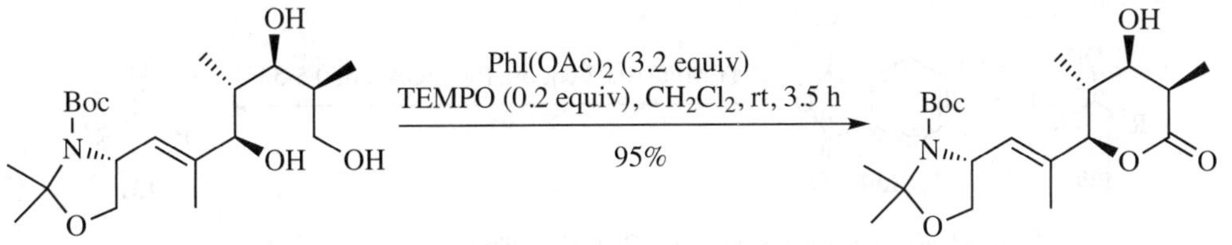

PhI(OAc)$_2$ (3.2 equiv)
TEMPO (0.2 equiv), CH$_2$Cl$_2$, rt, 3.5 h
95%

127 128

Scheme 3.54

128, is given in Scheme 3.54. Monitoring of this reaction revealed initial formation of the intermediate lactol species, which then undergoes further oxidation to the lactone [155]. A similar PhI(OAc)$_2$–TEMPO promoted γ-lactonization has been utilized in the asymmetric total synthesis of the antitumor agent (+)-eremantholide A [156].

An efficient and mild procedure has been described for the oxidation of different types of alcohols to carbonyl compounds using TEMPO as the catalyst and (dichloroiodo)benzene as a stoichiometric oxidant at 50 °C in chloroform solution in the presence of pyridine [157]. Under these conditions, 1,2-diols are oxidized to β-hydroxyketones or β-diketones depending upon the amount of PhICl$_2$ used. Interestingly, a competitive study has shown that this system preferentially oxidizes aliphatic secondary alcohols over aliphatic primary alcohols [157], while the PhI(OAc)$_2$–TEMPO system selectively converts primary alcohols into the corresponding aldehydes in the presence of secondary alcohols.

A similar TEMPO-catalyzed system for the oxidation of alcohols using 1-chloro-1,2-benziodoxol-3(1*H*)-one (**130**) (Section 2.1.8.1.1) as the terminal oxidant in ethyl acetate in the presence of pyridine at room temperature has been reported [158]. Various alcohols **129** can be oxidized to the corresponding carbonyl compounds in high to excellent yields under these conditions (Scheme 3.55). The oxidation of primary alcohols (**129,** R^2 = H) in this reaction proceeds generally faster compared to the secondary alcohols. Different heteroaromatic rings (e.g., pyridine, furan and thiophene) and carbon–carbon double bonds are well tolerated under the reaction conditions. Moreover, reagent **130** can be easily recycled from the reaction mixture and reused. The mechanism of this reaction includes initial homolytic cleavage of the I–Cl bond providing a chlorine atom and the iodanyl radical **132,** which is in equilibrium with the benzoyloxy radical **133.** TEMPO (**134**) is oxidized by the chlorine atom to oxoammonium salt **135,** which then oxidizes the alcohol **129** to the corresponding carbonyl compound **131** and is itself reduced to hydroxylamine **136.** The benzoyloxy radical **133** accomplishes the regeneration of TEMPO **134** from hydroxylamine **136,** giving rise to 2-iodobenzoic acid and the catalytic cycle is complete (Scheme 3.55) [158].

Only a few examples of uncatalyzed oxidation of alcohols with hypervalent iodine(III) reagents have been reported [159–162]. Vicinal diols can be cleaved to aldehydes in dichloromethane at room temperature by polymer-supported (diacetoxyiodo)benzene (Chapter 5) [159]. Substituted benzyl alcohols can be oxidized by [bis(trifluoroacetoxy)iodo]benzene in aqueous acetic acid to the corresponding benzaldehydes [160]. Benzylic alcohols can also be oxidized with [hydroxy(tosyloxy)iodo]benzene under solvent-free microwave irradiation conditions to afford the corresponding aldehydes or ketones in excellent yields [162]. Vicinal fullerene diol is oxidized to fullerene dione in 80% yield by PhI(OAc)$_2$ in benzene at 35 °C [161]. Oligomeric iodosylbenzene sulfate, (PhIO)$_3$·SO$_3$, can oxidize benzyl alcohol in aqueous acetonitrile at room temperature to afford benzaldehyde in 92% yield [163]. Alcohols can be selectively oxidized to the corresponding aldehydes or ketones with (PhIO)$_3$·SO$_3$ in water in the presence of β-cyclodextrin [164]. Transition metal catalyzed oxidations of alcohols using hypervalent iodine reagents as stoichiometric oxidants are discussed in Section 3.1.20.

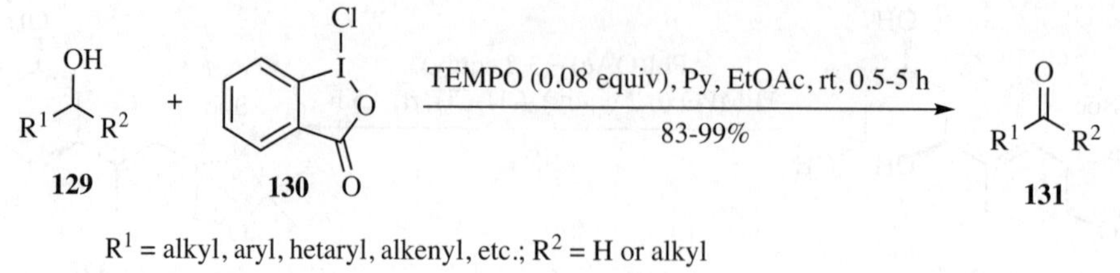

R^1 = alkyl, aryl, hetaryl, alkenyl, etc.; R^2 = H or alkyl

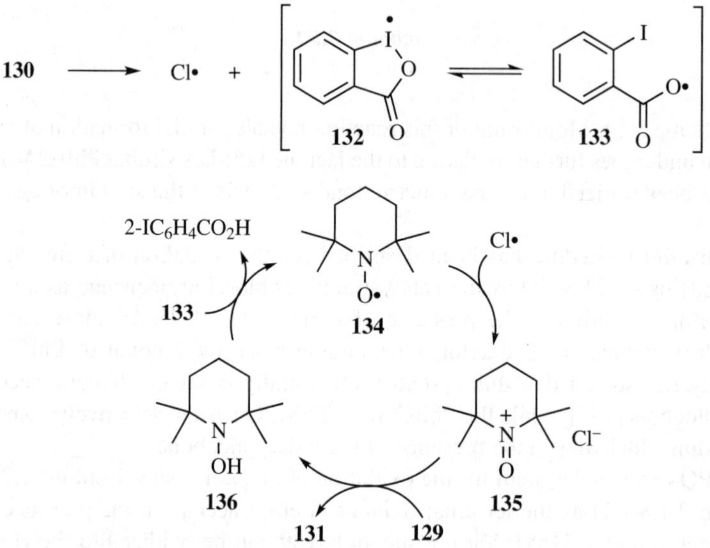

Scheme 3.55

3.1.6 Oxidative Functionalization of Carbonyl Compounds

Hypervalent iodine reagents are commonly used for oxidative α-functionalization of carbonyl compounds [165]. The α-hydroxycarbonyl group is of interest to synthetic organic chemists due to its ubiquity in nature, occurring in polyketide, terpenoid and alkaloid natural products. In the 1980s Moriarty and coworkers developed a particularly useful methodology for the oxidative α-hydroxylation of enolizable carbonyl compounds using iodosylbenzene, (diacetoxyiodo)benzene, or other hypervalent iodine oxidants [166]. The reagent system $PhI(OAc)_2$/KOH/MeOH is especially efficient for hydroxylation of enolizable ketones **137** via the initially formed α-hydroxydimethylacetals **138** (Scheme 3.56) [167–173]. This procedure has been further extended to the α-hydroxylation of carboxylic esters [167, 174].

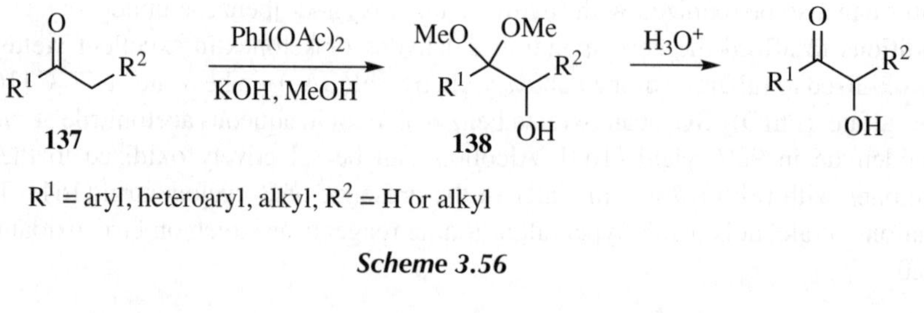

R^1 = aryl, heteroaryl, alkyl; R^2 = H or alkyl

Scheme 3.56

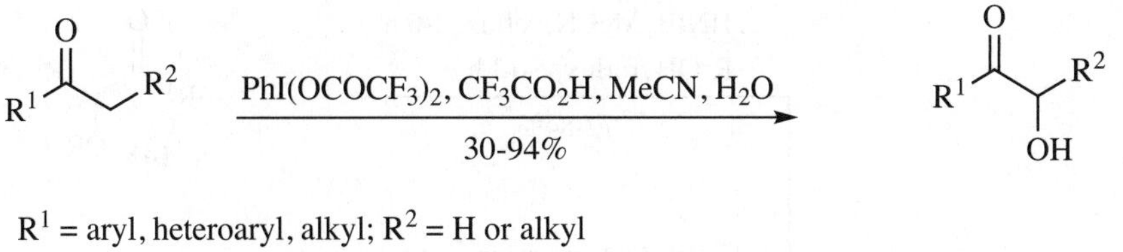

$PhI(OCOCF_3)_2$, CF_3CO_2H, MeCN, H_2O

30-94%

R^1 = aryl, heteroaryl, alkyl; R^2 = H or alkyl

Scheme 3.57

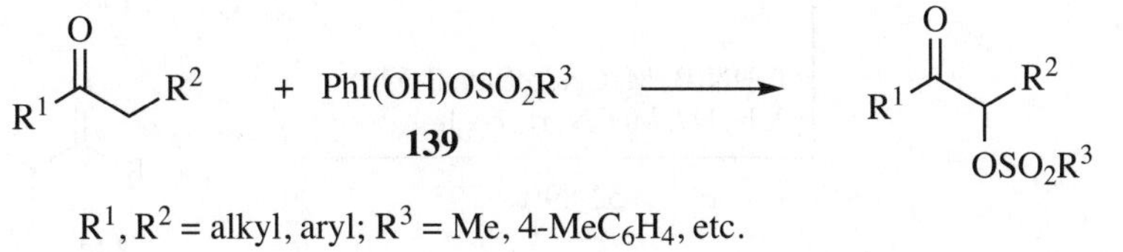

+ $PhI(OH)OSO_2R^3$ **139**

R^1, R^2 = alkyl, aryl; R^3 = Me, 4-MeC_6H_4, etc.

Scheme 3.58

Hydroxylation of some substituted acetophenones under these conditions (Scheme 3.56) is especially useful for the synthesis of various oxygen-containing heterocyclic compounds [175–179]. Applications of this methodology in organic synthesis, especially in the chemistry of heterocyclic compounds, have been summarized in several reviews [166, 180–182].

A similar hydroxylation of aromatic, heteroaromatic and aliphatic ketones can also be performed using [bis(trifluoroacetoxy)iodo]benzene under acidic conditions (Scheme 3.57) [183]. A plausible mechanism for this hydroxylation involves initial electrophilic addition of $PhI(OCOCF_3)_2$ to the enolized ketone and subsequent nucleophilic substitution in the iodonium intermediate.

The functionalization of carbonyl compounds at the α-carbon is the most typical reaction of [hydroxy(organosulfonyloxy)iodo]arenes **139** (Scheme 3.58) [184].

Of particular use is the reaction of [hydroxy(tosyloxy)iodo]benzene (HTIB, also known as Koser's reagent) with ketones leading to α-tosyloxyketones [185–187]. This is a highly chemoselective reaction; different functional groups, aromatic rings and carbon–carbon double bonds are well tolerated under the reaction conditions [188]. Scheme 3.59 shows a representative example of synthetic application of HTIB for the functionalization of the azabicyclic alkaloid anatoxin-a, which is one of the most potent nicotinic antagonists. Specifically, the reaction of *N*-Boc anatoxin-a **140** with HTIB is the method of choice for the preparation of the synthetically versatile α-tosyloxy ketone **141** (Scheme 3.59) [189].

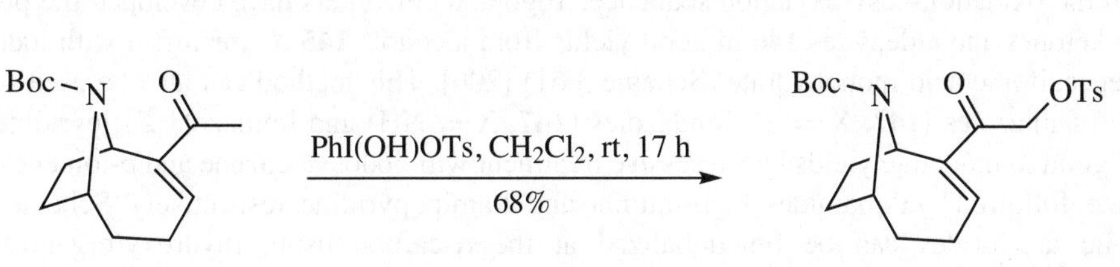

140 $PhI(OH)OTs$, CH_2Cl_2, rt, 17 h, 68% **141**

Scheme 3.59

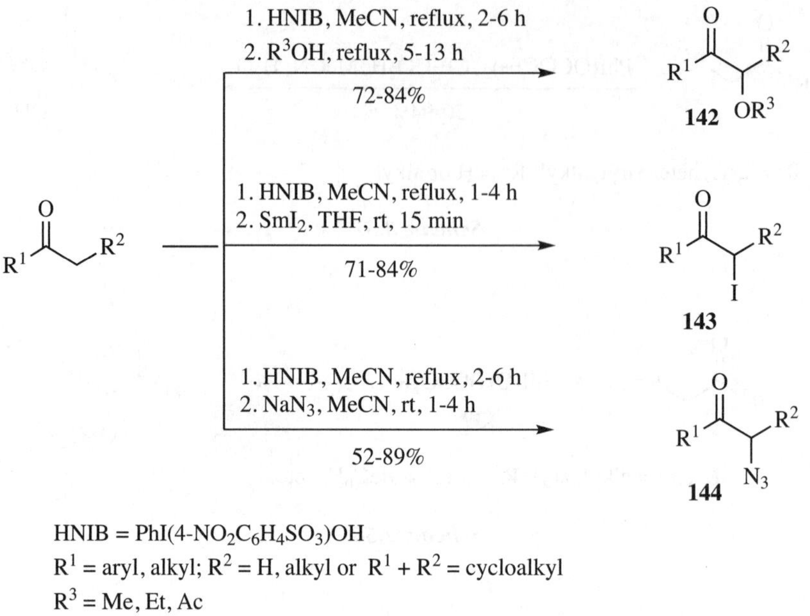

1. HNIB, MeCN, reflux, 2-6 h
2. R^3OH, reflux, 5-13 h
72-84%
142 ($R^1C(O)CH(OR^3)R^2$)

1. HNIB, MeCN, reflux, 1-4 h
2. SmI_2, THF, rt, 15 min
71-84%
143 ($R^1C(O)CH(I)R^2$)

1. HNIB, MeCN, reflux, 2-6 h
2. NaN_3, MeCN, rt, 1-4 h
52-89%
144 ($R^1C(O)CH(N_3)R^2$)

HNIB = $PhI(4\text{-}NO_2C_6H_4SO_3)OH$
R^1 = aryl, alkyl; R^2 = H, alkyl or $R^1 + R^2$ = cycloalkyl
R^3 = Me, Et, Ac

Scheme 3.60

This reaction of ketones with [hydroxy(organosulfonyloxy)iodo]arenes followed by treatment with an appropriate nucleophile *in situ* offers a convenient entry into various other α-substituted ketones, or can lead to various heterocycles via cyclization of the initially formed α-tosyloxyketones [176, 177, 184, 190, 191]. For example, the reaction of various ketones with [hydroxy(*p*-nitrobenzenesulfonyloxy)iodo]benzene (HNIB) and subsequent treatment with the appropriate nucleophile has been used for a one-pot preparation of secondary α-alkoxy or α-acetoxy ketones **142** [192], α-iodoketones **143** [118] and α-azidoketones **144** [193] in generally high yields (Scheme 3.60).

The tosyloxylation of suitable ketones followed by heterocyclization has been utilized in the syntheses of the following heterocyclic systems: 2-aroylbenzo[*b*]furans [194], 3-aryl-5,6-dihydroimidazo[2,1-*b*][1,3]-thiazoles [194], 6-arylimidazo[2,1-*b*]thiazoles [195], (1*S*,2*R*)-indene oxide [196], 2-mercaptothiazoles [197], triazolo-[3,4-*b*]-1,3,4-thiadiazines [198], dihydroindeno[1,2-*e*][1,2,4]triazolo[3,4-*b*][1,3,4]thiadiazines [199], furo[3,2-*c*]coumarins [200], 4,5-diarylisoxazoles [201], 2-substituted 4,5-diphenyloxazoles [202], quinoxaline [203], 3-carbomethoxy-4-arylfuran-2-(5*H*)-ones [204], thiazol-2(3*H*)-imine-linked glycoconjugates [205] and other important heterocycles.

Based on the oxidation–tosyloxylation sequence, Togo and coworkers have developed the preparation of α-tosyloxy ketones and aldehydes **146** in good yields from alcohols **145** by treatment with iodosylbenzene and *p*-toluenesulfonic acid monohydrate (Scheme 3.61) [206]. This method can also be used for the direct preparation of thiazoles (**147,** X = S), imidazoles (**147,** X = NH) and imidazo[1,2-*a*]pyridines **148** from alcohols in good to moderate yields by successive treatment with iodosylbenzene and *p*-toluenesulfonic acid monohydrate, followed by thioamides, benzamidine and 2-aminopyridine, respectively (Scheme 3.61) [206].

Carboxylic anhydrides can be functionalized at the α-carbon using [hydroxy(organosulfonyloxy)-iodo]arenes. Treatment of carboxylic anhydrides **149** with reagents **150** at about 100 °C followed by esterification of the reaction mixture with methanol affords 2-sulfonyloxycarboxylate esters **151** in moderate to good yields (Scheme 3.62) [207].

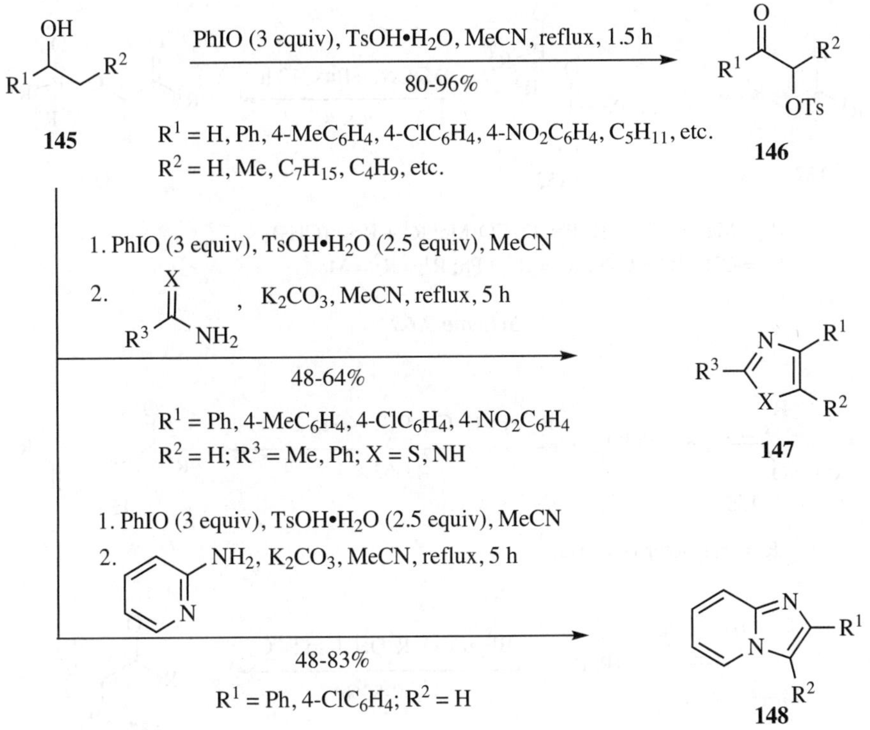

PhIO (3 equiv), $TsOH•H_2O$, MeCN, reflux, 1.5 h

80-96%

145 → **146**

R^1 = H, Ph, $4-MeC_6H_4$, $4-ClC_6H_4$, $4-NO_2C_6H_4$, C_5H_{11}, etc.
R^2 = H, Me, C_7H_{15}, C_4H_9, etc.

1. PhIO (3 equiv), $TsOH•H_2O$ (2.5 equiv), MeCN
2. $R^3C(=X)NH_2$, K_2CO_3, MeCN, reflux, 5 h

48-64%

R^1 = Ph, $4-MeC_6H_4$, $4-ClC_6H_4$, $4-NO_2C_6H_4$
R^2 = H; R^3 = Me, Ph; X = S, NH

147

1. PhIO (3 equiv), $TsOH•H_2O$ (2.5 equiv), MeCN
2. 2-aminopyridine (NH_2), K_2CO_3, MeCN, reflux, 5 h

48-83%

R^1 = Ph, $4-ClC_6H_4$; R^2 = H

148

Scheme 3.61

[Hydroxy(phosphoryloxy)iodo]benzenes **153** (Section 2.1.7) are useful reagents for the introduction of phosphonate or phosphinate groups at the α-position to ketone or ester carbonyl groups of carbonyl compounds **152** to produce the corresponding products **154** (Scheme 3.63) [208].

3.1.7 Oxidative Functionalization of Silyl Enol Ethers

Reactions of silyl enol ethers with iodosylbenzene in the presence of $BF_3{\cdot}Et_2O$ can be used to form new carbon–carbon bonds [209], prepare α-substituted ketones [210] and synthesize oxygen-containing

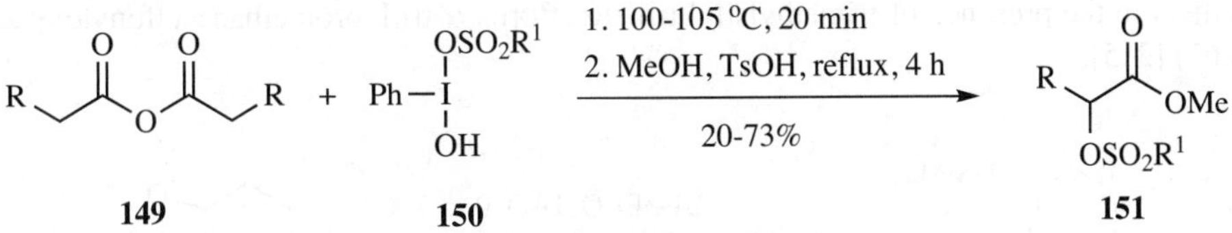

149 + **150** (Ph–I(OH)OSO_2R^1) → **151**

1. 100-105 °C, 20 min
2. MeOH, TsOH, reflux, 4 h

20-73%

R = Me, Et, Pr, Bu, C_8H_{17}, $C_{10}H_{21}$, MeOOC, Pr^i, Bu^i, $PhCH_2CH_2$
R^1 = $4-MeC_6H_4$, Me, $4-NO_2C_6H_4$, (+)-10-camphoryl

Scheme 3.62

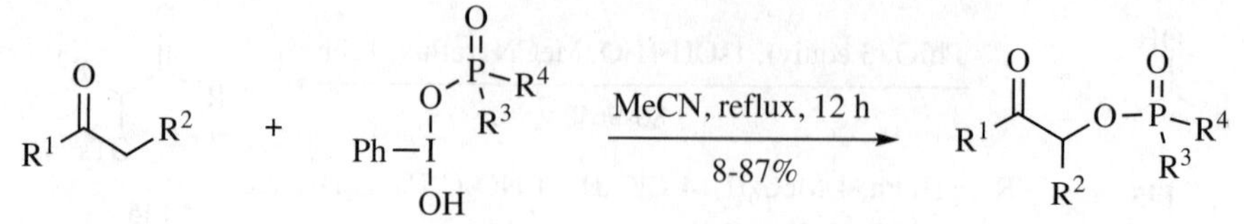

152 + 153 → 154 (MeCN, reflux, 12 h; 8-87%)

R^1 = Me, Ph; R^2 = H, PhCO, CO_2Me; $R^1 + R^2 = (CH_2)_4$
$R^3 = CH_3$, R^4 = OPh; $R^1 = R^2$ = Ph; $R^1 = R^2$ = Me

Scheme 3.63

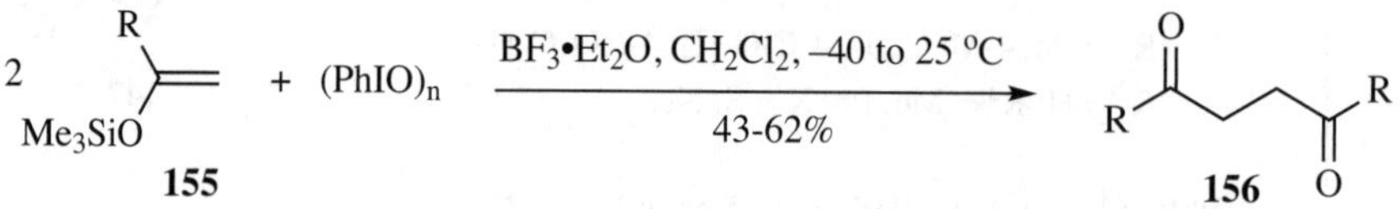

2 155 + $(PhIO)_n$ → 156 (BF_3•Et_2O, CH_2Cl_2, –40 to 25 °C; 43-62%)

R = aryl, heteroaryl, Bu^t

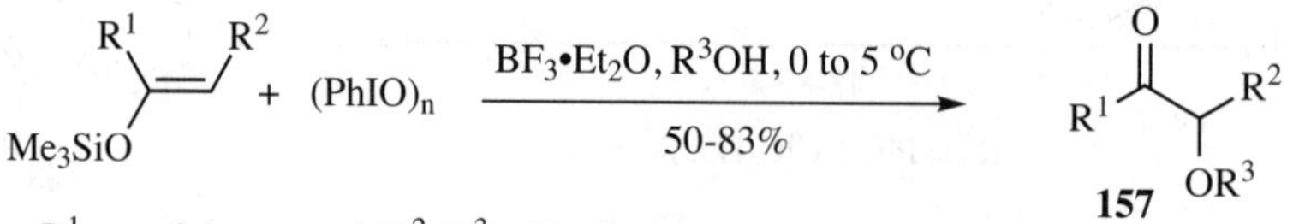

+ $(PhIO)_n$ → 157 (BF_3•Et_2O, R^3OH, 0 to 5 °C; 50-83%)

R^1 = aryl, heteroaryl; R^2, R^3 = H, alkyl

Scheme 3.64

heterocyclic compounds [176–178]. For example, the reaction of silyl enol ethers **155** with PhIO–BF_3 in non-nucleophilic solvents usually affords 1,4-butanediones **156** as major products [211], while in the presence of water or alcohols these reactants give α-hydroxy- or α-alkoxyketones **157** (Scheme 3.64) [212].

A similar intramolecular oxidation of silyl enol ether **158** affords oxygen heterocycle **159** (Scheme 3.65) [213].

The reaction of silyl enol ethers with PhIO–BF_3 in the presence of triethyl phosphite as an external nucleophile yields β-keto phosphonates **160** in good yield (Scheme 3.66) [214].

Various Lewis acids and strong Brønsted acids can be used instead of $BF_3 \cdot Et_2O$ to catalyze reactions of iodosylbenzene with unsaturated compounds. For example, the reaction of iodosylbenzene with silyl enol ethers in the presence of trimethylsilyl triflate affords α-trifluoromethanesulfonyloxyketones **161** (Scheme 3.67) [215].

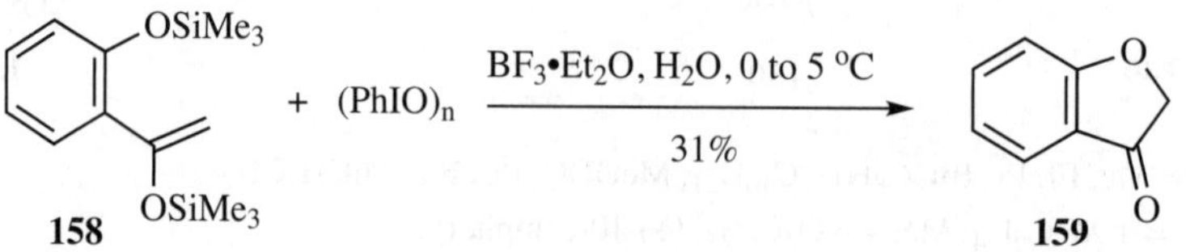

158 + $(PhIO)_n$ → 159 (BF_3•Et_2O, H_2O, 0 to 5 °C; 31%)

Scheme 3.65

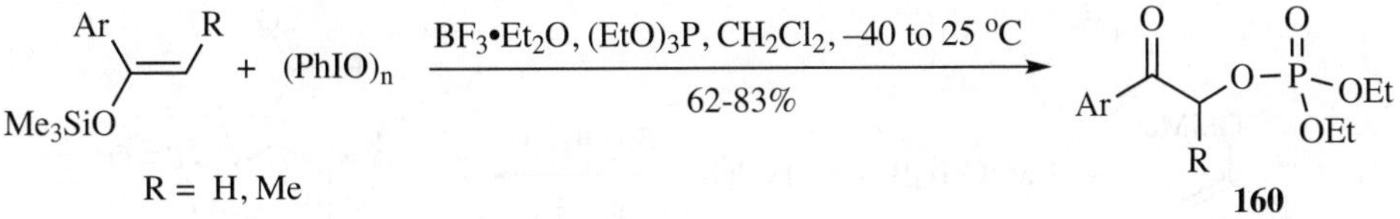

Ar R Me3SiO + (PhIO)n; R = H, Me; $BF_3 \cdot Et_2O$, $(EtO)_3P$, CH_2Cl_2, –40 to 25 °C; 62-83%; **160**

Scheme 3.66

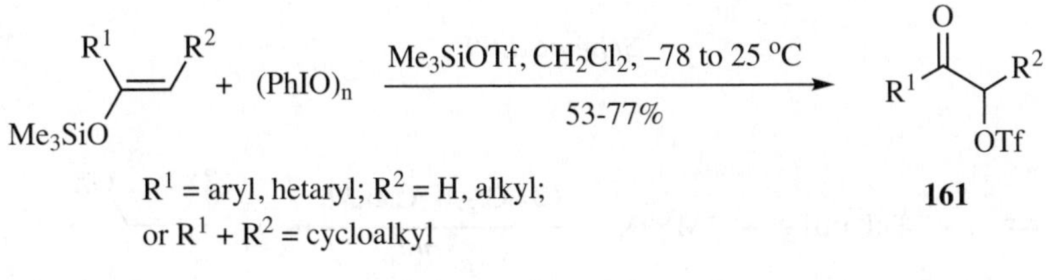

R^1 R^2 Me3SiO + (PhIO)n; Me_3SiOTf, CH_2Cl_2, –78 to 25 °C; 53-77%; **161**

R^1 = aryl, hetaryl; R^2 = H, alkyl;
or R^1 + R^2 = cycloalkyl

Scheme 3.67

Trimethylsilyl ketene acetals of esters **162** and lactones **164** react with iodosylbenzene in methanol to give the corresponding α-methoxylated esters **163** or α-methoxylated lactones **165** in good yields (Scheme 3.68) [216]. It is assumed that the actual oxidizing monomeric species in this reaction is $PhI(OMe)_2$.

In some reactions (difluoroiodo)arenes can be used as general oxidizing reagents. For example, Koser and coworkers applied a difluoroiodotoluene/phosphoric acid mixture as a reagent for direct conversion of silyl enol ethers into tris-ketol phosphates **166** (Scheme 3.69) [217].

3.1.8 Oxidation of Alkenes and Alkynes

Alkenes and alkynes can be oxidatively functionalized by electrophilic λ^3-iodanes, such as iodosylbenzene, [bis(acyloxy)iodo]arenes and organoiodine(III) derivatives of strong acids. Iodosylbenzene itself has a low reactivity to alkenes due to the polymeric structure. However, the relatively weak electrophilic reactivity of $(PhIO)_n$ can be increased considerably in the presence of $BF_3 \cdot Et_2O$ or other Lewis acids. This activation

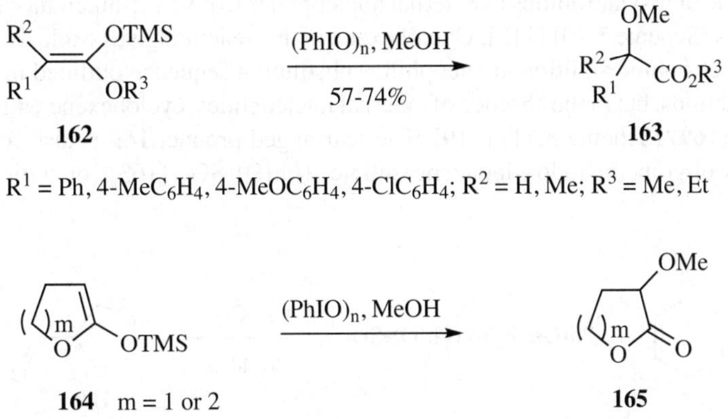

162 (PhIO)n, MeOH; 57-74%; **163**

R^1 = Ph, 4-MeC_6H_4, 4-$MeOC_6H_4$, 4-ClC_6H_4; R^2 = H, Me; R^3 = Me, Et

164 m = 1 or 2; (PhIO)n, MeOH; **165**

Scheme 3.68

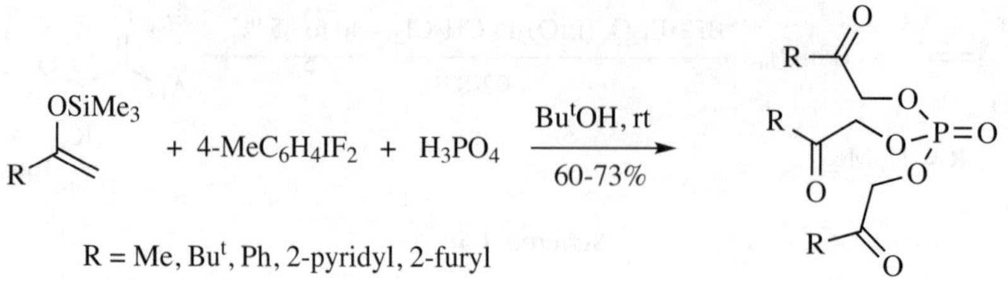

$R = Me, Bu^t, Ph, 2\text{-pyridyl}, 2\text{-furyl}$

166

Scheme 3.69

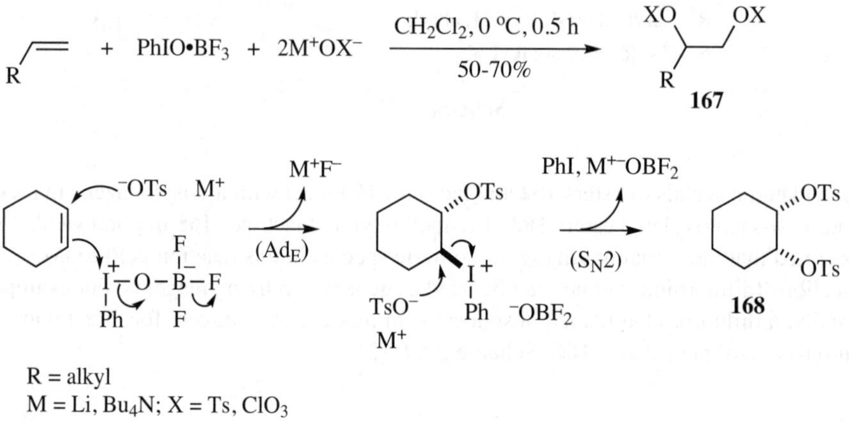

167

$R = alkyl$
$M = Li, Bu_4N; X = Ts, ClO_3$

168

Scheme 3.70

is usually explained by the formation of non-isolable, highly electrophilic complexes such as $PhI^+OBF_3^-$. Reactions of these complexes with unsaturated organic substrates can afford various products depending on the reaction conditions and structure of the organic substrate. Alkenes react with activated iodosylbenzene in the presence of perchlorate or tosylate anions as external nucleophiles to give 1,2-diperchlorates or 1,2-ditosylates **167** as major products (Scheme 3.70) [218]. Cyclohexene in this reaction gives exclusively *cis*-ditosylate **168** resulting from the electrophilic addition–nucleophilic substitution sequence outlined in Scheme 3.70.

Under similar conditions, but in the absence of external nucleophiles, cyclohexene with $PhI^+OBF_3^-$ affords formylcyclopentane (**169**) (Scheme 3.71) [219]. The rearranged product **169** is also formed in the reaction of cyclohexene with oligomeric iodosylbenzene sulfate, $(PhIO)_3SO_3$ [163], or with $(PhIO)_n$ in aqueous H_2SO_4 [220].

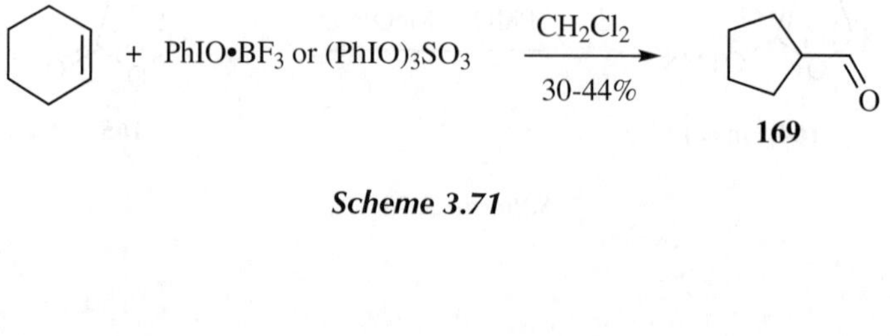

169

Scheme 3.71

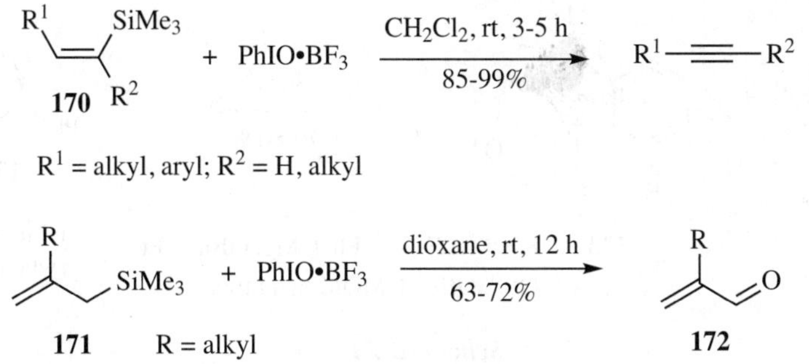

R^1, SiMe$_3$ (170), R^2 + PhIO•BF$_3$ → (CH_2Cl_2, rt, 3-5 h; 85-99%) R^1—≡—R^2

R^1 = alkyl, aryl; R^2 = H, alkyl

171 (R, SiMe$_3$) + PhIO•BF$_3$ → (dioxane, rt, 12 h; 63-72%) 172 (R, O)

R = alkyl

Scheme 3.72

Various products can be prepared by the reaction of $PhI^+OBF_3^-$ with unsaturated silylated substrates, such as silylalkenes and silylalkynes. (*E*)-Vinylsilanes react with $PhI^+OBF_3^-$ to give stable alkenyl(phenyl)iodonium tetrafluoroborates (Section 2.1.9.2), whereas (*Z*)-vinylsilanes **170** undergo dehydrosilylation to afford alkynes in high yields (Scheme 3.72) [221–223]. Allylsilanes **171** are oxidized with $PhI^+OBF_3^-$ to give conjugated enals **172** [224].

HTIB [PhI(OH)OTs] has high reactivity toward alkenes and alkynes. Reactions of HTIB with alkenes afford *vic*-ditosyloxyalkanes in moderate yield (Scheme 3.73) [225–227]. With cyclohexene and alkyl-substituted alkenes, this reaction proceeds as a stereoselective *syn*-addition, whereas phenyl-substituted alkenes and norbornene give products of skeletal rearrangements.

The reactions of HTIB with alkenes (Scheme 3.73) can be rationalized by a polar addition–substitution mechanism similar to the one shown in Scheme 3.70. The first step in this mechanism involves electrophilic *anti*-addition of the reagent to the double bond and the second step is nucleophilic substitution of the iodonium fragment by tosylate anion with inversion of configuration. Such a polar mechanism also explains the skeletal rearrangements in the reactions of HTIB with polycyclic alkenes [227], the participation of external nucleophiles [228] and the intramolecular participation of a nucleophilic functional group with the formation of lactones and other cyclic products [229–231]. An analogous reactivity pattern is also typical of [hydroxy(methanesulfonyloxy)iodo]benzene [232] and other [hydroxy(organosulfonyloxy)iodo]arenes.

Chiral [hydroxy(organosulfonyloxy)iodo]arenes **173** (Section 2.1.6.1) have been evaluated as enantioselective electrophilic reagents towards alkenes and ketones. Enantioselectivities as high as 65% have been achieved in the dioxytosylation of styrene to give products **174** (Scheme 3.74). The maximum selectivity is observed in the reactions of *ortho*-ethyl compounds **173,** with lower selectivity being observed for reagents **173** bearing both smaller and larger substituents R. X-Ray structure analysis and *ab initio* calculations have been used to develop a model to rationalize the stereoselectivities in the reactions of chiral hypervalent iodine reagents **173.** In this model, high enantiomeric excess in the reaction correlates with the relative population of a conformation in which a methyl group on the asymmetric carbon atom is in the axial position [233].

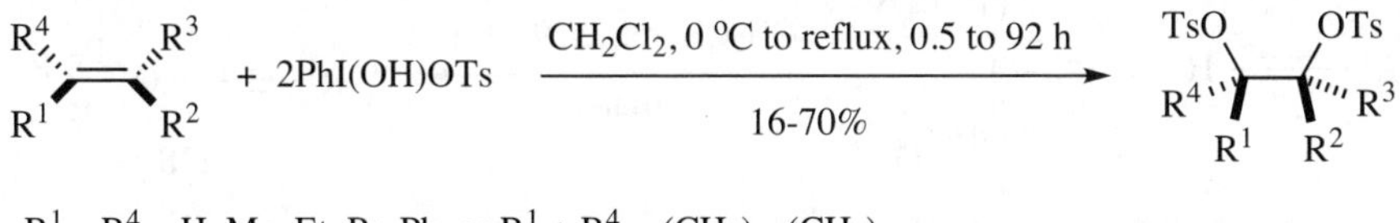

R^4, R^3, R^1, R^2 (alkene) + 2PhI(OH)OTs → (CH_2Cl_2, 0 °C to reflux, 0.5 to 92 h; 16-70%) TsO, OTs, R^4, R^3, R^1, R^2

$R^1 - R^4$ = H, Me, Et, Pr, Ph, or $R^1 + R^4$ = $(CH_2)_3$, $(CH_2)_4$

Scheme 3.73

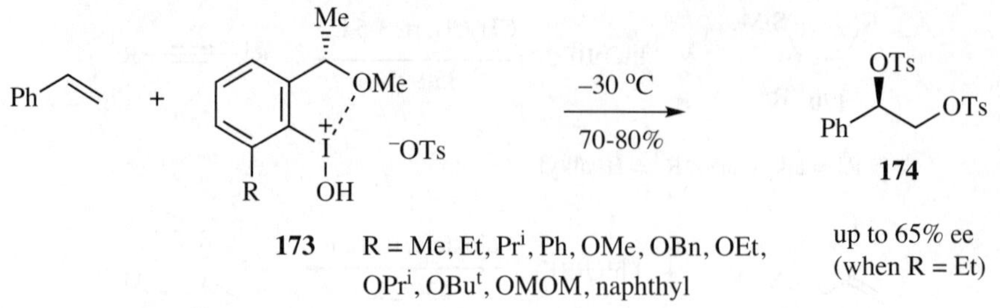

173 R = Me, Et, Pr^i, Ph, OMe, OBn, OEt, OPr^i, OBu^t, OMOM, naphthyl

–30 °C, 70-80%

174 up to 65% ee (when R = Et)

Scheme 3.74

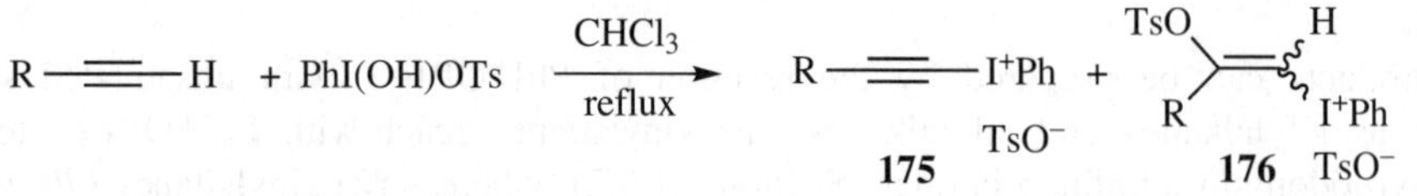

R—≡—H + PhI(OH)OTs → ($CHCl_3$, reflux) R—≡—I^+Ph TsO^- (175) + 176 TsO^-

Scheme 3.75

The reaction of HTIB with alkynes generally affords two products: alkynyl- (**175**) and alkenyl-(phenyl)iodonium (**176**) tosylates (Scheme 3.75) [234–238]. Alkynyl(phenyl)iodonium tosylates **175** are major products in the reactions of terminal alkynes bearing bulky substituents (R = *tert*-butyl, *sec*-butyl, cyclohexyl, aryl, etc.), while alkenyliodonium tosylates **176** are formed from non-sterically hindered terminal alkynes or from internal alkynes.

[Hydroxy(dialkylphosphoryloxy)iodo]benzenes **177** (Section 2.1.7) react with terminal alkynes under anhydrous conditions to afford alkynyl phosphates **178** (Scheme 3.76) via intermediate formation of the respective alkynyl(phenyl)iodonium phosphates [239].

μ-Oxo-bridged triflate **179** and perchlorate **180** (Section 2.1.7) react with alkenes to afford *vic*-diperchlorates or ditriflates; in the case of cyclohexene this reaction proceeds as a stereospecific *syn*-addition (Scheme 3.77) [218, 240–242]. Likewise, the reaction of alkenes and cycloalkenes with a reagent generated *in situ* from (diacetoxyiodo)benzene and magnesium perchlorate (Section 2.1.7) affords 1,2-diperchlorates **181** and **182** [243].

Adducts of iodosylbenzene with sulfur trioxide (**183** and **185**) (Section 2.1.7) react with alkenes under mild conditions to give the respective cyclic sulfate esters, such as **184, 186** and **187** (Scheme 3.78) [218, 244]. Likewise, phenyliodine(III) sulfate **185,** generated from iodosylbenzene and trimethylsilyl chlorosulfonate

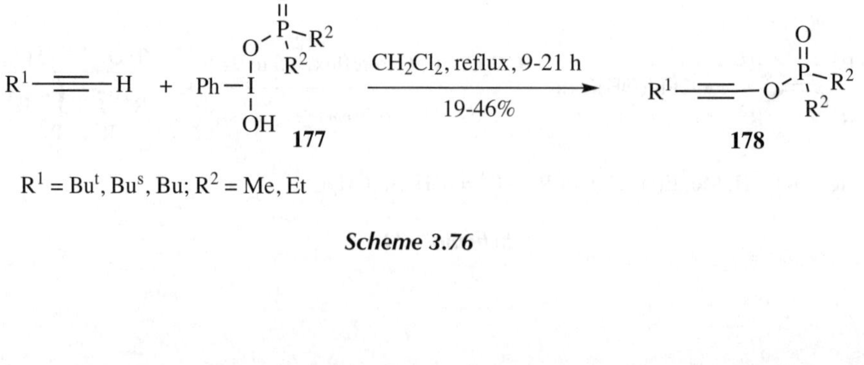

R^1—≡—H + 177 → (CH_2Cl_2, reflux, 9-21 h; 19-46%) 178

R^1 = Bu^t, Bu^s, Bu; R^2 = Me, Et

Scheme 3.76

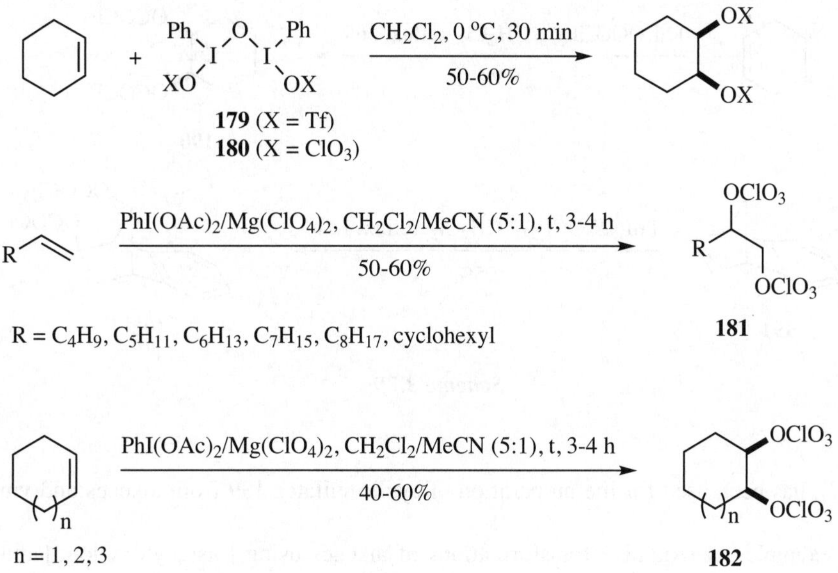

Ph, O, Ph, I, I, XO, OX

CH_2Cl_2, 0 °C, 30 min

50-60%

OX, OX

179 (X = Tf)
180 (X = ClO_3)

$PhI(OAc)_2/Mg(ClO_4)_2$, CH_2Cl_2/MeCN (5:1), t, 3-4 h

50-60%

R, $OClO_3$, $OClO_3$

R = C_4H_9, C_5H_{11}, C_6H_{13}, C_7H_{15}, C_8H_{17}, cyclohexyl

181

$PhI(OAc)_2/Mg(ClO_4)_2$, CH_2Cl_2/MeCN (5:1), t, 3-4 h

40-60%

n, $OClO_3$, $OClO_3$

n = 1, 2, 3

182

Scheme 3.77

R, Ph, O, Ph, I, I, O, O, S, O_2

CH_2Cl_2, rt, 5 h

80-85%

R, O, O, S, O_2

R = H, Bu

183

184

+ $PhIOSO_2O^-$

CH_2Cl_2, 0 °C

72%

185

O, O, S, O_2

186

O, SO_2, O

187

R^1, R^2 + **185**

CH_2Cl_2, −78 °C to rt

50-76%

R^1, R^2, O, O, S, O_2

188

R^1 = Me_3Si, Me_2PhSi, Bu; R^2 = H or Bu

189

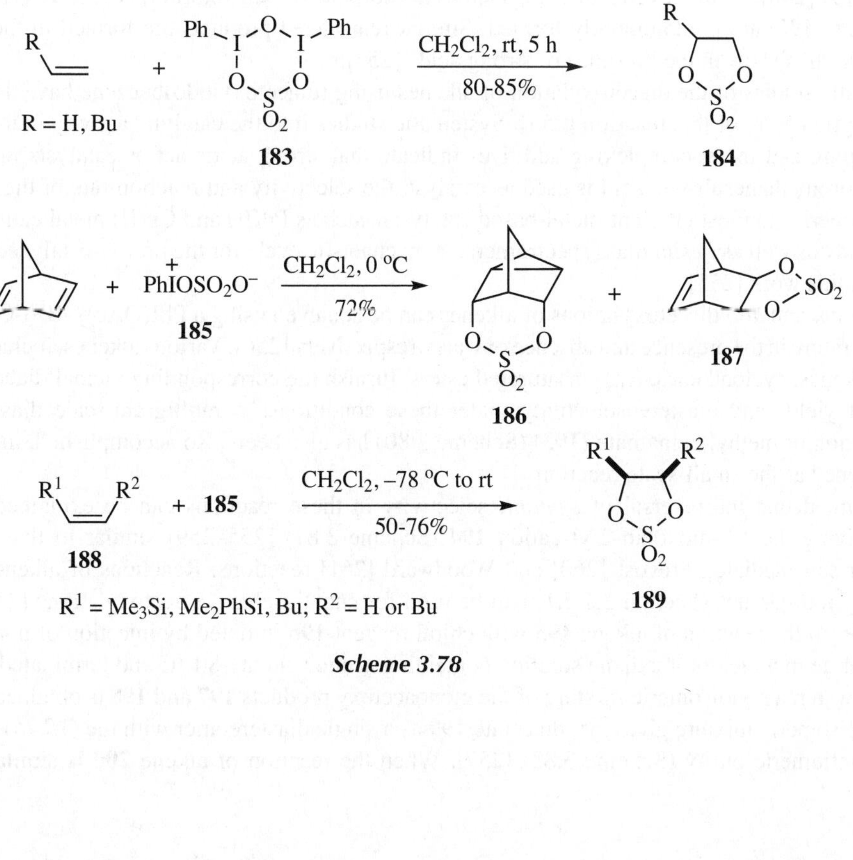

Scheme 3.78

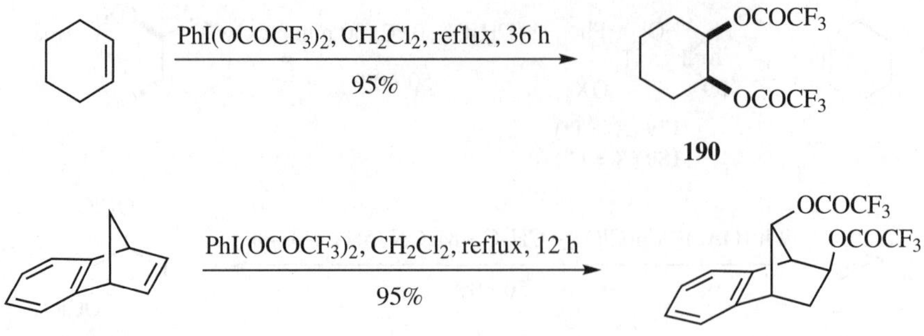

$PhI(OCOCF_3)_2$, CH_2Cl_2, reflux, 36 h
95%

OCOCF$_3$
OCOCF$_3$

190

$PhI(OCOCF_3)_2$, CH_2Cl_2, reflux, 12 h
95%

191

OCOCF$_3$
OCOCF$_3$

192

Scheme 3.79

(Section 2.1.7), has been used for the preparation of cyclic sulfates **189** from alkenes and vinylsilanes **188** [245, 246].

Numerous examples of oxidative transformations of alkenes using [bis(acyloxy)iodo]arenes have been reported [11, 247–253]. [Bis(trifluoroacetoxy)iodo]benzene reacts with alkenes in the absence of any additive or catalyst, affording vicinal bis(trifluoroacetates), which can be converted into the corresponding glycols or carbonyl compounds by hydrolysis [248, 253]. For example, cyclohexene reacts with $PhI(OCOCF_3)_2$ in dichloromethane under reflux conditions to give *cis*-1,2-bis(trifluoroacetate) **190** in almost quantitative yield (Scheme 3.79) [248]. With bicyclic alkenes, such as norbornene or benzonorbornadiene **191,** the rearranged products (e.g., **192**) are predominantly formed. Similar rearranged products are formed in the reactions of alkenes with $PhI(OAc)_2$ in the presence of strong acids [254].

Mechanistic studies of the diacetoxylation of alkenes using (diacetoxyiodo)benzene have demonstrated a protio-catalytic nature of this reaction [255]. Systematic studies into the catalytic activity in the presence of proton-trapping and metal-complexing additives indicate that strong acids act as catalysts in the reaction. When trifluoromethanesulfonic acid is used as catalyst, the selectivity and reaction rate of the conversion is similar or superior to most efficient metal-based catalysts, such as Pd(II) and Cu(II) metal cations. Based on a kinetic study as well as *in situ* mass spectrometry, a mechanistic cycle for the proton-catalyzed reaction was proposed in this work [255].

Selective *syn* and *anti* diacetoxylations of alkenes can be achieved using a $PhI(OAc)_2$–$BF_3{\cdot}OEt_2$ system at room temperature in the presence and absence of water, respectively [256]. Various alkenes, including styrenes, aliphatic alkenes, cycloalkenes, α,β-unsaturated esters, furnish the corresponding vicinal diacetates in good to excellent yields and diastereoselectivity under these conditions. A multigram-scale diastereoselective diacetoxylation of methyl cinnamate (**193**) (Scheme 3.80) has also been also accomplished, maintaining the same efficiency as the small-scale reaction.

It is assumed that the reversal of *syn*/*anti* selectivity in these reactions can be explained by a mechanism involving the 1,3-dioxolan-2-yl cation **194** (Scheme 3.81) [255–259] similar to that proposed for the I_2/silver salt mediated Prevost [260] and Woodward [261] reactions. Reactions of alkenes with chiral [bis(acyloxy)iodo]arenes (Section 2.1.5.1) can be used for enantioselective acetoxylations [257, 262, 263]. For example, in the reaction of alkene **195** with chiral reagent **196** initiated by injection of boron trifluoride diethyl etherate in a dichloromethane solution containing acetic acid at –80 °C and terminated at –40 °C by addition of water, a regioisomeric mixture of the monoacetoxy products **197** and **198** is obtained. Acetylation of this regioisomeric mixture gives *syn*-diacetate **199** as a single diastereomer with the (1*S*,2*S*) configuration in high enantiomeric purity (Scheme 3.82) [257]. When the reaction of alkene **200** is similarly started at

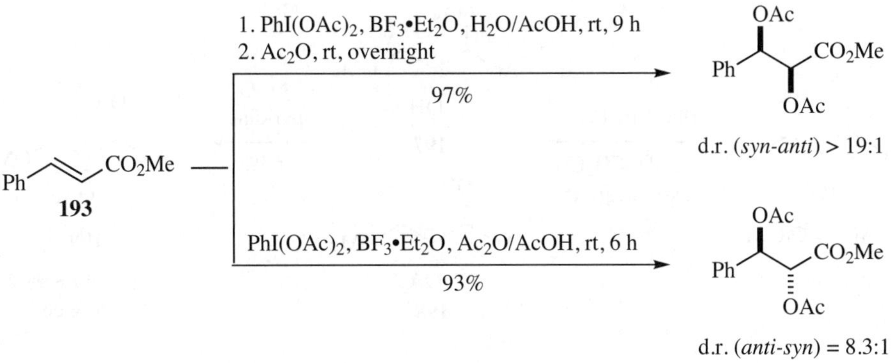

1. $PhI(OAc)_2$, $BF_3{\bullet}Et_2O$, $H_2O/AcOH$, rt, 9 h
2. Ac_2O, rt, overnight

97%

d.r. (*syn-anti*) > 19:1

193

$PhI(OAc)_2$, $BF_3{\bullet}Et_2O$, $Ac_2O/AcOH$, rt, 6 h

93%

d.r. (*anti-syn*) = 8.3:1

Scheme 3.80

−80 °C and the reaction mixture is allowed to warm to room temperature, *anti*-diacetate **201** with the (1*R*,2*S*) configuration is preferentially obtained.

The reversal of *syn*/*anti* selectivity in this reaction (Scheme 3.82) can be explained by a mechanism similar to that shown in Scheme 3.81. The chiral, non-racemic 1,3-dioxolan-2-yl cation intermediates **194** are initially generated during enantioselective dioxyacetylation of alkene with chiral [bis(acyloxy)iodo]arene **196.** Regioselective attack of a nucleophile toward the intermediate results in reversal of enantioselectivity of the dioxyacetylation [257].

The oxylactonization of *ortho*-alkenylbenzoates with lactate-derived optically active hypervalent iodine(III) reagents proceeds with a high degree of regio-, diastereo- and enantioselectivity leading to the asymmetric synthesis of 3-alkyl-4-oxyisochroman-1-ones [263]. A specific example – the enantioselective oxylactonization of substrate **202** with reagent **203** – is shown in Scheme 3.83.

Iodosylbenzene and [bis(acyloxy)iodo]arenes are useful reagents for nucleophilic epoxidation of electron-deficient alkenes, such as tetrasubstituted perfluoroalkenes [264] and α,β-unsaturated carbonyl compounds [265, 266]. In particular, iodosylbenzene reacts with enones **204** to furnish the corresponding epoxides **205** in generally high yields (Scheme 3.84) [265].

Likewise, [bis(acyloxy)iodo]arenes can be used as the oxidants in organocatalytic, asymmetric epoxidation of α,β-unsaturated aldehydes using chiral imidazolidinone catalyst **207** [266]. In a specific example, the

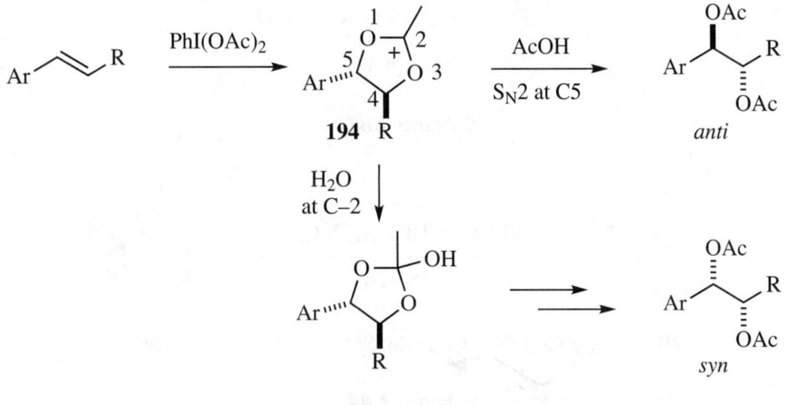

$PhI(OAc)_2$

194

AcOH

S_N2 at C5

anti

H_2O at C–2

syn

Scheme 3.81

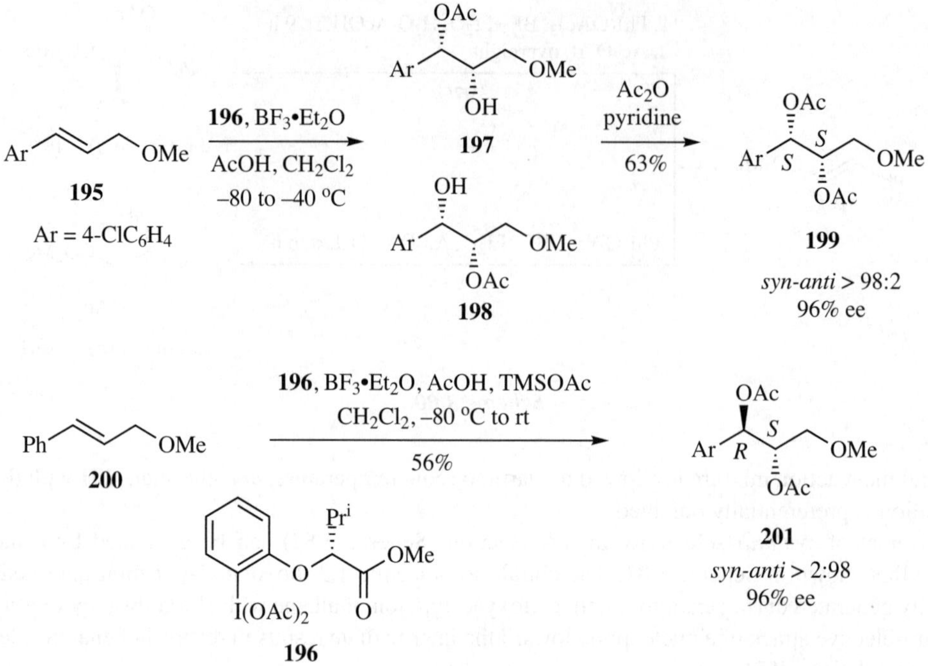

Ar⁄⁄OMe
195
Ar = 4-ClC$_6$H$_4$

196, BF$_3$•Et$_2$O
AcOH, CH$_2$Cl$_2$
–80 to –40 °C

OAc
Ar OMe
OH
197

OH
Ar OMe
OAc
198

Ac$_2$O
pyridine
63%

OAc
S
Ar S OMe
OAc
199
syn-anti > 98:2
96% ee

Ph OMe
200

196, BF$_3$•Et$_2$O, AcOH, TMSOAc
CH$_2$Cl$_2$, –80 °C to rt
56%

Pr^i
OMe
O
I(OAc)$_2$ O
196

OAc
S
Ar R OMe
OAc
201
syn-anti > 2:98
96% ee

Scheme 3.82

OMe
CO$_2$Me
202

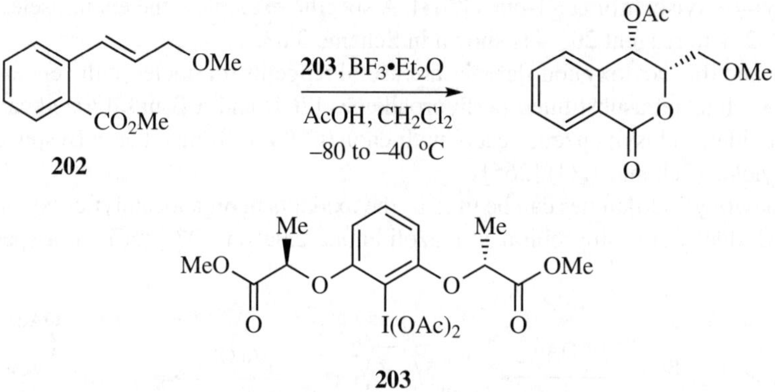

203, BF$_3$•Et$_2$O
AcOH, CH$_2$Cl$_2$
–80 to –40 °C

OAc
OMe
O
O

Me Me
MeO O O OMe
O I(OAc)$_2$ O
203

Scheme 3.83

O
R
204

(PhIO)$_n$, CHCl$_3$, rt, 2-18 h
45-91%

O
R
O
205

R = C(O)OEt, C(O)Me, SO$_2$Ph, CN

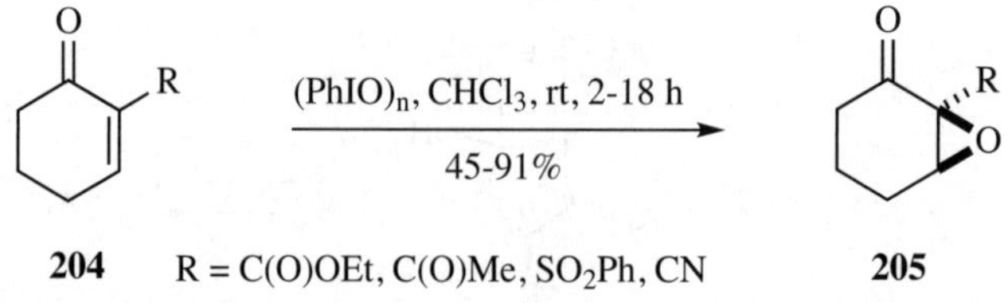

Scheme 3.84

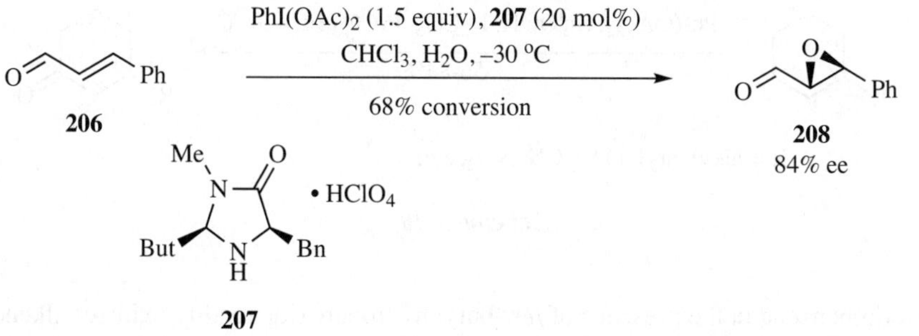

$PhI(OAc)_2$ (1.5 equiv), **207** (20 mol%)
$CHCl_3$, H_2O, –30 °C
68% conversion

206 **208** 84% ee

207 • $HClO_4$

Scheme 3.85

reaction of aldehyde **206** with (diacetoxyiodo)benzene affords epoxide **208** with good enantioselectivity (Scheme 3.85).

3.1.9 Oxidations at the Benzylic or Allylic Position

Oxidations of C—H bonds at the benzylic or allylic position using λ^3-iodanes can be achieved only in the presence of catalysts, such as, iodine, peroxides and transition metals (Section 3.1.20). For example, the oxidation of tetrahydroisoquinoline **209** by iodosylbenzene in the presence of tetrabutylammonium iodide proceeds at the benzylic position to give the lactam **210** (Scheme 3.86) [267].

Likewise, treatment of alkylbenzenes **211** with (diacetoxyiodo)benzene in the presence of catalytic amounts of molecular iodine and *p*-toluenesulfonamide or *p*-nitrobenzenesulfonamide in 1,2-dichloroethane at 60 °C gives the corresponding (α-acetoxy)alkylbenzenes **212** in generally good yields (Scheme 3.87) [268]. A plausible mechanism for this reaction involves the initial generation of $ArSO_2NH^\bullet$ radicals from $PhI(OAc)_2$, I_2 and $ArSO_2NH_2$, which further promote radical substitution at the benzylic carbon [268].

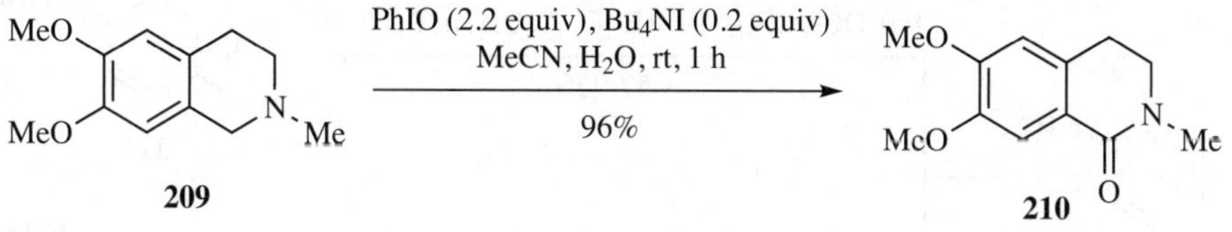

PhIO (2.2 equiv), Bu_4NI (0.2 equiv)
MeCN, H_2O, rt, 1 h
96%

209 **210**

Scheme 3.86

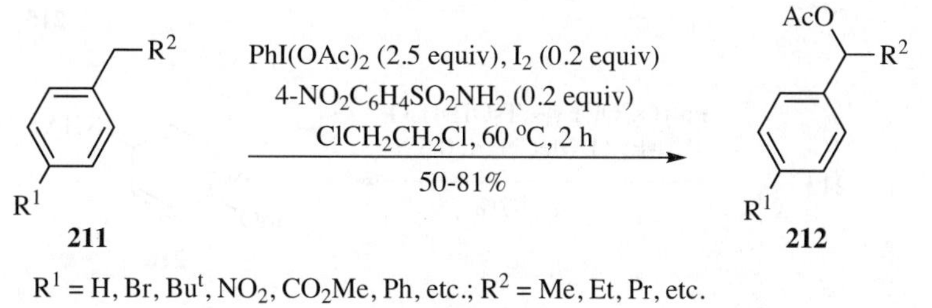

$PhI(OAc)_2$ (2.5 equiv), I_2 (0.2 equiv)
4-$NO_2C_6H_4SO_2NH_2$ (0.2 equiv)
$ClCH_2CH_2Cl$, 60 °C, 2 h
50-81%

211 **212**

R^1 = H, Br, Bu^t, NO_2, CO_2Me, Ph, etc.; R^2 = Me, Et, Pr, etc.

Scheme 3.87

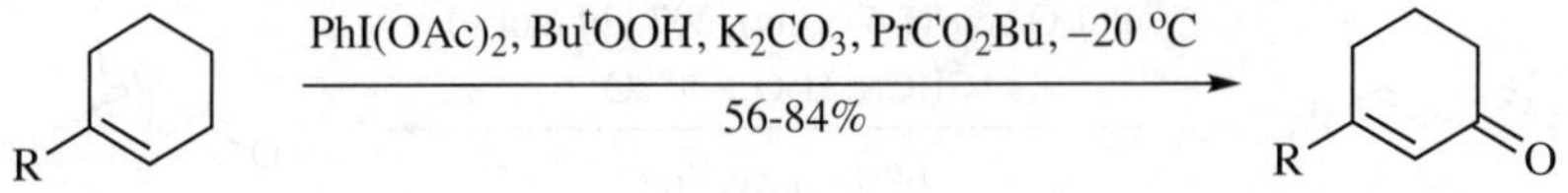

$PhI(OAc)_2$, Bu^tOOH, K_2CO_3, $PrCO_2Bu$, –20 °C

56-84%

R = alkyl, aryl, OAc, CN, NO_2, etc.

Scheme 3.88

(Diacetoxyiodo)benzene in the presence of *tert*-butyl hydroperoxide readily oxidizes alkenes at the allylic position (Scheme 3.88) [269]. This reaction proceeds via initial formation of the *tert*-butylperoxy radical and it can be extended to the oxidation of unactivated C—H bonds in alkyl esters and amides to give the corresponding keto compounds under mild conditions [270].

3.1.10 Oxidative Functionalization of Aromatic Compounds

The reactions of aromatic compounds with λ^3-iodanes usually afford products of oxidative dearomatization (Section 3.1.11) or oxidative coupling (Section 3.1.12). Only a few examples of non-catalytic oxygenation of aromatic C—H bonds have been reported in the literature. Direct acetoxylation and etherification of anilides can be achieved using [bis(trifluoroacetoxy)iodo]benzene in the presence of Lewis acids [271]. In particular, treatment of various anilides **213** with 1.5 equiv of $PhI(OCOCF_3)_2$ and 1.0 equiv of boron trifluoride etherate in acetic acid at room temperature affords the corresponding *para*-acetoxylated products **214** in good yields and with high regioselectivity (Scheme 3.89). Likewise, the reaction of anilides **213** with alcohols and 2.0 equiv of $PhI(OCOCF_3)_2$ in the presence of 2.0 equiv of $BF_3{\cdot}OEt_2$ provides the corresponding *para*-etherified products **215** in good yields [271]. The direct tosyloxylation of anilides **213** has been performed in a similar fashion by the treatment of anilides with $PhI(OCOCF_3)_2$ under mild conditions in the presence of $BF_3{\cdot}OEt_2$ and toluenesulfonic acid to give *para*-tosyloxylated products **216** with high regioselectivity [272].

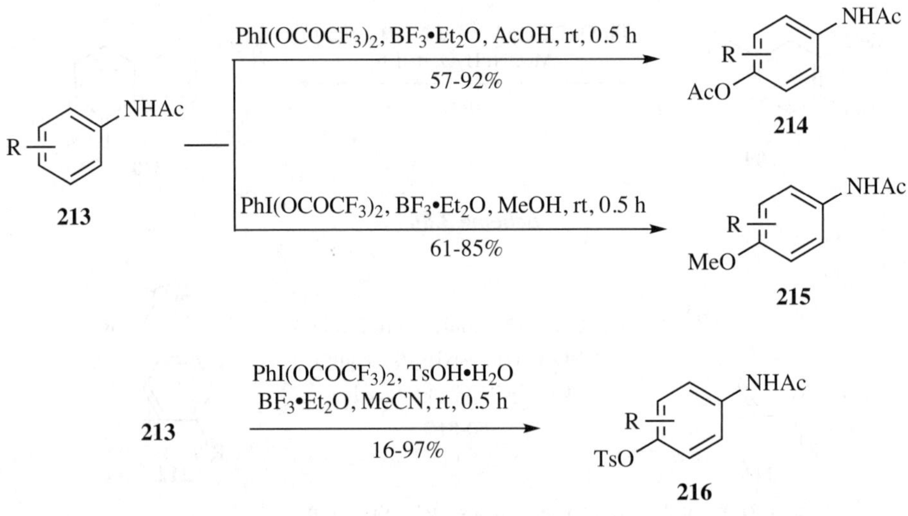

NHAc

R

213

$PhI(OCOCF_3)_2$, $BF_3{\bullet}Et_2O$, AcOH, rt, 0.5 h

57-92%

NHAc

R

AcO

214

$PhI(OCOCF_3)_2$, $BF_3{\bullet}Et_2O$, MeOH, rt, 0.5 h

61-85%

NHAc

R

MeO

215

213

$PhI(OCOCF_3)_2$, $TsOH{\bullet}H_2O$
$BF_3{\bullet}Et_2O$, MeCN, rt, 0.5 h

16-97%

NHAc

R

TsO

216

R = H, 2-Me, 3-Me, 2-OMe, 2-F, 2-Cl, 2-Br, 2-CO_2Me, 3-OAc, 2,5-Me_2, 3,5-Me_2, etc.

Scheme 3.89

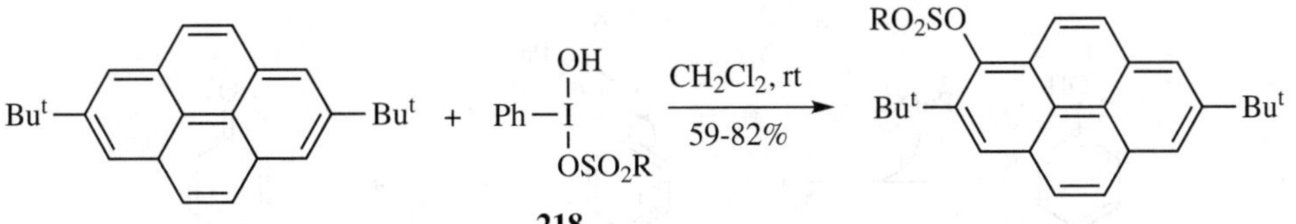

CH_2Cl_2, rt
59-82%

217 **218** **219**

R = Me, 4-MeC_6H_4, 2,4-$(NO_2)_2C_6H_3$, (1*R*)-10-camphoryl

Scheme 3.90

para-Triflates from anilides have been prepared similarly by direct oxidative triflation using $PhI(OCOCF_3)_2$ as the oxidant and AgOTf as the source of triflate anion [273].

[Hydroxy(organosulfonyloxy)iodo]benzenes can be used for the oxidative functionalization of arenes. Various polycyclic arenes, such as pyrene, anthracene, phenanthrene, perylene and others, undergo regioselective oxidative substitution reactions with [hydroxy(organosulfonyloxy)iodo]benzenes in dichloromethane at room temperature to give the corresponding aryl sulfonate esters in moderate to good yields [274]. For example, treatment of 2,7-di-*tert*-butylpyrene (**217**) with iodine(III) organosulfonate reagents **218,** containing tosylate, mesylate, (+)-10-camphorsulfonate and 2,4-dinitrobenzenesulfonate ligands, affords the respective organosulfonyloxy derivatives **219** in good yields and with high regioselectivity (Scheme 3.90) [274].

3.1.11 Oxidative Dearomatization of Phenols and Related Substrates

Hypervalent iodine compounds are commonly used as the reagents for various synthetically useful oxidative transformations of phenols and other electron-rich aromatic substrates. The oxidation of various *ortho*-substituted phenols or *o*- and *p*-hydroquinones with [bis(acyloxy)iodo]arenes usually affords the corresponding benzoquinones in excellent yields [275–282]. (Diacetoxyiodo)benzene is a reagent of choice for the oxidation of various substituted *o*- and *p*-hydroquinones to the corresponding benzoquinones [277–279]. The oxidation generally proceeds in a methanol solution at room temperature to give benzoquinones in almost quantitative yield [277]. This procedure has been utilized in organic synthesis; for example, the oxidation of phenol **220** with (diacetoxyiodo)benzene has been used for the preparation of quinone **221** (Scheme 3.91), which is a key intermediate in the synthesis of an important class of antitumor agents [278]. [Bis(trifluoroacetoxy)iodo]benzene can selectively oxidize polychlorinated phenols to the respective benzoquinones in aqueous media [280]. This reaction has been utilized in the sensitive electrochemical identification of pentachlorophenol, which is one of the most toxic polychlorinated phenols.

Of particular use are the reactions of oxidative dearomatization of 4- or 2-substituted phenols **222** and **225** with λ^3-iodanes in the presence of an external or internal nucleophile (Nu) leading to the respective

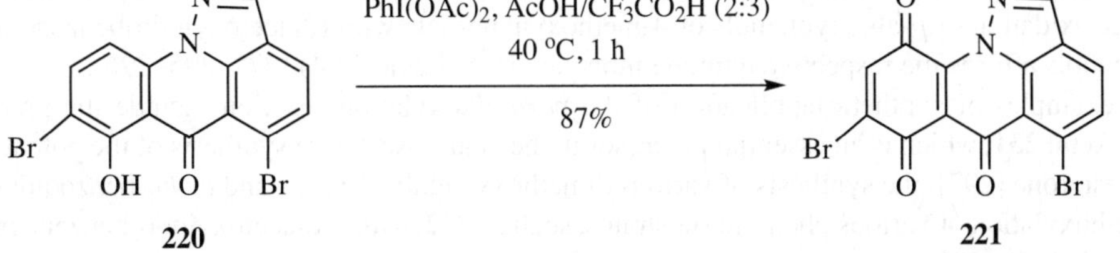

$PhI(OAc)_2$, $AcOH/CF_3CO_2H$ (2:3)
40 °C, 1 h
87%

220 **221**

Scheme 3.91

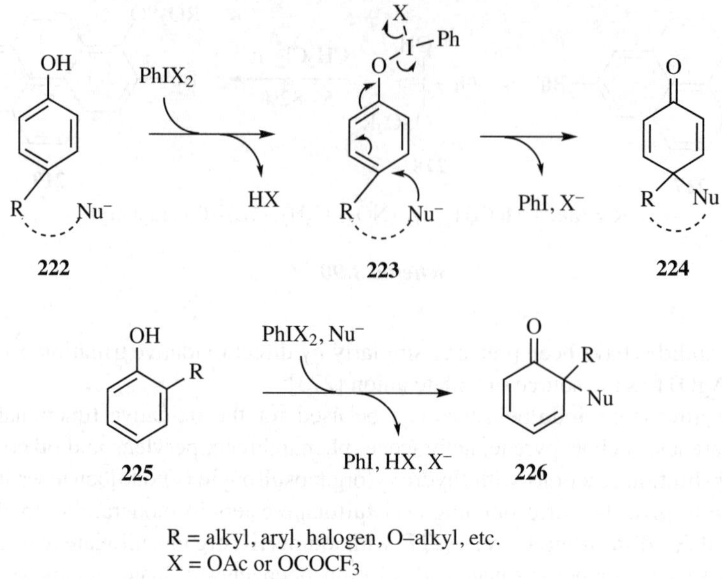

R = alkyl, aryl, halogen, O–alkyl, etc.
X = OAc or $OCOCF_3$

Scheme 3.92

cyclohexadienones **224** or **226** according to Scheme 3.92. The mechanism of this reaction most likely involves the initial formation of phenoxyiodine(III) species **223** followed by elimination of PhI and the generation of cationic phenoxenium intermediates, which finally combine with the nucleophile [283].

Various nucleophiles, such as water, alcohols, fluoride ion, carboxylic acids, amides, oximes and carbon nucleophiles, have been used successfully in these reactions (Scheme 3.92) in either an inter- or intramolecular mode. Synthetic applications of oxidative dearomatization reactions of phenols and related substrates have been summarized in several reviews [284–292].

3.1.11.1 Oxidative Dearomatization of 4-Substituted Phenols

Oxidative dearomatization of 4-substituted phenols **222** with [bis(acyloxy)iodo]arenes in the presence of an external nucleophile provides a convenient approach to various 3,3-disubstituted cyclohexadienones **224** according to Scheme 3.92. Several examples of this reaction are provided below in Schemes 3.93–3.97.

The oxidation of various substituted phenols **227** with [bis(trifluoroacetoxy)iodo]benzene in aqueous acetonitrile affords *p*-quinols **228** in moderate to good yields (Scheme 3.93) [293]. Even higher yields of *p*-quinols are obtained when trimethylsilyl ethers of phenols are used as starting material. In particular, it was shown that the oxidation of trimethylsilyl ethers **229** affords *p*-quinols **230** in greatly improved yields due to the minimization of oligomer side-product formation compared to the oxidation of free phenol [294].

A similar oxidation of *p*-alkoxyphenols or 4-methoxynaphthols with (diacetoxyiodo)benzene in the presence of alcohols affords the respective quinone monoketals (Scheme 3.94) [277, 295, 296].

Further examples of synthetic application of the *para*-alkoxylation reaction include the preparation of dimethoxy ketal **231,** which is an essential precursor in the enantioselective synthesis of the potent antifungal agent (–)-jesterone [297], the synthesis of various dimethoxy ketals of *para*- and *ortho*-benzoquinones [298] and the methoxylation of various phenolic substrates, such as **232,** using (diacetoxyiodo)benzene in methanol (Scheme 3.95) [299–301].

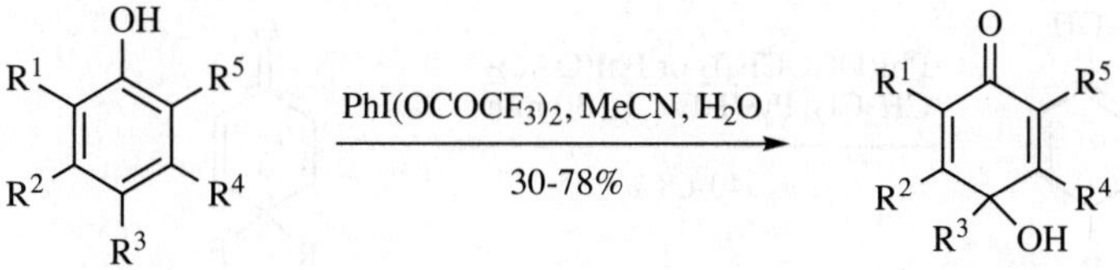

OH, R^1, R^2, R^3, R^4, R^5 — $PhI(OCOCF_3)_2$, MeCN, H_2O — 30-78% — O, R^1, R^2, R^3, R^4, R^5, OH

227 **228**

R^1, R^2, R^4, R^5 = H, alkyl, Br, CO_2Et; R^3 = Me, Bu^t, CH_2CO_2Et

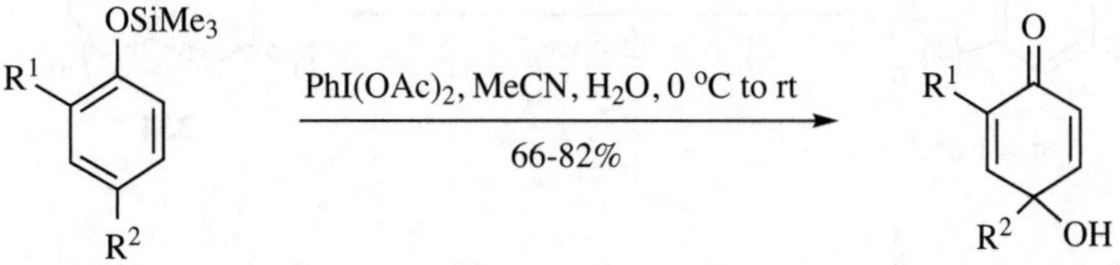

$OSiMe_3$, R^1, R^2 — $PhI(OAc)_2$, MeCN, H_2O, 0 °C to rt — 66-82% — O, R^1, R^2, OH

229 **230**

R^1 = H or Br; R^2 = Me, Ph, 2-FC_6H_4, 3-$NO_2C_6H_4$

Scheme 3.93

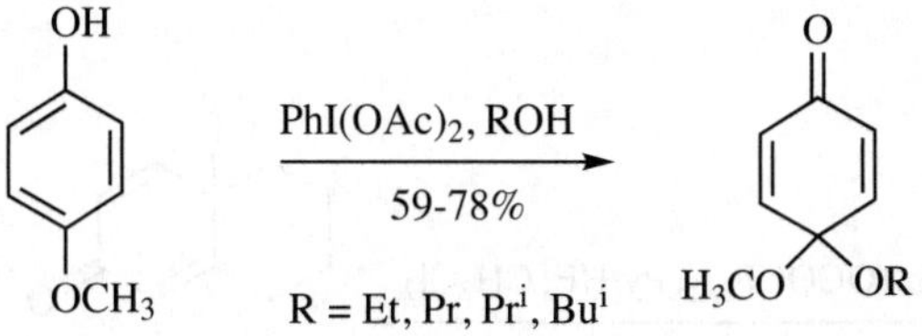

OH, OCH_3 — $PhI(OAc)_2$, ROH — 59-78% — O, H_3CO, OR

R = Et, Pr, Pr^i, Bu^i

Scheme 3.94

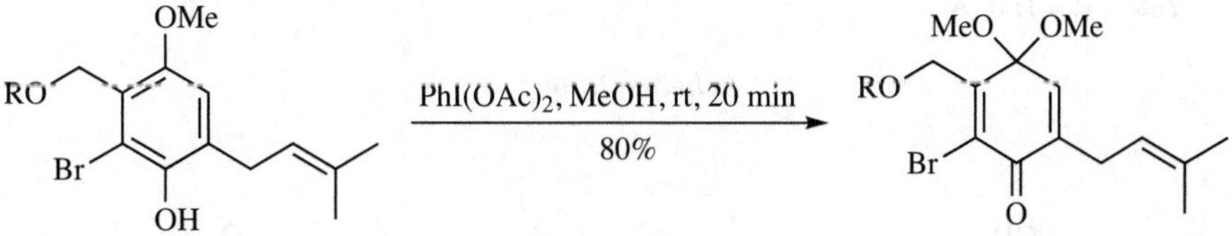

OMe, RO, Br, OH — $PhI(OAc)_2$, MeOH, rt, 20 min — 80% — MeO, OMe, RO, Br, O

R = Bu^tPh_2Si **231**

R, $(\ )_n$, OH — $PhI(OAc)_2$ (1equiv), MeOH, 0 °C, 1-1.5 h — 73-89% — R, MeO, $(\ )_n$, O

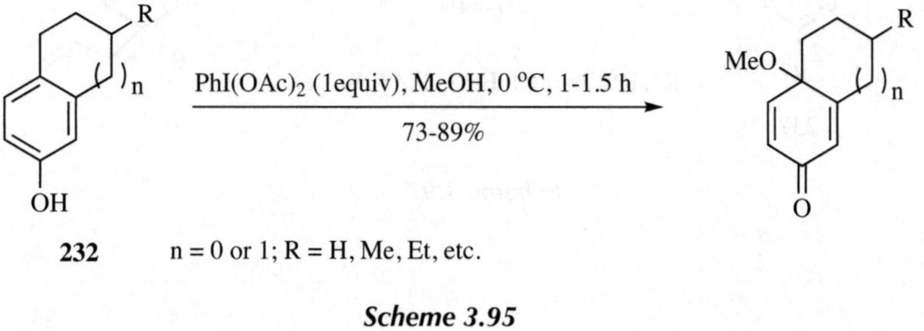

232 n = 0 or 1; R = H, Me, Et, etc.

Scheme 3.95

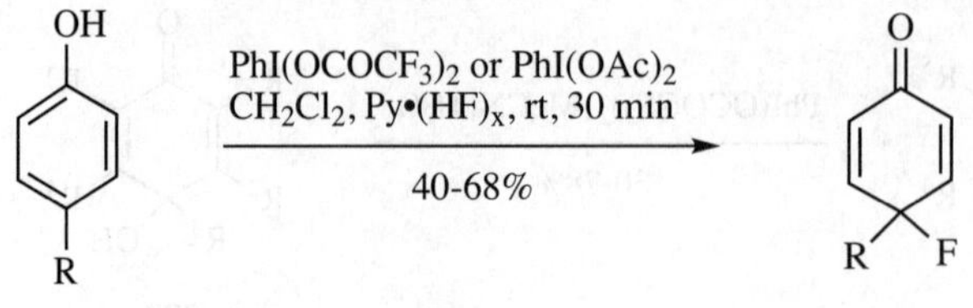

OH, R, O, R, F

$PhI(OCOCF_3)_2$ or $PhI(OAc)_2$
CH_2Cl_2, Py•(HF)$_x$, rt, 30 min
40-68%

233 R = Me, Et, CH_2CH_2Br, F, Cl, etc.

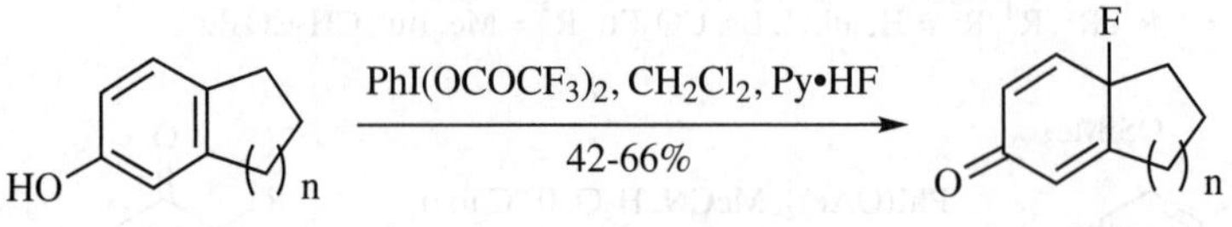

HO, n, F, O, n

$PhI(OCOCF_3)_2$, CH_2Cl_2, Py•HF
42-66%

n = 1 or 2

234

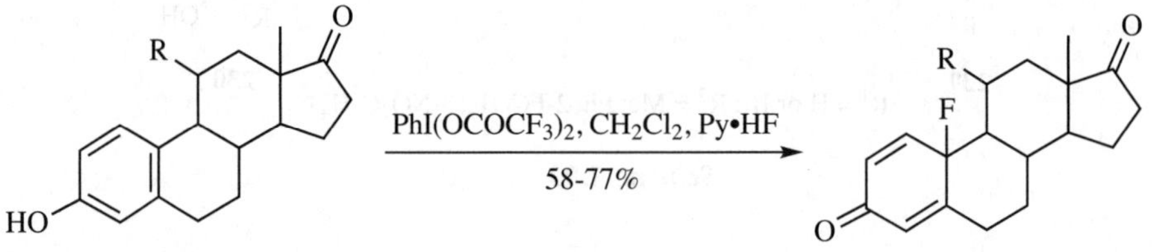

R, O, HO, R, F, O, O

$PhI(OCOCF_3)_2$, CH_2Cl_2, Py•HF
58-77%

R = H, OH

235

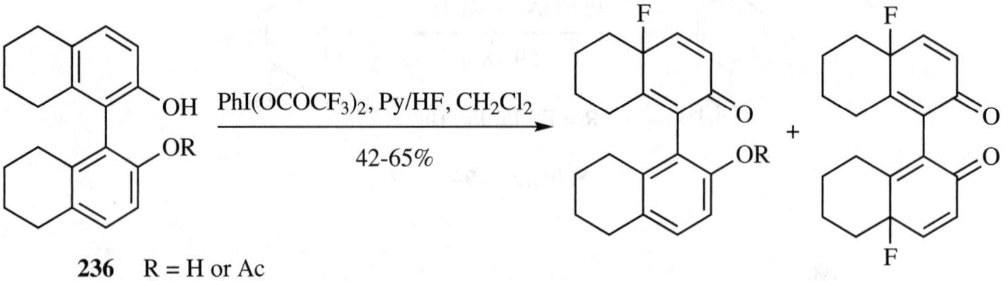

OH, OR, F, O, OR, +, F, O, O, F

$PhI(OCOCF_3)_2$, Py/HF, CH_2Cl_2
42-65%

236 R = H or Ac

Scheme 3.96

OH, R^1, R^2, R^3, O, R^1, R^2, R^3

$PhI(OAc)_2$, TMS
37-84%

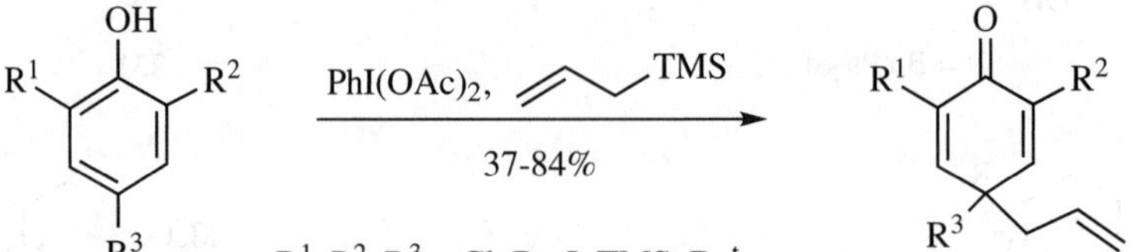

R^1, R^2, R^3 = Cl, Br, I, TMS, Bu^t, etc.

237

Scheme 3.97

The oxidation of 4-alkylphenols **233** with [bis(acyloxy)iodo]arenes in the presence of pyridinium polyhydrogen fluoride, Py·$(HF)_x$, as the source of fluoride anion results in nucleophilic *ipso*-fluorination (Scheme 3.96) [302–304]. This reaction has been used for the preparation of polycyclic 4-fluorocyclohexa-2,5-dienones **234** and **235** [302] and for the nucleophilic *para*-fluorination of tetrahydro-2-naphthol **236** [305].

Carbon nucleophiles can also be used as external nucleophiles, as illustrated by the oxidative allylation of phenols **237** (Scheme 3.97) [306].

The oxidation of phenolic substrates in the intramolecular mode has been widely exploited as a powerful synthetic tool for the construction of a spirodienone fragment. Numerous oxidative spirocyclizations of phenolic substrates containing an internal oxygen, nitrogen, or carbon nucleophile have been reported and utilized in natural product syntheses. Representative examples of spirocyclizations employing oxygen-based internal nucleophiles are shown below in Schemes 3.98–3.102.

Oxidative cyclizations of amides **238** and ketoximes **240** by the action of [bis(trifluoroacetoxy)iodo]benzene in non-nucleophilic solvents afford the respective spirohexadienones **239** and **241** in good yields (Scheme 3.98) [307–309].

The oxidatively induced cyclization of N-protected tyrosine **242** has been used as an approach to the spirocyclic core intermediate product **243** (Scheme 3.99), which is an important step in the total synthesis of the antitumor antibiotic aranorosin [310].

A similar oxidation of tyrosine derivative **244** with an excess of (diacetoxyiodo)benzene in cold acetonitrile followed by quenching with aqueous sodium bromide furnished brominated spirolactone **246** in 80% overall yield. A plausible mechanism for this reaction involves the iodonium derivative **245** as an intermediate product and subsequent substitution of each phenyliodonium group with a bromide anion (Scheme 3.100) [311].

The spirocyclic product **248** has been prepared by a hypervalent iodine-induced oxidation of catechol **247** in a key step of the enantiospecific synthesis of the antituberculosis marine sponge metabolite (+)-puupehenone (Scheme 3.101) [312]. A similar oxidative cyclodearomatization approach has been utilized in

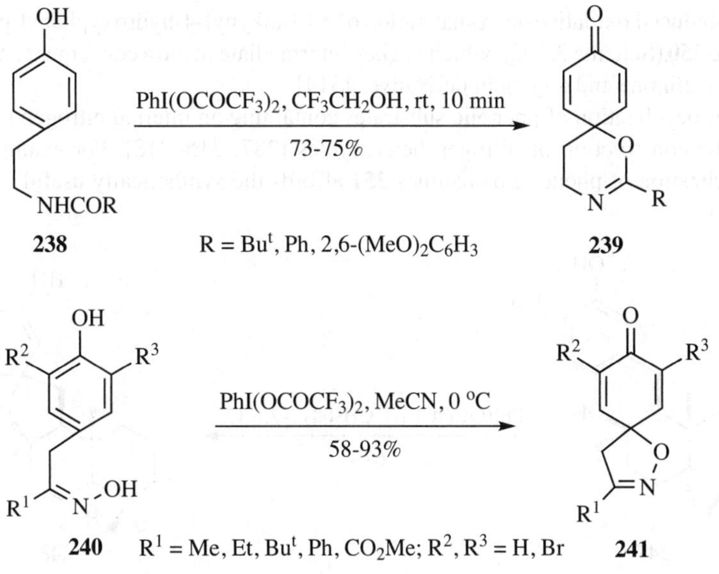

238 → **239**: $PhI(OCOCF_3)_2$, CF_3CH_2OH, rt, 10 min; 73-75%

R = Bu^t, Ph, 2,6-$(MeO)_2C_6H_3$

240 → **241**: $PhI(OCOCF_3)_2$, MeCN, 0 °C; 58-93%

240 R^1 = Me, Et, Bu^t, Ph, CO_2Me; R^2, R^3 = H, Br **241**

Scheme 3.98

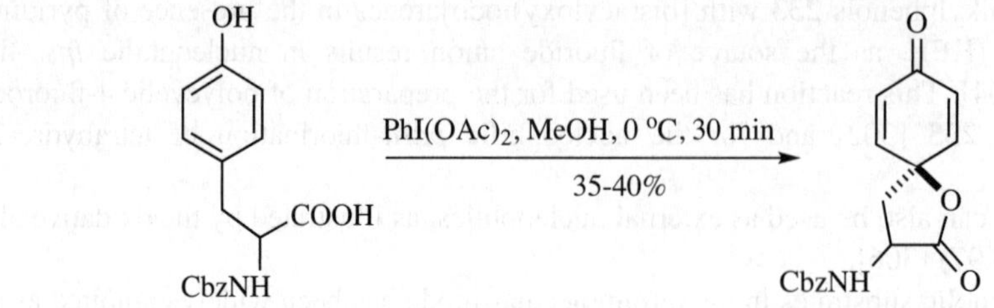

OH COOH CbzNH **242** $PhI(OAc)_2$, MeOH, 0 °C, 30 min 35-40% O O CbzNH O **243** Cbz = $PhCH_2O_2C–$

Scheme 3.99

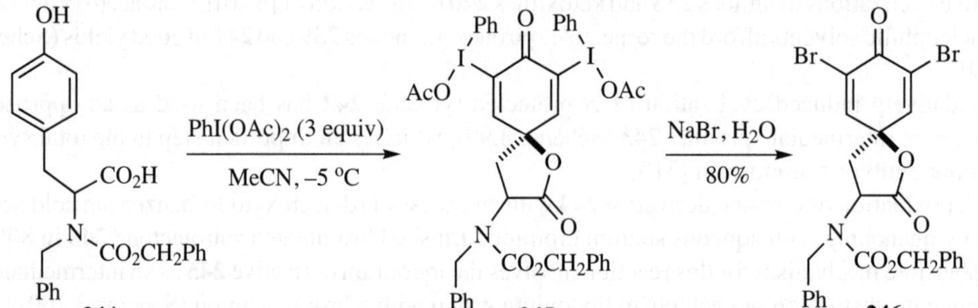

OH CO_2H N CO_2CH_2Ph Ph **244** $PhI(OAc)_2$ (3 equiv) MeCN, –5 °C Ph O Ph I I AcO OAc O O N CO_2CH_2Ph Ph **245** NaBr, H_2O 80% O Br Br O O N CO_2CH_2Ph Ph **246**

Scheme 3.100

the stereocontrolled synthesis of a complex pentacycle embodying the molecular architecture of the cortistatin class of natural products [313].

Iodosylbenzene-induced oxidative dearomatization of 3-(3-alkynyl-4-hydroxyphenyl)propanoic acid (**249**) affords spirolactone **250** (Scheme 3.102), which is a key intermediate in the a convergent and efficient synthetic approach to furoquinolinone and angelicin derivatives [314].

The oxidative spirocyclization of phenolic substrates containing an internal nitrogen nucleophile provides a useful tool for the construction of nitrogen heterocycles [287, 315–318]. For example, the hypervalent iodine-induced cyclization of phenolic oxazolines **251** affords the synthetically useful spirolactam products

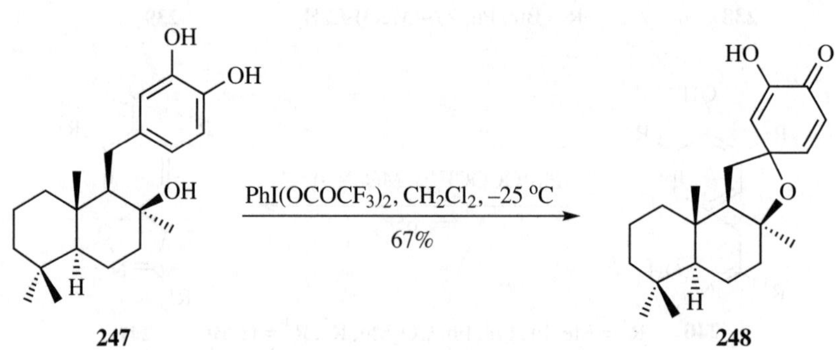

OH OH OH H **247** $PhI(OCOCF_3)_2$, CH_2Cl_2, –25 °C 67% HO O O H **248**

Scheme 3.101

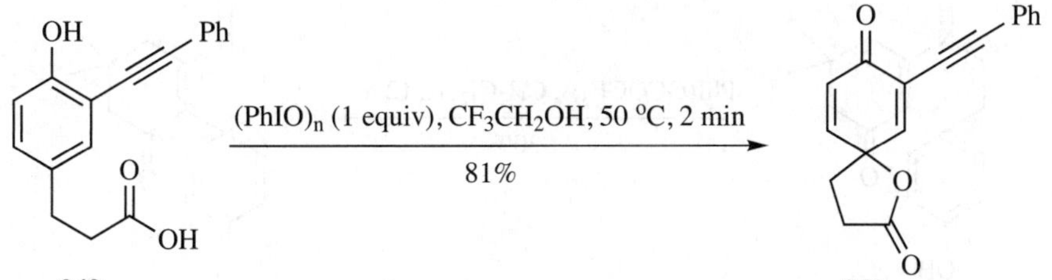

$(PhIO)_n$ (1 equiv), CF_3CH_2OH, 50 °C, 2 min
81%

249 → **250**

Scheme 3.102

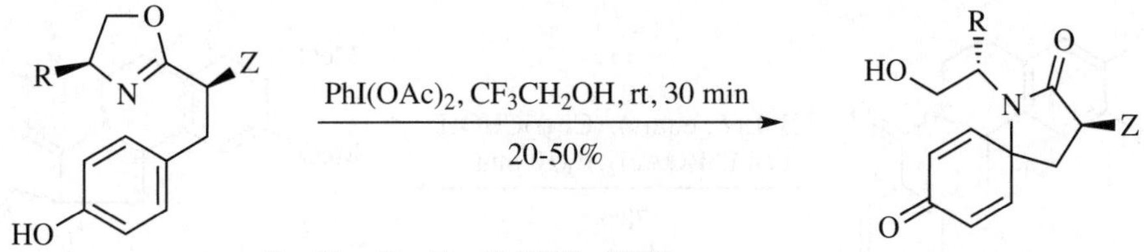

$PhI(OAc)_2$, CF_3CH_2OH, rt, 30 min
20-50%

R = H or Bn; Z = H, NHTs, NHBoc

251 → **252**

Scheme 3.103

252 (Scheme 3.103) [287, 315–318]. This methodology has been applied in the total synthesis of tricyclic azaspirane derivatives of tyrosine, FR901483 and TAN1251C [318,319].

Of particular interest are hypervalent iodine-induced cyclizations of phenolic precursors in which carbon–carbon bond formation is achieved. Representative examples of spirocyclizations employing carbon-based internal nucleophiles are shown below in Schemes 3.104–3.111.

Kita and coworkers have applied the oxidative coupling of various phenolic derivatives towards the synthesis of several pharmacologically interesting natural products [320–323]. For example, spirodienone compounds **254,** which are intermediates for the synthesis of an Amaryllidaceae alkaloid, (+)-maritidine, were selectively obtained by the reaction of phenolic precursor **253** and [bis(trifluoroacetoxy)iodo]benzene (Scheme 3.104) [323].

The analogous oxidation of phenolic enaminone derivatives **255** has been used to prepare spirocyclohexadienones **256** (Scheme 3.105) [324].

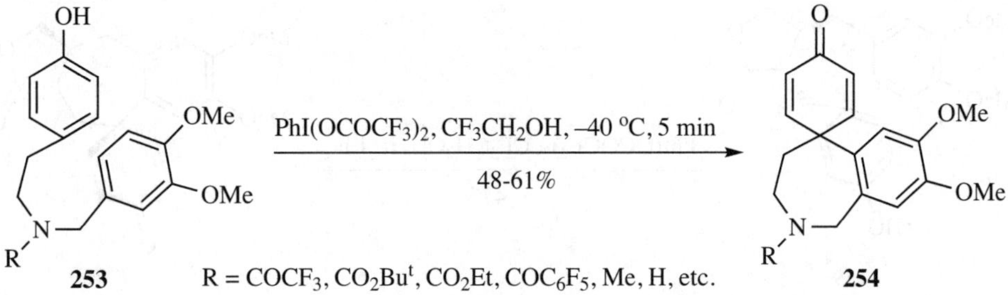

$PhI(OCOCF_3)_2$, CF_3CH_2OH, –40 °C, 5 min
48-61%

R = $COCF_3$, CO_2Bu^t, CO_2Et, COC_6F_5, Me, H, etc.

253 → **254**

Scheme 3.104

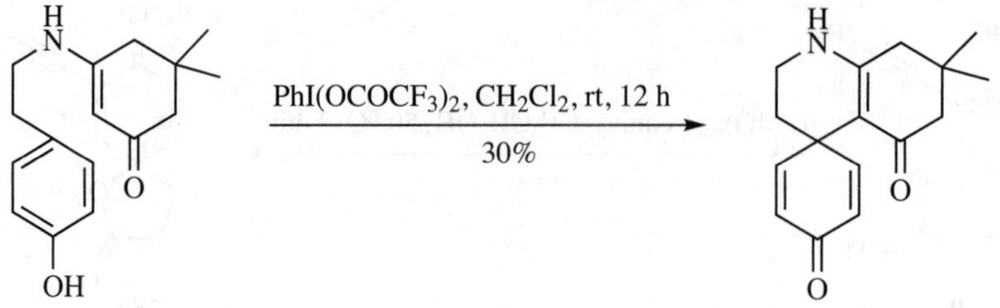

255 → 256: $PhI(OCOCF_3)_2$, CH_2Cl_2, rt, 12 h; 30%

Scheme 3.105

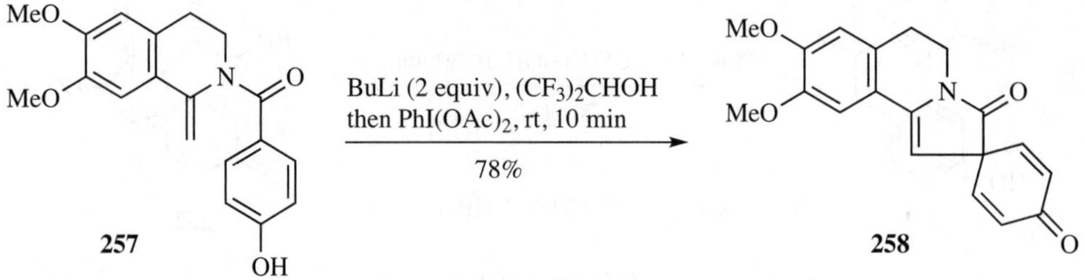

257 → 258: BuLi (2 equiv), $(CF_3)_2CHOH$ then $PhI(OAc)_2$, rt, 10 min; 78%

Scheme 3.106

The hypervalent iodine-promoted oxidation of enamide **257** in the presence of a base leads to the spiroenamide **258,** which is a key intermediate product in the total synthesis of annosqualine (Scheme 3.106) [325]. A similar oxidative cyclodearomatization approach has been utilized in the total synthesis of a proaporphine alkaloid, (±)-stepharine [326].

Treatment of the dibenzylbutyrolactone **259** with [bis(trifluoroacetoxy)iodo]benzene in trifluoroethanol for one hour gives as the major product spirodienone **260,** which has been postulated as an intermediate in the biosynthesis of tetrahydrodibenzocyclooctene lignans (Scheme 3.107) [327].

Canesi and coworkers have developed several synthetically useful tandem rearrangements on the basis of hypervalent iodine-promoted phenolic oxidation [328–331]. An oxidative Prins–pinacol tandem process mediated by a hypervalent iodine reagent allows the stereoselective transformation of simple phenols **261** into highly elaborate spirocyclic dienone cores **262** containing several quaternary carbon centers (Scheme 3.108).

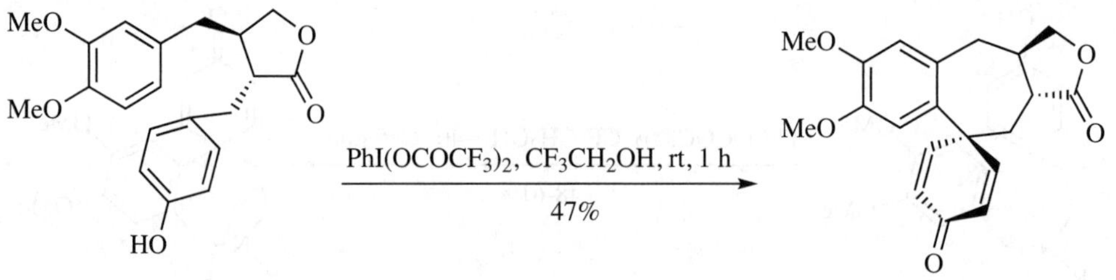

259 → 260: $PhI(OCOCF_3)_2$, CF_3CH_2OH, rt, 1 h; 47%

Scheme 3.107

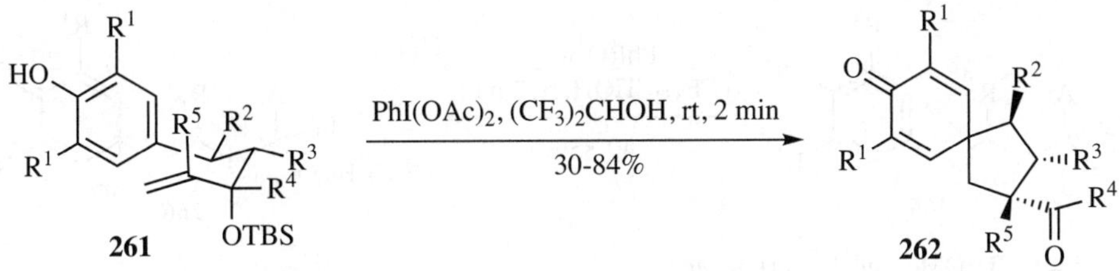

R^1 = H or Br; R^2 = H or Me; R^3 = H, Me, $CH_2CH{=}CH_2$
R^4 = Me, $CH{=}CH_2$, Ph; R^5 = H or Me
TBS = Bu^tMe_2Si

Scheme 3.108

This stereoselective process has been directly applied toward the formal synthesis of (–)-platensimycin, an important antibiotic agent [328].

Activation of phenol derivatives **263** with a hypervalent iodine reagent promotes the formation of bicyclic and tricyclic (**264**) products via a cationic cyclization process (Scheme 3.109). The method allows efficient one-step syntheses of scaffolds present in several natural products and occurs with total stereocontrol [329].

An oxidative *ipso*-rearrangement mediated by a hypervalent iodine reagent that enables rapid generation of a functionalized dienone system **266** containing a quaternary carbon center has been developed (Scheme 3.110) [330]. The process occurs through transfer of an aryl group from a silyl segment present on the lateral chain of the phenol derivative **265.** This transformation has been utilized in a total synthesis of an alkaloid sceletenone [330].

Oxidative 1,2- and 1,3-alkyl shifts mediated by a hypervalent iodine reagent using simple and inexpensive phenol derivatives **267** enable rapid construction of highly functionalized scaffolds containing a prochiral

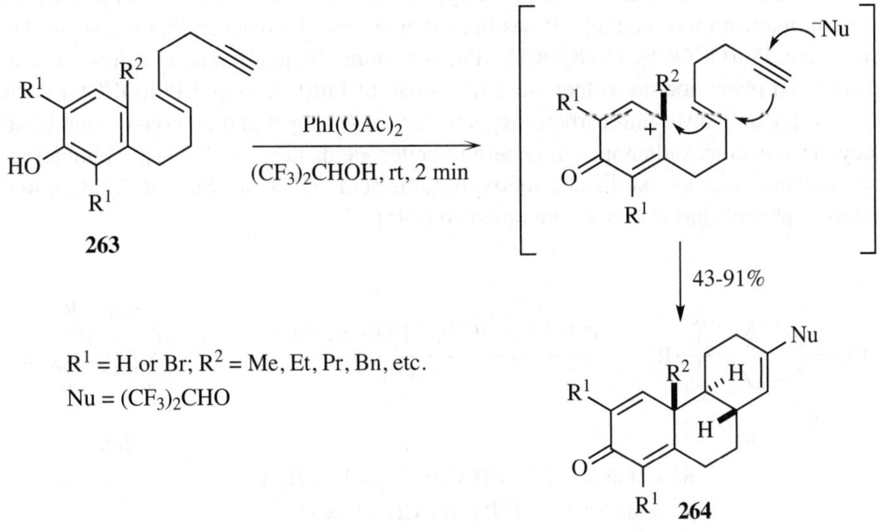

R^1 = H or Br; R^2 = Me, Et, Pr, Bn, etc.
Nu = $(CF_3)_2CHO$

Scheme 3.109

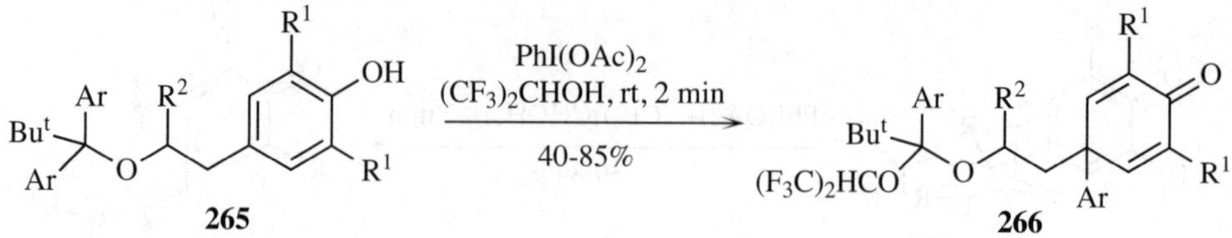

$PhI(OAc)_2$
$(CF_3)_2CHOH$, rt, 2 min
40-85%

265 → **266**

R^1 = H, Br, OMe, Bu^t; R^2 = H or Me
Ar = Ph, 4-$MeOC_6H_4$, 4-MeC_6H_4

Scheme 3.110

dienone system (**268**) (Scheme 3.111) [331]. An efficient enantioselective version of this process resulting in the formation of a challenging quaternary carbon center has also been developed. As an illustration of the synthetic potential of this method, the rapid synthesis of several functionalized polycyclic systems as well as a formal synthesis of acetylaspidoalbidine, a hexacyclic alkaloid belonging to the Aspidosperma family, has been described [331].

Additional examples of the hypervalent iodine-induced oxidative phenolic cyclizations include the following studies: the preparation of different heterocyclic rings such as dihydrofuranobenzofurans, tetrahydrofuranobenzofurans, tetrahydropyranofurans and dihydrobenzofurans by the treatment of various substituted phenols with (diacetoxyiodo)benzene in the presence of furan, allylsilanes, or cyclic enol ethers [332], the synthesis of galanthamine, a natural alkaloid isolated from the Amaryllidaceae family [333], the asymmetric total syntheses of the pentacyclic Stemona alkaloids tuberostemonine and didehydrotuberostemonine [334], the fully stereocontrolled total syntheses of (–)-cylindricine C and (–)-2-epicylindricine C [335], the asymmetric total synthesis of platensimycin [336], the total synthesis of a potent antitumor alkaloid, discorhabdin A [337], the total synthesis of the Amaryllidaceae alkaloid (+)-plicamine [338] and the development of a flow process for the multistep synthesis of the alkaloid natural product oxomaritidine [339].

(Diacetoxyiodo)benzene and [bis(trifluoroacetoxy)iodo]benzene are the most commonly used reagents for oxidative dearomatization of phenols. It has been demonstrated, however, that μ-oxo-bridged phenyliodine trifluoroacetate, $PhI(OCOCF_3)O(OCOCF_3)IPh$, is a more efficient oxidant in these reactions. The use of the μ-oxo-bridged phenyliodine trifluoroacetate instead of $PhI(OAc)_2$ and $PhI(OCOCF_3)_2$ in the oxidative cyclization of phenols involving carbon–oxygen, carbon–nitrogen and carbon–carbon bond formations affords spirocyclized cyclohexadienones in generally better yields [282, 340, 341]. Application of hypervalent iodine(V) oxidants, such as stabilized 2-iodoxybenzoic acid (SIBX, see Section 2.2.3.1), in the oxidative dearomatization of phenols has also been documented [291].

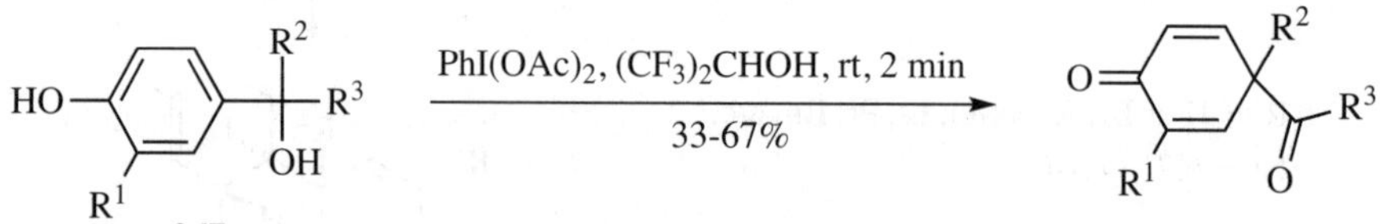

$PhI(OAc)_2$, $(CF_3)_2CHOH$, rt, 2 min
33-67%

267 → **268**

R^1 = H or Me; R^2 = $CH_2CH{=}CH_2$, $CH{=}CH_2$, Ph, Bu
R^3 = Me, Et, CH_2OTBS, Bu, $CH_2CH{=}CH_2$

Scheme 3.111

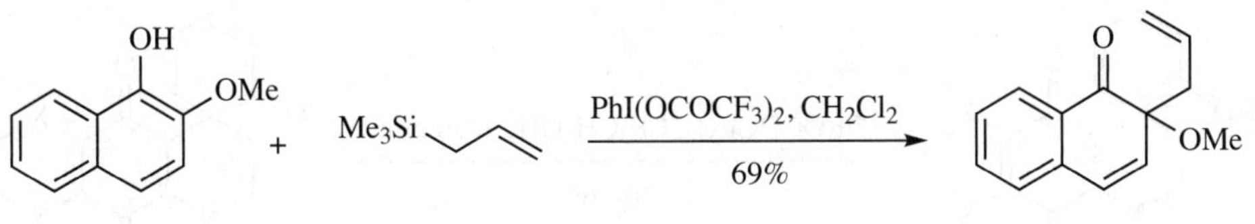

OH OMe + Me_3Si → $PhI(OCOCF_3)_2$, CH_2Cl_2 69% → O OMe

269 270

Scheme 3.112

3.1.11.2 Oxidative Dearomatization of 2-Substituted Phenols

Oxidative dearomatization of 2-substituted phenols **225** with [bis(acyloxy)iodo]arenes in the presence of external or internal nucleophile provides a convenient approach to 6,6-disubstituted cyclohexa-2,4-dienones **226** according to general Scheme 3.92. Specific examples of this reaction are provided below in Schemes 3.112–3.114.

Quideau and coworkers have developed a hypervalent iodine-mediated regioselective protocol for the oxidative dearomatization of 2-alkoxyarenols in the presence of external carbon-based nucleophiles, such as allylsilanes or silyl enol ethers [342–345]. For example, the oxidation of 2-alkoxynaphthol **269** with [bis(trifluoroacetoxy)iodo]benzene in the presence of allylsilane affords 2,4-cyclohexadienone derivative **270** (Scheme 3.112) [342].

This is a synthetically valuable process, as illustrated by the hypervalent iodine-mediated oxidative nucleophilic substitution of **269** with the silyl enol ether **271,** leading to the highly functionalized naphthoid cyclohexa-2,4-dienone **272** (Scheme 3.113), which is an important intermediate product in the synthesis of aquayamycin-type angucyclinones [343, 344].

Synthetic application of the oxidative dearomatization of *ortho*-substituted phenolic substrates in the intramolecular mode is exemplified by the preparation of azacarbocyclic spirodienones **274** from phenol derivatives **273** (Scheme 3.114) [322].

Hypervalent iodine induced oxidative dearomatization of *ortho*-substituted phenolic substrates in the intramolecular mode has been realized as an enantioselective reaction. In particular, Kita and coworkers have developed the enantioselective spirocyclization reaction of the *ortho*-substituted phenolic substrates **275** using chiral aryliodine(III) diacetate **276** having a rigid spirobiindane backbone (Scheme 3.115) [346]. Similar enantioselective oxidative spirocyclization reactions of the *ortho*-substituted phenolic substrates under catalytic conditions in the presence of chiral iodoarenes or chiral quaternary ammonium iodide catalysts are discussed in Sections 4.1.6 and 4.4.

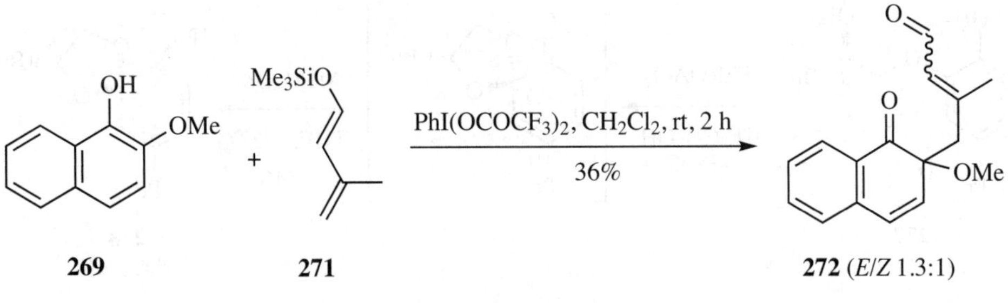

OH OMe + Me_3SiO → $PhI(OCOCF_3)_2$, CH_2Cl_2, rt, 2 h 36% → O O OMe

269 271 272 (*E*/*Z* 1.3:1)

Scheme 3.113

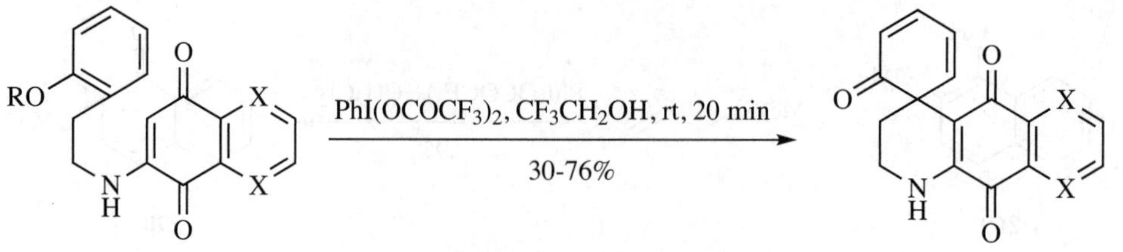

Scheme 3.114

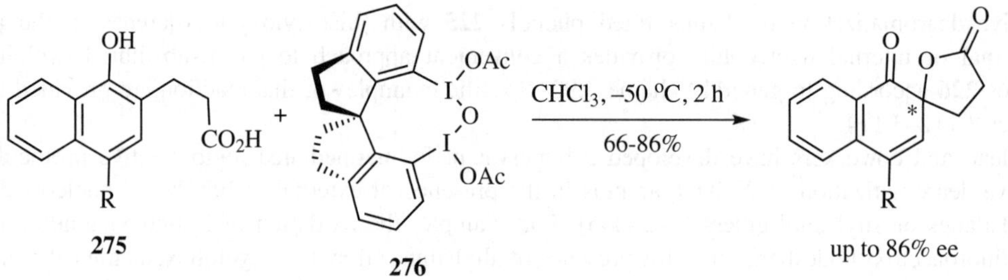

Scheme 3.115

A versatile chiral substrate **278** for asymmetric synthesis has been prepared through the hypervalent iodine induced spiroketalization of phenols **277** with a chiral substituted ethanol unit O-tethered to the *ortho* position (Scheme 3.116) [347]. This reaction has been successfully utilized in the asymmetric total synthesis of the natural product (+)-biscarvacrol.

The oxidative dearomatization of *ortho*-substituted phenols **225** leads to 6,6-disubstituted cyclohexa-2,4-dienones **226** (Scheme 3.92), which can be conveniently utilized *in situ* as dienes in the Diels–Alder cycloaddition reaction. When the oxidation of phenols is performed in the absence of an external dienophile, a dimerization via [4+2] cycloaddition often occurs spontaneously at ambient temperature to afford the corresponding dimers with an extraordinary level of regio-, site- and stereoselectivity [348–350]. A detailed experimental and theoretical investigation of such hypervalent iodine induced Diels–Alder cyclodimerizations

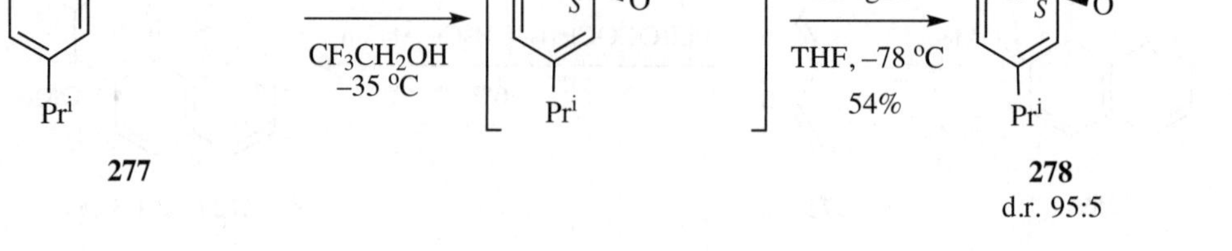

Scheme 3.116

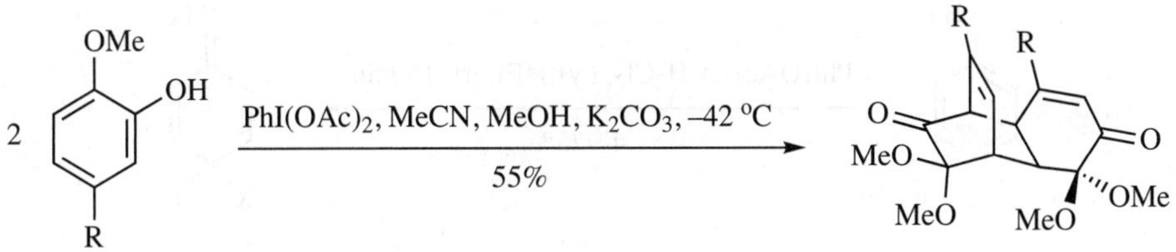

2 279 → 280: $PhI(OAc)_2$, MeCN, MeOH, K_2CO_3, –42 °C, 55%; R = CO_2Me

Scheme 3.117

has been published by Quideau and coauthors [349]. Scheme 3.117 shows a representative example of an oxidative Diels–Alder cyclodimerization of a phenolic substrate **279** to the dimer **280** [349].

If the oxidation is performed in the presence of an external dienophile, the respective products of [4+2] cycloaddition are formed [351–356]. Typical examples are illustrated by a one-pot synthesis of several silyl bicyclic alkenes **283** by intermolecular Diels–Alder reactions of 4-trimethylsilyl substituted masked *o*-benzoquinones **282** generated by oxidation of the corresponding 2-methoxyphenols **281** [351] and by the hypervalent iodine-mediated oxidative dearomatization/Diels–Alder cascade reaction of phenols **284** with allyl alcohol affording polycyclic acetals **285** (Scheme 3.118) [352]. This hypervalent iodine-promoted tandem phenolic oxidation/Diels–Alder reaction has been utilized in the stereoselective synthesis of the bacchopetiolone carbocyclic core [353].

3.1.11.3 *Oxidative Dearomatization of Anilines*

Aniline derivatives can be oxidatively dearomatized analogously to the phenol derivatives. In particular, the oxidative *ipso*-fluorination of *para*-substituted tosylated anilines **286** using hypervalent iodine reagents

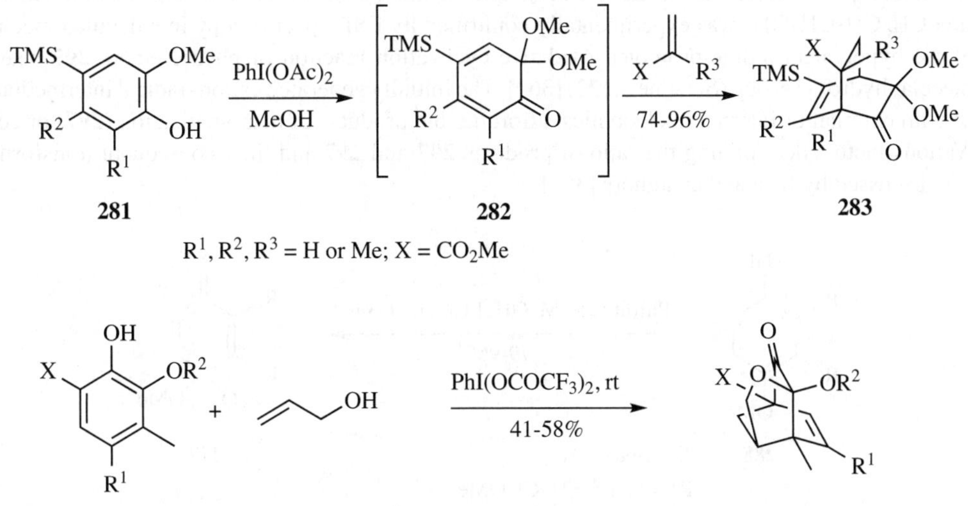

281 → [282] → 283: $PhI(OAc)_2$, MeOH; then dienophile, 74-96%; R^1, R^2, R^3 = H or Me; X = CO_2Me

284 + allyl alcohol → 285: $PhI(OCOCF_3)_2$, rt, 41-58%; X = Br or H; R^1 = CO_2Me or OTs; R^2 = Me or Bn

Scheme 3.118

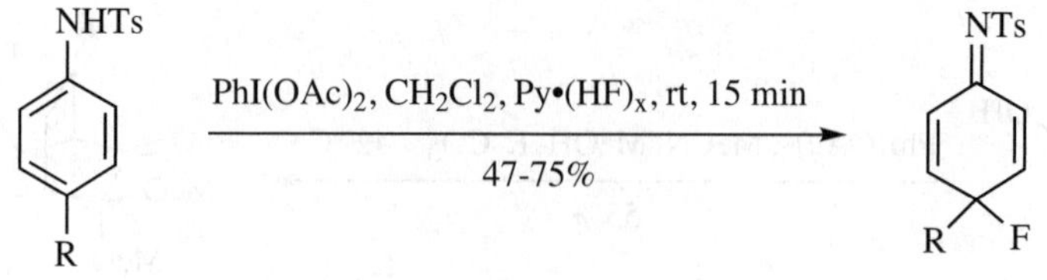

NHTs → NTs

$PhI(OAc)_2$, CH_2Cl_2, Py•$(HF)_x$, rt, 15 min

47-75%

286 R = Me, Et, F, Cl, etc. **287**

Scheme 3.119

in combination with pyridinium polyhydrogen fluoride affords 4-fluorocyclohexa-2,5-dienimines **287** in generally good yields (Scheme 3.119) [357].

Likewise, the oxidative dearomatization of *para*-methoxy substituted N-protected anilines **288** using (diacetoxyiodo)benzene in the presence of methanol gives *p*-quinone monoimide ketals **289** (Scheme 3.120) [358]. If the oxidation of aniline derivatives is performed in the presence of water, the final isolated products are the respective *p*-benzoquinones or *p*-benzoquinone monoketals resulting from the hydrolysis of initially formed monoimide ketals **289** [358, 359].

The oxidative dearomatization of *para*-substituted *o*-alkynylanilines **290** using (diacetoxyiodo)benzene affords 2-alkynyl cyclohexadienimines **291,** which can act as active substrates for reaction with electron-rich styrenes **292** in the presence of metal salts: the $Bi(OTf)_3$-catalyzed reactions give 3,4-dihydrocyclopenta[*c,d*]indoles **293** and the AgOTf-catalyzed reactions provide tricyclic pyrrole derivatives **294** (Scheme 3.121) [360].

3.1.12 Oxidative Coupling of Aromatic Substrates

Oxidative coupling typically occurs in the reactions of phenol ethers or other electron-rich aromatic substrates with [bis(acyloxy)iodo]arenes in polar, non-nucleophilic solvents, under conditions favorable for single-electron transfer (SET) reactions (Section 1.5.3). The formation of the cation–radical intermediates **296** (R^1–Nu = $CH_2CH_2CH_2OH$) was experimentally confirmed by ESR spectroscopy in a detailed mechanistic study of the hypervalent-iodine-promoted oxidative cyclization reaction of phenol ethers **295** bearing an intramolecular hydroxyl group (Scheme 3.122) [361]. The initially generated cation–radical intermediates **296** combine with external or internal nucleophiles, affording the products of dearomatization (**297**) or coupling (**298**). Various factors determining the ratio of products **297** and **297** and their consequent transformations have been discussed by Kita and coauthors [361].

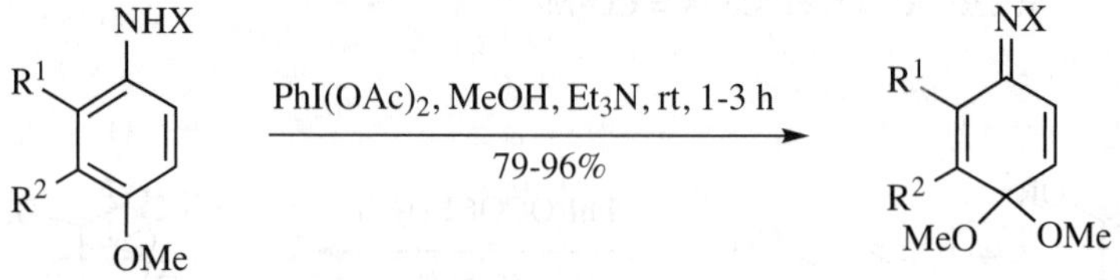

NHX, R^1, R^2, OMe → NX, R^1, R^2, MeO, OMe

$PhI(OAc)_2$, MeOH, Et_3N, rt, 1-3 h

79-96%

288 X = Boc or Ac **289**

R^1 = H, R^2 = H, Cl, OMe

or R^2 = H, R^1 = H, Me, OMe

Scheme 3.120

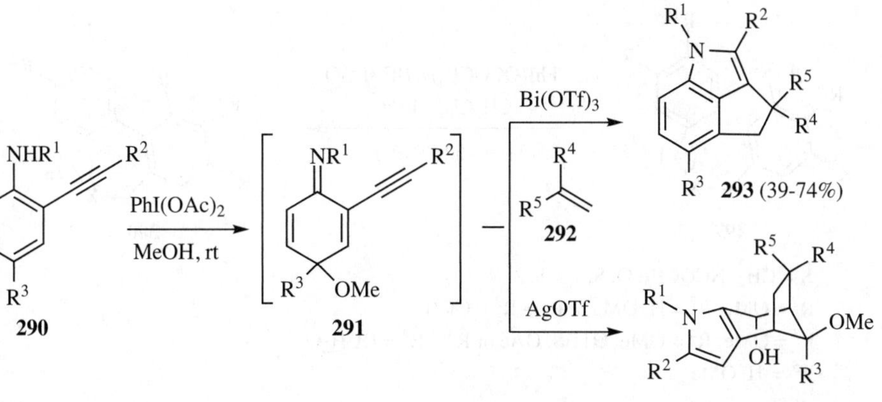

R^1 = Ts, $MeSO_2$
R^2 = Ph, 4-$MeOC_6H_4$, 4-ClC_6H_4, 2-thienyl, 2-pyridyl, cyclopropyl
R^3 = Me, Et, Bu
R^4 = Ph, 4-$MeOC_6H_4$, 4-$BnOC_6H_4$, 3,4-$(MeO)_2C_6H_3$, 4-Br-4,5-$(MeO)_2C_6H_2$, 4-MeC_6H_4
R^5 = H or Me

Scheme 3.121

The direct nucleophilic substitution of electron-rich phenol ethers using hypervalent iodine oxidants in the presence of Lewis acid or fluorinated alcohols and involving aromatic cation–radical intermediates was originally developed by Kita and coworkers in 1994 [362]. Since then this procedure with some variations has been extensively applied by Kita and other researchers for various oxidative transformations. In the intermolecular mode, this reaction (Scheme 3.122) has been utilized for the preparation of the products **298** from N_3^-, AcO^-, ArS^-, SCN^-, β-dicarbonyl compounds and other external nucleophiles [320]. The oxidative coupling reaction in the intramolecular mode provides a powerful synthetic tool for the preparation of various

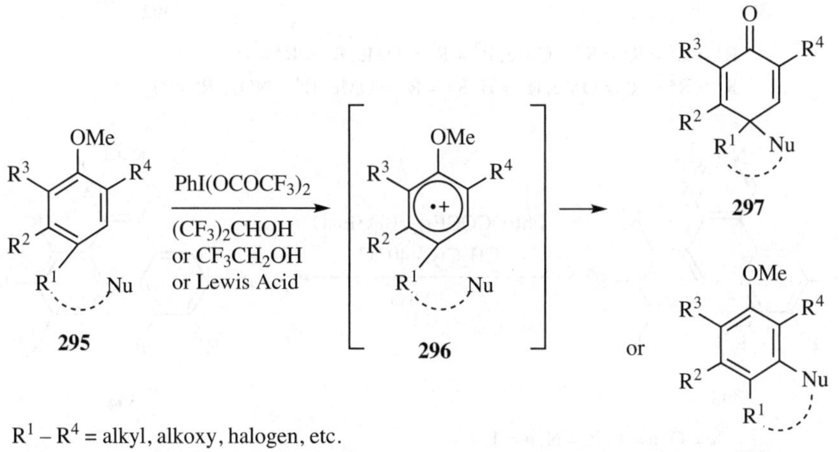

$R^1 - R^4$ = alkyl, alkoxy, halogen, etc.
Nu = external or internal nucleophilic group, including electron-rich aromatics

Scheme 3.122

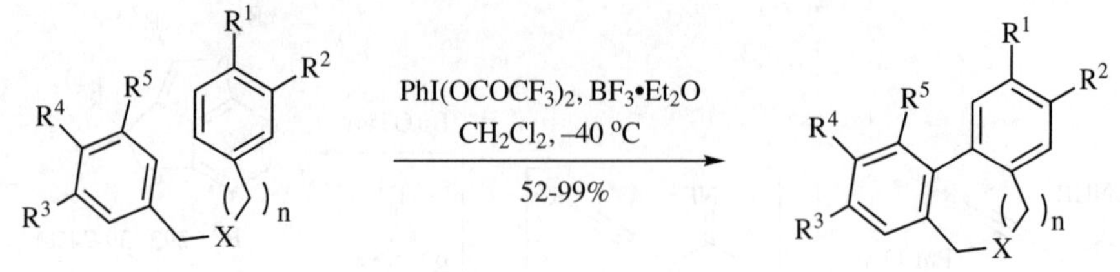

R^1 R^2 R^5 R^4 R^3 X n

$PhI(OCOCF_3)_2, BF_3{\cdot}Et_2O$
CH_2Cl_2, –40 °C
52-99%

299 **300**

$X = CH_2, NCOCF_3, O, S; n = 1, 2$
$R^1 = OMe; R^2 = H, OMe$ or $R^1 + R^2 = OCH_2O$
$R^3 = OMe; R^4 = OMe, OTBS, OAc$ or $R^3 + R^4 = OCH_2O$
$R^5 = H, OMe$

Scheme 3.123

carbocyclic and heterocyclic compounds [321, 363–369]. Specific examples of synthetic applications of this reaction are provided below in Schemes 3.123–3.129.

Numerous dibenzoheterocyclic compounds **300** have been prepared by the oxidation of phenol ether derivatives **299** with [bis(trifluoroacetoxy)iodo]benzene in the presence of $BF_3{\cdot}Et_2O$ in dichloromethane (Scheme 3.123) [363–365].

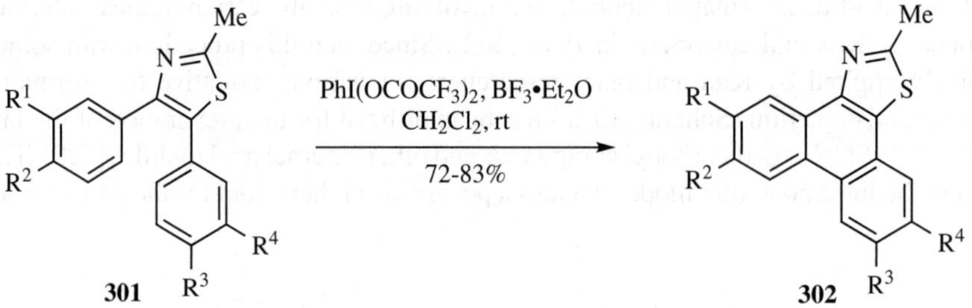

Me N S R^1 R^2 R^3 R^4

$PhI(OCOCF_3)_2, BF_3{\cdot}Et_2O$
CH_2Cl_2, rt
72-83%

301 **302**

$R^1 = R^2 = R^3 = R^4 = OMe$; $R^1 = R^2 = OMe, R^3 = R^4 = H$
$R^1 = R^2 = R^4 = OMe, R^3 = H$; $R^1 = R^2 = OMe, R^3 = NO_2, R^4 = H$

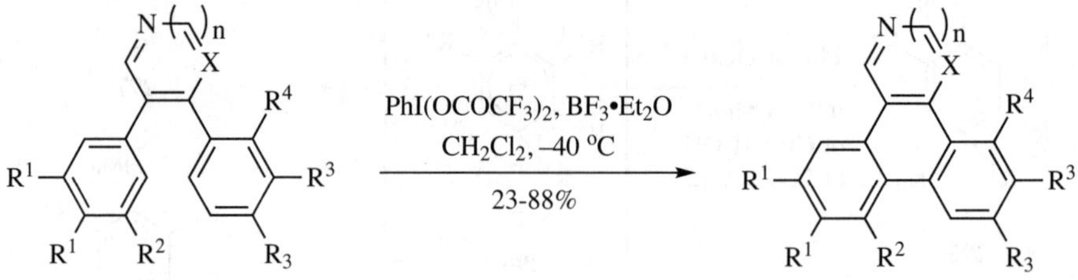

N X n R^1 R^2 R^3 R^4 R_3

$PhI(OCOCF_3)_2, BF_3{\cdot}Et_2O$
CH_2Cl_2, –40 °C
23-88%

303 **304**

$X = O, n = 0; X = N, n = 1$
$R^1, R^2, R^3, R^4 = H$ or OMe

Scheme 3.124

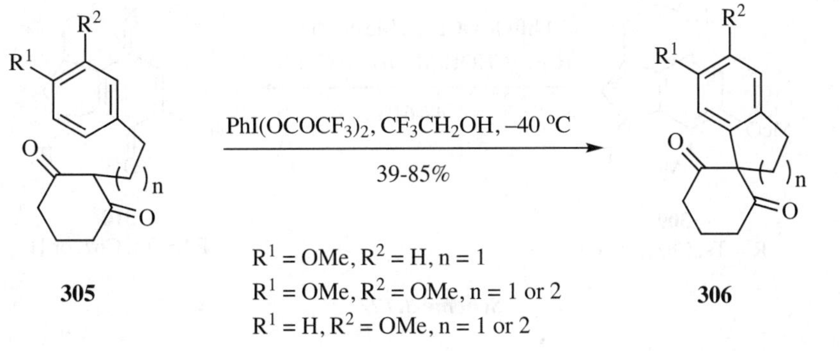

$PhI(OCOCF_3)_2$, CF_3CH_2OH, –40 °C
39-85%

305 R^1 = OMe, R^2 = H, n = 1
R^1 = OMe, R^2 = OMe, n = 1 or 2
R^1 = H, R^2 = OMe, n = 1 or 2 **306**

Scheme 3.125

Under similar conditions, the phenanthro-fused thiazoles **302,** isoxazoles **304** (X = O, $n = 0$) and pyrimidines **304** (X = N, $n = 1$) can be prepared by oxidative coupling of the respective phenol ethers **301** and **303** (Scheme 3.124) [368, 369].

The hypervalent iodine-induced direct intramolecular cyclization of α-(aryl)alkyl-β-dicarbonyl derivatives **305** affords biologically important spirobenzannulated compounds **306** (Scheme 3.125) [366].

The oxidation of phenol ethers containing the azido group as an internal nitrogen nucleophile provides a useful methodology for the construction of nitrogen heterocycles [370–372]. Kita and coworkers have reported an efficient synthesis of quinone imine ketals **308** from the substituted phenol ethers **307** bearing an alkyl azido side chain (Scheme 3.126) [371].

A similar intramolecular cyclization of 3-(azidoethyl)indole derivatives **309** provides an efficient route to the pyrroloiminoquinone system **310** (Scheme 3.127), which is an essential component of several recently isolated marine alkaloids such as makaluvamines, isobatzelline C and discorhabdins, which possess potent biological activities [372].

The oxidation of phenol ethers **311** bearing an alkyl sulfide side chain followed by treatment with aqueous methylamine selectively affords various dihydrobenzothiophenes **312** (Scheme 3.128) without yielding any sulfoxides as by-products [372]. This procedure has been applied in the total synthesis of the potent cytotoxic makaluvamine F, a sulfur-containing pyrroloiminoquinone marine product [373].

Additional examples of intramolecular oxidative coupling of phenolic ethers include the oxidative biaryl coupling of various N-substituted 1-benzyltetrahydroisoquinolines **313** to the corresponding aporphines **314** [374], the oxidative cyclization of 3,4-dimethoxyphenyl 3,4-dimethoxyphenylacetate (**315**) leading to the seven-membered lactone **316** [375] and the conversion of phenol ether derivatives **317** into the products of

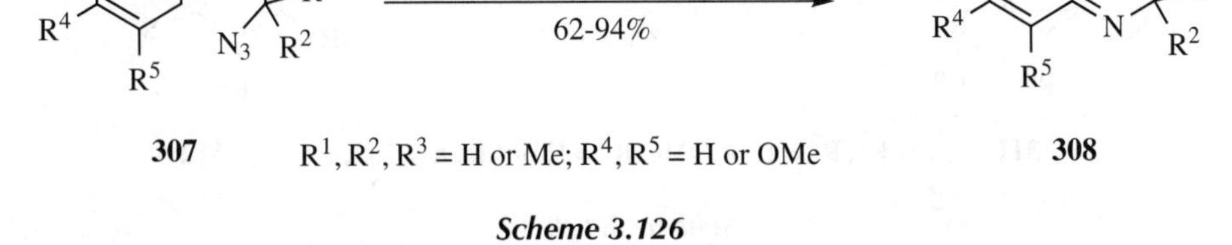

$PhI(OCOCF_3)_2$, Me_3SiOTf
$CF_3CH_2OH/MeOH$, 0 °C, 30 min
62-94%

307 R^1, R^2, R^3 = H or Me; R^4, R^5 = H or OMe **308**

Scheme 3.126

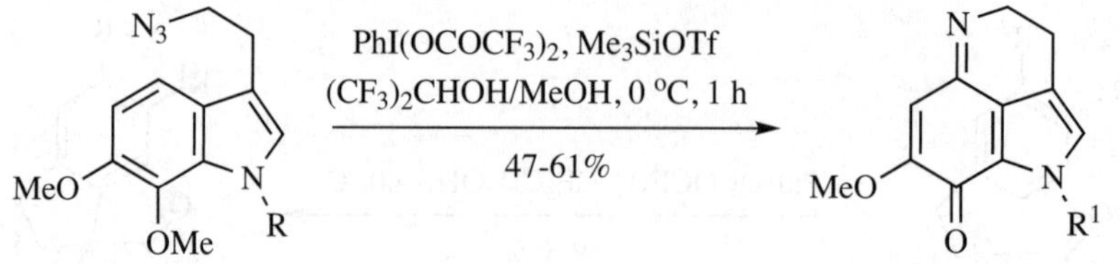

$$\mathbf{309} \xrightarrow[\text{47-61\%}]{\text{PhI(OCOCF}_3)_2,\ \text{Me}_3\text{SiOTf},\ (\text{CF}_3)_2\text{CHOH/MeOH},\ 0\ ^\circ\text{C},\ 1\ \text{h}} \mathbf{310}$$

309 R = Ts, Cbz, Ac, Bz

310 R1 = Ts, Cbz, or H

Scheme 3.127

intramolecular coupling **318** using a combination of [bis(trifluoroacetoxy)iodo]benzene and heteropoly acid (Scheme 3.129) [376]. A similar oxidative coupling reaction of benzyltetrahydroisoquinolines (laudanosine derivatives) using [bis(trifluoroacetoxy)iodo]benzene and heteropoly acid has been used in an efficient synthesis of morphinandienone alkaloids [377]. Catalytic versions of the oxidative coupling of phenolic ethers using iodoarenes as catalysts and *m*CPBA or hydrogen peroxide as stoichiometric oxidants have also been reported (Section 4.1).

Non-phenolic electron-rich aromatic substrates can also be oxidatively coupled using hypervalent iodine reagents (Schemes 3.130–3.133). Kita and coworkers reported a facile and efficient oxidative coupling reaction of alkylarenes **319** leading to alkylbiaryls **320** using a combination of [bis(trifluoroacetoxy)iodo]benzene and $BF_3{\cdot}OEt_2$ (Scheme 3.130) [378]. Similarly, multiply iodinated biaryls can be prepared in good yields by the [bis(trifluoroacetoxy)iodo]benzene-induced direct oxidative coupling reaction of the iodinated arenes [379].

Oxidation of N-aromatic methanesulfonamides **321** with (diacetoxyiodo)benzene in the presence of thiophene in trifluoroethanol or hexafluoroisopropanol (HFIP) affords the respective coupling products **322** in good yield (Scheme 3.131) [380]. The head-to-tail thiophene dimers **324** can be selectively prepared by the hypervalent iodine oxidation of 3-substituted thiophenes **323** [381, 382] and bipyrroles **326** can be regioselectively synthesized by oxidative dimerization of pyrroles **325** with [bis(trifluoroacetoxy)iodo]benzene in the presence of bromotrimethylsilane [383]. Likewise, bithiophenes **328** have been synthesized from 3,4-disubstituted thiophenes **327** using [hydroxy(tosyloxy)iodo]benzene in the presence of bromotrimethylsilane in hexafluoroisopropanol [384].

A direct coupling reaction of cycloalkenylsilanes **329** with a silylated nucleobase **330** promoted by (diacetoxyiodo)benzene in the presence of trimethylsilyl triflate in dichloromethane at room temperature has been reported (Scheme 3.132) [385]. This procedure was applied in the synthesis of a novel carbocyclic cytidine derivative having bis(hydroxymethyl)cyclohexene as a pseudo-sugar moiety, which was designed as a potential anti-HIV agent.

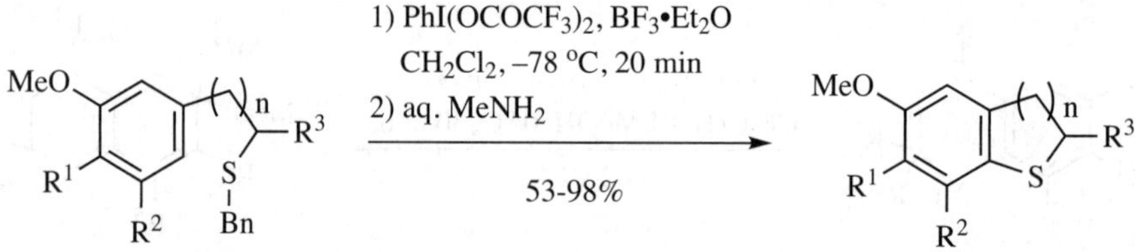

$$\mathbf{311} \xrightarrow[\text{53-98\%}]{\text{1) PhI(OCOCF}_3)_2,\ \text{BF}_3{\cdot}\text{Et}_2\text{O},\ \text{CH}_2\text{Cl}_2,\ -78\ ^\circ\text{C},\ 20\ \text{min};\ \text{2) aq. MeNH}_2} \mathbf{312}$$

R^1, R^2 = H or OMe; R^3 = H or Me; n = 1, 2, 3

Scheme 3.128

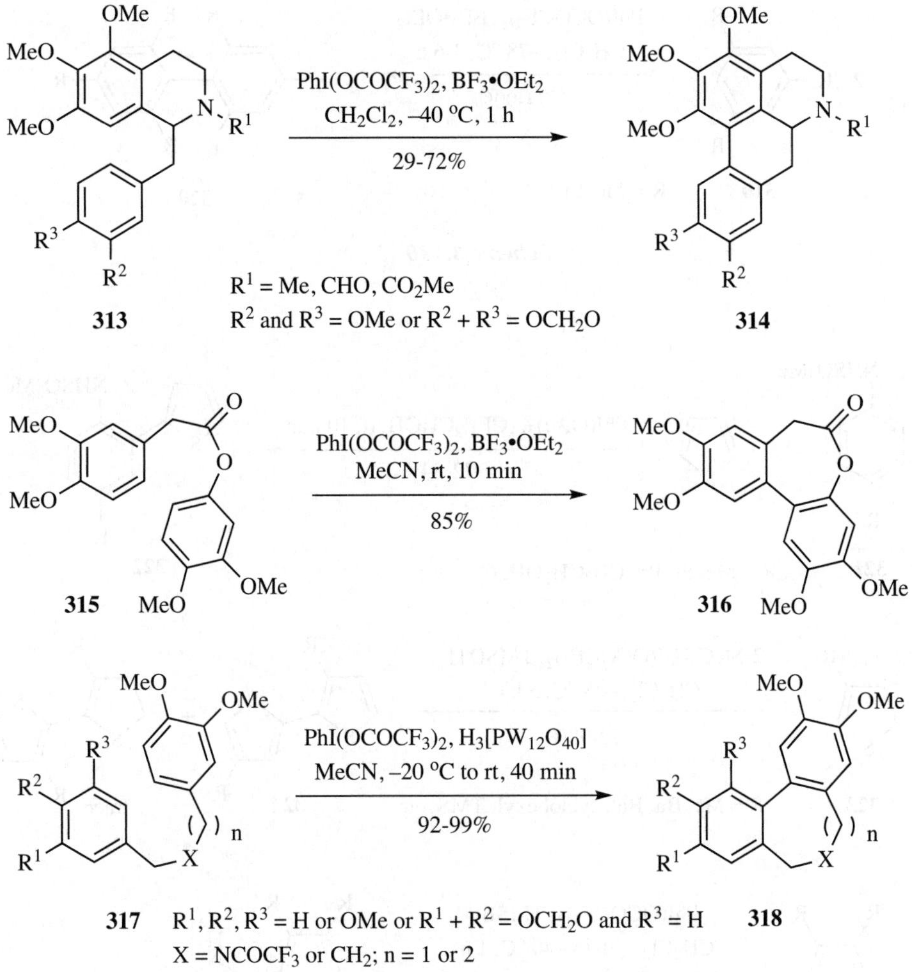

$PhI(OCOCF_3)_2$, $BF_3{\bullet}OEt_2$
CH_2Cl_2, –40 °C, 1 h
29-72%

313 R^1 = Me, CHO, CO_2Me
R^2 and R^3 = OMe or $R^2 + R^3 = OCH_2O$ **314**

$PhI(OCOCF_3)_2$, $BF_3{\bullet}OEt_2$
MeCN, rt, 10 min
85%

315 **316**

$PhI(OCOCF_3)_2$, $H_3[PW_{12}O_{40}]$
MeCN, –20 °C to rt, 40 min
92-99%

317 R^1, R^2, R^3 = H or OMe or $R^1 + R^2 = OCH_2O$ and R^3 = H **318**
X = $NCOCF_3$ or CH_2; n = 1 or 2

Scheme 3.129

[Bis(trifluoroacetoxy)iodo]benzene in conjunction with a Lewis acid promotes C–C coupling of Bodipy (4,4′-difluoro-4-bora-3*a*,4*a*-diaza-*s*-indacene) monomers **331** leading to mixtures of dimers **332** (when X = I or *p*-tolyl) (Scheme 3.133) and higher oligomers when X = H [386]. Bodipy dyes have attracted significant interest in recent years due to their outstanding optical properties.

3.1.13 Oxidative Cationic Cyclizations, Rearrangements and Fragmentations

Hypervalent iodine(III) compounds, such as [bis(trifluoroacetoxy)iodo]benzene, (diacetoxyiodo)benzene and [hydroxy(tosyloxy)iodo]benzene, are commonly used as reagents in various cationic cyclizations, rearrangements and fragmentations. Numerous examples of such reactions have been reported in the literature and summarized in the reviews dedicated to synthetic applications of hypervalent iodine compounds [4, 7, 10, 11, 180, 191, 387].

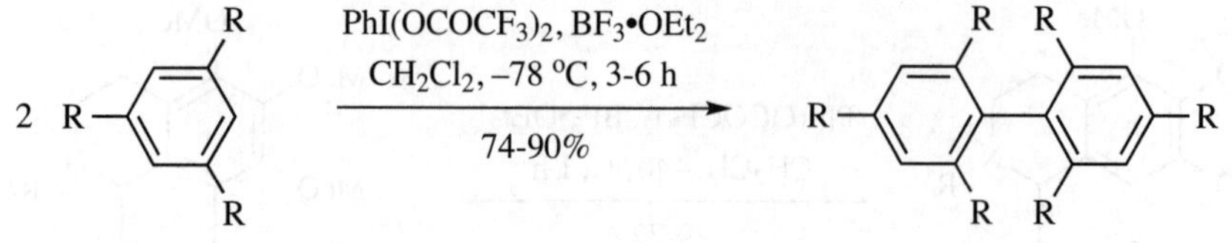

$PhI(OCOCF_3)_2$, BF_3•OEt_2
CH_2Cl_2, –78 °C, 3-6 h
74-90%

319 R = Me, Et or I **320**

Scheme 3.130

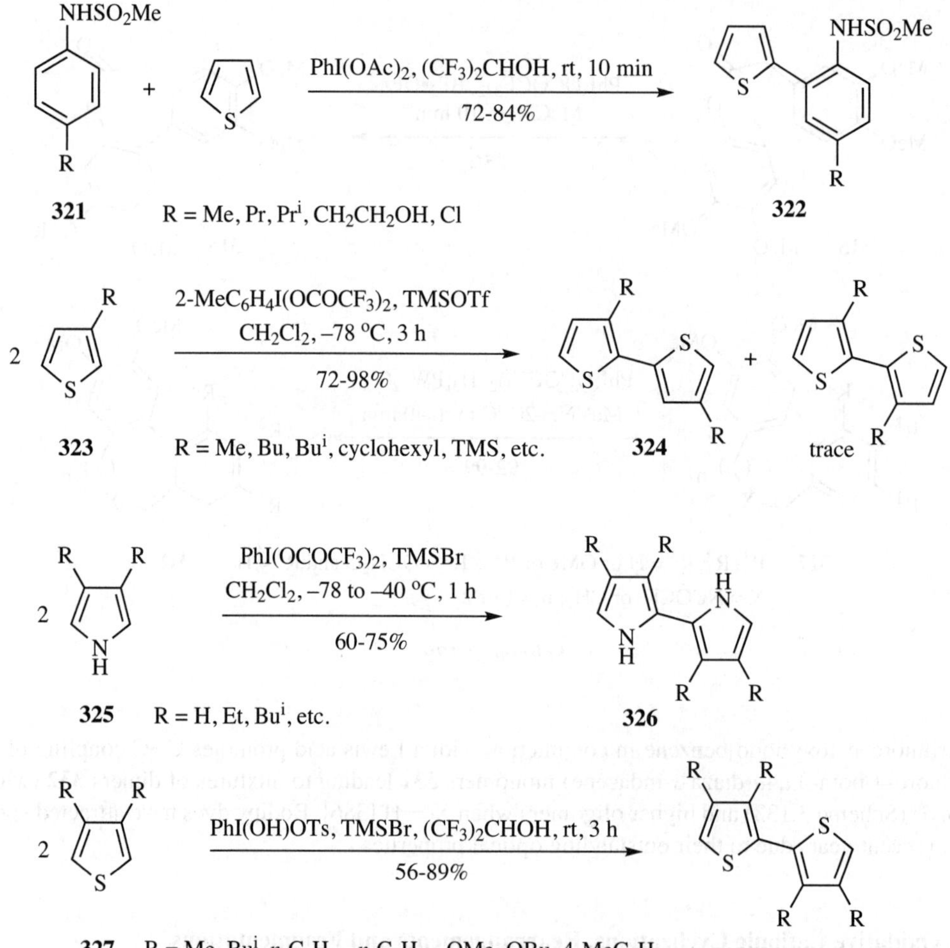

$NHSO_2Me$

$PhI(OAc)_2$, $(CF_3)_2CHOH$, rt, 10 min
72-84%

321 R = Me, Pr, Pr^i, CH_2CH_2OH, Cl **322**

$2\text{-}MeC_6H_4I(OCOCF_3)_2$, TMSOTf
CH_2Cl_2, –78 °C, 3 h
72-98%

323 R = Me, Bu, Bu^i, cyclohexyl, TMS, etc. **324** trace

$PhI(OCOCF_3)_2$, TMSBr
CH_2Cl_2, –78 to –40 °C, 1 h
60-75%

325 R = H, Et, Bu^i, etc. **326**

PhI(OH)OTs, TMSBr, $(CF_3)_2CHOH$, rt, 3 h
56-89%

327 R = Me, Bu^i, $n\text{-}C_6H_{13}$, $n\text{-}C_8H_{17}$, OMe, OBu, $4\text{-}MeC_6H_4$
or R + R = $O(CH_2)_2O$ **328**

Scheme 3.131

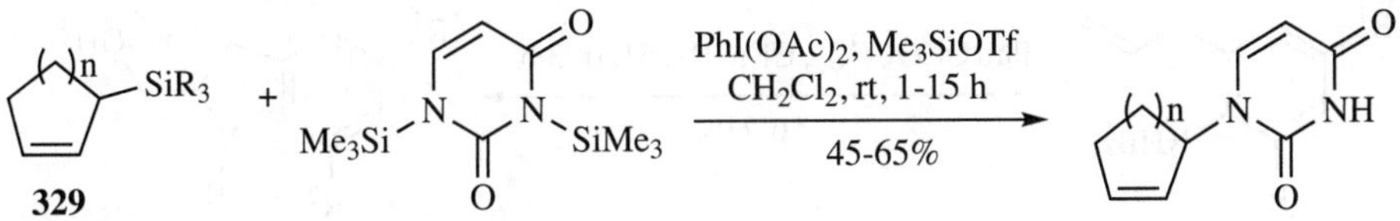

R = OEt or Me; n = 1 or 2

Scheme 3.132

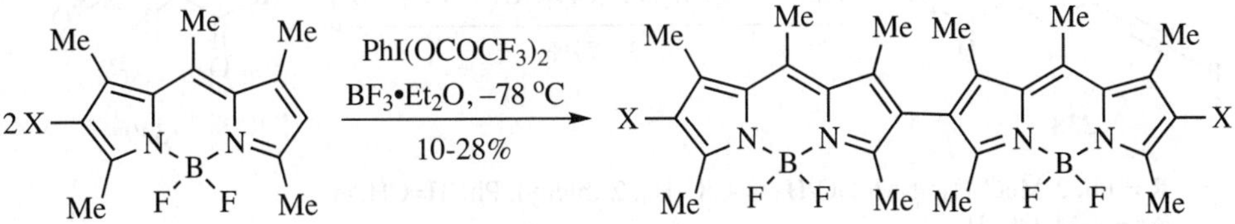

X = H, I, 4-MeC_6H_4

Scheme 3.133

3.1.13.1 Heterocyclizations

Cationic cyclizations, induced by hypervalent iodine reagents, are particularly useful in the synthesis of heterocycles. Tellitu and Dominguez have developed a series of [bis(trifluoroacetoxy)iodo]benzene-promoted intramolecular amidation reactions, generalized in Scheme 3.134, leading to various five, six and seven-membered heterocycles **335** [388, 389]. Experimental evidence supports the ionic mechanism of these reactions, involving *N*-acylnitrenium intermediates **334** generated in the initial reaction of the amide **333** with the hypervalent iodine reagent [390].

This methodology with some variations has been utilized in the synthesis of numerous heterocyclic systems, such as heterocycle-fused quinolinone derivatives [391], 1,4-benzodiazepin-2-ones [392], benzo-, naphtho- and heterocycle-fused pyrrolo[2,1-*c*][1,4]diazepines [393], quinolinone or pyrrolidinone derivatives [394], dibenzo[*a,c*]phenanthridines [395], thiazolo-fused quinolinones [396], isoindolinone and isoquinolin-2-one derivatives [397], indoline derivatives [398], 5-aroyl-pyrrolidinones [399, 400], indazolone derivatives [401, 402], substituted indolizidinones [403], 1-arylpyrrolopyrazinones [404], structurally diverse

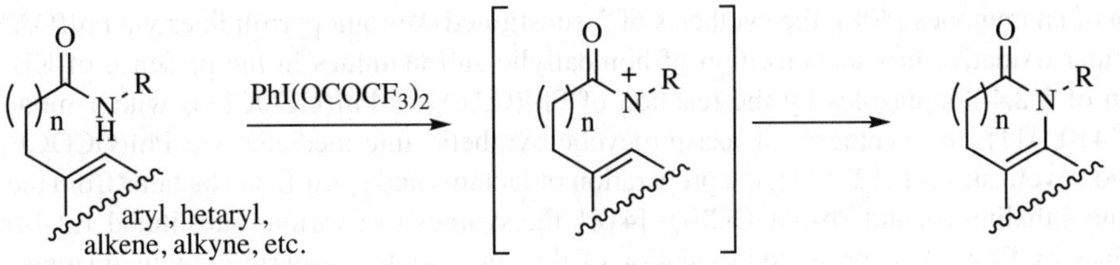

R = OMe, alkyl, Bn, Ph, Ts, Bz, etc.
n = 0, 1, 2

Scheme 3.134

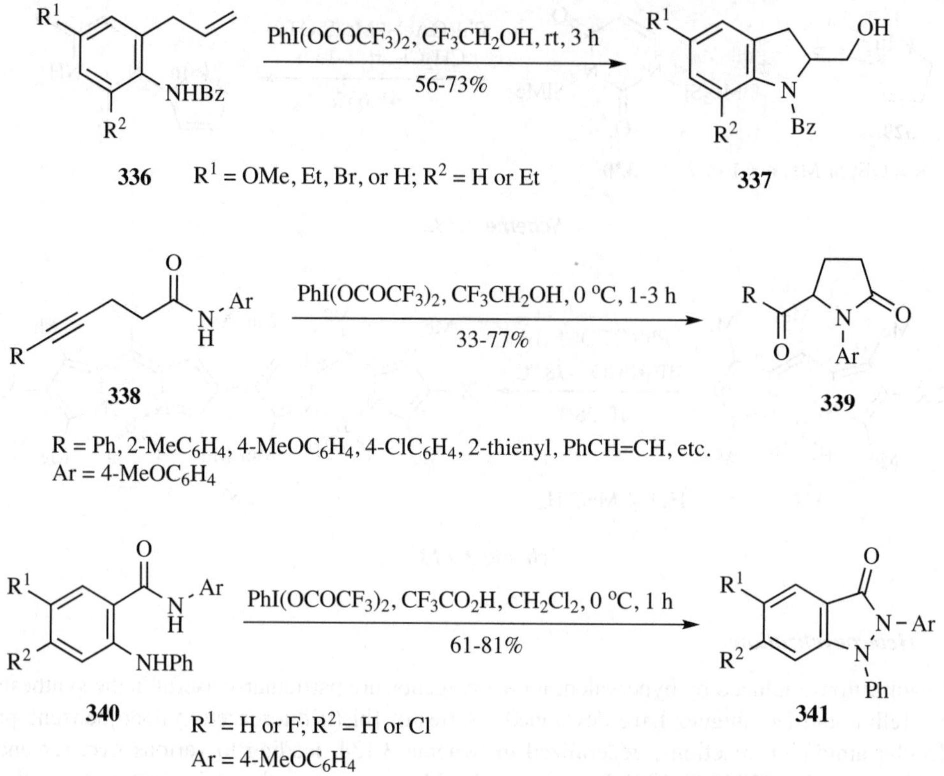

$PhI(OCOCF_3)_2$, CF_3CH_2OH, rt, 3 h
56-73%

336 R^1 = OMe, Et, Br, or H; R^2 = H or Et **337**

$PhI(OCOCF_3)_2$, CF_3CH_2OH, 0 °C, 1-3 h
33-77%

338 **339**

R = Ph, 2-MeC_6H_4, 4-$MeOC_6H_4$, 4-ClC_6H_4, 2-thienyl, PhCH=CH, etc.
Ar = 4-$MeOC_6H_4$

$PhI(OCOCF_3)_2$, CF_3CO_2H, CH_2Cl_2, 0 °C, 1 h
61-81%

340 R^1 = H or F; R^2 = H or Cl
Ar = 4-$MeOC_6H_4$ **341**

Scheme 3.135

pyrrolo(benzo)diazepines [405] and furopyrimidinones [406]. Representative examples of these cyclizations are shown in Scheme 3.135 and include the preparation of indoline derivatives **337** from anilides **336** [398], pyrrolidinones **339** from alkynylamides **338** [399,400] and indazol-3-ones **341** from anthranilamides **340** [401,402].

Similar hypervalent iodine-induced heterocyclizations of the appropriate amide or amine precursors have been used in numerous useful synthetic transformations, such as the synthesis of 1-arylcarbazoles by metal-free electrocyclization [407], the synthesis of highly substituted pyrrolin-4-ones via $PhI(OCOCF_3)_2$-mediated cyclization of enaminones [408], the synthesis of 2-substituted-4-bromopyrrolidines via $PhI(OAc)_2$-induced intramolecular oxidative bromocyclization of homoallylic sulfonamides in the presence of KBr [409], the preparation of 1,2,4-thiadiazoles by the reaction of $PhI(OAc)_2$ or $PhI(OCOCF_3)_2$ with 1-monosubstituted thioureas [410, 411], the synthesis of azaspirocyclic synthetic intermediates via $PhI(OCOCF_3)_2$-induced nitrenium ion cyclizations [412–417], the preparation of lactams and spiro-fused lactams from the reaction of *N*-acylaminophthalimides and $PhI(OCOCF_3)_2$ [418], the synthesis of various substituted 1,2,4-triazolo[4,3-*a*]pyrimidines by $PhI(OAc)_2$-promoted oxidation of the appropriate 2,4-pyrimidinylhydrazones [419–421], the synthesis of pyrrolidino[60]fullerene from the $PhI(OAc)_2$-promoted reaction between C_{60} and amino acid esters [422], the synthesis of 1,3,4-oxadiazoles from acylhydrazones by $PhI(OCOCF_3)_2$ oxidation [423–425], the synthesis of 1-aryl-4-methyl-1,2,4-triazolo[4,3-*a*]quinoxalines from arenecarboxaldehyde-3-methyl-2-quinoxalinylhydrazones [426, 427], the synthesis of various N-substituted indole derivatives via $PhI(OCOCF_3)_2$-mediated intramolecular cyclization of enamines [428] and the synthesis of enantiomerically

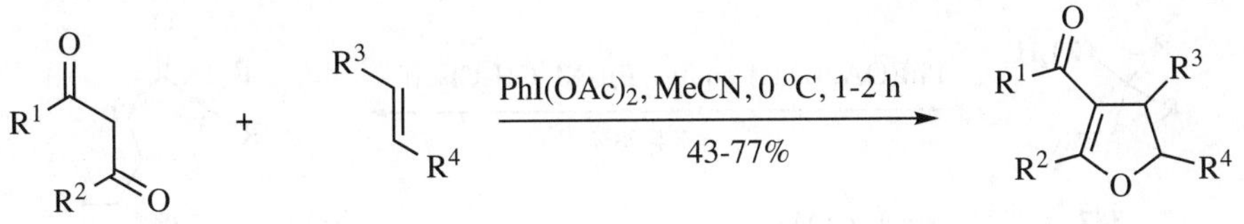

R^1 = Me, R^2 = Me
R^1 = Me, R^2 = OEt
R^1 + R^2 = $CH_2CMe_2CH_2$

R^3 and R^4 = H, alkyl, Ph, CH_2OAc, CH_2Cl, etc.

$PhI(OAc)_2$, MeCN, 0 °C, 1-2 h; 43-77%

342 **343** **344**

Scheme 3.136

pure 2-arylproline derivatives by intramolecular oxyamination of alkenes with ureas employing chiral lactic acid-based hypervalent iodine reagents [429].

Numerous hypervalent iodine-promoted cyclizations of non-amine substrates have also been reported. Several examples of oxidative cyclizations leading to the formation of oxygen heterocycles are shown in Schemes 3.136–3.138. In particular, the (diacetoxyiodo)benzene-mediated oxidative addition of 1,3-dicarbonyl compounds **342** to alkenes **343** allows an efficient one-pot synthesis of 2,3-dihydrofuran derivatives **344** (Scheme 3.136) [430]. Various alkenes and cycloalkenes bearing electron-withdrawing or electron-donating substituents can be used in this cyclization. A similar intramolecular cycloaddition/cycloisomerization of 2-propargyl-1,3-dicarbonyl compounds upon treatment with $PhI(OCOCF_3)_2$ in hexafluoroisopropanol affords 4,5-disubstituted furfuryl alcohols in high yields [431].

The lactonization of 4-phenyl-4-pentenoic acid (**345**) upon treatment with $PhI(OAc)_2$ has been reported (Scheme 3.137) [432]. The mechanism of this reaction includes electrophilic lactonization induced by the addition of the iodine(III) electrophile to the double bond of substrate **345** followed by 1,2-phenyl migration leading to the final rearranged lactone **346.**

The (diacetoxyiodo)benzene-promoted oxidative iodolactonization of pentenoic acids **347** in the presence of tetrabutylammonium iodide proceeds smoothly at room temperature to afford lactones **348** in high yields (Scheme 3.138) [433]. A catalytic version of this iodolactonization using iodobenzene as a catalyst and sodium perborate monohydrate as the stoichiometric oxidant has also been reported (Section 4.1).

Additional examples of synthetic applications of hypervalent iodine-induced heterocyclizations include the following: the metal-free one-pot synthesis of 2-acylbenzothiazoles by oxidative cyclization of multiform substrates [434], iodine(III)-mediated tandem oxidative cyclization for construction of 2-nitrobenzo[*b*]furans [435], hypervalent iodine mediated oxidative cyclization of *o*-hydroxystilbenes into benzo- and naphthofurans [436], $PhI(OCOCF_3)_2$-mediated synthesis of 3-hydroxy-2-oxindoles and spirooxindoles from anilides [437], synthesis of isoxazoles by hypervalent iodine-induced cycloaddition of nitrile oxides to alkynes [438],

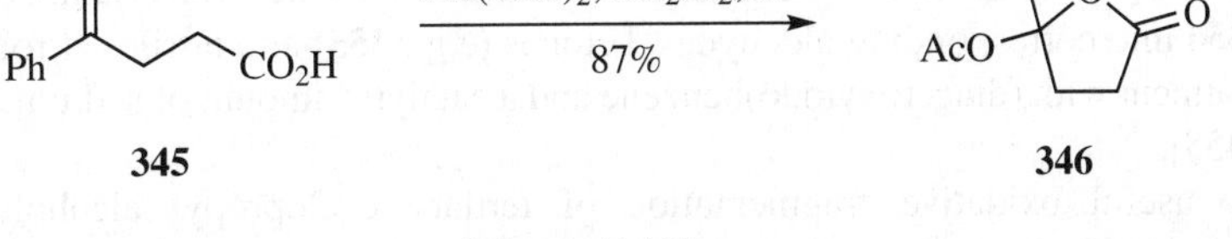

$PhI(OAc)_2$, CH_2Cl_2, rt; 87%

345 **346**

Scheme 3.137

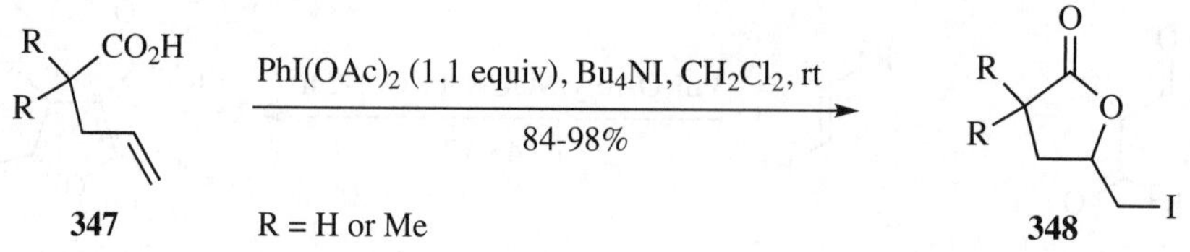

R, R, CO_2H (347) → $PhI(OAc)_2$ (1.1 equiv), Bu_4NI, CH_2Cl_2, rt; 84-98% → 348 (I)

R = H or Me

Scheme 3.138

metal-free synthesis of polysubstituted pyrroles by (diacetoxyiodo)benzene-mediated cascade reaction of 3-alkynyl amines [439], highly efficient synthesis of multisubstituted 2-acyl furans via $PhI(OCOCF_3)_2/I_2$-mediated oxidative cycloisomerization of *cis*-2-en-4-yn-1-ols [440], synthesis of carbazoles by intramolecular oxidative C–N bond formation [441], one-pot synthesis of [1,2,4]triazolo[4,3-*a*][1,4]benzodiazepine derivatives by oxidative cyclization reaction of 2-hydrazino-1,4-benzodiazepines with various aldehydes in presence of (diacetoxyiodo)benzene [442], preparation of benzopyrano- and furopyrano-2-isoxazoline derivatives from 2-allyloxybenzaldoximes by $PhI(OAc)_2$ oxidation [443], preparation of 2-(*N*-acylaminal) substituted tetrahydropyrans by $PhI(OAc)_2$-induced oxidative cyclization of hydroxy-substituted *N*-acyl enamines [444], synthesis of 2-substituted benzothiazoles via the oxidative cyclization of thiobenzamides [445], preparation of 2,3-diphenylquinoxaline-1-oxide from benzil-α-arylimino oximes using $PhI(OAc)_2$ [446], synthesis of 2,5-disubstituted-1,3,4-oxadiazoles via $PhI(OCOCF_3)_2$-mediated oxidative cyclization of aldazines [447], preparation of 2-substituted oxazolines from aldehydes and 2-amino alcohols using $PhI(OAc)_2$ as an oxidant [448], synthesis of 3,4-bis(1-phenyl-3-arylpyrazolyl)-1,2,5-oxadiazole-*N*-oxides by the $PhI(OAc)_2$ oxidation of pyrazole-4-carboxaldehyde oximes [449], synthesis of 2-arylbenzimidazoles from phenylenediamines and aldehydes via a one-step process using $PhI(OAc)_2$ as an oxidant [450], $PhI(OAc)_2$-mediated efficient synthesis of imidazoles from α-hydroxy ketones, aldehydes and ammonium acetate [451], preparation of dihydrooxazole derivatives by $PhI(OAc)_2$-promoted 1,3-dipolar cycloaddition reactions of phthalhydrazide [452], synthesis of 2,3-diarylbenzo[*b*]furans by $PhI(OCOCF_3)_2$-mediated oxidative coupling sequence [453] and the synthesis of *seco*-psymberin/irciniastatin A via a $PhI(OAc)_2$-mediated cascade cyclization reaction [454]. Some of these heterocyclizations can be performed under catalytic conditions in the presence of an iodoarene catalyst and peracids as stoichiometric oxidants (Section 4.1).

3.1.13.2 Fragmentations and Rearrangements

Numerous examples of hypervalent iodine-promoted fragmentations or rearrangements at electron-deficient centers have been reported. Several examples of oxidative fragmentations are shown below in Schemes 3.139–3.142. A mild and efficient fragmentation reaction of β-amino alcohols **349** and α-amino acids **350** upon treatment with [bis(trifluoroacetoxy)iodo]pentafluorobenzene leading to N,O-acetals **351** has been developed (Scheme 3.139). This method has been utilized in an improved synthesis of the key intermediate of discorhabdins [455,456].

Aryl-substituted aldehydes **352** can be cleaved to chain-shortened carbonyl compounds **353** and formaldehyde by iodosylbenzene in the presence of acids or Lewis acids (Scheme 3.140). Formaldehyde is further oxidized to CO and CO_2 under the reaction conditions [457]. Oxidative decarboxylation of 2-aryl-substituted carboxylic acids **354** into corresponding aldehydes, ketones (e.g., **355**) and nitriles at room temperature can be achieved by treatment with (diacetoxyiodo)benzene and a catalytic amount of sodium azide in acetonitrile (Scheme 3.141) [458].

A synthetically useful oxidative fragmentation of tertiary cyclopropyl alcohols (e.g., **356**) with [bis(trifluoroacetoxy)iodo]benzene, which produces alkenoic acids or esters, has been reported [459,460].

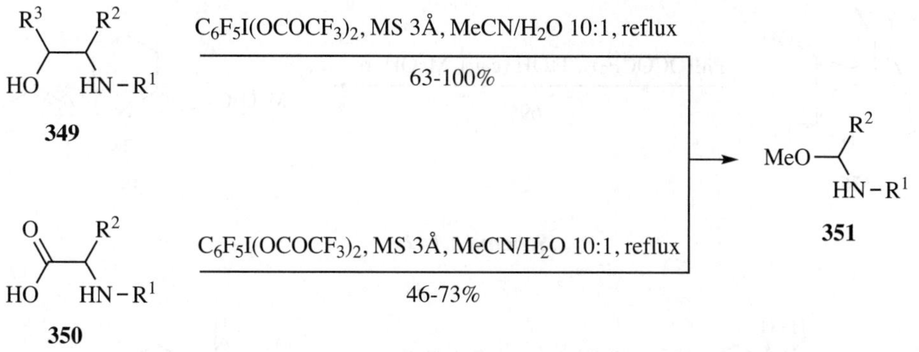

349

350

$C_6F_5I(OCOCF_3)_2$, MS 3Å, MeCN/H_2O 10:1, reflux

63-100%

$C_6F_5I(OCOCF_3)_2$, MS 3Å, MeCN/H_2O 10:1, reflux

46-73%

351

R^1 = Cbz or Fmoc; R^2 = H, Me, CO_2Me, etc; R^3 = H or Me

Scheme 3.139

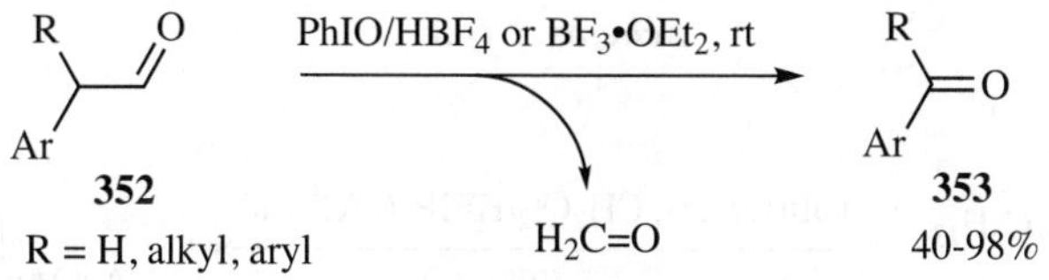

PhIO/HBF_4 or BF_3•OEt_2, rt

352

R = H, alkyl, aryl

$H_2C{=}O$

353

40-98%

Scheme 3.140

This fragmentation has been successfully employed for the preparation of product **357** (Scheme 3.142), which is the key precursor in an efficient asymmetric synthesis of the alkaloid (–)-pinidine [460].

Hypervalent-iodine-promoted rearrangements have been utilized in various ring-expansion reactions (Schemes 3.143–3.146 below). A (diacetoxyiodo)benzene-promoted oxidative rearrangement of *cis*- and *trans*-1,5-diazadecalins to macrocyclic lactams has been reported [461]. In a specific example, upon treatment with (diacetoxyiodo)benzene in aqueous NaOH, 1,5-diaza-*trans*-decalin (**358**) undergoes oxidation along with C—C bond cleavage to yield the ring-expanded bislactam **359** (Scheme 3.143) [461].

A stereoselective synthesis of five- to seven-membered cyclic ethers has been achieved by de-iodonative ring-enlargement of cyclic ethers having an iodoalkyl substituent. For example, the reaction of tetrahydrofuran derivative **360** with (diacetoxyiodo)toluene proceeds under mild conditions to afford ring-expanded product **361** (Scheme 3.144). The use of hexafluoroisopropanol (HFIP) as solvent in this reaction is critical [462].

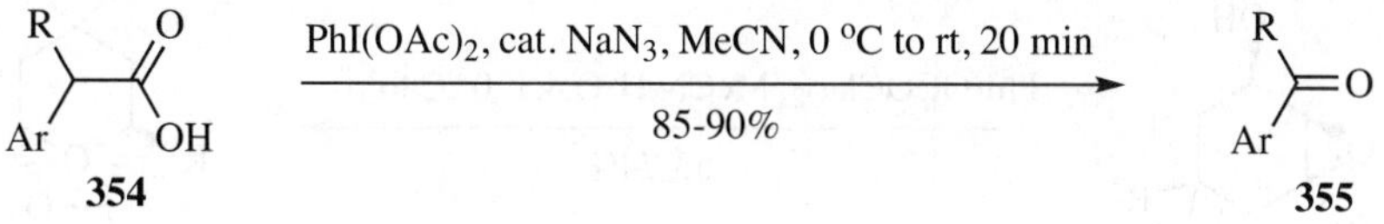

$PhI(OAc)_2$, cat. NaN_3, MeCN, 0 °C to rt, 20 min

85-90%

354

355

R = H or Me

Ar = Ph, 4-$MeOC_6H_4$, 2,5-$(MeO)_2C_6H_3$, 2-ClC_6H_4, 4-ClC_6H_4, 4-$EtOCOC_6H_4$, 4-$NO_2C_6H_4$, etc.

Scheme 3.141

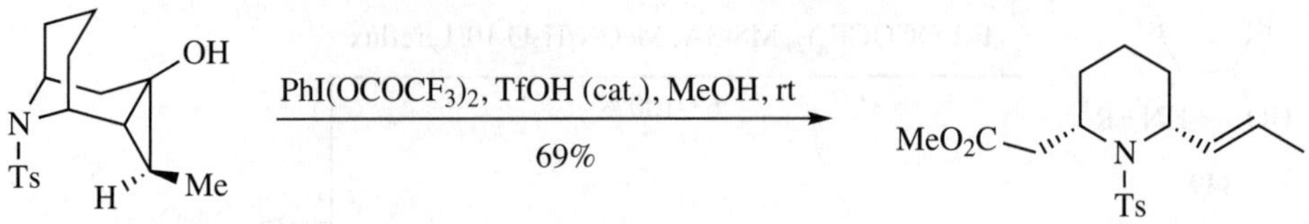

$PhI(OCOCF_3)_2$, TfOH (cat.), MeOH, rt
69%

356 **357**

Scheme 3.142

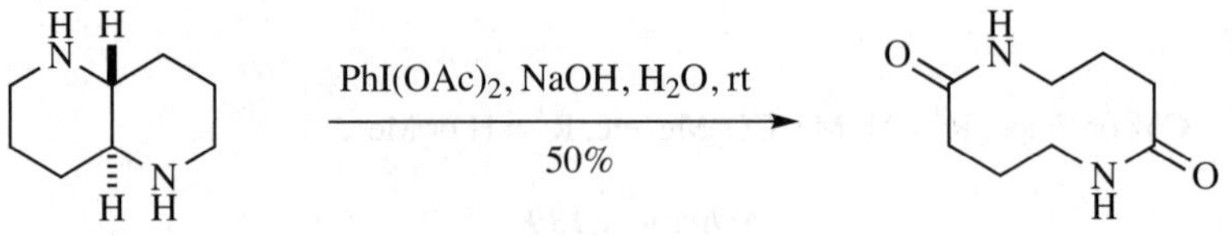

$PhI(OAc)_2$, NaOH, H_2O, rt
50%

358 **359**

Scheme 3.143

$TolI(OAc)_2$, CH_2Cl_2/HFIP, 0 °C, 1 h
76%

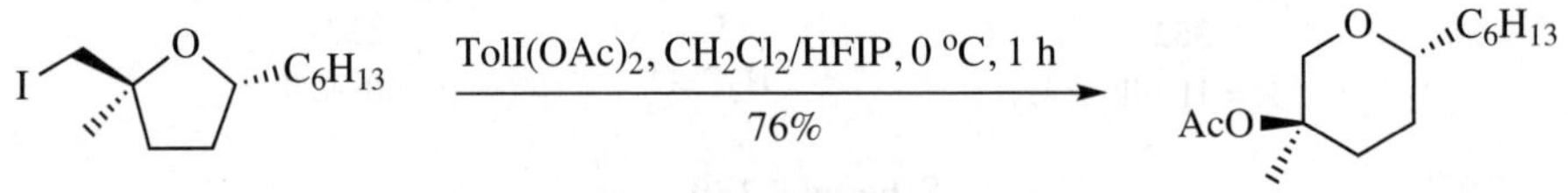

360 **361**

Scheme 3.144

A facile and efficient synthesis of lactols **363** via an oxidative rearrangement reaction of 2,3-epoxy alcohols **362** with [bis(trifluoroacetoxy)iodo]benzene has been reported (Scheme 3.145) [463–465]. This hypervalent-iodine-induced oxidative transformation has been utilized in the synthesis of several lactones and in the asymmetric synthesis of the marine γ-lactone metabolite (+)-tanikolide [463,464].

Reactions of 4-hydroxy-2-cyclobutenones **364** with (diacetoxyiodo)benzene in 1,2-dichloroethane at reflux afford 5-acetoxy-2(5*H*)-furanones **365** as rearranged products (Scheme 3.146) [466]. The formation of these products is explained by ring cleavage in the hypervalent iodine intermediate **366** followed by recyclization of the resulting acyl cation **367** with the carbonyl oxygen (Scheme 3.146). In a similar procedure,

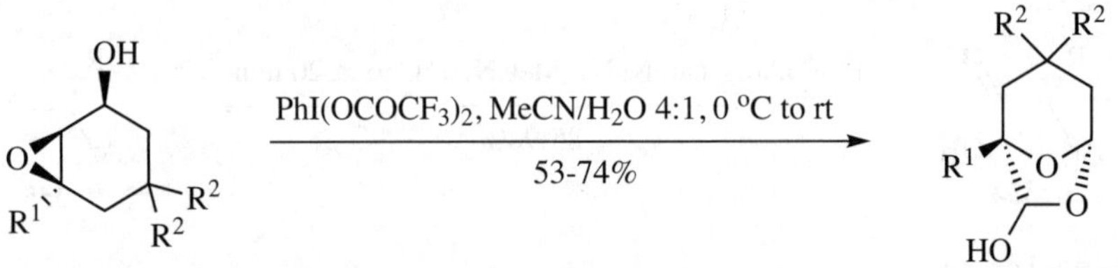

$PhI(OCOCF_3)_2$, MeCN/H_2O 4:1, 0 °C to rt
53-74%

362 R^1 = Me, Et, n-$C_{11}H_{23}$, $CH_2CH(CH_3)_2$, CH_2Ph
R^2 = H or Me **363**

Scheme 3.145

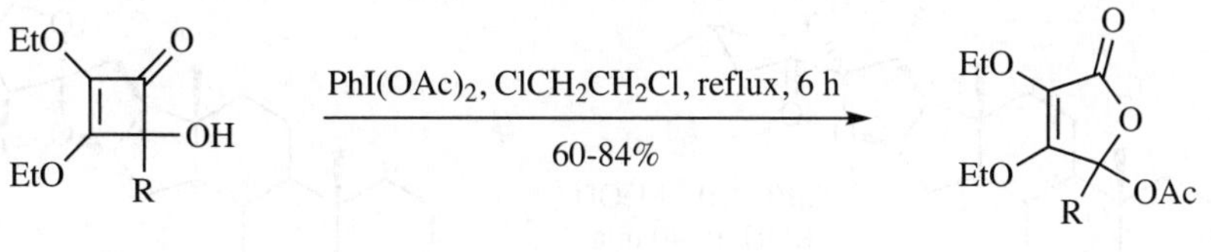

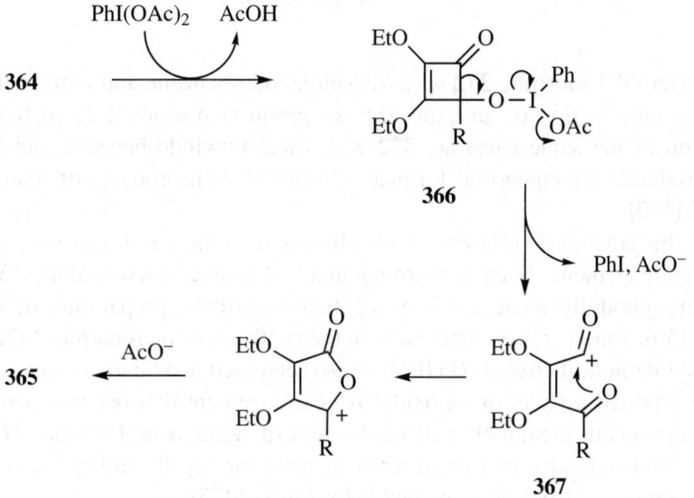

Scheme 3.146

5-methoxy-2(5*H*)-furanones are obtained in good yields by using methanol as both a solvent and a nucleophile [466].

A (diacetoxyiodo)benzene-induced domino reaction of the vicinal unsaturated diol **368** affords cyclic ene-acetal **369** (Scheme 3.147), which has been further utilized in the synthesis of a norsesquiterpene spirolactone/testosterone hybrid [467].

Iglesias-Arteaga and coworkers have reported several (diacetoxyiodo)benzene-promoted oxidative transformations of steroidal substrates (Schemes 3.148 and 3.149) [468–471]. In particular, the treatment of (25*R*)-3α-acetoxy-5β-spirostan-23-one (**370**) with (diacetoxyiodo)benzene in basic methanol leads to F-ring contraction via Favorskii rearrangement to afford product **371** (Scheme 3.148) [468].

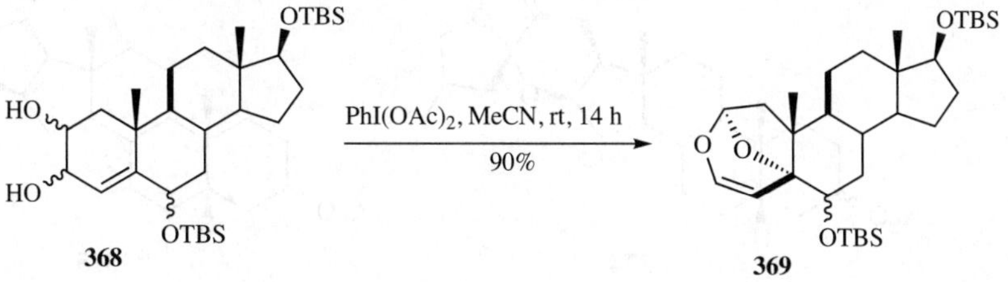

Scheme 3.147

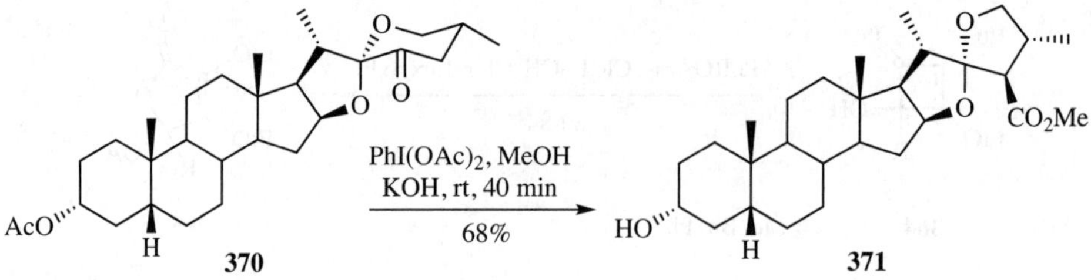

Scheme 3.148

Treatment of steroidal substrate **372** with (diacetoxyiodo)benzene and boron trifluoride etherate in acetic acid leads to the introduction of an axial acetoxy group at position C23 of the side chain [469], while a similar reaction of the same substrate **372** with (diacetoxyiodo)benzene and $BF_3{\cdot}OEt_2$ in formic acid unexpectedly produced the equatorial formate **373** mixed with products of rearrangement (**374** and **375**) (Scheme 3.149) [470].

Reactions of [bis(acyloxy)iodo]arenes with alkenes in some cases can give products of an oxidative, carbocationic rearrangement. Such a rearrangement of pentaalkoxychalcones **376** upon treatment with [bis(trifluoroacetoxy)iodo]benzene has been applied toward the preparation of the rearranged chalcones **377** (Scheme 3.150), which are key precursors in the synthesis of pterocarpins [472].

[Hydroxy(tosyloxy)iodo]benzene (HTIB) has also been used in oxidative rearrangements of alkenes. Justik and Koser have reported a study of an oxidative rearrangement that occurs upon treatment of arylalkenes **378** with HTIB in aqueous methanol to afford the corresponding α-aryl ketones **379** in generally high yields (Scheme 3.151). This oxidative rearrangement is general for acyclic and cyclic arylalkenes and permits the regioselective syntheses of isomeric α-phenyl ketone pairs [473].

A similar HTIB-induced oxidative rearrangement has been utilized in the regioselective synthesis of 6-prenylpolyhydroxyisoflavone (wighteone) [474] and in a diastereoselective total synthesis of (±)-indatraline

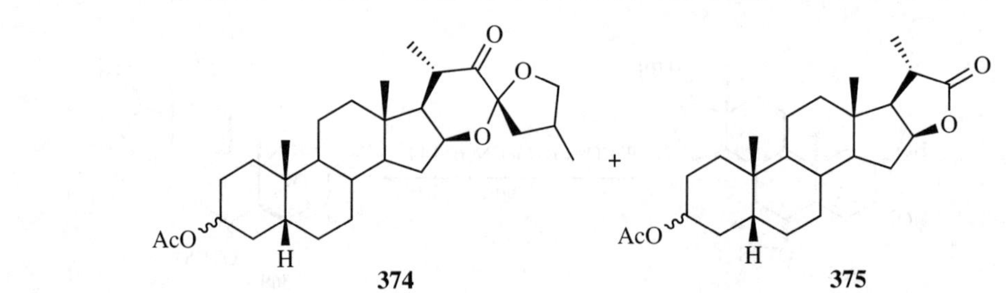

Scheme 3.149

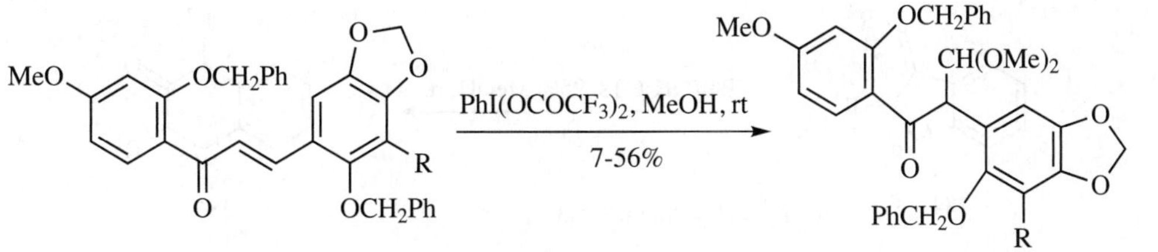

MeO, OCH_2Ph, O, O, R, OCH_2Ph — $PhI(OCOCF_3)_2$, MeOH, rt; 7-56% — MeO, OCH_2Ph, $CH(OMe)_2$, O, O, O, $PhCH_2O$, R

376 R = H, Br 377

Scheme 3.150

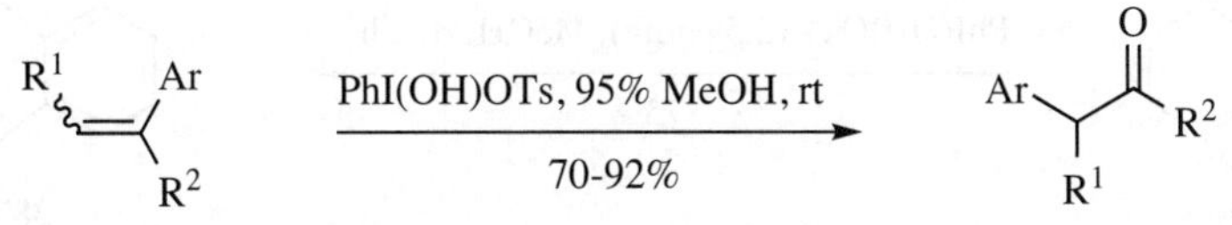

R^1, Ar, R^2 — PhI(OH)OTs, 95% MeOH, rt; 70-92% — O, Ar, R^2, R^1

378 R^1, R^2 = alkyl, aryl 379

Scheme 3.151

[475]. In particular, the key intermediate product **381** in the synthesis of wighteone was prepared by the oxidative rearrangement of 3′-iodotetraalkoxychalcone **380** [474] and the key step in the synthesis of (±)-indatraline involved the HTIB-promoted diastereoselective ring contraction of a 1,2-dihydronaphthalene **382** to construct the indane ring system **383** (Scheme 3.152) [475]. A similar oxidative rearrangement of 3-cinnamoyl-4-hydroxy-6-methyl-2*H*-pyran-2-ones with HTIB in dichloromethane followed by cyclization was used by Prakash and coworkers for the direct conversion of *o*-hydroxychalcones into isoflavone derivatives [476].

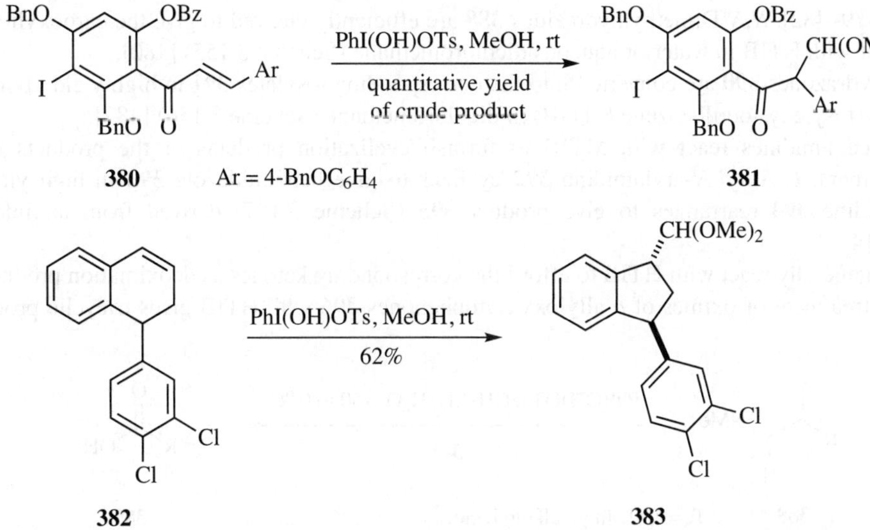

BnO, OBz, I, BnO, O, Ar — PhI(OH)OTs, MeOH, rt; quantitative yield of crude product — BnO, OBz, $CH(OMe)_2$, I, BnO, O, Ar

380 Ar = 4-$BnOC_6H_4$ 381

Cl, Cl — PhI(OH)OTs, MeOH, rt; 62% — $CH(OMe)_2$, Cl, Cl

382 383

Scheme 3.152

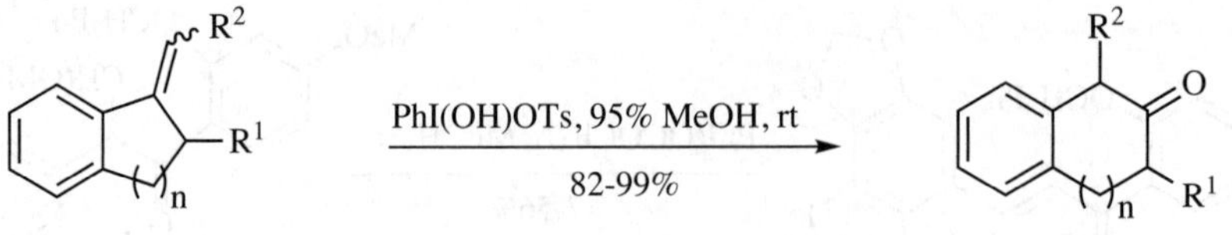

PhI(OH)OTs, 95% MeOH, rt

82-99%

384 R^1, R^2 = H or Me; n = 1-3 **385**

Scheme 3.153

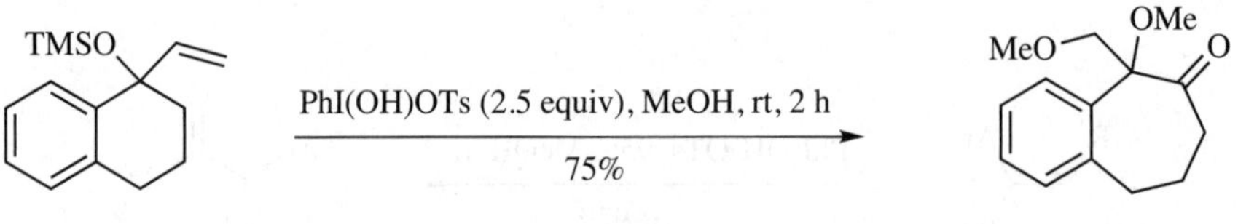

PhI(OH)OTs (2.5 equiv), MeOH, rt, 2 h

75%

386 **387**

Scheme 3.154

The HTIB-induced oxidative rearrangement of alkenes has been effectively used in ring expansion reactions (Schemes 3.153 and 3.154). Justik and Koser have investigated the oxidative ring expansions of alkylidenebenzocycloalkenes **384** to β-benzocycloalkenones **385** using HTIB in methanol (Scheme 3.153) [477]. This reaction allows the efficient conversion of alkenes **384,** which can be conveniently prepared from the respective α-benzocycloalkenones by Wittig olefination, into the homologous β-benzocycloalkenones **385** containing six-, seven- and eight-membered rings.

A similar HTIB-promoted ring expansion of 1-vinylcycloalkanol derivatives leading to seven- or eight-membered rings has been reported. In a specific example, the reaction of the unsaturated trimethylsilyl ether **386** with excess HTIB affords benzocycloheptanone derivative **387** in high yield (Scheme 3.154) [478].

HTIB can also be used in the oxidative rearrangements and fragmentations of various nitrogen-containing compounds [479–482]. *N,N*-Dimethylhydrazides **388** are efficiently cleaved to give the carboxylic acid **389** upon treatment with HTIB in water or aqueous dichloromethane (Scheme 3.155) [480].

Aromatic hydrazones **390** are converted into the corresponding tosylates **391** in high yield upon reaction with [methoxy(tosyloxy)iodo]benzene (MTIB) in dichloromethane (Scheme 3.156) [482].

N-Substituted amidines react with MTIB to furnish cyclization products or the products of oxidative rearrangement. *C*-Alkyl-*N*-arylamidine **392** cyclizes to give benzimidazole **393** in high yield, while *C,N*-diarylamidine **394** rearranges to give product **395** (Scheme 3.157) derived from an intermediate carbodiimide [479].

Ketoximes generally react with HTIB to afford the corresponding ketones as deoximation products [481]. However, the treatment of oximes of *o*-allyloxyacetophenones **396** with HTIB gives tricyclic products **397**

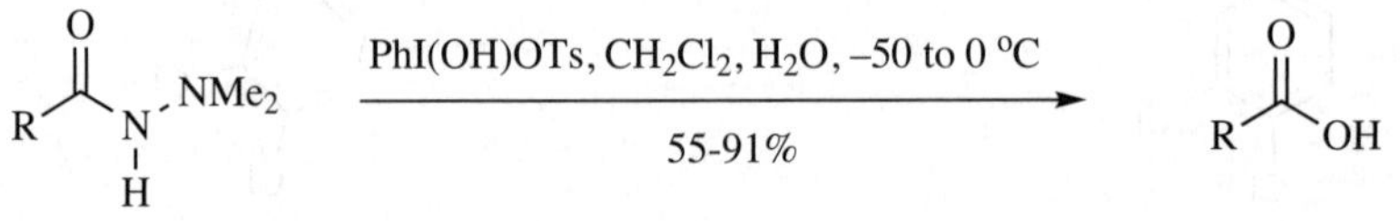

PhI(OH)OTs, CH_2Cl_2, H_2O, –50 to 0 °C

55-91%

388 R = aryl, alkyl, alkenyl, etc. **389**

Scheme 3.155

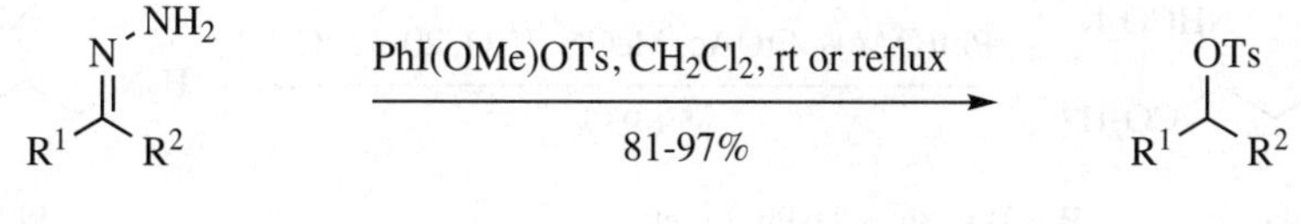

NH_2 N R^1 R^2

$PhI(OMe)OTs$, CH_2Cl_2, rt or reflux

81-97%

OTs R^1 R^2

390 R^1 = Ph, 3-$NO_2C_6H_4$, 4-CNC_6H_4, 4-$NO_2C_6H_4$
R^2 = H, Me, PhCO **391**

Scheme 3.156

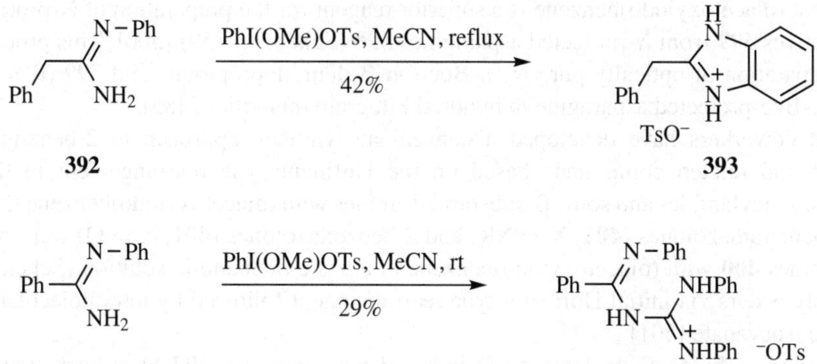

N−Ph Ph NH_2

$PhI(OMe)OTs$, MeCN, reflux

42%

H N + N H Ph TsO^-

392 **393**

N−Ph Ph NH_2

$PhI(OMe)OTs$, MeCN, rt

29%

N−Ph Ph NHPh HN + NHPh ^-OTs

394 **395**

Scheme 3.157

(Scheme 3.158) resulting from the intramolecular cyclization of an intermediate nitrosoalkene generated from the oxime and HTIB [481].

3.1.13.3 *Hofmann Rearrangement*

Organohypervalent iodine(III) compounds are particularly useful as the oxidants in the Hofmann-type degradation of aliphatic or aromatic carboxamides to the respective amines [483]. The most common reagents for Hofmann-type rearrangements include (diacetoxyiodo)benzene [484–488], [bis(trifluoroacetoxy)-iodo]benzene [489–494], [hydroxy(tosyloxy)]iodobenzene [495–499] and their recyclable analogs (Chapter 5). The catalytic version of the Hofmann rearrangement using aryl iodides and *m*-chloroperoxybenzoic acid as terminal oxidant has also been reported (Section 4.1).

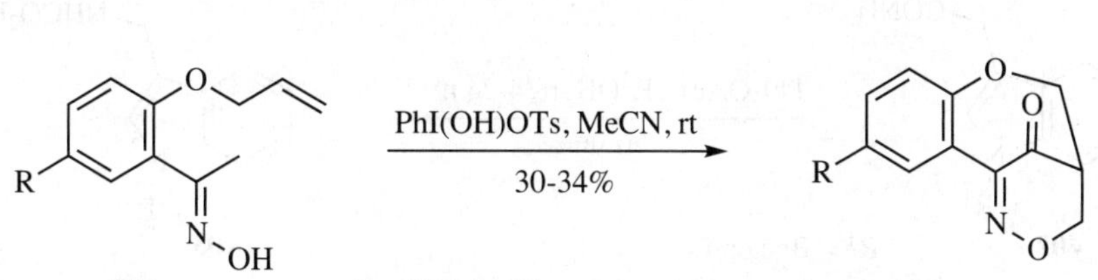

O R N OH

$PhI(OH)OTs$, MeCN, rt

30-34%

O O R N O

396 R = H, Me, Cl **397**

Scheme 3.158

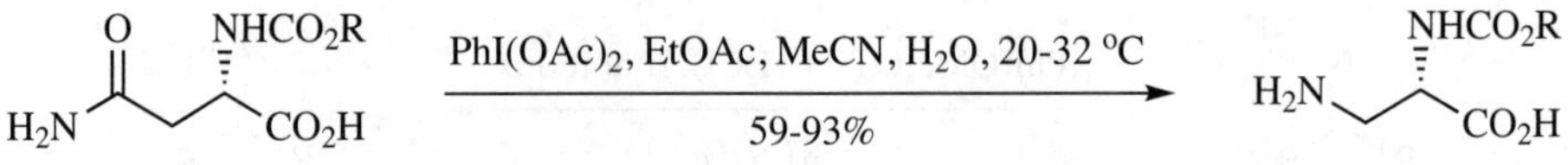

$PhI(OAc)_2$, EtOAc, MeCN, H_2O, 20-32 °C

59-93%

398 → 399

R = Bu, Bu^t, CH_2Ph, Et, etc.

Scheme 3.159

(Diacetoxyiodo)benzene is a preferred reagent for performing the Hofmann rearrangement on a large scale. A comparative study of various oxidants (hypochlorite, hypobromide, *N*-bromosuccinimide, etc.) has demonstrated that (diacetoxyiodo)benzene is a superior reagent for the preparation of N-protected β-amino-L-alanine derivatives **399** from *N*-protected asparagines **398** (Scheme 3.159) [500]. This procedure has been used for the preparation of optically pure N_α-*n*-Boc-L-α,β-diaminopropionic acid **399** (R = CO_2Bu) from the respective *N*-Boc-protected asparagine in hundred kilogram quantities [485].

Moriarty and coworkers have developed a convenient synthetic approach to 2-benzimidazolones, 2-benzoxazolones and related compounds based on the Hofmann-type rearrangement in the reaction of anthranilamides, salicylamides and some β-substituted amides with (diacetoxyiodo)benzene [501]. For example, various 2-benzimidazolones (**401,** X = NR) and 2-benzoxazolones (**401,** X = O) were prepared by the treatment of amides **400** with (diacetoxyiodo)benzene in a basic methanolic solution (Scheme 3.160). This reaction probably occurs via initial Hofmann-type rearrangement followed by intramolecular cyclization of the intermediate isocyanate [501].

A series of alkyl carbamates of 1-protected indole-3-methylamines **403** have been prepared from the corresponding acetamides **402** in good to excellent yields via (diacetoxyiodo)benzene-promoted Hofmann rearrangement (Scheme 3.161). This procedure has been further extended to the preparation of alkyl carbamates of thiophenylmethylamines and pyrrolylmethylamines [502].

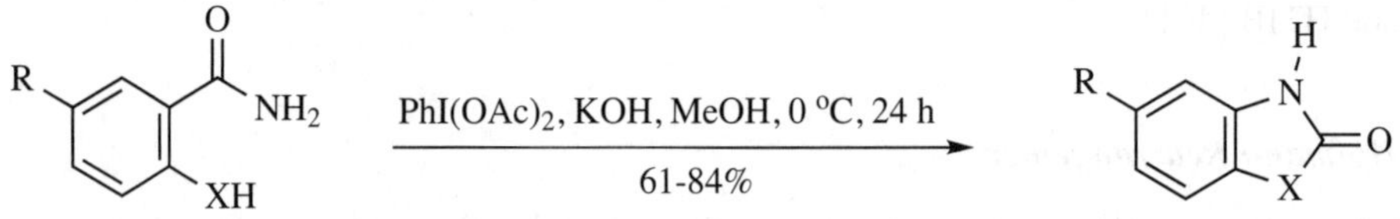

$PhI(OAc)_2$, KOH, MeOH, 0 °C, 24 h

61-84%

400 → 401

R = H or Cl

X = NH, NMe, NEt, NPr, NPr^i, NBu, or O

Scheme 3.160

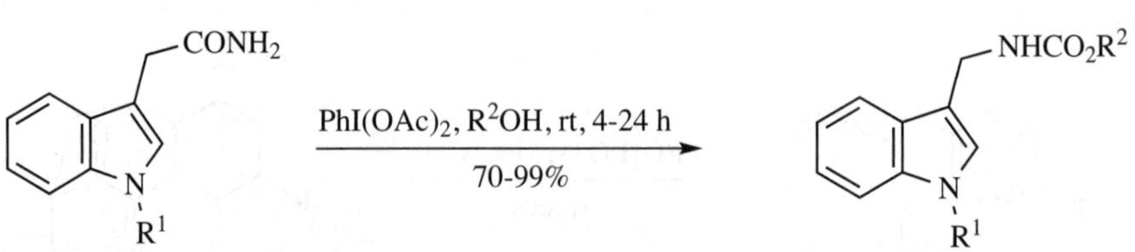

$PhI(OAc)_2$, R^2OH, rt, 4-24 h

70-99%

402 → 403

R^1 = Boc or Ts

R^2 = Me, Et, Pr^i, Bu^t, Bn

Scheme 3.161

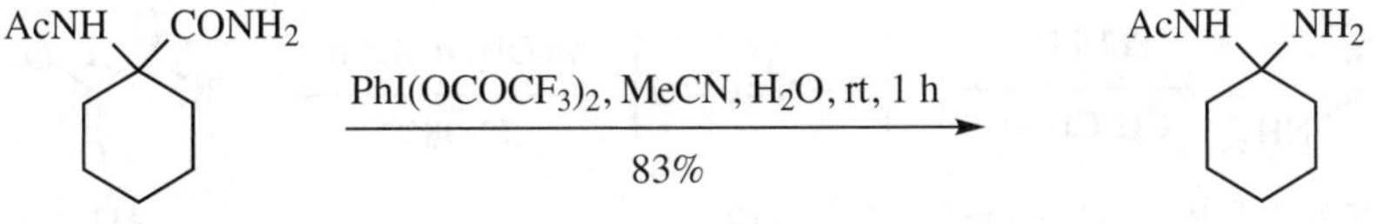

AcNH CONH$_2$ → PhI(OCOCF$_3$)$_2$, MeCN, H$_2$O, rt, 1 h; 83% → AcNH NH$_2$

404 405

Scheme 3.162

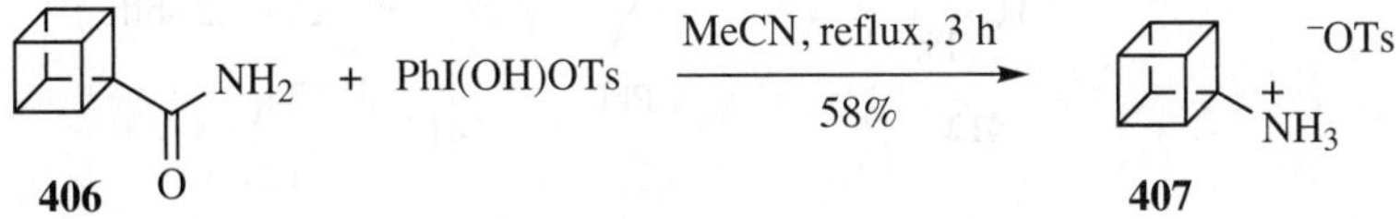

406 + PhI(OH)OTs → MeCN, reflux, 3 h; 58% → 407 ($^-$OTs, $\overset{+}{N}H_3$)

Scheme 3.163

[Bis(trifluoroacetoxy)iodo]benzene has also been used as a reagent for the Hofmann rearrangement, as illustrated by the conversion of amide **404** into the respective amine **405** (Scheme 3.162) [503]. A similar [bis(trifluoroacetoxy)iodo]benzene-induced Hofmann rearrangement has been used for the preparation of both enantiomers of *trans*-2-aminocyclohexanecarboxylic acid from *trans*-cyclohexane-1,2-dicarboxylic acid [492].

Similar to [bis(acyloxy)iodo]arenes, HTIB can serve as an efficient oxidant in Hofmann-type degradation of carboxamides to the respective amines [495–499]. In a representative example, HTIB has been used for the preparation of cubyl ammonium salt **407** from the respective carboxamide derivative **406** under mild reaction conditions (Scheme 3.163) [504].

An efficient method for the synthesis of 1,3-disubstituted ureas and carbamates from carboxamides by using iodosylbenzene as the oxidant in the presence of amines or alcohols has been described [505]. For example, carbamates **408** have been prepared by this procedure in high yields (Scheme 3.164). Various symmetric and asymmetric ureas and ureidopeptides can also be prepared in good yields by this method [505].

A very mild procedure for the Hofmann rearrangement of aromatic and aliphatic carboxamides **409** is based on the use of (tosylimino)phenyl-λ^3-iodane, PhINTs, as the oxidant (Scheme 3.165) [506]. Owing to the mild reaction conditions, this method is particularly useful for the Hofmann rearrangement of substituted benzamides **409** (R = aryl), which usually afford complex reaction mixtures with other hypervalent iodine oxidants. The mild reaction conditions and high selectivity in the reaction of carboxamides with PhINTs allow the isolation of the initially formed labile isocyanates **410,** or their subsequent conversion into stable carbamates **411** by treatment with alcohols. Based on the previously reported mechanistic studies of the Hofmann rearrangement using other hypervalent iodine reagents [489, 490, 496], it is assumed that the reaction

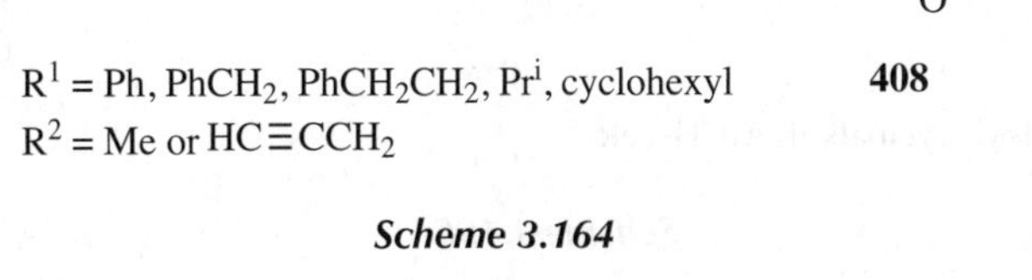

$R^1C(O)NH_2$ → (PhIO)$_n$, R^{2}OH, rt, 2 h; 53-91% → $R^1NHC(O)OR^2$

R^1 = Ph, PhCH$_2$, PhCH$_2$CH$_2$, Pr^i, cyclohexyl **408**

R^2 = Me or HC≡CCH$_2$

Scheme 3.164

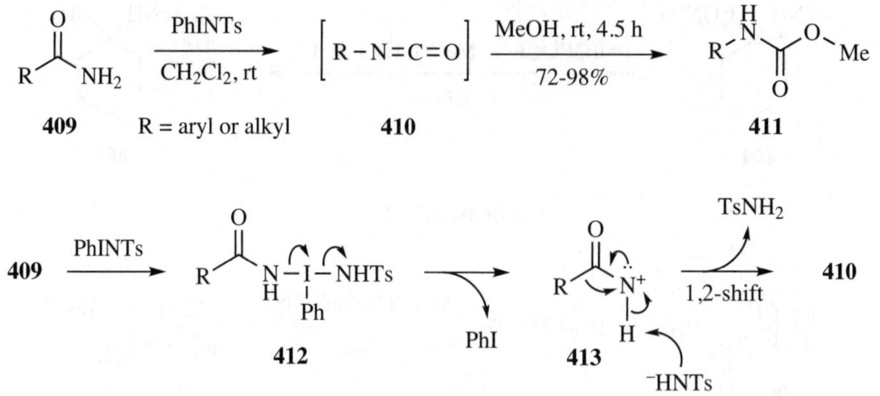

Scheme 3.165

starts from the formation of amidoiodane **412** (Scheme 3.165). Subsequently, the reductive elimination of iodobenzene and the 1,2-alkyl or -aryl shift to the electron-deficient nitrenium nitrogen atom in the intermediate **413** afford isocyanate **410.** Subsequent addition of an alcohol to isocyanate **410** gives the final carbamate **411** [506].

Alkylcarboxamides can be converted into the respective amines by Hofmann rearrangement using hypervalent iodine species generated *in situ* from iodobenzene and a terminal oxidant, such as Oxone® ($2KHSO_5 \cdot KHSO_4 \cdot K_2SO_4$) or *m*-chloroperoxybenzoic acid (*m*CPBA). In particular, a convenient experimental procedure for the preparation of alkylcarbamates using Oxone as the oxidant in the presence of iodobenzene in methanol (Scheme 3.166) has been developed [359].

Likewise, the Hofmann-type rearrangement of substituted phthalimides **414** or succinimides **416** using a hypervalent iodine(III) reagent generated *in situ* from iodobenzene and *m*CPBA in alcohol in the presence of a base affords anthranilic acid derivatives **415** or amino acid derivatives **417,** respectively (Scheme 3.167) [507].

3.1.14 Oxidations at Nitrogen, Sulfur and other Heteroatoms

Hypervalent iodine(III) compounds have found wide application for the oxidation of organic derivatives of nitrogen, sulfur, selenium, tellurium and other elements. Reactions of λ^3-iodanes with organonitrogen compounds leading to the electron-deficient nitrenium intermediates and followed by cyclizations and rearrangements (e.g., Hofmann rearrangement) are discussed in Section 3.1.13. Several other examples of oxidations at a nitrogen center are shown below in Schemes 3.168–3.170.

Primary aliphatic amines or benzylamines can be dehydrogenated by iodosylbenzene to nitriles **418** in dry dichloromethane at room temperature (Scheme 3.168) [508].

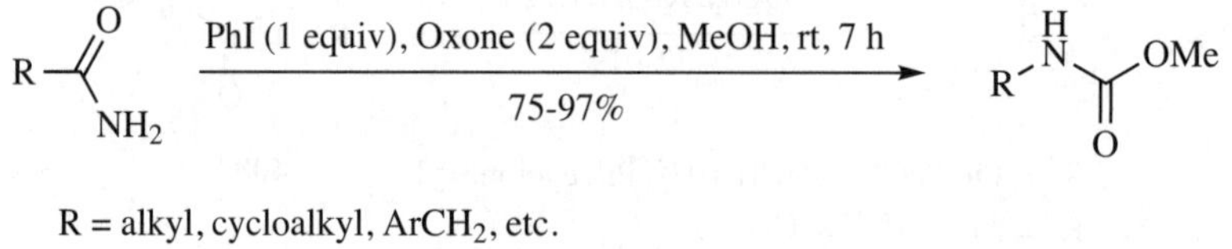

Scheme 3.166

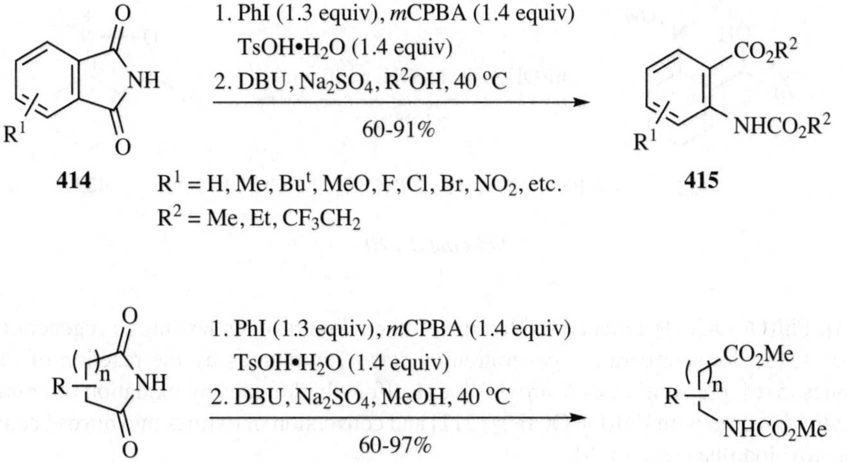

1. PhI (1.3 equiv), mCPBA (1.4 equiv) TsOH•H_2O (1.4 equiv)
2. DBU, Na_2SO_4, R^2OH, 40 °C
60-91%

414 → **415** (CO_2R^2, $NHCO_2R^2$)

R^1 = H, Me, Bu^t, MeO, F, Cl, Br, NO_2, etc.
R^2 = Me, Et, CF_3CH_2

1. PhI (1.3 equiv), mCPBA (1.4 equiv) TsOH•H_2O (1.4 equiv)
2. DBU, Na_2SO_4, MeOH, 40 °C
60-97%

416 → **417** (CO_2Me, $NHCO_2Me$)

n = 1 or 2; R= H, Me, Bu, Bn, $(CH_3)_2CHCH_2$, etc.

Scheme 3.167

RCH_2NH_2 + $(PhIO)_n$ → (CH_2Cl_2, rt, 3 days; 48-57%) R−C≡N **418**

R = aryl or alkyl

Scheme 3.168

Aldoximes **419** are selectively oxidized by (diacetoxyiodo)benzene in methanol containing a catalytic amount of trifluoroacetic acid to give nitrile oxides **420,** which can be trapped *in situ* with olefins in a bimolecular or an intramolecular mode to give the synthetically valuable isoxazoline products **421** (Scheme 3.169) [509]. A similar reaction of α-oxo-aldoximes with $PhI(OAc)_2$ affords α-oxo-nitrile oxides, which can be further trapped with norbornene or styrene [510].

Various isoxazoline N-oxide derivatives **423** can be synthesized by the oxidation of β-hydroxyketoximes **422** using [hydroxy(tosyloxy)iodo]benzene in methanol or water (Scheme 3.170) [511].

Additional examples of oxidations at a nitrogen center include the following: the $PhI(OAc)_2$-induced oxidation of aromatic amines to imines applied for deprotection of protected amino diols [512], N-acylation of 1,3-disubstituted thioureas using $PhI(OAc)_2$ [513], $PhI(OAc)_2$-promoted oxidation of 1,2-dicarbethoxyhydrazine to diethyl azodicarboxylate as a key step in an organocatalytic Mitsunobu

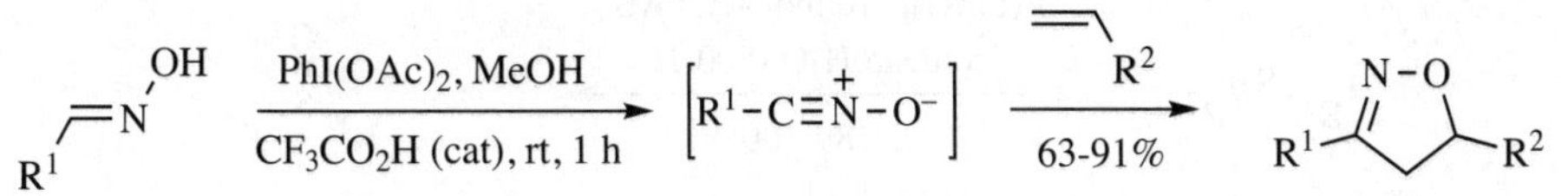

419 → ($PhI(OAc)_2$, MeOH; CF_3CO_2H (cat), rt, 1 h) [$R^1-C\equiv N^+-O^-$] **420** → (R^2-alkene; 63-91%) **421**

R^1 and R^2 = aryl or alkyl

Scheme 3.169

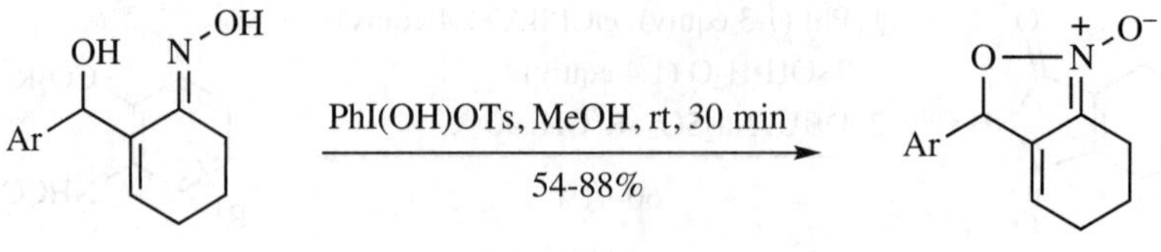

OH N OH / Ar — PhI(OH)OTs, MeOH, rt, 30 min; 54-88% — Ar, O–N$^+$–O$^-$

422 Ar = Ph, 2-$NO_2C_6H_4$, 2-BrC_6H_4, 2,4-$Cl_2C_6H_3$, etc. **423**

Scheme 3.170

reaction [514], $PhI(OCOCF_3)_2$-induced oxidations of phenylhydrazones leading to regeneration of the carbonyl function [515], low-temperature generation of diazo compounds by the reaction of $PhI(OCOCF_3)_2$ with hydrazones [516], preparation of *N*-aroyl-*N'*-arylsulfonylhydrazines by oxidation of aromatic aldehyde *N*-arylsulfonylhydrazones with $PhI(OCOCF_3)_2$ [517] and conversion of oximes into nitroso compounds using *p*-bromo(diacetoxyiodo)benzene [518].

Hypervalent iodine reagents are commonly used for the oxidation of organosulfur compounds. Applications of hypervalent iodine reagents for the preparation of sulfoxides, including enantioselective oxidations of organic sulfides, have been summarized in reviews [519, 520].

The oxidation of various sulfides **424** with iodosylbenzene in the presence of catalytic amounts of quaternary ammonium bromides affords the respective sulfoxides **425** in high yields (Scheme 3.171) [521]. The best catalytic effect in this reaction is observed when oxidation is carried out in a nonpolar solvent (toluene, hexane, dichloromethane) in the presence of trace amounts of water and 10 mol% of cetyltrimethylammonium bromide (CTAB). Iodosylbenzene can also be activated in the solid state by pulverization with natural clays or silica gels [522, 523]. The oxidation of various alkyl aryl sulfides with $(PhIO)_n$ supported on natural (montmorillonite, KSF and bentonite clays) as well as cation-exchanged K10-montmorillonite clays affords sulfoxides in excellent yields. A mechanism involving depolymerization of $(PhIO)_n$ by the acidic SiOH sites on the clay is proposed for this reaction [522]. Organic sulfides are also selectively oxidized to sulfoxides by the solid reagent system $PhI(OAc)_2$–alumina [524], or by $PhI(OAc)_2$ in water in the presence of KBr [525].

Oligomeric iodosylbenzene sulfate, $(PhIO)_3{\cdot}SO_3$, which can be formally considered as a partially depolymerized activated iodosylbenzene (Section 2.1.4), is a readily available, stable and water-soluble hypervalent iodine reagent that is useful for the oxidation of sulfides to sulfoxides. Furthermore, it can be conveniently generated *in situ* from $PhI(OAc)_2$ and $NaHSO_4$ and can be used without isolation for oxidations in aqueous solution or under solvent-free conditions. The reaction of $(PhIO)_3{\cdot}SO_3$ with organic sulfides at room temperature affords sulfoxides in high yields without overoxidation to sulfones and this reaction is compatible with the presence in the substrate molecule of hydroxy groups, double bonds, benzylic carbon atoms, carboxylate groups and various substituted aromatic rings [526].

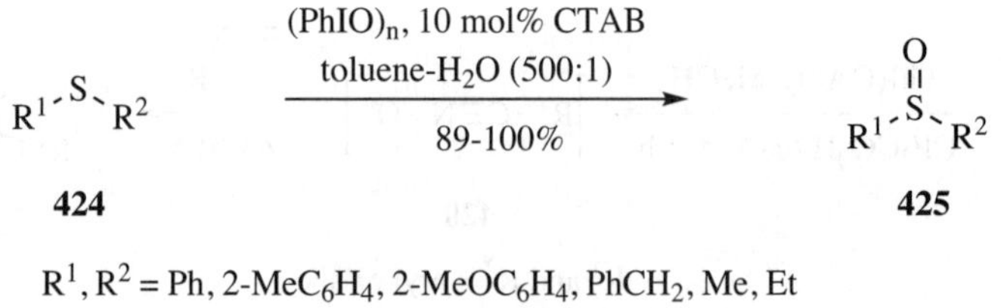

R^1–S–R^2 — $(PhIO)_n$, 10 mol% CTAB; toluene-H_2O (500:1); 89-100% — R^1–S(=O)–R^2

424 **425**

R^1, R^2 = Ph, 2-MeC_6H_4, 2-$MeOC_6H_4$, $PhCH_2$, Me, Et

Scheme 3.171

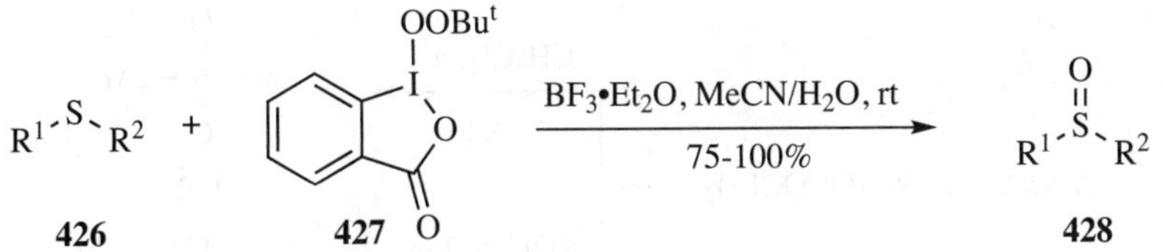

R^1–S–R^2 (**426**) + **427** (OOBu^t) $\xrightarrow[75\text{-}100\%]{BF_3{\bullet}Et_2O,\ MeCN/H_2O,\ rt}$ R^1–S(O)–R^2 (**428**)

R^1 = Bu, Bu^i, Bu^s, $PhCH_2$, $Me(CH_2)_4$, CH_2=$CHCH_2$, 4-MeC_6H_4, 4-$MeOC_6H_4$, 4-ClC_6H_4, Ph
R^2 = $PhCH_2$, Ph, Me, $CH_2P(O)(OEt)_2$

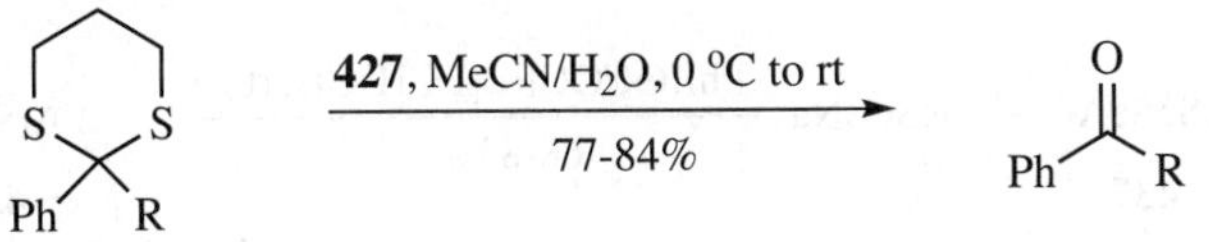

429 $\xrightarrow[77\text{-}84\%]{\mathbf{427},\ MeCN/H_2O,\ 0\ ^oC\ to\ rt}$ PhC(O)R (**430**)

R = Me, C_5H_{11}

Scheme 3.172

Alkylperoxybenziodoxoles (Section 2.1.8.1.4) are useful oxidizing reagents toward organic sulfides, selenides and phosphines [527–529]. Sulfides **426** can be oxidized with peroxybenziodoxole **427** under mild conditions to afford sulfoxides **428** in high yields (Scheme 3.172) [511]. A similar oxidation of dithioacetals **429** leads to regeneration of the parent carbonyl compound **430** and thus can be useful as a method for selective deprotection [511].

Organic sulfides are oxidized to sulfonyl chlorides by iodosylbenzene, activated by crushing and grinding with a HCl-treated silica gel [522,523]. Sulfides **431** with benzylic substituents, such as dibenzyl, alkyl benzyl and benzyl phenyl sulfides, are converted by this reaction into the corresponding sulfonyl chlorides **432** and **433** in high yields (Scheme 3.173). Other types of sulfides, such as dialkyl and alkyl phenyl sulfides, give sulfonyl chlorides only in moderate yield.

The reaction of diaryl disulfides **434** or thiophenols with $PhI(OCOCF_3)_2$ affords the corresponding thiosulfonic S-esters **435** in good yields under mild conditions (Scheme 3.174) [530]. A similar reaction of diaryl disulfides **434** with $PhI(OCOCF_3)_2$ in the presence of alcohols affords the respective arylsulfinic esters **436** [531]. Diselenides **437** in the presence of sodium organosulfinates are oxidized by $PhI(OCOCF_3)_2$ to the respective selenosulfonates **438** [532]. The oxidation of ditellurides **439** by $PhI(OCOCF_3)_2$ under similar conditions affords arenetellurinic mixed anhydrides **440** [533].

A similar oxidation of disulfides **441** with $PhI(OCOCF_3)_2$ in the presence of sodium trifluoromethanesulfinate provides a convenient synthetic approach to trifluoromethanethiosulfonates **442** (Scheme 3.175) [534].

R–S–CH_2Ph (**431**) $\xrightarrow[89\text{-}98\%]{(PhIO)_n,\ HCl\text{-}SiO_2,\ rt,\ 5\ min\ pulverization}$ RSO_2Cl (**432**) + $PhCH_2SO_2Cl$ (**433**)

R = $PhCH_2$, $PhCH_2CH_2$, Ph, CH_3, C_4H_9, C_8H_{17}, etc.

Scheme 3.173

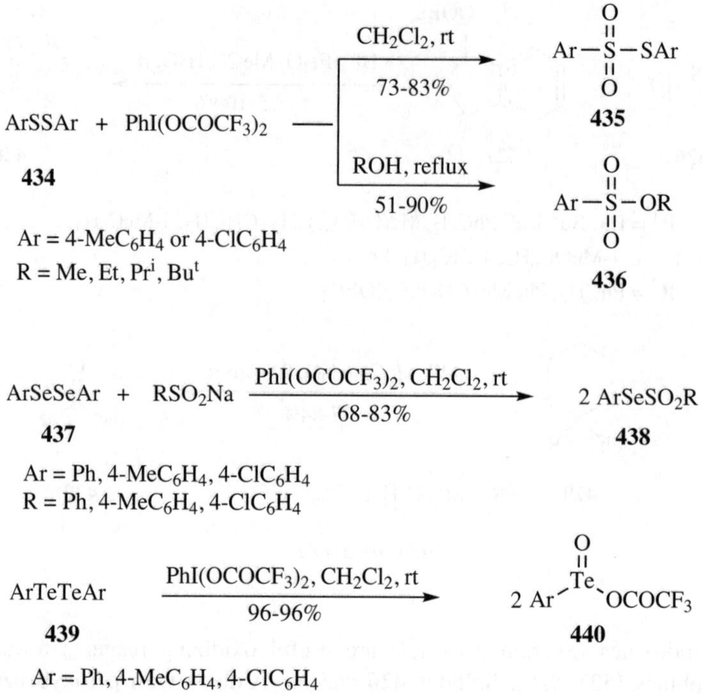

$$ArSSAr + PhI(OCOCF_3)_2 \xrightarrow[73\text{-}83\%]{CH_2Cl_2,\ rt} Ar{-}S(=O)_2{-}SAr$$

434 → 435

$$ArSSAr + PhI(OCOCF_3)_2 \xrightarrow[51\text{-}90\%]{ROH,\ reflux} Ar{-}S(=O)_2{-}OR$$

434 → 436

Ar = 4-MeC_6H_4 or 4-ClC_6H_4
R = Me, Et, Pr^i, Bu^t

$$ArSeSeAr + RSO_2Na \xrightarrow[68\text{-}83\%]{PhI(OCOCF_3)_2,\ CH_2Cl_2,\ rt} 2\ ArSeSO_2R$$

437 → 438

Ar = Ph, 4-MeC_6H_4, 4-ClC_6H_4
R = Ph, 4-MeC_6H_4, 4-ClC_6H_4

$$ArTeTeAr \xrightarrow[96\text{-}96\%]{PhI(OCOCF_3)_2,\ CH_2Cl_2,\ rt} 2\ Ar{-}Te(=O){-}OCOCF_3$$

439 → 440

Ar = Ph, 4-MeC_6H_4, 4-ClC_6H_4

Scheme 3.174

The analogous reaction of diselenides can be used for the preparation of trifluoromethaneselenosulfonates [534].

Enantioselective oxidation of sulfides **443** has been achieved by using chiral, non-racemic [menthyloxy(tosyloxy)iodo]benzenes derived from (1*S*,2*R*,5*S*)-(+)-menthol (structure **444**) and (1*R*,2*S*,5*R*)-(–)-menthol. Reaction of organic sulfides with these reagents in dichloromethane gives optically active (menthyloxy)sulfonium tosylates (e.g., **445**). For example, the treatment of methyl *p*-tolyl sulfide with (+)-**444** gave (+)-menthyloxylmethyl-*p*-tolylsulfonium tosylate **445** in 92% yield as a mixture (ca. 16% de) of diastereomers. The major diastereomer (+)-**445** was separated with 49% efficiency by recrystallization of the mixture from CH_2Cl_2/Et_2O and hydrolyzed in aqueous NaOH to optically pure (*S*)-(–)-methyl *p*-tolyl sulfoxide **446** (Scheme 3.176) [535].

The oxidation of alkyl and aryl sulfides **447** with [hydroxy((+)-10-camphorosulfonyloxy)iodo]benzene **448** as chiral oxidant has been reported (Scheme 3.177). This reaction affords the corresponding sulfoxides **449** in good yields but with low enantioselectivity [536].

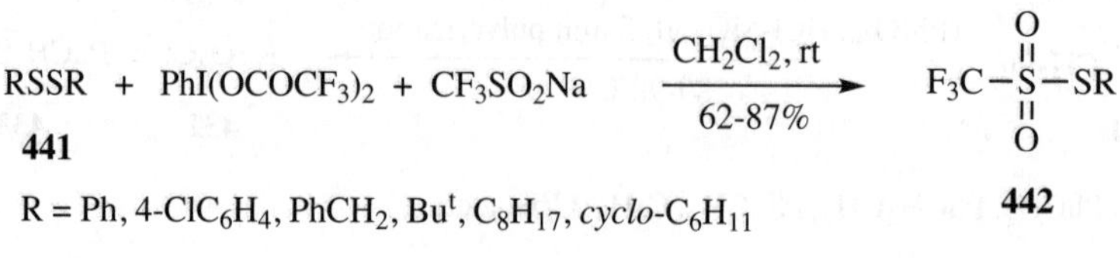

$$RSSR + PhI(OCOCF_3)_2 + CF_3SO_2Na \xrightarrow[62\text{-}87\%]{CH_2Cl_2,\ rt} F_3C{-}S(=O)_2{-}SR$$

441 → 442

R = Ph, 4-ClC_6H_4, $PhCH_2$, Bu^t, C_8H_{17}, *cyclo*-C_6H_{11}

Scheme 3.175

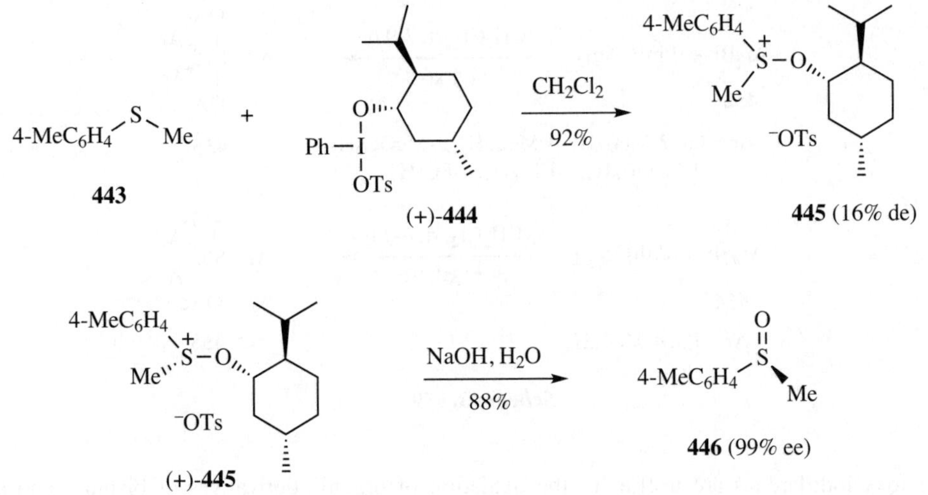

Scheme 3.176

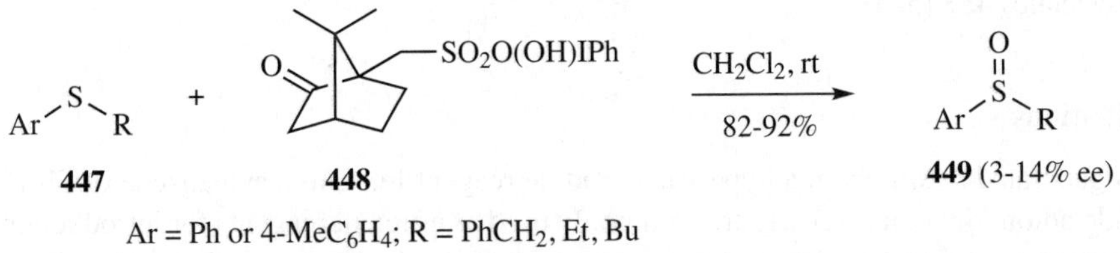

Scheme 3.177

For the enantioselective oxidations of organic sulfides to chiral sulfoxides using iodine(V) reagents, also see Section 3.2.

Thioacetals and thioketals **450** are efficiently cleaved to carbonyl compounds **451** with $PhI(OCOCF_3)_2$ or $PhI(OAc)_2$ under mild conditions (Scheme 3.178). This reaction is especially useful for the selective deprotection of either thioacetals or thioketals and is compatible with various other functional groups [537–541].

Hypervalent iodine compounds can serve as effective oxidants of trivalent phosphorus compounds. In particular, various organic phosphines are selectively oxidized by iodosylbenzene to the respective phosphine oxides in quantitative yield [542].

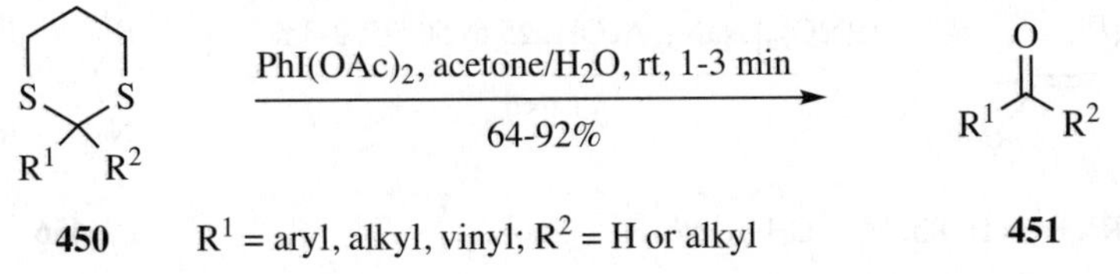

Scheme 3.178

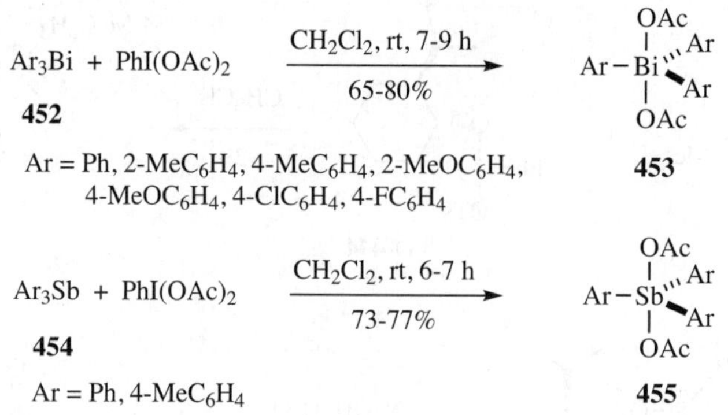

Ar_3Bi + $PhI(OAc)_2$ →(CH_2Cl_2, rt, 7-9 h; 65-80%) $Ar_3Bi(OAc)_2$

452 → **453**

Ar = Ph, 2-MeC_6H_4, 4-MeC_6H_4, 2-$MeOC_6H_4$, 4-$MeOC_6H_4$, 4-ClC_6H_4, 4-FC_6H_4

Ar_3Sb + $PhI(OAc)_2$ →(CH_2Cl_2, rt, 6-7 h; 73-77%) $Ar_3Sb(OAc)_2$

454 → **455**

Ar = Ph, 4-MeC_6H_4

Scheme 3.179

[Bis(acyloxy)iodo]arenes are useful for the oxidation of organic derivatives of bismuth and antimony [543, 544]. Triarylbismuthanes **452** react with (diacetoxyiodo)benzene in dichloromethane under mild, neutral conditions to afford pentavalent triarylbismuth diacetates **453,** which can be isolated in good yields (Scheme 3.179) [543]. Triarylstibines **454** react with $PhI(OAc)_2$ under similar conditions to afford triarylantimony(V) diacetates **455** [544].

3.1.15 Azidations

Azidoidanes generated *in situ* from a hypervalent iodine reagent [e.g., iodosylbenzene or $PhI(OAc)_2$] and a source of azide anion (Section 2.1.12.1) are commonly used as efficient reagents for introduction of the azido function into organic molecules. Zbiral and coworkers in the early 1970s first investigated reactions of the $PhI(OAc)_2/TMSN_3$ system with alkenes leading to diazides, α-azido ketones, or nitriles [545–547]. Later, in the 1980s, Moriarty and coworkers developed a convenient experimental procedure for the preparation of vicinal diazides **456** from alkenes by reaction with iodosylbenzene/NaN_3 in acetic acid (Scheme 3.180) [548].

This reaction (Scheme 3.180) is predominantly *anti* stereoselective; in several cases α-azido ketones has been isolated as by-products [548]. Steroids **457** under the same conditions give exclusively allylic azides **458** (Scheme 3.181) [549].

Ochiai and coworkers have developed a simple method for the synthesis of allyl azides **460** from allyltrimethylsilanes **459** using the iodosylbenzene/$TMSN_3/BF_3{\cdot}Et_2O$ combination (Scheme 3.182) [550].

Treatment of β-dicarbonyl compounds **461** with iodosylbenzene/$TMSN_3$ in chloroform at reflux affords azides **462** in moderate to good yield (Scheme 3.183) [551]. Under similar conditions, 2-(trimethylsiloxy)furan (**463**) gives azidofuranone **464** as the main product [552].

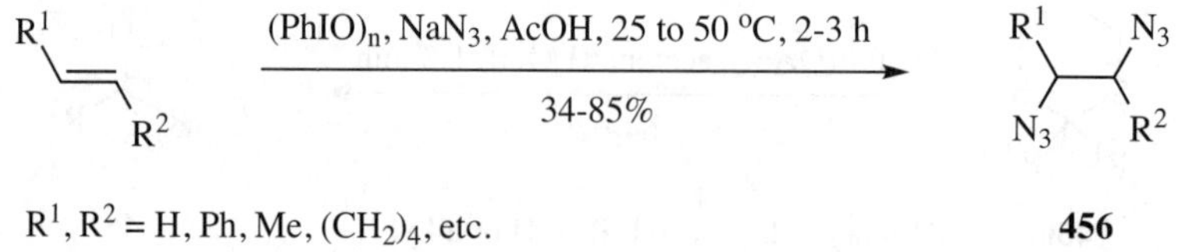

$R^1CH{=}CHR^2$ →($(PhIO)_n$, NaN_3, AcOH, 25 to 50 °C, 2-3 h; 34-85%) $R^1CH(N_3)CH(N_3)R^2$ **456**

R^1, R^2 = H, Ph, Me, $(CH_2)_4$, etc.

Scheme 3.180

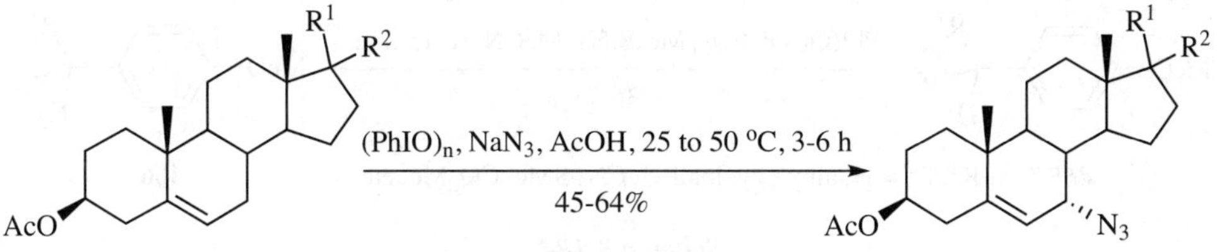

$(PhIO)_n$, NaN_3, AcOH, 25 to 50 °C, 3-6 h

45-64%

457 $R^1, R^2 = H, C_8H_{17}, COCH_3, OAc$ **458**

Scheme 3.181

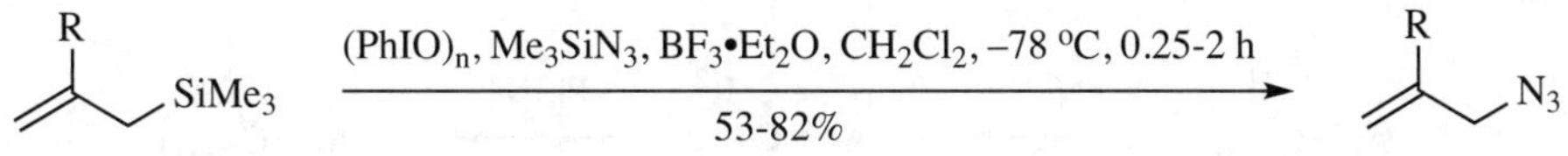

$(PhIO)_n$, Me_3SiN_3, $BF_3{\bullet}Et_2O$, CH_2Cl_2, –78 °C, 0.25-2 h

53-82%

459 R = alkyl, cycloalkyl **460**

Scheme 3.182

Aromatic compounds, such as anisoles and alkylbenzenes, can be azidated in the ring by treatment with $PhI(OCOCF_3)_2/TMSN_3$ in hexafluoroisopropanol (HFIP) [553]. It is assumed that this reaction does not involve azidoiodanes as reactive intermediates, but proceeds via cation–radicals generated by SET from the aromatic substrate to $PhI(OCOCF_3)_2$ (Section 3.1.12) [362]. It is possible, however, to change the course of this reaction by using acetonitrile instead of HFIP as the solvent. Treatment of *p*-alkylanisols **465** with $PhI(OCOCF_3)_2/TMSN_3$ under these conditions affords arylalkyl azides **466** in moderate yields (Scheme 3.184) [554].

A radical mechanism for this reaction (Scheme 3.184) involving the initial formation of azidoiodane **467** and its subsequent homolytic decomposition has been proposed (Scheme 3.185) [554]. Experimental evidence supporting the generation of azide radicals in the thermal decomposition of azidoiodanes formed from $PhI(OAc)_2/NaN_3$ has been reported by Fontana and coworkers [555].

Magnus and coworkers have reported several synthetically important azidations using the iodosylbenzene/$TMSN_3$ combination, which serves as the source of non isolable azidoiodanes $PhI(N_3)_2$

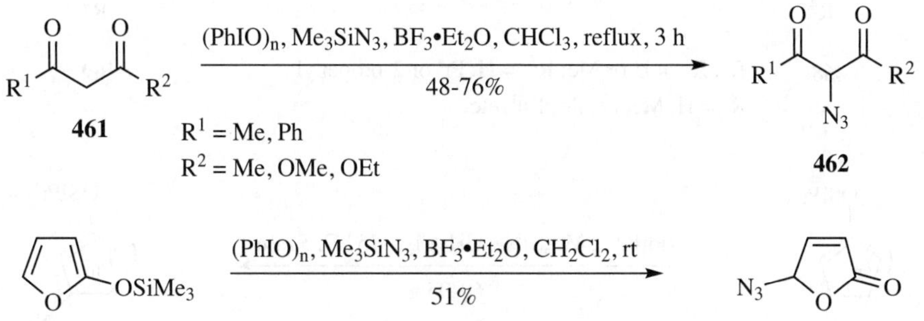

$(PhIO)_n$, Me_3SiN_3, $BF_3{\bullet}Et_2O$, $CHCl_3$, reflux, 3 h

48-76%

461 R^1 = Me, Ph

R^2 = Me, OMe, OEt **462**

$(PhIO)_n$, Me_3SiN_3, $BF_3{\bullet}Et_2O$, CH_2Cl_2, rt

51%

463 **464**

Scheme 3.183

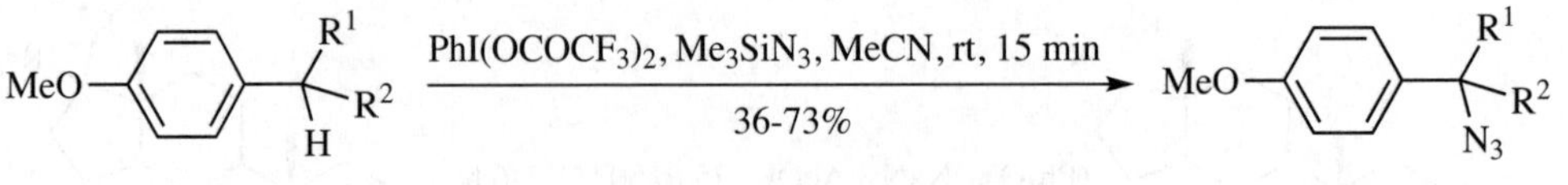

MeO–C$_6$H$_4$–CH(R^1)(R^2) (465) → PhI(OCOCF$_3$)$_2$, Me$_3$SiN$_3$, MeCN, rt, 15 min; 36-73% → MeO–C$_6$H$_4$–C(N$_3$)(R^1)(R^2) (466)

465 R^1, R^2 = H, alkyl, cycloalkyl, CN, SMe, CO_2Me, etc. 466

Scheme 3.184

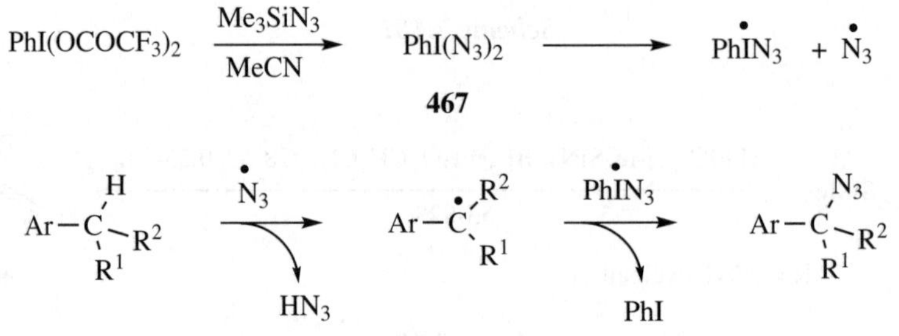

$PhI(OCOCF_3)_2$ → (Me_3SiN_3, MeCN) → $PhI(N_3)_2$ (467) → $Ph\dot{I}N_3$ + $\dot{N}_3$

Ar–CH(R^1)(R^2) → ($\dot{N}_3$, −HN$_3$) → Ar–$\dot{C}$(R^1)(R^2) → ($Ph\dot{I}N_3$, −PhI) → Ar–C(N$_3$)(R^1)(R^2)

Scheme 3.185

or $PhI(N_3)OTMS$ [556–568]. In particular, various triisopropylsilyl (TIPS) enol ethers **468** and **470** react with this reagent at –15 to –18 °C to give the β-azido adducts **469** and **471,** respectively, in excellent yields (Scheme 3.186) [562].

A similar reaction of silyl ether **472** with iodosylbenzene/TMSN$_3$ at lower temperatures and in the presence of catalytic amounts of the stable radical TEMPO stereoselectively affords vicinal *trans*-diazides **473** as the major products (Scheme 3.187) [562]. The effect of TEMPO on the outcome of this reaction has been explained by a change of mechanism from ionic dehydrogenation to a radical addition process in the presence of TEMPO [562].

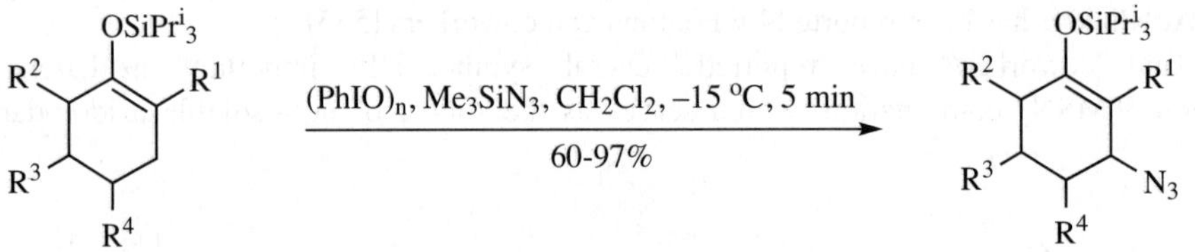

468 → $(PhIO)_n$, Me_3SiN_3, CH_2Cl_2, –15 °C, 5 min; 60-97% → 469

468 R^1, R^2 = H or Me; R^3 = H, Pr^i or 2-propenyl; R^4 = H, Me, Et, Pr^i, Bu^t, etc. 469

470 (OSiPr$^i_3$) → $(PhIO)_n$, Me_3SiN_3, CH_2Cl_2, –15 °C, 5 min; 60-95% → 471 (OSiPr$^i_3$, N$_3$)

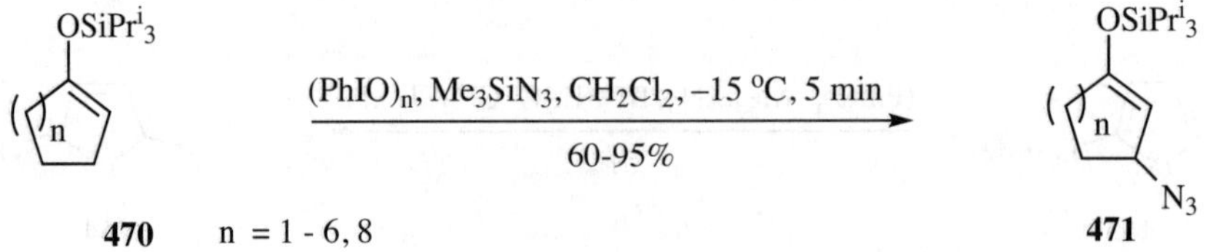

470 n = 1 - 6, 8 471

Scheme 3.186

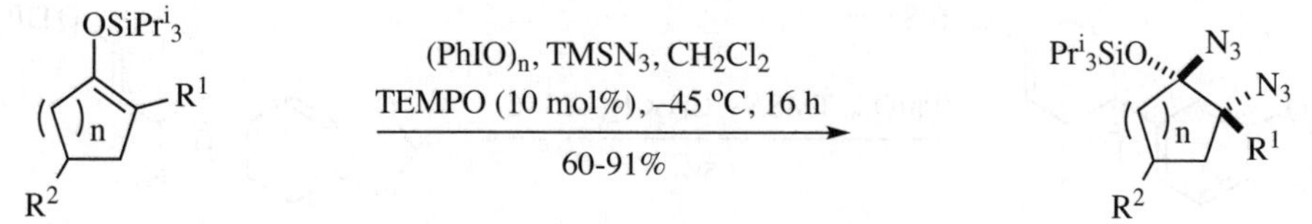

(PhIO)$_n$, TMSN$_3$, CH$_2$Cl$_2$
TEMPO (10 mol%), –45 °C, 16 h
60-91%

472 → **473**

n = 1 or 2; R^1 = H or Ph; R^2 = H, Bu^t, etc.
TEMPO = 2,2,6,6-tetramethylpiperidine-*N*-oxyl

Scheme 3.187

The β-azidation reaction of triisopropylsilyl enol ethers (Schemes 3.186 and 3.187) has been effectively utilized in organic synthesis [563–565]. Magnus and coworkers have developed a mechanistically different enone synthesis that involves treatment of β-azido TIPS enol ethers **469** and **471** with fluoride anion to effect desilylation and concomitant β-elimination to give an α,β-enone [563]. Alternatively, the β-azido group in **469** or **471** can be ionized with Me_3Al or Me_2AlCl and the intermediate enonium ion trapped by various nucleophiles, such as an allylstannane, electron-rich aromatics and trimethylsilyl enol ethers, to give various β-substituted TIPS enol ethers. Reduction of the β-azido TIPS enol ether provides access to the synthetically useful β-amino TIPS enol ethers [563].

The β-azidation of the TIPS derivative **474** has been utilized in the total synthesis of the antitumor alkaloid (+)-pancratistatin (**476**). Azide **475,** the key intermediate product in this synthetic scheme, was obtained in high yield as a mixture of *trans*- and *cis*-diastereomers (3.5 : 1 ratio) and was further converted into pancratistatin (**476**) in 13 steps (Scheme 3.188) [564, 565].

Likewise, the β-azidation reaction has been applied to the enantioselective synthesis of the core structure of lycorane Amaryllidaceae alkaloids. The key intermediate **478** in this synthesis was obtained by β-azidation of the TIPS derivative **477** as a mixture of *trans*- and *cis*-diastereomers (3.5 : 1 ratio) (Scheme 3.189) [566].

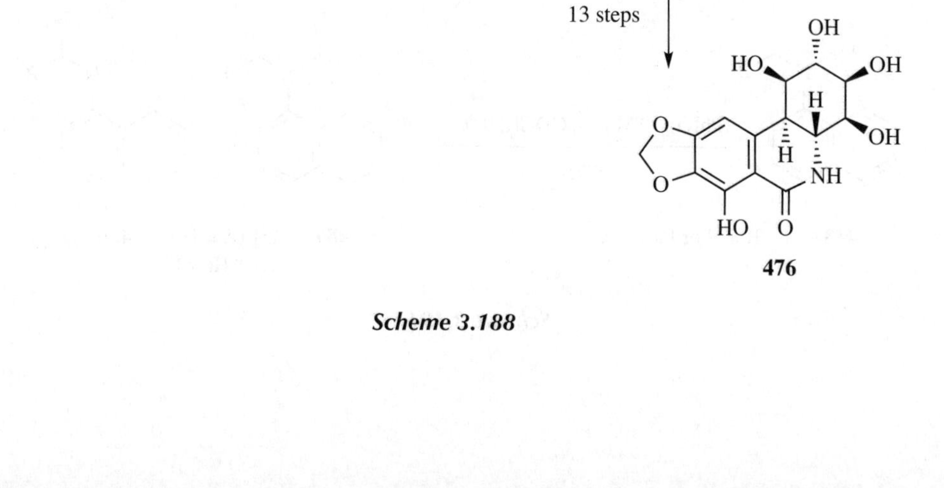

(PhIO)$_n$, TMSN$_3$, CH$_2$Cl$_2$, –15 °C
95%

474 → **475**

13 steps

476

Scheme 3.188

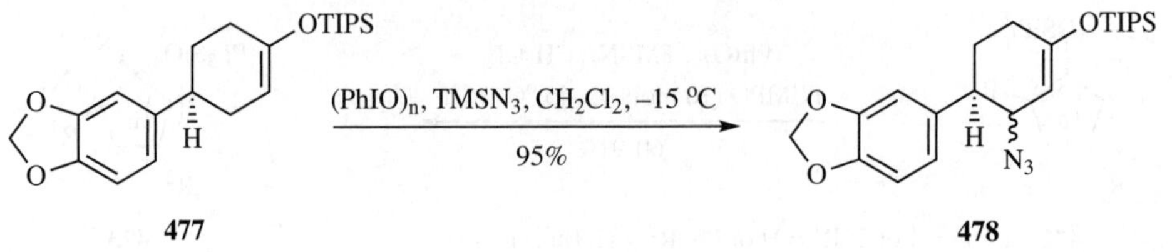

OTIPS (PhIO)$_n$, TMSN$_3$, CH$_2$Cl$_2$, –15 °C, 95%

477 → 478

Scheme 3.189

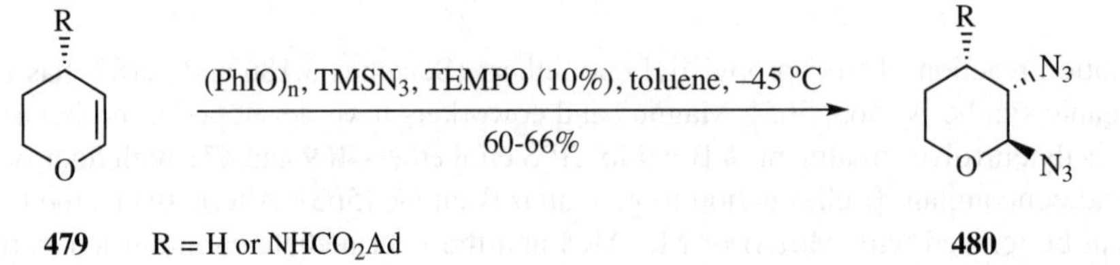

(PhIO)$_n$, TMSN$_3$, TEMPO (10%), toluene, –45 °C, 60-66%

479 R = H or NHCO$_2$Ad → 480

Scheme 3.190

The diazidation reaction leading to vicinal *trans*-diazides (Scheme 3.187) has also been utilized in organic synthesis [567]. Dihydropyrans **479** react with the (PhIO)$_n$/TMSN$_3$/TEMPO system to give 2,3-bis-azido adducts **480** (Scheme 3.190), which can be further elaborated into amino-pyrans [567].

Treatment of *N*,*N*-dimethylarylamines **481** with the iodosylbenzene/TMSN$_3$ reagent system results in the functionalization of one of the methyl groups to give *N*-azidomethyl derivatives **482** (Scheme 3.191). The reaction with an excess of the iodosylbenzene/TMSN$_3$ reagent (2.6–4 equiv) affords the respective bis(*N*,*N*-azidomethyl) derivatives. The azidation of unsymmetrical substrates **483** gives a mixture of products **484** and **485** [568].

Amides, carbamates and ureas can also be α-azidated under similar conditions; however, the yields of azides in this case are generally lower [557, 561]. For example, the piperidine derivatives **486** are converted

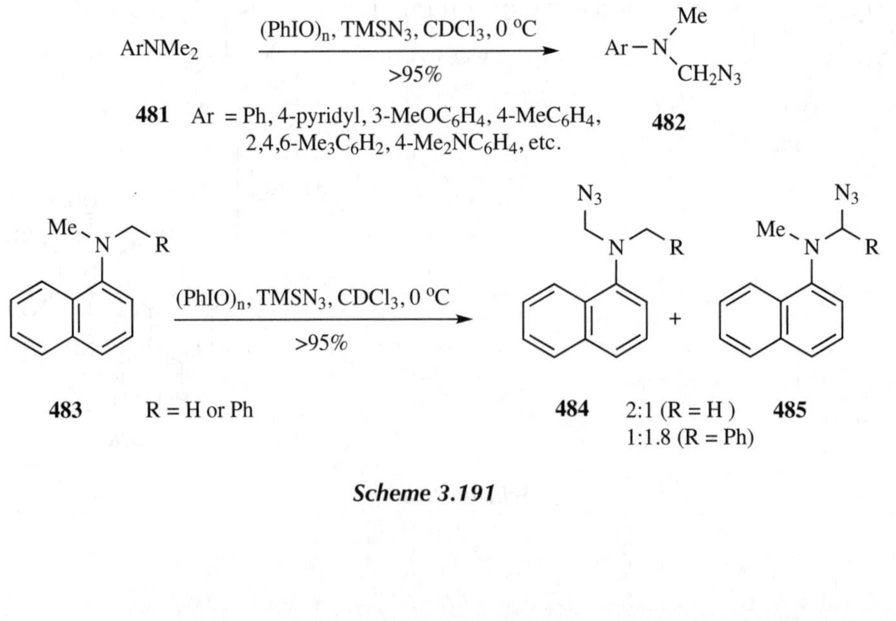

ArNMe$_2$ —(PhIO)$_n$, TMSN$_3$, CDCl$_3$, 0 °C; >95%→ Ar–N(Me)CH$_2$N$_3$

481 Ar = Ph, 4-pyridyl, 3-MeOC$_6$H$_4$, 4-MeC$_6$H$_4$, 2,4,6-Me$_3$C$_6$H$_2$, 4-Me$_2$NC$_6$H$_4$, etc. **482**

(PhIO)$_n$, TMSN$_3$, CDCl$_3$, 0 °C; >95%

483 R = H or Ph → **484** + **485** 2:1 (R = H) 1:1.8 (R = Ph)

Scheme 3.191

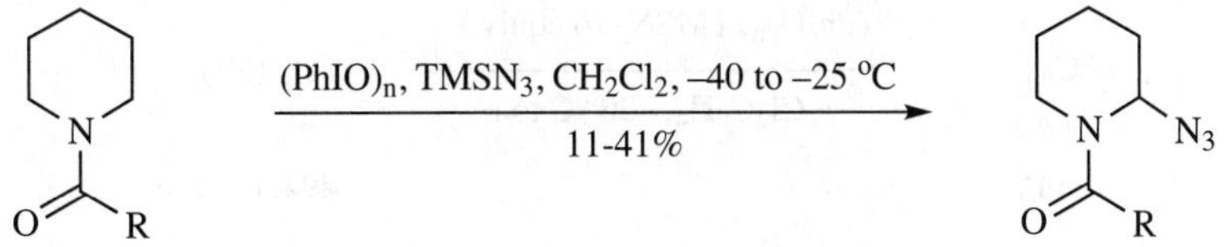

$(PhIO)_n$, $TMSN_3$, CH_2Cl_2, –40 to –25 °C
11-41%

486 R = Me, Bu^{t}O, Ph, Ph_2N, PhO, 4-$NO_2C_6H_4$, etc. **487**

Scheme 3.192

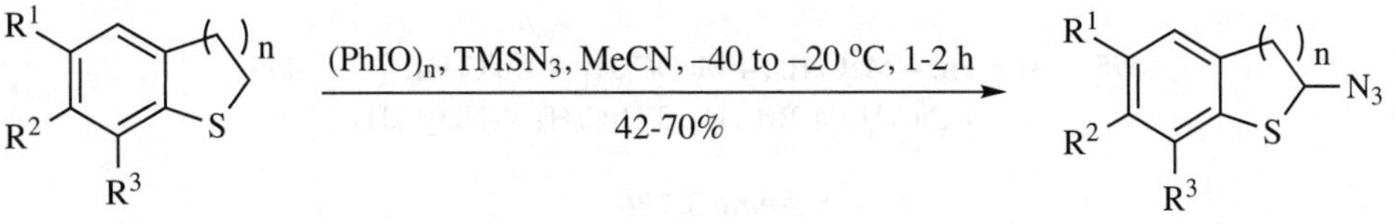

$(PhIO)_n$, $TMSN_3$, MeCN, –40 to –20 °C, 1-2 h
42-70%

488 R^1, R^2 = H, OMe, OAc
R^3 = H or OMe; n = 1, 2 **489**

Scheme 3.193

into azides **487** in moderate yield (Scheme 3.192) with some starting material remaining in the reaction mixture.

Kita and coworkers have reported the direct α-azidation of cyclic sulfides using the iodosylbenzene/$TMSN_3$ reagent system [569]. This method is applicable to substrates that are easily aromatized under oxidative conditions, such as monocyclic and bicyclic sulfides **488,** to give the corresponding α-azido sulfides **489** in moderate to good yields (Scheme 3.193).

A similar α-azidation of cyclic sulfide **490** leading to the α-azido sulfide **491** has been applied in the total synthesis of the strongly cytotoxic marine alkaloid (±)-makaluvamine F (**492**) (Scheme 3.194) [373, 570].

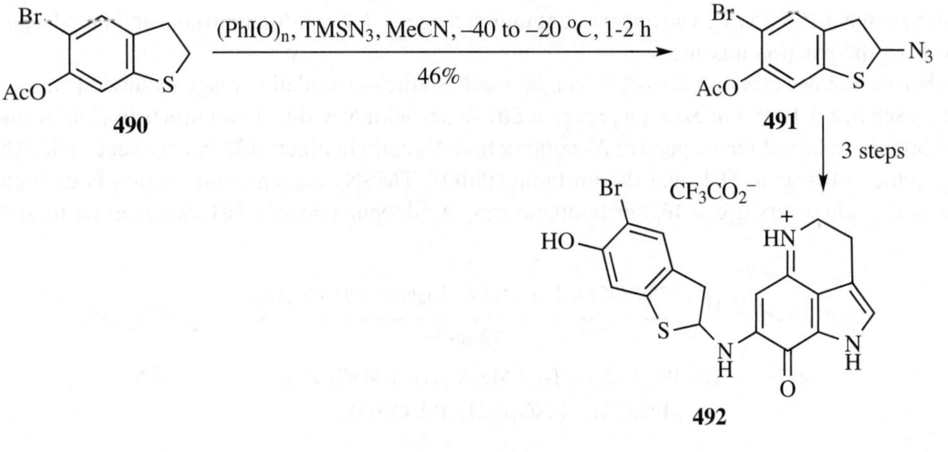

$(PhIO)_n$, $TMSN_3$, MeCN, –40 to –20 °C, 1-2 h
46%

490 **491**

3 steps

$CF_3CO_2^-$

492

Scheme 3.194

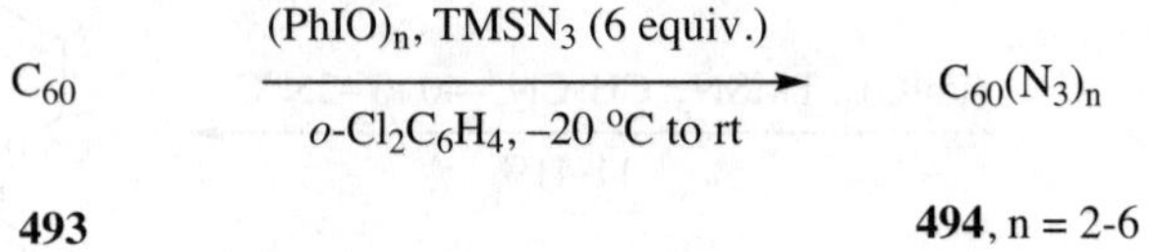

$$C_{60} \xrightarrow[o\text{-}Cl_2C_6H_4,\ -20\ ^\circ C \text{ to rt}]{(PhIO)_n,\ TMSN_3\ (6\ equiv.)} C_{60}(N_3)_n$$

493 → **494**, n = 2-6

Scheme 3.195

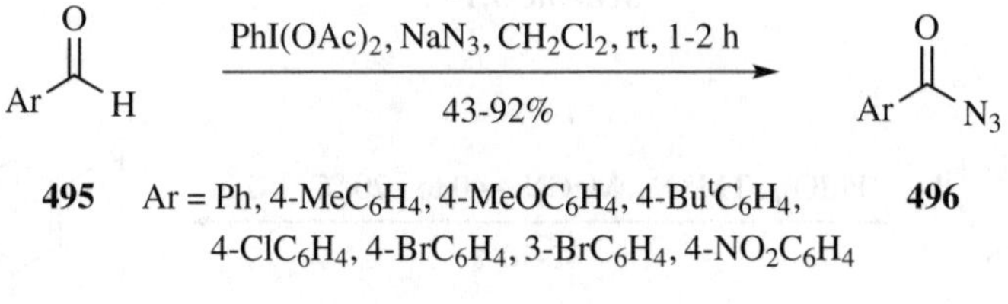

$$ArCHO \xrightarrow[43\text{-}92\%]{PhI(OAc)_2,\ NaN_3,\ CH_2Cl_2,\ rt,\ 1\text{-}2\ h} ArC(O)N_3$$

495 Ar = Ph, 4-MeC_6H_4, 4-$MeOC_6H_4$, 4-$Bu^tC_6H_4$, 4-ClC_6H_4, 4-BrC_6H_4, 3-BrC_6H_4, 4-$NO_2C_6H_4$ **496**

Scheme 3.196

Fullerene C_{60} (**493**) smoothly reacts with the iodosylbenzene/$TMSN_3$ reagent system under typical azidation conditions to form explosive polyazidofullerenes **494** (Scheme 3.195) [571].

Polystyrene can be directly azidated in 1,2-dichloroethane or chlorobenzene using the $PhI(OAc)_2/TMSN_3$ reagent combination at 0 °C for 4 h followed by heating to 50 °C for 2 h. The 2D NMR HMBC spectra indicate that the azido groups are attached to the polymer backbone and also possibly to the aromatic rings. The amount of introduced azido groups is approximately one in every eleven styrene units according to semiquantitative IR spectroscopy and elemental analysis [572].

The combination of (diacetoxyiodo)benzene and sodium azide, which presumably generates (diazidoiodo)benzene as the principal reagent, readily reacts with aryl aldehydes **495** to afford aroyl azides **496** in generally high yields (Scheme 3.196) [573].

Vinyl azides **498** can be prepared directly from α,β-unsaturated carboxylic acids **497** by treatment with [bis(trifluoroacetoxy)iodo]benzene and sodium azide in dichloromethane under phase-transfer conditions (Scheme 3.197) [574]. This method is also useful for the preparation of acyl azides [574].

A simple method for the preparation of α-azido ketones and esters **500** in good yields by direct azidation of carbonyl derivatives **499** at the α-carbon using 4,4′-bis-(dichloroiodo)biphenyl and sodium azide has been reported (Scheme 3.198) [575]. The hypervalent iodine reagent, 4,4′-bis-(dichloroiodo)biphenyl, can be easily recycled from the reaction mixture.

Azidobenziodoxoles (Section 2.1.8.1.5) can be used as efficient azidating reagents toward various organic substrates (Scheme 3.199). For example, reagent **501** reacts with *N,N*-dimethylanilines in dichloromethane at reflux in 30 min to afford the respective *N*-azidomethyl-*N*-methylanilines **502** in excellent yields [576]. The main advantage of reagent **501** over the unstable $(PhIO)_n/TMSN_3$ reagent combination is its high thermal stability, which allows its use at higher temperatures. Azidobenziodoxole **501** can even be used for direct

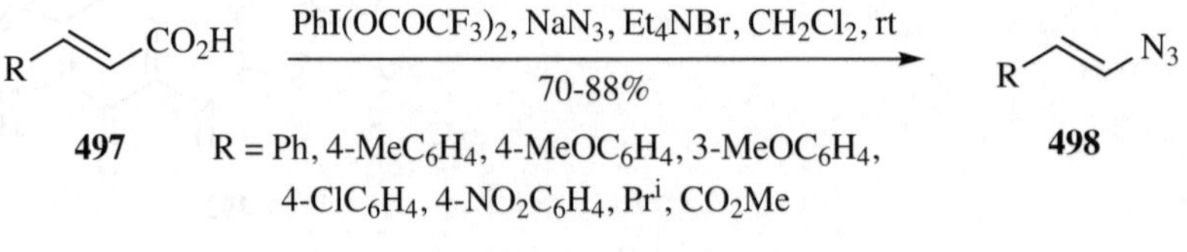

$$RCH{=}CHCO_2H \xrightarrow[70\text{-}88\%]{PhI(OCOCF_3)_2,\ NaN_3,\ Et_4NBr,\ CH_2Cl_2,\ rt} RCH{=}CHN_3$$

497 R = Ph, 4-MeC_6H_4, 4-$MeOC_6H_4$, 3-$MeOC_6H_4$, 4-ClC_6H_4, 4-$NO_2C_6H_4$, Pr^i, CO_2Me **498**

Scheme 3.197

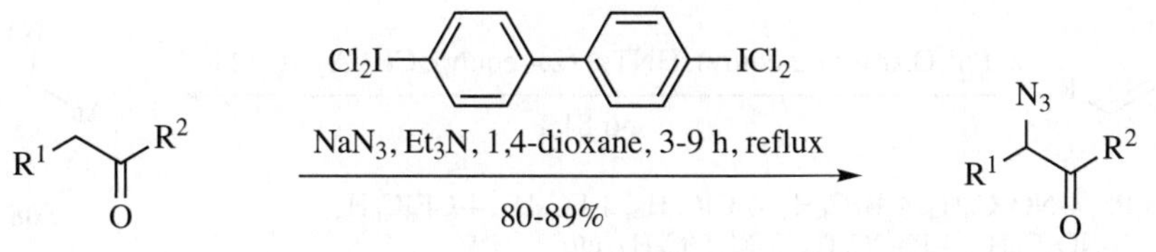

499 R^1 = H, Me, Et, Ph, 4-$MeOC_6H_4$, 4-ClC_6H_4, 4-$NO_2C_6H_4$, etc. **500**
R^2 = OMe, OEt, OBn, Ph, Me
or $R^1 + R^2 = (CH_2)_4$

Scheme 3.198

azidation of hydrocarbons at higher temperatures and in the presence of radical initiators (Scheme 3.199). Reagent **501** selectively reacts with isooctane upon reflux in 1,2-dichloroethane in the presence of catalytic amounts of benzoyl peroxide to afford tertiary azide **506.** Under similar conditions, reactions of azidobenziodoxole **501** with bicyclic and tricyclic hydrocarbons afford the respective alkyl azides **503–505.** Cyclohexene is azidated at the allylic position to form 3-azidocyclohexene **507** [576].

Azidations with azidobenziodoxoles can be effectively catalyzed by iron salts, such as iron(II) propionate, $Fe(O_2CEt)_2$. In the presence of chiral oxazoline ligands (boxmi ligands), the Fe-catalyzed azidations of

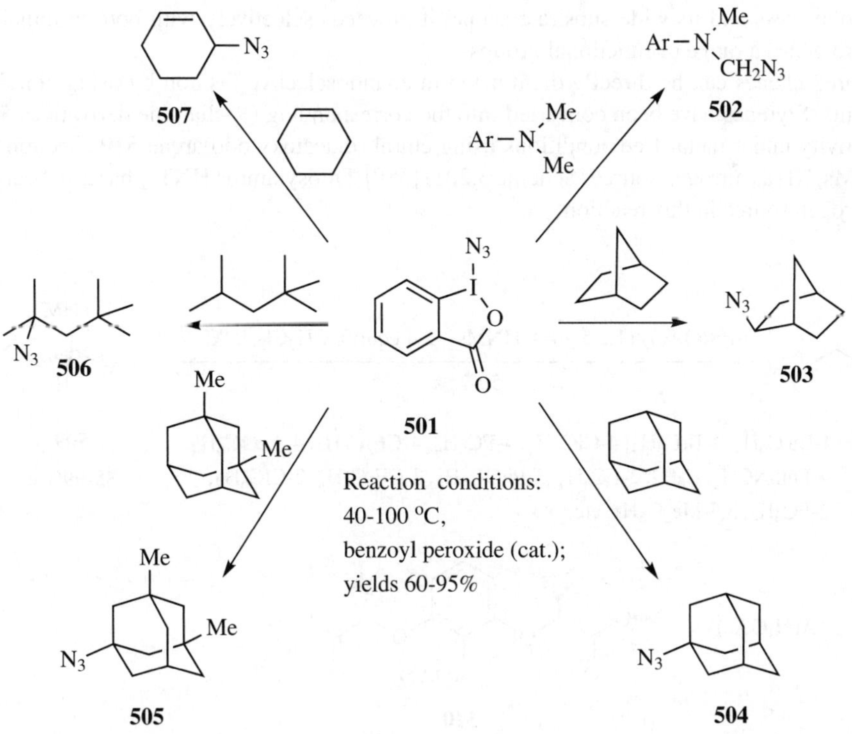

Scheme 3.199

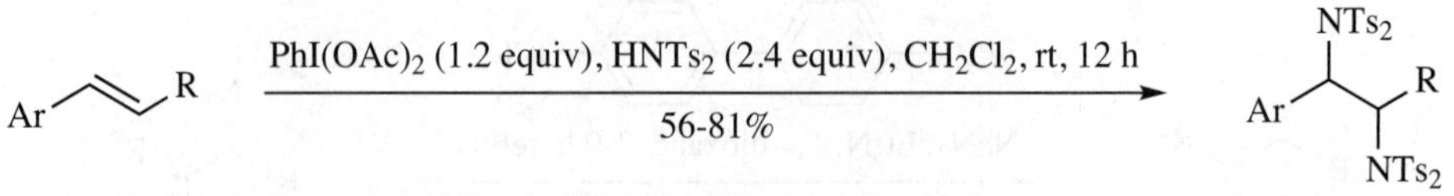

$$\mathrm{Ar{\diagup\!\!\!=\!\!\diagdown}R} \xrightarrow[\text{56-81\%}]{\text{PhI(OAc)}_2\text{ (1.2 equiv), HNTs}_2\text{ (2.4 equiv), CH}_2\text{Cl}_2\text{, rt, 12 h}} \mathrm{Ar\text{-}CH(NTs_2)\text{-}CH(NTs_2)\text{-}R}$$

508

Ar = Ph, 4-$NO_2C_6H_4$, 4-BrC_6H_4, 4-ClC_6H_4, 4-FC_6H_4, 4-$CF_3C_6H_4$, 3-$NO_2C_6H_4$, 3-$PhOC_6H_4$, 3-$MeOC_6H_4$, etc.
R = H, Me, Ph

Scheme 3.200

β-keto esters and oxindoles afford the respective α-azido-β-keto esters and 3-azidooxindoles with high enantioselectivity (up to 94% ee) [577].

3.1.16 Aminations

Direct introduction of the amino or amido functional group is an important synthetic goal that can be effectively achieved by using hypervalent iodine amides and imides (Section 2.1.12) as reagents. The amides, for example, $ArI(OAc)NTs_2$ or $ArI(NTs_2)_2$, can be conveniently generated *in situ* from [bis(acyloxy)iodo]arenes and ditosylamine $HNTs_2$ and used without isolation in reactions with alkenes [578, 579]. For example, numerous vicinal diamines **508** have been prepared by direct diamination of alkenes by the action of (diacetoxyiodo)benzene and bis-sulfonimides as nitrogen sources (Scheme 3.200) [578]. The reaction is characterized by its robustness and its wide substrate scope; it proceeds selectively with both terminal and internal alkenes and tolerates a range of functional groups.

Furthermore, alkenes can be directly diaminated in enantioselective fashion by using chiral hypervalent iodine reagents. Styrenes have been converted into the corresponding (*S*)-diamine derivatives **509** with high enantioselectivity under metal-free conditions using chiral (diacetoxyiodo)arene **510** (Section 2.1.5) as the oxidant and Ms_2NH as nitrogen source (Scheme 3.201) [580]. Ditosylamine $HNTs_2$ has also been successfully used as a nitrogen source in this reaction.

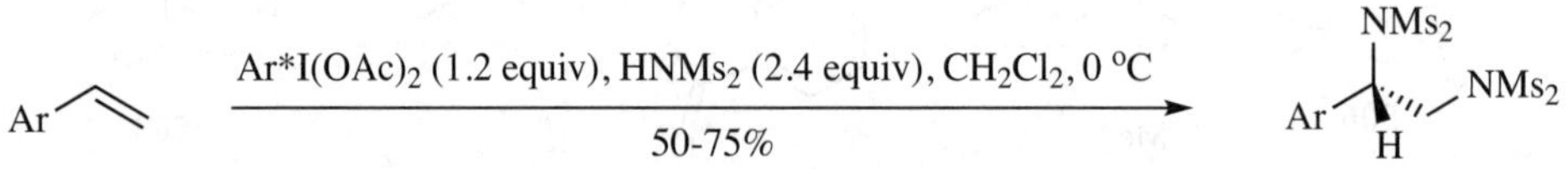

$$\mathrm{Ar{\diagup\!\!\!=}} \xrightarrow[\text{50-75\%}]{\text{Ar*I(OAc)}_2\text{ (1.2 equiv), HNMs}_2\text{ (2.4 equiv), CH}_2\text{Cl}_2\text{, 0 °C}} \mathrm{Ar\text{-}C(H)(NMs_2)\text{-}CH_2NMs_2}$$

509
85-99% ee

Ar = 4-$Bu^tC_6H_4$, 4-BrC_6H_4, 4-ClC_6H_4, 4-FC_6H_4, 4-$CF_3C_6H_4$, 4-$AcOC_6H_4$, 4-$PhthNCH_2$, C_6H_4, 3-FC_6H_4, 3-$PhOC_6H_4$, 3-$CF_3C_6H_4$, 2-ClC_6H_4, 2-FC_6H_4, 3,5-$Me_2C_6H_3$, etc.

$Ar^*I(OAc)_2$ = [2,6-bis(MeO₂C–CH(Me)–O)-C_6H_3–$I(OAc)_2$]

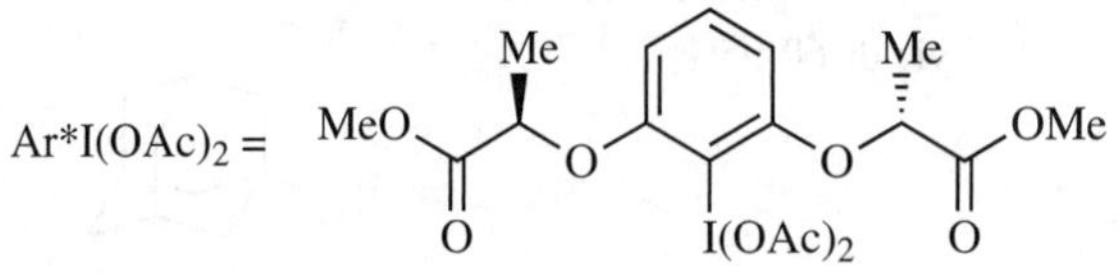

510

Scheme 3.201

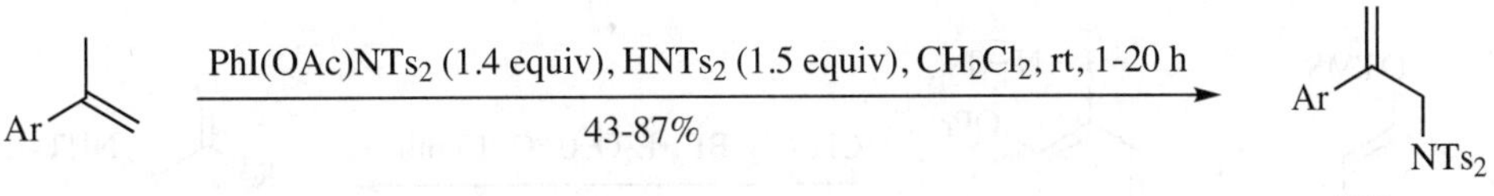

Ar (511) → [PhI(OAc)NTs$_2$ (1.4 equiv), HNTs$_2$ (1.5 equiv), CH_2Cl_2, rt, 1-20 h; 43-87%] → Ar, NTs_2 (512)

511 Ar = Ph, 4-$NO_2C_6H_4$, 4-MeC_6H_4, 4-$Bu^tC_6H_4$, 4-ClC_6H_4, 4-FC_6H_4, 4-$CF_3C_6H_4$, 3-MeC_6H_4, 3-BrC_6H_4, etc. 512

Scheme 3.202

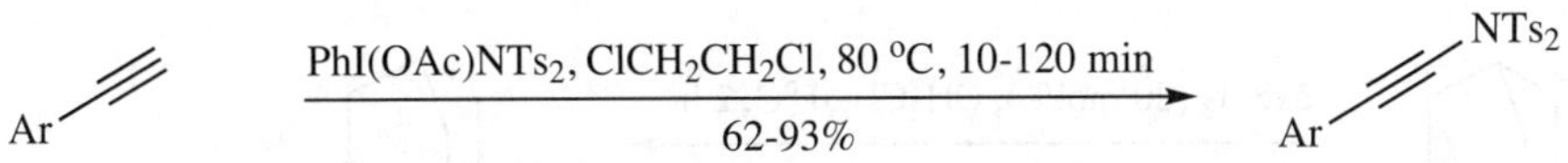

Ar (513) → [PhI(OAc)NTs$_2$, $ClCH_2CH_2Cl$, 80 °C, 10-120 min; 62-93%] → Ar, NTs_2 (514)

513 Ar = Ph, 4-$MeOC_6H_4$, 4-MeC_6H_4, 4-$Bu^tC_6H_4$, 4-BrC_6H_4, 2-MeC_6H_4, 3-$MeOC_6H_4$, 3-ClC_6H_4, 2,4-$F_2C_6H_3$, 1-naphthyl 514

Scheme 3.203

Hypervalent iodine amides have been used for direct intermolecular allylic amination. α-Methyl styrenes **511** are selectively aminated by hypervalent iodine amide PhI(OAc)NTs$_2$ in the presence of ditosylamine as a nitrogen source to give allylic bistosylamides **512** in generally high yields (Scheme 3.202) [581].

A direct metal-free amination of arylethynes by the reaction of terminal alkynes **513** with hypervalent iodine imide allows a simple, one-step synthesis of an important class of ynamides **514** (Scheme 3.203) [582].

Amidobenziodoxoles (Section 2.1.8.1.6) have been used as the amidating reagents toward polycyclic alkanes under radical conditions. For example, reagent **515** reacts with adamantane in chlorobenzene at 100–105 °C in the presence of a catalytic amount of benzoyl peroxide to afford 1-amidoadamantane **516** in moderate yield (Scheme 3.204) [583].

Imidoiodanes, ArINTs (Section 2.1.12.4), can be used for various amidations under transition metal catalysis (Section 3.1.21) [584–586] or under metal-free conditions [587, 588]. In particular, *o*-alkoxyphenyliminoiodane **518** readily reacts with silyl enol ethers **517** in the presence of BF_3·etherate to give products of α-tosylamination **519** in good yields (Scheme 3.205) [588]. Furthermore, reagent **518** in the presence of catalytic amounts of iodine readily reacts with adamantane to give the product of tosylamination (**520**) in excellent yield under very mild conditions. For comparison, PhINTs reacts with adamantane and iodine (0.2 equiv) in dichloromethane at room temperature in 2 h to afford 1-tosylaminoadamantane **520** in only 63% yield [589].

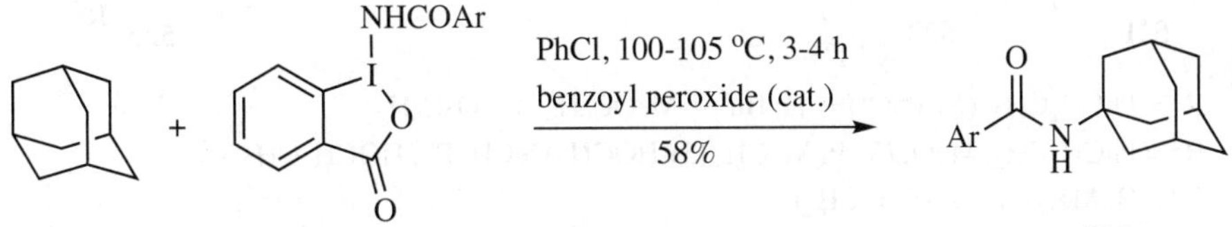

Adamantane + NHCOAr benziodoxolone (515) → [PhCl, 100-105 °C, 3-4 h; benzoyl peroxide (cat.); 58%] → Ar–C(=O)–NH–adamantyl (516)

515 Ar = 4-ClC_6H_4 516

Scheme 3.204

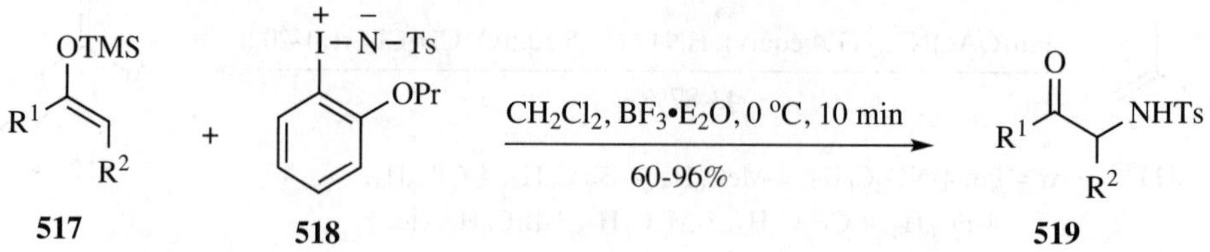

OTMS, R^1, R^2 (517) + 518 (I^+-N^--Ts, OPr) → CH_2Cl_2, $BF_3{\bullet}E_2O$, 0 °C, 10 min; 60-96% → 519 ($R^1C(O)CH(R^2)NHTs$)

R^1 = Ph, 4-$MeOC_6H_4$, 4-ClC_6H_4, 4-BrC_6H_4, 2-ClC_6H_4 and R^2 = H;
or R^1 = Pr and R^2 = Et; or R^1 + R^2 = $(CH_2)_4$

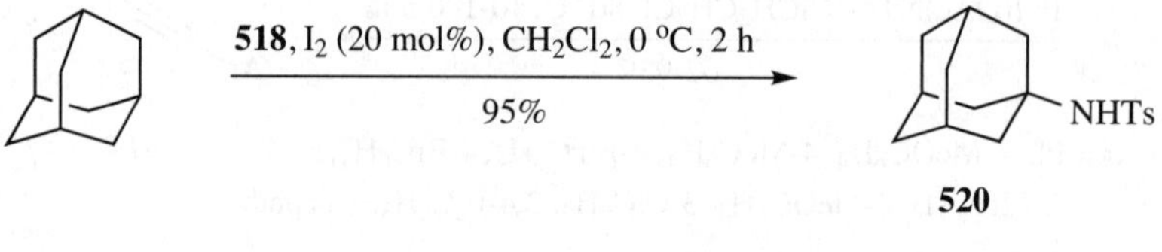

adamantane → **518**, I_2 (20 mol%), CH_2Cl_2, 0 °C, 2 h; 95% → NHTs (520)

Scheme 3.205

An oxidative amination of aldehydes using amines, *N*-hydroxysuccinimide and (diacetoxyiodo)benzene has been developed [590]. The method allows the coupling of a wide range of aldehydes **521** with amines **522** under mild reaction conditions, providing amides **523** in good to excellent yield (Scheme 3.206). A radical mechanism was proposed for this reaction based on an ESR study [590]. A similar protocol has been reported for a one-pot synthesis of a series of glycosyl carboxamides in 78–96% yield by the coupling of aldehydes with secondary amines using (diacetoxyiodo)benzene in the presence of ionic liquid at room temperature [591].

3.1.17 Thiocyanations and Arylselenations

Hypervalent iodine reagents in combination with a source of appropriate nucleophiles are commonly used to prepare products with new C—S and C—Se bonds. Moriarty and coworkers have developed convenient procedures for the thiocyanation of organic substrates using the combination of $PhICl_2$ with $Pb(SCN)_2$ [592–594]. Various enol silyl ethers **524,** ketene silyl acetals **526** and **528** and β-dicarbonyl compounds **530** can be effectively thiocyanated with this combination of reagents to produce the respective thiocyanato

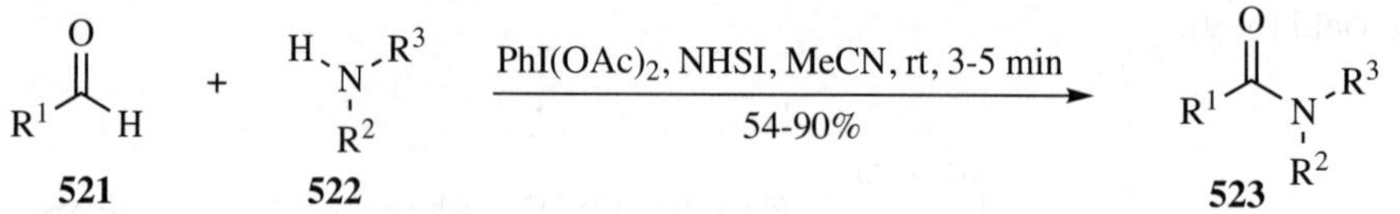

R^1CHO (521) + HNR^2R^3 (522) → $PhI(OAc)_2$, NHSI, MeCN, rt, 3-5 min; 54-90% → $R^1C(O)NR^2R^3$ (523)

R^1 = $PhCH_2CH_2$, (*E*)-PhCH=CH, Bu^t, 4-$MeOC_6H_4$, 4-$NO_2C_6H_4$
R^2 = $PhCH_2CH_2$, MeO, (*S*)-PhMeCH, (*S*)-$HOCH_2MeCH$, Ph, $HOCH_2CH_2$, etc.
R^3 = H, Me, or R^2 + R^3 = $(CH_2)_4$
NHSI = *N*-hydroxysuccinimide

Scheme 3.206

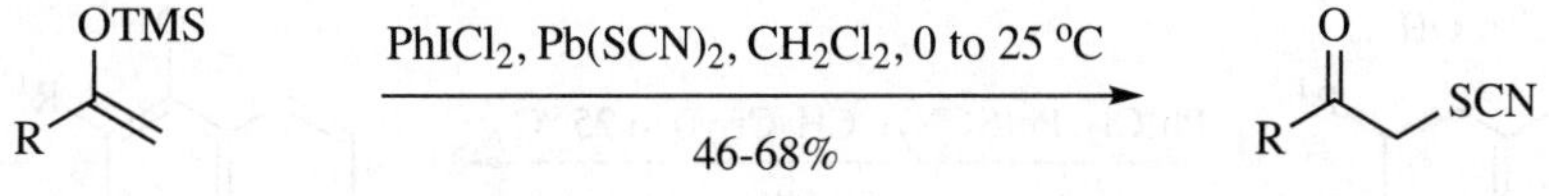

OTMS, R, **524** R = Ph, 2-furyl, 2-thienyl, 2-pyridyl, etc. → $PhICl_2$, $Pb(SCN)_2$, CH_2Cl_2, 0 to 25 °C, 46-68% → **525** (R–CO–CH₂SCN)

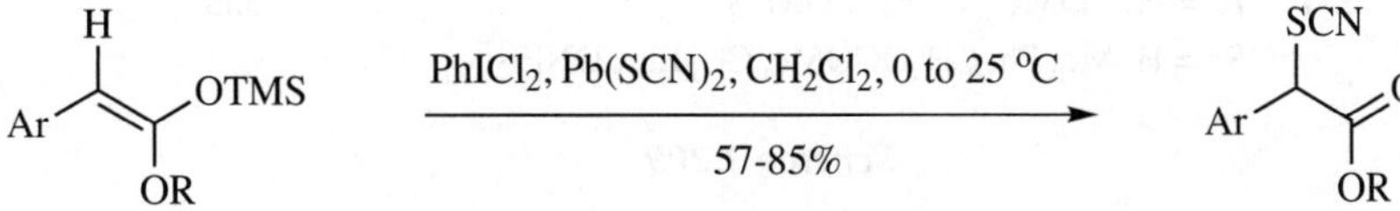

526 Ar = Ph, 4-ClC_6H_4, 4-MeC_6H_4, 4-$MeOC_6H_4$; R = Me, Et → $PhICl_2$, $Pb(SCN)_2$, CH_2Cl_2, 0 to 25 °C, 57-85% → **527**

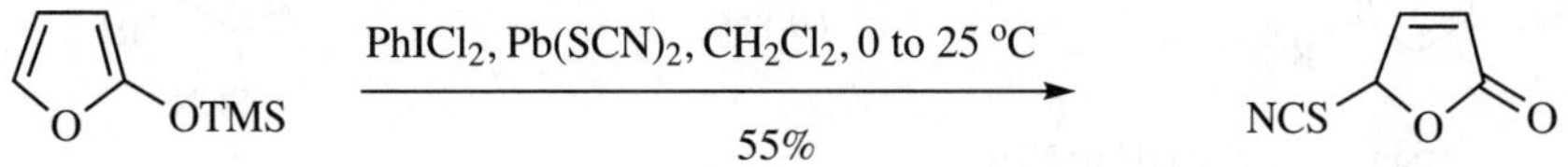

528 → $PhICl_2$, $Pb(SCN)_2$, CH_2Cl_2, 0 to 25 °C, 55% → **529**

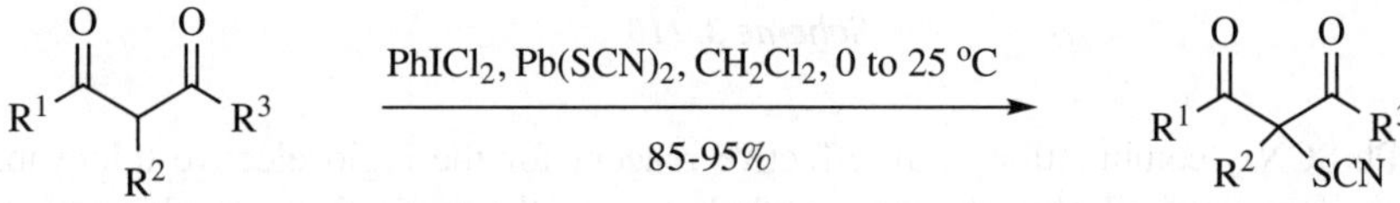

530 R^1, R^3 = Me or Ph; R^2 = H or Me → $PhICl_2$, $Pb(SCN)_2$, CH_2Cl_2, 0 to 25 °C, 85-95% → **531**

Scheme 3.207

derivatives of carbonyl and β-dicarbonyl compounds **525, 527, 529** and **531,** respectively (Scheme 3.207) [592, 593]. The mechanism of these reactions presumably involves the intermediate formation of unstable iodine(III) thiocyanate, $PhI(SCN)_2$.

Under similar conditions, various alkynes **532** are stereoselectively converted into (*E*)-1,2-dithiocyanated alkenes **533** in generally good yield and with less than 5% of the corresponding (*Z*) isomers (Scheme 3.208) [594].

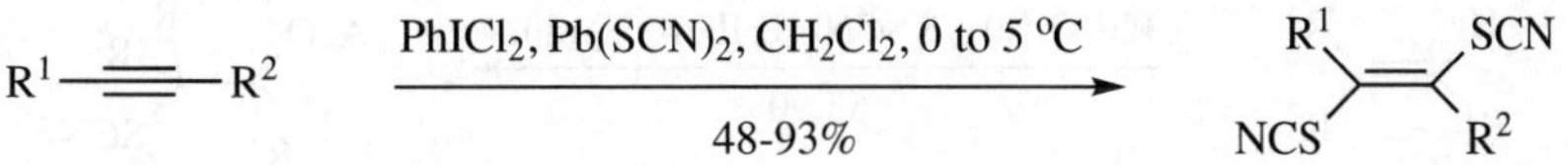

R^1—≡—R^2 **532** → $PhICl_2$, $Pb(SCN)_2$, CH_2Cl_2, 0 to 5 °C, 48-93% → **533**

R^1 = Ph, 4-MeC_6H_4, Pr, $CH_3(CH_2)_3$, $CH_3(CH_2)_5$, $CH_3(CH_2)_7$, $HO(CH_2)_4$, 1-cyclohexenyl
R^2 = H, Me, Ph, TMS

Scheme 3.208

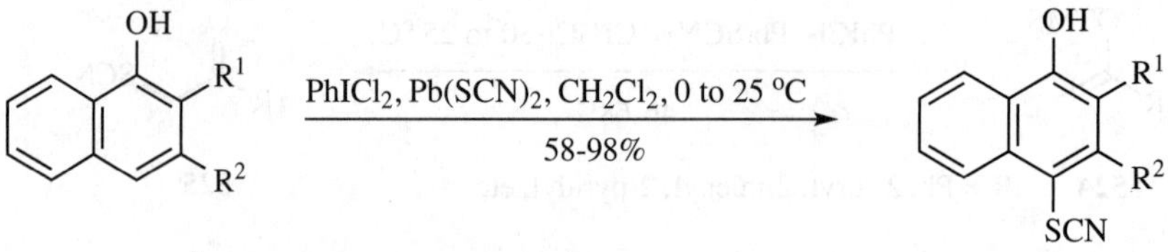

$$\text{534} \xrightarrow[\text{58-98\%}]{\text{PhICl}_2\text{, Pb(SCN)}_2\text{, CH}_2\text{Cl}_2\text{, 0 to 25 °C}} \text{535}$$

534 R^1 = H, COMe, CO_2Et, $CONEt_2$
R^2 = H, Me, Ph, CH_2OCOMe, CO_2Et, $CONEt_2$

535

Scheme 3.209

$$\text{536} \xrightarrow[\text{69-86\%}]{\text{PhICl}_2\text{, KSCN, CH}_2\text{Cl}_2\text{, 0 °C to rt, 50 min}} \text{537}$$

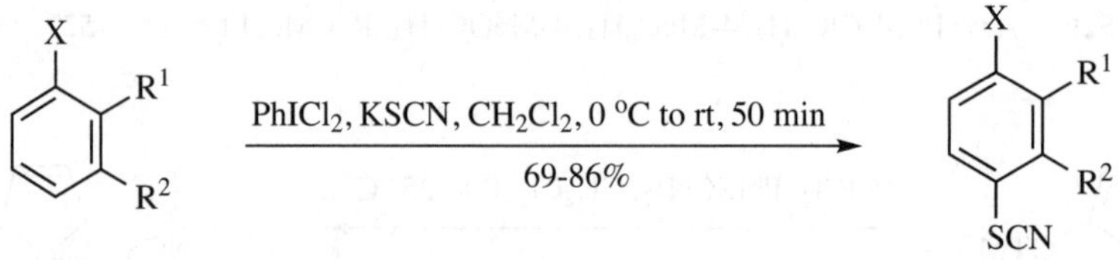

536 X = OH or NH_2
R^1 = H, OH or CO_2H
R^2 = H or OH

537

Scheme 3.210

The $PhICl_2/Pb(SCN)_2$ combination is an effective reagent for the regioselective thiocyanation of various types of *para*-unsubstituted phenols and naphthols to give the respective *para*-thiocyanatophenols and naphthols in good to quantitative yields [595]. Various functional groups, such as chloro, allyl, carbonyl, ester, amide and primary hydroxyl groups, have been shown to be compatible with this reaction. For example, the thiocyanation of naphthols **534** affords the respective thiocyanates **535** in generally high yield (Scheme 3.209) [595].

An improved method for the thiocyanation of 2-arylindan-1,3-diones, phenols and anilines employs a reagent combination of (dichloroiodo)benzene and potassium thiocyanate in dry dichloromethane [596]. For example, the *para*-unsubstituted phenols and anilines **536** are efficiently converted under these reaction conditions into the respective *p*-thiocyanato derivatives **537** in high yields (Scheme 3.210).

The combination of (diacetoxyiodo)benzene and thiocyanate anion in a polar, protic, non-nucleophilic solvent has been used for the oxidative functionalization of alkenes **538** to acetoxy thiocyanate derivatives **539** with *anti*-stereoselectivity and with good regioselectivity (Scheme 3.211) [597].

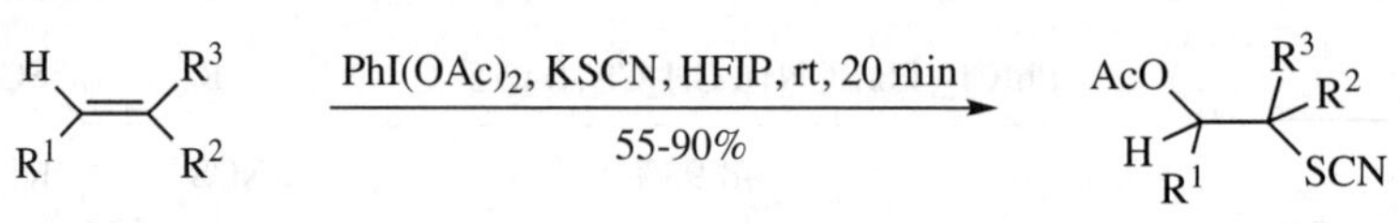

$$\text{538} \xrightarrow[\text{55-90\%}]{\text{PhI(OAc)}_2\text{, KSCN, HFIP, rt, 20 min}} \text{539}$$

538 **539**

R^1 = C_4H_9, C_6H_{13}, C_8H_{17}, *cyclo*-C_6H_{11}; R^2 and R^3 = H
R^1 = C_5H_{11}, R^2 = H, R^3 = Me or $R^1 + R^2$ = $(CH_2)_4$, $(CH_2)_5$, R^3 = H
HFIP = 1,1,1,3,3,3-hexafluoroisopropanol

Scheme 3.211

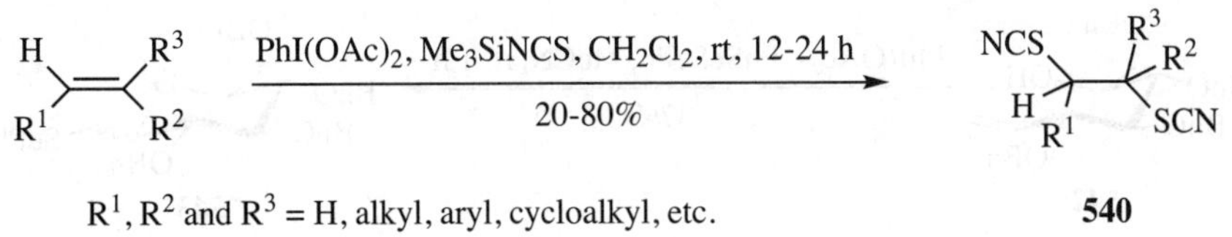

R^1, R^2 and R^3 = H, alkyl, aryl, cycloalkyl, etc.

PhI(OAc)$_2$, Me$_3$SiNCS, CH$_2$Cl$_2$, rt, 12-24 h; 20-80%

540

Scheme 3.212

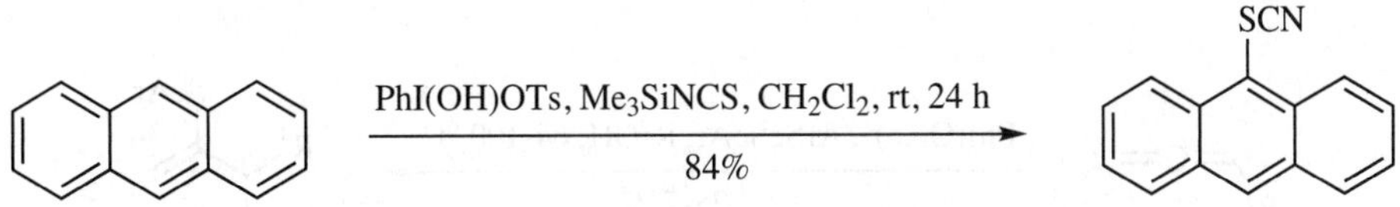

PhI(OH)OTs, Me$_3$SiNCS, CH$_2$Cl$_2$, rt, 24 h; 84%

Scheme 3.213

A similar reaction of (diacetoxyiodo)benzene with alkenes and trimethylsilyl isothiocyanate in dichloromethane affords 1,2-dithiocyanates **540** in moderate yield (Scheme 3.212). Cyclic alkenes, such as cyclohexene and 1-methylcyclohexene, react with this reagent system stereoselectively with the formation of the respective *trans*-adduct [598, 599].

The reaction of polycyclic aromatic hydrocarbons with [hydroxy(tosyloxy)iodo]benzene in the presence of trimethylsilyl isothiocyanate leads to the regioselective thiocyanation of an arene nucleus, as illustrated by the reaction of anthracene shown in Scheme 3.213 [274].

In the 1990s, Tingoli and coworkers developed a general approach to various arylselenated products by the reaction of unsaturated compounds with diaryl diselenides and (diacetoxyiodo)benzene [600–603]. Various phenylselenated products are formed in good yields from the reaction of alkenes with diphenyl diselenide and (diacetoxyiodo)benzene in acetonitrile. In particular, cyclohexene under these conditions stereoselectively affords *trans*-1-acetoxy-2-(phenylseleno)cyclohexane (**541**) in good yield (Scheme 3.214) [603].

Cyclic phenylselenated products are obtained when this reaction is applied to alkenes containing hydroxy, benzamido, enolizable ketones and carboxylic acids as remote functional groups. For example, the alkenol derivative **542** reacts with diphenyl diselenide and (diacetoxyiodo)benzene in acetonitrile to furnish C-glycoside **543** in moderate yield (Scheme 3.215) [603].

Several further modifications of this reaction have been reported more recently [250–252, 604]. In particular, a multicomponent reaction of allenes **544,** diaryl diselenides, (diacetoxyiodo)benzene and alcohols or acids affords 3-functionalized 2-arylselenyl-substituted allyl derivatives **545** in moderate yields (Scheme 3.216) [250].

Alkenes react with (diacetoxyiodo)benzene in the presence of diphenyl diselenide and sodium azide to produce vicinal azido selenides **546** in good yield (Scheme 3.217) [600]. This azidophenylselenation reaction

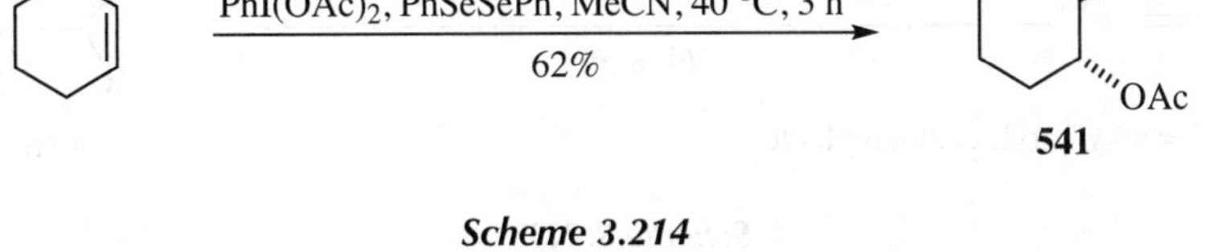

PhI(OAc)$_2$, PhSeSePh, MeCN, 40 °C, 3 h; 62%

541

Scheme 3.214

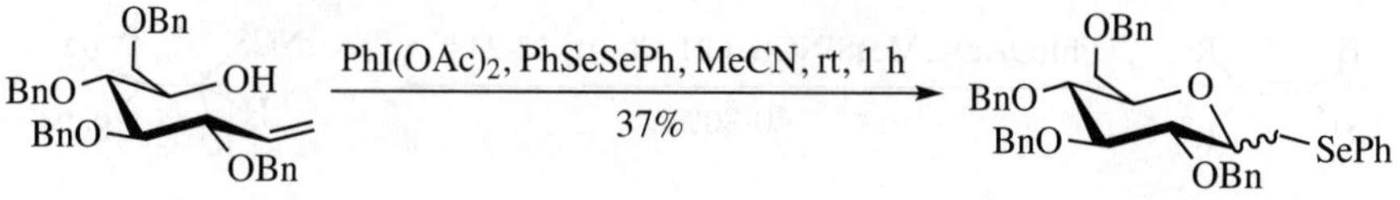

OBn BnO BnO OH OBn 542 — $PhI(OAc)_2$, PhSeSePh, MeCN, rt, 1 h, 37% → OBn BnO BnO O SePh OBn 543

3:1 mixture of diastereoisomers

Scheme 3.215

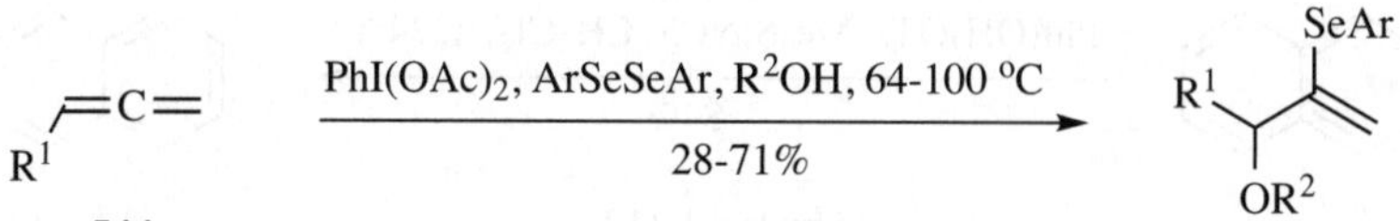

R^1 =C= 544 — $PhI(OAc)_2$, ArSeSeAr, R^2OH, 64-100 °C, 28-71% → SeAr R^1 OR^2 545

R^1 = Ph, 4-MeC_6H_4, 2-MeC_6H_4, 2,6-$Me_2C_6H_3$, α-$C_{10}H_7$
Ar = Ph or 4-MeC_6H_4; R^2 = Me, Et, Pr^i, Bu^t, Ac, C_3H_7CO

Scheme 3.216

proceeds with complete *anti*-Markovnikov regioselectivity, which has been explained by a radical process initiated by the azido radicals.

The *anti*-Markovnikov azidophenylselenation of protected galactals **547** under similar conditions proceeds as a stereospecific *anti*-addition to afford exclusively α-galacto isomer **548,** while the analogous glucals **549** give both *anti*- and *syn*-addition products **550** and **551** (Scheme 3.218) [605].

Nifantiev and coworkers have reported an improved preparative method for homogeneous azidophenylselenylation of glycols by the reaction with (diacetoxyiodo)benzene, diphenyldiselenide and trimethylsilyl azide [251]. In a representative example, the reaction of tri-*O*-benzyl-galactal **552** with $PhI(OAc)_2/Ph_2Se_2/TMSN_3$ in dichloromethane under mild conditions affords the corresponding selenoglycoside **553** in moderate yield (Scheme 3.219) [251]. Non-carbohydrate alkenes, such as styrene and substituted cyclopentenes, can also be azidophenylselenated under these conditions.

The selenodecarboxylation of cinnamic acid derivatives **554** with diaryldiselenides promoted by (diacetoxyiodo)benzene in acetonitrile affords vinyl selenides **555** in moderate yields (Scheme 3.220). A similar reaction of arylpropiolic acids gives the respective alkynyl selenides in 60–90% yields [604].

3.1.18 Radical Fragmentations, Rearrangements and Cyclizations

[Bis(acyloxy)iodo]arenes are commonly used as efficient initiators of radical processes. Under photochemical conditions or heating these reagents undergo decarboxylative decomposition generating alkyl

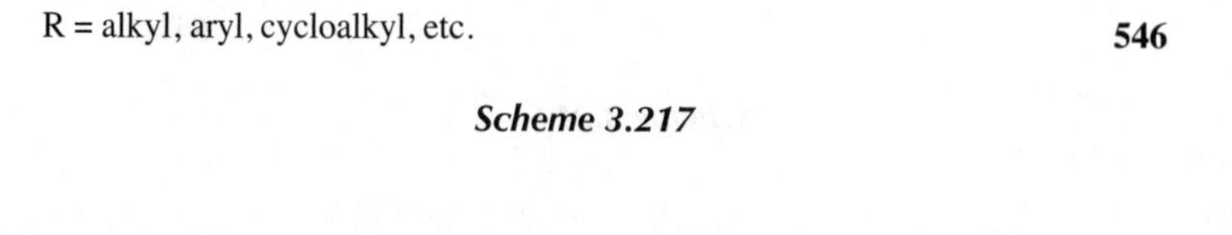

R — $PhI(OAc)_2$, NaN_3, PhSeSePh, CH_2Cl_2, rt, 10-12 h, 68-83% → PhSe R N_3 546

R = alkyl, aryl, cycloalkyl, etc.

Scheme 3.217

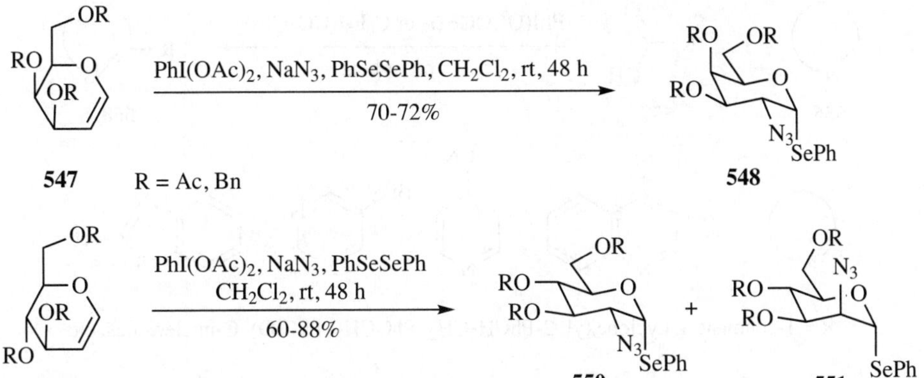

547 R = Ac, Bn

$PhI(OAc)_2$, NaN_3, PhSeSePh, CH_2Cl_2, rt, 48 h

70-72%

548

549 R = Ac, Bn

$PhI(OAc)_2$, NaN_3, PhSeSePh

CH_2Cl_2, rt, 48 h

60-88%

550 + 551

6:4, R = Ac

1:1, R = Bn

Scheme 3.218

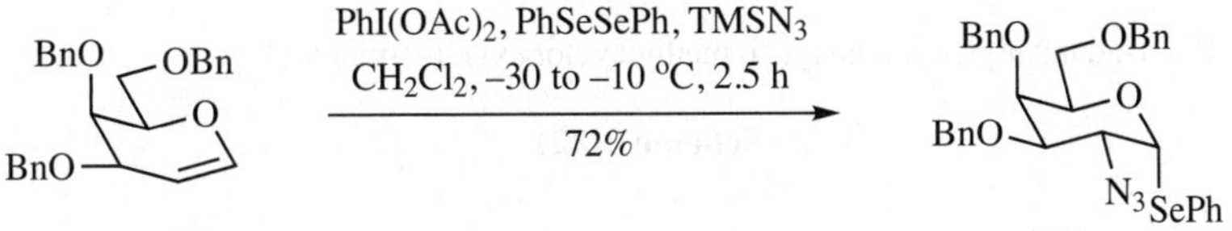

552

$PhI(OAc)_2$, PhSeSePh, $TMSN_3$

CH_2Cl_2, –30 to –10 °C, 2.5 h

72%

553

Scheme 3.219

radicals, which can be effectively trapped with various heteroaromatic bases or electron-deficient alkenes [606–611]. In particular, a convenient experimental procedure for radical alkylation of nitrogen heterocycles **556** to products **558** using carboxylic acids **557** in the presence of [bis(trifluoroacetoxy)iodo]benzene or [bis(trifluoroacetoxy)iodo]pentafluorobenzene has been developed (Scheme 3.221) [606, 608]. This reaction has been used for the preparation of C-nucleosides and their analogs [609]. This procedure has been further modified to allow the use of alcohols as the source of alkyl radicals. In this case, alcohols are first converted into the oxalic acid monoalkyl esters **559,** which are used as reagents in the radical alkylation of heteroaromatic bases as exemplified in Scheme 3.221 [607–609].

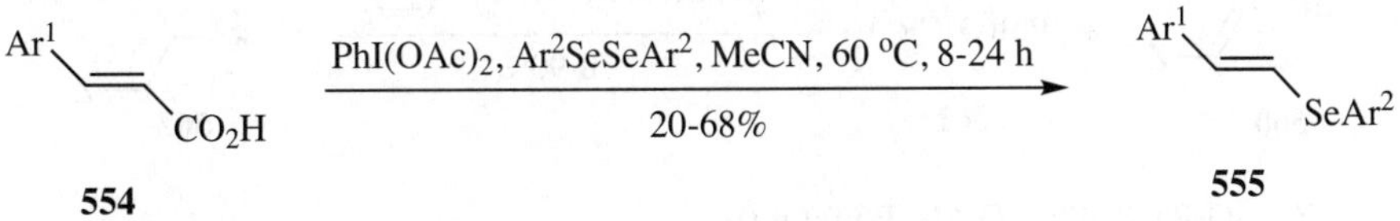

554

$PhI(OAc)_2$, $Ar^2SeSeAr^2$, MeCN, 60 °C, 8-24 h

20-68%

555

Ar^1 = 3,4-$(OCH_2O)C_6H_3$, 4-$MeOC_6H_4$, 3,4,5-$(MeO)_3C_6H_2$

Ar^2 = Ph or 4-MeC_6H_4

Scheme 3.220

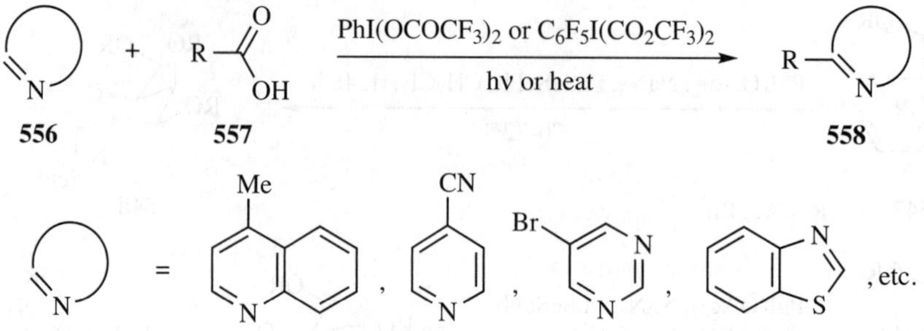

556 + 557 (R–CO₂H) → 558
$PhI(OCOCF_3)_2$ or $C_6F_5I(CO_2CF_3)_2$, hν or heat

R = 1-adamantyl, cyclohexyl, 2-$PhCH_2CH_2$, $PhOCH_2$, PhC(O), C-nucleosides, etc.

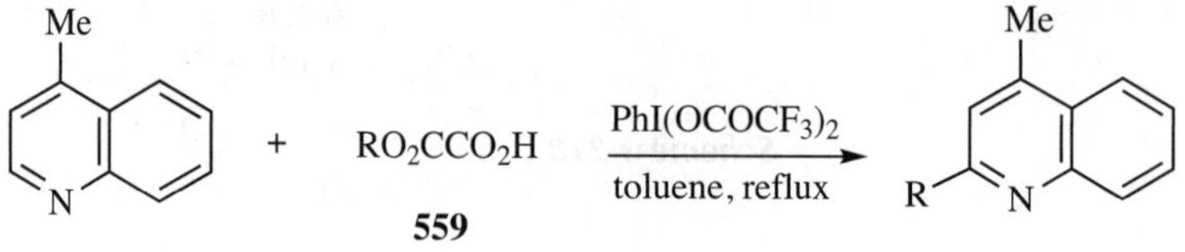

4-methylquinoline + RO_2CCO_2H (559) → 2-R-4-methylquinoline
$PhI(OCOCF_3)_2$, toluene, reflux

R = 1-adamantyl, cyclohexyl, 1-methylcyclohexyl, (–)-menthyl, etc.

Scheme 3.221

Electron-deficient alkenes can be alkylated under similar conditions. In this procedure, a mixture of alkene **560** and [bis(acyloxy)iodo]arenes **561** [prepared from $PhI(OAc)_2$ and the respective carboxylic acid or monoalkyl esters of oxalic acid] is irradiated with a high-pressure mercury lamp in dichloromethane in the presence of 1,4-cyclohexadiene to give a reductive addition product (**562**) (Scheme 3.222) [610,611].

Conjugate addition of radicals generated by decarboxylative fragmentation of (diacyloxyiodo)benzene **564** to dehydroamino acid derivatives **563** has been used in the synthesis of diaminopimelic acid analogues **565** (Scheme 3.223) [612].

Several useful synthetic methodologies are based on the generation of the oxygen-centered radicals from carboxylic acids and the (diacetoxyiodo)benzene–iodine system [613–617]. In particular, a direct conversion of 2-substituted benzoic acids **566** into lactones **567** via oxidative cyclization induced by [bis(acyloxy)iodo]arene/iodine has been reported (Scheme 3.224) [613,614].

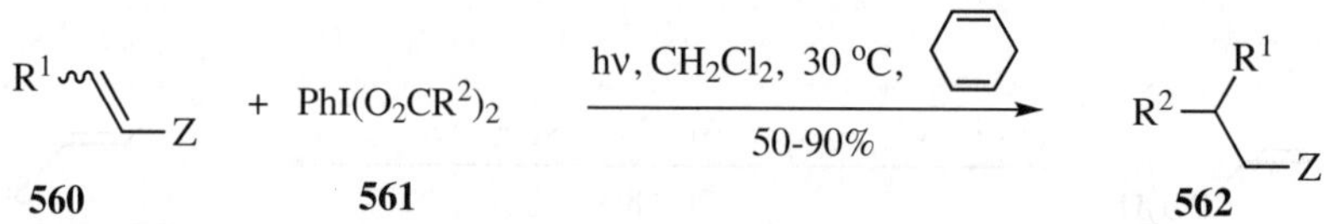

560 + $PhI(O_2CR^2)_2$ (561) → 562
hν, CH_2Cl_2, 30 °C, 1,4-cyclohexadiene; 50-90%

Z = SO_2Ph, SOPh, CO_2Me, $P(O)(OEt)_2$
R^1 = H, Me
R^2 = 1-adamantyl, cycloalkyl, 2-$PhCH_2CH_2$, C-nucleosides, etc.

Scheme 3.222

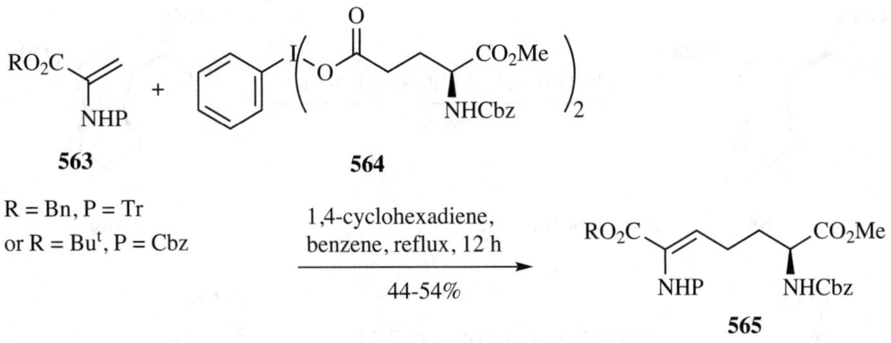

563 + 564

R = Bn, P = Tr
or R = Bu^t, P = Cbz

1,4-cyclohexadiene, benzene, reflux, 12 h
44-54%

565

Scheme 3.223

The reaction of carboxylic acids with the PhI(OAc)$_2$–iodine system may result in a decarboxylation leading to the intermediate formation of a carbon-centered radical, which can be further oxidized to a carbocation and trapped by a nucleophile. This process has been utilized in several syntheses [97, 615,616, 617]. In a typical example, the oxidative decarboxylation of uronic acid derivatives **568** in acetonitrile under mild conditions affords acetates **569** in good yields (Scheme 3.225) [615]. A similar oxidative decarboxylation has been be used for the synthesis of 2-substituted pyrrolidines **571** from the cyclic amino acid derivatives **570** [616,617].

Kita and coworkers have developed a simple and reliable method for the direct construction of biologically important aryl lactones **573** from carboxylic acids **572** using a combination of PhI(OAc)$_2$ with NaBr (Scheme 3.226). The mechanism of this reaction includes initial generation of carbonyloxy radical followed by intramolecular benzylic hydrogen abstraction and cyclization [618].

Adamantyl sulfides **576** have been prepared by radical decarboxylation of [bis(1-adamantanecarboxy)iodo]benzene **575** (Ar = Ph) in the presence of disulfides **574** (Scheme 3.227). A study of the reactivity of various [bis(1-adamantanecarboxy)iodo]arenes **575** in this reaction has shown that the introduction of strong electron-withdrawing groups, such as nitro and perfluoro, into the aromatic ring of **575** leads to a significant reduction of the yield of product **576,** which is explained by the lower reactivity of *p*-nitro and perfluorophenyl derivatives **575** in radical reactions due to the increased I—O bond strength in these compounds [619].

Useful synthetic methodologies are based on the cyclization, rearrangement, or fragmentation of the alkoxy radicals generated in the reaction of alcohols with [bis(acyloxy)iodo]arenes in the presence of iodine under photochemical conditions or in the absence of irradiation. Photolysis of PhI(OAc)$_2$ with cyclic alcohols in the presence of iodine leads to the generation of the respective alkoxy radical, which can undergo various

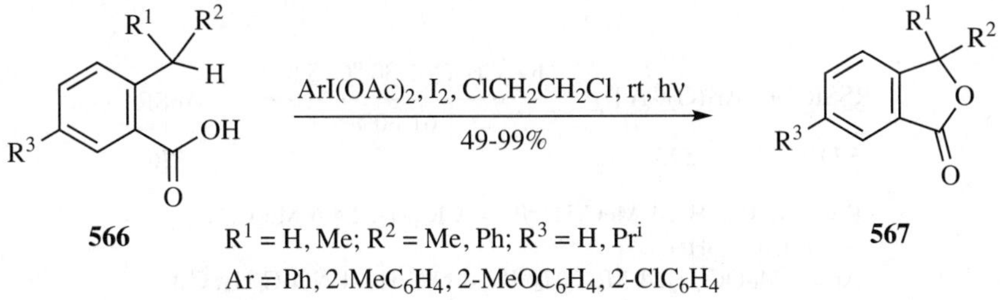

566

ArI(OAc)$_2$, I$_2$, ClCH$_2$CH$_2$Cl, rt, hν
49-99%

567

R^1 = H, Me; R^2 = Me, Ph; R^3 = H, Pr^i
Ar = Ph, 2-MeC$_6$H$_4$, 2-MeOC$_6$H$_4$, 2-ClC$_6$H$_4$

Scheme 3.224

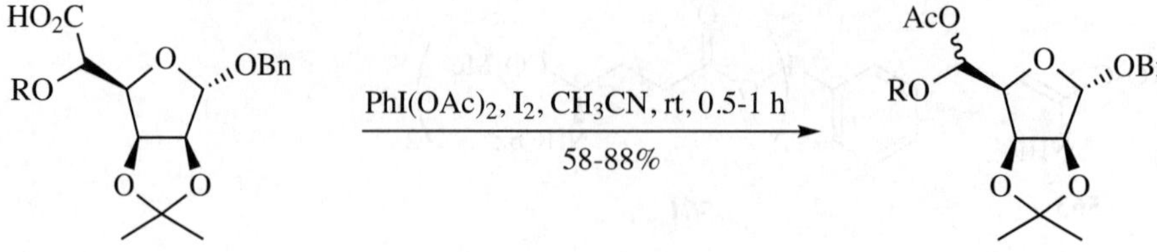

$PhI(OAc)_2$, I_2, CH_3CN, rt, 0.5-1 h; 58-88%

568 R = Me or Bn → 569

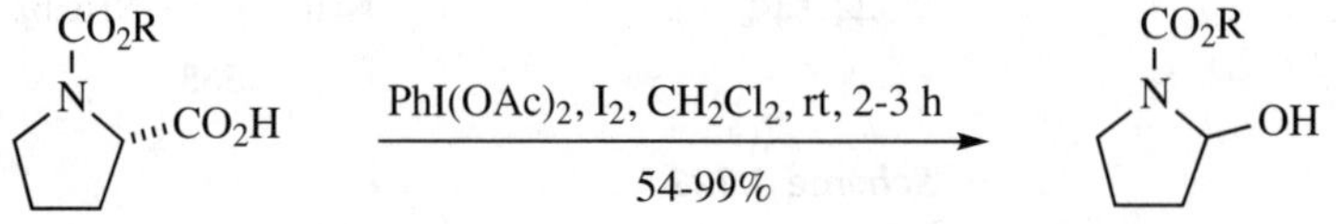

$PhI(OAc)_2$, I_2, CH_2Cl_2, rt, 2-3 h; 54-99%

570 R = Me, Ph, Bn → 571

Scheme 3.225

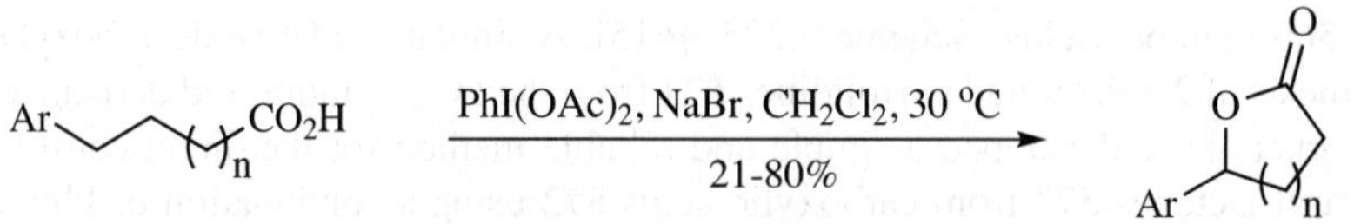

$PhI(OAc)_2$, NaBr, CH_2Cl_2, 30 °C; 21-80%

572 → 573

n = 1-3; Ar = Ph, 4-$MeOC_6H_4$, 4-$NO_2C_6H_4$, 4-FC_6H_4, etc.

Scheme 3.226

sequential reactions. This methodology has been utilized in several syntheses. Suarez and coworkers applied a sequential alkoxy radical fragmentation for the preparation of highly functionalized macrocycles and some steroidal derivatives [620–625]. For example, photolysis of steroidal hemiacetals, such as **577,** with stoichiometric amounts of $PhI(OAc)_2$ and iodine in the absence of oxygen affords medium-sized lactones in good yield as a result of alkoxy radical fragmentation (Scheme 3.228) [621].

When the reaction of cyclic alcohols **578** is conducted under an oxygen atmosphere, the initially produced C-radical traps a molecule of O_2, yielding peroxide **579** according to Scheme 3.229 [622]. This and similar sequential alkoxy radical fragmentations can be applied to the preparation of various synthetically interesting medium-sized ketones and lactones.

RSSR (574) + $ArI(O_2CAd)_2$ (575) → AdSR (576); hν, CH_2Cl_2, 30 °C, 3 h; 61-80%

R = 4-$MeOC_6H_4$, 4-MeC_6H_4, Ph, 4-ClC_6H_4, 2,4,6-$Me_3C_6H_2$, Bu, n-$C_{12}H_{25}$
Ar = 4-$MeOC_6H_4$, 4-MeC_6H_4, Ph, 4-ClC_6H_4, 4-$NO_2C_6H_4$, C_6F_5

Scheme 3.227

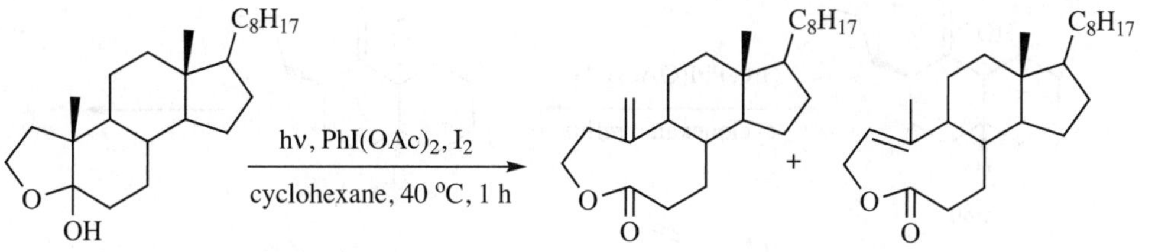

Scheme 3.228

Treatment of bicyclic dienol **580** with $PhI(OAc)_2/I_2$ in degassed cyclohexane under irradiation and reflux results in a cascade radical fragmentation–transannulation–cyclization sequence leading to ketone **581** in high yield (Scheme 3.230) [626].

Additional examples of this methodology include the synthesis of 1,1-difluoro-1-iodo alditols **583** [627], 2-azido-1,2-dideoxy-1-iodo-alditols **585** [628, 629] and chiral vinyl sulfones **587** [630] by fragmentation of carbohydrate anomeric alkoxy radicals generated from the respective carbohydrates **582, 584** and **586** (Scheme 3.231).

The methodology, based on generation of the alkoxy radicals, has also been used for the oxidative cyclization of various alcohols. For example, the irradiation of alcohols **588** with (diacetoxyiodo)benzene and iodine affords the chroman derivatives **589** in moderate to good yields (Scheme 3.232) [631].

Photolysis of $PhI(OAc)_2$ with tertiary allylic alcohols **590** in the presence of iodine affords α-iodoepoxides **591** (50–72% yield) as a result of alkoxy radical rearrangement (Scheme 3.233) [632].

Fragmentation of anomeric alkoxy radicals, generated from appropriate carbohydrates by a similar method, provides a convenient entry into useful chiral synthetic intermediates [633, 634]. Suarez and coworkers have reported the $PhI(OAc)_2$-initiated radical fragmentation of protected furanoses **592** leading to four-carbon chiral building blocks of the erythrose type **593** in good yield (Scheme 3.234) [633]. A similar fragmentation of pyranose derivatives **594** to the product **595** has also been reported. Interestingly, these reactions proceed smoothly at room temperature and do not require irradiation [634].

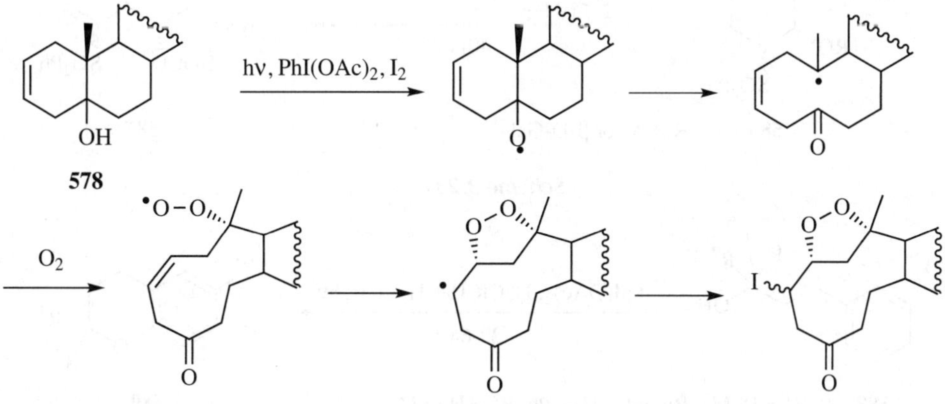

Scheme 3.229

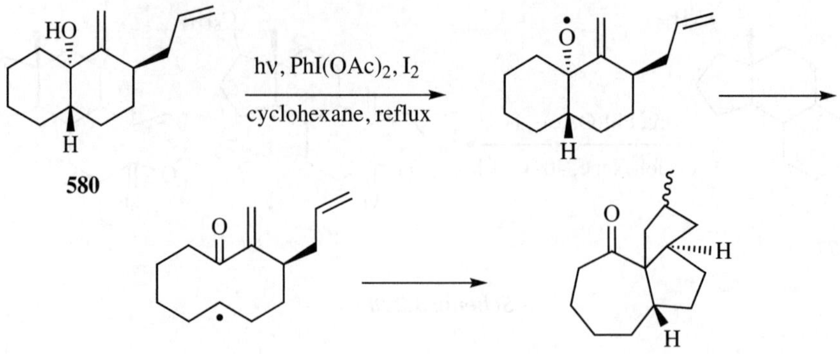

HO
H
580
hν, $PhI(OAc)_2$, I_2
cyclohexane, reflux
O
H
O
O
H
H
581

Scheme 3.230

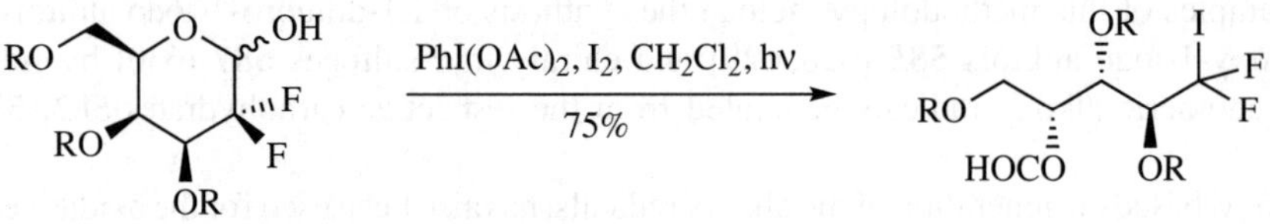

RO
O
OH
F
RO
F
OR
582
R = OAc or OMe
$PhI(OAc)_2$, I_2, CH_2Cl_2, hν
75%
OR
I
F
RO
F
HOCO
OR
583

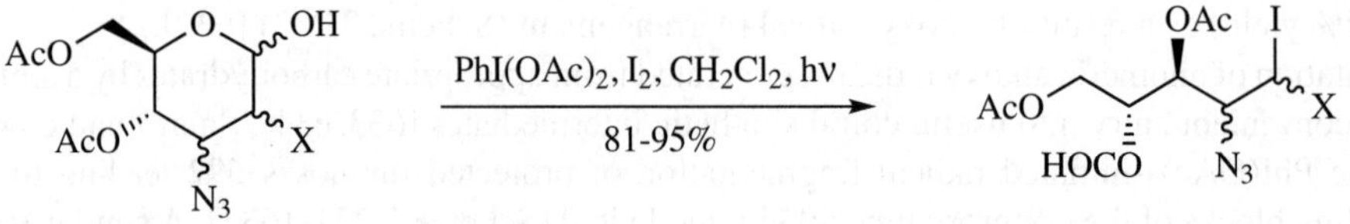

AcO
O
OH
AcO
X
N_3
584
X = H, F, Cl, Br, I
$PhI(OAc)_2$, I_2, CH_2Cl_2, hν
81-95%
OAc
I
AcO
X
HOCO
N_3
585

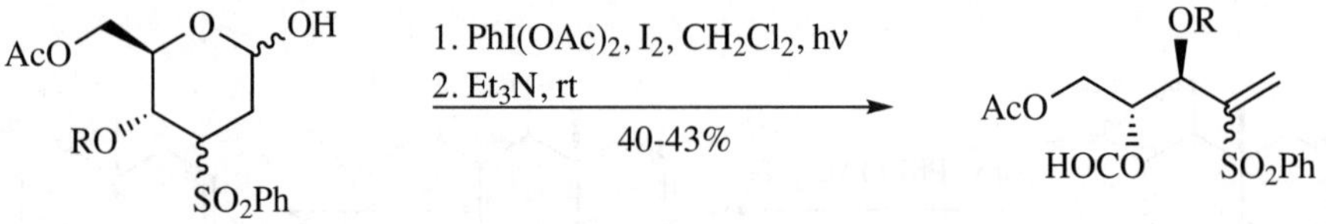

AcO
O
OH
RO
SO_2Ph
586
R = Ac or β-D-Gal
1. $PhI(OAc)_2$, I_2, CH_2Cl_2, hν
2. Et_3N, rt
40-43%
OR
AcO
HOCO
SO_2Ph
587

Scheme 3.231

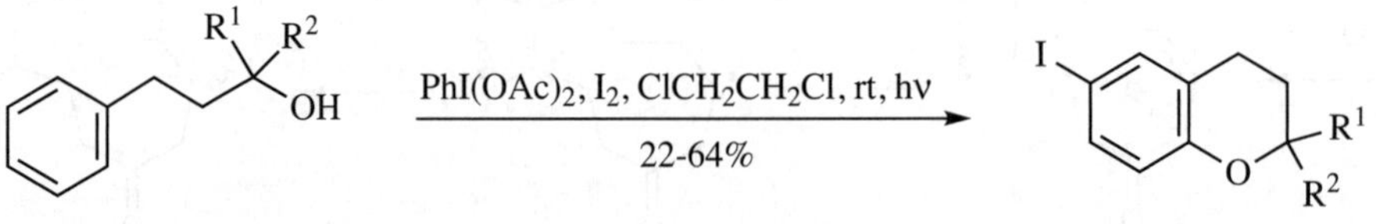

R^1
R^2
OH
588
R^1 = H, Me, Bu, n-$C_{13}H_{27}$, Ph; R^2 = H or Me
$PhI(OAc)_2$, I_2, $ClCH_2CH_2Cl$, rt, hν
22-64%
I
O
R^1
R^2
589

Scheme 3.232

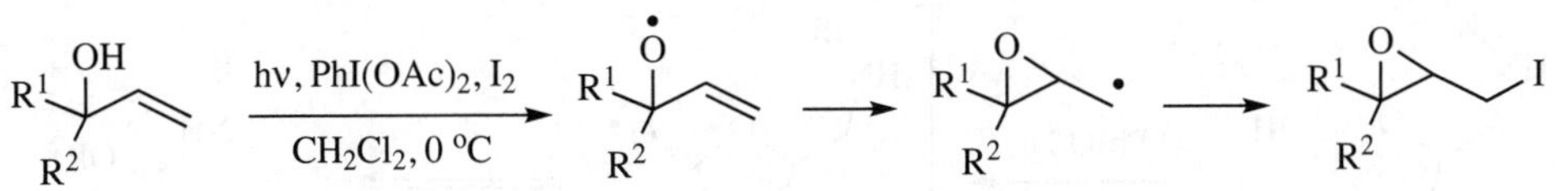

hν, $PhI(OAc)_2, I_2$

CH_2Cl_2, 0 °C

590 $R^1 = Et, n\text{-}C_5H_{11}; R^2 = Me, Et;$ or $R^1 + R^2 = (CH_2)_n$, n = 4, 5, 6, 7 **591**

Scheme 3.233

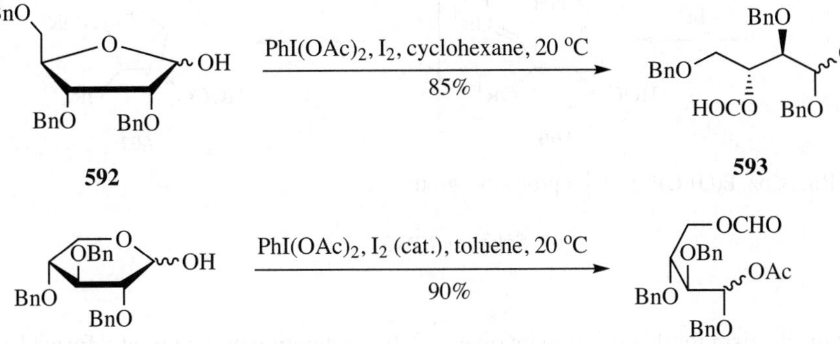

$PhI(OAc)_2, I_2$, cyclohexane, 20 °C

85%

592 **593**

$PhI(OAc)_2, I_2$ (cat.), toluene, 20 °C

90%

594 **595**

Scheme 3.234

Iodosylbenzene can also react with alcohols in the presence of iodine to form the respective alkoxy radicals, presumably through an alkyl hypoiodite intermediate [635]. Suarez and coworkers have developed a valuable synthetic methodology based on a sequential fragmentation of alkoxy radicals generated from alcohols, iodosylbenzene and iodine. This methodology has been applied in the synthesis of carbohydrate derivatives **596** and imino sugars **597** (Scheme 3.235) [636, 637].

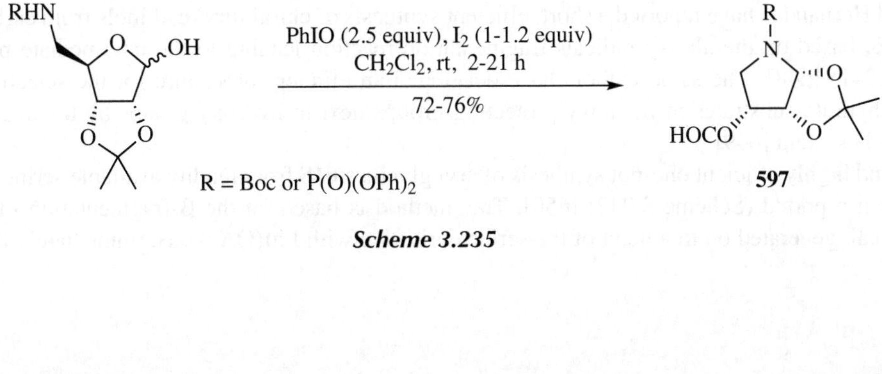

PhIO (2 equiv), I_2 (1 equiv)

CH_2Cl_2, rt, 90 min

90%

596

PhIO (2.5 equiv), I_2 (1-1.2 equiv)

CH_2Cl_2, rt, 2-21 h

72-76%

R = Boc or $P(O)(OPh)_2$ **597**

Scheme 3.235

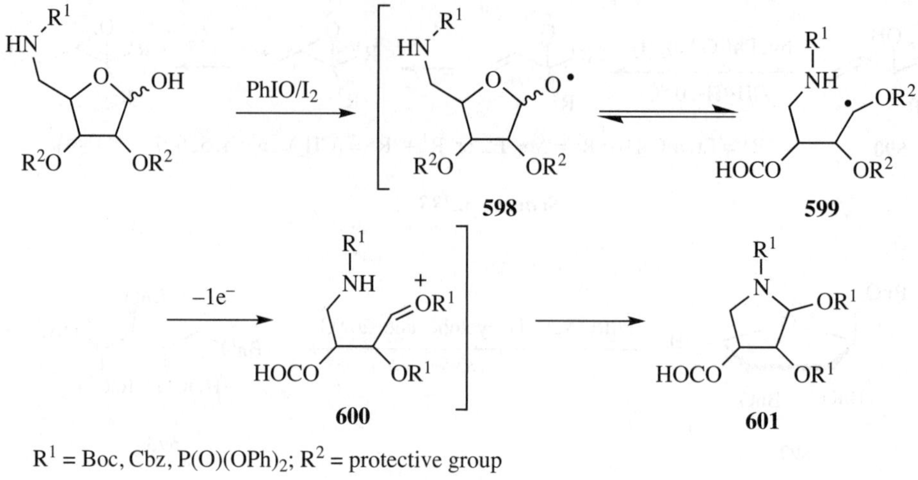

R^1 = Boc, Cbz, P(O)(OPh)$_2$; R^2 = protective group

Scheme 3.236

The proposed mechanism for this reaction involves a β-fragmentation of the initially formed alkoxy radical **598,** leading to a carbon-centered radical **599,** which is further oxidized by an excess of the reagent to the oxycarbenium ion **600** (Scheme 3.236). Intramolecular nucleophilic cyclization of **600** affords the final sugar derivative **601** [636].

A similar methodology has been applied in the synthesis of various useful chiral synthetic intermediates. Representative examples include the syntheses of alduronic acid lactone **603** [637,638], chiral dispiroacetals **605** and **606** [639] and α-iodoalkyl ester **608** from respective carbohydrate derivatives **602, 604** and **607** (Scheme 3.237) [640].

The oxidative cyclization of steroidal bromohydrins **609** selectively affords products **610** in high yields (Scheme 3.238) [641]. This reaction can be promoted by either photolysis or ultrasonic irradiation. The yields of products **610** are significantly better under the ultrasound-assisted conditions.

A similar oxidative cyclization initiated by the irradiation of a substrate in the presence of (diacetoxyiodo)benzene and iodine can be used for the deprotection of benzyl ethers (e.g., **611**) situated next to the hydroxyl in the α, β, or γ-position [642]. Depending on the substrate, the corresponding cyclic ethers of diols (such as **612**) can be isolated (Scheme 3.239).

The intramolecular hydrogen abstraction reactions promoted by alkoxy radicals in carbohydrates are particularly useful for the stereoselective synthesis of various polycyclic oxygen-containing ring systems [643–647]. This reaction can be illustrated by the intramolecular 1,8-hydrogen abstraction between glucopyranose units in disaccharide **613** promoted by alkoxy radicals and leading to the 1,3,5-trioxocane derivative **614** (Scheme 3.240) [644].

Boto and Hernandez have reported a short, efficient synthesis of chiral furyl carbinols from carbohydrates, such as **615,** based on the alkoxy radicals fragmentation reaction leading to the intermediate product **616** (Scheme 3.241) [648]. The same authors have developed an efficient procedure for the selective removal from carbohydrate substrates of methoxy protecting groups next to hydroxy groups by treatment with the PhI(OAc)$_2$–I$_2$ system [649].

A mild and highly efficient one-pot synthesis of aryl glycines **618** from readily available serine derivatives **617** has been reported (Scheme 3.242) [650]. This method is based on the β-fragmentation of a primary alkoxy radical, generated on treatment of the serine derivative with PhI(OAc)$_2$–I$_2$, immediately followed by

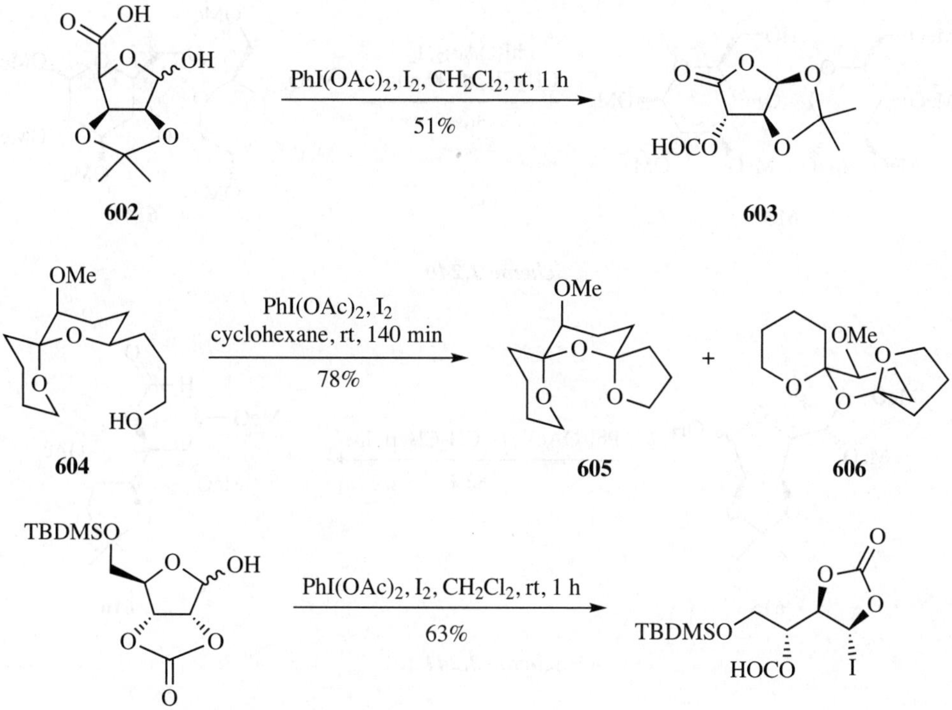

$PhI(OAc)_2$, I_2, CH_2Cl_2, rt, 1 h
51%

602 **603**

$PhI(OAc)_2$, I_2
cyclohexane, rt, 140 min
78%

604 **605** **606**

$PhI(OAc)_2$, I_2, CH_2Cl_2, rt, 1 h
63%

607 **608**

Scheme 3.237

$PhI(OAc)_2$, I_2, cyclohexane, 45 °C
35 KHz ultrasound, 50 min
86-99%

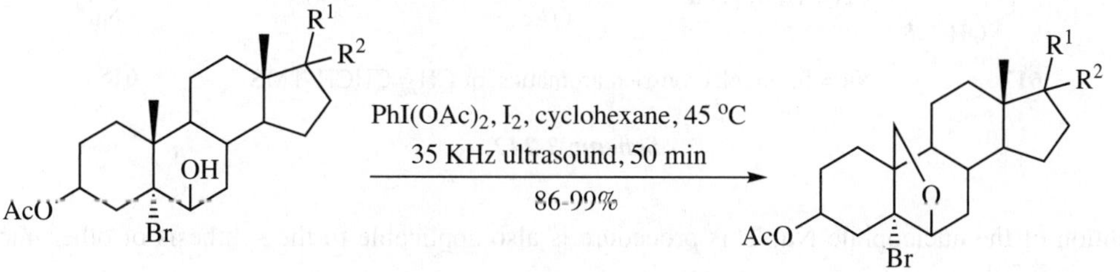

609 $R^1 = n\text{-}C_8H_{17}$, COMe, $COCH_2OAc$; $R^2 = H$ or $R^1 + R^2 = O$ **610**

Scheme 3.238

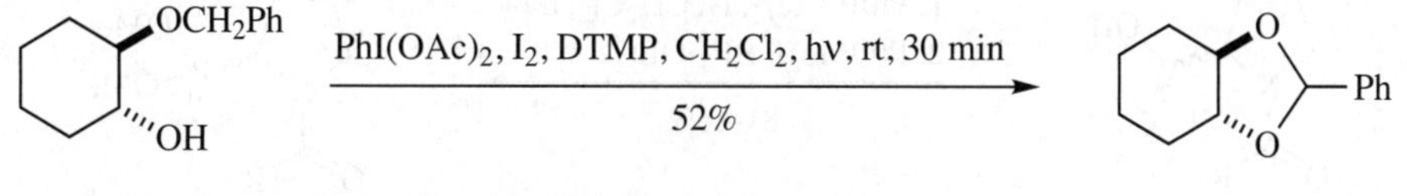

$PhI(OAc)_2$, I_2, DTMP, CH_2Cl_2, hν, rt, 30 min
52%

611 DTMP = 2,6-di-*tert*-butyl-4-methylpyridine **612**

Scheme 3.239

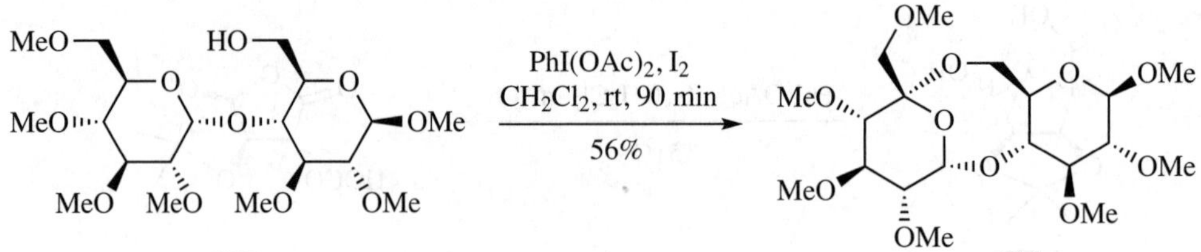

MeO HO O O MeO O O OMe MeO MeO MeO OMe

$PhI(OAc)_2$, I_2
CH_2Cl_2, rt, 90 min
56%

OMe O O OMe MeO O O OMe MeO OMe OMe

613 614

Scheme 3.240

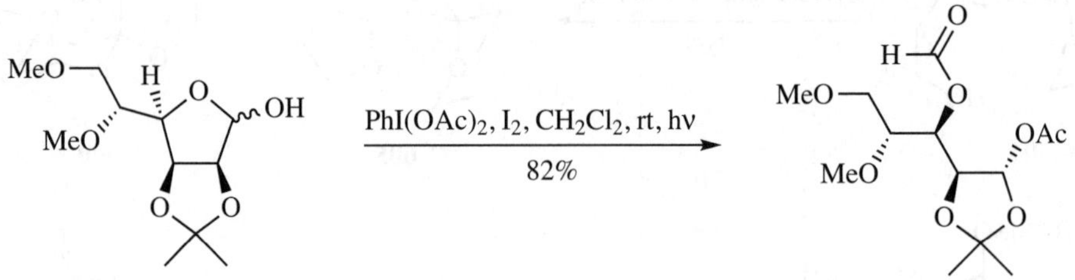

MeO H O OH MeO O O

$PhI(OAc)_2$, I_2, CH_2Cl_2, rt, hν
82%

O H O MeO O OAc MeO O O

615 616

Scheme 3.241

Ph O HN CO_2Me OH

$PhI(OAc)_2$, I_2
CH_2Cl_2, rt, 2.5 h

Ph O HN CO_2Me OAc

Nu, BF_3•OEt_2
63-90%

Ph O HN CO_2Me Nu

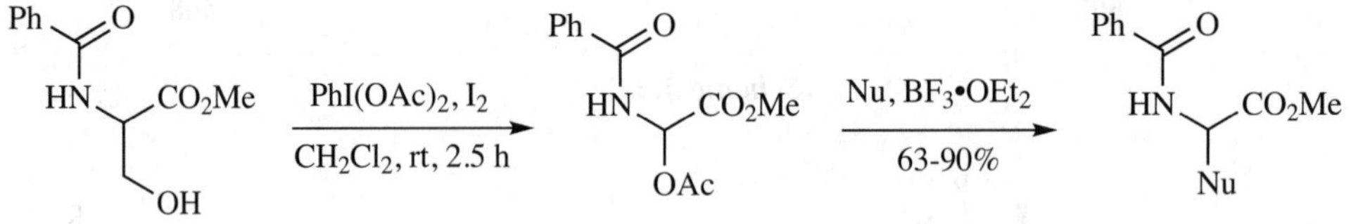

617 Nu = furan, elecron-rich aromatics, or CH_2=$CHCH_2TMS$ 618

Scheme 3.242

the addition of the nucleophile Nu. This procedure is also applicable to the synthesis of other uncommon amino acids [650].

The one-pot radical fragmentation–phosphorylation reaction of α-amino acids or β-amino alcohols (e.g., **619**) affords α-amino phosphonates **620** in good yields (Scheme 3.243). This reaction has been applied to the synthesis of biologically active phosphonates [651].

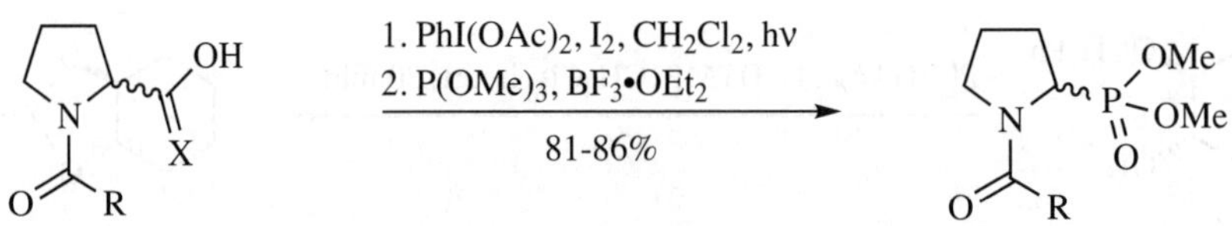

OH N X O R

1. $PhI(OAc)_2$, I_2, CH_2Cl_2, hν
2. $P(OMe)_3$, BF_3•OEt_2
81-86%

OMe N P OMe O O R

619 R = Ph or OMe; X = 2H or O 620

Scheme 3.243

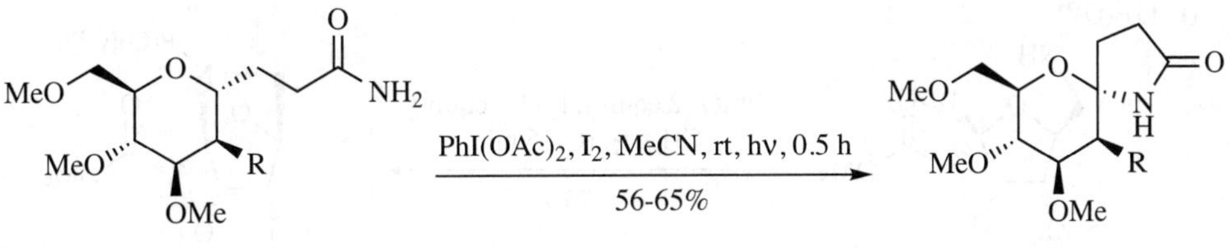

PhI(OAc)$_2$, I$_2$, MeCN, rt, hν, 0.5 h

56-65%

621 R = H or OMe **622**

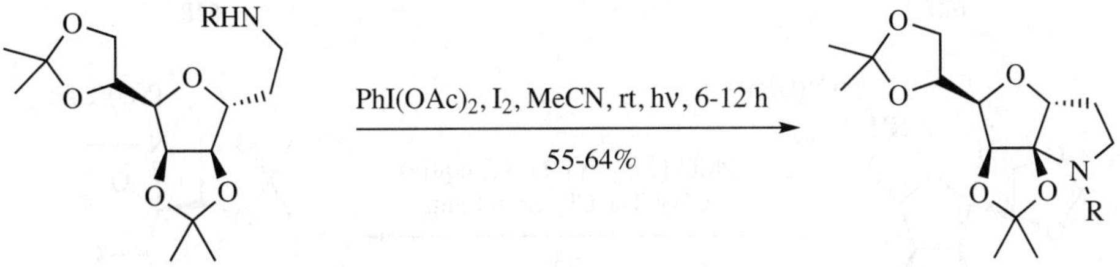

PhI(OAc)$_2$, I$_2$, MeCN, rt, hν, 6-12 h

55-64%

623 R = Boc or PO(OPh)$_2$ **624**

Scheme 3.244

Useful synthetic methodologies are based on the cyclization or rearrangement of the nitrogen-centered radicals generated in the reaction of the appropriate amides with (diacetoxyiodo)benzene in the presence of iodine [652–655]. Specific examples are illustrated by the synthesis of bicyclic spirolactams **622** from amides **621** [653] and preparation of the oxa-azabicyclic systems (e.g., **624**) by the intramolecular hydrogen atom transfer reaction promoted by carbamoyl and phosphoramidyl radicals generated from the appropriately substituted carbohydrates **623** (Scheme 3.244) [654].

Togo, Yokoyama and coworkers have developed a useful synthetic procedure for the preparation of nitrogen heterocycles based on the N-radical cyclization onto an aromatic ring [247, 656–658]. For example, various *N*-alkylsaccharins **626** can be conveniently prepared in moderate to good yields by the reaction of arenesulfonamides **625** with (diacetoxyiodo)benzene in the presence of iodine under irradiation with a tungsten lamp (Scheme 3.245) [656]. A similar procedure has been applied to the synthesis of 1,2,3,4-tetrahydroquinoline derivatives [247, 657] and 3,4-dihydro-2,1-benzothiazine 2,2-dioxides [658].

The methodology based on the generation of N-centered radicals from the reaction of amides with iodosylbenzene and iodine has been utilized in the synthesis of homochiral 7-oxa-2-azabicyclo[2.2.1]heptane

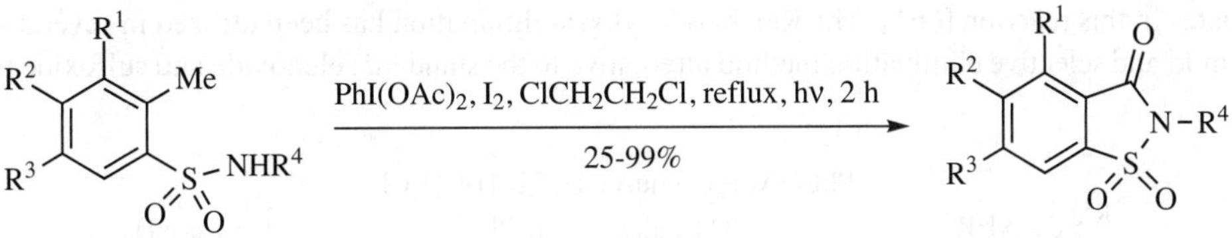

PhI(OAc)$_2$, I$_2$, ClCH$_2$CH$_2$Cl, reflux, hν, 2 h

25-99%

625 **626**

R^1, R^2 = H or Me; R^3 = H, Me, Br, Bu^t, CONEt$_2$, NO$_2$, SO$_2$CH$_3$; R^4 = H, Me, Et, Pr, Bu, CH$_2$Ph

Scheme 3.245

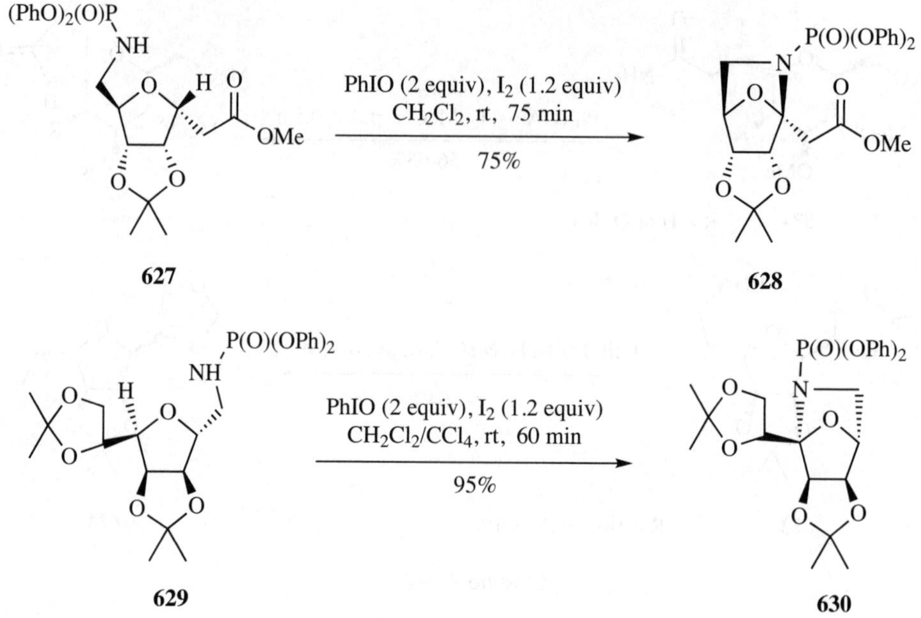

(PhO)$_2$(O)P NH O H O OMe PhIO (2 equiv), I_2 (1.2 equiv) CH_2Cl_2, rt, 75 min 75% P(O)(OPh)$_2$ N O O OMe O O O O

627 628

P(O)(OPh)$_2$ NH O H O O O O PhIO (2 equiv), I_2 (1.2 equiv) CH_2Cl_2/CCl_4, rt, 60 min 95% P(O)(OPh)$_2$ N O O O O O O

629 630

Scheme 3.246

derivatives **628** and **630** from the respective phosphoramidate derivatives of carbohydrates **627** and **629** (Scheme 3.246) [659].

Under conditions of ultrasonic irradiation in the presence of (diacetoxyiodo)benzene and iodine, *N*-alkylsulfonamides **631** are dealkylated to afford sulfonamides **632** in moderate to good yields (Scheme 3.247) [660].

3.1.19 Reactions via Alkyliodine(III) Intermediates

Alkyliodides can be readily oxidized with various oxidizing agents to afford either products of elimination or the products of oxidatively assisted nucleophilic substitution of iodine. Both pathways involve alkyliodine(III) derivatives as reactive intermediates. The elimination pathway generally occurs in reactions of alkyliodides with peracids in a non-nucleophilic solvent, such as dichloromethane [661–663]. Reich and Peake first demonstrated that the elimination proceeds with *syn*-stereochemistry and also proposed iodosylalkanes as reactive intermediates in this reaction [661]. The Reich iodosyl *syn*-elimination has been utilized in several syntheses as a very mild and selective elimination method alternative to the standard selenoxide and sulfoxide protocols.

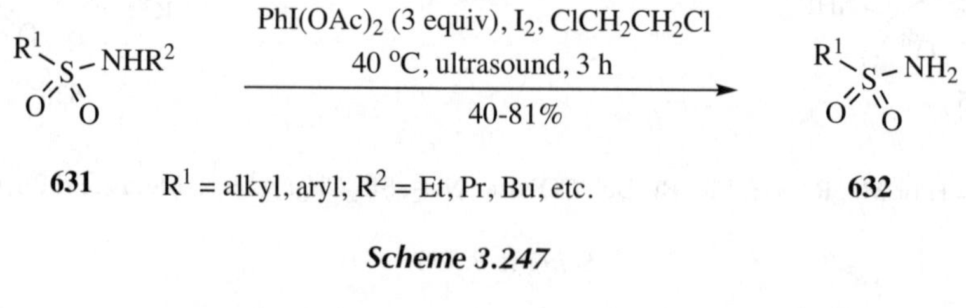

R^1–$S(O)_2$–NHR^2 PhI(OAc)$_2$ (3 equiv), I_2, $ClCH_2CH_2Cl$ 40 °C, ultrasound, 3 h 40-81% R^1–$S(O)_2$–NH_2

631 R^1 = alkyl, aryl; R^2 = Et, Pr, Bu, etc. **632**

Scheme 3.247

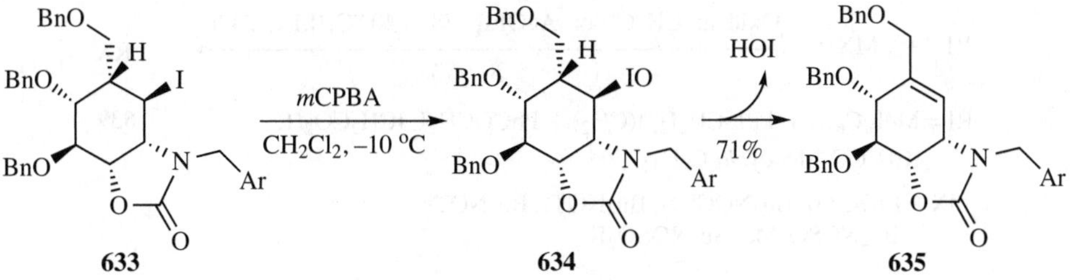

Scheme 3.248

The iodosyl elimination has been used in the preparation of unsaturated oxazolidinone **635** (Scheme 3.248), the key intermediate in the synthesis of valienamine [662]. This reaction proceeds via oxidation of iodide **633** to the intermediate iodosyl compound **634,** which spontaneously eliminates HOI to afford product **635.** The Reich iodosyl *syn*-elimination has also been used for the preparation of intermediate steroidal units of cephalostatin 7 [663].

The second pathway of oxidative deiodination of alkyliodides involves nucleophilic substitution of the oxidized iodine moiety. In fact, functionalities containing hypervalent iodine (e.g., $-IX_2$) belong to the best leaving groups [664]. For example, the normally unreactive 1-iodonorbornane can be readily transformed into 1-bromonorbornane by reaction with bromine via the intermediate $RIBr_2$ [664]. Likewise, the bridgehead 1-fluoronorbornanes **638** have been prepared by the analogous fluorodeiodination of 1-iodonorbornanes **636** by reaction with xenon difluoride (Scheme 3.249) [665]. The mechanism of this reaction involves initial formation of unstable difluoroiodides **637** and their subsequent decomposition with elimination of the IF fragment.

Even extremely weak nucleophiles, such as perchlorate and triflate anions, can participate in the oxidatively assisted nucleophilic substitution of iodine [666–668]. The oxidation of alkyliodides in the presence of lithium perchlorate or appropriate tetrabutylammonium salts in non-nucleophilic solvents affords the respective alkyl perchlorates or alkylsulfonates **639** as principal products (Scheme 3.250) [669].

The oxidatively assisted nucleophilic substitution of iodine with perchlorate anion in primary and secondary substrates proceeds according to an S_N2 mechanism as indicated by the stereochemistry of this reaction in the polycyclic substrates **640** leading to products **641** with inverted configuration (Scheme 3.251) [668].

Treatment of alkyliodides with peracids in the presence of water results in oxidatively assisted hydrolysis and the formation of alcohols [661]. This reaction can be used as a synthetic method for the mild hydrolysis of iodides; for example, iodide **642** undergoes facile stereoselective hydrolysis to the protected amino alcohol **643** under neutral conditions upon exposure to ten equivalents of buffered peroxytrifluoroacetic acid (Scheme 3.252) [670]. The oxidatively assisted hydrolysis of iodide **642** results in a complete inversion of configuration, which is consistent with the S_N2 mechanism of substitution in the hypervalent iodine intermediate [670].

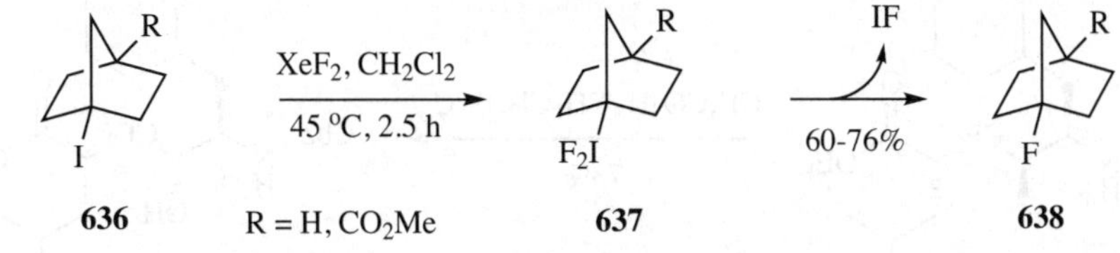

Scheme 3.249

$$RI + MX \xrightarrow[30\text{-}96\%]{\text{Oxidant, } CH_2Cl_2 \text{ or AcOEt, }-78 \text{ to } 40\,^\circ C,\ 0.1 \text{ to } 24 \text{ h}} RX$$

639

RI = MeI, $C_6H_{13}I$, Pr^iI, CH_2I_2, $I(CH_2)_6I$, $PhCOCH_2I$, ICH_2CO_2H, ICH_2CO_2Me, *cyclo*-$C_6H_{11}I$

MX = $LiOClO_3$, Bu_4NOClO_3, Bu_4NOTf, Bu_4NOTs, Bu_4NOSO_2Me, Bu_4NOSO_2F

Oxidant = Cl_2, Br_2, H_5IO_6, NO_2BF_4, *m*CPBA, $PhI(OCOCF_3)_2$, PhI(OH)OTs

Scheme 3.250

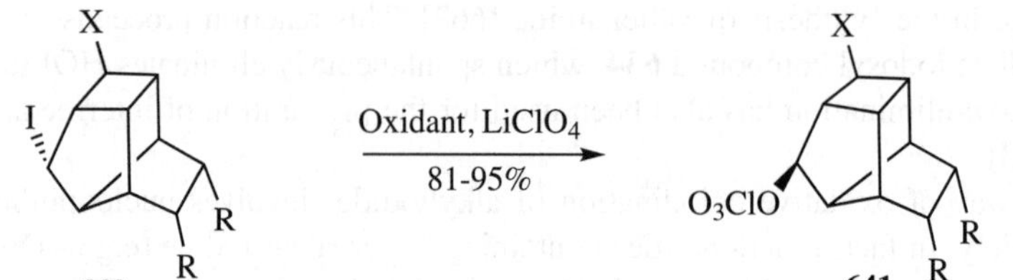

640 → (Oxidant, $LiClO_4$; 81-95%) → **641**

X = Br, 2,4-$(NO_2)_2C_6H_3S$, 2,4-$(NO_2)_2C_6H_3SO_2$, 2,4-$(NO_2)_2C_6H_3SO$
R = Me, CO_2Me
Oxidant = Cl_2, NO_2BF_4, $PhI(OCOCF_3)_2$

Scheme 3.251

3.1.20 Transition Metal Catalyzed Oxidations

Transition metals have a strong catalytic effect on oxidations with hypervalent iodine reagents. In 1979 Groves and coworkers found that iodosylbenzene is the most efficient source of oxygen in the oxygenation of hydrocarbons in the presence of iron(III) porphyrin complexes [671] and since then iodosylarenes and other hypervalent iodine reagents have been widely used as stoichiometric oxidants in reactions mimicking natural oxidations performed by the heme-containing cytochrome P-450 class of enzymes [672–686].

Iodosylbenzene has been used as an effective oxidant in hydrocarbon hydroxylation catalyzed by metalloporphyrins [687–696]. In particular, various iron(III) and manganese(III) porphyrins can be used as catalysts in hydroxylations of cyclohexane, cyclohexene, adamantane and aromatic hydrocarbons [687, 688, 692]. Breslow and coworkers have reported regioselective hydroxylations of several steroidal derivatives catalyzed by metalloporphyrins [689–691]. In a specific example androstanediol derivative **644** was

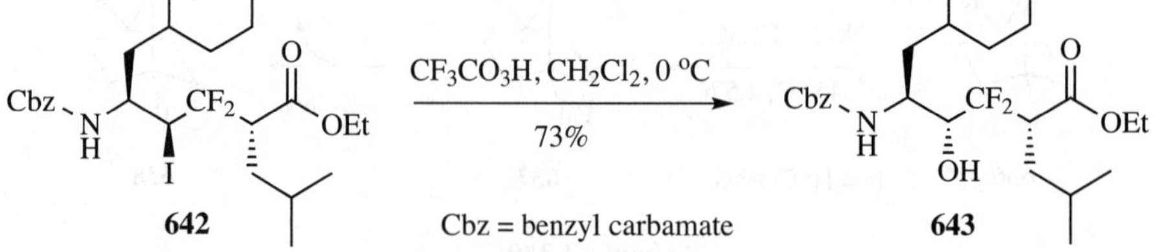

642 → (CF_3CO_3H, CH_2Cl_2, 0 °C; 73%) → **643**

Cbz = benzyl carbamate

Scheme 3.252

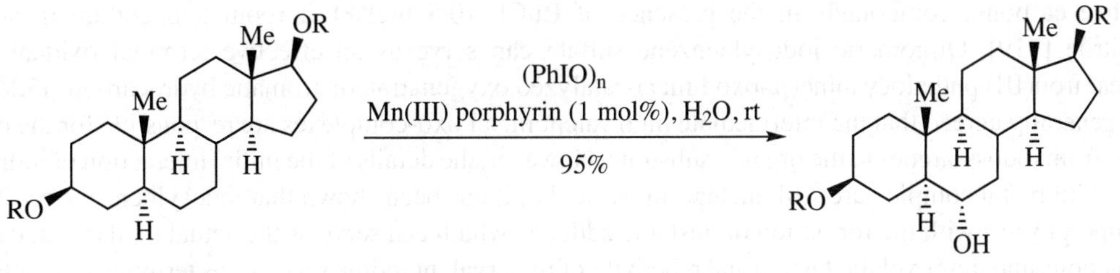

644 R = protective group

$(PhIO)_n$, Mn(III) porphyrin (1 mol%), H_2O, rt, 95%

Scheme 3.253

hydroxylated at the 6α carbon with complete positional selectivity in the presence of a manganese(III) porphyrin catalyst (Scheme 3.253) [689]. Presumably, this selective hydroxylation is directed by the geometry of the catalyst–substrate complex, as in the enzyme.

Transition metal complexes can effectively catalyze the epoxidation of alkenes using iodosylarenes as the source of oxygen atom [266, 697–701]. In particular, a highly enantioselective alkene epoxidation catalyzed by chiral non-racemic chromium salen complexes has been reported [701–703]. Iodosylbenzene was found to be the only applicable oxidant in these reactions and the highest ee (92%) was achieved in the epoxidation of (*E*)-β-methylstyrene **645** mediated by a chromium salen complex in stoichiometric mode and in the presence of Ph_3PO as a donor ligand (Scheme 3.254) [701]. Carrying out this reaction in catalytic mode (5–10 mol% of chromium complex), or the use of other substituted salen ligands, results in a slightly lower enantioselectivity.

Epoxidation of alkenes with iodosylbenzene can be effectively catalyzed by the analogous salen or chiral Schiff base complexes of manganese(III), ruthenium(II), or ruthenium(III). For example, the oxidation of indene with iodosylbenzene in the presence of (*R,S*)-Mn-salen complexes as catalysts affords the respective (1*S*,2*R*)-epoxyindane in good yield with 91–96% ee [704]. Additional examples include epoxidation of alkenes with iodosylbenzene catalyzed by various metalloporphyrins [705–709], corrole metal complexes, ruthenium-pyridinedicarboxylate complexes of terpyridine and chiral bis(oxazolinyl)pyridine [710, 711].

Iodosylbenzene in the presence of transition metal catalysts can effectively oxidize alcohols to carbonyl compounds [712–714] and organic sulfides to sulfoxides [715–717].

Iodosylarenes other than iodosylbenzene have also been used in the transition metal catalyzed oxidation reactions. The soluble *o*-(*tert*-butylsulfonyl)iodosylbenzene (Section 2.1.4) can serve as an alternative to iodosylbenzene in (porphyrin)manganese(III)-catalyzed alkene epoxidation reactions [718]. A convenient recyclable reagent, *m*-iodosylbenzoic acid, can selectively oxidize primary and secondary alcohols to the

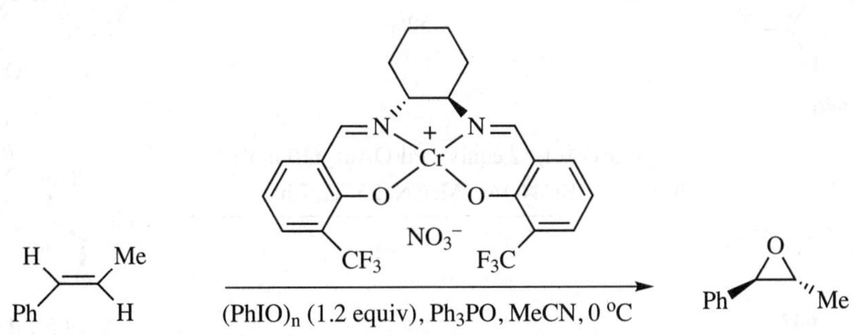

645

NO_3^-; $(PhIO)_n$ (1.2 equiv), Ph_3PO, MeCN, 0 °C

92% ee

Scheme 3.254

respective carbonyl compounds in the presence of $RuCl_3$ (0.5 mol%) at room temperature in aqueous acetonitrile [719]. Oligomeric iodosylbenzene sulfate can serve as an effective terminal oxidant in the binuclear iron(III) phthalocyanine(μ-oxodimer)-catalyzed oxygenation of aromatic hydrocarbons [720,721].

It is generally agreed that the intermediate high valent metal oxo-complexes are responsible for the oxygen transfer from iodosylarenes to the organic substrate. However, the details of the initial interaction of iodine(III) species with metal complex are still unclear. In particular, it has been shown that iodosylbenzene reacts with metalloporphyrins with the formation of unstable adducts, which can serve as the actual oxidants in catalytic oxygenations and thus explain the unusual reactivity of hypervalent iodine reagents as terminal oxidants [717, 722–725].

Various oxidations with [bis(acyloxy)iodo]arenes are also effectively catalyzed by transition metal salts and complexes [726]. (Diacetoxyiodo)benzene is occasionally used instead of iodosylbenzene as the terminal oxidant in biomimetic oxygenations catalyzed by metalloporphyrins and other transition metal complexes [727–729]. Primary and secondary alcohols can be selectively oxidized to the corresponding carbonyl compounds by $PhI(OAc)_2$ in the presence of transition metal catalysts, such as $RuCl_3$ [730–732], Ru(Pybox)(Pydic) complex [733], polymer-micelle incarcerated ruthenium catalysts [734], chiral-Mn(salen)-complexes [735, 736], Mn(TPP)CN/Im catalytic system [737] and (salen)Cr(III) complexes [738]. The epoxidation of alkenes, such as stilbenes, indene and 1-methylcyclohexene, using (diacetoxyiodo)benzene in the presence of chiral binaphthyl ruthenium(III) catalysts (5 mol%) has also been reported; however, the enantioselectivity of this reaction was low (4% ee) [739].

The mechanisms and applications of palladium-catalyzed reactions of (diacetoxyiodo)benzene and other hypervalent iodine reagents in synthetically useful organic transformations have been reviewed by Deprez and Sanford [740]. Particularly useful are the Pd-catalyzed oxidation reactions, including the oxidative functionalization of C—H bonds and the 1,2-aminooxygenation of olefinic substrates [740–756]. Representative examples of these catalytic oxidations are illustrated by the selective acetoxylation of C—H bonds adjacent to coordinating functional groups (e.g., pyridine in substrate **646**) [741] and by the $Pd(OAc)_2$-catalyzed intramolecular aminoacetoxylation in the reaction of γ-aminoolefins (e.g., cinnamyl alcohol derived tosyl carbamate **647**) with $PhI(OAc)_2$ (Scheme 3.255) [742]. A key mechanistic step in these catalytic transformations is the hypervalent iodine-promoted oxidation of Pd(II) to the Pd(IV) species, as proved by the isolation and X-ray structural identification of stable Pd(IV) complexes prepared by the reaction of $PhI(O_2CPh)_2$ with Pd(II) complexes containing chelating 2-phenylpyridine ligands [753]. Several examples of Pd-catalyzed chlorinations of organic substrates using (dichloroiodo)benzene have also been reported [754, 755].

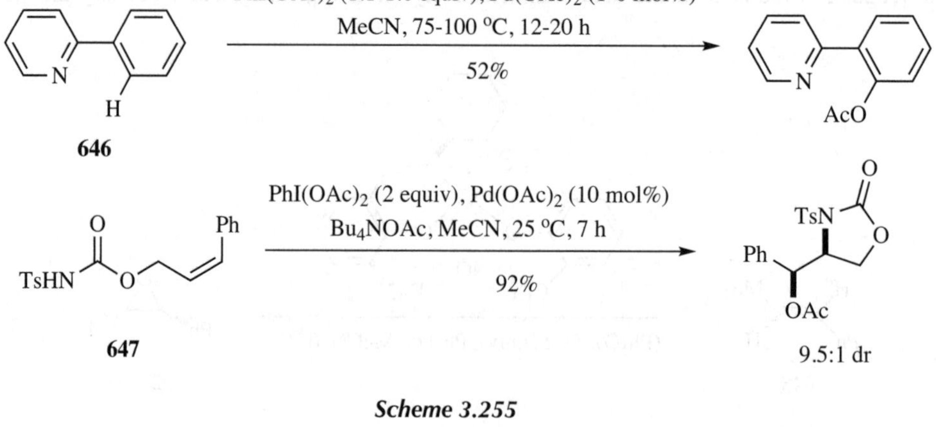

$PhI(OAc)_2$ (1.1-1.6 equiv), $Pd(OAc)_2$ (1-6 mol%)
MeCN, 75-100 °C, 12-20 h
52%

N H **646** → N AcO

$PhI(OAc)_2$ (2 equiv), $Pd(OAc)_2$ (10 mol%)
Bu_4NOAc, MeCN, 25 °C, 7 h
92%

O TsHN O Ph **647** → O TsN O Ph OAc 9.5:1 dr

Scheme 3.255

ArH $\xrightarrow[\text{35-82\%}]{\text{PhI(OAc)}_2\text{, AuCl}_3\text{ (2-5 mol\%), ClCH}_2\text{CH}_2\text{Cl, 110 °C}}$ ArOAc

648 ArH = 1,3,5-Me_3C_6H; 1,2,3,4,5-Me_5C_6H; 1,2,4,5-$Me_5C_6H_2$; $MeOC_6H_5$; 1,4-$(MeO)_2C_6H_4$; 2-Br,1,3-MeC_6H_3; 1-F,4-$MeOC_6H_4$, etc.

Scheme 3.256

Hypervalent iodine oxidations of organic substrates can be effectively catalyzed by salts and complexes of gold [757–760]. For example, direct acetoxylation of electron-rich aromatic compounds **648** can be performed with (diacetoxyiodo)benzene in the presence of catalytic amounts of $AuCl_3$ (Scheme 3.256) [757].

3.1.21 Transition Metal Catalyzed Aziridinations and Amidations

Imidoiodanes and especially *N*-tosyliminoiodanes, ArINTs (Section 2.1.12.4), have found broad synthetic application as useful nitrene precursors in transition metal catalyzed aziridination of alkenes and amidation of various organic substrates [584, 761]. Mansuy and coworkers in 1984 first reported the aziridination of alkenes with tosyliminoiodane PhINTs in the presence of iron- or manganese-porphyrins [762]. This reaction has a mechanism similar to the metal-catalyzed oxygen atom transfer reactions of iodosylbenzene (Section 3.1.20) and involves a metal–nitrene complex as the intermediate.

Significant recent interest in the transition metal catalyzed reactions of imidoiodanes was initiated in the 1990s by the pioneering works of Evans [586, 763, 764] and Jacobsen [765, 766] on the asymmetric aziridination of olefins using copper catalysts (2–10 mol%) with chiral dinitrogen ligands and PhINTs as the nitrene precursor. Since these initial publications, research activity in this area has surged and the copper-catalyzed aziridination of alkenes has been utilized in numerous syntheses. For example, Dodd and coworkers applied the Evans aziridination procedure to 2-substituted acrylates and cinnamates **649** [767] and to steroids **650** (Scheme 3.257) [768].

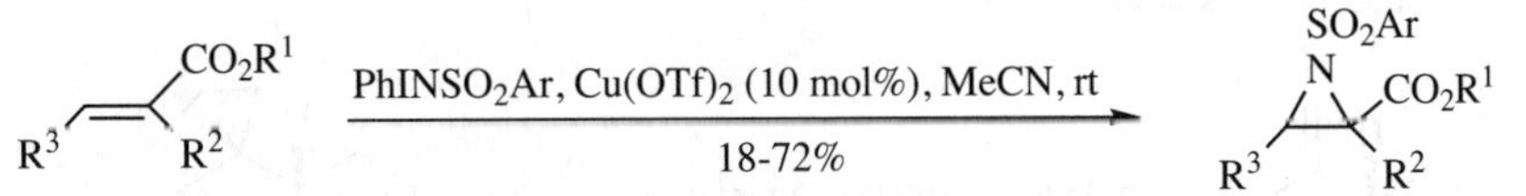

CO_2R^1, R^3, R^2 $\xrightarrow[\text{18-72\%}]{\text{PhINSO}_2\text{Ar, Cu(OTf)}_2\text{ (10 mol\%), MeCN, rt}}$ SO_2Ar, N, CO_2R^1, R^3, R^2

649 R^1 = Me, Et, Bu^t; R^2 and R^3 = H, Me, Ph
Ar = 4-MeC_6H_4, 4-$NO_2C_6H_4$

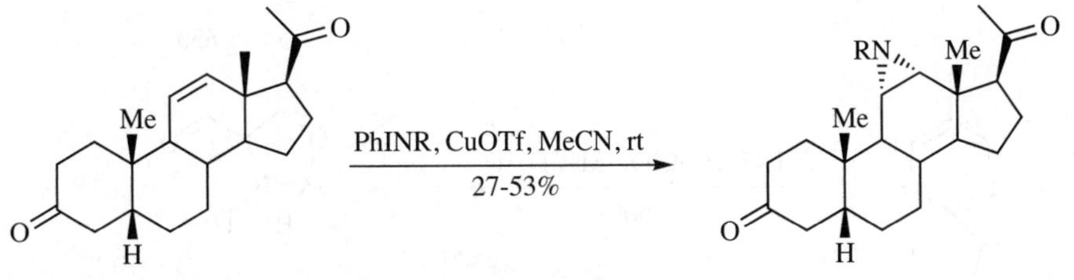

O, Me, H $\xrightarrow[\text{27-53\%}]{\text{PhINR, CuOTf, MeCN, rt}}$ RN, Me, O, Me, H, O

650 R = Ts, Ns, $Me_3SiCH_2CH_2SO_2$

Scheme 3.257

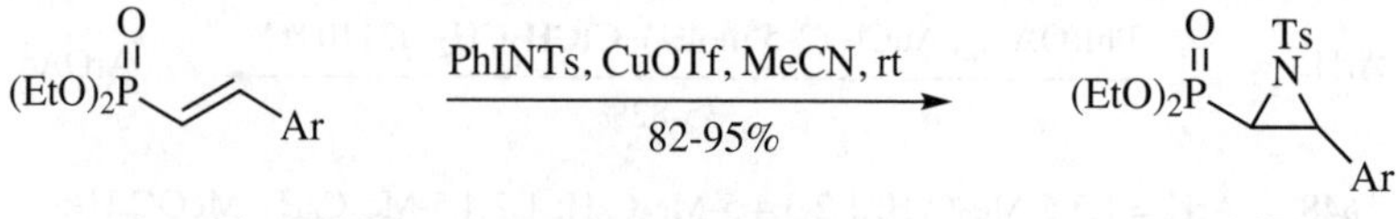

$(EtO)_2P(O)CH=CHAr$ —PhINTs, CuOTf, MeCN, rt; 82-95%→ **651**

Ar = Ph, 4-ClC$_6$H$_4$, 1-naphthyl, 2-naphthyl

Scheme 3.258

The copper-catalyzed aziridination of the appropriate olefinic substrates has been employed in the preparation of various 2-acylaziridines [769,770] and aziridinylphosphonates **651** (Scheme 3.258) [771].

A similar catalytic aziridination has also been applied for the functionalization of the optically active azoninones **652** [772], in the preparation of a key intermediate **653** in the total synthesis of kalihinane diterpenoids [773], in the synthesis of α-methylserinal derivatives **654** (Scheme 3.259) [774], in the preparation of 2,4-disubstituted *N*-tosylpyrrolidines [775] and in the synthesis of nosylaziridines [776].

Particularly important are enantioselective aziridinations of alkenes using PhINTs and copper catalysts with chiral dinitrogen ligands [777–781]. In a representative example, the PhINTs-promoted asymmetric aziridination of alkene **655** affords chiral aziridine **656** with excellent enantioselectivity (Scheme 3.260) [777].

Various chiral ligands or counter-anions, or complexes of other than copper transition metals, have been evaluated in these reactions. High enantioselectivity in the copper-catalyzed aziridination of styrene derivatives was observed in the presence of chiral biaryldiamines [782], chiral C$_2$-symmetric bisferrocenyldiamines

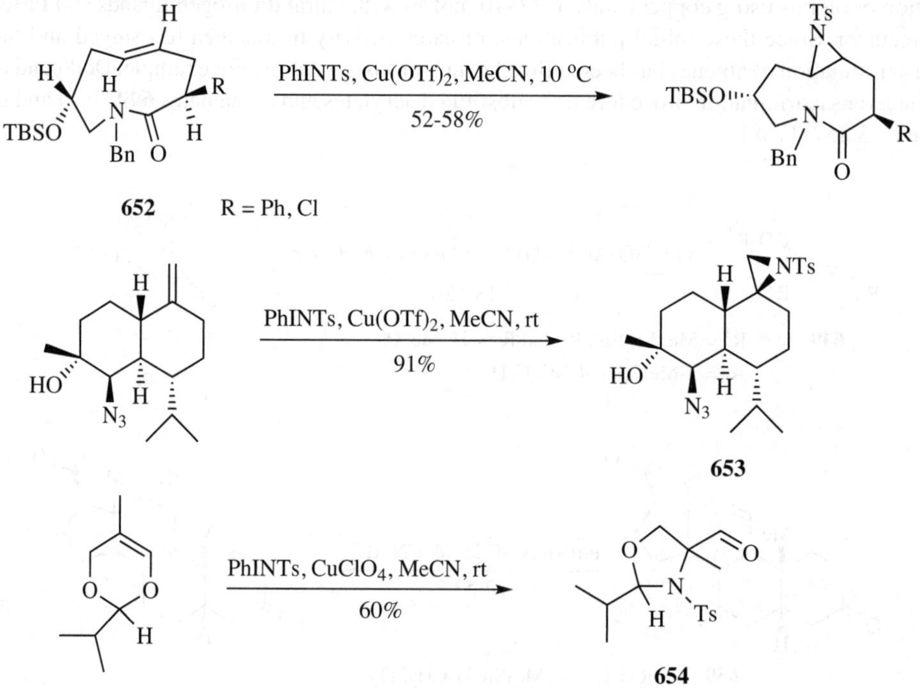

652 —PhINTs, Cu(OTf)$_2$, MeCN, 10 °C; 52-58%→ aziridine (R = Ph, Cl)

—PhINTs, Cu(OTf)$_2$, MeCN, rt; 91%→ **653**

—PhINTs, CuClO$_4$, MeCN, rt; 60%→ **654**

Scheme 3.259

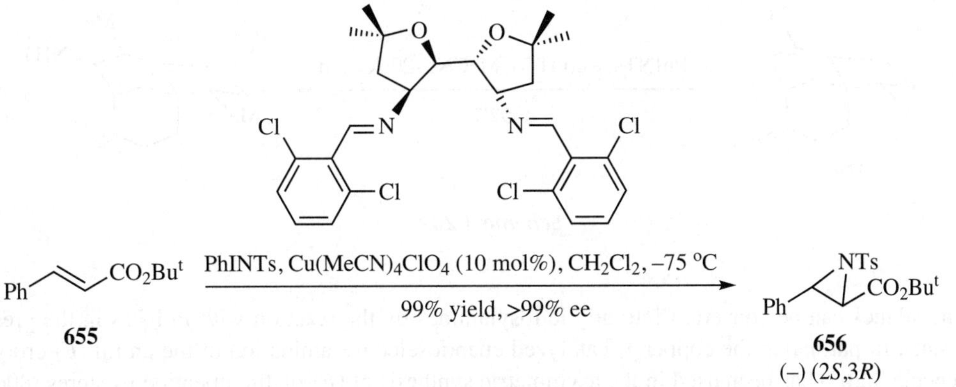

Ph CO$_2$Bu^t **655** — PhINTs, Cu(MeCN)$_4$ClO$_4$ (10 mol%), CH$_2$Cl$_2$, –75 °C; 99% yield, >99% ee → Ph NTs CO$_2$Bu^t **656** (–) (2*S*,3*R*)

Scheme 3.260

[783], chiral borate counter-anion [784], a phosphoramidite derived from (–)-(a*R*)-[1,1′-binaphthalene]-8,8′-diol [785], bis(oxazolines) on zeolite Y [786, 787], chiral tartrate-derived bis-oxazoline ligands [788] and C_2-symmetric bis(aziridine) ligands [789]. The highly enantioselective catalytic aziridination of styrenes was realized by using (salen)manganese(III) complexes [790], manganese and iron tetramethylchiroporphyrins [791] and the chiral rhodium(II) complexes [792–794]. An enhanced reactivity of PhINTs in the olefin aziridination reaction under achiral conditions was observed in the presence of the copper(II) complexes of pyridyl-appended diazacycloalkanes [795, 796], poly(pyrazolyl)borate–copper complexes [797], the copper(II) complexes of 1,4,7-triisopropyl-1,4,7-triazacyclononane [798], a Cu(I) complex of ferrocenyldiimine [799], bis(tosyl)imidoruthenium(VI) porphyrin complexes [800] and methyltrioxorhenium [801].

Mechanistic studies of copper-catalyzed aziridinations have demonstrated that copper nitrene species are the key intermediates in this reaction [802–804].

N-Tosyliminoiodanes, ArINTs, have found synthetic application as useful nitrene precursors in transition metal catalyzed amidation of saturated C—H bonds in various organic substrates. Breslow and coworkers have developed the regioselective amidation of steroidal derivatives catalyzed by metalloporphyrins [690, 691]. Specifically, the aromatic steroid equilenin acetate **657** undergoes regioselective and stereoselective amidation catalyzed by a manganese porphyrin using PhINTs as the nitrene donor (Scheme 3.261) [690].

Overman and Tomasi applied the copper-catalyzed amidation of compound **658** (Scheme 3.262) in the key step of the enantioselective total synthesis of the natural tetracyclic spermidine alkaloid (–)-hispidospermidin [805].

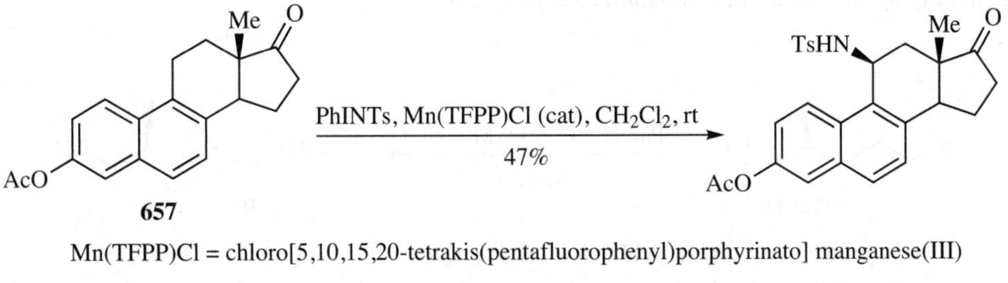

657 — PhINTs, Mn(TFPP)Cl (cat), CH$_2$Cl$_2$, rt; 47% → (TsHN product)

Mn(TFPP)Cl = chloro[5,10,15,20-tetrakis(pentafluorophenyl)porphyrinato] manganese(III)

Scheme 3.261

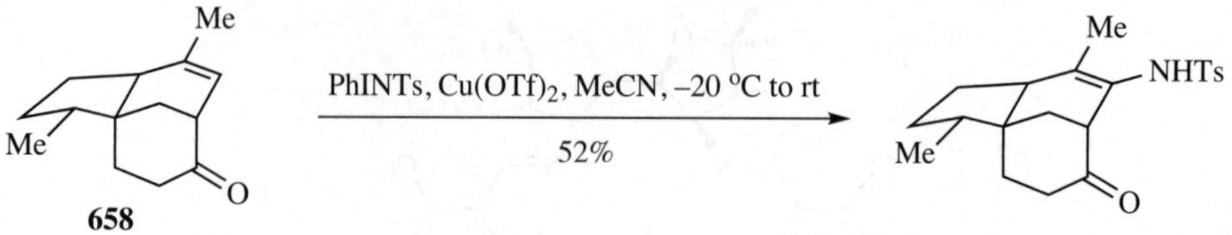

Scheme 3.262

Allylic silanes can be converted into allylic tosylamides by the reaction with PhINTs in the presence of copper salts. In particular, the copper(I)-catalyzed enantioselective amidation of the chiral (*E*)-crotylsilanes **659** (Scheme 3.263) has been used in the asymmetric synthesis of (*E*)-olefin dipeptide isosteres [806].

The amidation of saturated C—H bonds can be effectively catalyzed by ruthenium or manganese complexes. Unfunctionalized hydrocarbons, such as adamantane, cyclohexene, ethylbenzene, cumene, indane, tetralin, diphenylmethane and others, are selectively amidated with PhINTs in the presence of ruthenium or manganese porphyrins or the ruthenium cyclic amine complexes to afford N-substituted sulfonamides in 80–93% yields with high selectivity [807]. The enantioselective amidation of a C—H bond can be catalyzed by chiral (salen)manganese(III) complexes (e.g., **660**) [808], or by chiral ruthenium(II) and manganese(III) porphyrins (Scheme 3.264) [809].

The aziridination and amidation reactions of imidoiodanes can be efficiently catalyzed by Rh(II) complexes [810–815]. Dirhodium(II) tetrakis[*N*-tetrafluorophthaloyl-(*S*)-*tert*-leucinate], Rh_2(*S*-TFPTTL)$_4$, has been found to be an exceptionally efficient catalyst for enantioselective aminations of silyl enol ethers **661** with imidoiodane **662** to afford α-amido ketones **663** in high yields and with enantioselectivities of up to 95% ee (Scheme 3.265). The effectiveness of this catalytic protocol has been demonstrated by an asymmetric formal synthesis of (–)-metazocine [810]. This catalyst has also been used for the asymmetric synthesis of phenylglycine derivatives by enantioselective amidation of silylketene acetals with PhINTs [811].

Additional examples of C—H amidations using PhINTs as the nitrene precursor are represented by the following publications: the highly efficient Ru(II) porphyrin catalyzed C—H bond amidation of aldehydes [816,817], aromatic C—H amidation mediated by a diiron complex [818], gold-catalyzed nitrene insertion into aromatic and benzylic C—H bonds [819,820], silver-catalyzed intermolecular and intramolecular amidation of C—H bond in saturated hydrocarbons [821,822], α-amidation of cyclic ethers catalyzed by $Cu(OTf)_2$ [823], a mechanistic study of catalytic intermolecular amination of C—H bonds [824], nitrene insertion into the sp^3 C—H bonds of alkylarenes and cyclic ethers or the sp^2 C—H bonds of benzene using a copper homoscorpionate complex [825], Co(II)-catalyzed allylic amidation reactions [826], Ru(II) porphyrin-catalyzed amidation of aromatic heterocycles [827], non-heme iron-catalyzed amidation of aromatic substrates [828] and by the efficient stereoselective allylic C—H amination of terpenes and enol ethers involving the combination of a chiral aminating agent with a chiral rhodium catalyst [829].

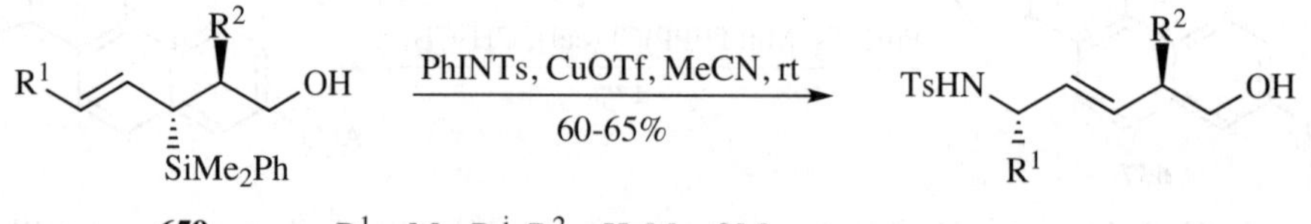

Scheme 3.263

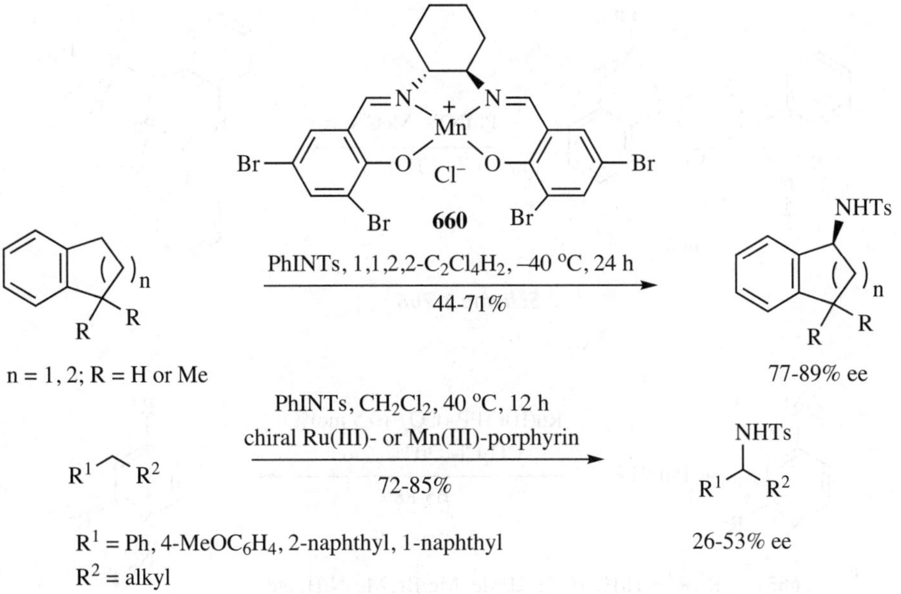

660

PhINTs, 1,1,2,2-$C_2Cl_4H_2$, –40 °C, 24 h

44-71%

n = 1, 2; R = H or Me

77-89% ee

PhINTs, CH_2Cl_2, 40 °C, 12 h
chiral Ru(III)- or Mn(III)-porphyrin

72-85%

R^1 = Ph, 4-$MeOC_6H_4$, 2-naphthyl, 1-naphthyl
R^2 = alkyl

26-53% ee

Scheme 3.264

Sanford and coworkers have investigated the carbon–nitrogen bond–forming reactions of palladacycles with aryliodonium imides [830]. In particular, palladium(II) complexes (e.g., **664**) containing bidentate cyclometalated chelating ligands react with PhINTs at room temperature to give products of insertion of the tosylimino group into the Pd—C bond (Scheme 3.266). This tosylimino insertion reaction has been applied to palladacycle complexes of azobenzene, benzo[*h*]quinoline and 8-ethylquinoline. The newly aminated organic ligands can be liberated from the metal center by protonolysis with a strong acid [830].

Imidoiodanes can be used to transfer the imido group to other elements under catalytic conditions. The imido group can be efficiently transferred to the sulfur atom in organic sulfides or sulfoxides [588, 831–835], or the nitrogen atom in aromatic nitrogen heterocycles using aryliodonium imides in the presence of copper, ruthenium, or iron complexes [836,837]. Specific examples are illustrated by the selective N-imidation of aromatic nitrogen heterocycles (e.g., **665**) catalyzed by carbonyl[*meso*-tetrakis(*p*-tolyl)porphyrinato]ruthenium(II), [Ru(II)(TPP)(CO)] [836] and the iron-catalyzed imidation of sulfoxides (e.g., **666**) and sulfides using imidoiodane **662** (Scheme 3.267) [831].

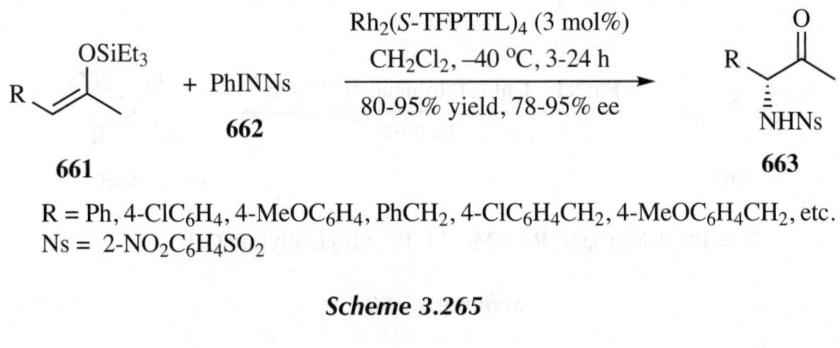

661 + PhINNs **662**

Rh_2(*S*-TFPTTL)$_4$ (3 mol%)
CH_2Cl_2, –40 °C, 3-24 h

80-95% yield, 78-95% ee

663

R = Ph, 4-ClC_6H_4, 4-$MeOC_6H_4$, $PhCH_2$, 4-$ClC_6H_4CH_2$, 4-$MeOC_6H_4CH_2$, etc.
Ns = 2-$NO_2C_6H_4SO_2$

Scheme 3.265

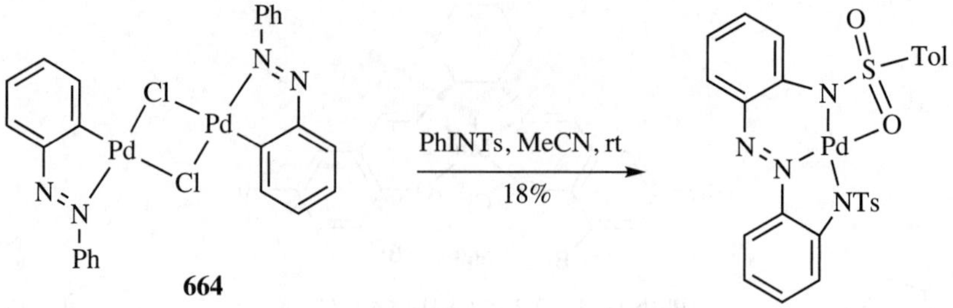

664

Scheme 3.266

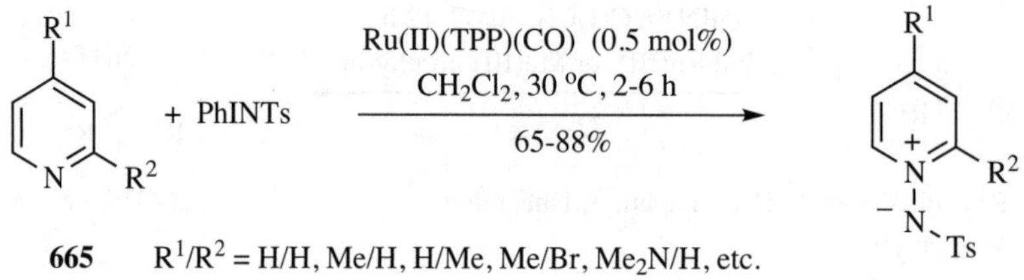

665 R^1/R^2 = H/H, Me/H, H/Me, Me/Br, Me_2N/H, etc.

666

662

Ns = 4-$NO_2C_6H_4SO_2$

Scheme 3.267

Likewise, the reaction of PhINTs with sulfoxides **667** in the presence of catalytic amounts of copper(I) triflate affords the corresponding *N*-tosylsulfoximides **668** in high yield (Scheme 3.268) [835]. The imidation of enantiomerically pure sulfoxides **667** allows stereoselective access to *N*-tosylsulfoximides **668** with complete retention of configuration at sulfur. A similar imidation procedure has been used for the preparation of chiral ferrocenylsulfoximides [838, 839].

Enantioselective imidation of alkyl aryl sulfides **669** can be achieved by using the chiral manganese(salen) complex shown in Scheme 3.269 as a catalyst [840, 841].

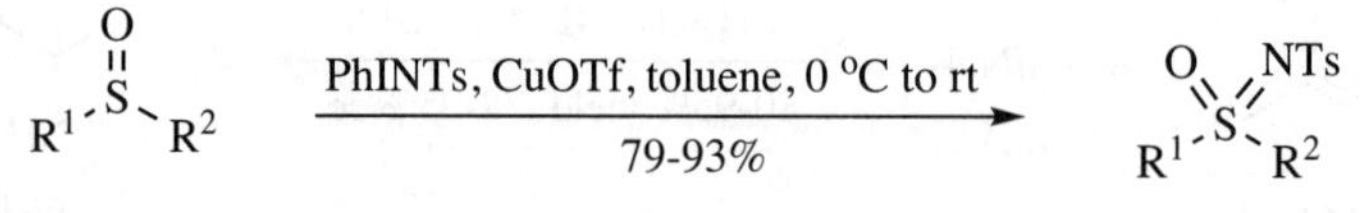

667 668

R^1 = Ph, 4-MeC_6H_4; R^2 = Me, Et, Pr^i, vinyl, allyl, $PhCH_2$

Scheme 3.268

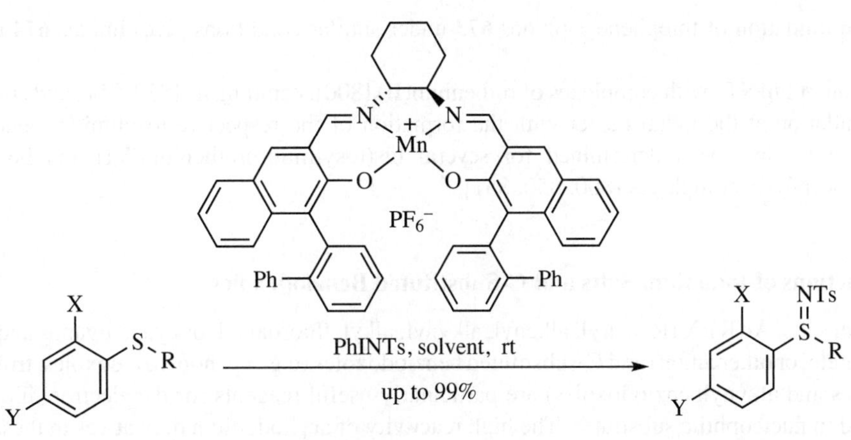

669 Solvent = PhCl, MeCN, PhCN
X = NO_2, H, Br; Y = H, NO_2; R = Me, Et, Bn

PhINTs, solvent, rt
up to 99%
up to 97% ee

Scheme 3.269

Similarly, a direct catalytic sulfimidation of sulfides or 1,3-dithianes with PhINTs using a catalytic amount of copper(I) triflate and a chiral 4,4′-disubstituted bis(oxazoline) as ligand affords the respective chiral monosulfimides in good yield and with moderate enantioselectivity of up to 40–71% ee [842, 843]. Under same conditions, prochiral selenides react with PhINTs in the presence of CuOTf and the chiral 4,4′-disubstituted 2,2′-bis(oxazoline) ligands to give the corresponding chiral selenimides with up to 64% yield and 36% ee [844, 845].

The reaction of 3,4-di-*tert*-butylthiophene (**670**) with PhINTs in the presence of copper(I) or copper(II) catalysts affords a mixture of imide **671** and diimide **672** as principal products (Scheme 3.270) [846–848].

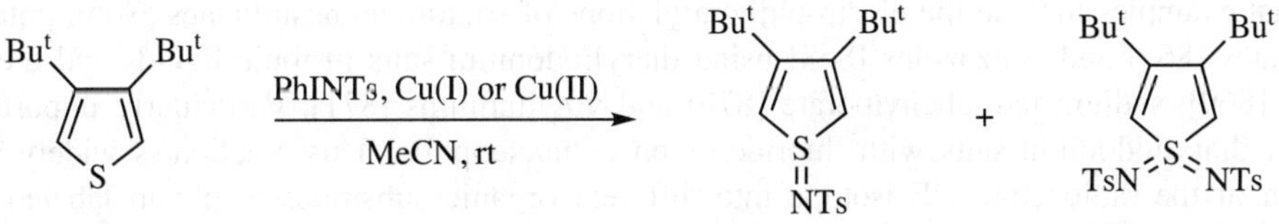

PhINTs, Cu(I) or Cu(II)
MeCN, rt

670 Cu(I) = $Cu(MeCN)_4PF_6$
Cu(II) = $Cu(OTf)_2$
671 (23-61%) **672** (7-18%)

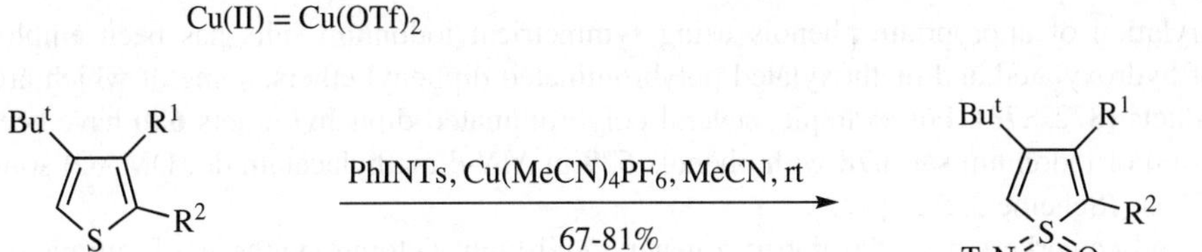

PhINTs, $Cu(MeCN)_4PF_6$, MeCN, rt
67-81%

673 $R^1 = Bu^t, R^2 = H$ or $R^1 = H, R^2 = Bu^t$ **674**

Scheme 3.270

Likewise, the imidation of thiophene 1-oxides **673** under similar conditions gives imides **674** in good yield [849].

The reaction of PhINTs with complexes of ruthenium(II) [800], osmium(II) [850, 851] and cobalt(III)[852] results in imidation at the metal center with the formation of the respective tosylimido-metal complexes. X-Ray structures have been determined for several bis(tosylimido)ruthenium(VI) and bis(tosylimido)-osmium(VI) porphyrin complexes [800, 850, 851].

3.1.22 Reactions of Iodonium Salts and C-Substituted Benziodoxoles

Aryliodonium salts, Ar(R)IX (R = aryl, alkenyl, alkynyl, alkyl, fluoroalkyl, or cyano ligands and X = triflate, tetrafluoroborate, or other anion) and C-substituted benziodoxoles (e.g., cyanobenziodoxoles, trifluoromethyl-benziodoxoles and alkynylbenziodoxoles) are particularly useful reagents for the electrophilic transfer of a carbon ligand to nucleophilic substrates. The high reactivity of aryliodonium derivatives in these reactions is explained by the "hyperleaving group ability" of the ArI group; for example, the leaving group ability of PhI is about a million times greater than that of triflate [853]. The chemistry of iodonium salts [854–856] and the atom-transfer reactions of benziodoxoles [857–859] have been summarized in several reviews.

3.1.22.1 Reactions of Diaryliodonium Salts

Diaryliodonium salts represent the most stable and well-investigated class of iodonium salts. Their chemistry was extensively covered in several reviews [854, 855, 860]. The most important and synthetically useful reactions of diaryliodonium salts include the following: the direct electrophilic arylation of various nucleophiles, transition metal mediated cross-coupling reactions and reactions involving generation and trapping of the benzyne intermediates.

Diaryliodonium salts have found synthetic application as arylating reagents in reactions with various organic substrates under polar, catalytic, or photochemical conditions. Typical examples of arylations of nucleophiles under polar, non-catalytic conditions are shown in Scheme 3.271 and include the reactions of diaryliodonium salts with thiosulfonate anions **675** [861], fluoride anion [862, 863], malonates **676** [864] and silyl enol ethers **677** [865].

Additional examples include the electrophilic arylations of sodium arenesulfinates [866], potassium carbonotrithioates [867] and benzazoles [868] using diaryliodonium salts in ionic liquids and the arylations of anilines [869], sodium tetraphenylborate [870] and vinylindiums [871]. Particularly important are the reactions of diaryliodonium salts with fluoride anion as nucleophile. This reaction is widely applied for introduction of the radioactive ^{18}F isotope into different organic substrates to obtain labeled agents for positron emission tomography. This topic is covered separately in Chapter 7 due to its importance and wide usage.

The O-arylation of appropriate phenols using symmetrical iodonium salts has been employed in the synthesis of hydroxylated and methoxylated polybrominated diphenyl ethers, some of which are related to natural products [872, 873]. For example, several polybrominated diphenyl ethers **680** have been prepared by the reaction of iodonium salt **678** with phenols **679** in *N,N*-dimethylacetamide (DMAC) solution in the presence of base (Scheme 3.272) [872].

Olofsson and coworkers have developed a general and high-yielding synthesis of various diaryl ethers **681** using the reaction of diaryliodonium salts with phenols under basic conditions (Scheme 3.273) [874]. The scope of products includes bulky *ortho*-substituted diaryl ethers, which are difficult to obtain by metal-catalyzed protocols. A similar procedure has been used for the metal-free synthesis of aryl esters from carboxylic acids and diaryliodonium salts [875].

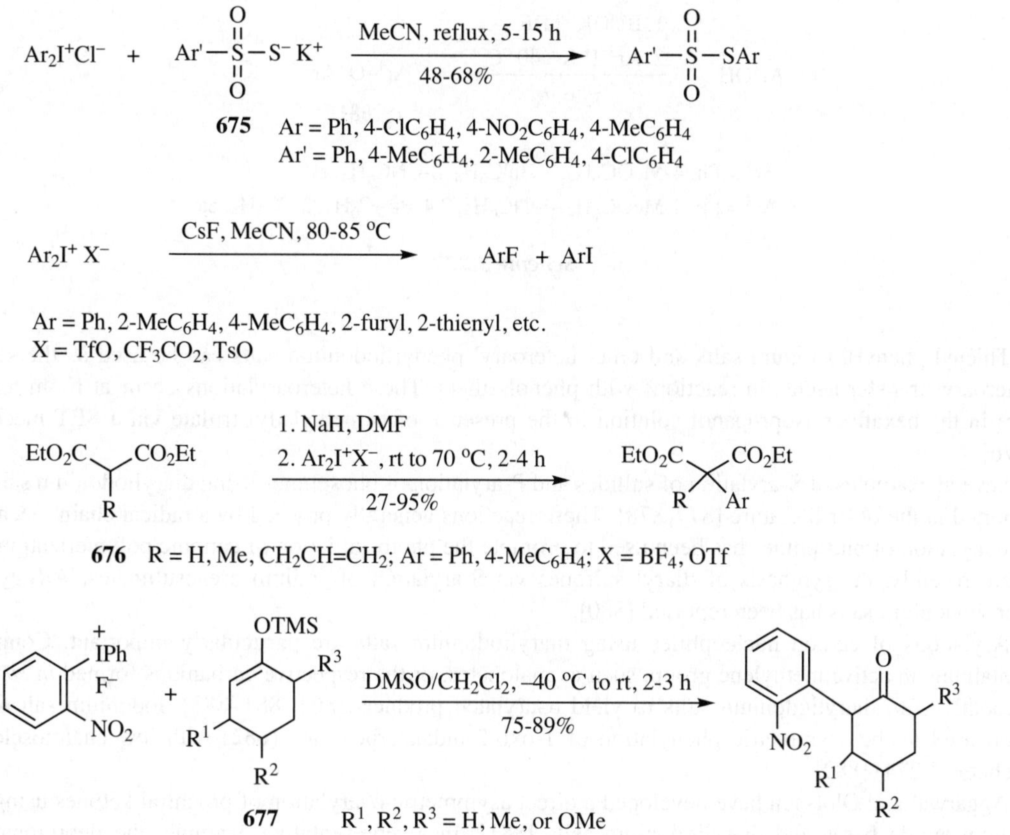

$Ar_2I^+Cl^-$ + $Ar'-S(O_2)-S^-K^+$ → (MeCN, reflux, 5-15 h; 48-68%) $Ar'-S(O_2)-SAr$

675 Ar = Ph, 4-ClC_6H_4, 4-$NO_2C_6H_4$, 4-MeC_6H_4
Ar' = Ph, 4-MeC_6H_4, 2-MeC_6H_4, 4-ClC_6H_4

$Ar_2I^+X^-$ → (CsF, MeCN, 80-85 °C) ArF + ArI

Ar = Ph, 2-MeC_6H_4, 4-MeC_6H_4, 2-furyl, 2-thienyl, etc.
X = TfO, CF_3CO_2, TsO

EtO_2C–CH(R)–CO_2Et → (1. NaH, DMF; 2. $Ar_2I^+X^-$, rt to 70 °C, 2-4 h; 27-95%) EtO_2C–C(R)(Ar)–CO_2Et

676 R = H, Me, $CH_2CH{=}CH_2$; Ar = Ph, 4-MeC_6H_4; X = BF_4, OTf

$^+$IPh, F^-, NO_2 + OTMS, R^3, R^1, R^2 → (DMSO/CH_2Cl_2, –40 °C to rt, 2-3 h; 75-89%) O, R^3, NO_2, R^1, R^2

677 R^1, R^2, R^3 = H, Me, or OMe

Scheme 3.271

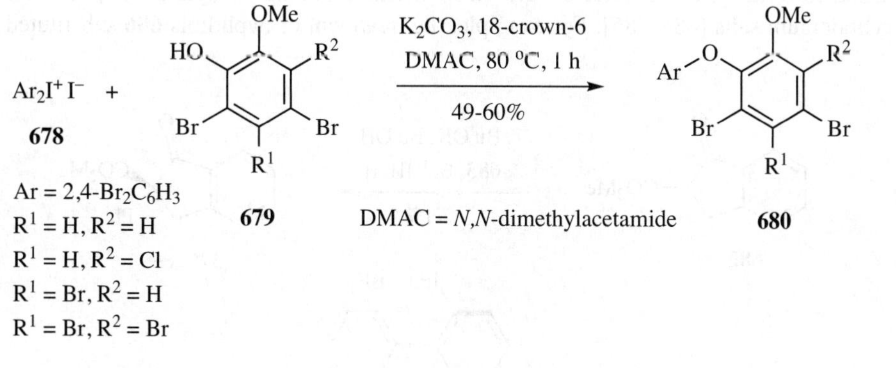

$Ar_2I^+I^-$ + (HO, OMe, R^2, Br, Br, R^1) → (K_2CO_3, 18-crown-6, DMAC, 80 °C, 1 h; 49-60%) (Ar–O, OMe, R^2, Br, Br, R^1)

678

Ar = 2,4-$Br_2C_6H_3$
R^1 = H, R^2 = H
R^1 = H, R^2 = Cl
R^1 = Br, R^2 = H
R^1 = Br, R^2 = Br

679 DMAC = *N,N*-dimethylacetamide **680**

Scheme 3.272

$$\mathrm{Ar^1OH} \xrightarrow[\text{72-98\%}]{\text{1. Bu}^t\text{OK, THF; 2. Ar}^2{}_2\text{I}^+\text{X}^-\text{, 40 °C}} \mathrm{Ar^1{-}O{-}Ar^2}\ \textbf{681}$$

Ar^1 = Ph, 4-$MeOC_6H_4$, 4-$Bu^tC_6H_4$, 3-CNC_6H_4, etc.
Ar^2 = Ph, 4-$MeOC_6H_4$, 4-ClC_6H_4, 2,4-$Bu^t_2C_6H_3$, 2-FC_6H_4, etc.

Scheme 3.273

Thienyl(phenyl)iodonium salts and other heteroaryl(phenyl)iodonium salts can be used as the selective heteroaryl transfer agents in reactions with phenol ethers. These heteroarylations occur at room temperature in the hexafluoroisopropanol solution in the presence of trimethylsilyl triflate via a SET mechanism [876].

Several examples of S-arylation of sulfides and P-arylation of phosphines using diaryliodonium salts were reported in the older literature [877, 878]. These reactions generally proceed by a radical chain mechanism. The arylation of phosphines has been used to promote the photo-initiation of cationic polymerization [879]. More recently, the synthesis of diaryl sulfones via S-arylation of sodium arenesulfinates, $ArSO_2Na$, by diaryliodonium salts has been reported [880].

Arylations of carbon nucleophiles using diaryliodonium salts are particularly important. Compounds containing an active methylene group, such as malonates, or the respective carbanions formed *in situ,* react smoothly with diaryliodonium salts to yield α-arylated products [864, 881–883]. Iodonium salt **683** has been used in the asymmetric phenylation of 1-oxo-2-indancarboxylate (**682**) with low enantioselectivity (Scheme 3.274) [882].

Aggarwal and Olofsson have developed a direct asymmetric α-arylation of prochiral ketones using chiral lithium amide bases and diaryliodonium salts [881]. In a representative example, the deprotonation of cyclohexanone derivative **684** using chiral Simpkins' (*R,R*)-base followed by reaction with the pyridyl iodonium salt gave the arylated product **685** in 94% ee (Scheme 3.275). This reaction has been employed in a short total synthesis of the alkaloid (–)-epibatidine [881].

Quideau and coworkers have reported a regioselective dearomatizing phenylation of phenols and naphthols using diaryliodonium salts [884, 885]. For example, the treatment of naphthols **686** substituted at the *ortho*

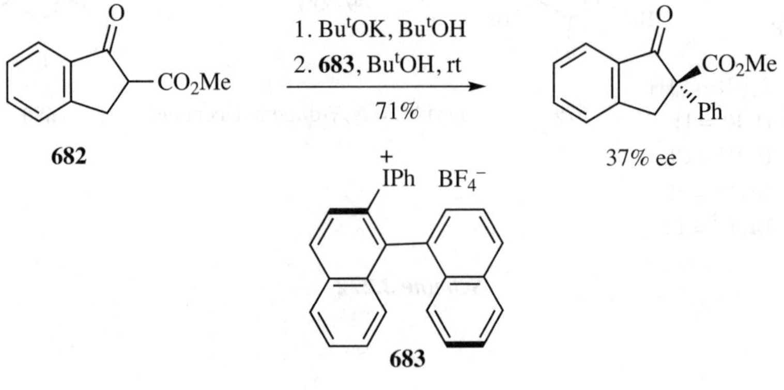

682 (–CO_2Me) → [1. Bu^tOK, Bu^tOH; 2. **683**, Bu^tOH, rt; 71%] → product (CO_2Me, Ph), 37% ee

683: $\overset{+}{I}Ph$ BF_4^-

Scheme 3.274

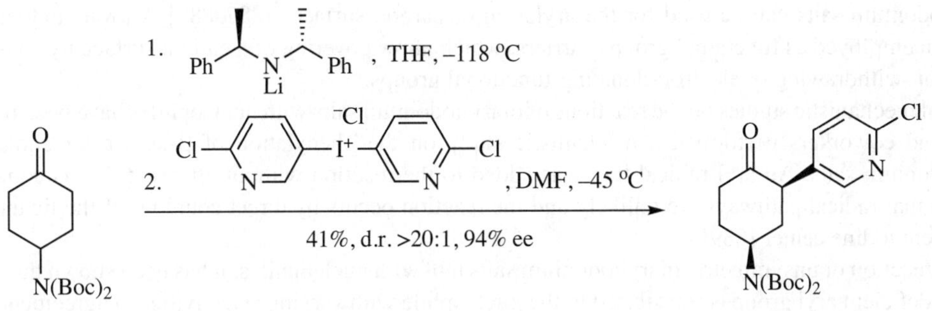

684 **685**

Scheme 3.275

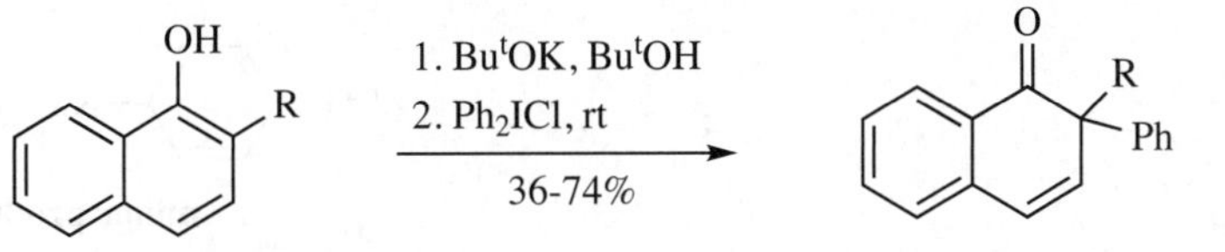

686 R = Me, OMe, NO **687**

Scheme 3.276

position by a small electron-donating group with diphenyliodonium chloride leads to regioselective *ortho*-phenylation to give products **687** (Scheme 3.276). The mechanism of this reaction involves a nonradical direct coupling of the ligands on the hypervalent iodine center [884]. The formation of phenol ethers due to the O-phenylation can also occur when the reaction of phenolate anion with diphenyliodonium chloride is carried out in a polar aprotic solvent such as dimethylformamide [884].

Indoles **688** and pyrroles can be efficiently arylated by aryliodonium salts at position 3 in the absence of metal catalysts to give products **689** (Scheme 3.277) [886]. The reaction of unsymmetrically substituted diaryliodonium salts results in the preferential transfer of the less sterically hindered aromatic ring.

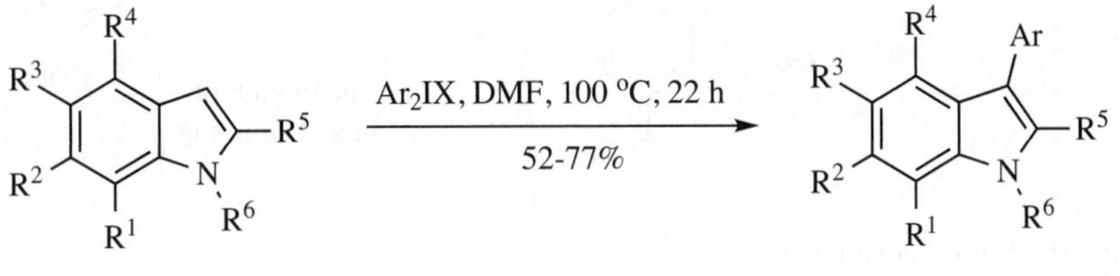

688 R^1 - R^4 = H, Me, Cl **689**

R^5 = H or Me; R^6 = H, Me, Pr, Bu, Bn

Ar = Ph, 4-$MeOC_6H_4$, 4-$Bu^tOC_6H_4$, etc.

Scheme 3.277

Aryliodonium salts can be used for the arylation of carbon surfaces [887, 888]. Various iodonium salts have been employed as functional groups carriers, which allow covering of a carbon surface by a wide range of electron-withdrawing or electron-donating functional groups.

Several mechanistic studies on the reactions of diaryliodonium salts with nucleophiles have been published. Ochiai and coworkers performed a mechanistic study on the phenylation of β-keto ester enolates with diaryliodonium salts. An aryl radical trap was added to the reaction without affecting the outcome, which indicates that radical pathways are unlikely and the reaction occurs by direct coupling of the ligands on the hypervalent iodine center [889].

In the reaction of unsymmetric diaryliodonium salts **690** with nucleophiles, it has been shown that the most electron-deficient aryl group is transferred to the nucleophile with varying selectivities in agreement with the mechanism outlined in Scheme 3.278. The initial ligand exchange leads to the hypervalent intermediates **691** and **692** with the electronegative ligand Nu occupying an axial position (Section 1.5.1). Rapid pseudorotation

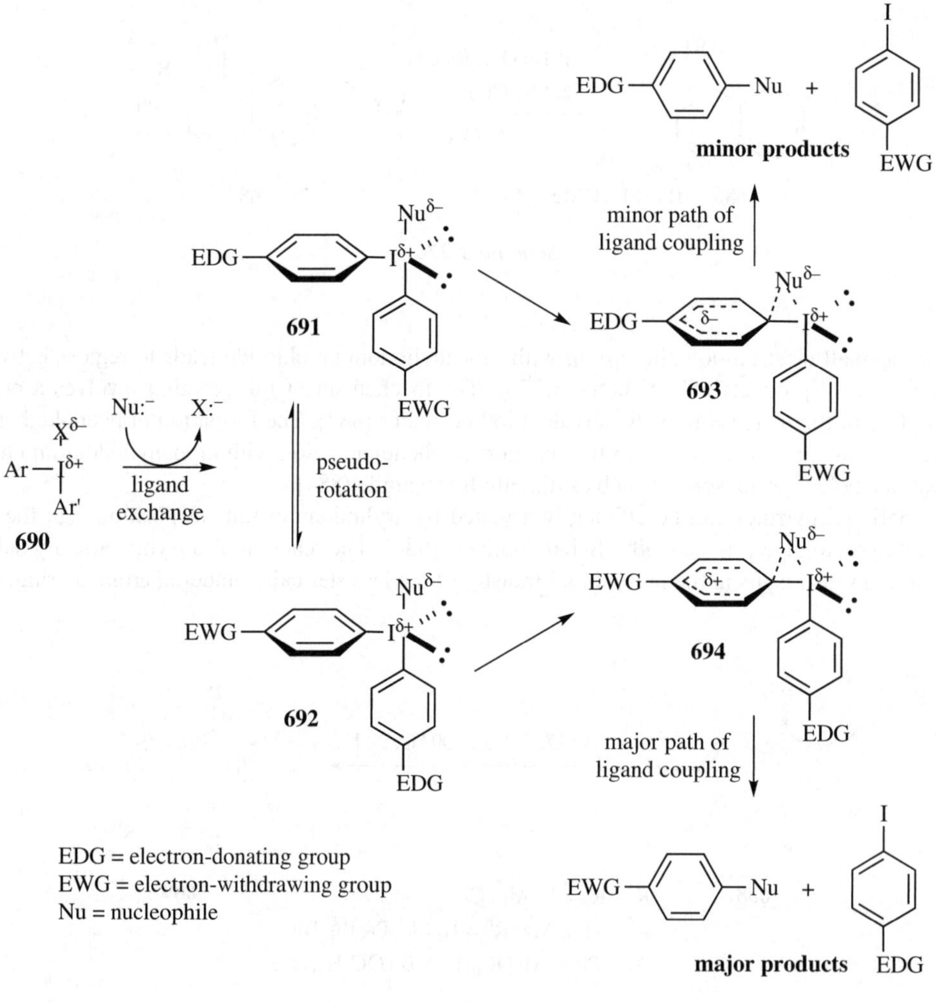

EDG = electron-donating group
EWG = electron-withdrawing group
Nu = nucleophile

Scheme 3.278

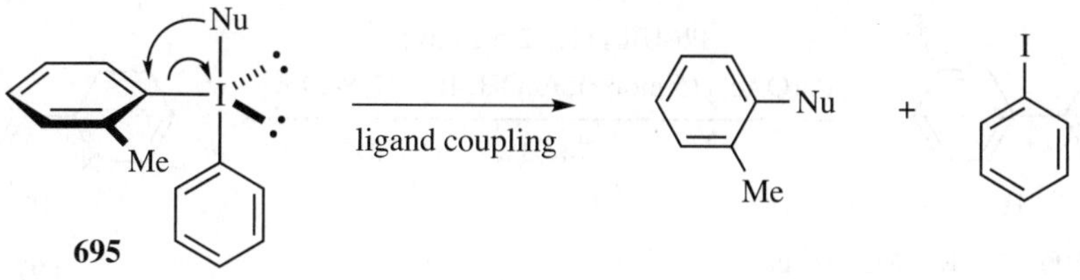

Scheme 3.279

occurs between intermediates **691** and **692,** which provides two different transition states, **693** and **694** [882, 890]. Of the two possible transition states for the subsequent ligand coupling, **694** is more favorable than **693,** because both the negative charge on the aromatic ring and the enhanced positive charge on the iodine(III) are stabilized by the substituents more effectively [889].

The so-called "ortho-effect" is observed in the reactions between a nucleophile and a diaryliodonium salt where one aryl ligand has an *ortho*-substituent, such as methyl. In these reactions the *ortho*-substituted aryl ligand is often coupled with the nucleophile, even if it is more electron rich [891, 892]. This has been explained by the predominant conformation **695** of the Ar(Ph)INu intermediate with the most bulky aryl ligand and the two lone pairs occupying the equatorial position for steric reasons (Scheme 3.279). Ligand coupling in the intermediate **695** leads to a reductive elimination of PhI and transfer of the nucleophile to the *ortho*-substituted aryl group situated in the equatorial position, even though it is the more electron-rich aromatic ring.

The mechanism of solvolysis of methoxy-substituted diaryliodonium tetrafluoroborates, $ArI^{+}Ph\ ^{-}BF_4$, in methanol and 2,2,2-trifluoroethanol has been investigated [893]. The solvolysis products include alkoxide substitution products (ArOR and PhOR) as well as iodoarenes (PhI and ArI). The ratios of products, ArOR/PhOR, range from 8 : 2 to 4 : 6. The results of this study provide experimental evidence against the formation of aryl cation under these conditions and support the pathways via ligand coupling or S_NAr2 mechanisms involving a solvent molecule as a nucleophile in the transition state [893]. If the reaction is performed in inert (not nucleophilic) solvent, various nucleophiles may be involved in the reaction with iodonium salt.

Arylations with diaryliodonium salts can be effectively catalyzed by transition metals. Diaryliodonium salts can serve as efficient reagents in the copper-catalyzed arylation of lithium enolates [894], α-aryl carbonyl compounds [895], thiophenes [896], 5-aryl-2*H*-tetrazole [897], uracil nucleosides [898], aniline and phenol derivatives [899] and alkenes [900]. Palladium salts and complexes are efficient catalysts in the cross-coupling reactions of diaryliodonium salts with various substrates, such as organoboron compounds [901], organostannanes [902], silanes [903], organolead triacetates [904], organobismuth(V) derivatives [905], carbon monoxide [906], allylic alcohols [907], functionalized allenes [908, 909], Grignard reagents [910], alkenes [911, 912], terminal alkynes [913], simple arenes [914] and arenecarboxylic acids via decarboxylative cross-coupling reaction [915]. Particularly interesting is the palladium-catalyzed directed C—H activation/phenylation of substituted 2-phenylpyridines and indoles with aryliodonium salts reported by Sanford and coworkers [916, 917]. In a representative example, 2-pyridyl substituted substrates **696** are selectively phenylated to the *ortho*-position, affording products **697** in good yields (Scheme 3.280). Preliminary mechanistic experiments have provided evidence in support of a rare Pd(II)/(IV) catalytic cycle for this transformation [917]. The preparation of stable triorganyl Pd(IV) complexes by the electrophilic arylation of palladium(II) bipyridine complexes using $Ph_2I^{+}TfO^{-}$ has been reported by Canty and coworkers [918]. It was also demonstrated that the carbamate (R_2NCO_2) function is an excellent directing group for palladium-catalyzed direct arylation reactions using diaryliodonium salts [919].

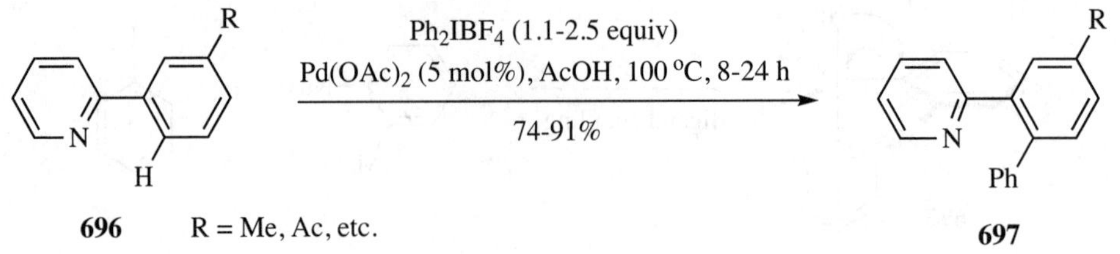

Ph_2IBF_4 (1.1-2.5 equiv)
$Pd(OAc)_2$ (5 mol%), AcOH, 100 °C, 8-24 h
74-91%

696 R = Me, Ac, etc. **697**

Scheme 3.280

McMillan and Allen have accomplished the enantioselective α-arylation of aldehydes using diaryliodonium salts and a combination of copper and organic amine catalysts. This asymmetric protocol has been applied to the rapid synthesis of (*S*)-ketoprofen, a commercially successful oral and topical analgesic [920].

Phenyl benziodoxolone **698** (Section 2.1.8.1.9) is a classical reagent that is commonly used to generate benzyne under heating (Scheme 3.281) [921–923].

In a more recent work, Kitamura and coworkers have developed more efficient benzyne precursors based on diaryliodonium salts [924–928]. Of particular use is phenyl[2-(trimethylsilyl)phenyl]iodonium triflate **699,** which is readily prepared by the reaction of 1,2-bis(trimethylsilyl)benzene with the $PhI(OAc)_2$/TfOH reagent system [924]. Treatment of reagent **699** with tetrabutylammonium fluoride in dichloromethane at room temperature generates benzyne, which can be trapped with a diene to afford the respective benzyne adducts in high yields [924]. Examples of synthetic applications of reagent **699** as benzyne precursor include O-arylation of carboxylic acids leading to aryl esters **700** [929], preparation of 2-aryl-substituted nitriles **702** by arylation of nitriles **701** via a benzyne reaction [930] and cycloaddition/elimination reaction of thiophene *S*-oxide **703** with benzyne leading to product **704** (Scheme 3.282) [931]. Reagent **699** has also been used in the synthesis of spiro(imidazolidine-2,3′-benzo[*b*]thiophene) by a one-pot reaction of benzyne, aryl isothiocyanates and N-heterocyclic carbenes [932] and for the preparation of benzo[*b*]seleno[2,3-*b*]pyridines by the reaction of acetic acid 2-selenoxo-2*H*-pyridin-1-yl esters with benzyne [933].

The efficient acylbenzyne precursors, [5-acyl-2-(trimethylsilyl)phenyl]iodonium triflates **705** have been prepared by reaction of the appropriate 1,2-bis(trimethylsilyl)benzenes with $PhI(OAc)_2$ in the presence of trifluoromethanesulfonic acid in dichloromethane at room temperature. Treatment of these reagents with Bu_4NF in dichloromethane generates acylbenzynes **706,** which can be trapped by furan to give adducts **707** in high yield (Scheme 3.283) [927].

The carborane analog of benzyne, 1,2-dehydrocarborane, can be generated similarly from phenyl[*o*-(trimethylsilyl)carboranyl]iodonium acetate by treatment with CsF in ether and trapped with dienes such as anthracene, naphthalene, norbornadiene and 2,5-dimethylfuran to give the respective 1,2-dehydrocarborane adducts in high yield [934].

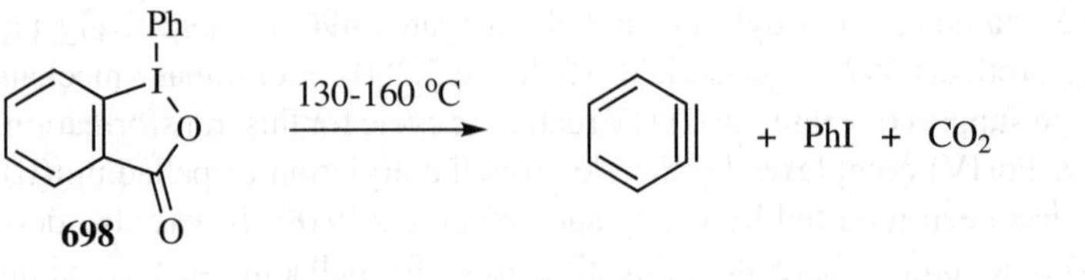

130-160 °C

698 + PhI + CO_2

Scheme 3.281

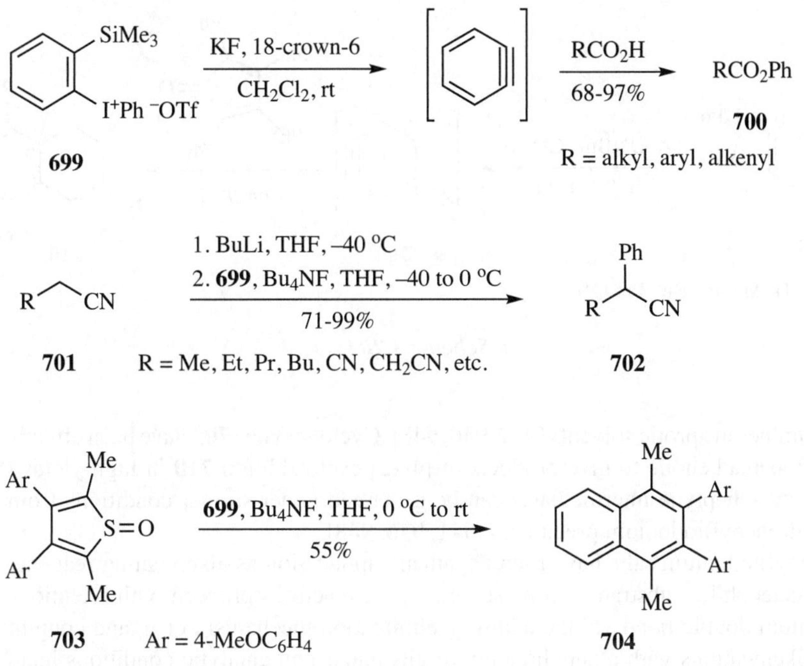

699: 2-(trimethylsilyl)phenyl(phenyl)iodonium triflate ($I^{+}Ph\ ^{-}OTf$) → KF, 18-crown-6, CH_2Cl_2, rt → [benzyne] → RCO_2H, 68-97% → RCO_2Ph **700**; R = alkyl, aryl, alkenyl

701 RCH_2CN → 1. BuLi, THF, –40 °C; 2. **699**, Bu_4NF, THF, –40 to 0 °C; 71-99% → $RCH(Ph)CN$ **702**; R = Me, Et, Pr, Bu, CN, CH_2CN, etc.

703 (Ar, Ar, Me, Me, S=O) → **699**, Bu_4NF, THF, 0 °C to rt; 55% → **704** (Me, Ar, Ar, Me); Ar = 4-$MeOC_6H_4$

Scheme 3.282

3.1.22.2 Reactions of Alkenyl(Aryl)Iodonium Salts

The chemistry of alkenyliodonium salts has been summarized in the reviews of Ochiai [935,936], Okuyama [937–939] and Zefirov [940]. Alkenyl(phenyl)iodonium salts are very reactive compounds because of the exceptional leaving group ability of the phenyliodonium moiety (10^{12} times greater than for iodine itself) combined with its high electron-withdrawing properties (the Hammett substituent constant σ_m for the PhI^{+} group is 1.35) [941]. Several research groups have been involved in mechanistic studies of the nucleophilic substitution in alkenyliodonium salts [942–949]. Several mechanisms, including S_N1, S_N2, ligand coupling and Michael addition–elimination, have been observed in these reactions. The mechanistic aspects of the reactions of vinylic iodonium salts with nucleophiles have been reviewed by Okuyama [937] and by Ochiai [935].

Cyclohexyne intermediates **709** have been observed as the result of β-elimination in the reactions of 1-cyclohexenyl(phenyl)iodonium salts **708** with mild bases, such as tetrabutylammonium acetate, fluoride ion,

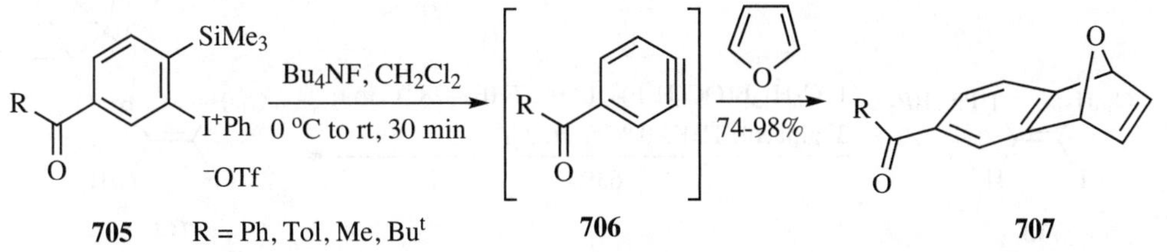

705 (RC(O)-aryl, $SiMe_3$, $I^{+}Ph\ ^{-}OTf$) → Bu_4NF, CH_2Cl_2, 0 °C to rt, 30 min → [**706**] → furan, 74-98% → **707**; R = Ph, Tol, Me, Bu[t]

Scheme 3.283

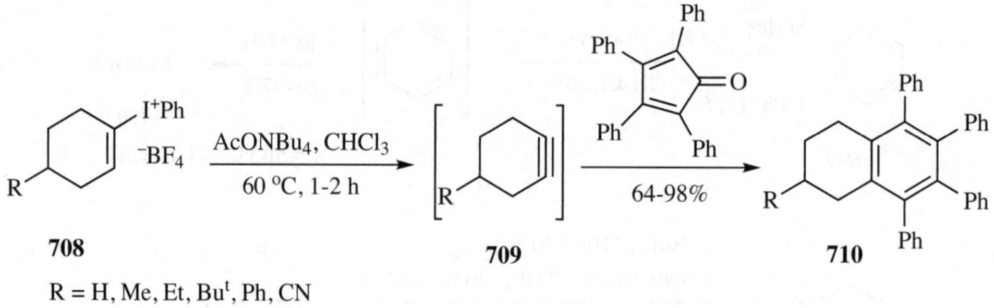

$I^{+}Ph$ $^{-}BF_4$ — $AcONBu_4$, $CHCl_3$, 60 °C, 1-2 h → [709] → (Ph, Ph, Ph, Ph, =O) 64-98% →

708 **709** **710**

R = H, Me, Et, Bu^t, Ph, CN

Scheme 3.284

alkoxides and amines in aprotic solvents [941, 946, 948]. Cyclohexynes **709** have been effectively trapped with tetraphenylcyclopentadienone to give products of [4+2] cycloaddition **710** in high yields (Scheme 3.284). The analogous cycloheptyne intermediates can be generated under similar conditions from the appropriate 1-cycloheptenyl(phenyl)iodonium precursors [941, 946, 948].

Alkenyl(phenyl)iodonium salts have found synthetic application as alkenylating reagents in the reactions with various nucleophilic substrates. In most cases these reactions proceed with retention of configuration at the carbon atom double bond via the addition–elimination mechanism or ligand coupling on the iodine. Examples of alkenylations with alkenyliodonium salts under non-catalytic conditions include the reactions with the following nucleophiles: thioamides [950], sodium dithiocarbamates and potassium carbonotrithioates [951], sodium tellurolates and selenolates [952, 953], potassium phosphorothioates, phosphorodithioates and phosphoroselenoates [954–956], group 15 element nucleophiles [957], formamides [958], tetrafluoroborate anion upon thermolysis leading to fluoroalkenes [959], potassium thiocyanate [960] and benzotriazole [961].

(*E*)- and (*Z*)-(Fluoroalkenyl)boronate derivatives **712** and **714** have been prepared stereospecifically by the reaction of (*E*)- or (*Z*)-(2-fluoroalkenyl)iodonium salts **711** and **713** with [bis(4-fluorophenoxy)]alkylboranes, followed by transesterification to pinacol esters (Scheme 3.285). The mechanism of this reaction involves

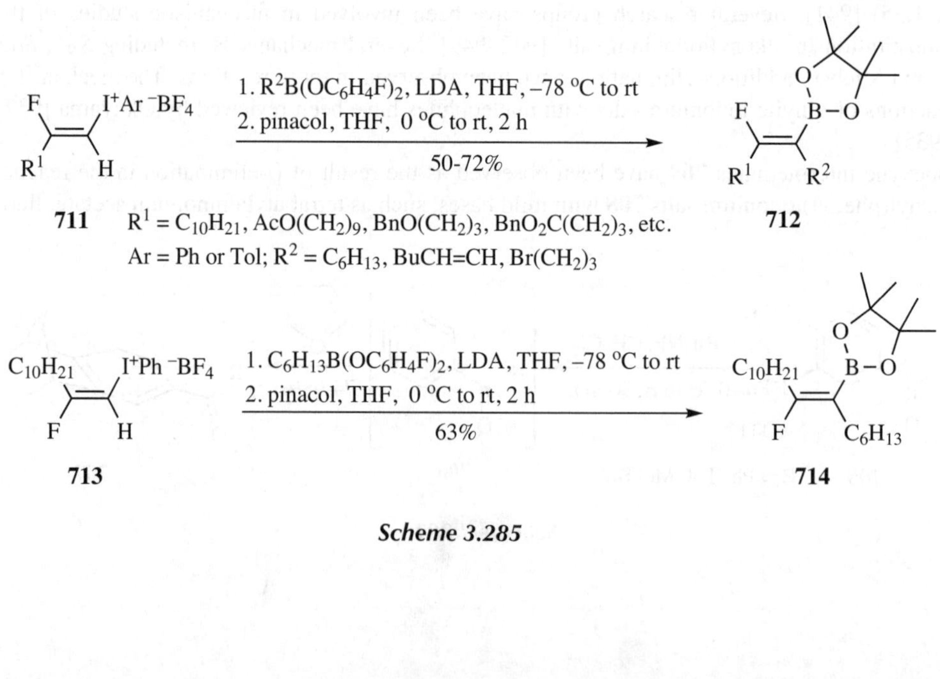

1. $R^2B(OC_6H_4F)_2$, LDA, THF, –78 °C to rt
2. pinacol, THF, 0 °C to rt, 2 h

50-72%

711 $R^1 = C_{10}H_{21}$, $AcO(CH_2)_9$, $BnO(CH_2)_3$, $BnO_2C(CH_2)_3$, etc.
Ar = Ph or Tol; $R^2 = C_6H_{13}$, BuCH=CH, $Br(CH_2)_3$

712

1. $C_6H_{13}B(OC_6H_4F)_2$, LDA, THF, –78 °C to rt
2. pinacol, THF, 0 °C to rt, 2 h

63%

713 **714**

Scheme 3.285

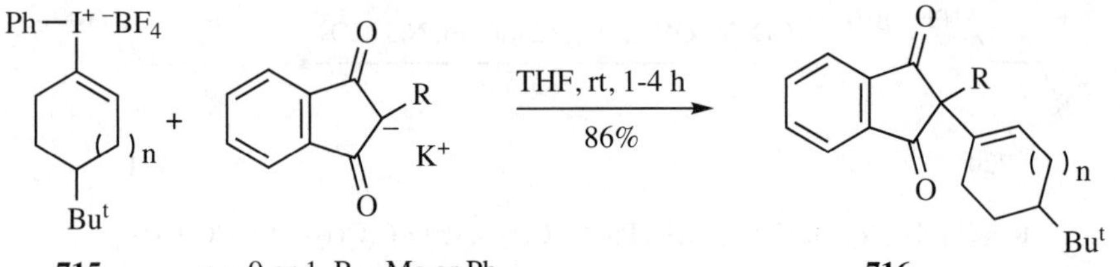

715 + n = 0 or 1, R = Me or Ph → THF, rt, 1-4 h, 86% → 716

Scheme 3.286

initial generation of 2-fluoroalkylideneiodonium ylide by the α-deprotonation of iodonium salts with lithium diisopropylamide (LDA) followed by its reaction with [bis(4-fluorophenoxy)]alkylboranes [962, 963].

Few examples of non-catalytic alkenylations of carbon nucleophiles have been published. For example, enolate anions derived from various 1,3-dicarbonyl compounds can be vinylated with cyclohexenyl or cyclopentenyl iodonium salts **715** to afford products **716** (Scheme 3.286) [964].

The selectivity of the alkenylation reactions and the yields of products can be significantly improved by carrying out the reaction of alkenyliodonium salts with carbon nucleophiles in the presence of transition metal compounds in stoichiometric or catalytic amounts. In the presence of a copper(I) catalyst iodonium salts selectively react with iodide anion [965, 966], organoborates [967], Grignard reagents [968] and terminal alkynes [969] to afford the respective products of cross-coupling in high yields with complete retention of geometry. An example of such a reaction is represented by the copper-mediated cross-coupling of *H*-phosphonates **718** with vinyliodonium salts **717** leading to 2-arylvinylphosphonates **719** under mild conditions (Scheme 3.287) [970].

Alkenyliodonium salts have been used as highly reactive reagents for Heck-type olefination [39, 971], Sonogashira-type coupling with alkynes [965, 972] and similar other palladium-catalyzed cross-coupling reactions [966, 973, 974]. In a specific example, (*Z*)-β-fluoro-α,β-unsaturated esters **721** were stereoselectively synthesized from (*Z*)-2-fluoro-1-alkenyliodonium salts **720** by the Pd-catalyzed methoxycarbonylation reaction (Scheme 3.288) [974]. This reaction proceeds at room temperature and is compatible with various functional groups on the substrate.

Reactions of alkenyliodonium salts with strong bases may lead to the generation of alkylidenecarbenes via a base-induced α-elimination. Alkylidenecarbenes generated by this method can undergo cyclization by a 1,5-C—H bond insertion, providing a useful route for the construction of substituted cyclopentenes [975–977]. For example, a synthesis of fluorocyclopentenes **723** by the reaction of (*Z*)-(2-fluoroalkenyl)iodonium salts **722** with potassium *tert*-butoxide has been developed (Scheme 3.289). The mechanism of this reaction

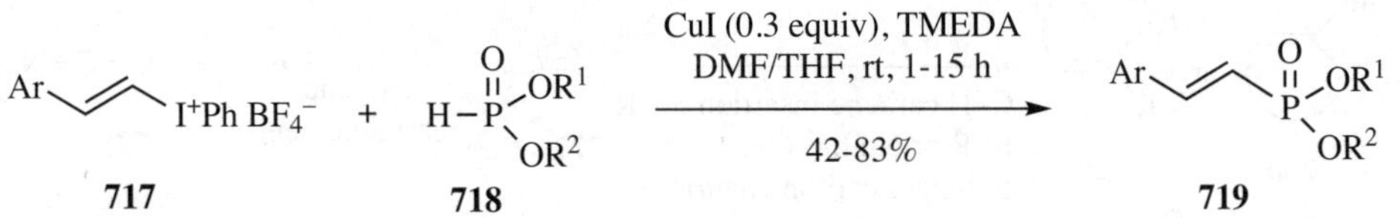

717 + 718 → CuI (0.3 equiv), TMEDA, DMF/THF, rt, 1-15 h, 42-83% → 719

Ar = Ph, 2-FC_6H_4, 2-MeC_6H_4, 2-$MeOC_6H_4$, 3-$MeOC_6H_4$, 4-$NO_2C_6H_4$, etc.
R^1, R^2 = Me, Et, Bu, Bn, Ph, etc.

Scheme 3.287

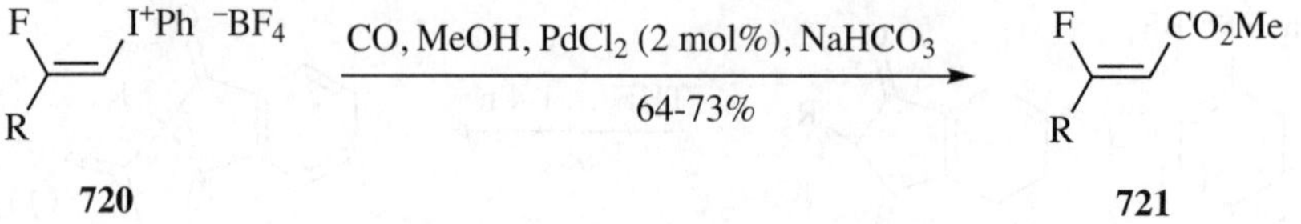

F, $I^+Ph\ ^-BF_4$, R — CO, MeOH, $PdCl_2$ (2 mol%), $NaHCO_3$, 64-73% → F, CO_2Me, R

720 → **721**

$R = C_{10}H_{21}$, (*cyclo*-C_6H_{11})CH_2, Ph, $Cl(CH_2)_9$, $Pr^iO_2C(CH_2)_8$, $Bu^tCO(CH_2)_8$

Scheme 3.288

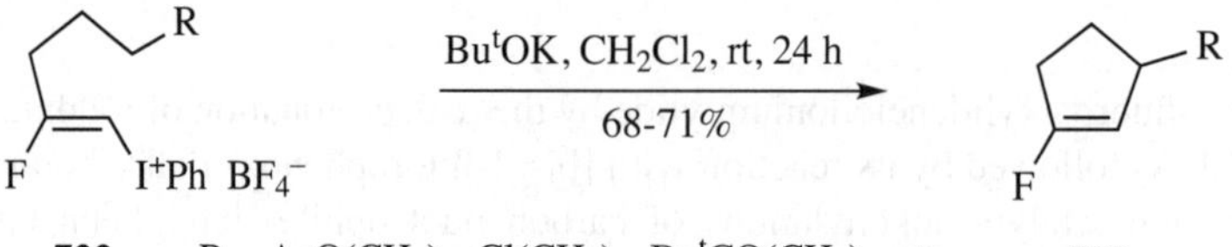

R, F, $I^+Ph\ BF_4^-$ — Bu^tOK, CH_2Cl_2, rt, 24 h, 68-71% → R, F

722 $R = AcO(CH_2)_6$, $Cl(CH_2)_6$, $Bu^tCO(CH_2)_5$, etc. **723**

Scheme 3.289

involves initial generation of (α-fluoroalkylidene)carbenes, which give fluorocyclopentenes via 1,5-C—H insertion [975].

3.1.22.3 Reactions of Alkynyl(aryl)iodonium Salts and Alkynylbenziodoxoles

The chemistry of alkynyliodonium salts has been summarized in several reviews [856, 978, 979]. Reactions of alkynyliodonium salts **724** with nucleophiles proceed via an addition–elimination mechanism involving alkylideneiodonium ylides **725** and alkylidene carbenes **727** as key intermediates (Scheme 3.290). Depending on the structure of the alkynyliodonium salt **724,** the specific reaction conditions and the nucleophile employed, this process can lead to the following products: β-functionalized alkenyliodonium salt **726** due to

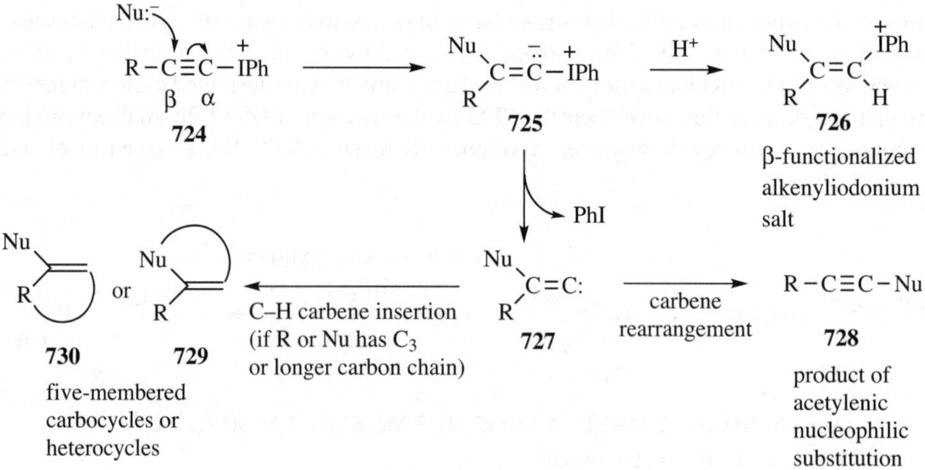

Nu:⁻ → $R-C\equiv C-\overset{+}{I}Ph$ (β, α) **724** → Nu(R)C=C̈⁻–$\overset{+}{I}Ph$ **725** —H^+→ Nu(R)C=C(H)$\overset{+}{I}Ph$ **726**

β-functionalized alkenyliodonium salt

725 —PhI→ Nu(R)C=C: **727**

727 —carbene rearrangement→ $R-C\equiv C-Nu$ **728**

product of acetylenic nucleophilic substitution

727 —C–H carbene insertion (if R or Nu has C_3 or longer carbon chain)→ **730** or **729**

five-membered carbocycles or heterocycles

Scheme 3.290

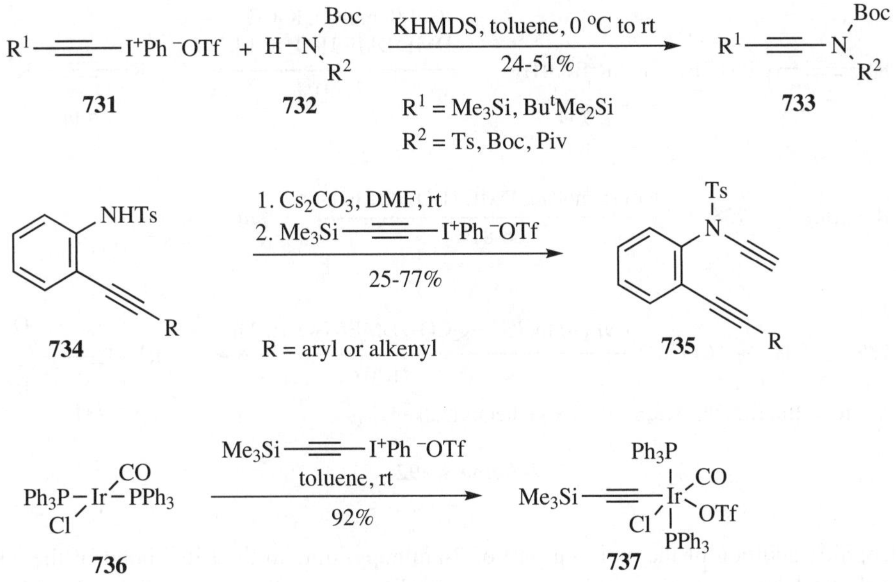

R^1—≡—$I^+Ph\ ^-OTf$ + H–N(Boc)(R^2) → KHMDS, toluene, 0 °C to rt; 24-51% → R^1—≡—N(Boc)(R^2)

731 **732** **733**

$R^1 = Me_3Si, Bu^tMe_2Si$

R^2 = Ts, Boc, Piv

NHTs, R → 1. Cs_2CO_3, DMF, rt; 2. Me_3Si—≡—$I^+Ph\ ^-OTf$; 25-77% → Ts, N, R

734 R = aryl or alkenyl **735**

CO, Ph_3P–Ir–PPh_3, Cl → Me_3Si—≡—$I^+Ph\ ^-OTf$; toluene, rt; 92% → Ph_3P, CO, Me_3Si—≡—Ir, Cl, OTf, PPh_3

736 **737**

Scheme 3.291

the protonation of ylide **725,** a substituted alkyne **728** due to the carbene rearrangement, or cyclic products **729** or **730** via intramolecular 1,5-carbene insertion [856]. These reaction pathways have been widely utilized in organic synthesis.

Alkynyl(phenyl)iodonium salts have found synthetic application for the preparation of various substituted alkynes by the reaction with appropriate nucleophiles, such as enolate anions [980, 981], selenide and telluride anions [982–984], dialkylphosphonate anions [985], benzotriazolate anion [986], imidazolate anion [987], N-functionalized amide anions [988–990] and transition metal complexes [991–993]. Scheme 3.291 shows several representative reactions: the preparation of *N*-alkynyl carbamates **733** by alkynylation of carbamates **732** using alkynyliodonium triflates **731** [989], synthesis of ynamides **735** by the alkynylation/desilylation of tosylanilides **734** using trimethylsilylethynyl(phenyl)iodonium triflate [990] and the preparation of Ir(III) σ-acetylide complex **737** by the alkynylation of Vaska's complex **736** [991].

Alkynyl(phenyl)iodonium salts can be coupled with organocopper reagents [994], or with organoboronic acids or organostannanes in the presence of Cu(I) catalysts [995, 996]. For example, the copper iodide-catalyzed cross-coupling and carbonylative coupling reactions of alkynyliodonium salts **738** with organoboronic acids **739** or organostannanes (R^2SnBu_3) under mild conditions afford acetylenes **740** and alkynyl ketones **741,** respectively, in high yields (Scheme 3.292) [996]. Interestingly, alkynyl(phenyl)-iodonium tetrafluoroborates **738** are more efficient in these coupling reactions than the corresponding iodonium triflates and tosylates.

Various five-membered heterocycles can be prepared by inter- or intramolecular addition/cyclizations of appropriate nucleophiles with alkynyliodonium salts via alkylidene carbene intermediates [856, 978, 979]. The intermolecular variant of this cyclization has been employed in the synthesis of 3-substituted-5,6-dihydroimidazo[2,1-*b*]thiazoles [997], 2-substituted imidazo[1,2-*a*]pyrimidines [998] and 2-substituted-imidazo[1,2-*a*]pyridines [999]. In a representative example, 2-substituted imidazo[1,2-*a*]pyridines **744** were synthesized in good yield by cyclocondensation of 2-aminopyridine (**742**) with alkynyl(phenyl)iodonium tosylates **743** under mild conditions (Scheme 3.293) [999]. The mechanism of this cyclization involves

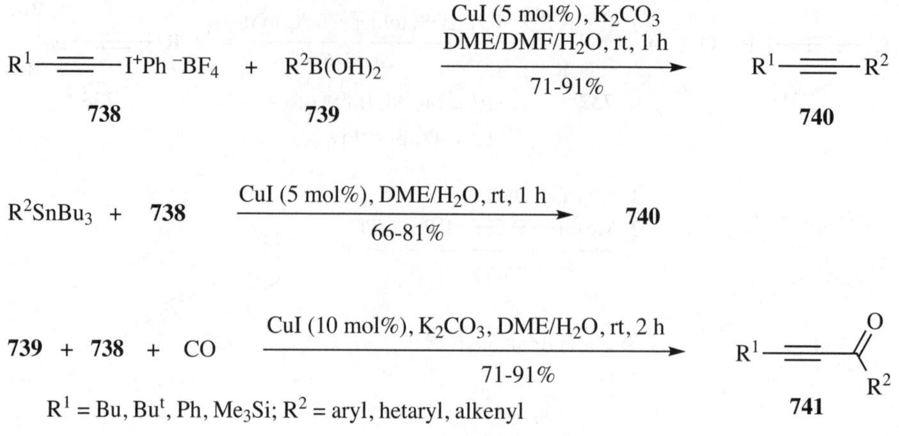

$R^1—≡—I^+Ph\ ^-BF_4$ + $R^2B(OH)_2$ $\xrightarrow[71\text{-}91\%]{CuI\ (5\ mol\%),\ K_2CO_3,\ DME/DMF/H_2O,\ rt,\ 1\ h}$ $R^1—≡—R^2$

738 **739** **740**

R^2SnBu_3 + **738** $\xrightarrow[66\text{-}81\%]{CuI\ (5\ mol\%),\ DME/H_2O,\ rt,\ 1\ h}$ **740**

739 + **738** + CO $\xrightarrow[71\text{-}91\%]{CuI\ (10\ mol\%),\ K_2CO_3,\ DME/H_2O,\ rt,\ 2\ h}$ $R^1—≡—C(O)R^2$ **741**

R^1 = Bu, Bu^t, Ph, Me_3Si; R^2 = aryl, hetaryl, alkenyl

Scheme 3.292

initial nucleophilic addition of the amino group of 2-aminopyridine to the triple bond of the alkynyliodonium salt followed by generation and subsequent cyclization of the intermediate alkylidene carbene (Scheme 3.290).

Ochiai and coworkers have investigated the mechanism for the one-pot synthesis of 2,4-disubstituted thiazoles **747** by cyclocondensation of alkynyliodonium salts **745** with thioureas or thioamides **746** (Scheme 3.294) [1000]. This reaction was originally reported by Wipf and Venkatraman in 1996 [1001]. Ochiai and coworkers have isolated and identified by X-ray analysis intermediate products **750** (as mesylate or tetrafluoroborate salts), which suggests a mechanism involving Michael addition of sulfur nucleophile **746** to alkynyliodonium salt **745** giving intermediate alkylideneiodonium ylide **748** followed by the 1,2-rearrangement of sulfenyl groups in the resulting alkylidene carbene **749** (Scheme 3.294) [1000].

The intramolecular variant of the alkylidene carbene cyclization is achieved by treating functionalized alkynyliodonium salts with a suitable nucleophile. These cyclizations are exemplified by the following works: the preparation of various functionalized 2,5-dihydrofurans by treatment of 3-alkoxy-1-alkynyl-(phenyl)iodonium triflates with sodium benzenesulfinate [1002], employment of the alkylidene carbene cyclization in the total syntheses of natural products agelastatin A and agelastatin B [1003] and preparation of the tricyclic core of (±)-halichlorine through the use of an alkynyliodonium salt/alkylidenecarbene/1,5 C–H insertion sequence [1004]. In particular, Wardrop and Fritz have employed the sodium benzenesulfinate-induced cyclization of alkynyliodonium triflate **751** for the preparation of dihydrofuran **752** (Scheme 3.295), which is a key intermediate product in the total synthesis of (±)-magnofargesin [1002].

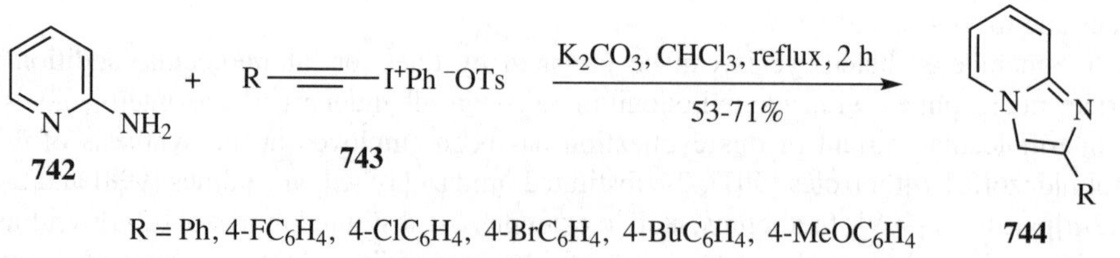

742 (2-aminopyridine, N, NH_2) + $R—≡—I^+Ph\ ^-OTs$ **743** $\xrightarrow[53\text{-}71\%]{K_2CO_3,\ CHCl_3,\ reflux,\ 2\ h}$ **744** (imidazopyridine, N, N, R)

R = Ph, $4\text{-}FC_6H_4$, $4\text{-}ClC_6H_4$, $4\text{-}BrC_6H_4$, $4\text{-}BuC_6H_4$, $4\text{-}MeOC_6H_4$

Scheme 3.293

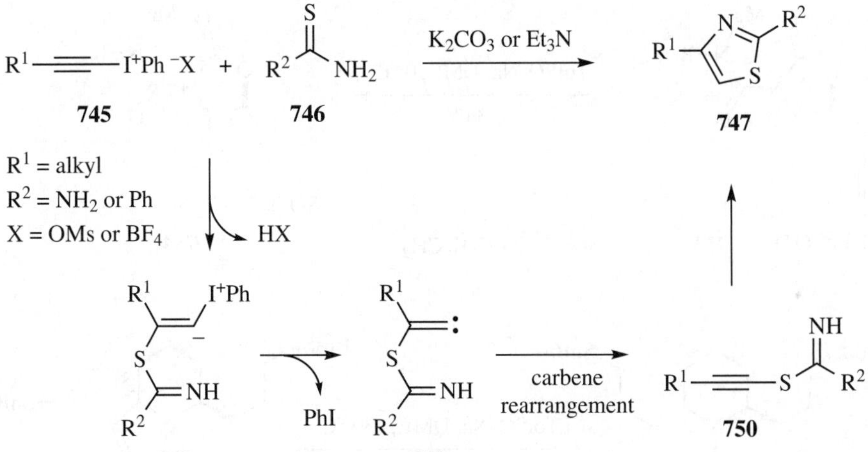

R1 = alkyl
R2 = NH2 or Ph
X = OMs or BF4

Scheme 3.294

Feldman and coworkers have employed the sodium *p*-toluenesulfinate-induced cyclizations of alkynyliodonium salts **753** and **755** in the preparation of compounds **754** and **756** (Scheme 3.296), which are key intermediates in the total syntheses of agelastatins [1003] and (±)-halichlorine, respectively [1004].

Waser and coworkers have demonstrated that triisopropylsilyl and trimethylsilyl substituted ethynylbenziodoxolones (Section 2.1.8.1.8) are excellent acetylene transfer reagents, both in metal-free and metal-catalyzed reactions [858, 1005–1012]. In particular, reagent **758** efficiently alkynylates β-dicarbonyl compounds and other C—H acidic substrates **757** in the presence of fluoride source in THF under mild conditions (Scheme 3.297) [1012]. Indoles (e.g., **759,** Scheme 3.297), pyrroles and thiophenes are directly alkynylated by reagent **760** in the presence of AuCl as catalyst [1005, 1006, 1008, 1010]. Additional examples include palladium-catalyzed intramolecular oxyalkynylation of non-activated olefins using triisopropylsilylethynylbenziodoxolone [1011], palladium-catalyzed alkynylations of *o*-allylphenols and carboxylic acids [858] and *para*-selective gold-catalyzed direct alkynylation of anilines [1009]. Aubineau and Cossy have found that trimethylsilyl substituted ethynylbenziodoxolone is a useful reagent for direct chemoselective alkynylation of *N*-sulfonylamides [1013].

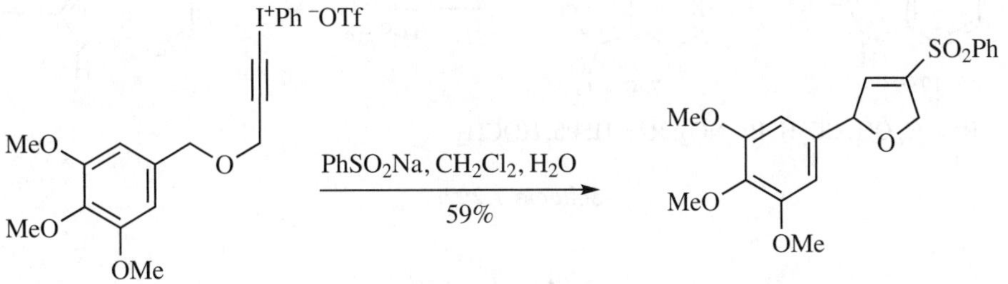

Scheme 3.295

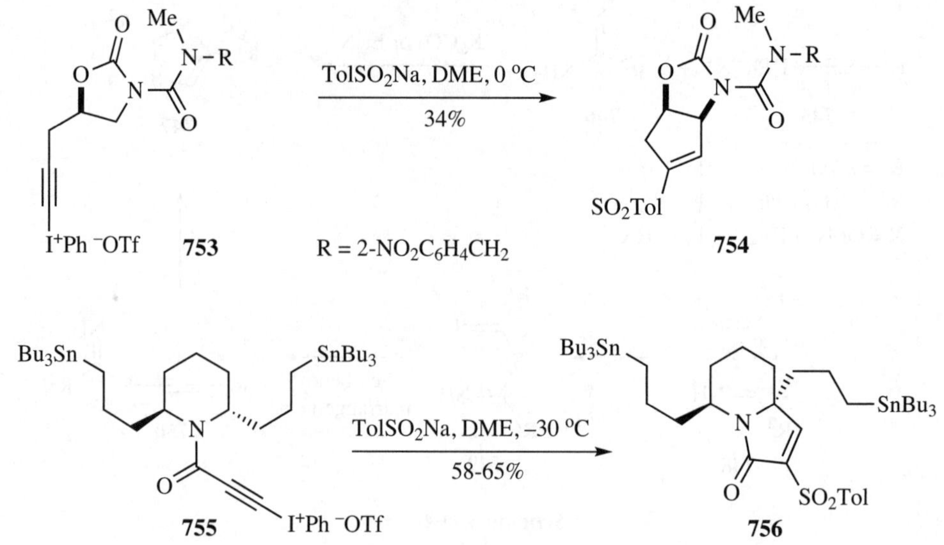

O Me N–R O N O I$^+$Ph $^-$OTf **753**

TolSO$_2$Na, DME, 0 °C
34%

O Me N–R O N O SO$_2$Tol **754**

R = 2-NO$_2$C$_6$H$_4$CH$_2$

Bu$_3$Sn SnBu$_3$ N O I$^+$Ph $^-$OTf **755**

TolSO$_2$Na, DME, –30 °C
58-65%

Bu$_3$Sn N SnBu$_3$ O SO$_2$Tol **756**

Scheme 3.296

EWG O OR^2 R^1 **757** + SiMe$_3$ I O O **758**

Bu$_4$NF, THF, –78 °C to rt
63-98%

EWG O OR^2 R^1

EWG = C(O)R, CN, NO$_2$; R^1 = Me, Et, PhCH$_2$, 4-BrC$_6$H$_4$CH$_2$, Ph, allyl; R^2 = Me, Et, Bu^t

R^1 R^2 N H **759** + SiPr$^i_3$ I O O **760**

AuCl (1-5 mol%)
Et$_2$O, 23 °C
67-93%

SiPr$^i_3$ R^1 R^2 N H

R^1 = H, OH, CO$_2$H, Br, NO$_2$; R^2 = H, Ph, HOCH$_2$

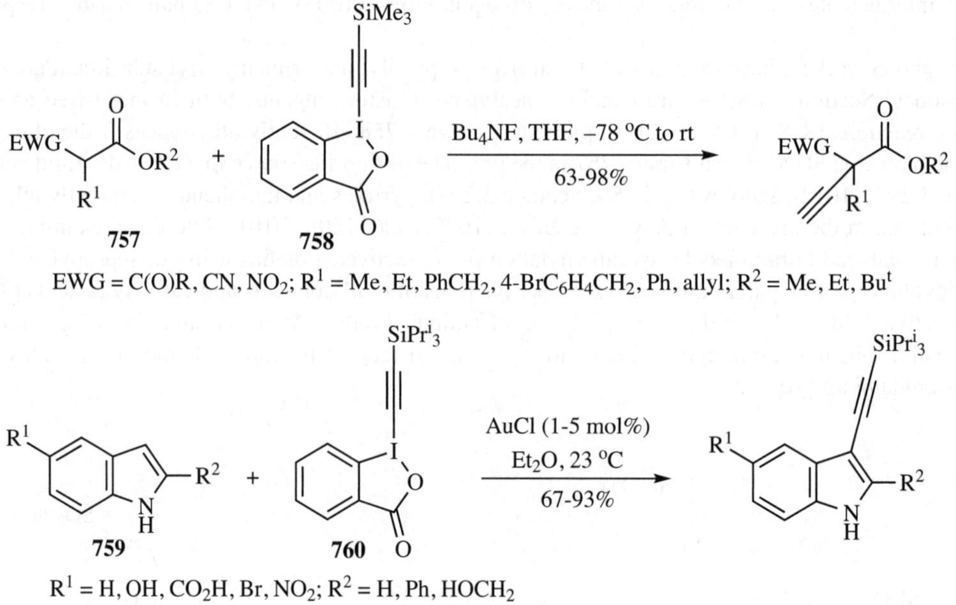

Scheme 3.297

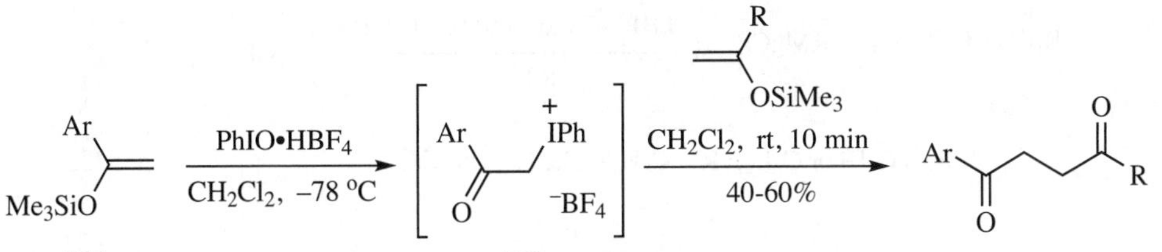

Ar, Me_3SiO **761** —$PhIO{\bullet}HBF_4$, CH_2Cl_2, –78 °C→ [Ar, $\overset{+}{I}Ph$, O, $^{-}BF_4$] **762** —R, $OSiMe_3$; CH_2Cl_2, rt, 10 min; 40-60%→ Ar, O, O, R

Ar = Ph, 4-MeC_6H_4, 4-ClC_6H_4, 4-$NO_2C_6H_4$, 4-$MeOC_6H_4$
R = alkyl or aryl

Scheme 3.298

3.1.22.4 Alkylations and Fluoroalkylations

The unstable β-oxoalkyl(phenyl)iodonium salts **762,** generated *in situ* by a low-temperature reaction of silyl enol ethers **761** with a complex of iodosylbenzene and tetrafluoroboric acid (Section 2.1.9.5), have been utilized in synthetically useful carbon–carbon bond forming reactions with various silyl enol ethers (Scheme 3.298) and other C–nucleophiles to afford the respective products of C—C bond formation [209, 1014, 1015].

The relatively stable (arylsulfonylmethyl)iodonium salts **763** (Section 2.1.9.5) are efficient electrophilic alkylating reagents towards various organic nucleophiles (thiophenolate anion, amines, pyridine, triphenyl phosphine and silyl enol ethers). All these reactions proceed under mild conditions and selectively afforded the appropriate product of alkylation along with iodobenzene as the by-product (Scheme 3.299) [1016].

Fluoroalkyl(aryl)iodonium salts are the most stable and practically important class of alkyl(aryl)iodonium derivatives. The application of such salts as electrophilic fluoroalkylating reagents was reviewed in 1996 by Umemoto [1017]. Perfluoroalkyl(phenyl)iodonium triflates (FITS reagents) **764** are efficient perfluoroalkylating reagents toward various nucleophilic substrates, such as arenes, carbanions, alkynes, alkenes, carbonyl compounds, amines, phosphines and sulfides [1017]. Scheme 3.300 shows several representative examples of electrophilic perfluoroalkylations using FITS reagents.

(Dihydroperfluoroalkyl)phenyliodonium triflates, $R_fCH_2I(Ph)OTf$, are electrophilic fluoroalkylating reagents with a reactivity pattern similar to FITS reagents **764** [1017]. The fluoroalkylation of amines is a particularly important reaction. For example, trifluoroethyl(phenyl)iodonium triflate **765** has been used for the N-trifluoroethylation of aminoalcohols (Scheme 3.301) [1018].

Likewise, fluoroalkyliodonium salt **766** (Section 2.1.9.5) is a useful reagent for fluoroalkylation of amino acids and peptides [1019–1024]. In particular, the reaction of iodonium salt **766** with the *tert*-butyl carboxyl ester of tyrosine (**767**) in the presence of collidine results in quantitative formation of the monoalkylation product **768** (Scheme 3.302) [1021, 1024]. Owing to this reactivity, iodonium salt **766** and other

$ArSO_2CH_2\overset{+}{I}Ph\ ^{-}OTf$ + Nu: —CH_2Cl_2, rt; 80-95%→ $ArSO_2CH_2Nu$

763 Ar = Ph or 4-MeC_6H_4
Nu: = PhS^-, PhO^-, R_3N, Ph_3P, silyl enol ethers, etc.

Scheme 3.299

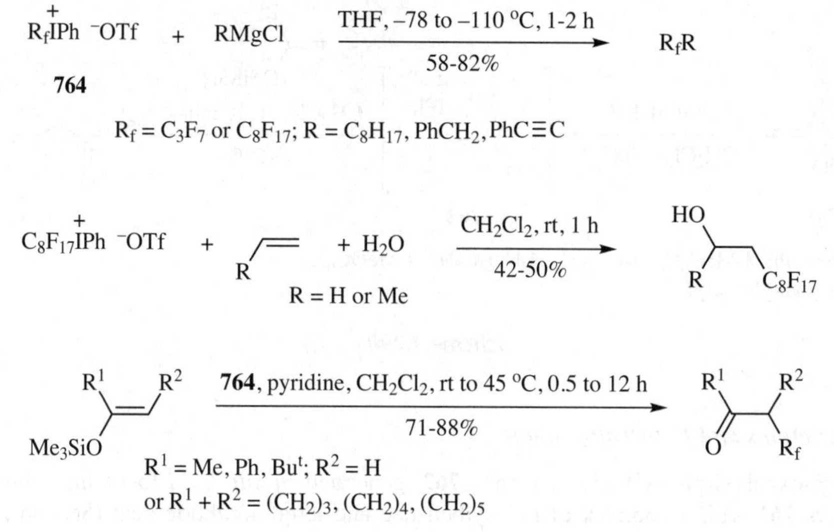

$R_f\overset{+}{I}Ph\ {}^{-}OTf$ + RMgCl $\xrightarrow[58\text{-}82\%]{THF,\ -78\ to\ -110\ ^\circ C,\ 1\text{-}2\ h}$ R_fR

764

$R_f = C_3F_7$ or C_8F_{17}; R = C_8H_{17}, $PhCH_2$, PhC≡C

$C_8F_{17}\overset{+}{I}Ph\ {}^{-}OTf$ + R–CH=CH$_2$ + H_2O $\xrightarrow[42\text{-}50\%]{CH_2Cl_2,\ rt,\ 1\ h}$ HO, R, C_8F_{17}

R = H or Me

R^1, R^2, Me_3SiO $\xrightarrow[71\text{-}88\%]{\mathbf{764},\ pyridine,\ CH_2Cl_2,\ rt\ to\ 45\ ^\circ C,\ 0.5\ to\ 12\ h}$ R^1, R^2, O, R_f

R^1 = Me, Ph, Bu^t; R^2 = H
or $R^1 + R^2 = (CH_2)_3, (CH_2)_4, (CH_2)_5$

Scheme 3.300

$CF_3CH_2\overset{+}{I}Ph\ {}^{-}OTf$ + (prolinol: N–H, OH) $\xrightarrow[88\%]{CH_2Cl_2,\ 25\ ^\circ C}$ (N–CH_2CF_3, OH)

765

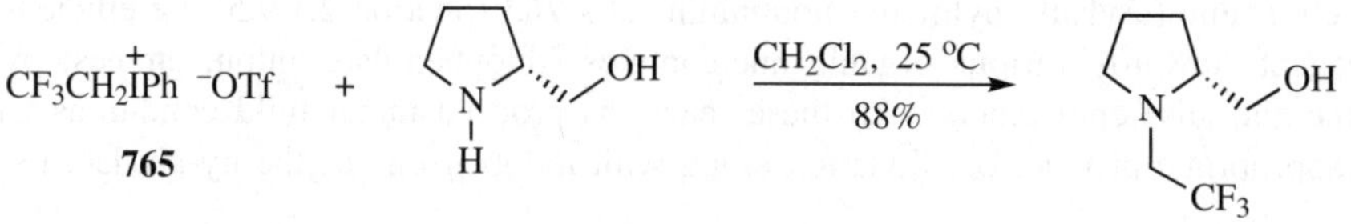

Scheme 3.301

(dihydroperfluoroalkyl)phenyliodonium triflimides can be used as fluorous capping reagents for facile purification of peptides synthesized on a solid phase [1021, 1024]. It has also been demonstrated that (dihydroperfluoroalkyl)phenyliodonium triflimide salts are useful for the regioselective N- or C-fluoroalkylation of imidazoles [1025].

Togni and coworkers have found that 1-trifluoromethylbenziodoxole **770** is a useful reagent for electrophilic trifluoromethylation of nucleophilic substrates. This reagent, in particular, reacts with β-ketoesters **769** under mild conditions in the presence of potassium carbonate to give α-trifluoromethylated product **771** in good yield (Scheme 3.303) [1026, 1027]. Likewise, this mild electrophilic trifluoromethylating

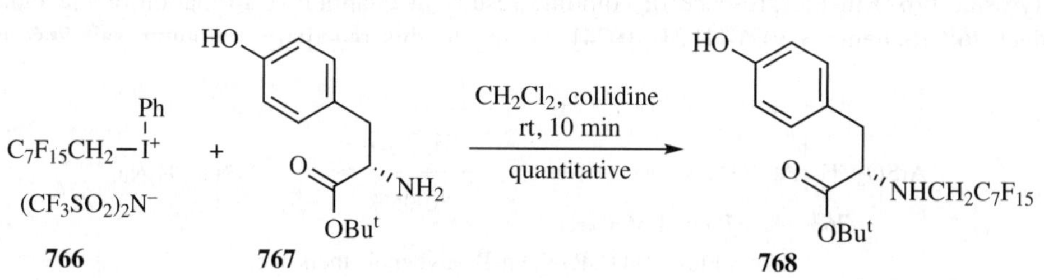

$C_7F_{15}CH_2$–I^+(Ph) $(CF_3SO_2)_2N^-$ **766** + (HO, O, NH_2, OBu^t) **767** $\xrightarrow[quantitative]{CH_2Cl_2,\ collidine,\ rt,\ 10\ min}$ (HO, O, $NHCH_2C_7F_{15}$, OBu^t) **768**

Scheme 3.302

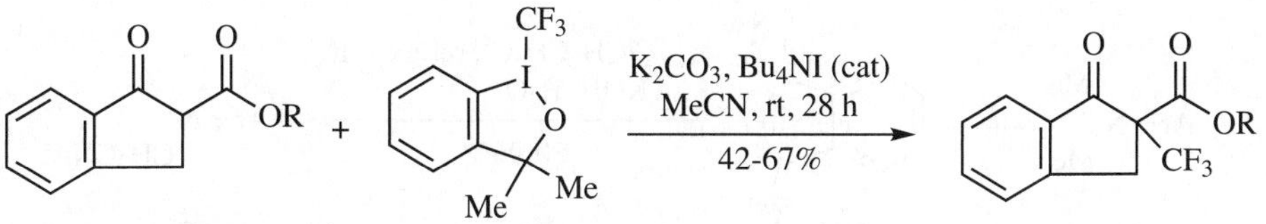

769 R = Me, Et, Bu^t + **770** → K_2CO_3, Bu_4NI (cat), MeCN, rt, 28 h, 42-67% → **771**

Scheme 3.303

reagent can be used to transfer a CF_3 group to other C-centered nucleophiles, such as α-nitro esters [1027], as well as to S-centered nucleophiles (thiols and S-hydrogen phosphorothioates) [1027–1029], O-centered nucleophiles (alcohols, phenols and triflate anions) [1030–1033], N-centered nucleophiles (azoles and nitriles) [1034, 1035] and P-centered nucleophiles (secondary or primary aryl- and alkylphosphines) [1036–1038].

MacMillan and coworkers have reported the combination of organo- and Lewis acid-catalysis for the asymmetric α-trifluoromethylation of aldehydes **772** using hypervalent iodine reagent **770** and imidazolidinone catalyst **773** (Scheme 3.304) [1039]. The use of Lewis acid is crucial to obtain the product in high yield. A range of Lewis acids have been tried in this reaction and CuCl gave the best yield and enantioselectivity. The proposed mechanism starts from the initial Lewis acid-mediated opening of the benziodoxole reagent **770** followed by a sequence of steps including the formation of chiral enamine intermediate from the aldehyde and imidazolidinone catalyst and trifluoromethylation of this intermediate by the activated iodine(III) reagent [1039].

The analog of **770** bearing a $PhSO_2CF_2$ substituent on iodine (Section 2.1.8.1.11) has been found to act as an electrophilic (phenylsulfonyl)difluoromethylating reagent for various S-nucleophiles under mild reaction conditions [1040].

3.1.22.5 Cyanation

The stable cyanobenziodoxole **775** can be used as an efficient cyanating reagent toward *N,N*-dialkylarylamines. In a typical example, reagent **775** reacts with *N,N*-dimethylanilines **774** in 1,2-dichloroethane at reflux to afford the respective *N*-cyanomethyl-*N*-methylanilines **776** in good yield (Scheme 3.305) [1041].

This procedure has been applied to the synthesis of *N*-cyanomethyl-*N*-cyclopropylamine, which is a possible metabolite of cyclopropylamine-derived drugs [1042].

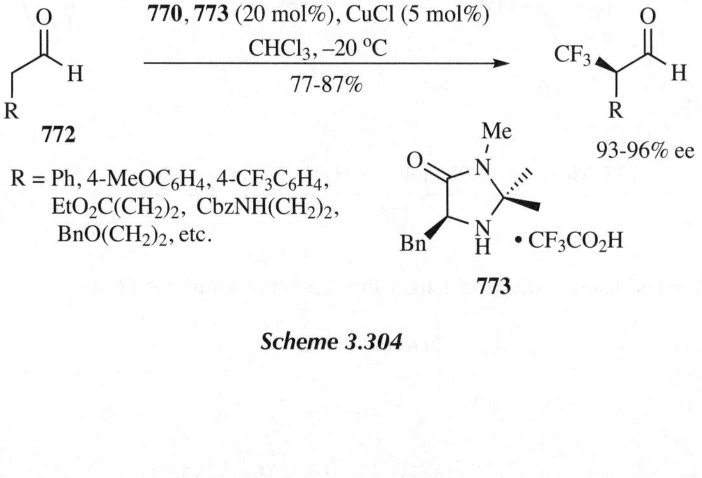

772 → **770**, **773** (20 mol%), CuCl (5 mol%), $CHCl_3$, –20 °C, 77-87% → product, 93-96% ee

R = Ph, 4-$MeOC_6H_4$, 4-$CF_3C_6H_4$, $EtO_2C(CH_2)_2$, $CbzNH(CH_2)_2$, $BnO(CH_2)_2$, etc.

773 • CF_3CO_2H

Scheme 3.304

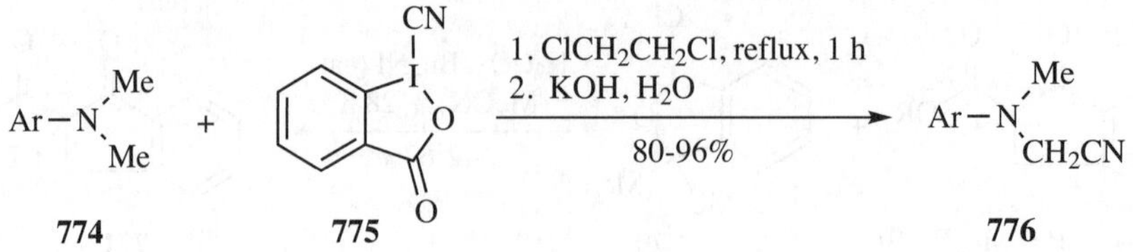

Ar = Ph, 4-BrC_6H_4, 4-MeC_6H_4, 1-naphthyl

Scheme 3.305

3.1.23 Reactions of Iodonium Ylides

Iodonium ylides can serve as convenient precursors to the respective carbene intermediates under thermal, photochemical, or catalytic conditions. A detailed discussion of the reaction mechanisms and synthetic applications of iodonium ylides as carbene precursors can be found in the 2004 review of Muller [1043].

Bis(methoxycarbonyl)(phenyliodinio)methanide (**778**), the most common iodonium ylide derived from malonate methyl ester, has found synthetic applications in the C—H insertion reactions [1044–1048] and the cyclopropanation of alkenes [1049–1055], including enantioselective cyclopropanations in the presence of chiral rhodium complexes [1056–1058]. Representative examples of these reactions are shown in Scheme 3.306 and include the BF_3-catalyzed bis(carbonyl)alkylation of 2-alkylthiophenes **777** [1045] and the optimized procedure for rhodium-catalyzed cyclopropanation of styrene **779** [1052].

A particularly useful reagent in these carbenoid reactions is the highly soluble and reactive iodonium ylide **780** derived from malonate methyl ester and bearing an *ortho* methoxy substituent on the phenyl ring [1059]. This reagent shows higher reactivity than common phenyliodonium ylides in the Rh-catalyzed cyclopropanation, C–H insertion and transylidation reactions under homogeneous conditions. Scheme 3.307 shows representative examples of the carbenoid reactions of ylide **780** [1059].

Another synthetically useful reagent of this type is 5,5-dimethyl-1,3-cyclohexanedione phenyliodonium ylide (**781**) (Scheme 3.308), a relatively stable iodonium ylide synthesized by condensation of $PhI(OAc)_2$ with dimedone under basic condition [1060, 1061]. Under catalytic, thermal, or photochemical conditions, ylide **781** serves as an excellent carbenoid precursor; the transfer of such a carbenoid moiety to a suitable

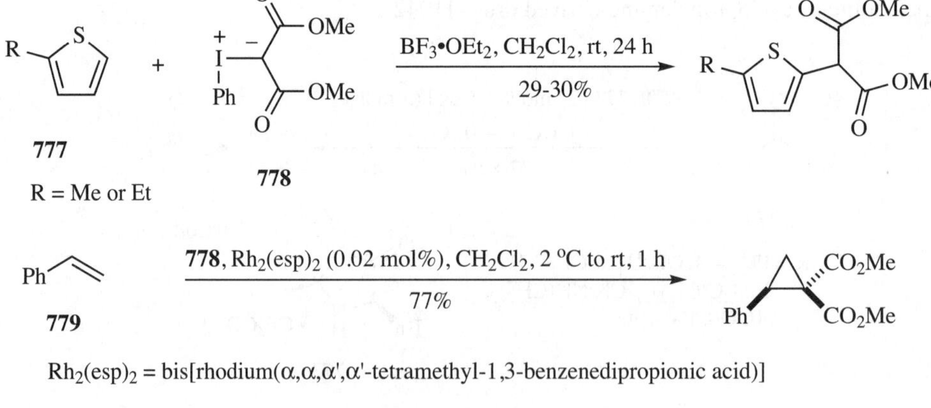

R = Me or Et

$Rh_2(esp)_2$ = bis[rhodium(α,α,α',α'-tetramethyl-1,3-benzenedipropionic acid)]

Scheme 3.306

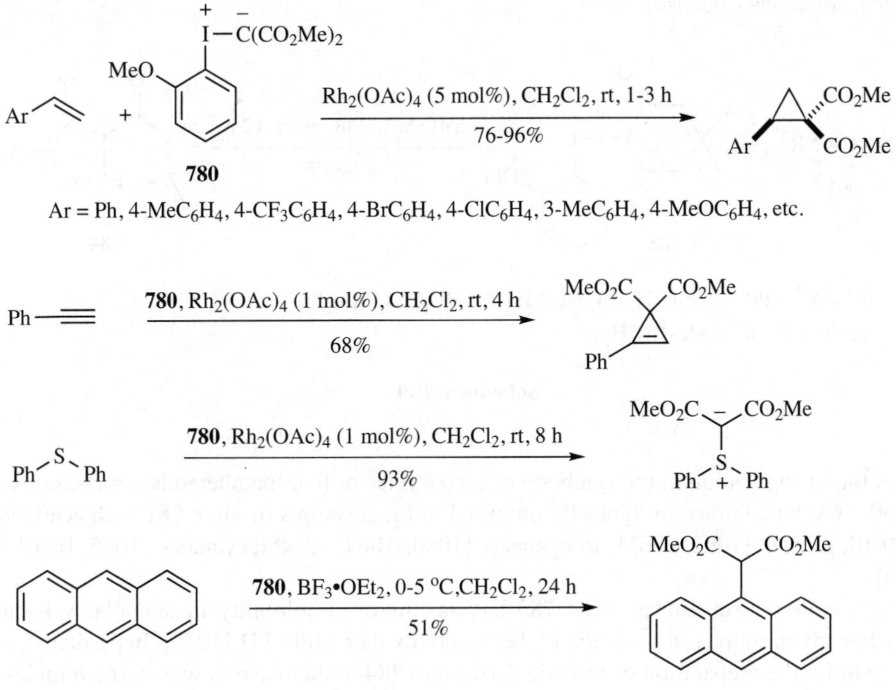

$I^{+}-C^{-}(CO_2Me)_2$

MeO

Ar + **780** → $Rh_2(OAc)_4$ (5 mol%), CH_2Cl_2, rt, 1-3 h; 76-96% → CO_2Me, CO_2Me, Ar

Ar = Ph, 4-MeC_6H_4, 4-$CF_3C_6H_4$, 4-BrC_6H_4, 4-ClC_6H_4, 3-MeC_6H_4, 4-$MeOC_6H_4$, etc.

Ph → **780**, $Rh_2(OAc)_4$ (1 mol%), CH_2Cl_2, rt, 4 h; 68% → MeO_2C, CO_2Me, Ph

Ph–S–Ph → **780**, $Rh_2(OAc)_4$ (1 mol%), CH_2Cl_2, rt, 8 h; 93% → MeO_2C – CO_2Me, $Ph-S^{+}-Ph$

780, BF_3•OEt_2, 0-5 °C, CH_2Cl_2, 24 h; 51% → MeO_2C, CO_2Me

Scheme 3.307

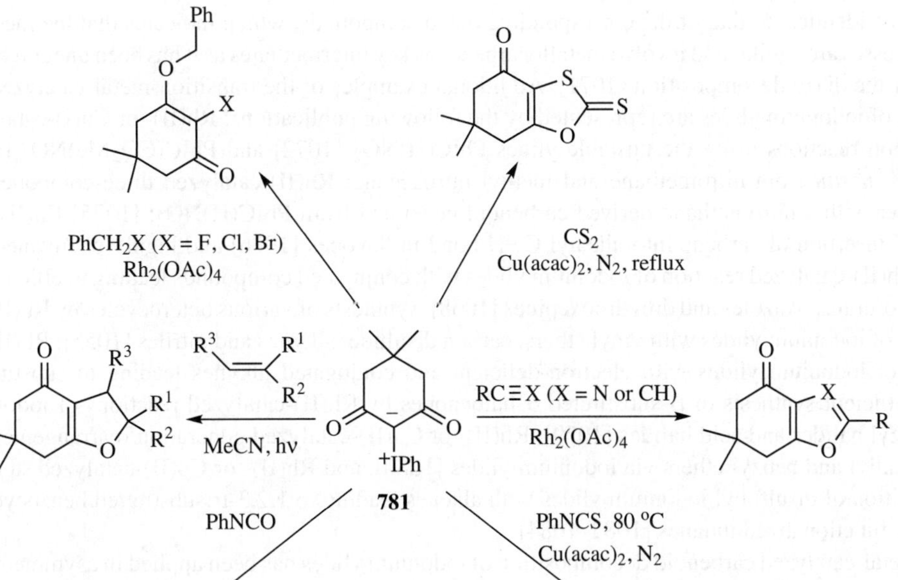

781 ($^{+}$IPh)

$PhCH_2X$ (X = F, Cl, Br), $Rh_2(OAc)_4$ → Ph, O, X, O

CS_2, $Cu(acac)_2$, N_2, reflux → O, S, S, O

R^3, R^1, R^2; MeCN, hν → O, R^3, R^1, R^2, O

RC≡X (X = N or CH), $Rh_2(OAc)_4$ → O, X, R, O

PhNCO → O, Ph, N, O, O

PhNCS, 80 °C, $Cu(acac)_2$, N_2 → O, S, =NPh, O

Scheme 3.308

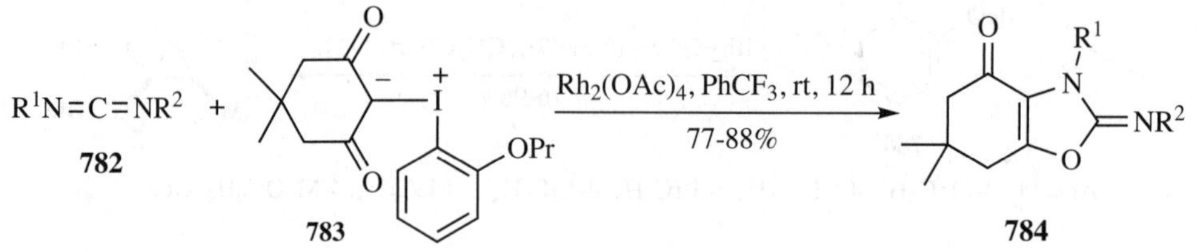

$R^1N{=}C{=}NR^2$ + **783** → ($Rh_2(OAc)_4$, $PhCF_3$, rt, 12 h; 77-88%) **784**

782 **783** **784**

$R^1 = R^2 = Pr^i$, *cyclo*-C_6H_{11}, 4-MeC_6H_4, 2,6-$Pr^i{}_2C_6H_3$, Me_3Si
or R^1 = Et, $R^2 = Me_2N(CH_2)_3$

Scheme 3.309

acceptor has found application in the synthesis of carbocycles or five-membered heterocycles as outlined in Scheme 3.308. Cycloadditions are typically observed in the reactions of ylide **781** with acetylenes [1062], ketenes [1060], nitriles [1062, 1063], isocyanates [1060, 1064], isothiocyanates [1065, 1066] and dienes [1067, 1068].

A modified, *o*-alkoxy-substituted ylide **783** has an improved solubility in nonpolar solvents, such as aromatic hydrocarbons and has a generally higher reactivity than ylide **781** [1069]. In particular, ylide **783** is a useful reagent for the preparation of oxazole derivatives **784** in the reaction with carbodiimides **782** under homogeneous conditions in the presence of Rh(II) or Cu(II) catalysts (Scheme 3.309) [1069].

The carbenoid reactions of iodonium ylides are effectively catalyzed by rhodium(II) or copper complexes [1043, 1058, 1070]. The product composition in the rhodium(II)-catalyzed reactions of iodonium ylides was found to be identical to that of the corresponding diazo compounds, which indicates that the mechanisms of both processes are similar and involve metallocarbenes as key intermediates as it has been unequivocally established for the diazo decomposition [1071]. Additional examples of the transition-metal-catalyzed carbenoid reactions of iodonium ylides are represented by the following publications: Rh(II)- or Cu(I)-catalyzed cyclopropanation reactions using the unstable ylides $PhIC(H)NO_2$ [1072] and $PhIC(CO_2Me)NO_2$ [1073, 1074] generated *in situ* from nitromethane and methyl nitroacetate; Rh(II)-catalyzed three-component coupling of an ether with a nitromethane-derived carbenoid generated from $PhIC(H)NO_2$; [1075] Rh(II)- or Cu(II)-catalyzed insertion of carbene into alkenyl C—H bond in flavones [1076] and highly phenylated ethylenes; [1077] Rh(II)-catalyzed reaction of iodonium ylides with conjugated compounds leading to efficient synthesis of dihydrofurans, oxazoles and dihydrooxepines [1068]; synthesis of various heterocycles by Rh(II)-catalyzed reactions of iodonium ylides with vinyl ethers, carbon disulfide, alkynes and nitriles [1055]; Rh(II)-catalyzed reaction of iodonium ylides with electron-deficient and conjugated alkynes leading to substituted furans [1078]; efficient synthesis of β-substituted α-haloenones by Rh(II)-catalyzed reactions of iodonium ylides with benzyl halides and acid halides [1079]; Rh(II)- or Cu(II)-catalyzed generation/rearrangement of onium ylides of allyl and benzyl ethers via iodonium ylides [1080]; and Rh(II)- or Cu(II)-catalyzed stereoselective cycloaddition of disulfonyl iodonium ylides with alkenes leading to 1,2,3-trisubstituted benzocyclopentenes [1081] or functionalized indanes [1082–1084].

The metal-catalyzed carbenoid decomposition of iodonium ylides has been applied in asymmetric reactions [1047, 1053, 1074, 1085]. For example, the copper(II)-catalyzed intramolecular C—H insertion of phenyliodonium ylide **785** in the presence of chiral ligands followed by hydrolysis and decarboxylation affords product **786** in moderate yield with up to 72% ee (Scheme 3.310) [1047].

Ochiai and coworkers have reported several useful reactions of the unstable monocarbonyl iodonium ylides **788**, which can be quantitatively generated from (*Z*)-(2-acetoxyvinyl)iodonium salts **787** via an ester exchange reaction with lithium ethoxide in THF at low temperature (Scheme 3.311) [1086–1088]. Ylide **788,** generated

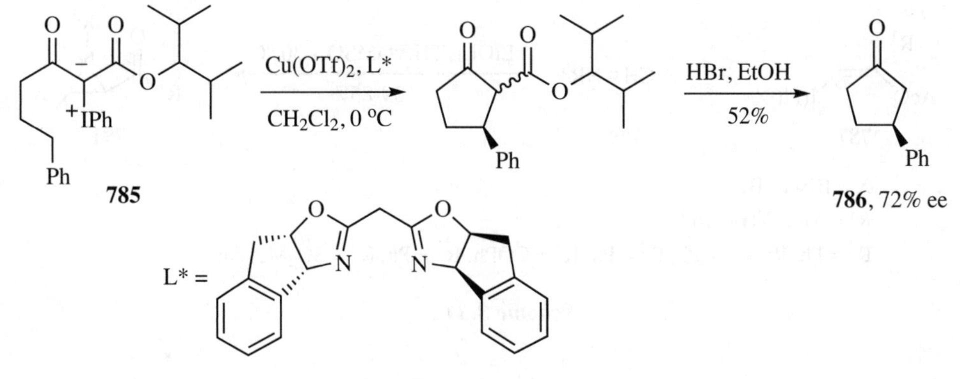

Cu(OTf)2, L*
CH2Cl2, 0 °C
HBr, EtOH
52%
+IPh
Ph
785
786, 72% ee
L* =

Scheme 3.310

in situ from iodonium salts **787,** reacts with aldehydes in THF/DMSO at low temperature to afford α,β-epoxy ketones **789** with predominant formation of the *trans* isomers. A Hammett reaction constant ($\rho = 2.95$) for this reaction indicates that monocarbonyl iodonium ylides **788** are moderately nucleophilic [1086].

Monocarbonyl iodonium ylides, generated *in situ* from iodonium salts **787,** undergo alkylidene transfer reactions to activated imines **790,** yielding 2-acylaziridines **791** in good yields (Scheme 3.312). The stereochemical outcome of this aziridination is dependent on both the activating groups of the imines and the reaction solvents; for example, aziridination of *N*-(2,4,6-trimethylbenzenesulfonyl)imines in THF affords *cis*-aziridines as a major product, while that of *N*-benzoylimines in THF/DMSO or THF gives the *trans* isomer stereoselectively [1087, 1088].

Treatment of iodonium tetrafluoroborates **792** with triethylamine in methanol in the presence of triphenylphosphine and aldehydes results in Wittig olefination to give products **793** (Scheme 3.313), which involves the intermediacy of monocarbonyl iodonium ylides **788** and their subsequent conversion into the respective phosphonium ylides upon the *in situ* reaction with Ph_3P [1089].

The interaction of monocarbonyl iodonium ylides, generated by the ester exchange of (*Z*)-(2-acetoxyvinyl)iodonium salts **792** with EtOLi, with organoboranes affords ketones **795,** probably via intermediate formation of the hitherto unknown α-boryl ketones **794** (Scheme 3.314) [1090].

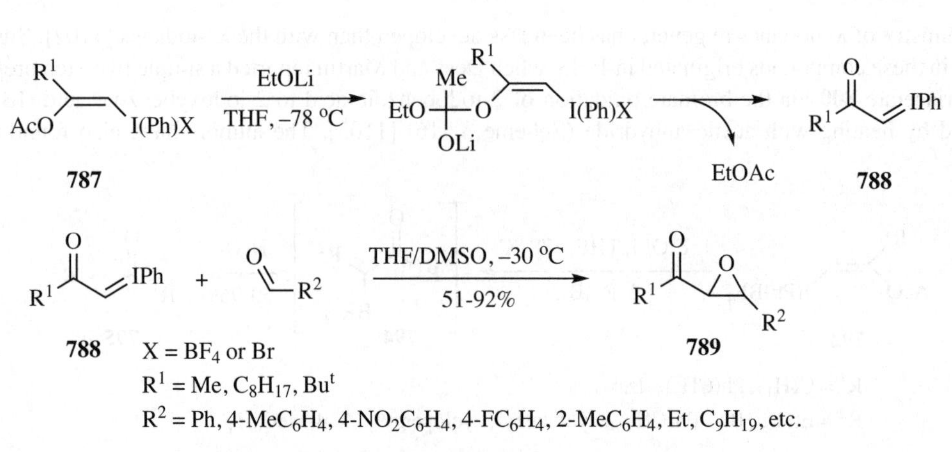

R1
AcO
I(Ph)X
EtOLi
THF, –78 °C
Me
EtO
O
OLi
I(Ph)X
EtOAc
R1
IPh
787
788
IPh
R2
THF/DMSO, –30 °C
51-92%
789

788 $X = BF_4$ or Br

$R^1 = Me, C_8H_{17}, Bu^t$

$R^2 = Ph, 4\text{-}MeC_6H_4, 4\text{-}NO_2C_6H_4, 4\text{-}FC_6H_4, 2\text{-}MeC_6H_4, Et, C_9H_{19}$, etc.

Scheme 3.311

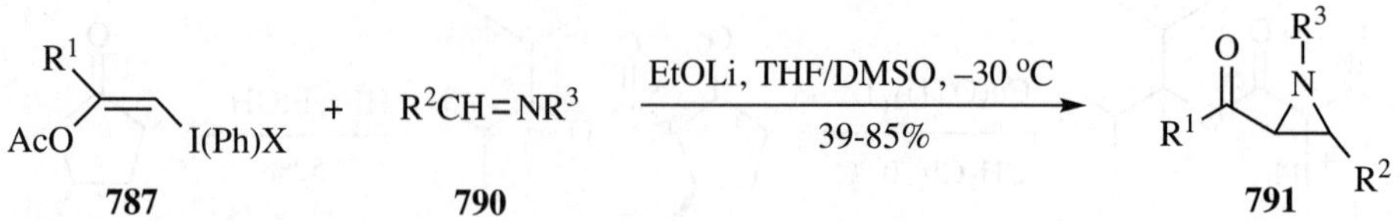

787 + $R^2CH{=}NR^3$ → (EtOLi, THF/DMSO, –30 °C; 39-85%) → 791

787: R^1, AcO, I(Ph)X; 790: $R^2CH{=}NR^3$; 791: R^1, O, N, R^3, R^2

X = BF_4 or Br
R^1 = Me, C_8H_{17}, Bu^t
R^2 = Ph, R^3 = SO_2Ph; R^2 = Ph, R^3 = COPh; R^2 = Ph, R^3 = SO_2Me; etc.

Scheme 3.312

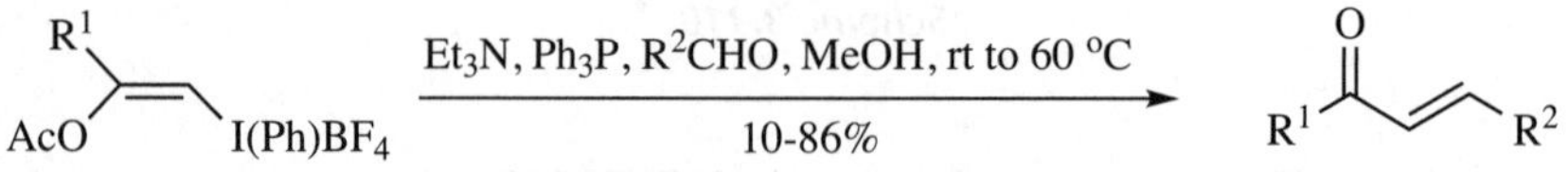

792 → (Et_3N, Ph_3P, R^2CHO, MeOH, rt to 60 °C; 10-86%) → 793

792: R^1, AcO, $I(Ph)BF_4$; 793: R^1, O, R^2

R^1 = Me, C_8H_{17}, $Ph(CH_2)_3$, Bu^t
R^2 = Ph, C_9H_{19}, 4-$MeOC_6H_4$, 4-MeC_6H_4, 4-ClC_6H_4, etc.

Scheme 3.313

The mixed phosphonium–iodonium ylides (Section 2.1.10.1), such as the tosylate **796,** represent a useful class of reagents that combine in one molecule the synthetic advantages of a phosphonium ylide and an iodonium salt [1091–1100]. Specifically, phosphorane-derived phenyliodonium tosylate **796** reacts with soft nucleophiles, such as iodide, bromide, benzenesulfinate and thiophenolate anions, to form selectively the respective α-functionalized phosphonium ylides **797** (Scheme 3.315), which can be further converted into alkenes (e.g., **798**) by the Wittig reaction with aldehydes [1092]. The analogous arsonium–iodonium ylides have a similar reactivity toward nucleophiles [1091, 1094, 1101].

3.2 Synthetic Applications of Iodine(V) Compounds

The chemistry of λ^5-iodanes in general has been less developed than with the λ^3-iodanes [1102]. Significant interest in these compounds originated in 1983, when Dess and Martin reported a simple two-step preparation of the triacetate **800** via the bromate oxidation of 2-iodobenzoic acid to 2-iodoxybenzoic acid (IBX, **799**) followed by heating with acetic anhydride (Scheme 3.316) [1103]. The authors have also found that the

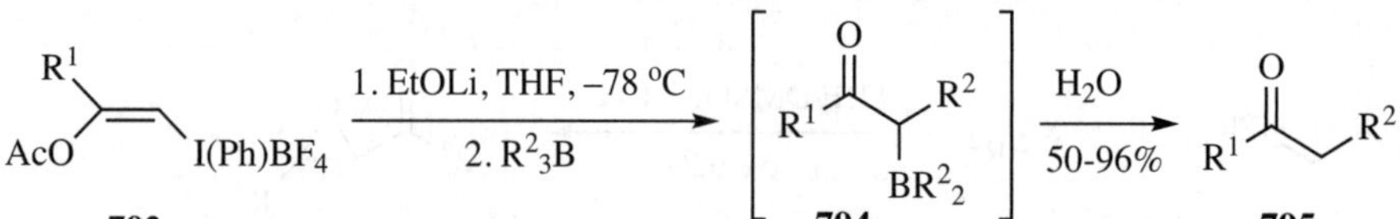

792 → (1. EtOLi, THF, –78 °C; 2. $R^2{}_3B$) → [794] → (H_2O; 50-96%) → 795

792: R^1, AcO, $I(Ph)BF_4$; 794: R^1, O, R^2, $BR^2{}_2$; 795: R^1, O, R^2

R^1 = C_8H_{17}, $Ph(CH_2)_3$, Bu^t
R^2 = Et, Bu, Bu^s, $Ph(CH_2)_3$, cyclohexyl, cyclopentyl, Ph, 4-MeC_6H_4

Scheme 3.314

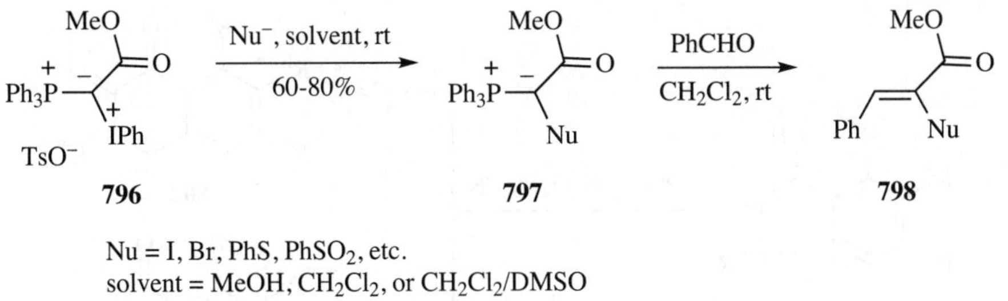

MeO, Ph_3P^+, O, I^+Ph, TsO^- **796** —(Nu^-, solvent, rt; 60-80%)→ MeO, Ph_3P^+, O, Nu **797** —(PhCHO; CH_2Cl_2, rt)→ MeO, O, Ph, Nu **798**

Nu = I, Br, PhS, $PhSO_2$, etc.
solvent = MeOH, CH_2Cl_2, or CH_2Cl_2/DMSO

Scheme 3.315

I, CO_2H —($KBrO_3$, H_2SO_4, 65 °C; 93%)→ **799** (IBX) —(Ac_2O, AcOH, 100 °C, 40 min; 93%)→ **800** (DMP)

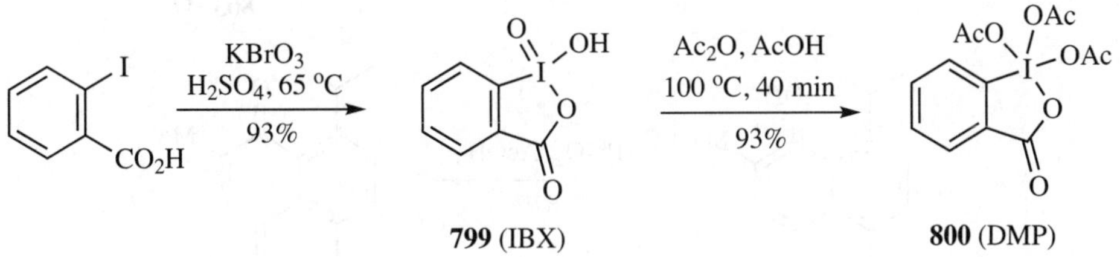

Scheme 3.316

triacetate **800,** which they referred to as periodinane, is a useful reagent for the facile and efficient oxidation of primary alcohols to aldehydes and secondary alcohols to ketones. Within a few years of publication of this paper, compound **800** found widespread application in organic synthesis under the name of the Dess–Martin periodinane (DMP).

Both IBX (**799**) and DMP (**800**) are now extensively employed in organic synthesis as mild and highly selective reagents for the oxidation of alcohols to carbonyl compounds, as well as for various other synthetically useful oxidative transformations [1102, 1104–1106]. For optimized experimental procedures for the preparation of IBX (Section 2.2.3.1) and DMP (Section 2.2.3.2), see Chapter 2.

Despite their importance, IBX and DMP are not perfect reagents and have some serious drawbacks. IBX is potentially explosive and it is insoluble in common organic solvents due to strong intermolecular secondary bonding, which creates a three-dimentional polymeric structure, while DMP is highly sensitive to moisture. In addition, IBX and DMP are not perfect with respect to the principles of green chemistry since they are normally used as non-recyclable, stoichiometric reagents in non-recyclable organic solvents, which have potentially damaging environmental effects. Several polymer-supported, recyclable IBX derivatives and analogs with improved properties have been introduced in the twenty-first century and utilized in organic synthesis (Chapter 5).

3.2.1 Noncyclic and Pseudocyclic Iodylarenes

Noncyclic iodylarenes have received only limited practical application because of their explosive properties and insolubility in organic solvents. In particular, iodylbenzene, $PhIO_2$, can oxidize alcohols to ketones [1107], or it can be used to generate alkoxy radicals from alcohols under photochemical conditions [1108]. Owing to its polymeric structure and insolubility in organic solvents, iodylbenzene has a relatively low reactivity as an oxidizing reagent and its reactions are usually conducted at high temperature or in the presence of a catalyst.

Several examples of synthetic application of iodylbenzene are illustrated below in Schemes 3.317–3.327. Specifically, iodylbenzene in an aqueous acetonitrile or acetic acid media oxidizes activated aromatic rings,

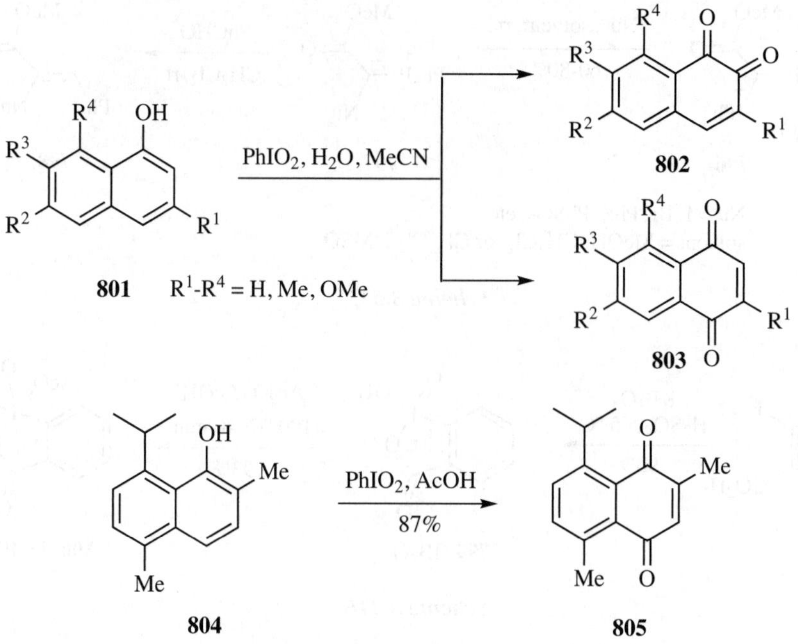

801 R^1-R^4 = H, Me, OMe

$PhIO_2$, H_2O, MeCN

802

803

804

$PhIO_2$, AcOH

87%

805

Scheme 3.317

yielding quinones or quinone imines [1109]. For example, substituted 1-naphthols **801** can be converted into corresponding 1,2- and 1,4-naphthoquinones **802** and **803** (Scheme 3.317) [1109]. This protocol was utilized in the synthesis of cadalenquinone (**805**), a naturally occurring sesquiterpene, starting from naphthol **804** [1110].

Several catalytic oxidative systems using iodylbenzene as a stoichiometric co-oxidant have been developed. Barton and coworkers have developed an efficient allylic oxidation protocol with 2-pyridineseleninic anhydride **806** (R = 2-Py) as the principal oxidant, generated *in situ* by oxidation of the corresponding diselenide **807** with iodylbenzene or 3-iodylbenzoic acid (Scheme 3.318) [1111]. This reaction proceeds in chlorobenzene at 100 °C within 2.5–3 h. Most likely, the initial oxidation of alkenes **808** leads to the formation of allylic alcohols, which undergo further oxidation into α,β-unsaturated ketones **809.** In contrast with the

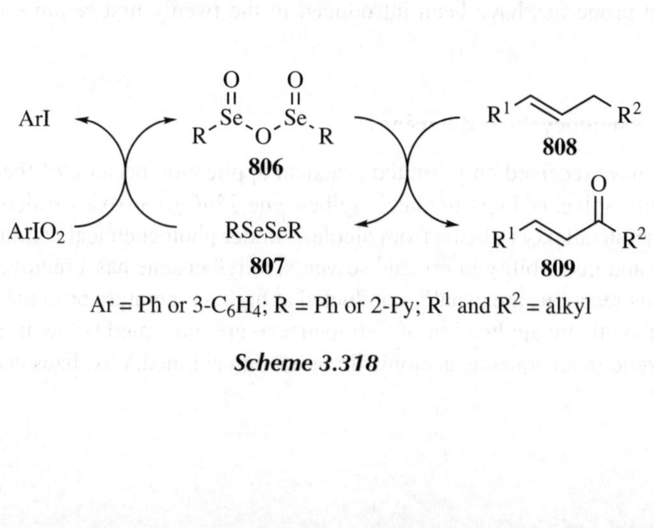

ArI

$ArIO_2$

806

RSeSeR

807

808

809

Ar = Ph or 3-C_6H_4; R = Ph or 2-Py; R^1 and R^2 = alkyl

Scheme 3.318

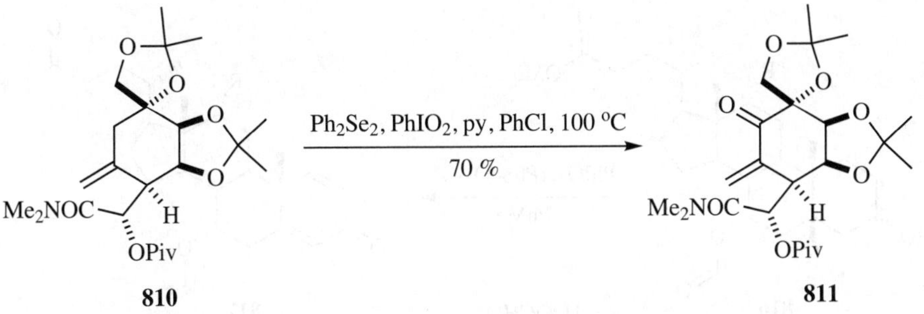

Ph2Se2, PhIO2, py, PhCl, 100 °C
70 %

810 811

Scheme 3.319

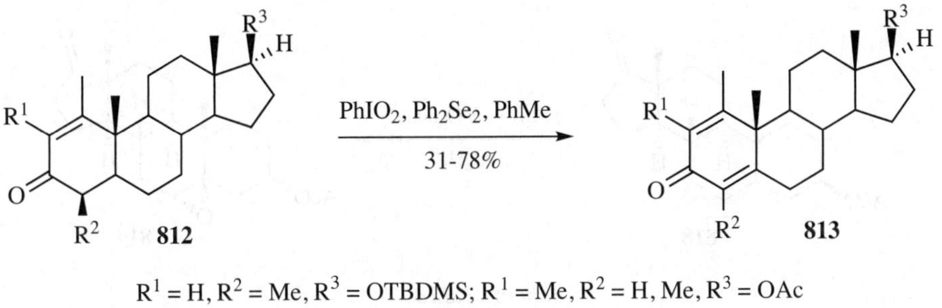

PhIO2, Ph2Se2, PhMe
31-78%

812 813

R^1 = H, R^2 = Me, R^3 = OTBDMS; R^1 = Me, R^2 = H, Me, R^3 = OAc

Scheme 3.320

classic allylic oxidation technique employing selenium dioxide, only a catalytic amount of the corresponding diselenide is required.

This convenient oxidation protocol has been used in several syntheses of complex organic molecules. In the stereoselective synthesis of (–)-tetrodotoxin by Du Bois and coworkers, the protected pentaol **810** was oxidized with $PhIO_2/Py_2Se_2$ to afford the unsaturated carbonyl compound **811** in a good yield (Scheme 3.319) [1112].

Based on this oxidizing system, a dehydrogenation protocol in the regioselective synthesis of ring A of polymethylated steroids has been developed. Intermediates **812** were converted into the corresponding 1,4-dienes **813,** which were key precursors to the target steroids (Scheme 3.320) [1113].

The procedure, employing 2-pyridyldiselenide, was used in the synthesis of tricyclo[5.4.0.0^{2,8}]undeca-3,5,9-triene, an interesting spiro compound with two mutually perpendicular π-systems [1114]. During this synthesis, the protected ketone **814** was oxidized to give the unsaturated ketone **815** in 51% yield (Scheme 3.321).

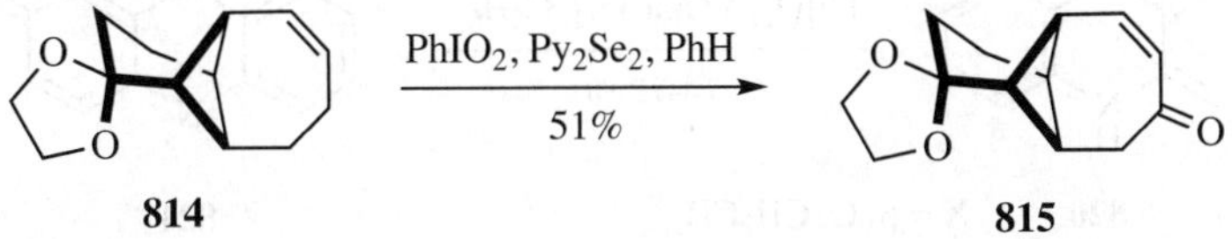

PhIO2, Py2Se2, PhH
51%

814 815

Scheme 3.321

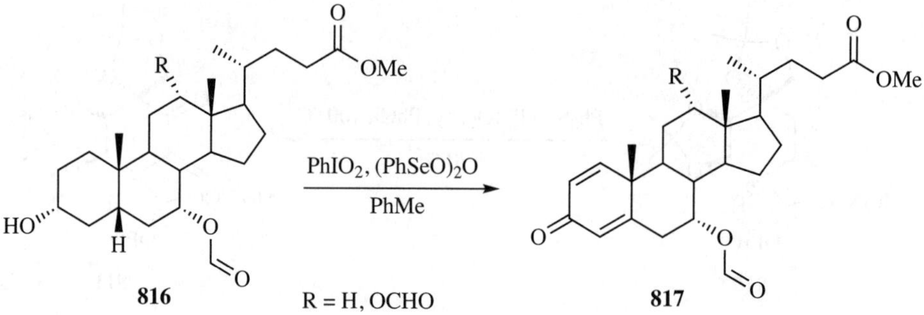

816 $\xrightarrow[\text{PhMe}]{\text{PhIO}_2,\ (\text{PhSeO})_2\text{O}}$ 817

R = H, OCHO

Scheme 3.322

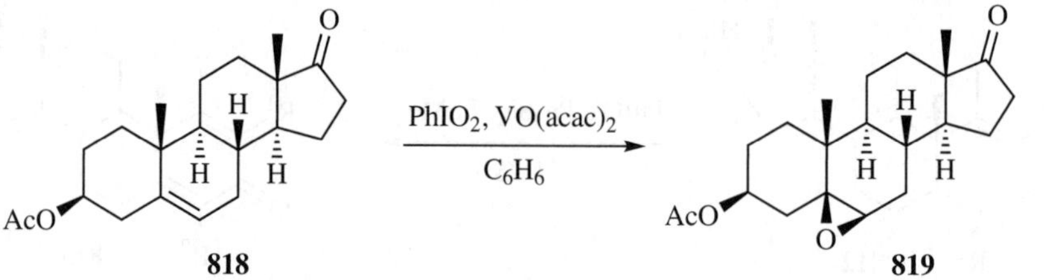

818 $\xrightarrow[\text{C}_6\text{H}_6]{\text{PhIO}_2,\ \text{VO(acac)}_2}$ 819

Scheme 3.323

A method for the preparation of allochenodeoxycholic and allocholic acids from the corresponding cholic acids has been reported. The key step in the synthesis is the oxidation–dehydrogenation of 3α-hydroxy-5β-bile acid formyl esters **816** to give oxodienes **817** (Scheme 3.322) [1115].

A series of oxidative transformations with iodylbenzene as the co-oxidant of vanadyl bis(acetylacetonate) have been reported [1116–1118]. In the presence of $VO(acac)_2$, iodylbenzene oxidizes Δ^5-steroids into epoxides; a radical mechanism was suggested for this reaction. Epoxidation of cholest-5-ene-3-one occurred with high α-selectivity, while the remaining substrates gave mainly β-epoxides. Oxidation of *trans*-dehydroepiandrosterone acetate (**818**) afforded epoxide **819** (Scheme 3.323) [1116].

A new route to quinone imines has been introduced based on this oxidizing system. The oxidation of the tricyclic scaffold **820** gives quinone imines **821** in moderate yields (Scheme 3.324) [1117].

Aryl sulfides **822** could be also converted by this reagent system into sulfoxides, sulfones and S-dealkylated products. Repeated treatment affords sulfones **823** in moderate yields (Scheme 3.325) [1118].

Kita and coworkers developed a catalytic asymmetric oxidation using iodoxybenzene in a cationic reversed micellar system in the presence of chiral tartaric acid derivatives. Under these conditions, sulfides **824**

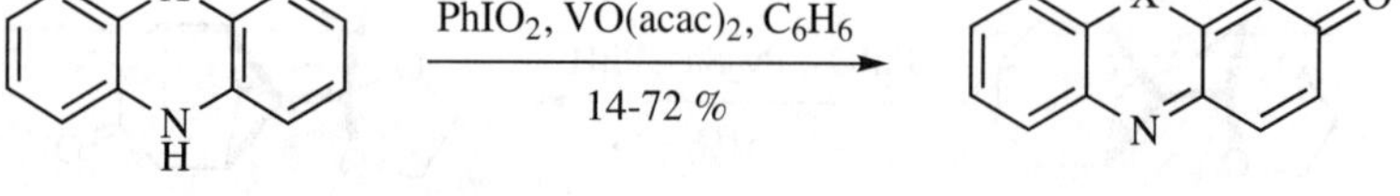

820 $\xrightarrow[\text{14-72 \%}]{\text{PhIO}_2,\ \text{VO(acac)}_2,\ \text{C}_6\text{H}_6}$ 821

X = S, O, CH_2CH_2

Scheme 3.324

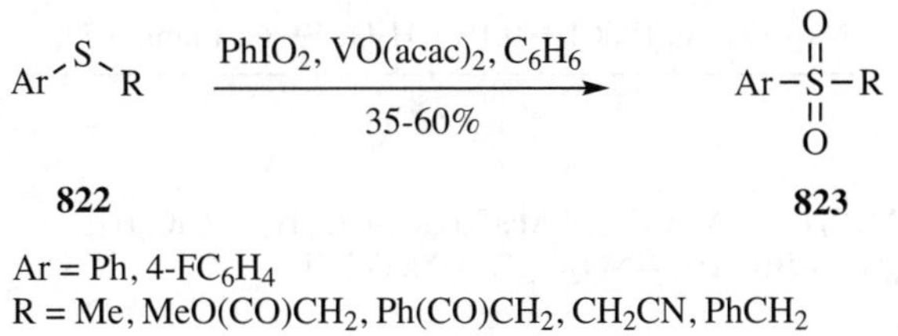

Ar–S–R (**822**) → ($PhIO_2$, $VO(acac)_2$, C_6H_6; 35-60%) → Ar–SO_2–R (**823**)

Ar = Ph, 4-FC_6H_4
R = Me, MeO(CO)CH_2, Ph(CO)CH_2, CH_2CN, $PhCH_2$

Scheme 3.325

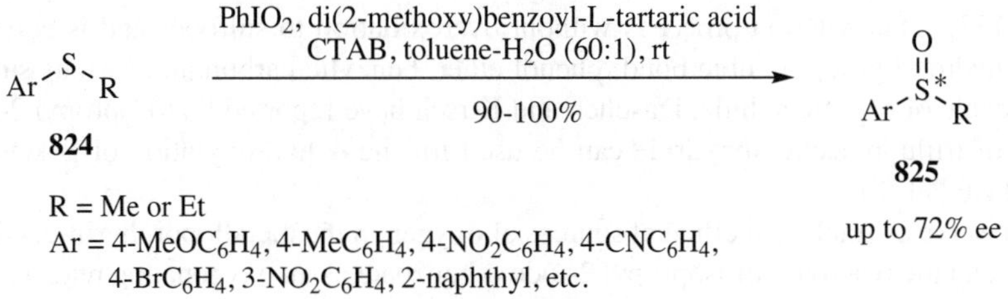

Ar–S–R (**824**) → ($PhIO_2$, di(2-methoxy)benzoyl-L-tartaric acid, CTAB, toluene-H_2O (60:1), rt; 90-100%) → Ar–S*(O)–R (**825**), up to 72% ee

R = Me or Et
Ar = 4-$MeOC_6H_4$, 4-MeC_6H_4, 4-$NO_2C_6H_4$, 4-CNC_6H_4, 4-BrC_6H_4, 3-$NO_2C_6H_4$, 2-naphthyl, etc.

Scheme 3.326

are oxidized to sulfoxides **825** in high chemical yield with moderate to good enantioselectivity (Scheme 3.326) [1119].

A purely water-based oxidation procedure has been developed by Kita using magnesium bromide as a catalyst instead of cetyltrimethylammonium bromide (CTAB). An asymmetric oxidizing reagent is formed by mixing (+)-dibenzoyl-D-tartaric acid, $MgBr_2$ and $PhIO_2$ in water for 5 min at room temperature. Treatment of 4-MeC_6H_4SMe with this oxidizing reagent at 0 °C for 24 h affords (*R*)-4-$MeC_6H_4S(O)Me$ in quantitative yield and 59% enantiomeric excess. Oxidation of 2-(phenylthio)ethanol (**826**) under these conditions gives sulfoxide **827,** leaving the primary hydroxyl group unaffected (Scheme 3.327) [1120].

Iodylarenes **828** react with CO in water in the presence of sodium tetrachloropalladate(II) and sodium carbonate at ambient temperature to furnish the corresponding carboxylic acid salts **829** (Scheme 3.328) [1121].

Iodylaryl derivatives bearing an appropriate substituent in the *ortho*-position to the iodine are characterized by the presence of a pseudocyclic structural moiety due to a strong intramolecular secondary bonding between the hypervalent iodine center and the oxygen atom in the *ortho*-substituent. Compared to the noncyclic aryliodyl derivatives, pseudocyclic iodine(V) compounds have much better solubility, which is explained by a partial disruption of their polymeric nature due to the redirection of secondary bonding [1122–1126]. Particularly useful reagents of this type are IBX esters (Section 2.2.2). IBX esters can oxidize alcohols to the respective aldehydes or ketones in the presence of trifluoroacetic acid or boron trifluoride etherate

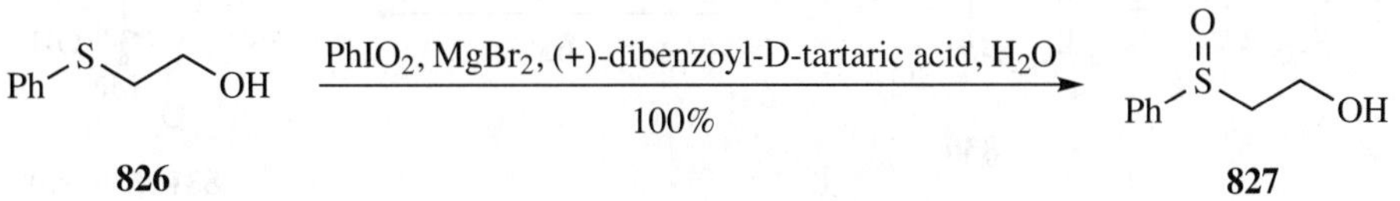

Ph–S–CH_2CH_2–OH (**826**) → ($PhIO_2$, $MgBr_2$, (+)-dibenzoyl-D-tartaric acid, H_2O; 100%) → Ph–S(O)–CH_2CH_2–OH (**827**)

Scheme 3.327

$$ArIO_2 + CO \xrightarrow[77\%]{Na_2CO_3,\ Na_2[PdCl_4]\ (0.1\%),\ H_2O,\ 40\ ^\circ C,\ 1\ atm,\ 6.5\ h} ArCOONa$$

828 **829**

Ar = Ph, 2-MeC_6H_4, 3-MeC_6H_4, 4-MeC_6H_4, 3-ClC_6H_4, 4-ClC_6H_4, 3-BrC_6H_4, 4-BrC_6H_4, 4-$NO_2C_6H_4$, 4-$MeOC_6H_4$

Scheme 3.328

[1123]. Isopropyl 2-iodoxybenzoate is a useful reagent for the clean, selective oxidation of organic sulfides to sulfoxides [1127]. This reaction proceeds without overoxidation to sulfones and is compatible with the presence of the hydroxy group, double bond, phenol ether, benzylic carbon and various substituted phenyl rings in the molecule of organic sulfide. Duschek and Kirsch have reported that isopropyl 2-iodoxybenzoate in the presence of trifluoroacetic anhydride can be used for the α-hydroxylation of β-keto esters at room temperature in THF [1128].

IBX esters can serve as stable and efficient sources of oxygen in the metalloporphyrin-catalyzed oxidations of hydrocarbons and the reactivity of isopropyl 2-iodoxybenzoate as an oxygenating reagent is similar to that of commonly used iodosylbenzene, which is a thermally unstable and potentially explosive compound [713, 1129, 1130].

The chiral, pseudocyclic 2-(*o*-iodoxyphenyl)-oxazolines **830** have been found to transform *ortho*-alkylphenols into *ortho*-quinol Diels–Alder dimers (e.g., **831**) with significant levels of asymmetric induction (Scheme 3.329) [1131].

3.2.2 2-Iodoxybenzoic Acid (IBX)

Applications of IBX in organic synthesis have been summarized in several comprehensive reviews [1105, 1106]. IBX is a particularly useful oxidant for the selective oxidation of alcohols to carbonyl compounds, even in complex molecules in the presence of other functional groups. Primary alcohols are oxidized by IBX in DMSO to the corresponding aldehydes at room temperature without overoxidation to the acids. The chiral primary alcohols are oxidized without epimerization and various functional groups like thioethers, amines, carboxylic acids, esters, carboxamides and both conjugated and isolated double bonds are compatible with IBX [1132, 1133]. Several representative examples of alcohol oxidations using IBX in DMSO are shown below in Schemes 3.330–3.335.

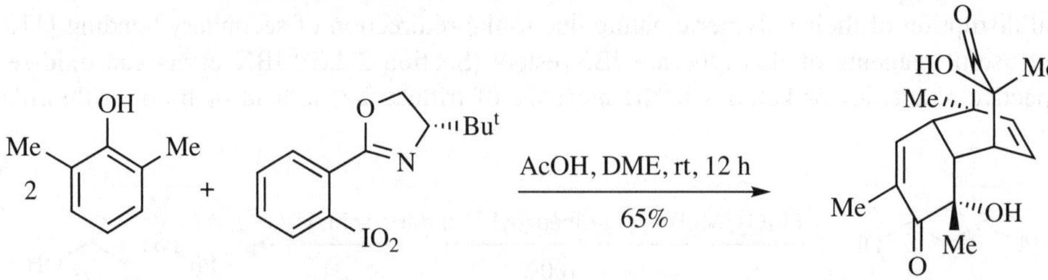

2 (2,6-dimethylphenol: OH, Me, Me) + **830** (IO_2, oxazoline, Bu^t) $\xrightarrow[65\%]{AcOH,\ DME,\ rt,\ 12\ h}$ **831** (63% ee)

Scheme 3.329

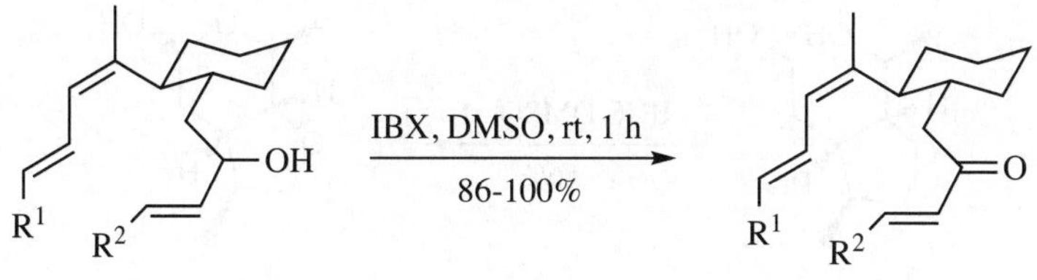

IBX, DMSO, rt, 1 h
86-100%

832 R^1 = H, SPh;
R^2 = H, SO_2Ph, CO_2Me, Ts

833

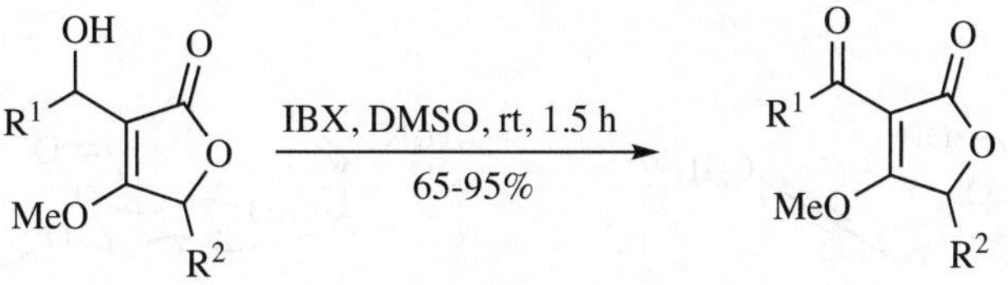

IBX, DMSO, rt, 1.5 h
65-95%

834 **835**

$R^1 = Bu^i, Bu^t$, Pr, Ph, $(CH_2)_2CH_2OTBDMS$; R^1 = H, Me

Scheme 3.330

Specifically, the allylic alcohols **832** are selectively oxidized by IBX to ketones **833** in high yield (Scheme 3.330) [1134]. The oxidation of alcohols **834** with IBX selectively affords 5-monosubstituted 3-acyl-4-*O*-methyl tetronates **835,** which are structurally similar to the tetrodecamycin antibiotics [1135].

The IBX oxidation of diol **836** has been utilized in the synthesis of the functionalized hexahydroanthracene dione **837** (Scheme 3.331), a model for the D ring of taxoids [1136].

Likewise, the total synthesis of the antifungal agent GM222712 was accomplished by a selective oxidation of diol **838** to hemiacetal **839** (Scheme 3.332) [1137].

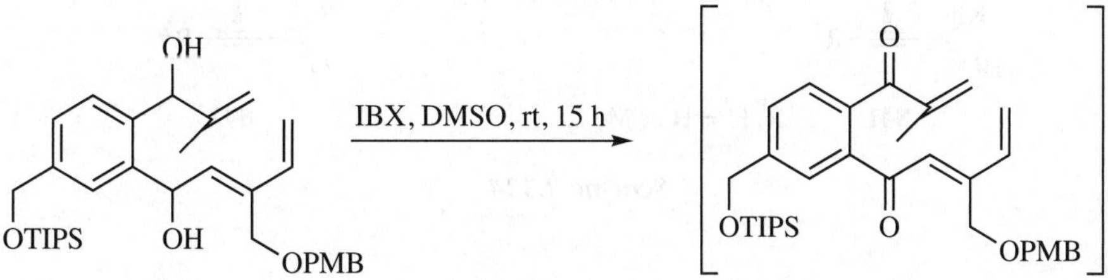

IBX, DMSO, rt, 15 h

836

[4+2]
55% overall

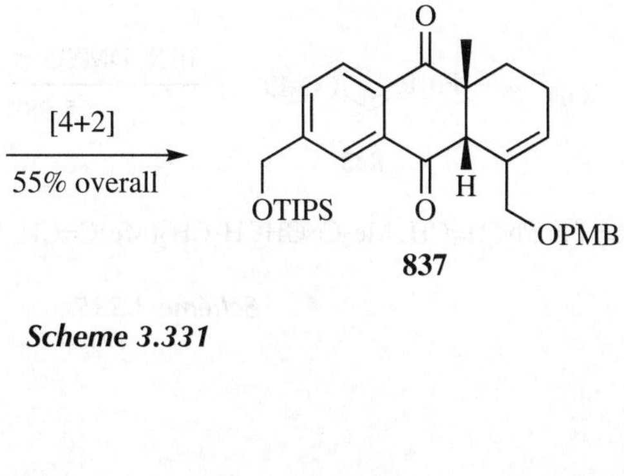

837

Scheme 3.331

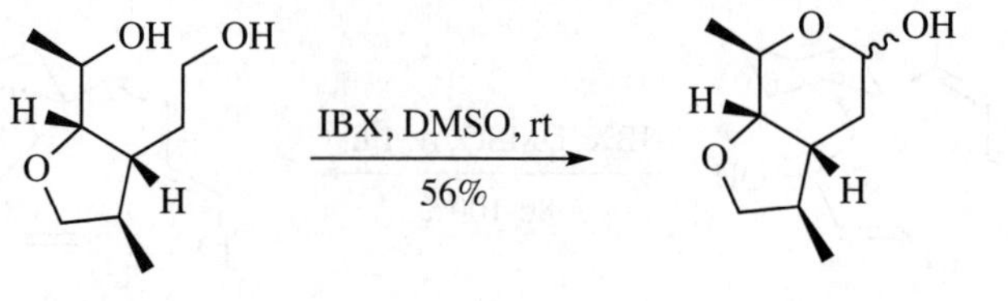

IBX, DMSO, rt
56%

838 **839**

Scheme 3.332

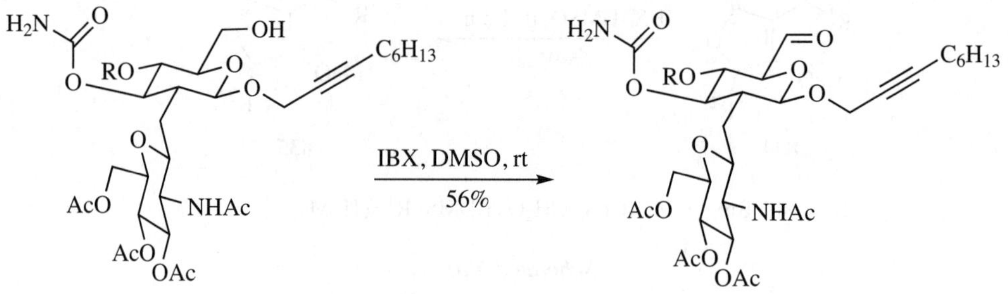

IBX, DMSO, rt
56%

840 R = TBDMS

Scheme 3.333

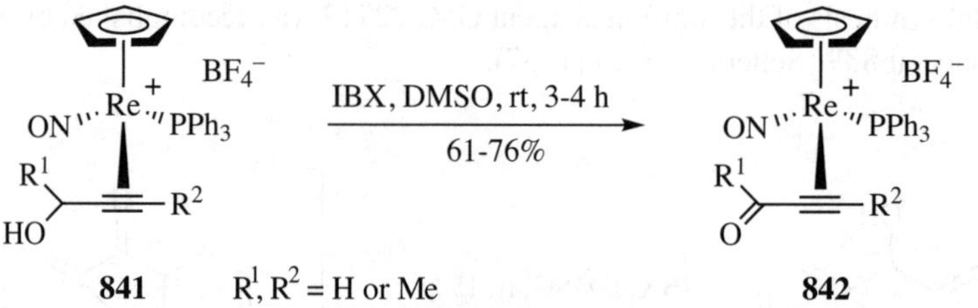

IBX, DMSO, rt, 3-4 h
61-76%

841 R^1, R^2 = H or Me **842**

Scheme 3.334

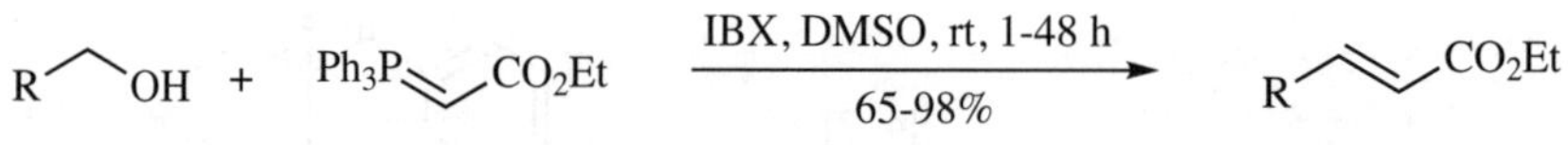

R⌒OH + Ph_3P=CHCO_2Et (**843**) → IBX, DMSO, rt, 1-48 h; 65-98% → R–CH=CH–CO_2Et (**844**)

R = Ph, PhCH=CH, Me_2C=CHCH_2CH_2(Me)C=CH, HC≡C, $C_5H_{11}C$≡C, etc.

Scheme 3.335

The IBX oxidation of carbohydrate **840** has been employed in synthetic studies of moenomycin A disaccharide analogs (Scheme 3.333) [1138].

The chiral rhenium complexes of allylic and propargylic alcohols **841** are selectively oxidized by IBX to the unsaturated carbonyl compounds **842** in good yields (Scheme 3.334) [1139].

A one-pot oxidation of benzylic, allylic and propargylic alcohols, as well as diols, with IBX in the presence of the stabilized Wittig ylide **843** affords α,β-unsaturated esters **844** in generally good yields (Scheme 3.335) [1140]. This is a useful one-pot procedure because the intermediate aldehydes are often unstable and difficult to isolate.

Selective oxidation of alcohols using IBX has been utilized in numerous syntheses, such as the total synthesis of (–)-decarbamoyloxysaxitoxin [1141], total synthesis of abyssomicin C and atrop-abyssomicin C [1142], stereoselective synthesis of pachastrissamine (jaspine B) [1143], syntheses of (±)-pterocarpans and isoflavones [1144], total synthesis of (±)-nitidanin [1145], total synthesis of lagunamycin [1146], synthesis of (–)-agelastatin [1147], syntheses of heliannuols B and D [1148], total syntheses of (–)-subincanadines A and B [1149], synthesis of marine sponge metabolite spiculoic acid A [1150] and in the total synthesis of a cyclic depsipeptide somamide A [1151]. Likewise, the oxidation of alcohols with IBX in DMSO has also been used in the development of a new silyl ether linker for solid-phase organic synthesis [1152], in the synthesis of optically pure highly functionalized tetrahydroisoquinolines [1153], in the kinetic study of organic reactions on polystyrene grafted microtubes [1154] and in the preparation of Fmoc-protected amino aldehydes from the corresponding alcohols [1155].

IBX is especially useful for the oxidation of 1,2-diols. Frigerio and Santagostino reported in 1994 that IBX, in contrast to DMP and iodylarenes, smoothly oxidizes 1,2-diols to α-ketols or α-diketones without cleaving the glycol C—C bond [1132]. More recently, Moorthy and coworkers have investigated the reactions of IBX with various vicinal diols and found that the oxidative cleavage of the C—C bond, as well as the oxidation to α-ketols or α-diketones, can occur in these reactions [1156]. In DMSO solutions, IBX oxidatively cleaves strained and sterically hindered *syn* 1,2-diols, while the non-hindered secondary glycols are oxidized to α-ketols or α-diketones. The use of trifluoroacetic acid as a solvent leads to efficient oxidative fragmentation of 1,2-diols of all types [1156]. The oxidation of 1,2-diols using IBX in DMSO has been utilized for the synthesis of α-ketols [1157–1159] or α-diketones [1160]. For example, in a key step of the total synthesis of the streptomyces maritimus metabolite wailupemycin B, IBX oxidation of the diol precursor **845** led to desired hydroxyketone **845** without any cleavage of the glycol C—C bond (Scheme 3.336) [1157].

The synthetic usefulness of IBX in general is significantly restricted by its low solubility in most organic solvents with the exception of DMSO. However, in several publications it has been shown that IBX can be used as effective oxidant in solvents other than DMSO [1161–1163]. More and Finney have found that primary and secondary alcohols can be oxidized into the corresponding aldehydes or ketones in excellent

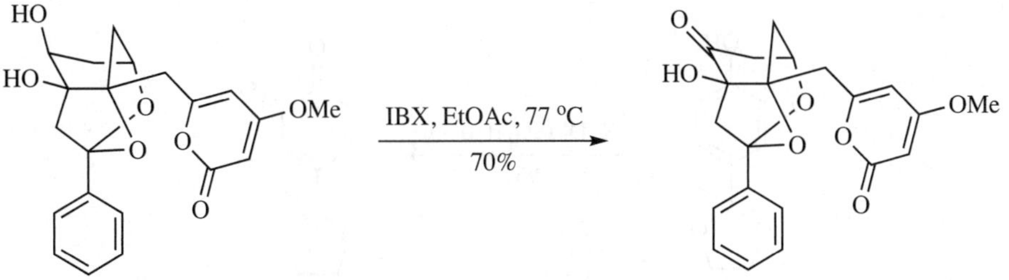

Scheme 3.336

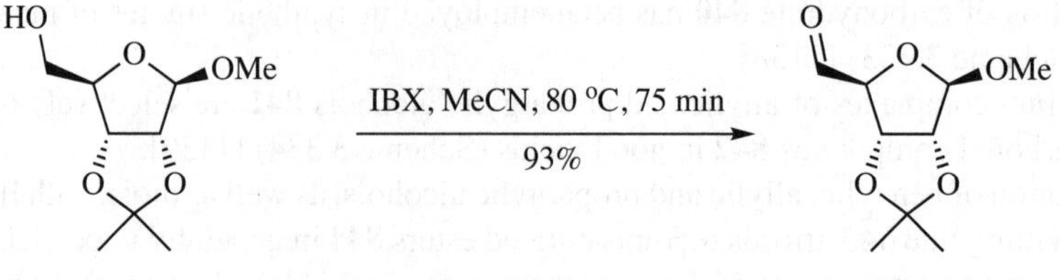

IBX, MeCN, 80 °C, 75 min
93%

847

Scheme 3.337

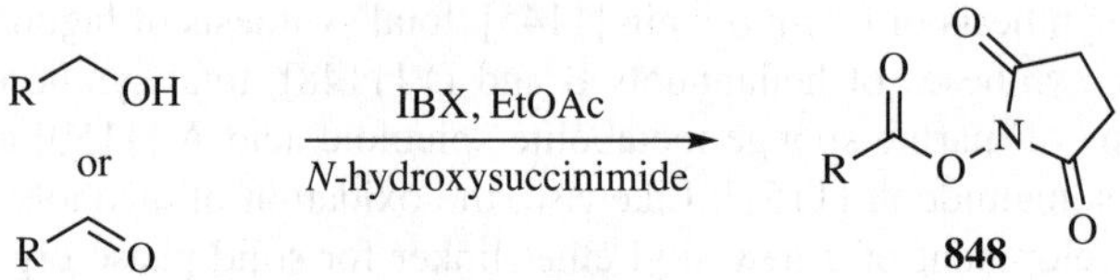

IBX, EtOAc
N-hydroxysuccinimide

848

Scheme 3.338

yields (90–100%) by heating a mixture of alcohol and IBX in common organic solvents [1161]. All reaction by-products can be completely removed by filtration. This method has been used for the efficient preparation of the ribosyl aldehyde **847** (Scheme 3.337), the key intermediate in the stereoselective synthesis of the core structure of the polyoxin and nikkomycin antibiotics [1162].

An IBX-mediated conversion of primary alcohols (or aldehydes) into *N*-hydroxysuccinimide esters **848** has been developed by Giannis and coworkers [1164]. The generality of this procedure was demonstrated on various aliphatic, allylic and benzylic alcohols (Scheme 3.338).

IBX in DMF or DMSO has been shown to be an excellent reagent for the oxidation of various phenols to *o*-quinones [1165]. This procedure was used for the oxidation of phenol **849** to quinone **850** (Scheme 3.339), the key intermediate in total synthesis of a novel cyclooxygenase inhibitor (±)-aiphanol [1166]. The same protocol was utilized in the synthesis of (±)-brazilin, a tinctorial compound found in the alcoholic extracts of trees collectively referred to as Brazil wood [1167].

The IBX-mediated oxygenative dearomatization of phenols leading to cyclohexa-2,4- or -2,5-dienone systems is a particularly useful synthetic transformation [292]. Representative examples include the use of IBX in key oxidation steps in the total synthesis of the resveratrol-derived polyphenol natural products

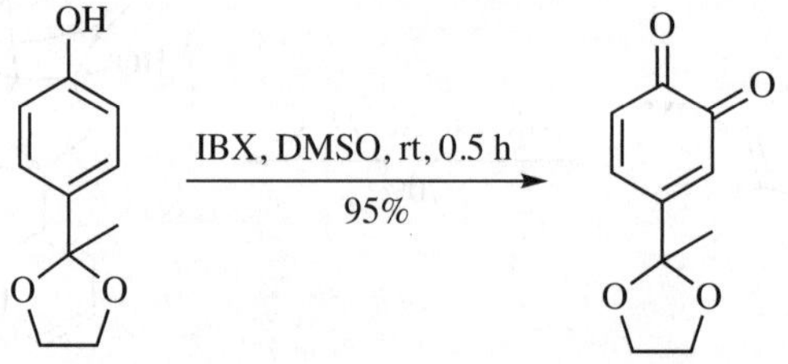

IBX, DMSO, rt, 0.5 h
95%

849 **850**

Scheme 3.339

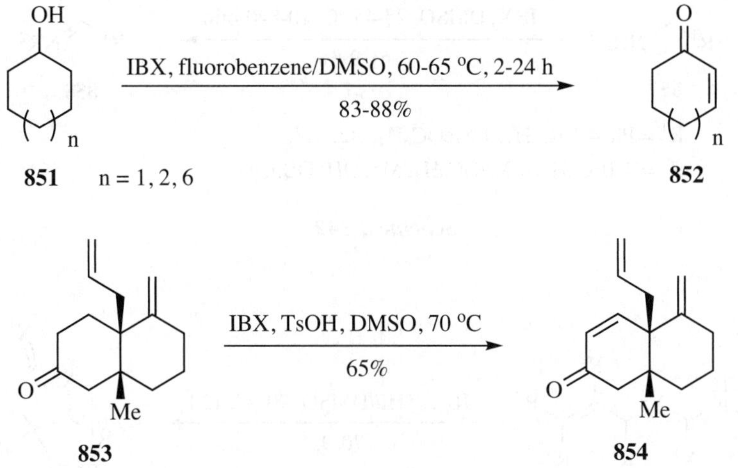

IBX, fluorobenzene/DMSO, 60-65 °C, 2-24 h
83-88%

851 n = 1, 2, 6 **852**

IBX, TsOH, DMSO, 70 °C
65%

853 **854**

Scheme 3.340

(–)-hopeanol and (–)-hopeahainol A [1168], the synthesis of carnosic acid and carnosol [1169] and the total synthesis of the bis-sesquiterpene (+)-aquaticol [1170].

The practical value of IBX as a reagent has been extended to various other synthetically useful oxidative transformations. In a series of papers, Nicolaou and coworkers have demonstrated the utility of IBX for the one-step synthesis of α,β-unsaturated carbonyl systems from saturated alcohols and carbonyl compounds [1171–1173], for the selective oxidation of the benzylic carbon [1174, 1175], for the oxidative cyclization of anilides and related compounds [1176–1179] and for the synthesis of amino sugars and libraries thereof [1179]. Specifically, alcohols, ketones and aldehydes are oxidized to the corresponding α,β-unsaturated species in one pot using IBX under mild conditions [1172]. For example, cycloalkanols **851** react with two equivalents of IBX in a 2 : 1 mixture of either fluorobenzene or toluene and DMSO under gentle heating to afford the corresponding α,β-unsaturated ketones **852** in good yields (Scheme 3.340) [1172]. A similar oxidative dehydrogenation of a cyclohexanone derivative **853** to the respective enone **854** has been employed in the total synthesis of (–)-anominine [1180].

IBX is an efficient and selective reagent for the oxidation of alkylarenes **855** at the benzylic positions to give ketones **856** (Scheme 3.341) [1174]. This reaction is quite general and can tolerate various substituents within the aromatic ring. Overoxidation to the corresponding carboxylic acids is not observed even in the presence of electron-rich substituents.

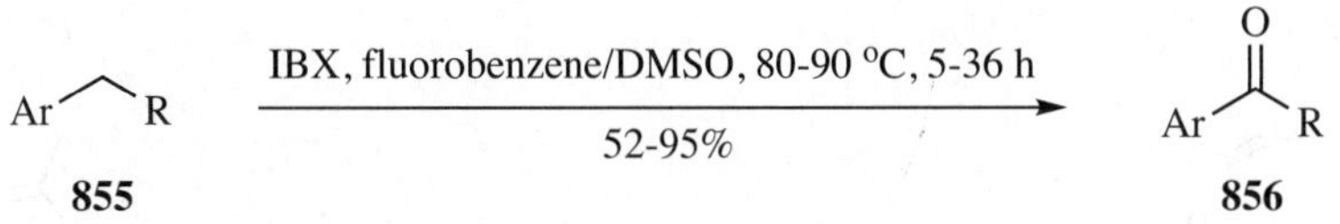

IBX, fluorobenzene/DMSO, 80-90 °C, 5-36 h
52-95%

855 **856**

Ar = Ph, 4-$Bu^tC_6H_4$, 2-MeC_6H_4, 3-IC_6H_4, 4-BrC_6H_4, 3,4-$(MeO)_2C_6H_3$,
2-PhC_6H_4, 4-(4-pyridyl)C_6H_4, etc.
R = H, C_3H_7, etc.

Scheme 3.341

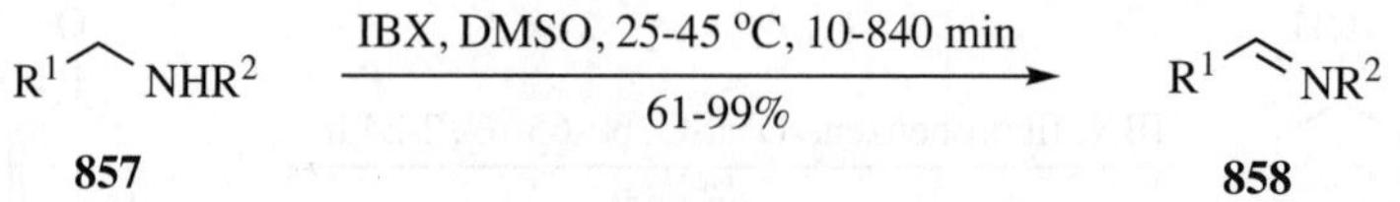

R^1 ⁀ NHR^2 (857) —[IBX, DMSO, 25-45 °C, 10-840 min; 61-99%]→ R^1=NR^2 (858)

R^1 = Ph, 4-BrC$_6$H$_4$, 4-MeOC$_6$H$_4$, etc.
R^2 = 4-BrC$_6$H$_4$, 4-MeOC$_6$H$_4$, Me, OH, OBn, etc.

Scheme 3.342

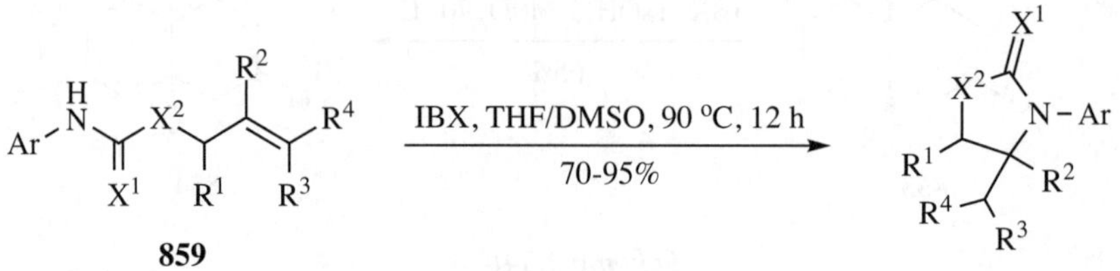

859 —[IBX, THF/DMSO, 90 °C, 12 h; 70-95%]→ 860

X^1 = O, S; X^2 = CH$_2$, O, N
Ar = Ph, 3-EtC$_6$H$_4$, 3-BrC$_6$H$_4$, 3-FC$_6$H$_4$, 4-EtC$_6$H$_4$, etc.
R^1 - R^4 = H, alkyl, cycloalkyl, etc.

Scheme 3.343

Similar to the oxidation of alcohols, secondary amines **857** can be oxidized with IBX in DMSO to yield the corresponding imines **858** in good to excellent yields (Scheme 3.342) [1175].

Various heterocycles **860** can be synthesized by the treatment of unsaturated aryl amides, carbamates, thiocarbamates and ureas **859** with IBX (Scheme 3.343) [1176, 1177]. The mechanism of this reaction has been investigated in detail [1178]. On the basis of solvent effects and D-labeling studies, it was proposed that the IBX-mediated cyclization of anilides in THF involves an initial single-electron transfer (SET) to a THF–IBX complex followed by deprotonation, radical cyclization and concluding termination by hydrogen abstraction from THF [1178]. A similar IBX-mediated cyclization has been applied in the synthetic protocol for the stereoselective preparation of amino sugars [1179].

Studer and Janza have developed a method for the generation of alkoxyamidyl radicals starting from the corresponding acylated alkoxyamines using IBX as a SET oxidant [1181]. For example, the stereoselective 5-*exo* cyclization of the respective N-centered radical generated from alkoxyamide **861** affords isoxazolidine **862** (Scheme 3.344) [1181].

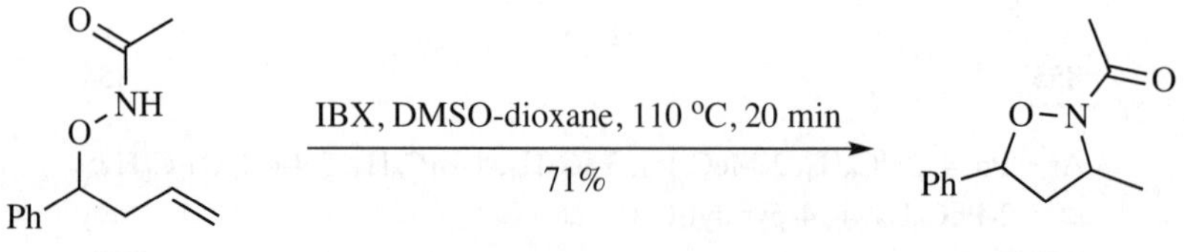

861 —[IBX, DMSO-dioxane, 110 °C, 20 min; 71%]→ 862

Scheme 3.344

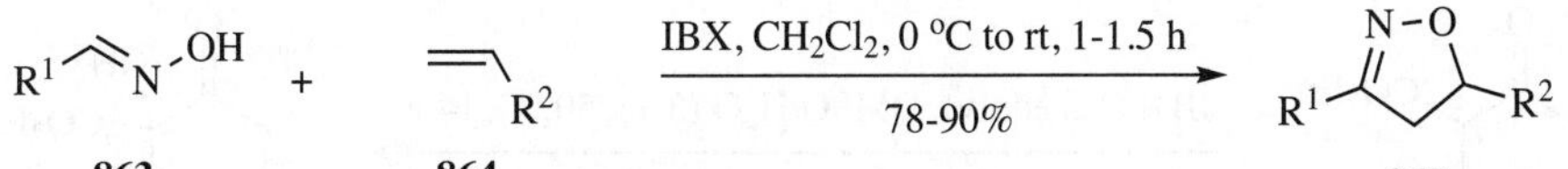

$R^1 = $ aryl, heteroaryl; $R^2 = CO_2Me$, CN, OAc

IBX, CH_2Cl_2, 0 °C to rt, 1-1.5 h; 78-90%

863 **864** **865**

Scheme 3.345

IBX/Bu_4NBr, MeCN, rt; 55-99%

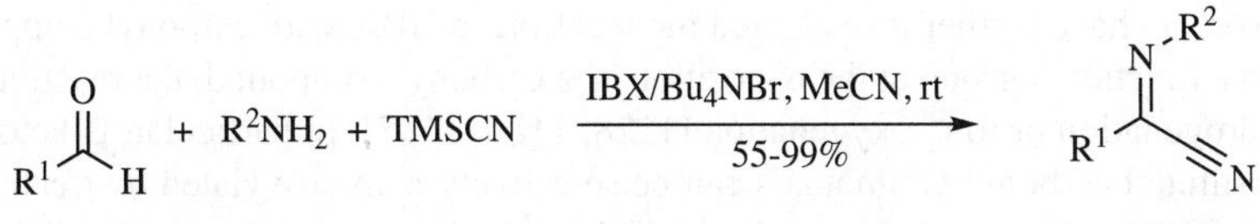

866

$R^1 = $ Ph, 4-$MeOC_6H_4$, 2,6-$Cl_2C_6H_3$, PhC=CH, $Ph(CH_2)_2$, Pr^i, etc.
$R^2 = Ph(CH_2)_2$, Bu^t, 4-$MeOC_6H_4$, Ph, etc.

Scheme 3.346

IBX has also been used for the preparation of the 3,5-disubstituted isoxazolines **865.** The oxidation of aldoximes **863** with IBX produces the respective nitrile oxides, which then undergo 1,3-dipolar addition with an alkene component **864** to give final products **865** (Scheme 3.345) [1182].

A one-pot three-component synthesis of α-iminonitriles **866** via an IBX/tetrabutylammonium bromide-mediated oxidative Strecker reaction has been developed (Scheme 3.346) [1183]. This methodology was employed in a two-step synthesis of indolizidines via a microwave-assisted intramolecular cycloaddition of α-iminonitriles.

The IBX-mediated oxidative Ugi-type multicomponent reaction of tetrahydroisoquinoline with isocyanides and carboxylic acids affords the nitrogen- and carbon-functionalized tetrahydroisoquinolines **867** in good to excellent yields [1184]. Likewise, the three-component Passerini reaction of an alcohol, carboxylic acid and an isonitrile in the presence of IBX affords the corresponding α-acyloxy carboxamides **868** in generally high yields (Scheme 3.347) [1185].

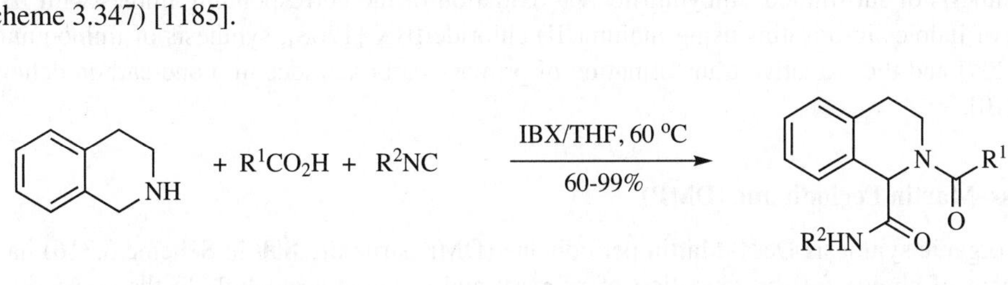

+ R^1CO_2H + R^2NC — IBX/THF, 60 °C; 60-99% → **867**

$R^1CH_2OH + R^2NC + R^3CO_2H$ — IBX/THF, 40 °C; 33-94% → **868**

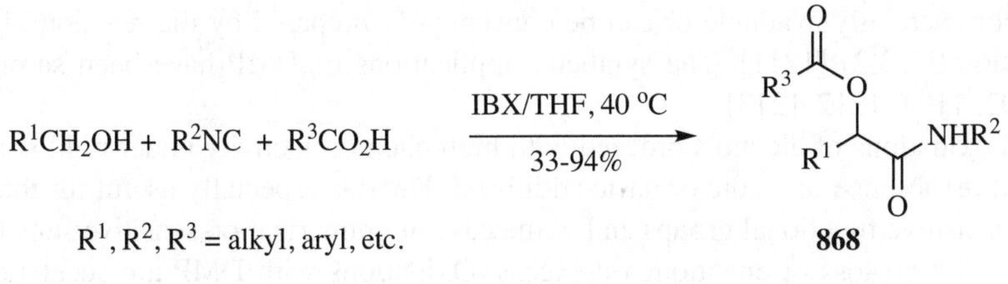

$R^1, R^2, R^3 = $ alkyl, aryl, etc.

Scheme 3.347

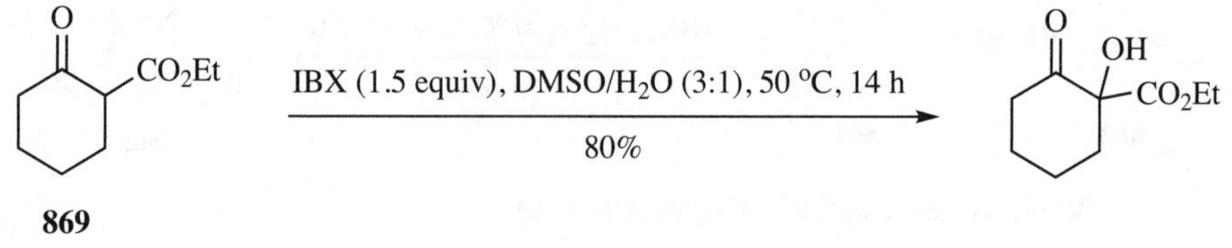

869

IBX (1.5 equiv), DMSO/H_2O (3:1), 50 °C, 14 h

80%

Scheme 3.348

Kirsch and coworkers have further investigated the reactions of IBX with carbonyl compounds and found that, depending on a functional group at the α-position of a carbonyl compound, the reaction may lead either to oxidative dehydrogenation or to α-oxygenation [1128, 1186, 1187]. In particular, β-keto esters and some other suitably substituted carbonyl compounds can be selectively α-hydroxylated by treatment with IBX in aqueous DMSO at 50 °C; a representative example of the α-hydroxylation reaction of β-keto ester (**869**) is shown in Scheme 3.348 [1128].

Additional representative examples of synthetic applications of IBX include the following oxidative transformations: the aromatization of tetrahydro-β-carbolines under mild conditions applied in a total synthesis of the marine indole alkaloid eudistomin U [1188], oxidation of glycosides to the respective 6-carbaldehydes used as precursors in the synthesis of amino-bridged oligosaccharides [1189], oxidation of amidoximes to carboxamides or nitriles with IBX or IBX/tetraethylammonium bromide [1190], aromatic hydroxylations of flavonoids [1191], hydroxylation of resveratrol diacyl derivatives [1192], synthesis of DOPA and DOPA peptides by oxidation of tyrosine residue [1193], oxidative preparation of γ-hydroxy-α-nitroolefins from α,β-epoxyketoximes [1194], aromatization of 1,4-dihydropyridines using IBX in water/acetone in the presence of β-cyclodextrin [1195], iodohydroxylation of alkenes and iodination of aromatics using IBX/I_2 in aqueous acetone [1196], conversion of alkenes and alkynes into α-iodo ketones using IBX/I_2 in water [1197], oxidation of primary amines to nitriles [1198,1199], oxidative cleavage of acetals using IBX/tetraethylammonium bromide in water [1200], one-pot synthesis of trifluoromethyl-containing pyrazoles via sequential $Yb(PFO)_3$-catalyzed three-component reaction and IBX-mediated oxidation [1201], oxidative thiocyanation of indoles, pyrrole and arylamines [1202], oxidative functionalization of Baylis–Hillman adducts [1203–1205], construction of multisubstituted 2-acyl furans by the IBX-mediated cascade oxidation/cyclization of *cis*-2-en-4-yn-1-ols [1206], one-pot synthesis of substituted salicylnitriles via oxidation of the corresponding imines with IBX [1207], conversion of indoles into isatins using indium(III) chloride/IBX [1208], synthesis of iminoquinones from anilines [1209] and the oxidative transformation of primary carboxamides into one-carbon dehomologated nitriles [1210].

3.2.3 Dess–Martin Periodinane (DMP)

In modern organic synthesis Dess–Martin periodinane (DMP, structure **800** in Scheme 3.316) has emerged as the reagent of choice for the oxidation of primary and secondary alcohols to the respective carbonyl compounds. DMP is commercially available or can be conveniently prepared by the reaction of IBX with acetic anhydride (Section 2.2.3.2) [1211]. The synthetic applications of DMP have been summarized in several overviews [1102, 1104, 1212, 1213].

The DMP-promoted oxidations of alcohols proceed with high chemoselectivity under mild reaction conditions (room temperature, absence of acidic or basic additives). DMP is especially useful for the oxidation of alcohols containing sensitive functional groups and in the case of epimerization sensitive substrates DMP allows clean oxidation with no loss of enantiomeric excess. Oxidations with DMP are accelerated by the addition of water to the reaction mixture immediately before or during the reaction [1214].

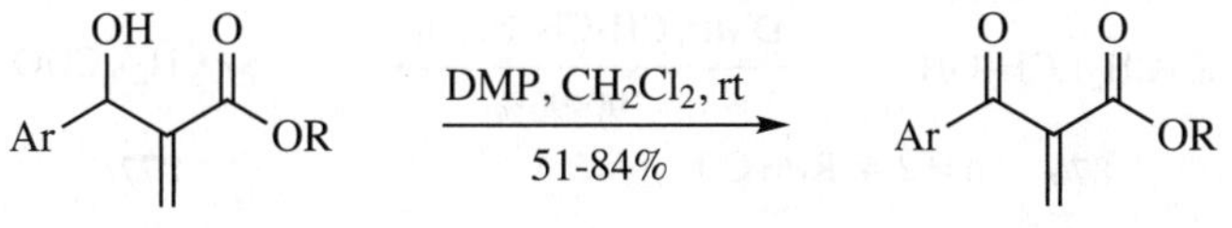

Ar = Ph, 2-ClC_6H_4, 4-MeC_6H_4, 3,4-$F_2C_6H_3$, 4-FC_6H_4, 4-$CF_3C_6H_4$
R = Me, Et

DMP, CH_2Cl_2, rt
51-84%

870 **871**

Scheme 3.349

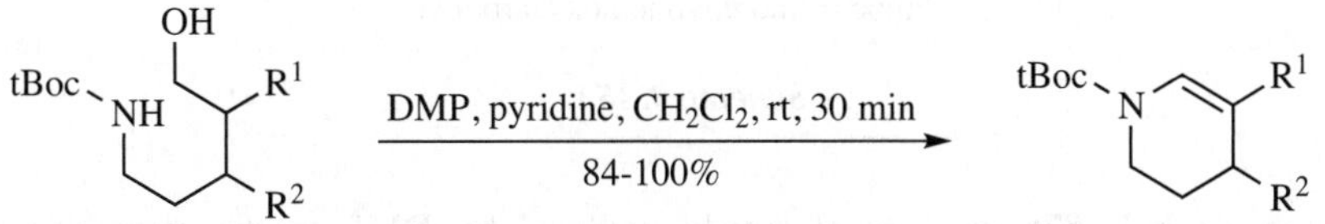

DMP, pyridine, CH_2Cl_2, rt, 30 min
84-100%

872 **873**

R^1 = H, Me, $CH_2OSi(Ph)_2Bu^t$
R^2 = H, OCH_2OMe, $OCH_2OCH_2CH_2SiMe_3$

Scheme 3.350

In numerous synthetic studies it has been demonstrated that DMP can be used for a selective oxidation of alcohols containing sensitive functional groups, such as unsaturated alcohols [297, 1215–1218], carbohydrates and polyhydroxy derivatives [1216, 1219–1221], silyl ethers [1222, 1223], amines and amides [1224–1227], various nucleoside derivatives [1228–1231], selenides [1232], tellurides [1233], phosphine oxides [1234], homoallylic and homopropargylic alcohols [1235], fluoroalcohols [1236–1239] and boronate esters [1240]. Several representative examples of these oxidations are shown below in Schemes 3.349–3.354. Specifically, the functionalized allylic alcohols **870,** the Baylis–Hillman adducts of aryl aldehydes and alkyl acrylates, are efficiently oxidized with DMP to the corresponding α-methylene-β-keto esters **871** (Scheme 3.349) [1217]. The attempted Swern oxidation of the same adducts **870** resulted in substitution of the allylic hydroxyl group by chloride.

Cyclic enecarbamates **873** have been prepared in excellent yields by the oxidation of ω-hydroxycarbamates **872** with DMP followed by cyclocondensation–dehydration of the intermediate aminoaldehydes (Scheme 3.350) [1227].

α-Hydroxyboronates **874** have been selectively oxidized to acylboronates **875** through the Dess–Martin oxidation (Scheme 3.351) [1240].

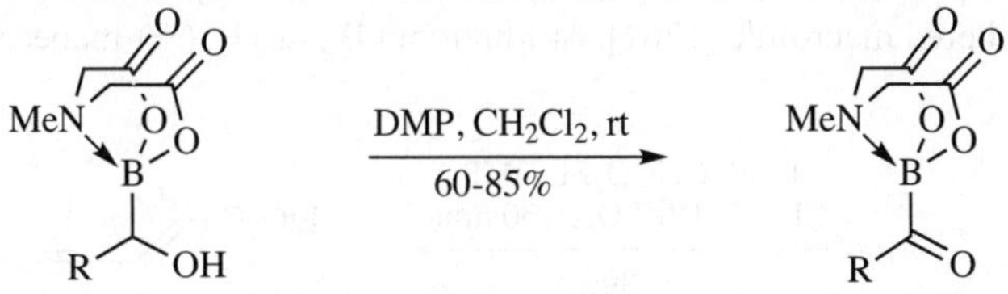

DMP, CH_2Cl_2, rt
60-85%

874 R = alkyl, aryl **875**

Scheme 3.351

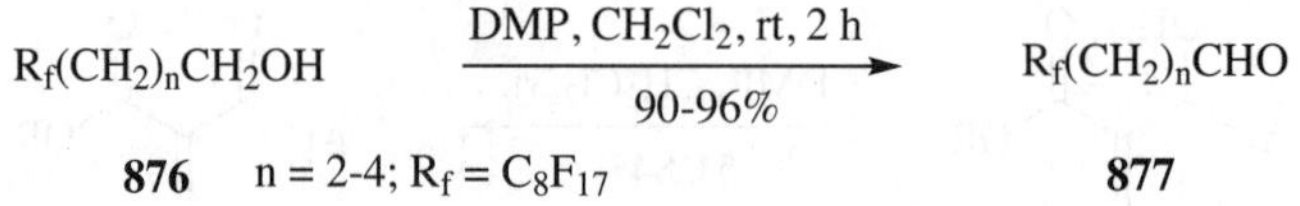

$R_f(CH_2)_nCH_2OH$ —[DMP, CH_2Cl_2, rt, 2 h; 90-96%]→ $R_f(CH_2)_nCHO$

876 n = 2-4; $R_f = C_8F_{17}$ **877**

Scheme 3.352

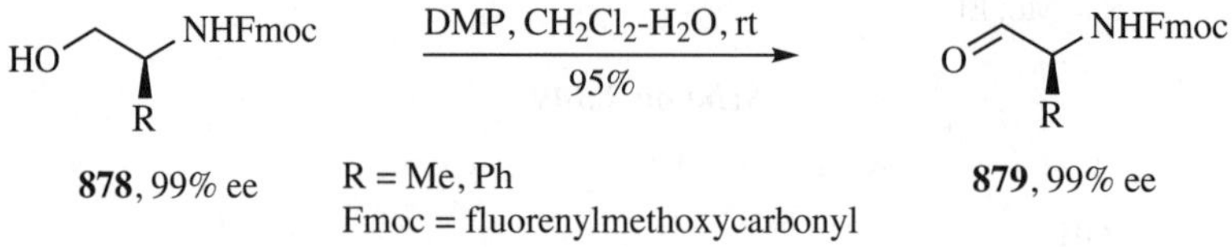

HO–CH₂–CH(R)–NHFmoc —[DMP, CH_2Cl_2-H_2O, rt; 95%]→ O=CH–CH(R)–NHFmoc

878, 99% ee R = Me, Ph **879**, 99% ee

Fmoc = fluorenylmethoxycarbonyl

Scheme 3.353

Polyfluorinated alcohols **876** can be selectively oxidized by DMP to the respective aldehydes **877** (Scheme 3.352) without the formation of dehydrofluorinated by-products [1238, 1239].

DMP is especially useful for the oxidation of the optically active, epimerization-sensitive substrates without loss of enantiomeric purity [1224, 1241, 1242]. In a typical example, DMP was found to be a superior oxidant for the efficient, epimerization-free synthesis of optically active N-protected α-amino aldehydes **879** from the corresponding N-protected β-amino alcohols **878** (Scheme 3.353) [1224]. In contrast, the Swern oxidation of amino alcohols **878** afforded products **879** of only 50–68% ee.

Primary alcohols can be oxidized with DMP in the presence of stabilized phosphonium ylides to afford the respective α,β-unsaturated esters in one pot [1232, 1243, 1244]. This is a useful procedure, especially when the intermediate aldehydes are unstable and difficult to isolate. In a representative example, a highly unstable dialdehyde, 2-butynedial, was generated by the oxidation of propargylic diol **880** with DMP and trapped by phosphonium ylide *in situ* to afford the adduct **881** as a 4 : 1 mixture of *trans-trans* and *trans-cis* isomers (Scheme 3.354) [674].

The DMP oxidation of 1,2-diols generally cleaves the glycol C—C bond, as illustrated by the synthesis of tricyclic enol ether **883** from diol **882** via tandem 1,2-diol cleavage–intramolecular cycloaddition (Scheme 3.355) [1220].

Because of the unique oxidizing properties and convenience of use, DMP has been widely employed in the synthesis of biologically important natural products. Representative examples include the use of DMP in key oxidation steps of the following total syntheses: (±)-deoxypreussomerin A [1245], racemic brevioxime [1246], erythromycin B [1247], (+)-cephalostatin 7 [1248], (+)-cephalostatin 12 [1248], (+)-ritterazine K [1248], fredericamycin A [1249], angucytcline antibiotics [1250], tricyclic β-lactam antibiotics [1251], ent-hyperforin [1252], (–)-spirotryprostatin B [1253], (+)-peloruside A [1254], (+)-ambruticin S [1255], (±)-platensimycin [1256], (–)-pseudolaric acid B [1257], azadirachtin [1258], salvinorin A [1259], amphidinol 3 [1260], FD-891 16-membered macrolide [1261], (+)-bretonin B [1262], (±)-maoecrystal V [1263], resolvin

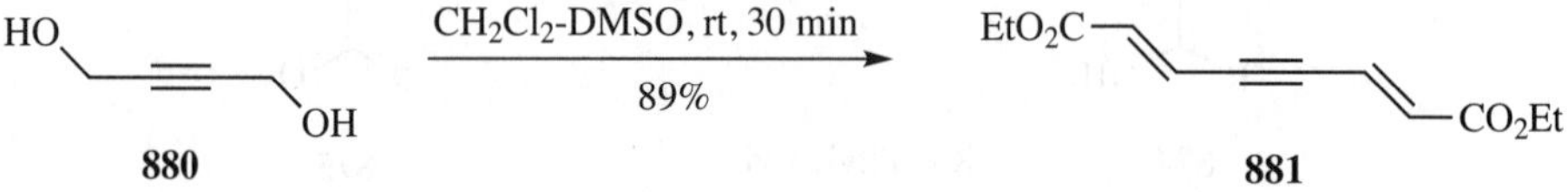

HOCH₂–C≡C–CH₂OH (**880**) —[Ph_3P=CHCO₂Et, DMP; CH_2Cl_2-DMSO, rt, 30 min; 89%]→ EtO_2C–CH=CH–C≡C–CH=CH–CO_2Et (**881**)

Scheme 3.354

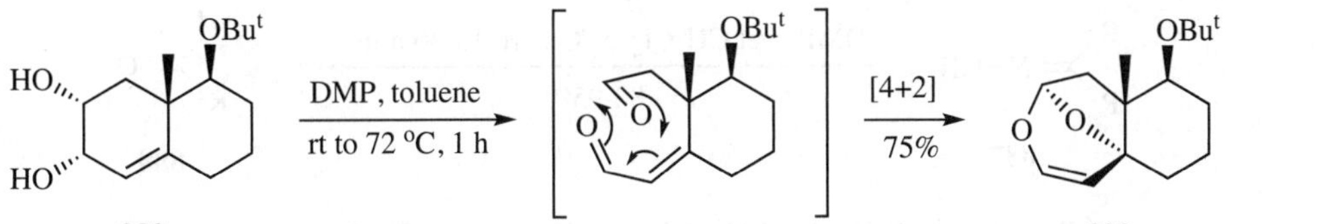

Scheme 3.355

D1 [1264], (–)-tirandamycin C [1265], gambieric acid A [1266], (+)-sieboldine A [1267], ripostatin B [1268], 15-deoxyripostatin A [1269], spirastrellolide A methyl ester [1270], (–)-fusarisetin A [1271], halichondrin C [1272], 16-membered macrolide FD-891 [1261] and numerous other synthetic works.

The unique oxidizing properties of DMP are best illustrated by its numerous applications in the total synthesis of the CP-molecules, lead structures for cardiovascular and anticancer drugs, published by Nicolaou and coworkers in 2002 [1273–1275]. For example, in the course of this synthetic study, a hindered secondary alcohol **884** was oxidized with DMP to give stable diol **886** via intermediate formation of hemiketal **885** (Scheme 3.356) [1274].

The practical value of DMP as a reagent has been extended to various other synthetically useful oxidative transformations, such as the dehydration of primary alcohols under extraordinarily mild conditions [1276], synthesis of various polycyclic heterocycles via the oxidative cascade cyclization of anilides with pendant double bonds [1277], one-pot oxidative allylation of Morita–Baylis–Hillman adducts with allyltrimethylsilane promoted by DMP/$BF_3 \cdot OEt_2$ [1278], synthesis of 2-amino-1,4-benzoquinone-4-phenylimides from anilines via DMP oxidation [1279], α-tosyloxylation of ketones using DMP and *p*-toluenesulfonic acid [1280] and the DMP-mediated oxidative aromatization of 1,3,5-trisubstituted pyrazolines [1281].

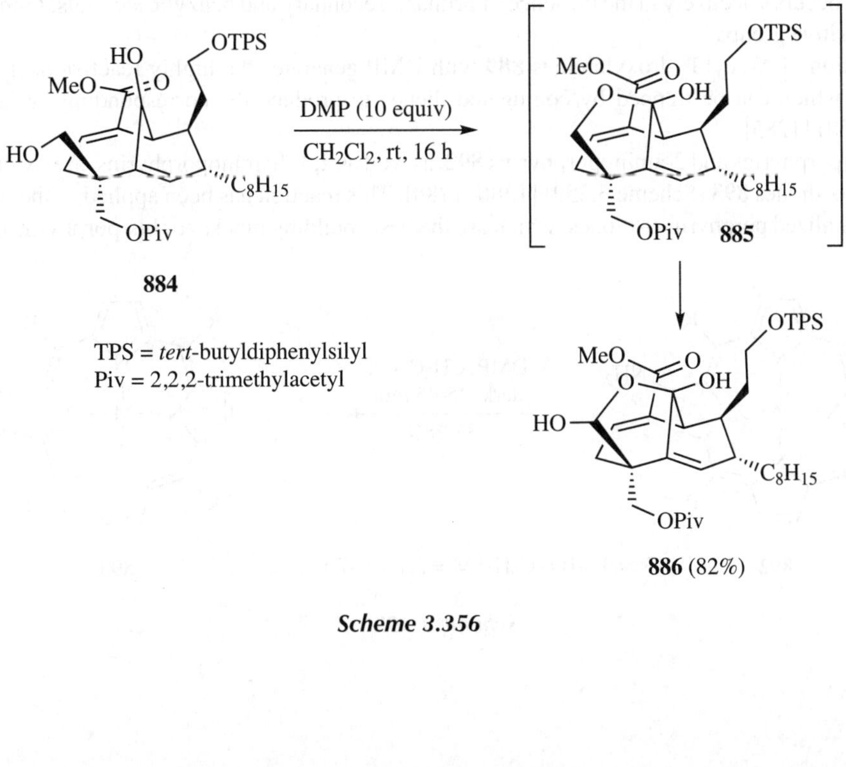

TPS = *tert*-butyldiphenylsilyl
Piv = 2,2,2-trimethylacetyl

Scheme 3.356

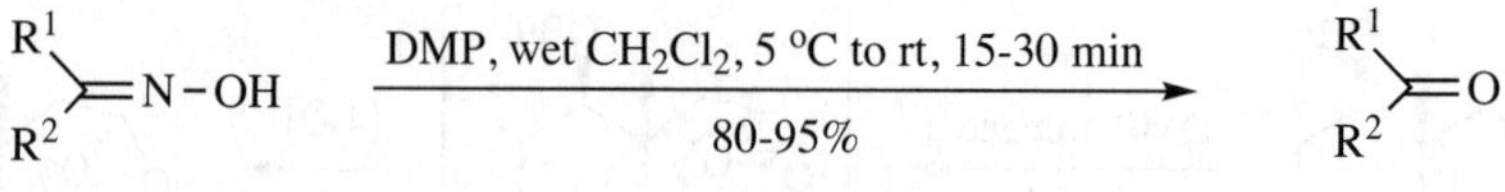

$R^1R^2C{=}N{-}OH$ (**887**) → DMP, wet CH_2Cl_2, 5 °C to rt, 15-30 min; 80-95% → $R^1R^2C{=}O$ (**888**)

R^1 = Ph, 4-ClC_6H_4, 4-FC_6H_4, 4-$MeOC_6H_4$, 4-$NO_2C_6H_4$, 2-furyl, PhCH=CH, 4-$Me_2NC_6H_4$, C_5H_{11}, C_7H_{15}, C_9H_{19}, PhC(O), Ph_2CHCH_2
R^2 = H, PhHC=CH, CO_2Me, Me, $(CH_2)_2CO_2H$, NH_2

Scheme 3.357

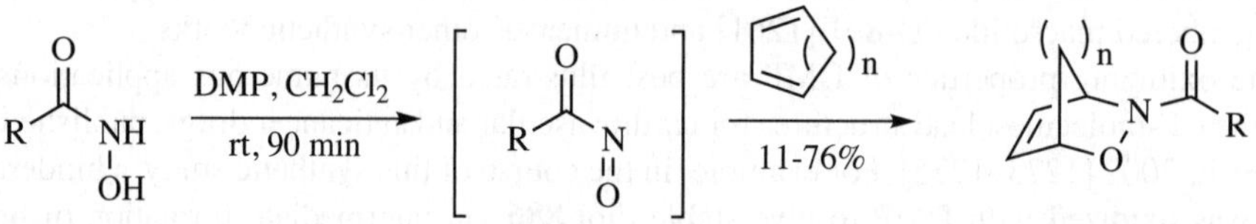

889 → DMP, CH_2Cl_2, rt, 90 min → [**890**] → diene, 11-76% → **891**

R = Ph, Me, OBu^t, OBn, NH_2; n = 1 or 2

Scheme 3.358

DMP can be used as an efficient and selective reagent for the oxidative cleavage of oximes [1282–1284] and tosylhydrazones [1284] to yield the corresponding carbonyl compounds under mild conditions in high yields. In a specific example, DMP oxidatively deoximates aldoximes or ketoximes **887** to give the respective carbonyl compounds **888** in excellent yields, smoothly in a short time and under mild conditions (Scheme 3.357) [1283]. Deoximation occurs selectively in the presence of primary, secondary and benzylic alcohols, *O*-methyl oximes and acid–sensitive groups.

The oxidation of *N*-acyl hydroxylamines **889** with DMP generates the highly reactive acyl nitroso compounds **890,** which can be trapped by conjugated dienes to produce the corresponding cycloadducts **891** (Scheme 3.358) [1285].

2-Hydroxyporphyrins and 2-aminoporphyrins **892,** as well as 2,3-diaminoporphyrins, are oxidized by DMP to porphyrin-α-diones **893** (Scheme 3.359) [1286–1288]. This reaction has been applied to the preparation of *meso*-functionalized porphyrin-α-diones, which are the basic building blocks for bis-porphyrin arrays [1287].

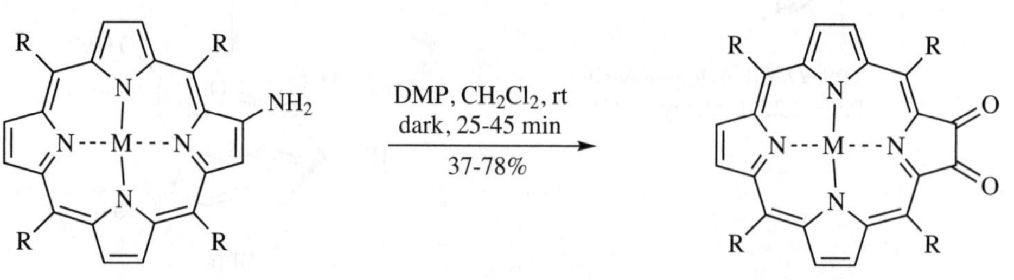

892 → DMP, CH_2Cl_2, rt, dark, 25-45 min; 37-78% → **893**

R = 3,5-$Bu^t_2C_6H_3$; M = 2H, Cu, Zn

Scheme 3.359

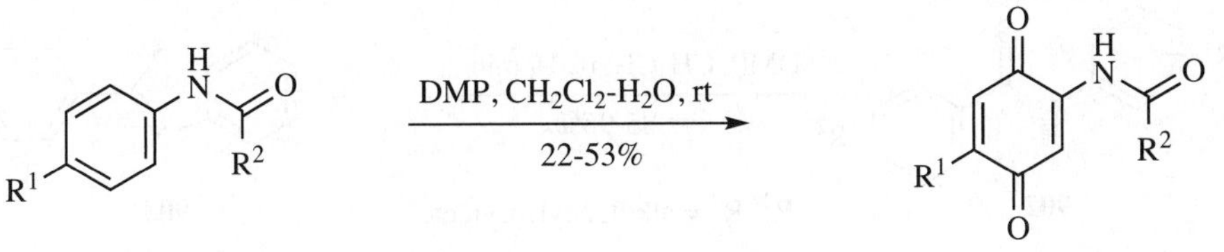

DMP, CH_2Cl_2-H_2O, rt
22-53%

894 R^1 = H, Et, Bu^t, Ph, OMe, F, Cl, Br, I
R^2 = Me, Ph, Bu^t, Pr^i **895**

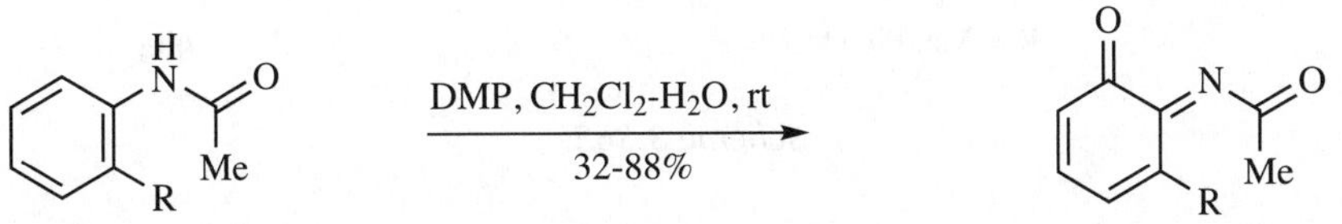

DMP, CH_2Cl_2-H_2O, rt
32-88%

896 R = Et, Bu^t, Ph, OMe, Cl, Br, I **897**

Scheme 3.360

In a series of publications, Nicolaou and coworkers have demonstrated the utility of DMP for the selective oxidation of 4-substituted anilides **894** to *p*-quinones **895** and 2-substituted anilides **896** to *o*-azaquinones **897** (Scheme 3.360) [1289–1291]. The first process was applied to the short, efficient total synthesis of epoxyquinomycin B [1290], while the second type of oxidation allowed rapid access to complex analogs of pseudopterosin and elisabethin natural products [1291].

Anilides with pendant double bonds **898** undergo DMP-induced stereoselective oxidative cyclization to give complex and diverse natural product-like polycycles **899** (Scheme 3.361) [1176, 1277]. This oxidative cyclization is proposed to occur by the initial *ortho* directed oxidation of anilide **898** to give an *ortho*-hydroxylated benzene ring that is further oxidized to the quinone imine; intramolecular Diels–Alder cyclization of the

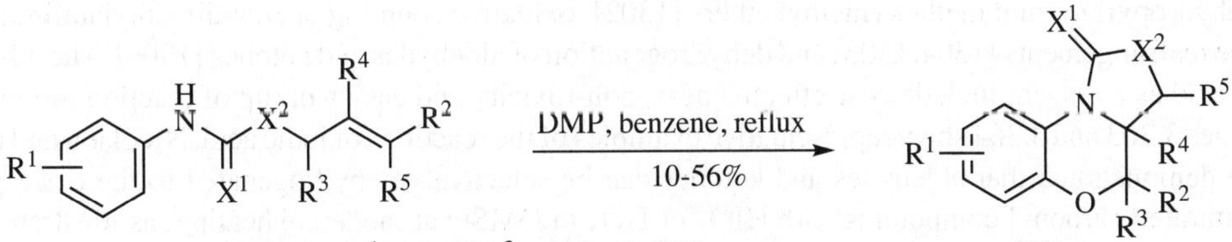

DMP, benzene, reflux
10-56%

898 X^1 = O, S; X^2 = CH_2, O, NH
R^1 = H, 3-F, 3-NO_2, 3-Br, 4-Et, etc.
R^2 - R^5 = H, alkyl, cycloalkyl, etc. **899**

DMP, benzene, 80 °C, 1 h
42-50%

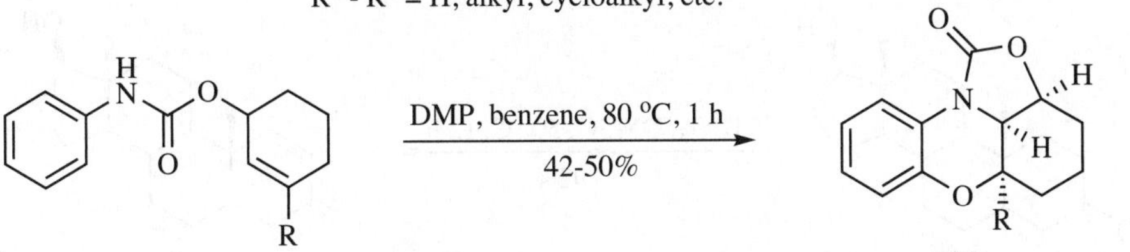

900 R = H, Me **901**

Scheme 3.361

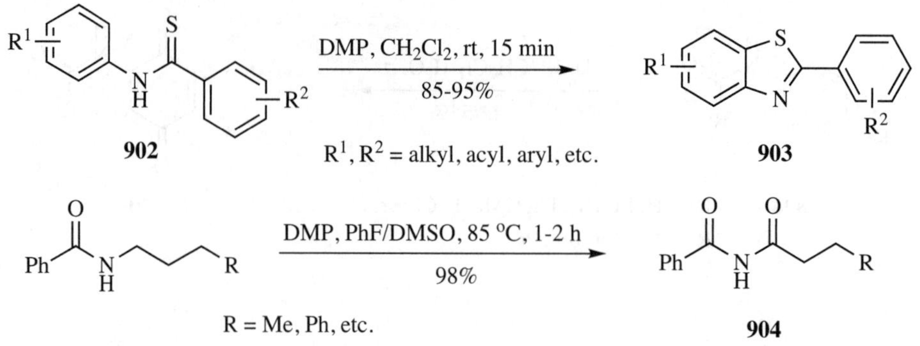

Scheme 3.362

quinone imine with the pendant alkene gives the final product **899** [1176]. A specific example of the oxidation of carbamates **900** leading to the benzomorpholine derivatives **901** is shown in Scheme 3.361.

Additional examples of the DMP-mediated oxidations of nitrogen substrates include the synthesis of 2-substituted benzothiazoles **903** via oxidative cyclization of thioanilides **902** [445] and the synthesis of imides (e.g., **904**), N-acyl vinylogous carbamates and ureas and nitriles by the oxidation of amides and amines with DMP (Scheme 3.362) [1292].

3.2.4 Inorganic Iodine(V) Reagents

Iodic acid (HIO_3), iodine pentoxide (I_2O_5) and inorganic iodate salts are commercially available, general-purpose oxidants, which are commonly used in organic synthesis. Applications of iodic acid in organic chemistry were summarized by Choghamarani in 2006 [1293]. Iodic acid has been used as a reagent in numerous organic reactions, such as oxidative iodination of aromatic compounds [1294–1296], oxidation of sulfides [1297, 1298], oxidation of aromatic amines to quinones [1299], deprotection of ketoximes and aromatic aldoximes [1300], deprotection of thioacetals and thioketals [1301], oxidative deprotection of trimethylsilyl, tetrahydropyranyl and methoxymethyl ethers [1302], oxidative coupling of *N,N*-dimethylanilines [1303], oxidative rearrangements [1304, 1305] and dehydrogenation of aldehydes and ketones [1306]. The advantages of iodic acid as a reagent include cost-effectiveness, non-toxicity and easy workup of reaction mixtures.

Schemes 3.363 and 3.364 show representative examples of the reactions of iodic acid. Nicolaou and coworkers have demonstrated that aldehydes and ketones can be selectively dehydrogenated to the corresponding 1,3-unsaturated carbonyl compounds with HIO_3 or I_2O_5 in DMSO at moderate heating, as illustrated by the reaction of steroidal substrate **905** (Scheme 3.363) [1306].

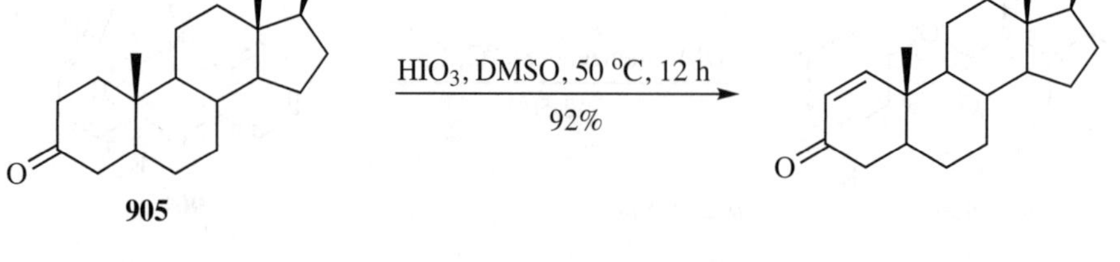

Scheme 3.363

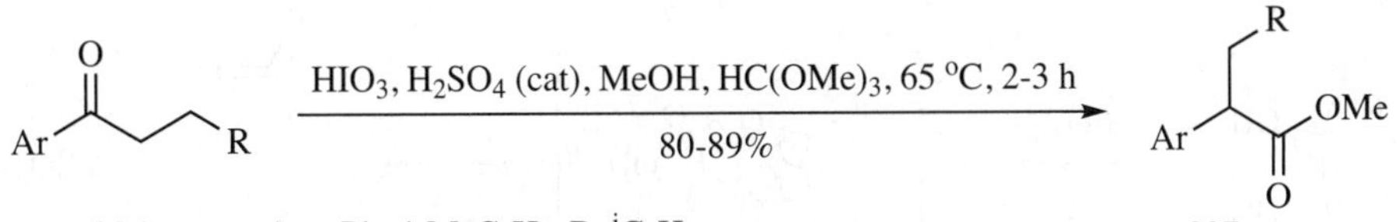

HIO_3, H_2SO_4 (cat), MeOH, $HC(OMe)_3$, 65 °C, 2-3 h

80-89%

906 Ar = Ph, 4-MeC_6H_4, $Bu^iC_6H_4$
R = H, Me, Et, CO_2H

907

Scheme 3.364

IF_5-pyridine-HF, CH_2Cl_2, rt, 24 h

89%

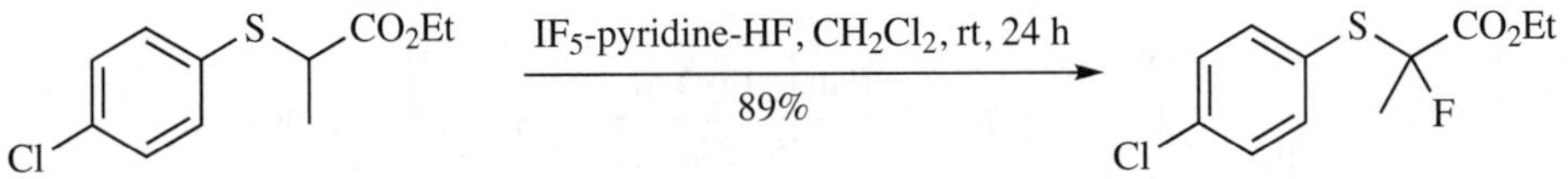

Scheme 3.365

Methyl esters of α-arylalkanoic acids **907** can be prepared by oxidative rearrangement of ketones **906** via a 1,2-aryl shift using iodic acid in methanol in the presence of trimethyl orthoformate and sulfuric acid (Scheme 3.364) [1304].

Iodine pentoxide can be used as a mild oxidant with generally similar reactivity to HIO_3. Representative examples of synthetic applications of I_2O_5 include α-thiocyanation of ketones using ammonium thiocyanate and iodine pentoxide [1307], thiocyanation of aromatic and heteroaromatic compounds using ammonium thiocyanate and iodine pentoxide [1308], oxidation of electron-rich alcohols in water using I_2O_5 or HIO_3 in the presence of KBr [1309] and oxidative decarboxylation of propiolic acids using the combination of iodine and I_2O_5 in methanol [1310].

Iodine pentafluoride, IF_5, has found some synthetic application as a powerful fluorinating reagent [1311, 1312]. For example, IF_5, in combination with pyridine and HF, can be used as a fluorination reagent for the introduction of fluorine atoms to the α-position in sulfides (Scheme 3.365) [1313]. A similar fluorination of sulfides using IF_5 without pyridine and HF affords a mixture of polyfluorinated products [1314].

3.3 Synthetic Applications of Iodine(VII) Compounds

Periodic acid, $HIO_4 \cdot 2H_2O$ or H_5IO_6 and periodate salts (e.g., sodium metaperiodate, $NaIO_4$) are common, commercially available, powerful oxidants, which are widely used in oxidation or oxidative cleavage reactions for many types of organic substrates [1315]. Periodic acid is soluble in ethereal organic solvents as well as in water, allowing a wider scope of use compared to sodium periodate.

The glycol-cleavage oxidation reaction (Scheme 3.366) is the major area of application of periodic acid, which is particularly important in carbohydrate chemistry [1316, 1317].

This reaction can be performed in non-aqueous solvents, such as diethyl ether or THF and the workup generally requires a simple filtration of iodic acid followed by evaporation of the solvent. Application of this methodology is exemplified by a convenient preparation of alkyl glyoxylates **908** in high yield (Scheme 3.367) [1318].

Periodic acid can also oxidatively cleave epoxides via a mechanism analogous to the glycol-cleavage shown in Scheme 3.366. The cleavage of terminal epoxides affords a single aldehyde, while cyclic substrates produce dialdehydes, as illustrated by the reaction of substrate **909** (Scheme 3.368) [1319]. The

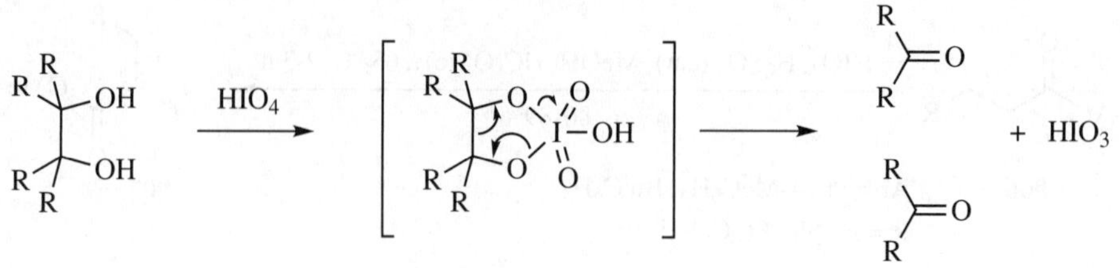

Scheme 3.366

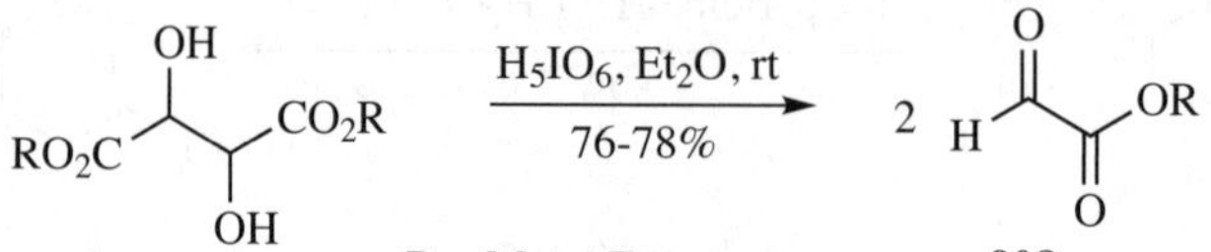

Scheme 3.367

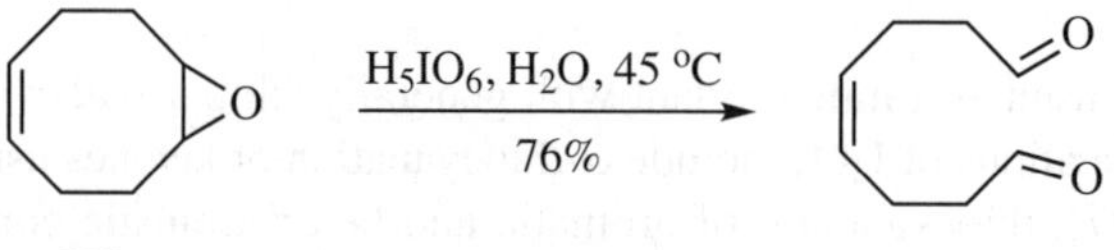

Scheme 3.368

reaction can be performed in water or aqueous organic solvents and it is compatible with various functional groups [1315].

Periodic acid can be applied as the stoichiometric oxidant in several transition metal catalyzed oxidations. Particularly useful are the chromium-catalyzed oxidations. In the presence of catalytic chromoyl diacetate, tertiary C—H bonds are oxidized to produce tertiary alcohols in moderate yields with retention of the original C—H stereochemistry, as exemplified in Scheme 3.369 [1320].

Secondary alcohols are oxidized by H_5IO_6 in the presence of various chromium catalysts to ketones [1321–1325], while primary alcohols can be oxidized to aldehydes or to carboxylic acids depending on the catalyst. Primary alcohols in the presence of pyridinium chlorochromate (PCC)[1322] or chromium (III) acetylacetonate, $Cr(acac)_3$ [1321] are oxidized to aldehydes or ketones in excellent yields, while the use of CrO_3 [1326, 1327] or pyridinium fluorochromate [1323] as catalysts results in the oxidation to carboxylic acids. The periodic acid promoted oxidation of primary and secondary alcohols to carbonyl compounds can also be catalyzed by Cu(II) derivatives [1328, 1329], by bromide anion [1330] and by TEMPO [1331].

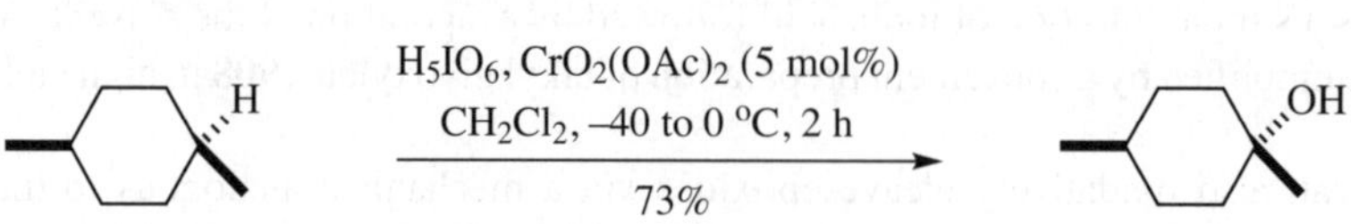

Scheme 3.369

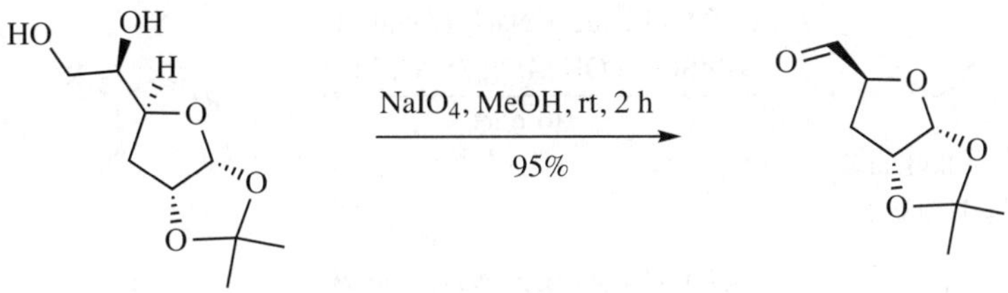

HO OH H O O O → $NaIO_4$, MeOH, rt, 2 h 95% → O O O O

Scheme 3.370

Chromium(VI) oxide is also an efficient catalyst for oxidation at the benzylic position with periodic acid as the terminal oxidant in acetonitrile. Substituted toluenes with an electron-withdrawing group at the 4- or 3-position and diarylmethanes such as Ph_2CH_2 and fluorene are oxidized to the respective substituted benzoic acids and ketones in excellent yields [1332]. Periodic acid in the presence of catalytic CrO_3 can be used for the oxidation of arenes, such as naphthalenes and anthracene, to the corresponding quinones; for example, 2-methylnaphthalene is oxidized to 2-methyl-1,4-naphthoquinone (vitamin K_3) by this catalytic system in high yield and regioselectivity [1333]. Sulfides can be oxidized by periodic acid to sulfoxides in the presence of catalytic $FeCl_3$ [1334], or to sulfones in the presence of CrO_3 [1335].

Additional examples of synthetic application of periodic acid as an oxidant include the oxidative iodination of aromatic compounds [1336–1341], iodohydrin formation by treatment of alkenes with periodic acid and sodium bisulfate [1342], oxidative cleavage of protecting groups (e.g., cyclic acetals, oxathioacetals and dithioacetals) [1315, 1343], conversion of ketone and aldehyde oximes into the corresponding carbonyl compounds [1344], oxidative cleavage of tetrahydrofuran-substituted alcohols to γ-lactones in the presence of catalytic PCC [1345] and direct synthesis of nitriles from alcohols or aldehydes using H_5IO_6/KI in aqueous ammonia [1346].

Sodium metaperiodate, $NaIO_4$, is a common, commercially available oxidant widely used in organic synthesis. As with periodic acid, sodium metaperiodate can be used for the glycol-cleavage oxidation reaction, which is particularly important in carbohydrate chemistry. Scheme 3.370 shows a representative example of a glycol-cleavage oxidation with $NaIO_4$; this reaction has been used in the synthesis of (2*S*,4*S*)-4-hydroxyproline from D-glucose [1347].

The glycol-cleavage oxidation has been utilized in numerous synthetic works, for example, the total syntheses of dipiperidine alkaloids virgidivarine and virgiboidine [1348], total synthesis of resolvin E2 [1349], synthesis of α-substituted oxazolochlorin aminals or acetals from *meso*-tetraaryldihydroxychlorins [1350], asymmetric synthesis of 1-(2- and 3-haloalkyl)azetidin-2-ones [1351], synthesis of 2-hydroxy-1,4-oxazin-3-ones through ring transformation of 3-hydroxy-4-(1,2-dihydroxyethyl)-β-lactams [1352], preparation of 1-O-protected (*R*)- and (*S*)-glycerols from L- and D-arabinose [1353], synthesis of unnatural glucose from cycloheptatriene [1354] and the synthesis of enantiomeric 2,3-disubstituted 5-norbornenes from D-mannitol [1355].

The combination of hydroxylamine hydrochloride, $NH_2OH{\cdot}HCl$ and sodium metaperiodate in dichloromethane at room temperature can be used as a mild oxidizing agent for selective oxidation of alcohols to carbonyl compounds [1356]. Various aliphatic, benzylic and heteroaryl substituted alcohols are oxidized to produce the corresponding carbonyl compounds in high yields; primary alcohols give the corresponding aldehydes without any noticeable further oxidation to acids. It is assumed that I_2 and NO, produced by the initial oxidation of $NH_2OH{\cdot}HCl$ with $NaIO_4$, act as the actual oxidants of alcohols under these conditions [1356].

A combination of $NaIO_4$ and KI in aqueous NH_3 converts alcohols into nitriles in moderate to good yield [1357]. This transformation, proceeds via an *in situ* oxidation–imination–aldimine oxidation sequence.

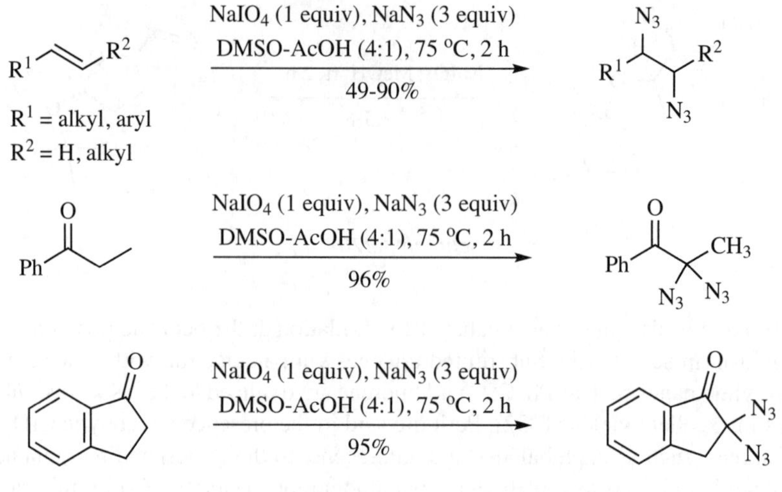

$NaIO_4$ (1 equiv), NaN_3 (3 equiv)
DMSO-AcOH (4:1), 75 °C, 2 h
49-90%

R^1 = alkyl, aryl
R^2 = H, alkyl

$NaIO_4$ (1 equiv), NaN_3 (3 equiv)
DMSO-AcOH (4:1), 75 °C, 2 h
96%

$NaIO_4$ (1 equiv), NaN_3 (3 equiv)
DMSO-AcOH (4:1), 75 °C, 2 h
95%

Scheme 3.371

The combination of $NaIO_4$, KI and NaN_3 is an efficient, simple and inexpensive reagent system for the β-azidoiodination of alkenes [1358]. This reaction proceeds in an anti-Markovnikov fashion to give β-iodo azides in excellent yields. Likewise, the $NaIO_4$–NaN_3 combination has been found to be an excellent reagent system suitable for the direct diazidation of styrenes, alkenes, benzylic alcohols and aryl ketones to produce the corresponding vicinal and geminal diazides, respectively, in high yields under mild reaction conditions (Scheme 3.371) [1359].

Sodium metaperiodate can be used for the oxidation of dihydrazones of α-diketones **910** to acetylenes **911** in high yields under mild condition (Scheme 3.372) [1360]. This procedure is also suitable for the deprotection of monohydrazones of aldehydes and ketones. This mild and efficient procedure is applicable to substrates with either electron-withdrawing or electron-donating substituents [1360].

McElwee-White and Gerack have developed a metal-free procedure for carbonylation of benzylamines **912** in methanol in the presence of $NaIO_4$, producing formamide derivatives **913** in good to excellent yields (Scheme 3.373) [1361]. Secondary amines can also be formylated under these conditions to give respective formamides in 51–58% yields. Labeling experiments have established that CO is the source of the formyl carbonyl, while its hydrogen is derived from the protic solvent.

Sodium metaperiodate has been applied as the stoichiometric oxidant in numerous transition metal catalyzed oxidations. Of particular use is a one-pot oxidative cleavage of olefins to aldehydes by the OsO_4–$NaIO_4$ catalytic system, as exemplified in Scheme 3.374 [1362]. This oxidative cleavage, with some modifications,

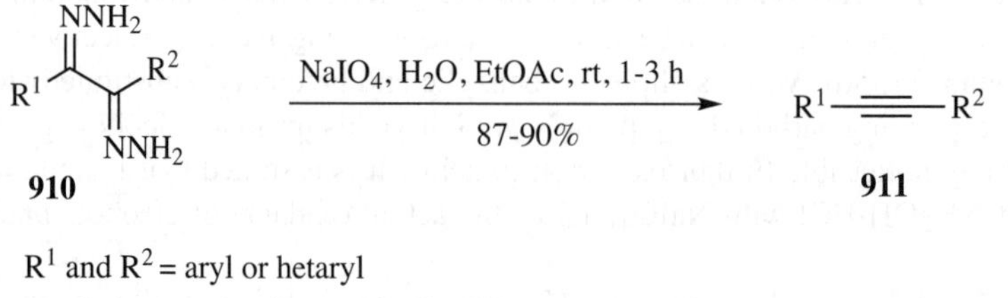

$NaIO_4$, H_2O, EtOAc, rt, 1-3 h
87-90%

910 → **911** (R^1—≡—R^2)

R^1 and R^2 = aryl or hetaryl

Scheme 3.372

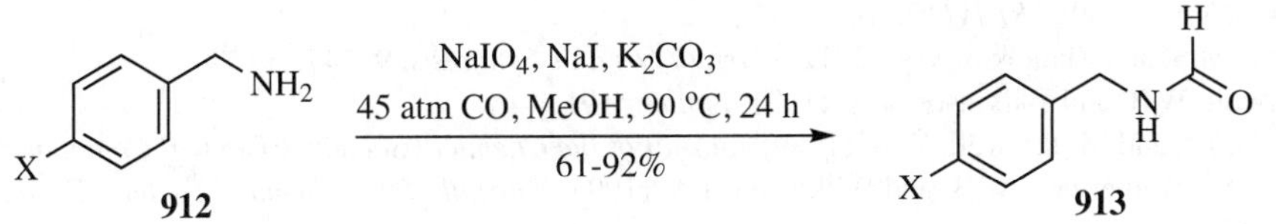

$NaIO_4$, NaI, K_2CO_3
45 atm CO, MeOH, 90 °C, 24 h
61-92%

912 → 913

X = H, OMe, Br, Cl

Scheme 3.373

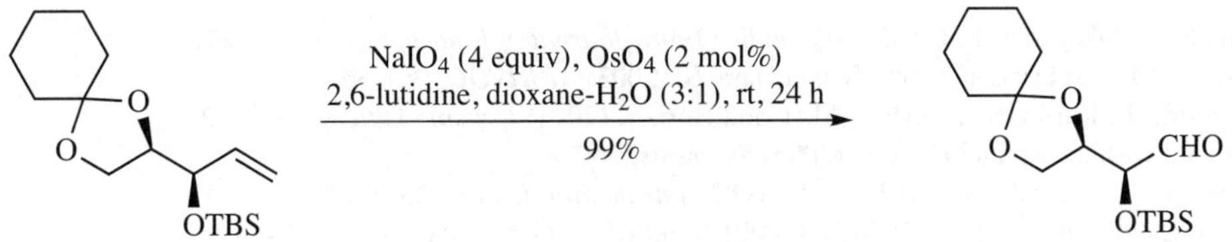

$NaIO_4$ (4 equiv), OsO_4 (2 mol%)
2,6-lutidine, dioxane-H_2O (3:1), rt, 24 h
99%

OTBS → CHO, OTBS

Scheme 3.374

has been utilized in numerous synthetic works [1363–1369]. A similar oxidative cleavage of olefins can be achieved by using catalytic RuO_4 (generated *in situ* from RuO_2 or $RuCl_3$) and $NaIO_4$ as the stoichiometric oxidant [1370].

Additional examples of the synthetic application of sodium metaperiodate as an oxidant in transition metal catalyzed oxidations include the hydroxylation of alkanes with $NaIO_4$ catalyzed by tetrakis(*p*-inophenyl)porphyrinatomanganese(III) chloride [1371], the use of $NaIO_4$ as a mild and efficient terminal oxidant for C—H oxidations with Cp*Ir (Cp* = C_5Me_5) precatalysts [1372], oxidation reactions of olefins with $NaIO_4$ using iron(III) *meso*-tetraarylporphyrins as the catalysts [1373], asymmetric epoxidation of unfunctionalized olefins with $NaIO_4$ using axially coordinated chiral salen Mn(III) complexes as the catalysts [1374], epoxidation of alkenes with $NaIO_4$ using multiwall carbon nanotube supported manganese(III) tetraphenylporphyrin [1375], oxidation of 2-imidazolines to 2-imidazoles with $NaIO_4$ catalyzed by manganese(III) tetraphenylporphyrin [1376], oxidation of 2-imidazolines to 2-imidazoles with $NaIO_4$ catalyzed by polystyrene-bound manganese(III) porphyrin [1377], oxidation of 2-imidazolines to 2-imidazoles with $NaIO_4$ catalyzed by Mn(salophen)Cl [1378] and numerous other works.

References

1. Varvoglis, A. (1992) *The Organic Chemistry of Polycoordinated Iodine*, VCH Publishers, Inc., New York.
2. Varvoglis, A. (1997) *Hypervalent Iodine in Organic Synthesis*, Academic Press, London.
3. Wirth, T. (ed.) (2003) *Hypervalent Iodine Chemistry: Modern Developments in Organic Synthesis*, Topics in Current Chemistry, vol. **224**, Springer, Berlin.
4. Silva, J.L.F. and Olofsson, B. (2011) *Natural Product Reports*, **28**, 1722.
5. Varvoglis, A. (1997) *Tetrahedron*, **53**, 1179.
6. Wirth, T. and Hirt, U.H. (1999) *Synthesis*, 1271.
7. Wirth, T. (2005) *Angewandte Chemie, International Edition*, **44**, 3656.
8. Stang, P.J. and Zhdankin, V.V. (1996) *Chemical Reviews*, **96**, 1123.
9. Zhdankin, V.V. and Stang, P.J. (2002) *Chemical Reviews*, **102**, 2523.
10. Zhdankin, V.V. and Stang, P.J. (2008) *Chemical Reviews*, **108**, 5299.

11. Zhdankin, V.V. (2009) *ARKIVOC*, (i), 1.
12. Yusubov, M.S. and Zhdankin, V.V. (2012) *Current Organic Synthesis*, **9**, 247.
13. Motherwell, W.B. and Wilkinson, J.A. (1991) *Synlett*, 191.
14. Edmunds, J.J. and Motherwell, W.B. (1989) *Journal of the Chemical Society, Chemical Communications*, 881.
15. Caddick, S., Motherwell, W.B. and Wilkinson, J.A. (1991) *Journal of the Chemical Society, Chemical Communications*, 674.
16. Edmunds, J.J. and Motherwell, W.B. (1989) *Journal of the Chemical Society, Chemical Communications*, 1348.
17. Lemal, D.M. (2001) in *e-EROS Encyclopedia of Reagents for Organic Synthesis* (editor in chief D. Crich), John Wiley & Sons, Inc., Hoboken, doi: 10.1002/047084289X.
18. Hara, S., Sekiguchi, M., Ohmori, A., *et al.* (1996) *Journal of the Chemical Society, Chemical Communications*, 1899.
19. Hara, S., Hatakeyama, T., Chen, S.-Q., *et al.* (1998) *Journal of Fluorine Chemistry*, **87**, 189.
20. Yoshida, M., Fujikawa, K., Sato, S. and Hara, S. (2003) *ARKIVOC*, (vi), 36.
21. Kitamura, T., Kuriki, S., Morshed, M.H. and Hori, Y. (2011) *Organic Letters*, **13**, 2392.
22. Sato, S., Yoshida, M. and Hara, S. (2005) *Synthesis*, 2602.
23. Tsushima, T., Kawada, K. and Tsuji, T. (1982) *Tetrahedron Letters*, **23**, 1165.
24. Greaney, M.F. and Motherwell, W.B. (2000) *Tetrahedron Letters*, **41**, 4467.
25. Greaney, M.F. and Motherwell, W.B. (2000) *Tetrahedron Letters*, **41**, 4463.
26. Caddick, S., Gazzard, L., Motherwell, W.B. and Wilkinson, J.A. (1996) *Tetrahedron*, **52**, 149.
27. Fuchigami, T. and Fujita, T. (1994) *Journal of Organic Chemistry*, **59**, 7190.
28. Motherwell, W.B., Greaney, M.F., Edmunds, J.J. and Steed, J.W. (2002) *Journal of the Chemical Society, Perkin Transactions* **1**, 2816.
29. Motherwell, W.B., Greaney, M.F. and Tocher, D.A. (2002) *Journal of the Chemical Society, Perkin Transactions* **1**, 2809.
30. Arrica, M.A. and Wirth, T. (2005) *European Journal of Organic Chemistry*, 395.
31. Inagaki, T., Nakamura, Y., Sawaguchi, M., *et al.* (2003) *Tetrahedron Letters*, **44**, 4117.
32. Fujita, T. and Fuchigami, T. (1996) *Tetrahedron Letters*, **37**, 4725.
33. Patrick, T.B., Scheibel, J.J., Hall, W.E. and Lee, Y.H. (1980) *Journal of Organic Chemistry*, **45**, 4492.
34. Hara, S., Nagahigashi, J., Ishi-i, K., *et al.* (1998) *Tetrahedron Letters*, **39**, 2589.
35. Hara, S., Nakahigashi, J., Ishi-i, K., *et al.* (1998) *Synlett*, 495.
36. Kasumov, T.M., Pirguliyev, N.S., Brel, V.K., *et al.* (1997) *Tetrahedron*, **53**, 13139.
37. Hara, S., Yoshida, M., Fukuhara, T. and Yoneda, N. (1998) *Journal of the Chemical Society, Chemical Communications*, 965.
38. Hara, S., Yamamoto, K., Yoshida, M., *et al.* (1999) *Tetrahedron Letters*, **40**, 7815.
39. Yoshida, M., Hara, S., Fukuhara, T. and Yoneda, N. (2000) *Tetrahedron Letters*, **41**, 3887.
40. Conte, P., Panunzi, B. and Tingoli, M. (2005) *Tetrahedron Letters*, **47**, 273.
41. Panunzi, B., Picardi, A. and Tingoli, M. (2004) *Synlett*, 2339.
42. Ochiai, M., Hirobe, M., Yoshimura, A., *et al.* (2007) *Organic Letters*, **9**, 3335.
43. Tian, T., Zhong, W.-H., Meng, S., *et al.* (2013) *Journal of Organic Chemistry*, **78**, 728.
44. Mylonas, V.E., Sigalas, M.P., Katsoulos, G.A., *et al.* (1994) *Journal of the Chemical Society, Perkin Transactions* 2, 1691.
45. Varvoglis, A. (1984) *Synthesis*, 709.
46. Merkushev, E.B. (1987) *Russian Chemical Reviews*, **56**, 826.
47. Moskovkina, T.V. and Vysotskii, V.I. (1991) *Zhurnal Organicheskoi Khimii*, **27**, 833.
48. Mikolajczyk, M., Midura, W.H., Grzejszczak, S., *et al.* (1994) *Tetrahedron*, **50**, 8053.
49. Masson, S. and Thuillier, A. (1969) *Bulletin de la Societe Chimique de France*, 4368.
50. Andreev, V.A., Anfilogova, S.N., Pekkh, T.I. and Belikova, N.A. (1993) *Zhurnal Organicheskoi Khimii*, **29**, 142.
51. Marchand, A.P., Sorokin, V.D., Rajagopal, D. and Bott, S.G. (1994) *Tetrahedron*, **50**, 9933.
52. Yu, J. and Zhang, C. (2009) *Synthesis*, 2324.
53. Ibrahim, H., Kleinbeck, F. and Togni, A. (2004) *Helvetica Chimica Acta*, **87**, 605.
54. Yusubov, M.S., Drygunova, L.A. and Zhdankin, V.V. (2004) *Synthesis*, 2289.

55. Yusubov, M.S., Yusubova, R.J., Filimonov, V.D. and Chi, K.-W. (2004) *Synthetic Communications*, **34**, 443.
56. Yusubov, M.S., Drygunova, L.A., Tkachev, A.V. and Zhdankin, V.V. (2005) *ARKIVOC*, (iv), 179.
57. Nicolaou, K.C., Simmons, N.L., Ying, Y., *et al.* (2011) *Journal of the American Chemical Society*, **133**, 8134.s
58. Zanka, A., Takeuchi, H. and Kubota, A. (1998) *Organic Process Research & Development*, **2**, 270.
59. Jin, L.-M., Yin, J.-J., Chen, L., *et al.* (2005) *Synlett*, 2893.
60. Ito, S. and Turnbull, K. (1996) *Synthetic Communications*, **26**, 1441.
61. Zhdankin, V.V., Hanson, K.J., Koposov, A.E., *et al.* (2001) *Mendeleev Communications*, 51.
62. Salamant, W. and Hulme, C. (2006) *Tetrahedron Letters*, **47**, 605.
63. Kitamura, T., Tazawa, Y., Morshed, M.H. and Kobayashi, S. (2012) *Synthesis*, 1159.
64. Drabowicz, J., Lyzwa, P., Luczak, J., *et al.* (1997) *Phosphorus, Sulfur and Silicon and the Related Elements*, **120–121**, 425.
65. Grebe, J., Schlecht, S., Weller, F., *et al.* (1999) *Zeitschrift für Anorganische und Allgemeine Chemie*, **625**, 633.
66. Burford, N., Clyburne, J.A.C., Gates, D.P., *et al.* (1994) *Journal of the Chemical Society, Dalton Transactions*, 997.
67. Poli, R. and Kelland, M.A. (1991) *Journal of Organometallic Chemistry*, **419**, 127.
68. Poli, R., Gordon, J.C., Desai, J.U. and Rheingold, A.L. (1991) *Journal of the Chemical Society, Chemical Communications*, 1518.
69. Gordon, J.C., Lee, V.T. and Poli, R. (1993) *Inorganic Chemistry*, **32**, 4460.
70. Filippou, A.C., Portius, P. and Jankowski, C. (2001) *Journal of Organometallic Chemistry*, **617–618**, 656.
71. Amor, F., Royo, P., Spaniol, T.P. and Okuda, J. (2000) *Journal of Organometallic Chemistry*, **604**, 126.
72. Cotton, F.A., Maloney, D.J. and Su, J. (1995) *Inorganica Chimica Acta*, **236**, 21.
73. Filippou, A.C., Winter, J.G., Kociok-Koehn, G., *et al.* (1999) *Organometallics*, **18**, 2649.
74. Bastian, M., Morales, D., Poli, R., *et al.* (2002) *Journal of Organometallic Chemistry*, **654**, 109.
75. Cook, T.R., Esswein, A.J. and Nocera, D.G. (2007) *Journal of the American Chemical Society*, **129**, 10094.
76. Whitfield, S.R. and Sanford, M.S. (2007) *Journal of the American Chemical Society*, **129**, 15142.
77. Sharma, M., Canty, A.J., Gardiner, M.G. and Jones, R.C. (2011) *Journal of Organometallic Chemistry*, **696**, 1441.
78. McCall, A.S., Wang, H., Desper, J.M. and Kraft, S. (2011) *Journal of the American Chemical Society*, **133**, 1832.
79. Khusniyarov, M.M., Harms, K. and Sundermeyer, J. (2006) *Journal of Fluorine Chemistry*, **127**, 200.
80. Hayton, T.W., Legzdins, P. and Patrick, B.O. (2002) *Inorganic Chemistry*, **41**, 5388.
81. Crispini, A., Ghedini, M. and Neve, F. (1993) *Inorganica Chimica Acta*, **209**, 235.
82. Hofer, M. and Nevado, C. (2012) *European Journal of Inorganic Chemistry*, **2012**, 1338.
83. Hartung, J. (2007) *Science of Synthesis*, **35**, 19.
84. Banks, D.F., Huyser, E.S. and Kleinberg, J. (1964) *Journal of Organic Chemistry*, **29**, 3692.
85. Hess, W.W., Huyser, E.S. and Kleinberg, J. (1964) *Journal of Organic Chemistry*, **29**, 1106.
86. Kajigaeshi, S., Kakinami, T., Moriwaki, M., *et al.* (1988) *Tetrahedron Letters*, **29**, 5783.
87. Troshin, P.A., Popkov, O. and Lyubovskaya, R.N. (2003) *Fullerenes, Nanotubes, and Carbon Nanostructures*, **11**, 165.
88. Troyanov, S.I., Shustova, N.B., Popov, A.A. and Sidorov, L.N. (2005) *Russian Chemical Bulletin*, **54**, 1656.
89. Amey, R.L. and Martin, J.C. (1979) *Journal of Organic Chemistry*, **44**, 1779.
90. Braddock, D.C., Cansell, G., Hermitage, S.A. and White, A.J.P. (2006) *Chemical Communications*, 1442.
91. Evans, P.A. and Brandt, T.A. (1996) *Tetrahedron Letters*, **37**, 6443.
92. Evans, P.A. and Brandt, T.A. (1997) *Journal of Organic Chemistry*, **62**, 5321.
93. Bovonsombat, P. and McNelis, E. (1993) *Synthesis*, 237.
94. Bovonsombat, P., Angara, G.J. and McNelis, E. (1992) *Synlett*, 131.
95. Bovonsombat, P., Djuardi, E. and Mc Nelis, E. (1994) *Tetrahedron Letters*, **35**, 2841.
96. Xia, J.-J., Wu, X.-L. and Wang, G.-W. (2008) *ARKIVOC*, (xvi), 22.
97. Camps, P., Lukach, A.E., Pujol, X. and Vazquez, S. (2000) *Tetrahedron*, **56**, 2703.
98. Nocquet-Thibault, S., Retailleau, P., Cariou, K. and Dodd, R.H. (2013) *Organic Letters*, **15**, 1842.
99. Merkushev, E.B. (1988) *Synthesis*, 923.
100. Kryska, A. and Skulski, L. (1999) *Journal of Chemical Research. Synopses*, 590.
101. Kryska, A. and Skulski, L. (1999) *Journal of Chemical Research. Synopses*, 2501.

102. Karade, N.N., Tiwari, G.B., Huple, D.B. and Siddiqui, T.A.J. (2006) *Journal of Chemical Research*, 366.
103. Barluenga, J., Gonzalez-Bobes, F. and Gonzalez, J.M. (2002) *Angewandte Chemie, International Edition*, **41**, 2556.
104. Togo, H., Nabana, T. and Yamaguchi, K. (2000) *Journal of Organic Chemistry*, **65**, 8391.
105. Togo, H., Nogami, G. and Yokoyama, M. (1998) *Synlett*, 534.
106. Togo, H., Abe, S., Nogami, G. and Yokoyama, M. (1999) *Bulletin of the Chemical Society of Japan*, **72**, 2351.
107. Panunzi, B., Rotiroti, L. and Tingoli, M. (2003) *Tetrahedron Letters*, **44**, 8753.
108. Benhida, R., Blanchard, P. and Fourrey, J.-L. (1998) *Tetrahedron Letters*, **39**, 6849.
109. Cheng, D.-P., Chen, Z.-C. and Zheng, Q.-G. (2003) *Synthetic Communications*, **33**, 2671.
110. Comins, D.L., Kuethe, J.T., Miller, T.M., *et al.* (2005) *Journal of Organic Chemistry*, **70**, 5221.
111. Muraki, T., Togo, H. and Yokoyama, M. (1998) *Synlett*, 286.
112. Muraki, T., Togo, H. and Yokoyama, M. (1999) *Journal of Organic Chemistry*, **64**, 2883.
113. Hashem, A., Jung, A., Ries, M. and Kirschning, A. (1998) *Synlett*, 195.
114. Kirschning, A., Kunst, E., Ries, M., *et al.* (2003) *ARKIVOC*, (vi), 145.
115. Gottam, H. and Vinod, T.K. (2011) *Journal of Organic Chemistry*, **76**, 974.
116. Koposov, A.Y., Boyarskikh, V.V. and Zhdankin, V.V. (2004) *Organic Letters*, **6**, 3613.
117. Muraki, T., Yokoyama, M. and Togo, H. (2000) *Journal of Organic Chemistry*, **65**, 4679.
118. Lee, J.C. and Jin, Y.S. (1999) *Synthetic Communications*, **29**, 2769.
119. Yusubov, M.S., Yusubova, R.Y., Funk, T.V., *et al.* (2010) *Synthesis*, 3681.
120. Iglesias-Arteaga, M.A., Castellanos, E. and Juaristi, E. (2003) *Tetrahedron: Asymmetry*, **14**, 577.
121. Iglesias-Arteaga, M.A., Avila-Ortiz, C.G. and Juaristi, E. (2002) *Tetrahedron Letters*, **43**, 5297.
122. Boto, A., Hernandez, R., De Leon, Y., *et al.* (2004) *Tetrahedron Letters*, **45**, 6841.
123. Diaz-Sanchez, B.R., Iglesias-Arteaga, M.A., Melgar-Fernandez, R. and Juaristi, E. (2007) *Journal of Organic Chemistry*, **72**, 4822.
124. Yan, J., Li, J. and Cheng, D. (2007) *Synlett*, 2442.
125. Angara, G.J. and McNelis, E. (1991) *Tetrahedron Letters*, **32**, 2099.
126. Bovonsombat, P., Angara, G.J. and NcNelis, E. (1994) *Tetrahedron Letters*, **35**, 6787.
127. Angara, G.J., Bovonsombat, P. and McNelis, E. (1992) *Tetrahedron Letters*, **33**, 2285.
128. Bovonsombat, P. and McNelis, E. (1992) *Tetrahedron Letters*, **33**, 7705.
129. Bovonsombat, P. and McNelis, E. (1993) *Tetrahedron*, **49**, 1525.
130. Bovonsombat, P. and Mc Nelis, E. (1994) *Tetrahedron Letters*, **35**, 6431.
131. Bovonsombat, P. and McNelis, E. (1993) *Tetrahedron Letters*, **34**, 4277.
132. Djuardi, E., Bovonsombat, P. and Mc Nelis, E. (1994) *Tetrahedron*, **50**, 11793.
133. Bovonsombat, P. and McNelis, E. (1995) *Synthetic Communications*, **25**, 1223.
134. Herault, X. and Mc Nelis, E. (1996) *Tetrahedron*, **52**, 10267.
135. Djuardi, E. and McNelis, E. (1999) *Tetrahedron Letters*, **40**, 7193.
136. Tohma, H., Takizawa, S., Maegawa, T. and Kita, Y. (2000) *Angewandte Chemie, International Edition*, **39**, 1306.
137. Tohma, H., Maegawa, T., Takizawa, S. and Kita, Y. (2002) *Advanced Synthesis & Catalysis*, **344**, 328.
138. Tohma, H., Maegawa, T. and Kita, Y. (2003) *Synlett*, 723.
139. Dohi, T., Maruyama, A., Yoshimura, M., *et al.* (2005) *Chemical Communications*, 2205.
140. Karade, N.N., Shirodkar, S.G., Dhoot, B.M. and Waghmare, P.B. (2005) *Journal of Chemical Research*, 274.
141. Karade, N.N., Tiwari, G.B. and Huple, D.B. (2005) *Synlett*, 2039.
142. Karade, N.N., Budhewar, V.H., Katkar, A.N. and Tiwari, G.B. (2006) *ARKIVOC*, (xi), 162.
143. De Mico, A., Margarita, R., Parlanti, L., *et al.* (1997) *Journal of Organic Chemistry*, **62**, 6974.
144. Moroda, A. and Togo, H. (2006) *Tetrahedron*, **62**, 12408.
145. Piancatelli, G., Leonelli, F., Do, N. and Ragan, J. (2006) *Organic Syntheses*, **83**, 18.
146. Pozzi, G., Quici, S. and Shepperson, I. (2002) *Tetrahedron Letters*, **43**, 6141.
147. Vatele, J.-M. (2006) *Tetrahedron Letters*, **47**, 715.
148. Vugts, D.J., Veum, L., al-Mafraji, K., *et al.* (2006) *European Journal of Organic Chemistry*, 1672.
149. Herrerias, C.I., Zhang, T.Y. and Li, C.-J. (2006) *Tetrahedron Letters*, **47**, 13.
150. But, T.Y.S., Tashino, Y., Togo, H. and Toy, P.H. (2005) *Organic & Biomolecular Chemistry*, **3**, 970.

151. Pozzi, G., Cavazzini, M., Quici, S., *et al.* (2004) *Organic Letters*, **6**, 441.
152. Holczknecht, O., Cavazzini, M., Quici, S., *et al.* (2005) *Advanced Synthesis & Catalysis*, **347**, 677.
153. Pozzi, G., Cavazzini, M., Holczknecht, O., *et al.* (2004) *Tetrahedron Letters*, **45**, 4249.
154. Qian, W., Jin, E., Bao, W. and Zhang, Y. (2006) *Tetrahedron*, **62**, 556.
155. Hansen, T.M., Florence, G.J., Lugo-Mas, P., *et al.* (2002) *Tetrahedron Letters*, **44**, 57.
156. Li, Y. and Hale, K.J. (2007) *Organic Letters*, **9**, 1267.
157. Zhao, X.-F. and Zhang, C. (2007) *Synthesis*, 551.
158. Li, X.-Q. and Zhang, C. (2009) *Synthesis*, 1163.
159. Chen, F.-E., Xie, B., Zhang, P., *et al.* (2007) *Synlett*, 619.
160. Kansara, A., Sharma, P.K. and Banerji, K.K. (2004) *Journal of Chemical Research*, 581.
161. Huang, S., Wang, F., Gan, L., *et al.* (2006) *Organic Letters*, **8**, 277.
162. Lee, J.C., Lee, J.Y. and Lee, S.J. (2004) *Tetrahedron Letters*, **45**, 4939.
163. Koposov, A.Y., Netzel, B.C., Yusubov, M.S., *et al.* (2007) *European Journal of Organic Chemistry*, 4475.
164. Zhu, C. and Wei, Y. (2011) *Catalysis Letters*, **141**, 582.
165. Merritt, E.A. and Olofsson, B. (2011) *Synthesis*, 517.
166. Moriarty, R.M. and Prakash, O. (1986) *Accounts of Chemical Research*, **19**, 244.
167. Moriarty, R.M. and Hu, H. (1981) *Tetrahedron Letters*, **22**, 2747.
168. Moriarty, R.M., Prakash, O., Karalis, P. and Prakash, I. (1984) *Tetrahedron Letters*, **25**, 4745.
169. Moriarty, R.M., Hu, H. and Gupta, S.C. (1981) *Tetrahedron Letters*, **22**, 1283.
170. Moriarty, R.M. and Hou, K.C. (1984) *Tetrahedron Letters*, **25**, 691.
171. Prakash, O., Tanwar, M.P., Goyal, S. and Pahuja, S. (1992) *Tetrahedron Letters*, **33**, 6519.
172. Prakash, O., Kumar, D., Saini, R.K. and Singh, S.P. (1994) *Synthetic Communications*, **24**, 2167.
173. Prakash, O. and Mendiratta, S. (1992) *Synthetic Communications*, **22**, 327.
174. Moriarty, R.M., Hou, K.C., Prakash, I. and Arora, S.K. (1986) *Organic Syntheses*, **64**, 138.
175. Moriarty, R.M. and Prakash, O. (1985) *Journal of Organic Chemistry*, **50**, 151.
176. Prakash, O., Saini, N. and Sharma, P.K. (1994) *Synlett*, 221.
177. Prakash, O., Saini, N. and Sharma, P. (1994) *Heterocycles*, **38**, 409.
178. Prakash, O. and Singh, S.P. (1994) *Aldrichimica Acta*, **27**, 15.
179. Prakash, O. (1995) *Aldrichimica Acta*, **28**, 63.
180. Moriarty, R.M. (2005) *Journal of Organic Chemistry*, **70**, 2893.
181. Moriarty, R.M. and Prakash, O. (1999) *Organic Reactions*, **54**, 273.
182. Prakash, O., Saini, N., Tanwar, M.P. and Moriarty, R.M. (1995) *Contemporary Organic Synthesis*, **2**, 121.
183. Moriarty, R.M., Berglund, B.A. and Penmasta, R. (1992) *Tetrahedron Letters*, **33**, 6065.
184. Koser, G.F. (2001) *Aldrichimica Acta*, **34**, 89.
185. Koser, G.F., Relenyi, A.G., Kalos, A.N., *et al.* (1982) *Journal of Organic Chemistry*, **47**, 2487.
186. Moriarty, R.M., Penmasta, R., Awasthi, A.K., *et al.* (1989) *Journal of Organic Chemistry*, **54**, 1101.
187. Tuncay, A., Dustman, J.A., Fisher, G., *et al.* (1992) *Tetrahedron Letters*, **33**, 7647.
188. Prakash, O., Sharma, N. and Ranjan, P. (2010) *Synthetic Communications*, **40**, 2875.
189. Magnus, N.A., Ducry, L., Rolland, V., *et al.* (1997) *Journal of the Chemical Society, Perkin Transactions* **1**, 2313.
190. Moriarty, R.M., Vaid, R.K. and Koser, G.F. (1990) *Synlett*, 365.
191. Koser, G.F. (2004) *Advances in Heterocyclic Chemistry*, **86**, 225.
192. Lee, J.C. and Hong, T. (1997) *Synthetic Communications*, **27**, 4085.
193. Lee, J.C., Kim, S. and Shin, W.C. (2000) *Synthetic Communications*, **30**, 4271.
194. Varma, R.S., Kumar, D. and Liesen, P.J. (1998) *Journal of the Chemical Society, Perkin Transactions* **1**, 4093.
195. Aggarwal, R. and Sumran, G. (2006) *Synthetic Communications*, **36**, 875.
196. Choi, O.K. and Cho, B.T. (2001) *Tetrahedron: Asymmetry*, **12**, 903.
197. Zhang, P.-F. and Chen, Z.-C. (2001) *Synthetic Communications*, **31**, 415.
198. Singh, S.P., Naithani, R., Aggarwal, R. and Prakash, O. (1998) *Synthetic Communications*, **28**, 3133.
199. Prakash, O., Aneja, D.K., Wadhwa, D., *et al.* (2012) *Journal of Heterocyclic Chemistry*, **49**, 566.
200. Prakash, O., Wadhwa, D., Hussain, K. and Kumar, R. (2012) *Synthetic Communications*, **42**, 2947.
201. Kamal, R., Sharma, D., Wadhwa, D. and Prakash, O. (2012) *Synlett*, 93.

202. Lee, J.C., Seo, J.-W. and Baek, J.W. (2007) *Synthetic Communications*, **37**, 2159.
203. Xu, X., Deng, Y., Li, X., *et al.* (2011) *Journal of Chemical Research*, **35**, 605.
204. Karade, N.N., Gampawar, S.V., Kondre, J.M. and Shinde, S.V. (2008) *Tetrahedron Letters*, **49**, 4402.
205. Zhou, G.-B., Guan, Y.-Q., Shen, C., *et al.* (2008) *Synthesis*, 1994.
206. Ueno, M., Nabana, T. and Togo, H. (2003) *Journal of Organic Chemistry*, **68**, 6424.
207. Goff, J.M., Justik, M.W. and Koser, G.F. (2001) *Tetrahedron Letters*, **42**, 5597.
208. Moriarty, R.M., Condeiu, C., Tao, A. and Prakash, O. (1997) *Tetrahedron Letters*, **38**, 2401.
209. Moriarty, R.M. and Vaid, R.K. (1990) *Synthesis*, 431.
210. Moriarty, R.M., Prakash, O., Duncan, M.P., *et al.* (1987) *Journal of Organic Chemistry*, **52**, 150.
211. Moriarty, R., Prakash, O. and Duncan, M.P. (1987) *Journal of the Chemical Society, Perkin Transactions* **1**, 559.
212. Moriarty, R.M., Duncan, M.P. and Prakash, O. (1987) *Journal of the Chemical Society, Perkin Transactions* **1**, 1781.
213. Moriarty, R.M., Prakash, O. and Duncan, M.P. (1986) *Synthetic Communications*, **16**, 1239.
214. Kim, D.Y., Mang, J.Y. and Oh, D.Y. (1994) *Synthetic Communications*, **24**, 629.
215. Moriarty, R.M., Epa, W.R., Penmasta, R. and Awasthi, A.K. (1989) *Tetrahedron Letters*, **30**, 667.
216. Moriarty, R.M., Rani, N., Condeiu, C., *et al.* (1997) *Synthetic Communications*, **27**, 3273.
217. Koser, G.F., Chen, K., Huang, Y. and Summers, C.A. (1994) *Journal of the Chemical Society, Perkin Transactions* **1**, 1375.
218. Zefirov, N.S., Zhdankin, V.V., Dan'kov, Y.V., *et al.* (1986) *Tetrahedron Letters*, **27**, 3971.
219. Zefirov, N.S., Caple, R., Palyulin, V.A., *et al.* (1988) *Izvestiya Akademii Nauk SSSR, Seriya Khimicheskaya*, 1452.
220. Moriarty, R.M., Prakash, O., Duncan, M.P., *et al.* (1996) *Journal of Chemical Research (S)*, 432.
221. Ochiai, M., Sumi, K., Takaoka, Y., *et al.* (1988) *Tetrahedron*, **44**, 4095.
222. Ochiai, M., Oshima, K. and Masaki, Y. (1991) *Journal of the Chemical Society, Chemical Communications*, 869.
223. Ochiai, M., Sumi, K., Nagao, Y., *et al.* (1985) *Journal of the Chemical Society, Chemical Communications*, 697.
224. Ochiai, M., Fujita, E., Arimoto, M. and Yamaguchi, H. (1983) *Tetrahedron Letters*, **24**, 777.
225. Koser, G.F., Rebrovic, L. and Wettach, R.H. (1981) *Journal of Organic Chemistry*, **46**, 4324.
226. Rebrovic, L. and Koser, G.F. (1984) *Journal of Organic Chemistry*, **49**, 2462.
227. Zefirov, N.S., Zhdankin, V.V., Dan'kov, Y.V., *et al.* (1986) *Doklady Akademii Nauk SSSR*, **288**, 385.
228. Zefirov, N.S., Zhdankin, V.V., Dan'kov, Y.V., *et al.* (1984) *Zhurnal Organicheskoi Khimii*, **20**, 444.
229. Shah, M., Tashner, M.J., Koser, G.F., *et al.* (1986) *Tetrahedron Letters*, **27**, 5437.
230. Kim, H.J. and Schlecht, M.F. (1987) *Tetrahedron Letters*, **28**, 5229.
231. Schaumann, E. and Kirschning, A. (1990) *Journal of the Chemical Society, Perkin Transactions* **1**, 1481.
232. Zefirov, N.S., Zhdankin, V.V., Dan'kov, Y.V., *et al.* (1985) *Zhurnal Organicheskoi Khimii*, **21**, 2461.
233. Hirt, U.H., Schuster, M.F.H., French, A.N., *et al.* (2001) *European Journal of Organic Chemistry*, 1569.
234. Rebrovic, L. and Koser, G.F. (1984) *Journal of Organic Chemistry*, **49**, 4700.
235. Margida, A.J. and Koser, G.F. (1984) *Journal of Organic Chemistry*, **49**, 4703.
236. Lodaya, J.S. and Koser, G.F. (1990) *Journal of Organic Chemistry*, **55**, 1513.
237. Stang, P.J. and Surber, B.W. (1985) *Journal of the American Chemical Society*, **107**, 1452.
238. Stang, P.J., Surber, B.W., Chen, Z.C., *et al.* (1987) *Journal of the American Chemical Society*, **109**, 228.
239. Stang, P.J., Kitamura, T., Boehshar, M. and Wingert, H. (1989) *Journal of the American Chemical Society*, **111**, 2225.
240. Hembre, R.T., Scott, C.P. and Norton, J.R. (1987) *Journal of Organic Chemistry*, **52**, 3650.
241. Yang, Y., Diederich, F. and Valentine, J.S. (1991) *Journal of the American Chemical Society*, **113**, 7195.
242. Zefirov, N.S., Zhdankin, V.V., Dan'kov, Y.V. and Koz'min, A.S. (1984) *Zhurnal Organicheskoi Khimii*, **20**, 446.
243. De Mico, A., Margarita, R., Parlanti, L., *et al.* (1997) *Tetrahedron*, **53**, 16877.
244. Zefirov, N.S., Sorokin, V.D., Zhdankin, V.V. and Koz'min, A.S. (1986) *Russian Journal of Organic Chemistry*, **22**, 450.
245. Bassindale, A.R., Katampe, I. and Taylor, P.G. (2000) *Canadian Journal of Chemistry*, **78**, 1479.
246. Bassindale, A.R., Katampe, I., Maesano, M.G., *et al.* (1999) *Tetrahedron Letters*, **40**, 7417.
247. Togo, H., Hoshina, Y. and Yokoyama, M. (1996) *Tetrahedron Letters*, **37**, 6129.
248. Celik, M., Alp, C., Coskun, B., *et al.* (2006) *Tetrahedron Letters*, **47**, 3659.

249. Alvarez, H.M., Barbosa, D.P., Fricks, A.T., *et al.* (2006) *Organic Process Research & Development*, **10**, 941.
250. Yu, L., Chen, B. and Huang, X. (2007) *Tetrahedron Letters*, **48**, 925.
251. Mironov, Y.V., Sherman, A.A. and Nifantiev, N.E. (2004) *Tetrahedron Letters*, **45**, 9107.
252. Shi, M., Wang, B.-Y. and Li, J. (2005) *European Journal of Organic Chemistry*, 759.
253. Tellitu, I. and Dominguez, E. (2008) *Tetrahedron*, **64**, 2465.
254. Yusubov, M.S., Zholobova, G.A., Filimonova, I.L. and Chi, K.-W. (2004) *Russian Chemical Bulletin*, **53**, 1735.
255. Kang, Y.-B. and Gade, L.H. (2011) *Journal of the American Chemical Society*, **133**, 3658.
256. Zhong, W., Yang, J., Meng, X. and Li, Z. (2011) *Journal of Organic Chemistry*, **76**, 9997.
257. Fujita, M., Wakita, M. and Sugimura, T. (2011) *Chemical Communications*, **47**, 3983.
258. Fujita, M., Suzawa, H., Sugimura, T. and Okuyama, T. (2008) *Tetrahedron Letters*, **49**, 3326.
259. Fujita, M., Ookubo, Y. and Sugimura, T. (2009) *Tetrahedron Letters*, **50**, 1298.
260. Prevost, C. (1933) *Comptes Rendus*, **196**, 1129.
261. Woodward, R.B. and Brutcher, F.V. Jr. (1958) *Journal of the American Chemical Society*, **80**, 209.
262. Fujita, M., Okuno, S., Lee, H.J., *et al.* (2007) *Tetrahedron Letters*, **48**, 8691.
263. Fujita, M., Yoshida, Y., Miyata, K., *et al.* (2010) *Angewandte Chemie, International Edition*, **49**, 7068.
264. Ono, T. and Henderson, P. (2002) *Tetrahedron Letters*, **43**, 7961.
265. McQuaid, K.M. and Pettus, T.R.R. (2004) *Synlett*, 2403.
266. Lee, S. and MacMillan, D.W.C. (2006) *Tetrahedron*, **62**, 11413.
267. Huang, W.-J., Singh, O.V., Chen, C.-H., *et al.* (2002) *Helvetica Chimica Acta*, **85**, 1069.
268. Baba, H., Moriyama, K. and Togo, H. (2011) *Tetrahedron Letters*, **52**, 4303.
269. Zhao, Y. and Yeung, Y.-Y. (2010) *Organic Letters*, **12**, 2128.
270. Zhao, Y., Yim, W.-L., Tan, C.K. and Yeung, Y.-Y. (2011) *Organic Letters*, **13**, 4308.
271. Liu, H., Wang, X. and Gu, Y. (2011) *Organic & Biomolecular Chemistry*, **9**, 1614.
272. Liu, H., Xie, Y. and Gu, Y. (2011) *Tetrahedron Letters*, **52**, 4324.
273. Pialat, A., Liegault, B. and Taillefer, M. (2013) *Organic Letters*, **15**, 1764.
274. Koser, G.F., Telu, S. and Laali, K.K. (2006) *Tetrahedron Letters*, **47**, 7011.
275. Ley, S.V., Thomas, A.W. and Finch, H. (1999) *Journal of the Chemical Society, Perkin Transactions* **1**, 669.
276. Ficht, S., Mulbaier, M. and Giannis, A. (2001) *Tetrahedron*, **57**, 4863.
277. Pelter, A. and Elgendy, S.M.A. (1993) *Journal of the Chemical Society, Perkin Transactions* **1**, 1891.
278. Kato, N., Sugaya, T., Mimura, T., *et al.* (1997) *Synthesis*, 625.
279. Rocaboy, C. and Gladysz, J.A. (2003) *Chemistry–A European Journal*, **9**, 88.
280. Saby, C. and Luong, J.H.T. (1997) *Chemical Communications*, 1197.
281. Kinugawa, M., Masuda, Y., Arai, H., *et al.* (1996) *Synthesis*, 633.
282. Dohi, T., Nakae, T., Takenaga, N., *et al.* (2012) *Synthesis*, 1183.
283. Kurti, L., Herczegh, P., Visy, J., *et al.* (1999) *Journal of the Chemical Society, Perkin Transactions* **1**, 379.
284. Moriarty, R.M. and Prakash, O. (2001) *Organic Reactions*, **57**, 327.
285. Moore, J.D. and Hanson, P.R. (2002) *Chemtracts*, **15**, 74.
286. Quideau, S., Pouysegu, L. and Deffieux, D. (2004) *Current Organic Chemistry*, **8**, 113.
287. Ciufolini, M.A., Braun, N.A., Canesi, S., *et al.* (2007) *Synthesis*, 3759.
288. Rodriguez, S. and Wipf, P. (2004) *Synthesis*, 2767.
289. Quideau, S., Pouysegu, L. and Deffieux, D. (2008) *Synlett*, 467.
290. Liang, H. and Ciufolini, M.A. (2011) *Angewandte Chemie, International Edition*, **50**, 11849.
291. Pouysegu, L., Sylla, T., Garnier, T., *et al.* (2010) *Tetrahedron*, **66**, 5908.
292. Pouysegu, L., Deffieux, D. and Quideau, S. (2010) *Tetrahedron*, **66**, 2235.
293. McKillop, A., McLaren, L. and Taylor, R.J.K. (1994) *Journal of the Chemical Society, Perkin Transactions* **1**, 2047.
294. Felpin, F.-X. (2007) *Tetrahedron Letters*, **48**, 409.
295. Fleck, A.E., Hobart, J.A. and Morrow, G.W. (1992) *Synthetic Communications*, **22**, 179.
296. Mitchell, A.S. and Russell, R.A. (1993) *Tetrahedron Letters*, **34**, 545.
297. Hu, Y., Li, C., Kulkarni, B.A., *et al.* (2001) *Organic Letters*, **3**, 1649.
298. Mitchell, A.S. and Russell, R.A. (1997) *Tetrahedron*, **53**, 4387.

299. Roy, H., Sarkar, M. and Mal, D. (2005) *Synthetic Communications*, **35**, 2183.
300. Dey, S. and Mal, D. (2005) *Tetrahedron Letters*, **46**, 5483.
301. Venkateswarlu, R., Kamakshi, C., Subhash, P.V., *et al.* (2006) *Tetrahedron*, **62**, 4463.
302. Karam, O., Jacquesy, J.-C. and Jouannetaud, M.-P. (1994) *Tetrahedron Letters*, **35**, 2541.
303. Karam, O., Martin, A., Jouannetaud, M.-P. and Jacquesy, J.-C. (1999) *Tetrahedron Letters*, **40**, 4183.
304. Karam, O., Martin-Mingot, A., Jouannetaud, M.-P., *et al.* (2004) *Tetrahedron*, **60**, 6629.
305. Martin, A., Jouannetaud, M.-P., Jacquesy, J.-C. and Cousson, A. (1996) *Tetrahedron Letters*, **37**, 7735.
306. Sabot, C., Commare, B., Duceppe, M.-A., *et al.* (2008) *Synlett*, 3226.
307. Kita, Y., Tohma, H., Kikuchi, K., *et al.* (1991) *Journal of Organic Chemistry*, **56**, 435.
308. Murakata, M., Yamada, K. and Hoshino, O. (1994) *Journal of the Chemical Society, Chemical Communications*, 443.
309. Kacan, M., Koyuncu, D. and McKillop, A. (1993) *Journal of the Chemical Society, Perkin Transactions* **1**, 1771.
310. Wipf, P., Kim, Y. and Fritch, P.C. (1993) *Journal of Organic Chemistry*, **58**, 7195.
311. Hara, H., Inoue, T., Nakamura, H., *et al.* (1992) *Tetrahedron Letters*, **33**, 6491.
312. Quideau, S., Lebon, M. and Lamidey, A.-M. (2002) *Organic Letters*, **4**, 3975.
313. Frie, J.L., Jeffrey, C.S. and Sorensen, E.J. (2009) *Organic Letters*, **11**, 5394.
314. Ye, Y., Zhang, L. and Fan, R. (2012) *Organic Letters*, **14**, 2114.
315. Braun, N.A., Ousmer, M., Bray, J.D., *et al.* (2000) *Journal of Organic Chemistry*, **65**, 4397.
316. Braun, N.A., Bray, J.D. and Ciufolini, M.A. (1999) *Tetrahedron Letters*, **40**, 4985.
317. Braun, N.A., Ciufolini, M.A., Peters, K. and Peters, E.-M. (1998) *Tetrahedron Letters*, **39**, 4667.
318. Ousmer, M., Braun, N.A., Bavoux, C., *et al.* (2001) *Journal of the American Chemical Society*, **123**, 7534.
319. Ousmer, M., Braun, N.A. and Ciufolini, M.A. (2001) *Organic Letters*, **3**, 765.
320. Kita, Y., Takada, T. and Tohma, H. (1996) *Pure and Applied Chemistry*, **68**, 627.
321. Kita, Y., Arisawa, M., Gyoten, M., *et al.* (1998) *Journal of Organic Chemistry*, **63**, 6625.
322. Kita, Y., Takada, T., Ibaraki, M., *et al.* (1996) *Journal of Organic Chemistry*, **61**, 223.
323. Kita, Y., Takada, T., Gyoten, M., *et al.* (1996) *Journal of Organic Chemistry*, **61**, 5857.
324. Asmanidou, A., Papoutsis, I., Spyroudis, S. and Varvoglis, A. (2000) *Molecules*, **5**, 874.
325. Shigehisa, H., Takayama, J. and Honda, T. (2006) *Tetrahedron Letters*, **47**, 7301.
326. Honda, T. and Shigehisa, H. (2006) *Organic Letters*, **8**, 657.
327. Ward, R.S., Pelter, A. and Abd-El-Ghani, A. (1996) *Tetrahedron*, **52**, 1303.
328. Beaulieu, M.-A., Guerard, K.C., Maertens, G., *et al.* (2011) *Journal of Organic Chemistry*, **76**, 9460.
329. Desjardins, S., Andrez, J.-C. and Canesi, S. (2011) *Organic Letters*, **13**, 3406.
330. Jacquemot, G. and Canesi, S. (2012) *Journal of Organic Chemistry*, **77**, 7588.
331. Guerard, K.C., Guerinot, A., Bouchard-Aubin, C., *et al.* (2012) *Journal of Organic Chemistry*, **77**, 2121.
332. Berard, D., Giroux, M.-A., Racicot, L., *et al.* (2008) *Tetrahedron*, **64**, 7537.
333. Krikorian, D., Tarpanov, V., Parushev, S. and Mechkarova, P. (2000) *Synthetic Communications*, **30**, 2833.
334. Wipf, P. and Spencer, S.R. (2005) *Journal of the American Chemical Society*, **127**, 225.
335. Canesi, S., Bouchu, D. and Ciufolini, M.A. (2004) *Angewandte Chemie, International Edition*, **43**, 4336.
336. Nicolaou, K.C., Edmonds, D.J., Li, A. and Tria, G.S. (2007) *Angewandte Chemie, International Edition*, **46**, 3942.
337. Tohma, H., Harayama, Y., Hashizume, M., *et al.* (2003) *Journal of the American Chemical Society*, **125**, 11235.
338. Baxendale, I.R., Ley, S.V., Nessi, M. and Piutti, C. (2002) *Tetrahedron*, **58**, 6285.
339. Baxendale, I.R., Deeley, J., Griffiths-Jones, C.M., *et al.* (2006) *Chemical Communications*, 2566.
340. Dohi, T., Uchiyama, T., Yamashita, D., *et al.* (2011) *Tetrahedron Letters*, **52**, 2212.
341. Takenaga, N., Uchiyama, T., Kato, D., *et al.* (2011) *Heterocycles*, **82**, 1327.
342. Quideau, S., Looney, M.A. and Pouysegu, L. (1999) *Organic Letters*, **1**, 1651.
343. Lebrasseur, N., Fan, G.-J. and Quideau, S. (2004) *ARKIVOC*, (xiii), 5.
344. Lebrasseur, N., Fan, G.-J., Oxoby, M., *et al.* (2005) *Tetrahedron*, **61**, 1551.
345. Quideau, S., Pouysegu, L., Oxoby, M. and Looney, M.A. (2001) *Tetrahedron*, **57**, 319.
346. Dohi, T., Maruyama, A., Takenage, N., *et al.* (2008) *Angewandte Chemie, International Edition*, **47**, 3787.
347. Pouysegu, L., Chassaing, S., Dejugnac, D., *et al.* (2008) *Angewandte Chemie, International Edition*, **47**, 3552.
348. Quideau, S., Pouysegu, L., Deffieux, D., *et al.* (2003) *ARKIVOC*, (vi), 106.

349. Gagnepain, J., Mereau, R., Dejugnac, D., *et al.* (2007) *Tetrahedron*, **63**, 6493.
350. Quideau, S., Looney, M.A., Pouysegu, L., *et al.* (1999) *Tetrahedron Letters*, **40**, 615.
351. Lai, C.-H., Lin, P.-Y., Peddinti, R.K. and Liao, C.-C. (2002) *Synlett*, 1520.
352. Cook, S.P. and Danishefsky, S.J. (2006) *Organic Letters*, **8**, 5693.
353. Berube, A., Drutu, I. and Wood, J.L. (2006) *Organic Letters*, **8**, 5421.
354. Chittimalla, S.K. and Liao, C.-C. (2002) *Synlett*, 565.
355. Hou, H.-F., Peddinti, R.K. and Liao, C.-C. (2002) *Organic Letters*, **4**, 2477.
356. Hsieh, M.-F., Rao, P.D. and Liao, C.-C. (1999) *Chemical Communications*, 1441.
357. Basset, L., Martin-Mingot, A., Jouannetaud, M.-P. and Jacquesy, J.-C. (2008) *Tetrahedron Letters*, **49**, 1551.
358. Bodipati, N. and Peddinti, R.K. (2012) *Organic & Biomolecular Chemistry*, **10**, 4549.
359. Zagulyaeva, A.A., Banek, C.T., Yusubov, M.S. and Zhdankin, V.V. (2010) *Organic Letters*, **12**, 4644.
360. Wang, L. and Fan, R. (2012) *Organic Letters*, **14**, 3596.
361. Hata, K., Hamamoto, H., Shiozaki, Y., *et al.* (2007) *Tetrahedron*, **63**, 4052.
362. Kita, Y., Tohma, H., Hatanaka, K., *et al.* (1994) *Journal of the American Chemical Society*, **116**, 3684.
363. Tohma, H., Morioka, H., Takizawa, S., *et al.* (2001) *Tetrahedron*, **57**, 345.
364. Kita, Y., Gyoten, M., Ohtsubo, M., *et al.* (1996) *Chemical Communications*, 1481.
365. Takada, T., Arisawa, M., Gyoten, M., *et al.* (1998) *Journal of Organic Chemistry*, **63**, 7698.
366. Arisawa, M., Ramesh, N.G., Nakajima, M., *et al.* (2001) *Journal of Organic Chemistry*, **66**, 59.
367. Hamamoto, H., Hata, K., Nambu, H., *et al.* (2004) *Tetrahedron Letters*, **45**, 2293.
368. Olivera, R., SanMartin, R., Pascual, S., *et al.* (1999) *Tetrahedron Letters*, **40**, 3479.
369. Moreno, I., Tellitu, I., SanMartin, R., *et al.* (1999) *Tetrahedron Letters*, **40**, 5067.
370. Kita, Y., Egi, M., Okajima, A., *et al.* (1996) *Journal of the Chemical Society, Chemical Communications*, 1491.
371. Kita, Y., Egi, M., Ohtsubo, M., *et al.* (1999) *Chemical & Pharmaceutical Bulletin*, **47**, 241.
372. Kita, Y., Watanabe, H., Egi, M., *et al.* (1998) *Journal of the Chemical Society, Perkin Transactions* **1**, 635.
373. Kita, Y., Egi, M. and Tohma, H. (1999) *Journal of the Chemical Society, Chemical Communications*, 143.
374. Pingaew, R. and Ruchirawat, S. (2007) *Synlett*, 2363.
375. Taylor, S.R., Ung, A.T., Pyne, S.G., *et al.* (2007) *Tetrahedron*, **63**, 11377.
376. Hamamoto, H., Anilkumar, G., Tohma, H. and Kita, Y. (2002) *Chemical Communications*, 450.
377. Hamamoto, H., Shiozaki, Y., Nambu, H., *et al.* (2004) *Chemistry–A European Journal*, **10**, 4977.
378. Tohma, H., Iwata, M., Maegawa, T. and Kita, Y. (2002) *Tetrahedron Letters*, **43**, 9241.
379. Mirk, D., Willner, A., Froehlich, R. and Waldvogel, S.R. (2004) *Advanced Synthesis & Catalysis*, **346**, 675.
380. Jean, A., Cantat, J., Berard, D., *et al.* (2007) *Organic Letters*, **9**, 2553.
381. Dohi, T., Morimoto, K., Kiyono, Y., *et al.* (2005) *Chemical Communications*, 2930.
382. Tohma, H., Iwata, M., Maegawa, T., *et al.* (2003) *Organic & Biomolecular Chemistry*, **1**, 1647.
383. Dohi, T., Morimoto, K., Maruyama, A. and Kita, Y. (2006) *Organic Letters*, **8**, 2007.
384. Morimoto, K., Nakae, T., Yamaoka, N., *et al.* (2011) *European Journal of Organic Chemistry*, 6326.
385. Yoshimura, Y., Ohta, M., Imahori, T., *et al.* (2008) *Organic Letters*, **10**, 3449.
386. Rihn, S., Erdem, M., De Nicola, A., *et al.* (2011) *Organic Letters*, **13**, 1916.
387. Zhdankin, V.V. (2007) *Science of Synthesis*, **31a**, 161.
388. Tellitu, I. and Dominguez, E. (2011) *Trends Heterocyclic Chem.*, **15**, 23.
389. Tellitu, I. and Dominguez, E. (2012) *Synlett*, **23**, 2165.
390. Tellitu, I., Urrejola, A., Serna, S., *et al.* (2007) *European Journal of Organic Chemistry*, 437.
391. Herrero, M.T., Tellitu, I., Dominguez, E., *et al.* (2002) *Tetrahedron*, **58**, 8581.
392. Herrero, M.T., Tellitu, I., Dominguez, E., *et al.* (2002) *Tetrahedron Letters*, **43**, 8273.
393. Correa, A., Tellitu, I., Dominguez, E., *et al.* (2005) *Journal of Organic Chemistry*, **70**, 2256.
394. Serna, S., Tellitu, I., Dominguez, E., *et al.* (2004) *Tetrahedron*, **60**, 6533.
395. Churruca, F., SanMartin, R., Carril, M., *et al.* (2005) *Journal of Organic Chemistry*, **70**, 3178.
396. Herrero, M.T., Tellitu, I., Hernandez, S., *et al.* (2002) *ARKIVOC*, (v), 31.
397. Serna, S., Tellitu, I., Dominguez, E., *et al.* (2003) *Tetrahedron Letters*, **44**, 3483.
398. Correa, A., Tellitu, I., Dominguez, E. and SanMartin, R. (2006) *Journal of Organic Chemistry*, **71**, 8316.
399. Serna, S., Tellitu, I., Dominguez, E., *et al.* (2005) *Organic Letters*, **7**, 3073.

400. Tellitu, I., Serna, S., Herrero, M.T., *et al.* (2007) *Journal of Organic Chemistry*, **72**, 1526.
401. Correa, A., Tellitu, I., Dominguez, E. and SanMartin, R. (2006) *Journal of Organic Chemistry*, **71**, 3501.
402. Correa, A., Tellitu, I., Dominguez, E. and SanMartin, R. (2006) *Tetrahedron*, **62**, 11100.
403. Pardo, L.M., Tellitu, I. and Dominguez, E. (2012) *Tetrahedron*, **68**, 3692.
404. Pardo, L.M., Tellitu, I. and Dominguez, E. (2010) *Synthesis*, 971.
405. Pardo, L.M., Tellitu, I. and Dominguez, E. (2010) *Tetrahedron*, **66**, 5811.
406. Couto, I., Tellitu, I. and Dominguez, E. (2010) *Journal of Organic Chemistry*, **75**, 7954.
407. Samanta, R., Kulikov, K., Strohmann, C. and Antonchick, A.P. (2012) *Synthesis*, 2325.
408. Huang, J., Liang, Y., Pan, W., *et al.* (2007) *Organic Letters*, **9**, 5345.
409. Fan, R., Wen, F., Qin, L., *et al.* (2007) *Tetrahedron Letters*, **48**, 7444.
410. Mamaeva, E.A. and Bakibaev, A.A. (2003) *Tetrahedron*, **59**, 7521.
411. Cheng, D.-P. and Chen, Z.-C. (2002) *Synthetic Communications*, **32**, 2155.
412. Wardrop, D.J., Zhang, W. and Landrie, C.L. (2004) *Tetrahedron Letters*, **45**, 4229.
413. Wardrop, D.J. and Burge, M.S. (2004) *Chemical Communications*, 1230.
414. Wardrop, D.J., Landrie, C.L. and Ortiz, J.A. (2003) *Synlett*, 1352.
415. Wardrop, D.J., Burge, M.S., Zhang, W. and Ortiz, J.A. (2003) *Tetrahedron Letters*, **44**, 2587.
416. Wardrop, D.J. and Basak, A. (2001) *Organic Letters*, **3**, 1053.
417. Wardrop, D.J. and Burge, M.S. (2005) *Journal of Organic Chemistry*, **70**, 10271.
418. Kikugawa, Y., Nagashima, A., Sakamoto, T., *et al.* (2003) *Journal of Organic Chemistry*, **68**, 6739.
419. Prakash, O., Bhardwaj, V., Kumar, R., *et al.* (2004) *European Journal of Medicinal Chemistry*, **39**, 1073.
420. Prakash, O., Kumar, R., Sharma, D., *et al.* (2006) *Heteroatom Chemistry*, **17**, 653.
421. Prakash, O., Kumar, R., Kumar, R., *et al.* (2007) *European Journal of Medicinal Chemistry*, **42**, 868.
422. Zhang, X., Gan, L., Huang, S. and Shi, Y. (2004) *Journal of Organic Chemistry*, **69**, 5800.
423. Rao, V.S. and Sekhar, K.C. (2004) *Synthetic Communications*, **34**, 2153.
424. Shang, Z. (2006) *Synthetic Communications*, **36**, 2927.
425. Somogyi, L. (2007) *Journal of Heterocyclic Chemistry*, **44**, 1235.
426. Kumar, D., Chandra Sekhar, K.V.G., Dhillon, H., *et al.* (2004) *Green Chemistry*, **6**, 156.
427. Aggarwal, R. and Sumran, G. (2006) *Synthetic Communications*, **36**, 1873.
428. Du, Y., Liu, R., Linn, G. and Zhao, K. (2006) *Organic Letters*, **8**, 5919.
429. Farid, U. and Wirth, T. (2012) *Angewandte Chemie, International Edition*, **51**, 3462.
430. Karade, N.N., Shirodkar, S.G., Patil, M.N., *et al.* (2003) *Tetrahedron Letters*, **44**, 6729.
431. Saito, A., Anzai, T., Matsumoto, A. and Hanzawa, Y. (2011) *Tetrahedron Letters*, **52**, 4658.
432. Boye, A.C., Meyer, D., Ingison, C.K., *et al.* (2003) *Organic Letters*, **5**, 2157.
433. Liu, H. and Tan, C.-H. (2007) *Tetrahedron Letters*, **48**, 8220.
434. Zhu, Y.-p., Jia, F.-c., Liu, M.-c. and Wu, A.-x. (2012) *Organic Letters*, **14**, 4414.
435. Lu, S.-C., Zheng, P.-R. and Liu, G. (2012) *Journal of Organic Chemistry*, **77**, 7711.
436. Singh, F.V. and Wirth, T. (2012) *Synthesis*, 1171.
437. Wang, J., Yuan, Y., Xiong, R., *et al.* (2012) *Organic Letters*, **14**, 2210.
438. Jawalekar, A.M., Reubsaet, E., Rutjes, F.P.J.T. and van Delft, F.L. (2011) *Chemical Communications*, 3198.
439. Mo, D.-L., Ding, C.-H., Dai, L.-X. and Hou, X.-L. (2011) *Chemistry – An Asian Journal*, **6**, 3200.
440. Du, X., Chen, H., Chen, Y., *et al.* (2011) *Synlett*, 1010.
441. Cho, S.H., Yoon, J. and Chang, S. (2011) *Journal of the American Chemical Society*, **133**, 5996.
442. Patil, S.S., Jadhav, R.P., Patil, S.V. and Bobade, V.D. (2010) *Journal of Chemical and Pharmaceutical Research*, **2**, 52.
443. Das, B., Holla, H., Mahender, G., *et al.* (2005) *Synthesis*, 1572.
444. Huang, X., Shao, N., Palani, A. and Aslanian, R. (2007) *Tetrahedron Letters*, **48**, 1967.
445. Bose, D.S. and Idrees, M. (2006) *Journal of Organic Chemistry*, **71**, 8261.
446. Aggarwal, R., Sumran, G., Saini, A. and Singh, S.P. (2006) *Tetrahedron Letters*, **47**, 4969.
447. Shang, Z., Reiner, J., Chang, J. and Zhao, K. (2005) *Tetrahedron Letters*, **46**, 2701.
448. Karade, N.N., Tiwari, G.B. and Gampawar, S.V. (2007) *Synlett*, 1921.
449. Prakash, O. and Pannu, K. (2007) *ARKIVOC*, (xiii), 28.

450. Du, L.-H. and Wang, Y.-G. (2007) *Synthesis*, 675.
451. Das, B., Srinivas, Y., Holla, H., *et al.* (2007) *Chemistry Letters*, **36**, 1270.
452. Liu, L.-P., Lu, J.-M. and Shi, M. (2007) *Organic Letters*, **9**, 1303.
453. Churruca, F., SanMartin, R., Tellitu, I. and Dominguez, E. (2005) *European Journal of Organic Chemistry*, 2481.
454. Huang, X., Shao, N., Palani, A., *et al.* (2008) *Tetrahedron Letters*, **49**, 3592.
455. Harayama, Y., Yoshida, M., Kamimura, D. and Kita, Y. (2005) *Chemical Communications*, 1764.
456. Harayama, Y., Yoshida, M., Kamimura, D., *et al.* (2006) *Chemistry–A European Journal*, **12**, 4893.
457. Havare, N. and Plattner, D.A. (2012) *Organic Letters*, **14**, 5078.
458. Telvekar, V.N. and Sasane, K.A. (2010) *Synlett*, 2778.
459. Kirihara, M., Yokoyama, S., Kakuda, H. and Momose, T. (1998) *Tetrahedron*, **54**, 13943.
460. Kirihara, M., Nishio, T., Yokoyama, S., *et al.* (1999) *Tetrahedron*, **55**, 2911.
461. Li, X., Xu, Z., DiMauro, E.F. and Kozlowski, M.C. (2002) *Tetrahedron Letters*, **43**, 3747.
462. Abo, T., Sawaguchi, M., Senboku, H. and Hara, S. (2005) *Molecules*, **10**, 183.
463. Fujioka, H., Matsuda, S., Horai, M., *et al.* (2007) *Chemistry–A European Journal*, **13**, 5238.
464. Kita, Y., Matsuda, S., Fujii, E., *et al.* (2005) *Angewandte Chemie, International Edition*, **44**, 5857.
465. Kita, Y., Matsuda, S., Fujii, E., *et al.* (2005) *Heterocycles*, **66**, 309.
466. Ohno, M., Oguri, I. and Eguchi, S. (1999) *Journal of Organic Chemistry*, **64**, 8995.
467. Chanu, A., Safir, I., Basak, R., *et al.* (2007) *European Journal of Organic Chemistry*, 4305.
468. Iglesias-Arteaga, M.A. and Velazquez-Huerta, G.A. (2005) *Tetrahedron Letters*, **46**, 6897.
469. Iglesias-Arteaga, M.A. and Arcos-Ramos, R.O. (2006) *Tetrahedron Letters*, **47**, 8029.
470. Iglesias-Arteaga, M.A., Arcos-Ramos, R.O. and Mendez-Stivalet, J.M. (2007) *Tetrahedron Letters*, **48**, 7485.
471. Sanchez-Flores, J., Pelayo-Gonzalez, V.G., Romero-Avila, M., *et al.* (2013) *Steroids*, **78**, 234.
472. Miki, Y., Fujita, R. and Matsushita, K.-i. (1998) *Journal of the Chemical Society, Perkin Transactions* **1**, 2533.
473. Justik, M.W. and Koser, G.F. (2004) *Tetrahedron Letters*, **45**, 6159.
474. Hossain, M.M., Tokuoka, T., Yamashita, K., *et al.* (2006) *Synthetic Communications*, **36**, 1201.
475. Silva, L.F., Jr., Siqueira, F.A., Pedrozo, E.C., *et al.* (2007) *Organic Letters*, **9**, 1433.
476. Prakash, O., Kumar, A., Sadana, A.K. and Singh, S.P. (2006) *Synthesis*, 21.
477. Justik, M.W. and Koser, G.F. (2005) *Molecules*, **10**, 217.
478. Silva, L.F., Jr., Vasconcelos, R.S. and Nogueira, M.A. (2008) *Organic Letters*, **10**, 1017.
479. Ramsden, C.A. and Rose, H.L. (1997) *Journal of the Chemical Society, Perkin Transactions 1*, 2319.
480. Wuts, P.G.M. and Goble, M.P. (2000) *Organic Letters*, **2**, 2139.
481. De, S.K. and Mallik, A.K. (1998) *Tetrahedron Letters*, **39**, 2389.
482. Ramsden, C.A. and Rose, H.L. (1997) *Synlett*, 27.
483. Gribble, G.W. (2009) in *Name Reactions for Homologations* (ed. J.J. Li), John Wiley & Sons, Inc., Hoboken, New Jersey, p. 164.
484. Moriarty, R.M., Chany, C.J. II, Vaid, R.K., *et al.* (1993) *Journal of Organic Chemistry*, **58**, 2478.
485. Zhang, L.-h., Chung, J.C., Costello, T.D., *et al.* (1997) *Journal of Organic Chemistry*, **62**, 2466.
486. Andruszkiewicz, R. and Rozkiewicz, D. (2004) *Synthetic Communications*, **34**, 1049.
487. Okamoto, N., Miwa, Y., Minami, H., *et al.* (2009) *Angewandte Chemie, International Edition*, **48**, 9693.
488. Landsberg, D. and Kalesse, M. (2010) *Synlett*, 1104.
489. Loudon, G.M., Radhakrishna, A.S., Almond, M.R., *et al.* (1984) *Journal of Organic Chemistry*, **49**, 4272.
490. Boutin, R.H. and Loudon, G.M. (1984) *Journal of Organic Chemistry*, **49**, 4277.
491. Davies, S.G. and Dixon, D.J. (2002) *Journal of the Chemical Society, Perkin Transactions* **1**, 1869.
492. Berkessel, A., Glaubitz, K. and Lex, J. (2002) *European Journal of Organic Chemistry*, 2948.
493. Hernandez, E., Velez, J.M. and Vlaar, C.P. (2007) *Tetrahedron Letters*, **48**, 8972.
494. Erdelmeier, I., Tailhan-Lomont, C. and Yadan, J.-C. (2000) *Journal of Organic Chemistry*, **65**, 8152.
495. Lazbin, I.M. and Koser, G.F. (1986) *Journal of Organic Chemistry*, **51**, 2669.
496. Lazbin, I.M. and Koser, G.F. (1987) *Journal of Organic Chemistry*, **52**, 476.
497. Della, E.W. and Head, N.J. (1995) *Journal of Organic Chemistry*, **60**, 5303.
498. Moriarty, R.M., Enache, L.A., Zhao, L., *et al.* (1998) *Journal of Medicinal Chemistry*, **41**, 468.
499. Liu, S.J., Zhang, J.Z., Tian, G.R. and Liu, P. (2005) *Synthetic Communications*, **36**, 823.

500. Zhang, L.-h., Kauffman, G.S., Pesti, J.A. and Yin, J. (1997) *Journal of Organic Chemistry*, **62**, 6918.
501. Prakash, O., Batra, H., Kaur, H., *et al.* (2001) *Synthesis*, 541.
502. Song, H., Chen, W., Wang, Y. and Qin, Y. (2005) *Synthetic Communications*, **35**, 2735.
503. Davis, M.C., Stasko, D. and Chapman, R.D. (2003) *Synthetic Communications*, **33**, 2677.
504. Moriarty, R.M., Khosrowshahi, J.S., Awasthi, A.K. and Penmasta, R. (1988) *Synthetic Communications*, **18**, 1179.
505. Liu, P., Wang, Z. and Hu, X. (2012) *European Journal of Organic Chemistry*, 1994.
506. Yoshimura, A., Luedtke, M.W. and Zhdankin, V.V. (2012) *Journal of Organic Chemistry*, **77**, 2087.
507. Moriyama, K., Ishida, K. and Togo, H. (2012) *Organic Letters*, **14**, 946.
508. Moriarty, R.M., Vaid, R.K., Duncan, M.P., *et al.* (1988) *Tetrahedron Letters*, **29**, 6913.
509. Mendelsohn, B.A., Lee, S., Kim, S., *et al.* (2009) *Organic Letters*, **11**, 1539.
510. Jen, T., Mendelsohn, B.A. and Ciufolini, M.A. (2011) *Journal of Organic Chemistry*, **76**, 728.
511. Raihan, M.J., Kavala, V., Habib, P.M., *et al.* (2011) *Journal of Organic Chemistry*, **76**, 424.
512. Momiyama, N., Yamamoto, Y. and Yamamoto, H. (2007) *Journal of the American Chemical Society*, **129**, 1190.
513. Singh, C.B., Ghosh, H., Murru, S. and Patel, B.K. (2008) *Journal of Organic Chemistry*, **73**, 2924.
514. But, T.Y.S. and Toy, P.H. (2006) *Journal of the American Chemical Society*, **128**, 9636.
515. Kansara, A., Sharma, P.K. and Banerji, K.K. (2004) *Journal of Chemical Research*, 315.
516. ter Wiel, M.K.J., Vicario, J., Davey, S.G., *et al.* (2005) *Organic & Biomolecular Chemistry*, **3**, 28.
517. Shang, Z., Reiner, J. and Zhao, K. (2006) *Synthetic Communications*, **36**, 1529.
518. Zhutov, E.V., Skornyakov, Y.V., Proskurina, M.V. and Zefirov, N.S. (2003) *Russian Journal of Organic Chemistry*, **39**, 1672.
519. Stingl, K.A. and Tsogoeva, S.B. (2010) *Tetrahedron: Asymmetry*, **21**, 1055.
520. O'Mahony, G.E., Kelly, P., Lawrence, S.E. and Maguire, A.R. (2011) *ARKIVOC*, (i), 1.
521. Tohma, H., Takizawa, S., Watanabe, H. and Kita, Y. (1998) *Tetrahedron Letters*, **39**, 4547.
522. Kannan, P., Sevvel, R., Rajagopal, S., *et al.* (1997) *Tetrahedron*, **53**, 7635.
523. Sohmiya, H., Kimura, T., Fujita, M. and Ando, T. (1998) *Tetrahedron*, **54**, 13737.
524. Varma, R.S., Saini, R.K. and Dahiya, R. (1998) *Journal of Chemical Research. Synopses*, 120.
525. Tohma, H., Maegawa, T. and Kita, Y. (2003) *ARKIVOC*, (vi), 62.
526. Yusubov, M.S., Yusubova, R.Y., Funk, T.V., *et al.* (2009) *Synthesis*, 2505.
527. Ochiai, M., Ito, T., Masaki, Y. and Shiro, M. (1992) *Journal of the American Chemical Society*, **114**, 6269.
528. Ochiai, M., Nakanishi, A. and Ito, T. (1997) *Journal of Organic Chemistry*, **62**, 4253.
529. Ochiai, M., Kajishima, D. and Sueda, T. (1999) *Tetrahedron Letters*, **40**, 5541.
530. Xia, M. and Chen, Z.-C. (1997) *Synthetic Communications*, **27**, 1301.
531. Xia, M. and Chen, Z.-C. (1997) *Synthetic Communications*, **27**, 1321.
532. Chen, D.-W. and Chen, Z.-C. (1994) *Tetrahedron Letters*, **35**, 7637.
533. Chen, D.-W. and Chen, Z.-C. (1995) *Synthetic Communications*, **25**, 1605.
534. Billard, T. and Langlois, B.R. (1997) *Journal of Fluorine Chemistry*, **84**, 63.
535. Ray, D.G. III and Koser, G.F. (1990) *Journal of the American Chemical Society*, **112**, 5672.
536. Xia, M. and Chen, Z.-C. (1997) *Synthetic Communications*, **27**, 1315.
537. Chen, L.-C. and Wang, H.-M. (1999) *Organic Preparations and Procedures International*, **31**, 562.
538. Shi, X.-X. and Wu, Q.-Q. (2000) *Synthetic Communications*, **30**, 4081.
539. Burghardt, T.E. (2005) *Journal of Sulfur Chemistry*, **26**, 411.
540. Wu, Y., Shen, X., Yang, Y.-Q., *et al.* (2004) *Journal of Organic Chemistry*, **69**, 3857.
541. Fleming, F.F., Funk, L., Altundas, R. and Tu, Y. (2001) *Journal of Organic Chemistry*, **66**, 6502.
542. Foss, V.L., Veits, Y.A., Lermontov, S.A. and Lutsenko, I.F. (1978) *Zhurnal obshchei khimii*, **48**, 1713.
543. Combes, S. and Finet, J.-P. (1998) *Tetrahedron*, **54**, 4313.
544. Kang, S.-K., Ryu, H.-C. and Lee, S.-W. (2000) *Journal of Organometallic Chemistry*, **610**, 38.
545. Cech, F. and Zbiral, E. (1975) *Tetrahedron*, **31**, 605.
546. Zbiral, E. and Nestler, G. (1970) *Tetrahedron*, **26**, 2945.
547. Ehrenfreund, J. and Zbiral, E. (1973) *Liebigs Annalen der Chemie*, 290.
548. Moriarty, R.M. and Khosrowshahi, J.S. (1986) *Tetrahedron Letters*, **27**, 2809.
549. Moriarty, R.M. and Khosrowshahi, J.S. (1987) *Synthetic Communications*, **17**, 89.

550. Arimoto, M., Yamaguchi, H., Fujita, E., *et al.* (1987) *Tetrahedron Letters*, **28**, 6289.
551. Moriarty, R.M., Vaid, R.K., Ravikumar, V.T., *et al.* (1988) *Tetrahedron*, **44**, 1603.
552. Moriarty, R.M., Vaid, R.K., Hopkins, T.E., *et al.* (1989) *Tetrahedron Letters*, **30**, 3019.
553. Kita, Y., Tohma, H., Inagaki, M., *et al.* (1991) *Tetrahedron Letters*, **32**, 4321.
554. Kita, Y., Tohma, H., Takada, T., *et al.* (1994) *Synlett*, 427.
555. Fontana, F., Minisci, F., Yan, Y.M. and Zhao, L. (1993) *Tetrahedron Letters*, **34**, 2517.
556. Magnus, P., Lacour, J. and Weber, W. (1993) *Journal of the American Chemical Society*, **115**, 9347.
557. Magnus, P., Hulme, C. and Weber, W. (1994) *Journal of the American Chemical Society*, **116**, 4501.
558. Magnus, P. and Lacour, J. (1992) *Journal of the American Chemical Society*, **114**, 3993.
559. Magnus, P., Evans, A. and Lacour, J. (1992) *Tetrahedron Letters*, **33**, 2933.
560. Magnus, P., Roe, M.B. and Hulme, C. (1995) *Journal of the Chemical Society, Chemical Communications*, 263.
561. Magnus, P. and Hulme, C. (1994) *Tetrahedron Letters*, **35**, 8097.
562. Magnus, P., Lacour, J., Evans, P.A., *et al.* (1996) *Journal of the American Chemical Society*, **118**, 3406.
563. Magnus, P., Lacour, J., Evans, P.A., *et al.* (1998) *Journal of the American Chemical Society*, **120**, 12486.
564. Magnus, P. and Sebhat, I.K. (1998) *Journal of the American Chemical Society*, **120**, 5341.
565. Magnus, P. and Sebhat, I.K. (1998) *Tetrahedron*, **54**, 15509.
566. Magnus, P., Bailey, J.M. and Porter, M.J. (1999) *Tetrahedron*, **55**, 13927.
567. Magnus, P. and Roe, M.B. (1996) *Tetrahedron Letters*, **37**, 303.
568. Magnus, P., Lacour, J. and Weber, W. (1998) *Synthesis*, 547.
569. Tohma, H., Egi, M., Ohtsubo, M., *et al.* (1998) *Journal of the Chemical Society, Chemical Communications*, 173.
570. Kita, Y., Egi, M., Takada, T. and Tohma, H. (1999) *Synthesis*, 885.
571. Zhdankin, V.V., Kuehl, C.J., Arif, A.M. and Stang, P.J. (1996) *Mendeleev Communications*, 50.
572. Tsarevsky, N.V. (2010) *Journal of Polymer Science Part A: Polymer Chemistry*, **48**, 966.
573. Chen, D.J. and Chen, Z.C. (2000) *Tetrahedron Letters*, **41**, 7361.
574. Telvekar, V.N., Takale, B.S. and Bachhav, H.M. (2009) *Tetrahedron Letters*, **50**, 5056.
575. Telvekar, V.N. and Patile, H.V. (2011) *Synthetic Communications*, **41**, 131.
576. Zhdankin, V.V., Krasutsky, A.P., Kuehl, C.J., *et al.* (1996) *Journal of the American Chemical Society*, **118**, 5192.
577. Deng, Q.-H., Bleith, T., Wadepohl, H. and Gade, L.H. (2013) *Journal of the American Chemical Society*, **135**, 5356.
578. Souto, J.A., Gonzalez, Y., Iglesias, A., *et al.* (2012) *Chemistry – An Asian Journal*, **7**, 1103.
579. Roben, C., Souto, J.A., Escudero-Adan, E.C. and Muniz, K. (2013) *Organic Letters*, **15**, 1008.
580. Roeben, C., Souto, J.A., Gonzalez, Y., *et al.* (2011) *Angewandte Chemie, International Edition*, **50**, 9478.
581. Souto, J.A., Zian, D. and Muniz, K. (2012) *Journal of the American Chemical Society*, **134**, 7242.
582. Souto, J.A., Becker, P., Iglesias, A. and Muniz, K. (2012) *Journal of the American Chemical Society*, **134**, 15505.
583. Zhdankin, V.V., McSherry, M., Mismash, B., *et al.* (1997) *Tetrahedron Letters*, **38**, 21.
584. Chang, J.W.W., Ton, T.M.U. and Chan, P.W.H. (2011) *The Chemical Record*, **11**, 331.
585. Johannsen, M. and Jorgensen, K.A. (1998) *Chemical Reviews*, **98**, 1689.
586. Evans, D.A., Bilodeau, M.T. and Faul, M.M. (1994) *Journal of the American Chemical Society*, **116**, 2742.
587. Lim, B.-W. and Ahn, K.-H. (1996) *Synthetic Communications*, **26**, 3407.
588. Yoshimura, A., Nemykin, V.N. and Zhdankin, V.V. (2011) *Chemistry–A European Journal*, **17**, 10538.
589. Lamar, A.A. and Nicholas, K.M. (2010) *Journal of Organic Chemistry*, **75**, 7644.
590. Yao, H., Tang, Y. and Yamamoto, K. (2012) *Tetrahedron Letters*, **53**, 5094.
591. Prasad, V., Kale, R.R., Mishra, B.B., *et al.* (2012) *Organic Letters*, **14**, 2936.
592. Prakash, O., Rani, N., Sharma, V. and Moriarty, R.M. (1997) *Synlett*, 1255.
593. Prakash, O., Kaur, H., Batra, H., *et al.* (2001) *Journal of Organic Chemistry*, **66**, 2019.
594. Prakash, O., Sharma, V., Batra, H. and Moriarty, R.M. (2001) *Tetrahedron Letters*, **42**, 553.
595. Kita, Y., Takeda, Y., Okuno, T., *et al.* (1997) *Chemical & Pharmaceutical Bulletin*, **45**, 1887.
596. Prakash, O., Kaur, H., Pundeer, R., *et al.* (2003) *Synthetic Communications*, **33**, 4037.
597. De Mico, A., Margarita, R., Mariani, A. and Piancatelli, G. (1997) *Journal of the Chemical Society, Chemical Communications*, 1237.
598. Bruno, M., Margarita, R., Parlanti, L., *et al.* (1998) *Tetrahedron Letters*, **39**, 3847.

599. Margarita, R., Mercanti, C., Parlanti, L. and Piancatelli, G. (2000) *European Journal of Organic Chemistry*, 1865.
600. Tingoli, M., Tiecco, M., Chianelli, D., *et al.* (1991) *Journal of Organic Chemistry*, **56**, 6809.
601. Tingoli, M., Tiecco, M., Testaferri, L. and Balducci, R. (1993) *Synlett*, 211.
602. Tingoli, M., Tiecco, M., Testaferri, L. and Temperini, A. (1994) *Journal of the Chemical Society, Chemical Communications*, 1883.
603. Tingoli, M., Tiecco, M., Testaferri, L. and Temperini, A. (1998) *Synthetic Communications*, **28**, 1769.
604. Das, J.P., Roy, U.K. and Roy, S. (2005) *Organometallics*, **24**, 6136.
605. Czernecki, S. and Randriamandimby, D. (1993) *Tetrahedron Letters*, **34**, 7915.
606. Togo, H., Aoki, M. and Yokoyama, M. (1991) *Tetrahedron Letters*, **32**, 6559.
607. Togo, H., Aoki, M. and Yokoyama, M. (1991) *Chemistry Letters*, 1691.
608. Togo, H., Aoki, M., Kuramochi, T. and Yokoyama, M. (1993) *Journal of the Chemical Society, Perkin Transactions* **1**, 2417.
609. Vismara, E., Torri, G., Pastori, N. and Marchiandi, M. (1992) *Tetrahedron Letters*, **33**, 7575.
610. Togo, H., Aoki, M. and Yokoyama, M. (1992) *Chemistry Letters*, 2169.
611. Togo, H., Aoki, M. and Yokoyama, M. (1993) *Tetrahedron*, **49**, 8241.
612. Sutherland, A. and Vederas, J.C. (2002) *Chemical Communications*, 224.
613. Muraki, T., Togo, H. and Yokoyama, M. (1999) *Journal of the Chemical Society, Perkin Transactions* **1**, 1713.
614. Togo, H., Muraki, T., Hoshina, Y., *et al.* (1997) *Journal of the Chemical Society, Perkin Transactions* **1**, 787.
615. Francisco, C.G., Gonzalez, C.C. and Suarez, E. (1997) *Tetrahedron Letters*, **38**, 4141.
616. Boto, A., Hernandez, R. and Suarez, E. (2000) *Journal of Organic Chemistry*, **65**, 4930.
617. Boto, A., Hernandez, R. and Suarez, E. (2000) *Tetrahedron Letters*, **41**, 2495.
618. Dohi, T., Takenaga, N., Goto, A., *et al.* (2007) *Organic Letters*, **9**, 3129.
619. Togo, H., Muraki, T. and Yokoyama, M. (1995) *Synthesis*, 155.
620. Boto, A., Betancor, C. and Suarez, E. (1994) *Tetrahedron Letters*, **35**, 6933.
621. Arencibia, M.T., Freire, R., Perales, A., *et al.* (1991) *Journal of the Chemical Society, Perkin Transactions* **1**, 3349.
622. Boto, A., Betancor, C., Prange, T. and Suarez, E. (1992) *Tetrahedron Letters*, **33**, 6687.
623. Boto, A., Bentancor, C. and Suarez, E. (1994) *Tetrahedron Letters*, **35**, 5509.
624. Boto, A., Betancor, C., Prange, T. and Suarez, E. (1994) *Journal of Organic Chemistry*, **59**, 4393.
625. Boto, A., Betancor, C., Hernandez, R., *et al.* (1993) *Tetrahedron Letters*, **34**, 4865.
626. Mowbray, C.E. and Pattenden, G. (1993) *Tetrahedron Letters*, **34**, 127.
627. Francisco, C.G., Gonzalez, C.C., Paz, N.R. and Suarez, E. (2003) *Organic Letters*, **5**, 4171.
628. Alonso-Cruz, C.R., Kennedy, A.R., Rodriguez, M.S. and Suarez, E. (2003) *Organic Letters*, **5**, 3729.
629. Alonso-Cruz, C.R., Kennedy, A.R., Rodriguez, M.S. and Suarez, E. (2007) *Tetrahedron Letters*, **48**, 7207.
630. Alonso-Cruz, C.R., Leon, E.I., Ortiz-Lopez, F.J., *et al.* (2005) *Tetrahedron Letters*, **46**, 5265.
631. Muraki, T., Togo, H. and Yokoyama, M. (1996) *Tetrahedron Letters*, **37**, 2441.
632. Galatsis, P. and Millan, S.D. (1991) *Tetrahedron Letters*, **32**, 7493.
633. De Armas, P., Francisco, C.G. and Suarez, E. (1992) *Angewandte Chemie, International Edition*, **31**, 772.
634. Inanaga, J., Sugimoto, Y., Yokoyama, Y. and Hanamoto, T. (1992) *Tetrahedron Letters*, **33**, 8109.
635. Courtneidge, J.L., Lusztyk, J. and Page, D. (1994) *Tetrahedron Letters*, **35**, 1003.
636. Francisco, C.G., Freire, R., Gonzalez, C.C., *et al.* (2001) *Journal of Organic Chemistry*, **66**, 1861.
637. Francisco, C.G., Gonzalez, C.C. and Suarez, E. (1996) *Tetrahedron Letters*, **37**, 1687.
638. Francisco, C.G., Gonzalez Martin, C. and Suarez, E. (1998) *Journal of Organic Chemistry*, **63**, 2099.
639. Dorta, R.L., Martin, A., Salazar, J.A., *et al.* (1998) *Journal of Organic Chemistry*, **63**, 2251.
640. Francisco, C.G., Gonzalez Martin, C. and Suarez, E. (1998) *Journal of Organic Chemistry*, **63**, 8092.
641. Costa, S.C.P., Moreno, M.J.S.M., Sa e Melo, M.L. and Neves, A.S.C. (1999) *Tetrahedron Letters*, **40**, 8711.
642. Madsen, J., Viuf, C. and Bols, M. (2000) *Chemistry–A European Journal*, **6**, 1140.
643. Francisco, C.G., Herrera, A.J. and Suarez, E. (2002) *Journal of Organic Chemistry*, **67**, 7439.
644. Francisco, C.G., Herrera, A.J., Kennedy, A.R., *et al.* (2002) *Angewandte Chemie, International Edition*, **41**, 856.
645. Francisco, C.G., Freire, R., Herrera, A.J., *et al.* (2007) *Tetrahedron*, **63**, 8910.
646. Francisco, C.G., Freire, R., Herrera, A.J., *et al.* (2002) *Organic Letters*, **4**, 1959.
647. Martin, A., Quintanal, L.M. and Suarez, E. (2007) *Tetrahedron Letters*, **48**, 5507.

648. Boto, A., Hernandez, D. and Hernandez, R. (2007) *Organic Letters*, **9**, 1721.
649. Boto, A., Hernandez, D., Hernandez, R. and Suarez, E. (2004) *Organic Letters*, **6**, 3785.
650. Boto, A., Hernandez, R., Montoya, A. and Suarez, E. (2002) *Tetrahedron Letters*, **43**, 8269.
651. Boto, A., Gallardo, J.A., Hernandez, R. and Saavedra, C.J. (2005) *Tetrahedron Letters*, **46**, 7807.
652. Francisco, C.G., Herrera, A.J. and Suarez, E. (2003) *Journal of Organic Chemistry*, **68**, 1012.
653. Martin, A., Perez-Martin, I. and Suarez, E. (2005) *Organic Letters*, **7**, 2027.
654. Francisco, C.G., Herrera, A.J., Martin, A., *et al.* (2007) *Tetrahedron Letters*, **48**, 6384.
655. Fan, R., Pu, D., Wen, F. and Wu, J. (2007) *Journal of Organic Chemistry*, **72**, 8994.
656. Katohgi, M., Togo, H., Yamaguchi, K. and Yokoyama, M. (1999) *Tetrahedron*, **55**, 14885.
657. Togo, H., Hoshina, Y., Muraki, T., *et al.* (1998) *Journal of Organic Chemistry*, **63**, 5193.
658. Togo, H., Harada, Y. and Yokoyama, M. (2000) *Journal of Organic Chemistry*, **65**, 926.
659. Francisco, C.G., Herrera, A.J. and Suarez, E. (2000) *Tetrahedron: Asymmetry*, **11**, 3879.
660. Katohgi, M. and Togo, H. (2001) *Tetrahedron*, **57**, 7481.
661. Reich, H.J. and Peake, S.L. (1978) *Journal of the American Chemical Society*, **100**, 4888.
662. Knapp, S., Naughton, A.B.J. and Dhar, T.G.M. (1992) *Tetrahedron Letters*, **33**, 1025.
663. Kim, S. and Fuchs, P.L. (1994) *Tetrahedron Letters*, **35**, 7163.
664. Wiberg, K.B., Pratt, W.E. and Matturro, M.G. (1982) *Journal of Organic Chemistry*, **47**, 2720.
665. Della, E.W. and Head, N.J. (1992) *Journal of Organic Chemistry*, **57**, 2850.
666. Zefirov, N.S., Zhdankin, V.V. and Koz'min, A.S. (1982) *Izvestiya Akademii Nauk SSSR, Seriya Khimicheskaya*, 1676.
667. Zefirov, N.S., Zhdankin, V.V., Sorokin, V.D., *et al.* (1982) *Zhurnal Organicheskoi Khimii*, **18**, 2608.
668. Zefirov, N.S., Zhdankin, V.V. and Koz'min, A.S. (1986) *Tetrahedron Letters*, **27**, 1845.
669. Zefirov, N.S., Zhdankin, V.V., Makhon'kova, G.V., *et al.* (1985) *Journal of Organic Chemistry*, **50**, 1872.
670. Damon, D.B. and Hoover, D.J. (1990) *Journal of the American Chemical Society*, **112**, 6439.
671. Groves, J.T., Nemo, T.E. and Myers, R.S. (1979) *Journal of the American Chemical Society*, **101**, 1032.
672. Sheldon, R.A. (ed.) (1994) *Metalloporphyrins in Catalytic Oxidations*, M. Dekker, New York.
673. Ortiz de Montellano, P.R. (ed.) (2005) *Cytochrome P450: Structure, Mechanism, and Biochemistry*, Kluwer Academic/Plenum Publishers, New York.
674. Meunier, B. (1992) *Chemical Reviews*, **92**, 1411.
675. Rose, E., Andrioletti, B., Zrig, S. and Quelquejeu-Etheve, M. (2005) *Chemical Society Reviews*, **34**, 573.
676. Simonneaux, G. and Tagliatesta, P. (2004) *Journal of Porphyrins and Phthalocyanines*, **8**, 1166.
677. Bernadou, J. and Meunier, B. (2004) *Advanced Synthesis & Catalysis*, **346**, 171.
678. Vinhado, F.S., Martins, P.R. and Iamamoto, Y. (2002) *Current Topics in Catalysis*, **3**, 199.
679. Meunier, B., Robert, A., Pratviel, G. and Bernadou, J. (2000) in *The Porphyrin Handbook* (ed. K.M. Kadish, K.M. Smith and R. Guilard), vol. 4, Academic Press, San Deigo, p. 119.
680. Groves, J.T., Shalyaev, K. and Lee, J. (2000) in *The Porphyrin Handbook* (ed. K.M. Kadish, K.M. Smith and R. Guilard), vol. 4, Academic Press, San Deigo, p. 17.
681. Ji, L., Peng, X. and Huang, J. (2002) *Progress in Natural Science*, **12**, 321.
682. Groves, J.T. (2000) *Journal of Porphyrins and Phthalocyanines*, **4**, 350.
683. Kimura, M. and Shirai, H. (2003) in *The Porphyrin Handbook* (ed. K.M. Kadish, K.M. Smith and R. Guilard), vol. 19, Academic Press, San Deigo, p. 151.
684. Moro-oka, Y. (1998) *Catalysis Today*, **45**, 3.
685. Moro-oka, Y. and Akita, M. (1998) *Catalysis Today*, **41**, 327.
686. Noyori, R. (1994) *Asymmetric Catalysis in Organic Synthesis*, John Wiley & Sons, Inc., New York.
687. Mukerjee, S., Stassinopoulos, A. and Caradonna, J.P. (1997) *Journal of the American Chemical Society*, **119**, 8097.
688. das Dores Assis, M. and Lindsay Smith, J.R.L. (1998) *Journal of the Chemical Society, Perkin Transactions* **2**, 2221.
689. Breslow, R., Gabriele, B. and Yang, J. (1998) *Tetrahedron Letters*, **39**, 2887.
690. Yang, J., Weinberg, R. and Breslow, R. (2000) *Chemical Communications*, 531.
691. Yang, J. and Breslow, R. (2000) *Tetrahedron Letters*, **41**, 8063.

692. Guo, C.C., Song, J.X., Chen, X.B. and Jiang, G.F. (2000) *Journal of Molecular Catalysis A: Chemical*, **157**, 31.
693. Kang, M.-J., Song, W.J., Han, A.-R., *et al.* (2007) *Journal of Organic Chemistry*, **72**, 6301.
694. de Visser, S.P., Oh, K., Han, A.-R. and Nam, W. (2007) *Inorganic Chemistry*, **46**, 4632.
695. Song, W.J., Seo, M.S., DeBeer George, S., *et al.* (2007) *Journal of the American Chemical Society*, **129**, 1268.
696. Silva, G.d.F., Carvalho da Silva, D., Guimaraes, A.S., *et al.* (2007) *Journal of Molecular Catalysis A: Chemical*, **266**, 274.
697. Murakami, Y. and Konishi, K. (2007) *Journal of the American Chemical Society*, **129**, 14401.
698. Santos, M.M.C., Silva, A.M.S., Cavaleiro, J.A.S., *et al.* (2007) *European Journal of Organic Chemistry*, 2877.
699. Babakhania, R., Bahadoran, F. and Safari, N. (2007) *Journal of Porphyrins and Phthalocyanines*, **11**, 95.
700. Pouralimardan, O., Chamayou, A.-C., Janiak, C. and Hosseini-Monfared, H. (2007) *Inorganica Chimica Acta*, **360**, 1599.
701. Daly, A.M., Renehan, M.F. and Gilheany, D.G. (2001) *Organic Letters*, **3**, 663.
702. Daly, A.M., Dalton, C.T., Renehan, M.F. and Gilheany, D.G. (1999) *Tetrahedron Letters*, **40**, 3617.
703. Ryan, K.M., Bousquet, C. and Gilheany, D.G. (1999) *Tetrahedron Letters*, **40**, 3613.
704. Nishida, T., Miyafuji, A., Ito, Y.N. and Katsuki, T. (2000) *Tetrahedron Letters*, **41**, 7053.
705. Lai, T.-S., Kwong, H.-L., Zhang, R. and Che, C.-M. (1998) *Journal of the Chemical Society, Dalton Transactions*, 3559.
706. Reginato, G., Di Bari, L., Salvadori, P. and Guilard, R. (2000) *European Journal of Organic Chemistry*, 1165.
707. Porhiel, E., Bondon, A. and Leroy, J. (2000) *European Journal of Inorganic Chemistry*, 1097.
708. Baciocchi, E., Boschi, T., Cassioli, L., *et al.* (1999) *European Journal of Organic Chemistry*, 3281.
709. Sacco, H.C., Iamamoto, Y. and Lindsay Smith, J.R. (2001) *Journal of the Chemical Society, Perkin Transactions* **2**, 181.
710. Gross, Z., Simkhovich, L. and Galili, N. (1999) *Chemical Communications*, 599.
711. Nishiyama, H., Shimada, T., Itoh, H., *et al.* (1997) *Chemical Communications*, 1863.
712. Yang, Z.W., Kang, Q.X., Quan, F. and Lei, Z.Q. (2007) *Journal of Molecular Catalysis A: Chemical*, **261**, 190.
713. Geraskin, I.M., Luedtke, M.W., Neu, H.M., *et al.* (2008) *Tetrahedron Letters*, **49**, 7410.
714. Vatele, J.-M. (2006) *Synlett*, 2055.
715. Wang, S.H., Mandimutsira, B.S., Todd, R., *et al.* (2004) *Journal of the American Chemical Society*, **126**, 18.
716. Ferrand, Y., Daviaud, R., Le Maux, P. and Simonneaux, G. (2006) *Tetrahedron: Asymmetry*, **17**, 952.
717. Bryliakov, K.P. and Talsi, E.P. (2007) *Chemistry–A European Journal*, **13**, 8045.
718. Collman, J.P., Zeng, L., Wang, H.J.H., *et al.* (2006) *European Journal of Organic Chemistry*, 2707.
719. Yusubov, M.S., Gilmkhanova, M.P., Zhdankin, V.V. and Kirschning, A. (2007) *Synlett*, 563.
720. Neu, H.M., Yusubov, M.S., Zhdankin, V.V. and Nemykin, V.N. (2009) *Advanced Synthesis & Catalysis*, **351**, 3168.
721. Yusubov, M.S., Nemykin, V.N. and Zhdankin, V.V. (2010) *Tetrahedron*, **66**, 5745.
722. Jin, R., Cho, C.S., Jiang, L.H. and Shim, S.C. (1997) *Journal of Molecular Catalysis A: Chemical*, **116**, 343.
723. Collman, J.P., Chien, A.S., Eberspacher, T.A. and Brauman, J.I. (2000) *Journal of the American Chemical Society*, **122**, 11098.
724. Feichtinger, D. and Plattner, D.A. (2000) *Journal of the Chemical Society, Perkin Transactions* **2**, 1023.
725. Bryliakov, K.P. and Talsi, E.P. (2004) *Angewandte Chemie, International Edition*, **43**, 5228.
726. Dumitru, I. (2011) *Synlett*, 432.
727. Young, K.J.H., Mironov, O.A. and Periana, R.A. (2007) *Organometallics*, **26**, 2137.
728. Li, Z. and Xia, C.-G. (2004) *Journal of Molecular Catalysis A: Chemical*, **214**, 95.
729. In, J.-H., Park, S.-E., Song, R. and Nam, W. (2003) *Inorganica Chimica Acta*, **343**, 373.
730. Kirschning, A., Yusubov, M.S., Yusubova, R.Y., *et al.* (2007) *Beilstein Journal of Organic Chemistry*, **3**, 19.
731. Yusubov, M.S., Chi, K.-W., Park, J.Y., *et al.* (2006) *Tetrahedron Letters*, **47**, 6305.
732. Kunst, E., Gallier, F., Dujardin, G., *et al.* (2007) *Organic Letters*, **9**, 5199.
733. Iwasa, S., Morita, K., Tajima, K., *et al.* (2002) *Chemistry Letters*, 284.
734. Miyamura, H., Akiyama, R., Ishida, T., *et al.* (2005) *Tetrahedron*, **61**, 12177.
735. Sun, W., Wang, H., Xia, C., *et al.* (2003) *Angewandte Chemie, International Edition*, **42**, 1042.
736. Li, Z., Tang, Z.H., Hu, X.X. and Xia, C.G. (2005) *Chemistry–A European Journal*, **11**, 1210.

737. Karimipour, G.R., Shadegan, H.A. and Ahmadpour, R. (2007) *Journal of Chemical Research*, 252.
738. Adam, W., Hajra, S., Herderich, M. and Saha-Moeller, C.R. (2000) *Organic Letters*, **2**, 2773.
739. Provins, L. and Murahashi, S.-I. (2007) *ARKIVOC*, (x), 107.
740. Deprez, N.R. and Sanford, M.S. (2007) *Inorganic Chemistry*, **46**, 1924.
741. Dick, A.R., Hull, K.L. and Sanford, M.S. (2004) *Journal of the American Chemical Society*, **126**, 2300.
742. Alexanian, E.J., Lee, C. and Sorensen, E.J. (2005) *Journal of the American Chemical Society*, **127**, 7690.
743. Desai, L.V., Hull, K.L. and Sanford, M.S. (2004) *Journal of the American Chemical Society*, **126**, 9542.
744. Kalyani, D., Dick, A.R., Anani, W.Q. and Sanford, M.S. (2006) *Organic Letters*, **8**, 2523.
745. Wang, D.-H., Hao, X.-S., Wu, D.-F. and Yu, J.-Q. (2006) *Organic Letters*, **8**, 3387.
746. Welbes, L.L., Lyons, T.W., Cychosz, K.A. and Sanford, M.S. (2007) *Journal of the American Chemical Society*, **129**, 5836.
747. Desai, L.V. and Sanford, M.S. (2007) *Angewandte Chemie, International Edition*, **46**, 5737.
748. Liu, G. and Stahl, S.S. (2006) *Journal of the American Chemical Society*, **128**, 7179.
749. Streuff, J., Hoevelmann, C.H., Nieger, M. and Muniz, K. (2005) *Journal of the American Chemical Society*, **127**, 14586.
750. Muniz, K., Hoevelmann, C.H. and Streuff, J. (2008) *Journal of the American Chemical Society*, **130**, 763.
751. Giri, R., Chen, X. and Yu, J.-Q. (2005) *Angewandte Chemie, International Edition*, **44**, 2112.
752. Daugulis, O. and Zaitsev, V.G. (2005) *Angewandte Chemie, International Edition*, **44**, 4046.
753. Dick, A.R., Kampf, J.W. and Sanford, M.S. (2005) *Journal of the American Chemical Society*, **127**, 12790.
754. Kalyani, D. and Sanford, M.S. (2008) *Journal of the American Chemical Society*, **130**, 2150.
755. Kalyani, D., Dick, A.R., Anani, W.Q. and Sanford, M.S. (2006) *Tetrahedron*, **62**, 11483.
756. Emmert, M.H., Cook, A.K., Xie, Y.J. and Sanford, M.S. (2011) *Angewandte Chemie, International Edition*, **50**, 9409.
757. Qiu, D., Zheng, Z.-T., Mo, F.-Y., *et al.* (2011) *Organic Letters*, **13**, 4988.
758. Ball, L.T., Lloyd-Jones, G.C. and Russell, C.A. (2012) *Chemistry–A European Journal*, **18**, 2931.
759. Pradal, A., Toullec, P.Y. and Michelet, V. (2011) *Organic Letters*, **13**, 6086.
760. Pradal, A., dit Bel, P.F., Toullec, P.Y. and Michelet, V. (2012) *Synthesis*, 2463.
761. Dauban, P. and Dodd, R.H. (2003) *Synlett*, 1571.
762. Mansuy, D., Mahy, J.P., Dureault, A., *et al.* (1984) *Journal of the Chemical Society, Chemical Communications*, 1161.
763. Evans, D.A., Faul, M.M. and Bilodeau, M.T. (1991) *Journal of Organic Chemistry*, **56**, 6744.
764. Evans, D.A., Faul, M.M., Bilodeau, M.T., *et al.* (1993) *Journal of the American Chemical Society*, **115**, 5328.
765. Li, Z., Quan, R.W. and Jacobsen, E.N. (1995) *Journal of the American Chemical Society*, **117**, 5889.
766. Li, Z., Conser, K.R. and Jacobsen, E.N. (1993) *Journal of the American Chemical Society*, **115**, 5326.
767. Dauban, P. and Dodd, R.H. (1998) *Tetrahedron Letters*, **39**, 5739.
768. Di Chenna, P.H., Dauban, P., Ghini, A., *et al.* (2000) *Tetrahedron Letters*, **41**, 7041.
769. Molander, G.A. and Stengel, P.J. (1997) *Tetrahedron*, **53**, 8887.
770. Pak, C.S., Kim, T.H. and Ha, S.J. (1998) *Journal of Organic Chemistry*, **63**, 10006.
771. Kim, D.Y. and Rhie, D.Y. (1997) *Tetrahedron*, **53**, 13603.
772. Sudau, A., Munch, W., Bats, J.W. and Nubbemeyer, U. (2001) *Chemistry–A European Journal*, **7**, 611.
773. White, R.D. and Wood, J.L. (2001) *Organic Letters*, **3**, 1825.
774. Flock, S. and Frauenrath, H. (2001) *Synlett*, 839.
775. Gupta, V., Besev, M. and Engman, L. (1998) *Tetrahedron Letters*, **39**, 2429.
776. Maligres, P.E., See, M.M., Askin, D. and Reider, P.J. (1997) *Tetrahedron Letters*, **38**, 5253.
777. Wang, X. and Ding, K. (2006) *Chemistry–A European Journal*, **12**, 4568.
778. Ma, L., Du, D.-M. and Xu, J. (2006) *Chirality*, **18**, 575.
779. Ma, L., Du, D.-M. and Xu, J. (2005) *Journal of Organic Chemistry*, **70**, 10155.
780. Gullick, J., Taylor, S., Ryan, D., *et al.* (2003) *Chemical Communications*, 2808.
781. Gillespie, K.M., Sanders, C.J., O'Shaughnessy, P., *et al.* (2002) *Journal of Organic Chemistry*, **67**, 3450.
782. Sanders, C.J., Gillespie, K.M., Bell, D. and Scott, P. (2000) *Journal of the American Chemical Society*, **122**, 7132.
783. Cho, D.-J., Jeon, S.-J., Kim, H.-S., *et al.* (1999) *Tetrahedron: Asymmetry*, **10**, 3833.

784. Llewellyn, D.B., Adamson, D. and Arndtsen, B.A. (2000) *Organic Letters*, **2**, 4165.
785. Muller, P., Nury, P. and Bernardinelli, G. (2000) *Helvetica Chimica Acta*, **83**, 843.
786. Langham, C., Piaggio, P., Bethell, D., *et al.* (1998) *Chemical Communications*, 1601.
787. Langham, C., Taylor, S., Bethell, D., *et al.* (1999) *Journal of the Chemical Society, Perkin Transactions* **2**, 1043.
788. Harm, A.M., Knight, J.G. and Stemp, G. (1996) *Tetrahedron Letters*, **37**, 6189.
789. Tanner, D., Johansson, F., Harden, A. and Andersson, P.G. (1998) *Tetrahedron*, **54**, 15731.
790. Nishikori, H. and Katsuki, T. (1996) *Tetrahedron Letters*, **37**, 9245.
791. Simonato, J.-P., Pecaut, J., Marchon, J.-C. and Robert Scheidt, W. (1999) *Chemical Communications*, 989.
792. Mueller, P., Baud, C. and Jacquier, Y. (1996) *Tetrahedron*, **52**, 1543.
793. Mueller, P., Baud, C. and Naegeli, I. (1998) *Journal of Physical Organic Chemistry*, **11**, 597.
794. Muller, P., Baud, C. and Jacquier, Y. (1998) *Canadian Journal of Chemistry*, **76**, 738.
795. Halfen, J.A., Fox, D.C., Mehn, M.P. and Que, L. Jr. (2001) *Inorganic Chemistry*, **40**, 5060.
796. Halfen, J.A., Uhan, J.M., Fox, D.C., *et al.* (2000) *Inorganic Chemistry*, **39**, 4913.
797. Handy, S.T. and Czopp, M. (2001) *Organic Letters*, **3**, 1423.
798. Halfen, J.A., Hallman, J.K., Schultz, J.A. and Emerson, J.P. (1999) *Organometallics*, **18**, 5435.
799. Cho, D.-J., Jeon, S.-J., Kim, H.-S. and Kim, T.-J. (1998) *Synlett*, 617.
800. Au, S.-M., Fung, W.-H., Cheng, M.-C., *et al.* (1997) *Chemical Communications*, 1655.
801. Jeon, H.-J. and Nguyen, S.T. (2001) *Chemical Communications*, 235.
802. Diaz-Requejo, M.M., Perez, P.J., Brookhart, M. and Templeton, J.L. (1997) *Organometallics*, **16**, 4399.
803. Brandt, P., Soedergren, M.J., Andersson, P.G. and Norrby, P.-O. (2000) *Journal of the American Chemical Society*, **122**, 8013.
804. Gillespie, K.M., Crust, E.J., Deeth, R.J. and Scott, P. (2001) *Chemical Communications*, 785.
805. Overman, L.E. and Tomasi, A.L. (1998) *Journal of the American Chemical Society*, **120**, 4039.
806. Masse, C.E., Knight, B.S., Stavropoulos, P. and Panek, J.S. (1997) *Journal of the American Chemical Society*, **119**, 6040.
807. Au, S.-M., Huang, J.-S., Che, C.-M. and Yu, W.-Y. (2000) *Journal of Organic Chemistry*, **65**, 7858.
808. Kohmura, Y. and Katsuki, T. (2001) *Tetrahedron Letters*, **42**, 3339.
809. Zhou, X.-G., Yu, X.-Q., Huang, J.-S. and Che, C.-M. (1999) *Chemical Communications*, 2377.
810. Anada, M., Tanaka, M., Washio, T., *et al.* (2007) *Organic Letters*, **9**, 4559.
811. Tanaka, M., Kurosaki, Y., Washio, T., *et al.* (2007) *Tetrahedron Letters*, **48**, 8799.
812. Liang, J.-L., Yuan, S.-X., Chan, P.W.H. and Che, C.-M. (2002) *Organic Letters*, **4**, 4507.
813. Okamura, H. and Bolm, C. (2004) *Organic Letters*, **6**, 1305.
814. Knapp, S. and Yu, Y. (2007) *Organic Letters*, **9**, 1359.
815. Guthikonda, K., Wehn, P.M., Caliando, B.J. and Du Bois, J. (2006) *Tetrahedron*, **62**, 11331.
816. Chang, J.W.W. and Chan, P.W.H. (2008) *Angewandte Chemie, International Edition*, **47**, 1138.
817. Chang, J.W.W., Ton, T.M.U., Tania, S., *et al.* (2010) *Chemical Communications*, **46**, 922.
818. Avenier, F., Goure, E., Dubourdeaux, P., *et al.* (2008) *Angewandte Chemie, International Edition*, **47**, 715.
819. Li, Z., Capretto, D.A., Rahaman, R.O. and He, C. (2007) *Journal of the American Chemical Society*, **129**, 12058.
820. Li, C. and Zhang, L. (2011) *Organic Letters*, **13**, 1738.
821. Li, Z., Capretto, D.A., Rahaman, R. and He, C. (2007) *Angewandte Chemie, International Edition*, **46**, 5184.
822. Llaveria, J., Beltran, A., Diaz-Requejo, M.M., *et al.* (2010) *Angewandte Chemie, International Edition*, **49**, 7092.
823. He, L., Yu, J., Zhang, J. and Yu, X.-Q. (2007) *Organic Letters*, **9**, 2277.
824. Fiori, K.W. and Du Bois, J. (2007) *Journal of the American Chemical Society*, **129**, 562.
825. Fructos, M.R., Trofimenko, S., Diaz-Requejo, M.M. and Perez, P.J. (2006) *Journal of the American Chemical Society*, **128**, 11784.
826. Caselli, A., Gallo, E., Ragaini, F., *et al.* (2005) *Journal of Organometallic Chemistry*, **690**, 2142.
827. He, L., Chan, P.W.H., Tsui, W.-M., *et al.* (2004) *Organic Letters*, **6**, 2405.
828. Jensen, M.P., Mehn, M.P. and Que, L. Jr. (2003) *Angewandte Chemie, International Edition*, **42**, 4357.
829. Lescot, C., Darses, B., Collet, F., *et al.* (2012) *Journal of Organic Chemistry*, **77**, 7232.
830. Dick, A.R., Remy, M.S., Kampf, J.W. and Sanford, M.S. (2007) *Organometallics*, **26**, 1365.
831. Mancheno, O.G. and Bolm, C. (2006) *Organic Letters*, **8**, 2349.

832. Mancheno, O.G. and Bolm, C. (2007) *Chemistry–A European Journal*, **13**, 6674.
833. Leca, D., Song, K., Amatore, M., *et al.* (2004) *Chemistry–A European Journal*, **10**, 906.
834. Archibald, S.J., Boa, A.N. and Pesa, N. (2003) *Chemical Communications*, 1736.
835. Muller, J.F.K. and Vogt, P. (1998) *Tetrahedron Letters*, **39**, 4805.
836. Jiang, Y., Zhou, G.-C., He, G.-L., *et al.* (2007) *Synthesis*, 1459.
837. Jain, S.L., Sharma, V.B. and Sain, B. (2003) *Tetrahedron Letters*, **44**, 4385.
838. Bolm, C., Muniz, K., Aguilar, N., *et al.* (1999) *Synthesis*, 1251.
839. Bolm, C., Kesselgruber, M., Muniz, K. and Raabe, G. (2000) *Organometallics*, **19**, 1648.
840. Nishikori, H., Ohta, C., Oberlin, E., *et al.* (1999) *Tetrahedron*, **55**, 13937.
841. Ohta, C. and Katsuki, T. (2001) *Tetrahedron Letters*, **42**, 3885.
842. Miyake, Y., Takada, H., Ohe, K. and Uemura, S. (1998) *Journal of the Chemical Society, Perkin Transactions* **1**, 2373.
843. Takada, H., Nishibayashi, Y., Ohe, K., *et al.* (1997) *Journal of Organic Chemistry*, **62**, 6512.
844. Takada, H., Oda, M., Miyake, Y., *et al.* (1998) *Chemical Communications*, 1557.
845. Miyake, Y., Oda, M., Oyamada, A., *et al.* (2000) *Journal of Organometallic Chemistry*, **611**, 475.
846. Otani, T., Sugihara, Y., Ishii, A. and Nakayama, J. (2000) *Tetrahedron Letters*, **41**, 8461.
847. Otani, T., Sugihara, Y., Ishii, A. and Nakayama, J. (1999) *Tetrahedron Letters*, **40**, 5549.
848. Nakayama, J., Otani, T., Sugihara, Y., *et al.* (2001) *Heteroatom Chemistry*, **12**, 333.
849. Nakayama, J., Sano, Y., Sugihara, Y. and Ishii, A. (1999) *Tetrahedron Letters*, **40**, 3785.
850. Au, S.-M., Fung, W.-H., Huang, J.-S., *et al.* (1998) *Inorganic Chemistry*, **37**, 6564.
851. Li, Y., Huang, J.-S., Zhou, Z.-Y. and Che, C.-M. (2001) *Journal of the American Chemical Society*, **123**, 4843.
852. Nomura, M., Yagisawa, T., Takayama, C., *et al.* (2000) *Journal of Organometallic Chemistry*, **611**, 376.
853. Okuyama, T., Takino, T., Sueda, T. and Ochiai, M. (1995) *Journal of the American Chemical Society*, **117**, 3360.
854. Merritt, E.A. and Olofsson, B. (2009) *Angewandte Chemie, International Edition*, **48**, 9052.
855. Yusubov, M.S., Maskaev, A.V. and Zhdankin, V.V. (2011) *ARKIVOC*, (i), 370.
856. Zhdankin, V.V. and Stang, P.J. (1998) *Tetrahedron*, **54**, 10927.
857. Zhdankin, V.V. (2005) *Current Organic Synthesis*, **2**, 121.
858. Brand, J.P., Gonzalez, D.F., Nicolai, S. and Waser, J. (2011) *Chemical Communications*, 102.
859. Zhdankin, V.V. (1997) *Reviews on Heteroatom Chemistry*, **17**, 133.
860. Grushin, V.V. (2000) *Chemical Society Reviews*, **29**, 315.
861. Xia, M. and Chen, Z.-C. (1997) *Synthetic Communications*, **27**, 1309.
862. Martin-Santamaria, S., Carroll, M.A., Carroll, C.M., *et al.* (2000) *Chemical Communications*, 649.
863. Shah, A., Pike, V.W. and Widdowson, D.A. (1998) *Journal of the Chemical Society, Perkin Transactions* **1**, 2043.
864. Oh, C.H., Kim, J.S. and Jung, H.H. (1999) *Journal of Organic Chemistry*, **64**, 1338.
865. Iwama, T., Birman, V.B., Kozmin, S.A. and Rawal, V.H. (1999) *Organic Letters*, **1**, 673.
866. Wang, F.-Y., Chen, Z.-C. and Zheng, Q.-G. (2003) *Journal of Chemical Research (S)*, 620.
867. Wang, F.-Y., Chen, Z.-C. and Zheng, Q.-G. (2003) *Journal of Chemical Research (S)*, 810.
868. Wang, F.-Y., Chen, Z.-C. and Zheng, Q.-G. (2004) *Journal of Chemical Research*, 206.
869. Carroll, M.A. and Wood, R.A. (2007) *Tetrahedron*, **63**, 11349.
870. Yan, J., Hu, W. and Rao, G. (2006) *Synthesis*, 943.
871. Xue, Z., Yang, D. and Wang, C. (2006) *Journal of Organometallic Chemistry*, **691**, 247.
872. Marsh, G., Stenutz, R. and Bergman, A. (2003) *European Journal of Organic Chemistry*, 2566.
873. Couladouros, E.A., Moutsos, V.I. and Pitsinos, E.N. (2003) *ARKIVOC*, (xv), 92.
874. Jalalian, N., Ishikawa, E.E., Silva, L.F. and Olofsson, B. (2011) *Organic Letters*, **13**, 1552.
875. Petersen, T.B., Khan, R. and Olofsson, B. (2011) *Organic Letters*, **13**, 3462.
876. Dohi, T., Ito, M., Yamaoka, N., *et al.* (2010) *Angewandte Chemie, International Edition*, **49**, 3334.
877. Ptitsyna, O.A., Pudeeva, M.E. and Reutov, O.A. (1965) *Doklady Akademii Nauk SSSR*, **165**, 582.
878. Kampmeier, J.A. and Nalli, T.W. (1993) *Journal of Organic Chemistry*, **58**, 943.
879. Muneer, R. and Nalli, T.W. (1998) *Macromolecules*, **31**, 7976.
880. Umierski, N. and Manolikakes, G. (2013) *Organic Letters*, **15**, 188.
881. Aggarwal, V.K. and Olofsson, B. (2005) *Angewandte Chemie, International Edition*, **44**, 5516.

882. Ochiai, M., Kitagawa, Y., Takayama, N., *et al.* (1999) *Journal of the American Chemical Society*, **121**, 9233.
883. Norrby, P.-O., Petersen, T.B., Bielawski, M. and Olofsson, B. (2010) *Chemistry–A European Journal*, **16**, 8251.
884. Ozanne-Beaudenon, A. and Quideau, S. (2005) *Angewandte Chemie, International Edition*, **44**, 7065.
885. Quideau, S., Pouysegu, L., Ozanne, A. and Gagnepain, J. (2005) *Molecules*, **10**, 201.
886. Ackermann, L., Dell'Acqua, M., Fenner, S., *et al.* (2011) *Organic Letters*, **13**, 2358.
887. Vase, K.H., Holm, A.H., Norrman, K., *et al.* (2007) *Langmuir*, **23**, 3786.
888. Weissmann, M., Baranton, S. and Coutanceau, C. (2010) *Langmuir*, **26**, 15002.
889. Ochiai, M., Kitagawa, Y. and Toyonari, M. (2003) *ARKIVOC*, (vi), 43.
890. Ochiai, M., Takaoka, Y., Masaki, Y., *et al.* (1990) *Journal of the American Chemical Society*, **112**, 5677.
891. Lancer, K.M. and Wiegand, G.H. (1976) *Journal of Organic Chemistry*, **41**, 3360.
892. Yamada, Y. and Okawara, M. (1972) *Bulletin of the Chemical Society of Japan*, **45**, 1860.
893. Fujita, M., Mishima, E. and Okuyama, T. (2007) *Journal of Physical Organic Chemistry*, **20**, 241.
894. Ryan, J.H. and Stang, P.J. (1997) *Tetrahedron Letters*, **38**, 5061.
895. Duong, H.A., Gilligan, R.E., Cooke, M.L., *et al.* (2011) *Angewandte Chemie, International Edition*, **50**, 463.
896. Zhang, B.-X., Nuka, T., Fujiwara, Y., *et al.* (2004) *Heterocycles*, **64**, 199.
897. Zhou, T. and Chen, Z.-C. (2004) *Journal of Chemical Research*, 404.
898. Zhou, T., Li, T.-C. and Chen, Z.-C. (2005) *Helvetica Chimica Acta*, **88**, 290.
899. Ciana, C.-L., Phipps, R.J., Brandt, J.R., *et al.* (2011) *Angewandte Chemie, International Edition*, **50**, 458.
900. Phipps, R.J., McMurray, L., Ritter, S., *et al.* (2012) *Journal of the American Chemical Society*, **134**, 10773.
901. Bumagin, N.A. and Tsarev, D.A. (1998) *Tetrahedron Letters*, **39**, 8155.
902. Al-Qahtani, M.H. and Pike, V.W. (2000) *Journal of the Chemical Society, Perkin Transactions 1*, 1033.
903. Kang, S.-K., Yamaguchi, T., Ho, P.-S., *et al.* (1997) *Tetrahedron Letters*, **38**, 1947.
904. Kang, S.-K., Choi, S.-C. and Baik, T.-G. (1999) *Synthetic Communications*, **29**, 2493.
905. Kang, S.-K., Ryu, H.-C. and Kim, J.-W. (2001) *Synthetic Communications*, **31**, 1021.
906. Kang, S.-K., Yamaguchi, T., Ho, P.-S., *et al.* (1998) *Journal of the Chemical Society, Perkin Transactions* **1**, 841.
907. Kang, S.-K., Lee, H.-W., Jang, S.-B., *et al.* (1996) *Journal of Organic Chemistry*, **61**, 2604.
908. Kang, S.-K., Ha, Y.-H. and Yang, H.-Y. (2002) *Journal of Chemical Research (S)*, 282.
909. Kang, S.-K., Baik, T.-G. and Hur, Y. (1999) *Tetrahedron*, **55**, 6863.
910. Wang, L. and Chen, Z.-C. (2000) *Synthetic Communications*, **30**, 3607.
911. Liang, Y., Luo, S., Liu, C., *et al.* (2000) *Tetrahedron*, **56**, 2961.
912. Zhu, M., Song, Y. and Cao, Y. (2007) *Synthesis*, 853.
913. Radhakrishnan, U. and Stang, P.J. (2001) *Organic Letters*, **3**, 859.
914. Storr, T.E. and Greaney, M.F. (2013) *Organic Letters*, **15**, 1410.
915. Becht, J.-M. and Le Drian, C. (2008) *Organic Letters*, **10**, 3161.
916. Deprez, N.R., Kalyani, D., Krause, A. and Sanford, M.S. (2006) *Journal of the American Chemical Society*, **128**, 4972.
917. Kalyani, D., Deprez, N.R., Desai, L.V. and Sanford, M.S. (2005) *Journal of the American Chemical Society*, **127**, 7330.
918. Canty, A.J., Patel, J., Rodemann, T., *et al.* (2004) *Organometallics*, **23**, 3466.
919. Bedford, R.B., Webster, R.L. and Mitchell, C.J. (2009) *Organic & Biomolecular Chemistry*, **7**, 4853.
920. Allen, A.E. and MacMillan, D.W.C. (2011) *Journal of the American Chemical Society*, **133**, 4260.
921. Fieser, L.F. and Haddadin, M.J. (1966) *Organic Syntheses*, **46**, 107.
922. Beringer, F.M. and Lillien, I. (1960) *Journal of the American Chemical Society*, **82**, 725.
923. Beringer, F.M. and Huang, S.J. (1964) *Journal of Organic Chemistry*, **29**, 445.
924. Kitamura, T., Yamane, M., Inoue, K., *et al.* (1999) *Journal of the American Chemical Society*, **121**, 11674.
925. Kitamura, T., Meng, Z. and Fujiwara, Y. (2000) *Tetrahedron Letters*, **41**, 6611.
926. Kitamura, T., Wasai, K., Todaka, M. and Fujiwara, Y. (1999) *Synlett*, 731.
927. Kitamura, T., Aoki, Y., Isshiki, S., *et al.* (2006) *Tetrahedron Letters*, **47**, 1709.
928. Kitamura, T., Abe, T., Fujiwara, Y. and Yamaji, T. (2003) *Synthesis*, 213.
929. Xue, J. and Huang, X. (2007) *Synthetic Communications*, **37**, 2179.
930. Kamila, S., Koh, B. and Biehl, E.R. (2006) *Synthetic Communications*, **36**, 3493.

931. Thiemann, T., Fujii, H., Ohira, D., *et al.* (2003) *New Journal of Chemistry*, **27**, 1377.
932. Xue, J., Yang, Y. and Huang, X. (2007) *Synlett*, 1533.
933. Rao, U.N., Sathunuru, R., Maguire, J.A. and Biehl, E. (2004) *Journal of Heterocyclic Chemistry*, **41**, 13.
934. Lee, T., Jeon, J., Song, K.H., *et al.* (2004) *Journal of the Chemical Society, Dalton Transactions*, 933.
935. Ochiai, M. (2000) *Journal of Organometallic Chemistry*, **611**, 494.
936. Ochiai, M. (1999) In *Chemistry of Hypervalent Compounds* (ed. K.y. Akiba), VCH Publishers, New York, p. 359.
937. Okuyama, T. (2002) *Accounts of Chemical Research*, **35**, 12.
938. Okuyama, T. and Fujita, M. (2005) *Russian Journal of Organic Chemistry*, **41**, 1245.
939. Okuyama, T. and Fujita, M. (2007) *ACS Symposium Series*, **965**, 68.
940. Pirkuliev, N.S., Brel, V.K. and Zefirov, N.S. (2000) *Russian Chemical Reviews*, **69**, 105.
941. Okuyama, T. and Fujita, M. (2005) *Accounts of Chemical Research*, **38**, 679.
942. Fujita, M., Kim, W.H., Sakanishi, Y., *et al.* (2004) *Journal of the American Chemical Society*, **126**, 7548.
943. Fujita, M., Sakanishi, Y., Nishii, M., *et al.* (2002) *Journal of Organic Chemistry*, **67**, 8130.
944. Fujita, M., Kim, W.H., Fujiwara, K. and Okuyama, T. (2005) *Journal of Organic Chemistry*, **70**, 480.
945. Hinkle, R.J. and Mikowski, A.M. (2003) *ARKIVOC*, (vi), 201.
946. Fujita, M., Sakanishi, Y., Nishii, M. and Okuyama, T. (2002) *Journal of Organic Chemistry*, **67**, 8138.
947. Slegt, M., Gronheid, R., Van der Vlugt, D., *et al.* (2006) *Journal of Organic Chemistry*, **71**, 2227.
948. Fujita, M., Sakanishi, Y., Kim, W.H. and Okuyama, T. (2002) *Chemistry Letters*, 908.
949. Fujita, M., Ihara, K., Kim, W.H. and Okuyama, T. (2003) *Bulletin of the Chemical Society of Japan*, **76**, 1849.
950. Ochiai, M., Yamamoto, S., Suefuji, T. and Chen, D.-W. (2001) *Organic Letters*, **3**, 2753.
951. Yan, J. and Chen, Z.-C. (2000) *Synthetic Communications*, **30**, 3897.
952. Yan, J. and Chen, Z.-C. (2000) *Synthetic Communications*, **30**, 2359.
953. Yan, J. and Chen, Z.-C. (2000) *Synthetic Communications*, **30**, 1009.
954. Yan, J. and Chen, Z.-C. (1999) *Tetrahedron Letters*, **40**, 5757.
955. Yan, J. and Chen, Z.-C. (1999) *Synthetic Communications*, **29**, 3605.
956. Yan, J. and Chen, Z.-C. (1999) *Synthetic Communications*, **29**, 3275.
957. Ochiai, M., Sueda, T., Noda, R. and Shiro, M. (1999) *Journal of Organic Chemistry*, **64**, 8563.
958. Ochiai, M., Yamamoto, S. and Sato, K. (1999) *Chemical Communications*, 1363.
959. Okuyama, T., Fujita, M., Gronheid, R. and Lodder, G. (2000) *Tetrahedron Letters*, **41**, 5125.
960. Yan, J., Jin, H. and Chen, Z. (2007) *Journal of Chemical Research*, 233.
961. Zhang, P.-F. and Chen, Z.-C. (2002) *Journal of Chemical Research (S)*, 388.
962. Guan, T., Yoshida, M. and Hara, S. (2007) *Journal of Organic Chemistry*, **72**, 9617.
963. Hara, S., Guan, T. and Yoshida, M. (2006) *Organic Letters*, **8**, 2639.
964. Ochiai, M., Shu, T., Nagaoka, T. and Kitagawa, Y. (1997) *Journal of Organic Chemistry*, **62**, 2130.
965. Yoshida, M. and Hara, S. (2003) *Organic Letters*, **5**, 573.
966. Yoshida, M., Komata, A. and Hara, S. (2006) *Tetrahedron*, **62**, 8636.
967. Kang, S.-K., Yamaguchi, T., Kim, T.-H. and Ho, P.-S. (1996) *Journal of Organic Chemistry*, **61**, 9082.
968. Huang, X. and Xu, X.-H. (1998) *Journal of the Chemical Society, Perkin Transactions* **1**, 3321.
969. Kang, S.-K., Yoon, S.-K. and Kim, Y.-M. (2001) *Organic Letters*, **3**, 2697.
970. Thielges, S., Bisseret, P. and Eustache, J. (2005) *Organic Letters*, **7**, 681.
971. Moriarty, R.M., Epa, W.R. and Awasthi, A.K. (1991) *Journal of the American Chemical Society*, **113**, 6315.
972. Yoshida, M., Yoshikawa, S., Fukuhara, T., *et al.* (2001) *Tetrahedron*, **57**, 7143.
973. Yoshida, M., Kawakami, K. and Hara, S. (2004) *Synthesis*, 2821.
974. Yoshida, M., Komata, A. and Hara, S. (2004) *Journal of Fluorine Chemistry*, **125**, 527.
975. Guan, T., Takemura, K., Senboku, H., *et al.* (2007) *Tetrahedron Letters*, **49**, 76.
976. Ochiai, M., Uemura, K. and Masaki, Y. (1993) *Journal of the American Chemical Society*, **115**, 2528.
977. Ochiai, M., Kunishima, M., Tani, S. and Nagao, Y. (1991) *Journal of the American Chemical Society*, **113**, 3135.
978. Feldman, K.S. (2003) *ARKIVOC*, (vi), 179.
979. Feldman, K.S. (2004) In *Strategies and Tactics in Organic Synthesis*, vol. **4** (ed. M. Harmata), Elsevier, London, p. 133.
980. Bachi, M.D., Bar-Ner, N. and Crittell, C.M., (1991) *Journal of Organic Chemistry*, **56**, 3912.

981. Ochiai, M., Ito, T., Takaoka, Y., *et al.* (1990) *Journal of the Chemical Society, Chemical Communications*, 118.
982. Stang, P.J. and Murch, P. (1997) *Synthesis*, 1378.
983. Zhang, J.-L. and Chen, Z.-C. (1997) *Synthetic Communications*, **27**, 3757.
984. Zhang, J.-L. and Chen, Z.-C. (1997) *Synthetic Communications*, **27**, 3881.
985. Zhang, J.-L. and Chen, Z.-C. (1998) *Synthetic Communications*, **28**, 175.
986. Kitamura, T., Tashi, N., Tsuda, K., *et al.* (2000) *Heterocycles*, **52**, 303.
987. Kerwin, S.M. and Nadipuram, A. (2004) *Synlett*, 1404.
988. Witulski, B. and Stengel, T. (1999) *Angewandte Chemie, International Edition*, **38**, 2426.
989. Hashmi, A.S.K., Salathe, R. and Frey, W. (2007) *Synlett*, 1763.
990. Martinez-Esperon, M.F., Rodriguez, D., Castedo, L. and Saa, C. (2005) *Organic Letters*, **7**, 2213.
991. Bykowski, D., McDonald, R. and Tykwinski, R.R. (2003) *ARKIVOC*, (vi), 21.
992. Canty, A.J. and Rodemann, T. (2003) *Inorganic Chemistry Communications*, **6**, 1382.
993. Canty, A.J., Watson, R.P., Karpiniec, S.S., *et al.* (2008) *Organometallics*, **27**, 3203.
994. Kitamura, T., Lee, C.H., Taniguchi, Y., *et al.* (1997) *Journal of the American Chemical Society*, **119**, 619.
995. Yang, D.-Y., He, J. and Miao, S. (2003) *Synthetic Communications*, **33**, 2695.
996. Yu, C.-M., Kweon, J.-H., Ho, P.-S., *et al.* (2005) *Synlett*, 2631.
997. Liu, Z., Chen, Z.-C. and Zheng, Q.-G. (2003) *Journal of Chemical Research (S)*, 715.
998. Liu, Z., Chen, Z.-C. and Zheng, Q.-G. (2003) *Journal of Heterocyclic Chemistry*, **40**, 909.
999. Liu, Z., Chen, Z.-C. and Zheng, Q.-G. (2004) *Synthetic Communications*, **34**, 361.
1000. Miyamoto, K., Nishi, Y. and Ochiai, M. (2005) *Angewandte Chemie, International Edition*, **44**, 6896.
1001. Wipf, P. and Venkatraman, S. (1996) *Journal of Organic Chemistry*, **61**, 8004.
1002. Wardrop, D.J. and Fritz, J. (2006) *Organic Letters*, **8**, 3659.
1003. Feldman, K.S., Saunders, J.C. and Wrobleski, M.L. (2002) *Journal of Organic Chemistry*, **67**, 7096.
1004. Feldman, K.S., Perkins, A.L. and Masters, K.M. (2004) *Journal of Organic Chemistry*, **69**, 7928.
1005. Brand, J.P., Charpentier, J. and Waser, J. (2009) *Angewandte Chemie, International Edition*, **48**, 9346.
1006. Brand, J.P., Chevalley, C. and Waser, J. (2011) *Beilstein Journal of Organic Chemistry*, **7**, 565.
1007. Brand, J.P. and Waser, J. (2012) *Chemical Society Reviews*, **41**, 4165.
1008. Brand, J.P. and Waser, J. (2012) *Synthesis*, 1155.
1009. Brand, J.P. and Waser, J. (2012) *Organic Letters*, **14**, 744.
1010. Brand, J.P., Chevalley, C., Scopelliti, R. and Waser, J. (2012) *Chemistry–A European Journal*, **18**, 5655.
1011. Nicolai, S., Erard, S., Gonzalez, D.F. and Waser, J. (2010) *Organic Letters*, **12**, 384.
1012. Fernandez Gonzalez, D., Brand, J.P. and Waser, J. (2010) *Chemistry–A European Journal*, **16**, 9457.
1013. Aubineau, T. and Cossy, J. (2013) *Chemical Communications*, 3303.
1014. Zhdankin, V.V., Tykwinski, R., Caple, R., *et al.* (1988) *Tetrahedron Letters*, **29**, 3703.
1015. Zhdankin, V.V., Tykwinski, R., Mullikin, M., *et al.* (1989) *Journal of Organic Chemistry*, **54**, 2605.
1016. Zhdankin, V.V., Erickson, S.A. and Hanson, K.J. (1997) *Journal of the American Chemical Society*, **119**, 4775.
1017. Umemoto, T. (1996) *Chemical Reviews*, **96**, 1757.
1018. Montanari, V. and Resnati, G. (1994) *Tetrahedron Letters*, **35**, 8015.
1019. DesMarteau, D.D. and Montanari, V. (2000) *Chemistry Letters*, 1052.
1020. DesMarteau, D.D. and Montanari, V. (2001) *Journal of Fluorine Chemistry*, **109**, 19.
1021. Montanari, V. and Kumar, K. (2006) *Journal of Fluorine Chemistry*, **127**, 565.
1022. Montanari, V. and Kumar, K. (2006) *European Journal of Organic Chemistry*, 874.
1023. DesMarteau, D.D. and Montanari, V. (1998) *Chemical Communications*, 2241.
1024. Montanari, V. and Kumar, K. (2004) *Journal of the American Chemical Society*, **126**, 9528.
1025. Lu, C., VanDerveer, D. and DesMarteau, D.D. (2008) *Organic Letters*, **10**, 5565.
1026. Eisenberger, P., Gischig, S. and Togni, A. (2006) *Chemistry–A European Journal*, **12**, 2579.
1027. Kieltsch, I., Eisenberger, P. and Togni, A. (2007) *Angewandte Chemie, International Edition*, **46**, 754.
1028. Capone, S., Kieltsch, I., Flogel, O., *et al.* (2008) *Helvetica Chimica Acta*, **91**, 2035.
1029. Santschi, N. and Togni, A. (2011) *Journal of Organic Chemistry*, **76**, 4189.
1030. Stanek, K., Koller, R. and Togni, A. (2008) *Journal of Organic Chemistry*, **73**, 7678.
1031. Koller, R., Huchet, Q., Battaglia, P., *et al.* (2009) *Chemical Communications*, 5993.

1032. Fantasia, S., Welch, J.M. and Togni, A. (2010) *Journal of Organic Chemistry*, **75**, 1779.
1033. Koller, R., Stanek, K., Stolz, D., *et al.* (2009) *Angewandte Chemie, International Edition*, **48**, 4332.
1034. Niedermann, K., Frueh, N., Senn, R., *et al.* (2012) *Angewandte Chemie, International Edition*, **51**, 6511.
1035. Niedermann, K., Frueh, N., Vinogradova, E., *et al.* (2011) *Angewandte Chemie, International Edition*, **50**, 1059.
1036. Eisenberger, P., Kieltsch, I., Armanino, N. and Togni, A. (2008) *Chemical Communications*, 1575.
1037. Armanino, N., Koller, R. and Togni, A. (2010) *Organometallics*, **29**, 1771.
1038. Santschi, N., Geissbuehler, P. and Togni, A. (2012) *Journal of Fluorine Chemistry*, **135**, 83.
1039. Allen, A.E. and MacMillan, D.W.C. (2010) *Journal of the American Chemical Society*, **132**, 4986.
1040. Zhang, W., Zhu, J. and Hu, J. (2008) *Tetrahedron Letters*, **49**, 5006.
1041. Zhdankin, V.V., Kuehl, C.J., Krasutsky, A.P., *et al.* (1995) *Tetrahedron Letters*, **36**, 7975.
1042. Shaffer, C.L., Morton, M.D. and Hanzlik, R.P. (2001) *Journal of the American Chemical Society*, **123**, 349.
1043. Muller, P. (2004) *Accounts of Chemical Research*, **37**, 243.
1044. Batsila, C., Gogonas, E.P., Kostakis, G. and Hadjiarapoglou, L.P. (2003) *Organic Letters*, **5**, 1511.
1045. Telu, S., Durmus, S. and Koser, G.F. (2007) *Tetrahedron Letters*, **48**, 1863.
1046. Muller, P. and Bolea, C. (2001) *Molecules*, **6**, 258.
1047. Muller, P. and Bolea, C. (2002) *Helvetica Chimica Acta*, **85**, 483.
1048. Muller, P. and Tohill, S. (2000) *Tetrahedron*, **56**, 1725.
1049. Moriarty, R.M., Prakash, O., Vaid, R.K. and Zhao, L. (1989) *Journal of the American Chemical Society*, **111**, 6443.
1050. Camacho, M.B., Clark, A.E., Liebrecht, T.A. and DeLuca, J.P. (2000) *Journal of the American Chemical Society*, **122**, 5210.
1051. Goudreau, S.R., Marcoux, D. and Charette, A.B. (2009) *Journal of Organic Chemistry*, **74**, 470.
1052. Goudreau, S.R., Marcoux, D., Charette, A.B. and Hughes, D. (2010) *Organic Syntheses*, **87**, 115.
1053. Ghanem, A., Lacrampe, F. and Schurig, V. (2005) *Helvetica Chimica Acta*, **88**, 216.
1054. Georgakopoulou, G., Kalogiros, C. and Hadjiarapoglou, L.P. (2001) *Synlett*, 1843.
1055. Batsila, C., Kostakis, G. and Hadjiarapoglou, L.P. (2002) *Tetrahedron Letters*, **43**, 5997.
1056. Ghanem, A., Aboul-Enein Hassan, Y. and Muller, P. (2005) *Chirality*, **17**, 44.
1057. Muller, P. and Bolea, C. (2001) *Helvetica Chimica Acta*, **84**, 1093.
1058. Muller, P., Allenbach, Y.F., Chappellet, S. and Ghanem, A. (2006) *Synthesis*, 1689.
1059. Zhu, C., Yoshimura, A., Ji, L., *et al.* (2012) *Organic Letters*, **14**, 3170.
1060. Koser, G.F. and Yu, S.-M. (1975) *Journal of Organic Chemistry*, **40**, 1166.
1061. Moriarty, R.M., Tyagi, S., Ivanov, D. and Constantinescu, M. (2008) *Journal of the American Chemical Society*, **130**, 7564.
1062. Gogonas, E.P. and Hadjiarapoglou, L.P. (2000) *Tetrahedron Letters*, **41**, 9299.
1063. Asouti, A. and Hadjiarapoglou, L.P. (1998) *Tetrahedron Letters*, **39**, 9073.
1064. Lee, Y.R., Suk, J.Y. and Kim, B.S. (1999) *Tetrahedron Letters*, **40**, 6603.
1065. Koser, G.F. and Yu, S.-M. (1976) *Journal of Organic Chemistry*, **41**, 125.
1066. Hadjiarapoglou, L.P. (1987) *Tetrahedron Letters*, **28**, 4449.
1067. Pirrung, M.C., Zhang, J., Lackey, K., *et al.* (1995) *Journal of Organic Chemistry*, **60**, 2112.
1068. Lee, Y.R., Yoon, S.H., Seo, Y. and Kim, B.S. (2004) *Synthesis*, 2787.
1069. Zhu, C., Yoshimura, A., Solntsev, P., *et al.* (2012) *Chemical Communications*, 10108.
1070. Kirmse, W. (2005) *European Journal of Organic Chemistry*, 237.
1071. Mueller, P. and Fernandez, D. (1995) *Helvetica Chimica Acta*, **78**, 947.
1072. Bonge, H.T. and Hansen, T. (2007) *Synlett*, 55.
1073. Wurz, R.P. and Charette, A.B. (2003) *Organic Letters*, **5**, 2327.
1074. Moreau, B. and Charette, A.B. (2005) *Journal of the American Chemical Society*, **127**, 18014.
1075. Bonge, H.T. and Hansen, T. (2008) *Tetrahedron Letters*, **49**, 57.
1076. Adam, W., Gogonas, E.P. and Hadjiarapoglou, L.P. (2003) *Tetrahedron*, **59**, 7929.
1077. Adam, W., Gogonas, E.P. and Hadjiarapoglou, L.P. (2003) *European Journal of Organic Chemistry*, 1064.
1078. Lee, Y.R. and Yoon, S.H. (2006) *Synthetic Communications*, **36**, 1941.
1079. Lee, Y.R. and Jung, Y.U. (2002) *Journal of the Chemical Society, Perkin Transactions* **1**, 1309.

1080. Murphy, G.K. and West, F.G. (2006) *Organic Letters*, **8**, 4359.
1081. Adam, W., Gogonas, E.P. and Hadjiarapoglou, L.P. (2003) *Journal of Organic Chemistry*, **68**, 9155.
1082. Adam, W., Gogonas, E.P. and Hadjiarapoglou, L.P. (2003) *Synlett*, 1165.
1083. Adam, W., Gogonas, E.P., Nyxas, I.A. and Hadjiarapoglou, L.P. (2007) *Synthesis*, 3211.
1084. Adam, W., Bosio, S.G., Gogonas, E.P. and Hadjiarapoglou, L.P. (2002) *Synthesis*, 2084.
1085. Mueller, P., Allenbach, Y.F., Ferri, M. and Bernardinelli, G. (2003) *ARKIVOC*, (vii), 80.
1086. Ochiai, M., Kitagawa, Y. and Yamamoto, S. (1997) *Journal of the American Chemical Society*, **119**, 11598.
1087. Ochiai, M. and Kitagawa, Y. (1998) *Tetrahedron Letters*, **39**, 5569.
1088. Ochiai, M. and Kitagawa, Y. (1999) *Journal of Organic Chemistry*, **64**, 3181.
1089. Ochiai, M., Nishi, Y., Nishitani, J., *et al.* (2000) *Chemical Communications*, 1157.
1090. Ochiai, M., Tuchimoto, Y. and Higashiura, N. (2004) *Organic Letters*, **6**, 1505.
1091. Huang, Z., Yu, X. and Huang, X. (2002) *Journal of Organic Chemistry*, **67**, 8261.
1092. Zhdankin, V.V., Maydanovych, O., Herschbach, J., *et al.* (2003) *Journal of Organic Chemistry*, **68**, 1018.
1093. Zhdankin, V.V., Maydanovych, O., Herschbach, J., *et al.* (2002) *Tetrahedron Letters*, **43**, 2359.
1094. Huang, Z.-Z., Yu, X.-C. and Huang, X. (2002) *Tetrahedron Letters*, **43**, 6823.
1095. Matveeva, E.D., Podrugina, T.A., Grishin, Y.K., *et al.* (2007) *Russian Journal of Organic Chemistry*, **43**, 201.
1096. Matveeva, E.D., Podrugina, T.A., Pavlova, A.S., *et al.* (2009) *Journal of Organic Chemistry*, **74**, 9428.
1097. Matveeva, E.D., Podrugina, T.A., Pavlova, A.S., *et al.* (2009) *European Journal of Organic Chemistry*, 2323.
1098. Matveeva, E.D., Podrugina, T.A., Taranova, M.A., *et al.* (2011) *Journal of Organic Chemistry*, **76**, 566.
1099. Matveeva, E.D., Podrugina, T.A., Taranova, M.A., *et al.* (2012) *Journal of Organic Chemistry*, **77**, 5770.
1100. Matveeva, E.D., Gleiter, R. and Zefirov, N.S. (2010) *Russian Chemical Bulletin*, **59**, 488.
1101. Deng, G. (2002) *Journal of Chemical Research (S)*, 558.
1102. Ladziata, U. and Zhdankin, V.V. (2006) *ARKIVOC*, (ix), 26.
1103. Dess, D.B. and Martin, J.C. (1983) *Journal of Organic Chemistry*, **48**, 4155.
1104. Zhdankin, V.V. (2011) *Journal of Organic Chemistry*, **76**, 1185.
1105. Satam, V., Harad, A., Rajule, R. and Pati, H. (2010) *Tetrahedron*, **66**, 7659.
1106. Duschek, A. and Kirsch, S.F. (2011) *Angewandte Chemie, International Edition*, **50**, 1524.
1107. Barton, D.H.R., Godfrey, C.R.A., Morzycki, J.W., *et al.* (1982) *Tetrahedron Letters*, **23**, 957.
1108. De Armas, P., Concepcion, J.I., Francisco, C.G., *et al.* (1989) *Journal of the Chemical Society, Perkin Transactions* **1**, 405.
1109. Murali, D. and Rao, G.S.K. (1987) *Indian Journal of Chemistry, Section B*, **26**, 668.
1110. Reddy, N.K. and Rao, G.S.K. (1987) *Indian Journal of Chemistry, Section B*, **26B**, 920.
1111. Barton, D.H.R. and Crich, D. (1985) *Tetrahedron*, **41**, 4359.
1112. Hinman, A. and Du Bois, J. (2003) *Journal of the American Chemical Society*, **125**, 11510.
1113. Kuenzer, H., Sauer, G. and Wiechert, R. (1989) *Tetrahedron*, **45**, 6409.
1114. Gleiter, R. and Mueller, G. (1988) *Journal of Organic Chemistry*, **53**, 3912.
1115. Iida, T., Nishida, S., Chang, F.C., *et al.* (1993) *Chemical & Pharmaceutical Bulletin*, **41**, 763.
1116. Barret, R., Sabot, J.P., Pautet, F., *et al.* (1989) *Oxidation Communication*, **12**, 55.
1117. Barret, R. and Daudon, M. (1990) *Synthetic Communications*, **20**, 1543.
1118. Barret, R., Pautet, F., Bordat, P., *et al.* (1989) *Phosphorus, Sulfur Silicon and the Related Elements*, **45**, 31.
1119. Tohma, H., Takizawa, S., Watanabe, H., *et al.* (1999) *Journal of Organic Chemistry*, **64**, 3519.
1120. Tohma, H., Takizawa, S., Morioka, H., *et al.* (2000) *Chemical & Pharmaceutical Bulletin*, **48**, 445.
1121. Grushin, V.V. and Alper, H. (1993) *Journal of Organic Chemistry*, **58**, 4794.
1122. Macikenas, D., Skrzypczak-Jankun, E. and Protasiewicz, J.D. (2000) *Angewandte Chemie, International Edition*, **39**, 2007.
1123. Zhdankin, V.V., Koposov, A.Y., Litvinov, D.N., *et al.* (2005) *Journal of Organic Chemistry*, **70**, 6484.
1124. Zhdankin, V.V., Koposov, A.Y., Netzel, B.C., *et al.* (2003) *Angewandte Chemie, International Edition*, **42**, 2194.
1125. Ladziata, U., Koposov, A.Y., Lo, K.Y., *et al.* (2005) *Angewandte Chemie, International Edition*, **44**, 7127.
1126. Ladziata, U. and Zhdankin, V.V. (2007) *Synlett*, 527.
1127. Koposov, A.Y. and Zhdankin, V.V. (2005) *Synthesis*, 22.
1128. Duschek, A. and Kirsch, S.F. (2009) *Chemistry–A European Journal*, **15**, 10713.

1129. Geraskin, I.M., Pavlova, O., Neu, H.M., *et al.* (2009) *Advanced Synthesis & Catalysis*, **351**, 733.
1130. Ye, W., Ho, D.M., Friedle, S., *et al.* (2012) *Inorganic Chemistry*, **51**, 5006.
1131. Boppisetti, J.K. and Birman, V.B. (2009) *Organic Letters*, **11**, 1221.
1132. Frigerio, M. and Santagostino, M. (1994) *Tetrahedron Letters*, **35**, 8019.
1133. Frigerio, M., Santagostino, M., Sputore, S. and Palmisano, G. (1995) *Journal of Organic Chemistry*, **60**, 7272.
1134. Zoller, T., Breuilles, P., Uguen, D., *et al.* (1999) *Tetrahedron Letters*, **40**, 6253.
1135. Paintner, F.F., Allmendinger, L. and Bauschke, G. (2001) *Synthesis*, 2113.
1136. Martin, C., Macintosh, N., Lamb, N. and Fallis, A.G. (2001) *Organic Letters*, **3**, 1021.
1137. Bueno, J.M., Coteron, J.M., Chiara, J.L., *et al.* (2000) *Tetrahedron Letters*, **41**, 4379.
1138. Weigelt, D., Krahmer, R., Bruschke, K., *et al.* (1999) *Tetrahedron*, **55**, 687.
1139. Legoupy, S., Crevisy, C., Guillemin, J.-C. and Gree, R. (1998) *Journal of Organometallic Chemistry*, **567**, 75.
1140. Maiti, A. and Yadav, J.S. (2001) *Synthetic Communications*, **31**, 1499.
1141. Iwamoto, O., Koshino, H., Hashizume, D. and Nagasawa, K. (2007) *Angewandte Chemie, International Edition*, **46**, 8625.
1142. Nicolaou, K.C. and Harrison, S.T. (2006) *Angewandte Chemie, International Edition*, **45**, 3256.
1143. Venkatesan, K. and Srinivasan, K.V. (2008) *Tetrahedron: Asymmetry*, **19**, 209.
1144. Skouta, R. and Li, C.-J. (2007) *Tetrahedron Letters*, **48**, 8343.
1145. Kuboki, A., Yamamoto, T., Taira, M., *et al.* (2007) *Tetrahedron Letters*, **48**, 771.
1146. Hosokawa, S., Kuroda, S., Imamura, K. and Tatsuta, K. (2006) *Tetrahedron Letters*, **47**, 6183.
1147. Ichikawa, Y., Yamaoka, T., Nakano, K. and Kotsuki, H. (2007) *Organic Letters*, **9**, 2989.
1148. Zhang, J., Wang, X., Wang, W., *et al.* (2007) *Tetrahedron*, **63**, 6990.
1149. Suzuki, K. and Takayama, H. (2006) *Organic Letters*, **8**, 4605.
1150. Kirkham, J.E.D., Lee, V. and Baldwin, J.E. (2006) *Chemical Communications*, 2863.
1151. Yokokawa, F. and Shioiri, T. (2002) *Tetrahedron Letters*, **43**, 8673.
1152. Boehm, T.L. and Showalter, H.D.H. (1996) *Journal of Organic Chemistry*, **61**, 6498.
1153. Kaluza, Z., Mostowicz, D., Dolega, G. and Wojcik, R. (2008) *Tetrahedron*, **64**, 2321.
1154. Li, W., Czarnik, A.W., Lillig, J. and Xiao, X.-Y. (2000) *Journal of Combinatorial Chemistry*, **2**, 224.
1155. Chen, J.J. and Aduda, V. (2007) *Synthetic Communications*, **37**, 3493.
1156. Moorthy, J.N., Singhal, N. and Senapati, K. (2007) *Organic & Biomolecular Chemistry*, **5**, 767.
1157. Kirsch, S. and Bach, T. (2003) *Angewandte Chemie, International Edition*, **42**, 4685.
1158. Corey, E.J. and Palani, A. (1995) *Tetrahedron Letters*, **36**, 3485.
1159. Corey, E.J. and Palani, A. (1995) *Tetrahedron Letters*, **36**, 7945.
1160. De Munari, S., Frigerio, M. and Santagostino, M. (1996) *Journal of Organic Chemistry*, **61**, 9272.
1161. More, J.D. and Finney, N.S. (2002) *Organic Letters*, **4**, 3001.
1162. More, J.D. and Finney, N.S. (2003) *Synlett*, 1307.
1163. Liu, Z., Chen, Z.-C., Zheng, Q.-G. (2003) *Organic Letters*, **5**, 3321.
1164. Schulze, A. and Giannis, A. (2004) *Advanced Synthesis & Catalysis*, **346**, 252.
1165. Magdziak, D., Rodriguez, A.A., Van De Water, R.W. and Pettus, T.R.R. (2002) *Organic Letters*, **4**, 285.
1166. Kuboki, A., Yamamoto, T. and Ohira, S. (2003) *Chemistry Letters*, **32**, 420.
1167. Huang, Y., Zhang, J. and Pettus, T.R.R. (2005) *Organic Letters*, **7**, 5841.
1168. Nicolaou, K.C., Kang, Q., Wu, T.R., *et al.* (2010) *Journal of the American Chemical Society*, **132**, 7540.
1169. Tada, M., Ohkanda, T. and Kurabe, J. (2010) *Chemical & Pharmaceutical Bulletin*, **58**, 27.
1170. Gagnepain, J., Castet, F. and Quideau, S. (2007) *Angewandte Chemie, International Edition*, **46**, 1533.
1171. Nicolaou, K.C., Zhong, Y.L. and Baran, P.S. (2000) *Journal of the American Chemical Society*, **122**, 7596.
1172. Nicolaou, K.C., Montagnon, T., Baran, P.S. and Zhong, Y.L. (2002) *Journal of the American Chemical Society*, **124**, 2245.
1173. Nicolaou, K.C., Baran, P.S. and Zhong, Y.-L. (2000) *Journal of the American Chemical Society*, **122**, 10246.
1174. Nicolaou, K.C., Baran, P.S. and Zhong, Y.-L. (2001) *Journal of the American Chemical Society*, **123**, 3183.
1175. Nicolaou, K.C., Mathison, C.J.N. and Montagnon, T. (2004) *Journal of the American Chemical Society*, **126**, 5192.
1176. Nicolaou, K.C., Zhong, Y.-L. and Baran, P.S. (2000) *Angewandte Chemie, International Edition*, **39**, 622.

1177. Nicolaou, K.C., Baran, P.S., Zhong, Y.L., *et al.* (2002) *Journal of the American Chemical Society*, **124**, 2233.
1178. Nicolaou, K.C., Baran, P.S., Kranich, R., *et al.* (2001) *Angewandte Chemie, International Edition*, **40**, 202.
1179. Nicolaou, K.C., Baran, P.S., Zhong, Y.-L. and Vega, J.A. (2000) *Angewandte Chemie, International Edition*, **39**, 2525.
1180. Bradshaw, B., Etxebarria-Jardi, G. and Bonjoch, J. (2010) *Journal of the American Chemical Society*, **132**, 5966.
1181. Janza, B. and Studer, A. (2005) *Journal of Organic Chemistry*, **70**, 6991.
1182. Das, B., Holla, H., Mahender, G., *et al.* (2004) *Tetrahedron Letters*, **45**, 7347.
1183. Fontaine, P., Chiaroni, A., Masson, G. and Zhu, J. (2008) *Organic Letters*, **10**, 1509.
1184. Ngouansavanh, T. and Zhu, J. (2006) *Angewandte Chemie, International Edition*, **45**, 3495.
1185. Ngouansavanh, T. and Zhu, J. (2007) *Angewandte Chemie, International Edition*, **46**, 5775.
1186. Kirsch, S.F. (2005) *Journal of Organic Chemistry*, **70**, 10210.
1187. Crone, B. and Kirsch, S.F. (2006) *Chemical Communications*, 764.
1188. Panarese, J.D. and Waters, S.P. (2010) *Organic Letters*, **12**, 4086.
1189. Neumann, J. and Thiem, J. (2010) *European Journal of Organic Chemistry*, 900.
1190. Deshmukh, S.S., Huddar, S.N., Bhalerao, D.S. and Akamanchi, K.G. (2010) *ARKIVOC*, (ii), 118.
1191. Barontini, M., Bernini, R., Crisante, F. and Fabrizi, G. (2010) *Tetrahedron*, **66**, 6047.
1192. Bernini, R., Barontini, M. and Spatafora, C. (2009) *Molecules*, **14**, 4669.
1193. Bernini, R., Barontini, M., Crisante, F., *et al.* (2009) *Tetrahedron Letters*, **50**, 6519.
1194. Souto, A., Rodriguez, J. and Jimenez, C. (2009) *Tetrahedron Letters*, **50**, 7395.
1195. Chen, J.-M. and Zeng, X.-M. (2009) *Synthetic Communications*, **39**, 3521.
1196. Moorthy, J.N., Senapati, K. and Kumar, S. (2009) *Journal of Organic Chemistry*, **74**, 6287.
1197. Yadav, J.S., Subba Reddy, B.V., Singh, A.P. and Basak, A.K. (2008) *Tetrahedron Letters*, **49**, 5880.
1198. Drouet, F., Fontaine, P., Masson, G. and Zhu, J. (2009) *Synthesis*, 1370.
1199. Chiampanichayakul, S., Pohmakotr, M., Reutrakul, V., *et al.* (2008) *Synthesis*, 2045.
1200. Kuhakarn, C., Panchan, W., Chiampanichayakul, S., *et al.* (2009) *Synthesis*, 929.
1201. Shen, L., Zhang, J., Cao, S., *et al.* (2008) *Synlett*, 3058.
1202. Yadav, J.S., Reddy, B.V.S. and Krishna, B.B.M. (2008) *Synthesis*, 3779.
1203. Yadav, L.D.S. and Awasthi, C. (2009) *Tetrahedron Letters*, **50**, 715.
1204. Yadav, J.S., Reddy, B.V.S., Singh, A.P. and Basak, A.K. (2007) *Tetrahedron Letters*, **48**, 7546.
1205. Yadav, J.S., Reddy, B.V.S., Singh, A.P. and Basak, A.K. (2007) *Tetrahedron Letters*, **48**, 4169.
1206. Du, X., Chen, H. and Liu, Y. (2008) *Chemistry–A European Journal*, **14**, 9495.
1207. Anwar, H.F. and Hansen, T.V. (2008) *Tetrahedron Letters*, **49**, 4443.
1208. Yadav, J.S., Reddy, B.V.S., Reddy, C.S. and Krishna, A.D. (2007) *Synthesis*, 693.
1209. Ma, H.C. and Jiang, X.Z. (2007) *Synthesis*, 412.
1210. Bhalerao, D.S., Mahajan, U.S., Chaudhari, K.H. and Akamanchi, K.G. (2007) *Journal of Organic Chemistry*, **72**, 662.
1211. Boeckman, R.K., Shao, P. and Mullins, J.J. (2000) *Organic Syntheses*, **77**, 141.
1212. Speicher, A., Bomm, V. and Eicher, T. (1996) *Journal für Praktische Chemie*, **338**, 588.
1213. Chaudhari, S.S. (2000) *Synlett*, 278.
1214. Meyer, S.D. and Schreiber, S.L. (1994) *Journal of Organic Chemistry*, **59**, 7549.
1215. Satoh, T., Nakamura, A., Iriuchijima, A., *et al.* (2001) *Tetrahedron*, **57**, 9689.
1216. Roels, J. and Metz, P. (2001) *Synlett*, 789.
1217. Lawrence, N.J., Crump, J.P., McGown, A.T. and Hadfield, J.A. (2001) *Tetrahedron Letters*, **42**, 3939.
1218. Wang, Z., Gu, Y., Zapata, A.J. and Hammond, G.B. (2001) *Journal of Fluorine Chemistry*, **107**, 127.
1219. Paquette, L.A. and Tae, J. (2001) *Journal of the American Chemical Society*, **123**, 4974.
1220. Candela Lena, J.I., Martin Hernando, J.I., Rico Ferreira, M.d.R., *et al.* (2001) *Synlett*, 597.
1221. Paquette, L.A., Tae, J., Branan, B.M., *et al.* (2000) *Journal of Organic Chemistry*, **65**, 9172.
1222. Heck, R., Henderson, A.P., Kohler, B., *et al.* (2001) *European Journal of Organic Chemistry*, 2623.
1223. Clive, D.L.J. and Zhang, J. (1999) *Tetrahedron*, **55**, 12059.
1224. Myers, A.G., Zhong, B., Movassaghi, M., *et al.* (2000) *Tetrahedron Letters*, **41**, 1359.
1225. Bonnet, D., Joly, P., Gras-Masse, H. and Melnyk, O. (2001) *Tetrahedron Letters*, **42**, 1875.

1226. Smith, A.B. 3rd, Liu, H. and Hirschmann, R. (2000) *Organic Letters*, **2**, 2037.
1227. Yu, C. and Hu, L. (2001) *Tetrahedron Letters*, **42**, 5167.
1228. Harry-O'kuru, R.E., Smith, J.M. and Wolfe, M.S. (1997) *Journal of Organic Chemistry*, **62**, 1754.
1229. Robins, M.J., Guo, Z. and Wnuk, S.F. (1997) *Journal of the American Chemical Society*, **119**, 3637.
1230. Wnuk, S.F., Yuan, C.-S., Borchardt, R.T., *et al.* (1997) *Journal of Medicinal Chemistry*, **40**, 1608.
1231. Robins, M.J., Sarker, S., Samano, V. and Wnuk, S.F. (1997) *Tetrahedron*, **53**, 447.
1232. Kumamoto, H., Ogamino, J., Tanaka, H., *et al.* (2001) *Tetrahedron*, **57**, 3331.
1233. Rahmeier, L.H.S. and Comasseto, J.V. (1997) *Organometallics*, **16**, 651.
1234. Gueguen, C., O'Brien, P., Powell, H.R., *et al.* (1998) *Journal of the Chemical Society, Perkin Transactions* **1**, 3405.
1235. Wavrin, L. and Viala, J. (2002) *Synthesis*, 326.
1236. Tanaka, Y., Ishihara, T. and Konno, T. (2012) *Journal of Fluorine Chemistry*, **137**, 99.
1237. Cox, R.J., Hadfield, A.T. and Mayo-Martin, M.B. (2001) *Chemical Communications*, 1710.
1238. Rocaboy, C., Bauer, W. and Gladysz, J.A. (2000) *European Journal of Organic Chemistry*, 2621.
1239. Leveque, L., Le Blanc, M. and Pastor, R. (1998) *Tetrahedron Letters*, **39**, 8857.
1240. He, Z., Trinchera, P., Adachi, S., *et al.* (2012) *Angewandte Chemie, International Edition*, **51**, 11092.
1241. Davis, F.A., Srirajan, V. and Titus, D.D. (1999) *Journal of Organic Chemistry*, **64**, 6931.
1242. Botuha, C., Haddad, M. and Larcheveque, M. (1998) *Tetrahedron: Asymmetry*, **9**, 1929.
1243. Barrett, A.G.M., Hamprecht, D. and Ohkubo, M. (1997) *Journal of Organic Chemistry*, **62**, 9376.
1244. Clive, D.L.J. and Yeh, V.S.C. (1999) *Tetrahedron Letters*, **40**, 8503.
1245. Wipf, P. and Jung, J.-K. (1998) *Journal of Organic Chemistry*, **63**, 3530.
1246. Clive, D.L.J. and Hisaindee, S. (1999) *Chemical Communications*, 2251.
1247. Paterson, I., Florence, G.J., Gerlach, K., *et al.* (2001) *Journal of the American Chemical Society*, **123**, 9535.
1248. Jeong, J.U., Guo, C. and Fuchs, P.L. (1999) *Journal of the American Chemical Society*, **121**, 2071.
1249. Kita, Y., Higuchi, K., Yoshida, Y., *et al.* (2001) *Journal of the American Chemical Society*, **123**, 3214.
1250. Larsen, D.S. and O'Shea, M.D. (1996) *Journal of Organic Chemistry*, **61**, 5681.
1251. Niu, C., Pettersson, T. and Miller, M.J. (1996) *Journal of Organic Chemistry*, **61**, 1014.
1252. Shimizu, Y., Shi, S.-L., Usuda, H., *et al.* (2010) *Tetrahedron*, **66**, 6569.
1253. Overman, L.E. and Rosen, M.D. (2010) *Tetrahedron*, **66**, 6514.
1254. McGowan, M.A., Stevenson, C.P., Schiffler, M.A. and Jacobsen, E.N. (2010) *Angewandte Chemie, International Edition*, **49**, 6147.
1255. Hanessian, S., Focken, T., Mi, X., *et al.* (2010) *Journal of Organic Chemistry*, **75**, 5601.
1256. Nicolaou, K.C., Tang, Y. and Wang, J. (2007) *Chemical Communications*, 1922.
1257. Trost, B.M., Waser, J. and Meyer, A. (2007) *Journal of the American Chemical Society*, **129**, 14556.
1258. Veitch, G.E., Beckmann, E., Burke, B.J., *et al.* (2007) *Angewandte Chemie, International Edition*, **46**, 7633.
1259. Scheerer, J.R., Lawrence, J.F., Wang, G.C. and Evans, D.A. (2007) *Journal of the American Chemical Society*, **129**, 8968.
1260. de Vicente, J., Huckins, J.R. and Rychnovsky, S.D. (2006) *Angewandte Chemie, International Edition*, **45**, 7258.
1261. Yadav, J.S., Das, S.K. and Sabitha, G. (2012) *Journal of Organic Chemistry*, **77**, 11109.
1262. Neubauer, T., Kammerer-Pentier, C. and Bach, T. (2012) *Chemical Communications*, 11629.
1263. Peng, F. and Danishefsky, S.J. (2012) *Journal of the American Chemical Society*, **134**, 18860.
1264. Rodriguez, A.R. and Spur, B.W. (2012) *Tetrahedron Letters*, **53**, 6990.
1265. Yadav, J.S., Dhara, S., Hossain, S.S. and Mohapatra, D.K. (2012) *Journal of Organic Chemistry*, **77**, 9628.
1266. Fuwa, H., Ishigai, K., Hashizume, K. and Sasaki, M. (2012) *Journal of the American Chemical Society*, **134**, 11984.
1267. Canham, S.M., France, D.J. and Overman, L.E. (2013) *Journal of Organic Chemistry*, **78**, 9.
1268. Glaus, F. and Altmann, K.-H. (2012) *Angewandte Chemie, International Edition*, **51**, 3405.
1269. Tang, W. and Prusov, E.V. (2012) *Angewandte Chemie, International Edition*, **51**, 3401.
1270. Paterson, I., Maltas, P., Dalby, S.M., *et al.* (2012) *Angewandte Chemie, International Edition*, **51**, 2749.
1271. Deng, J., Zhu, B., Lu, Z., *et al.* (2012) *Journal of the American Chemical Society*, **134**, 920.
1272. Yamamoto, A., Ueda, A., Bremond, P., *et al.* (2012) *Journal of the American Chemical Society*, **134**, 893.
1273. Nicolaou, K.C., Jung, J., Yoon, W.H., *et al.* (2002) *Journal of the American Chemical Society*, **124**, 2183.

1274. Nicolaou, K.C., Baran, P.S., Zhong, Y.L., *et al.* (2002) *Journal of the American Chemical Society*, **124**, 2190.
1275. Nicolaou, K.C., Zhong, Y.L., Baran, P.S., *et al.* (2002) *Journal of the American Chemical Society*, **124**, 2202.
1276. Andreou, T., Bures, J. and Vilarrasa, J. (2010) *Tetrahedron Letters*, **51**, 1863.
1277. Nicolaou, K.C., Baran, P.S., Zhong, Y.L. and Sugita, K. (2002) *Journal of the American Chemical Society*, **124**, 2212.
1278. Yadav, J.S., Reddy, B.V.S., Singh, *et al.* (2008) *Synthesis*, 469.
1279. Ma, H.C. and Jiang, X.Z. (2007) *Synlett*, 1679.
1280. Mahajan, U.S. and Akamanchi, K.G. (2008) *Synlett*, 987.
1281. Gamapwar, S.V., Tale, N.P. and Karade, N.N. (2012) *Synthetic Communications*, **42**, 2617.
1282. Chaudhari, S.S. and Akamanchi, K.G. (1998) *Tetrahedron Letters*, **39**, 3209.
1283. Chaudhari, S.S. and Akamanchi, K.G. (1999) *Synthesis*, 760.
1284. Bose, D.S. and Narsaiah, A.V. (1999) *Synthetic Communications*, **29**, 937.
1285. Jenkins, N.E., Ware, R.W. Jr., Atkinson, R.N. and King, S.B. (2000) *Synthetic Communications*, **30**, 947.
1286. Beavington, R., Rees, P.A. and Burn, P.L. (1998) *Journal of the Chemical Society, Perkin Transactions* **1**, 2847.
1287. Beavington, R. and Burn, P.L. (1999) *Journal of the Chemical Society, Perkin Transactions* **1**, 583.
1288. Promarak, V. and Burn, P.L. (2001) *Journal of the Chemical Society, Perkin Transactions* **1**, 14.
1289. Nicolaou, K.C., Sugita, K., Baran, P.S., *et al.* (2001) *Angewandte Chemie, International Edition*, **40**, 207.
1290. Nicolaou, K.C., Sugita, K., Baran, P.S. and Zhong, Y.L. (2002) *Journal of the American Chemical Society*, **124**, 2221.
1291. Nicolaou, K.C., Zhong, Y.-L., Baran, P.S. and Sugita, K. (2001) *Angewandte Chemie, International Edition*, **40**, 2145.
1292. Nicolaou, K.C. and Mathison, C.J.N. (2005) *Angewandte Chemie, International Edition*, **44**, 5992.
1293. Choghamarani, A.G. (2006) *Synlett*, 2347.
1294. Patil, B.R., Bhusare, S.R., Pawar, R.P. and Vibhute, Y.B. (2005) *Tetrahedron Letters*, **46**, 7179.
1295. Krassowska-Swiebocka, B., Prokopienko, G. and Skulski, L. (2005) *Molecules*, **10**, 394.
1296. Wirth, H.O., Konigstein, O. and Kern, W. (1960) *Justus Liebigs Annalen der Chemie*, **634**, 84.
1297. Lakouraj, M.M., Tajbakhsh, M., Shirini, F. and Tamami, M.V.A. (2005) *Synthetic Communications*, **35**, 775.
1298. Salgaonkar, P., Shukla, V. and Akamanchi, K. (2005) *Synthetic Communications*, **35**, 2805.
1299. Hashemi, M.M. and Akhbari, M. (2003) *Monatshefte für Chemie*, **134**, 1561.
1300. Chandrasekhar, S. and Gopalaiah, K. (2002) *Tetrahedron Letters*, **43**, 4023.
1301. Lakouraj, M., Tajbakhsh, M., Shirini, F. and Tamami, M. (2005) *Phosphorus, Sulfur, and Silicon and the Related Elements*, **180**, 2423.
1302. Shirini, F., Khademian, M. and Abedini, M. (2008) *ARKIVOC*, (xv), 71.
1303. Huddar, S.N., Mahajan, U.S. and Akamanchi, K.G. (2010) *Chemistry Letters*, **39**, 808.
1304. Huddar, S.N., Deshmukh, S.S. and Akamanchi, K.G. (2011) *ARKIVOC*, (v), 67.
1305. Loskutov, V.A., Gatilov, Y.V. and Shteingarts, V.D. (2010) *Tetrahedron Letters*, **51**, 6396.
1306. Nicolaou, K.C., Montagnon, T. and Baran, P.S. (2002) *Angewandte Chemie, International Edition*, **41**, 1386.
1307. Wu, L., Yang, X. and Yan, F. (2011) *Journal of Sulfur Chemistry*, **32**, 105.
1308. Wu, J., Wu, G. and Wu, L. (2008) *Synthetic Communications*, **38**, 2367.
1309. Liu, Z.-Q., Zhao, Y., Luo, H., *et al.* (2007) *Tetrahedron Letters*, **48**, 3017.
1310. Cohen, M.J. and McNelis, E. (1984) *Journal of Organic Chemistry*, **49**, 515.
1311. Hara, S. (2001) in *e-EROS Encyclopedia of Reagents for Organic Synthesis* (editor in chief D. Crich), John Wiley & Sons, Inc., Hoboken, doi: 10.1002/047084289X.
1312. Hara, S. (2006) *Advances in Organic Synthesis*, **2**, 49.
1313. Hara, S., Monoi, M., Umemura, R. and Fuse, C. (2012) *Tetrahedron*, **68**, 10145.
1314. Fukuhara, T. and Hara, S. (2010) *Journal of Organic Chemistry*, **75**, 7393.
1315. Stengel, J.H. and McMills, M.C. (2001) in *e-EROS Encyclopedia of Reagents for Organic Synthesis* (editor in chief D. Crich), John Wiley & Sons, Inc., Hoboken, doi: 10.1002/047084289X.
1316. Kristiansen, K.A., Potthast, A. and Christensen, B.E. (2010) *Carbohydrate Research*, **345**, 1264.
1317. Perlin, A.S. (2006) *Advances in Carbohydrate Chemistry and Biochemistry*, **60**, 183.
1318. Kelly, T.R., Schmidt, T.E. and Haggerty, J.G. (1972) *Synthesis*, 544.

1319. Nagarkatti, J.P. and Ashley, K.R. (1973) *Tetrahedron Letters*, **14** (46), 4599.
1320. Lee, S. and Fuchs, P.L. (2002) *Journal of the American Chemical Society*, **124**, 13978.
1321. Xu, L. and Trudell, M.L. (2003) *Tetrahedron Letters*, **44**, 2553.
1322. Hunsen, M. (2005) *Tetrahedron Letters*, **46**, 1651.
1323. Hunsen, M. (2005) *Journal of Fluorine Chemistry*, **126**, 1356.
1324. Zhang, S., Xu, L. and Trudell, M.L. (2005) *Synthesis*, 1757.
1325. Asadolah, K., Heravi, M.M., Hekmatshoar, R. and Majedi, S. (2007) *Molecules*, **12**, 958.
1326. Zhao, M., Li, J., Song, Z., *et al.* (1998) *Tetrahedron Letters*, **39**, 5323.
1327. Bekish, A.V. (2012) *Tetrahedron Letters*, **53**, 3082.
1328. Ramakrishna, D. and Bhat, B.R. (2011) *Inorganic Chemistry Communications*, **14**, 690.
1329. Babu, S.G., Priyadarsini, P.A. and Karvembu, R. (2011) *Applied Catalysis A: General*, **392**, 218.
1330. Zolfigol, M.A., Shirini, F., Chehardoli, G. and Kolvari, E. (2007) *Journal of Molecular Catalysis A: Chemical*, **265**, 272.
1331. Kim, S.S. and Nehru, K. (2002) *Synlett*, 616.
1332. Yamazaki, S. (1999) *Organic Letters*, **1**, 2129.
1333. Yamazaki, S. (2001) *Tetrahedron Letters*, **42**, 3355.
1334. Kim, S.S., Nehru, K., Kim, S.S., *et al.* (2002) *Synthesis*, 2484.
1335. Xu, L., Cheng, J. and Trudell, M.L. (2003) *Journal of Organic Chemistry*, **68**, 5388.
1336. Suzuki, H., Nakamura, K. and Goto, R. (1966) *Bulletin of the Chemical Society of Japan*, **39**, 128.
1337. Kraszkiewicz, L., Sosnowski, M. and Skulski, L. (2006) *Synthesis*, 1195.
1338. Sosnowski, M. and Skulski, L. (2005) *Molecules*, **10**, 401.
1339. Sosnowski, M., Skulski, L. and Wolowik, K. (2004) *Molecules*, **9**, 617.
1340. Kraszkiewicz, L., Sosnowski, M. and Skulski, L. (2004) *Tetrahedron*, **60**, 9113.
1341. Lulinski, P. and Skulski, L. (2000) *Bulletin of the Chemical Society of Japan*, **73**, 951.
1342. Masuda, H., Takase, K., Nishio, M., *et al.* (1994) *Journal of Organic Chemistry*, **59**, 5550.
1343. Shi, X.-X., Khanapure, S.P. and Rokach, J. (1996) *Tetrahedron Letters*, **37**, 4331.
1344. Li, Z., Ding, R.-B., Xing, Y.-L., Shi, S.-Y. (2005) *Synthetic Communications*, **35**, 2515.
1345. Roth, S. and Stark, C.B.W. (2008) *Chemical Communications*, 6411.
1346. Ghorbani-Choghamarani, A., Zolfigol, M.A., Hajjami, M. and Sardari, S. (2013) *Synthetic Communications*, **43**, 52.
1347. Mereyala, H.B., Pathuri, G. and Nagarapu, L. (2012) *Synthetic Communications*, **42**, 1278.
1348. Kress, S., Weckesser, J., Schulz, S.R. and Blechert, S. (2013) *European Journal of Organic Chemistry*, **2013**, 1346.
1349. Kosaki, Y., Ogawa, N. and Kobayashi, Y. (2010) *Tetrahedron Letters*, **51**, 1856.
1350. Ogikubo, J., Worlinsky, J.L., Fu, Y.-J. and Bruckner, C. (2013) *Tetrahedron Letters*, **54**, 1707.
1351. Van Brabandt, W., Vanwalleghem, M., D'Hooghe, M. and De Kimpe, N. (2006) *Journal of Organic Chemistry*, **71**, 7083.
1352. Mollet, K., Goossens, H., Piens, N., *et al.* (2013) *Chemistry–A European Journal*, **19**, 3383.
1353. Doboszewski, B. and Herdewijn, P. (2011) *Tetrahedron Letters*, **52**, 3853.
1354. Johnson, C.R., Golebiowski, A. and Steensma, D.H. (1992) *Journal of the American Chemical Society*, **114**, 9414.
1355. Takano, S., Kurotaki, A. and Ogasawara, K. (1987) *Synthesis*, 1075.
1356. Majee, A., Kundu, S.K., Santra, S. and Hajra, A. (2012) *Tetrahedron Letters*, **53**, 4433.
1357. Zolfigol, M.A., Hajjami, M. and Ghorbani-Choghamarani, A. (2011) *Bulletin of Korean Chemical Society*, **32**, 4191.
1358. Chouthaiwale, P.V., Karabal, P.U., Suryavanshi, G. and Sudalai, A. (2010) *Synthesis*, 3879.
1359. Kamble, D.A., Karabal, P.U., Chouthaiwale, P.V. and Sudalai, A. (2012) *Tetrahedron Letters*, **53**, 4195.
1360. Takale, B.S. and Telvekar, V.N. (2010) *Chemistry Letters*, **39**, 1279.
1361. Gerack, C.J. and McElwee-White, L. (2012) *Chemical Communications*, 11310.
1362. Yu, W., Mei, Y., Kang, Y., *et al.* (2004) *Organic Letters*, **6**, 3217.
1363. Nielsen, T.E. and Meldal, M. (2005) *Organic Letters*, **7**, 2695.
1364. Nielsen, T.E., Le Quement, S.T. and Meldal, M. (2007) *Organic Letters*, **9**, 2469.
1365. Chang, M.-Y., Tai, H.-Y., Chen, Y.-L. and Hsu, R.-T. (2012) *Tetrahedron*, **68**, 7941.

1366. Chang, M.-Y., Chan, C.-K., Lin, S.-Y. and Hsu, R.-T. (2012) *Tetrahedron*, **68**, 10272.
1367. Chang, M.-Y., Wu, M.-H., Lee, N.-C. and Lee, M.-F. (2012) *Tetrahedron Letters*, **53**, 2125.
1368. Kim, S., Chung, J. and Kim, B.M. (2011) *Tetrahedron Letters*, **52**, 1363.
1369. Hung, K.-y., Harris, P.W.R. and Brimble, M.A. (2010) *Journal of Organic Chemistry*, **75**, 8728.
1370. Rup, S., Sindt, M. and Oget, N. (2010) *Tetrahedron Letters*, **51**, 3123.
1371. Araghi, M. and Bokaei, F. (2013) *Polyhedron*, **53**, 15.
1372. Zhou, M., Hintermair, U., Hashiguchi, B.G., *et al.* (2013) *Organometallics*, **32**, 957.
1373. Zakavi, S. and Mokary Yazdeli, T. (2013) *Journal of Molecular Catalysis A: Chemical*, **367**, 108.
1374. Huang, J., Fu, X., Wang, G., *et al.* (2012) *Dalton Transactions*, **41**, 10661.
1375. Zakeri, M., Moghadam, M., Mohammadpoor-Baltork, I., *et al.* (2012) *Journal of Coordination Chemistry*, **65**, 1144.
1376. Kargar, H., Moghadam, M., Mirkhani, V., *et al.* (2011) *Bioorganic & Medicinal Chemistry Letters*, **21**, 2146.
1377. Kargar, H., Moghadam, M., Mirkhani, V., *et al.* (2011) *Polyhedron*, **30**, 1463.
1378. Kargar, H. (2011) *Inorganic Chemistry Communications*, **14**, 863.

4

Hypervalent Iodine Catalysis

It has been recognized for many years that there are similarities between hypervalent iodine compounds and transition metal organic complexes. The hypervalent bond is highly polarized and it is longer and weaker than a regular covalent bond and the presence of hypervalent bonding leads to special structural features and reactivity pattern characteristic of polyvalent iodine compounds (Chapter 1). The reactions of hypervalent molecules are commonly discussed in terms of oxidative addition, reductive elimination, ligand exchange and ligand coupling, which are typical of transition metal chemistry [1]. However, catalytic reactions, typical of transition metals, remained unknown for hypervalent iodine compounds until the beginning of twenty-first century.

In 2005, Kita and Ochiai independently reported the catalytic use of aryl iodides, in the presence of stoichiometric *m*-chloroperoxybenzoic acid (*m*CPBA), to perform oxidative dearomatization and biarylation of phenolic substrates [2], or α-acetoxylation of carbonyl compounds [3]. These reactions involved selective generation of the highly reactive hypervalent iodine(III) species *in situ* from aryl iodide and terminal oxidant. The first examples of catalytic application of the iodine(V) species in the oxidation of alcohols using Oxone® ($2KHSO_5{\cdot}KHSO_4{\cdot}K_2SO_4$) as a stoichiometric oxidant at 70 °C were independently reported by the groups of Vinod [4] in 2005 and Giannis [5] in 2006. These initial reports were quickly followed by the discovery of numerous other reactions based on the iodine(I)/iodine(III), iodine(I)/iodine(V) and iodide anion/hypoiodite catalytic cycles. Several examples of reactions catalyzed by hypervalent iodine species in the presence of a co-catalyst (e.g., TEMPO, ruthenium salts, or metalloporphyrins) have also been reported. While chemical reactions catalyzed by iodine species were discovered only in 2005, the electrochemical generation of iodine(III) species *in situ* from catalytic amounts of iodoarenes (0.05–0.2 equiv) and the use of these species as the in-cell mediators in electrochemical fluorination reactions, have been known since 1994 (Section 3.1.1) [6]. Reactions catalyzed by hypervalent iodine species have been overviewed in several review articles [7–15].

4.1 Catalytic Cycles Based on Iodine(III) Species

Catalytic reactions of this type usually involve the reoxidation of iodoarene to aryliodine(III) species *in situ* using oxidants such as peroxycarboxylic acids, hydrogen peroxide, sodium perborate, or Oxone at room temperature. The choice of oxidant is critically important; the oxidant must not react with the substrate, as the substrate should only be oxidized by the hypervalent iodine species. The stoichiometric oxidant has to be carefully selected to achieve the re-oxidation of the iodine compound under homogeneous reaction

Hypervalent Iodine Chemistry: Preparation, Structure and Synthetic Applications of Polyvalent Iodine Compounds, First Edition. Viktor V. Zhdankin.

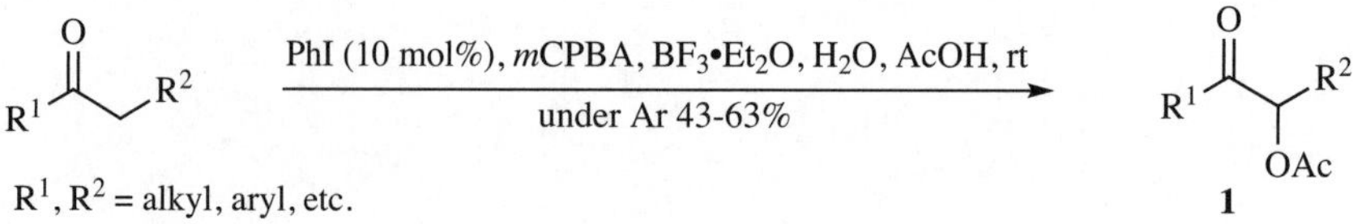

R^1, R^2 = alkyl, aryl, etc.

PhI (10 mol%), *m*CPBA, BF_3•Et_2O, H_2O, AcOH, rt under Ar 43-63%

Scheme 4.1

conditions. The nature of the aryl iodide is also important. Most commonly, iodobenzene is used as the catalyst; however, numerous other iodoarenes have also been tested in these reactions. Particularly important are the enantioselective reactions catalyzed by chiral iodoarenes [11, 16].

4.1.1 Oxidative α-Functionalization of Carbonyl Compounds

In 2005 Ochiai and coworkers reported the first iodobenzene-catalyzed reaction, a catalytic variant of α-acetoxylation of ketones based on the *in situ* generation of (diacetoxyiodo)benzene from iodobenzene using *m*-chloroperoxybenzoic acid (*m*CPBA) as a terminal oxidant [3]. In a typical example, the oxidation of a ketone with *m*CPBA (2 equiv) in acetic acid in the presence of a catalytic amount of iodobenzene (0.1 equiv), $BF_3{\cdot}OEt_2$ (3 equiv) and water (5 equiv) at room temperature under argon affords the respective α-acetoxy-ketone **1** in a moderate yield (Scheme 4.1). 4-Iodotoluene and 4-chloroiodobenzene can also serve as catalysts in the α-acetoxylation of ketones under these reaction conditions; however, the use of iodobenzene results in the highest yields [3]. The use of at least 10 mol% iodobenzene in this reaction is necessary. When smaller amounts are used, the reaction slows and Baeyer–Villiger oxidation products resulting from a direct reaction of *m*CPBA and the ketone are observed [3].

Scheme 4.2 shows a catalytic cycle for this oxidation. Boron trifluoride etherate accelerates the initial oxidation of iodobenzene to (diacyloxyiodo)benzene by *m*CPBA in the presence of acetic acid. Ligand

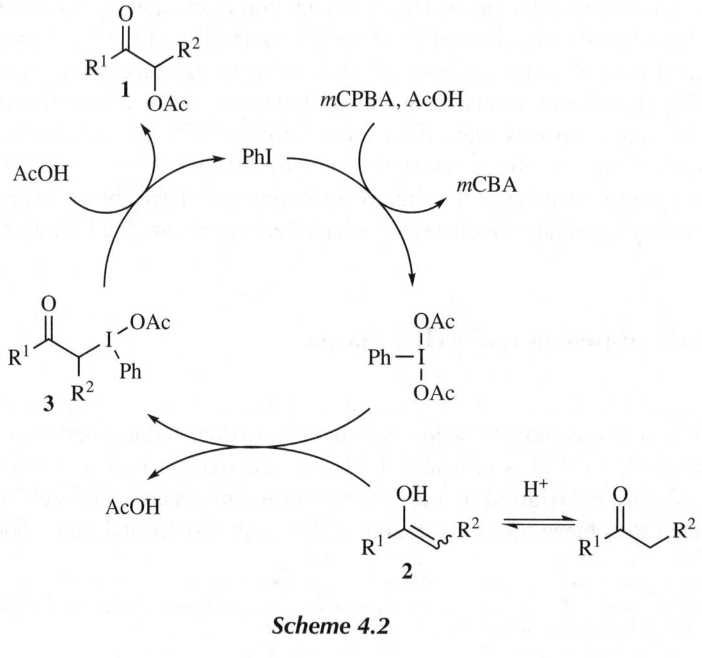

*m*CPBA, AcOH

PhI

AcOH

*m*CBA

AcOH

H^+

Scheme 4.2

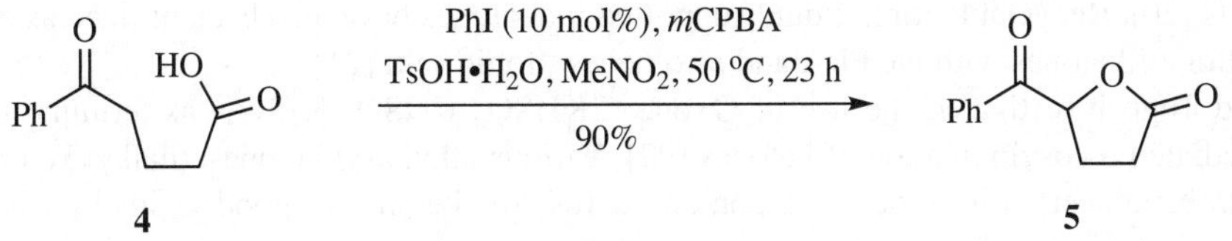

PhI (10 mol%), *m*CPBA
TsOH•H_2O, $MeNO_2$, 50 °C, 23 h
90%

4 → **5**

Scheme 4.3

exchange of $PhI(OAc)_2$ with enol **2** derived from a ketone produces an alkyliodonium intermediate (**3**), which on S_N2 displacement by acetic acid affords an α-acetoxy-ketone (**1**) with liberation of iodobenzene [3].

A study of the catalytic α-acetoxylation reaction of acetophenone by electrospray ionization tandem mass spectrometry (ESI-MS/MS) has confirmed the mechanism shown in Scheme 4.2. In particular, the trivalent iodine species was detected when iodobenzene and *m*CPBA in acetic acid were mixed, which indicated the facile oxidation of a catalytic amount of PhI to the iodine(III) species by *m*CPBA. Most importantly, the protonated alkyliodonium intermediate **3** (R^1 = Ph, R^2 = H) was observed at *m/z* 383 from the reaction solution and this ion gave the protonated α-acetoxylation product **1** at *m/z* 179 in MS/MS by an intramolecular reductive elimination of PhI [17].

Based on Ochiai's procedure for α-acetoxylation of ketones, Ishihara and coworkers have developed the hypervalent iodine-catalyzed oxylactonization of ketocarboxylic acids to ketolactones [18]. Optimized reaction conditions consist of the treatment of a ketocarboxylic acid with iodobenzene (10 mol%), *p*-toluenesulfonic acid monohydrate (20 mol%) and *m*CPBA as a stoichiometric oxidant in nitromethane solution; Scheme 4.3 shows as a representative example the cyclization of ketocarboxylic acid **4** to ketolactone **5.**

As a further extension of the Ochiai's procedure for α-acetoxylation, the catalytic procedure for α-tosyloxylation of ketones using *m*CPBA as stoichiometric oxidant and iodoarenes as catalysts in the presence of *p*-toluenesulfonic acid has been developed. Various α-tosyloxyketones **7** can be efficiently prepared in high yields from the reaction of ketones **6** with *m*CPBA (1.1 equiv) and *p*-toluenesulfonic acid (1.1 equiv) in the presence of a catalytic amount of iodobenzene with moderate warming (Scheme 4.4) [19]. The mechanism of this reaction involves initial oxidation of PhI by *m*CPBA in the presence of *p*-toluenesulfonic acid to generate [hydroxy(tosyloxy)iodo]benzene *in situ,* which then reacts with the enol form of ketone to give the α-tosyloxyketone.

Further modification of this reaction (Scheme 4.4) involves the use of polystyrene-supported iodobenzene as a recyclable catalyst, which can be recovered by simple filtration of the reaction mixture and reused (Section 5.5) [20]. Alcohols can be used instead of ketones for the preparation of α-tosyloxyketones **7;** in this case an excess of *m*CPBA (2.1 equiv) is employed in the presence of KBr (0.1 equiv) and PhI or poly(4-iodostyrene)

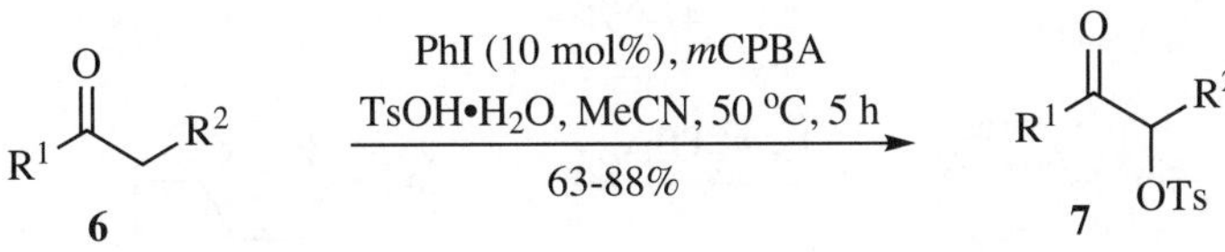

PhI (10 mol%), *m*CPBA
TsOH•H_2O, MeCN, 50 °C, 5 h
63-88%

6 → **7**

R^1 = Ph, 4-MeC_6H_4, 4-ClC_6H_4, 4-$NO_2C_6H_4$, Et, C_5H_{11}
R^2 = H, Me, Bu, C_7H_{15}
or $R^1 + R^2 = (CH_2)_5$

Scheme 4.4

as the catalysts [20]. Recyclable ionic-liquid supported iodoarenes have also been utilized as catalysts in the α-tosyloxylation of ketones with *m*CPBA and *p*-toluenesulfonic acid [21].

Tanaka and Togo investigated the use of Oxone ($2KHSO_5{\cdot}KHSO_4{\cdot}K_2SO_4$) as terminal oxidant in the iodoarene-mediated α-tosyloxylation of ketones [22]. Various alkyl aryl ketones, dialkyl ketones and cycloheptanone can be converted into the corresponding α-tosyloxyketones in good yields by using Oxone and *p*-toluenesulfonic acid monohydrate in the presence of *p*-iodotoluene in acetonitrile. In these reactions, *p*-iodotoluene, $4\text{-}MeC_6H_4I$, works as the catalyst and *p*-[hydroxy(tosyloxy)iodo]toluene, $4\text{-}MeC_6H_4I(OH)OTs$, is formed *in situ* as the reactive species for the α-tosyloxylation of ketones. However, one equivalent of *p*-iodotoluene was required to obtain α-tosyloxyketones in good yields [22]. The catalytic efficiency of this reaction is low, because *p*-iodotoluene is partially oxidized by Oxone to the iodine(V) species, which are not active in the α-tosyloxylation of ketones.

Togo and coworkers have found that alkyl aryl ketones and dialkyl ketones could be converted into the corresponding α-tosyloxyketones in generally low yields by the reaction with *m*CPBA and *p*-toluenesulfonic acid monohydrate in the presence of a catalytic amount of molecular iodine in a mixture of acetonitrile and 2,2,2-trifluoroethanol (method A, Scheme 4.5) [23]. The same conversion of ketones into the corresponding α-tosyloxyketones could be smoothly carried out by the reaction with *m*CPBA and $TsOH{\cdot}H_2O$ in the presence of catalytic amounts of iodine and *tert*-butylbenzene in a mixture of acetonitrile and 2,2,2-trifluoroethanol (method B). In these reactions, *p*-iodotoluene **8** (method A) and 4-*tert*-butyl-1-iodobenzene **9** (method B) are formed at first and are then converted into the corresponding [hydroxy(tosyloxy)iodo]arenes, ArI(OH)OTs,

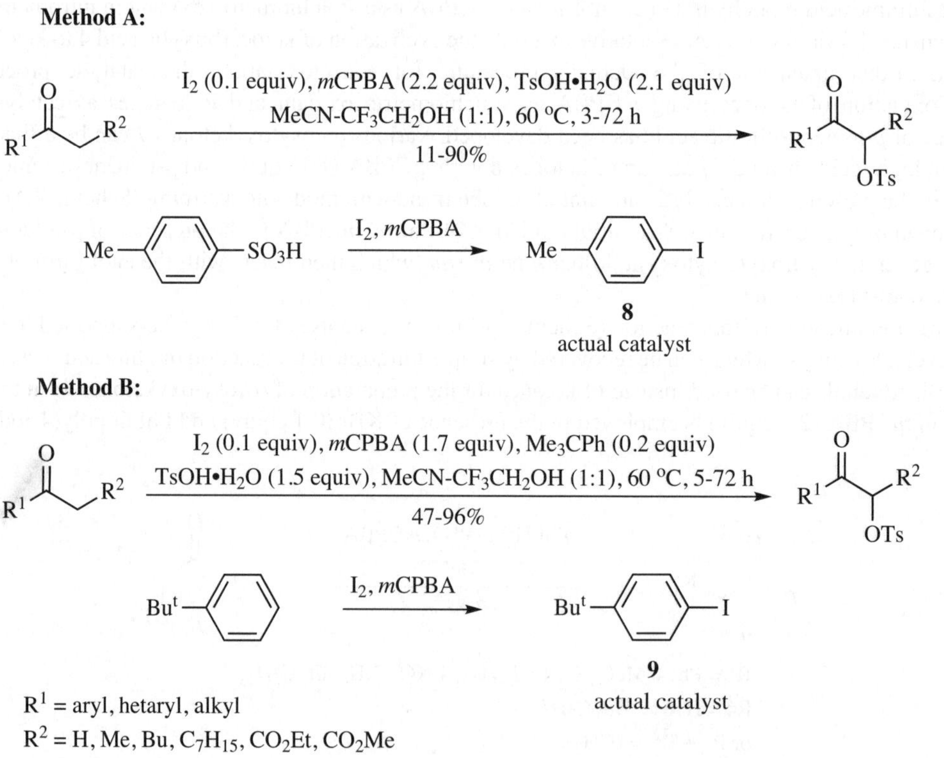

Method A:

$R^1C(O)CH_2R^2$ → $R^1C(O)CH(OTs)R^2$

I_2 (0.1 equiv), *m*CPBA (2.2 equiv), $TsOH{\bullet}H_2O$ (2.1 equiv)
MeCN-CF_3CH_2OH (1:1), 60 °C, 3-72 h
11-90%

Me–C_6H_4–SO_3H → (I_2, *m*CPBA) → Me–C_6H_4–I

8
actual catalyst

Method B:

$R^1C(O)CH_2R^2$ → $R^1C(O)CH(OTs)R^2$

I_2 (0.1 equiv), *m*CPBA (1.7 equiv), Me_3CPh (0.2 equiv)
$TsOH{\bullet}H_2O$ (1.5 equiv), MeCN-CF_3CH_2OH (1:1), 60 °C, 5-72 h
47-96%

Bu^t–C_6H_5 → (I_2, *m*CPBA) → Bu^t–C_6H_4–I

9
actual catalyst

R^1 = aryl, hetaryl, alkyl
R^2 = H, Me, Bu, C_7H_{15}, CO_2Et, CO_2Me

Scheme 4.5

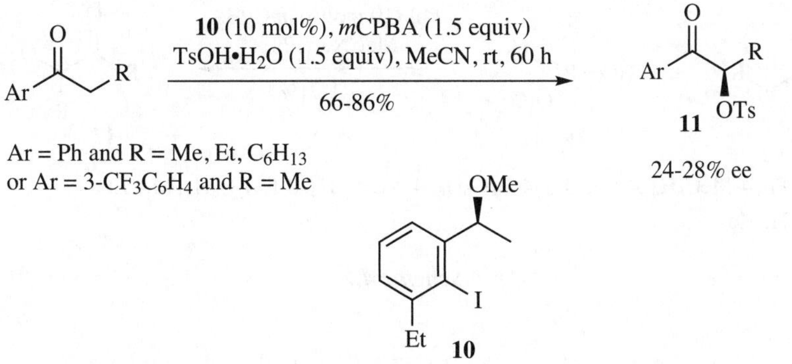

10 (10 mol%), mCPBA (1.5 equiv)
TsOH•H_2O (1.5 equiv), MeCN, rt, 60 h
66-86%

Ar = Ph and R = Me, Et, C_6H_{13}
or Ar = 3-$CF_3C_6H_4$ and R = Me

11
24-28% ee

OMe
I
Et **10**

Scheme 4.6

by reaction with mCPBA and TsOH·H_2O. [Hydroxy(tosyloxy)iodo]arenes work as the actual reagents for the α-tosyloxylation of ketones in the catalytic cycle. A similar procedure for the α-tosyloxylation of ketones employs Oxone (2 equiv) as the oxidant and I_2 (0.7 equiv) as the catalyst [24].

Several research groups have investigated the enantioselective α-oxytosylation of ketones catalyzed by chiral iodoarenes. The first reaction of this type, catalyzed by the enantiopure iodoarene **10** and resulting in the enantioenriched α-tosyloxyketones **11** (Scheme 4.6), was reported in 2007 by Wirth and coworkers [25]. The authors have tested numerous chiral iodoarenes as catalysts, but in most cases the enantioselectivity of this reaction was very low [25–27].

Higher levels of enantioselectivities in the α-oxytosylation of ketones were achieved in several recent works. Chi Zhang and coworkers have evaluated spirobiindane-based chiral iodoarenes as catalyst and were able to obtain α-tosyloxylated ketones in up to 58% enantiomeric excess using catalyst **12** (Figure 4.1) [28]. Moran and Rodriguez have prepared several chiral aryl iodides (e.g., structures **13** and **14**, Figure 4.1) and

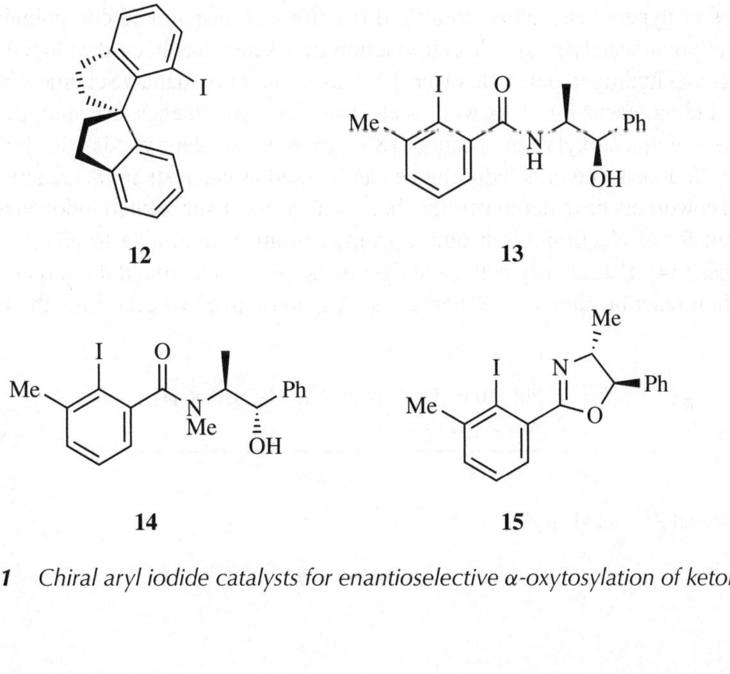

12 **13** **14** **15**

Figure 4.1 *Chiral aryl iodide catalysts for enantioselective α-oxytosylation of ketones.*

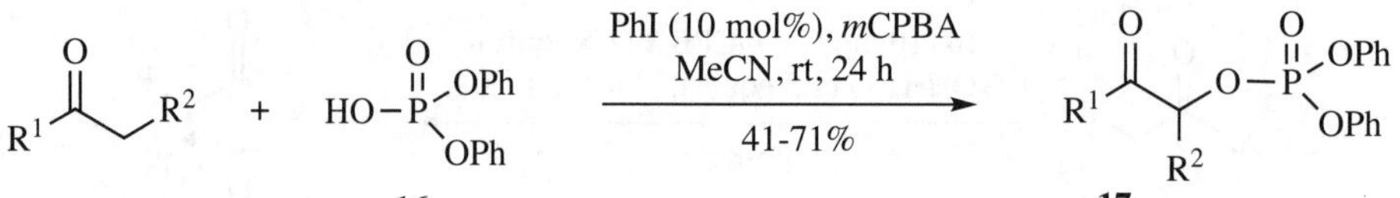

$R^1C(O)CH_2R^2$ + $HO{-}P(O)(OPh)_2$ (16) → PhI (10 mol%), mCPBA, MeCN, rt, 24 h, 41-71% → $R^1C(O)CH(R^2){-}O{-}P(O)(OPh)_2$ (17)

R^1 = Ph, 4-MeC_6H_4, 4-ClC_6H_4, 4-BrC_6H_4, 4-$NO_2C_6H_4$, 3-$NO_2C_6H_4$, Me, Et
R^2 = H, Me

Scheme 4.7

assessed them as catalysts in the enantioselective α-oxytosylation of propiophenone and in the oxidative cyclization of 5-oxo-5-phenylpentanoic acid to 5-benzoyldihydrofuran-2(3*H*)-one (Scheme 4.3) [29]. The highest enantioselectivities obtained were 18% for the α-oxytosylation using catalyst **13** and 51% ee for the oxidative cyclization using catalyst **14.** Legault and coworkers have developed a family of iodooxazoline catalysts (e.g., structure **15,** Figure 4.1) for the iodine(III)-mediated α-tosyloxylation of ketone derivatives [30]. The use of catalyst **15** (10 mol%) in dichloromethane solution allows the best levels of enantioselectivity to be achieved for this reaction (up to 54% ee). A drastic enhancement in catalytic activity was observed due to the introduction of steric hindrance *ortho* to the iodine atom of the catalyst (e.g., methyl group in catalyst **15**) [31].

Analogously to the α-oxytosylation, an effective catalytic method for the α-phosphoryloxylation of ketones has been developed [32]. When ketones are treated with phosphates **16** in the presence of iodobenzene as the catalyst and *m*CPBA as the terminal oxidant in acetonitrile at room temperature, the α-phosphoryloxylation of ketones takes place easily and the corresponding keto phosphates **17** are obtained in moderate to good yields (Scheme 4.7) [32].

4.1.2 Oxidative Functionalization of Alkenes and Alkynes

Several examples of hypervalent iodine-catalyzed reactions of unsaturated compounds have been reported. A method for the organocatalytic *syn* diacetoxylation of alkenes has been developed using aryl iodides as efficient catalysts and hydrogen peroxide or *m*CPBA as terminal oxidants (Scheme 4.8) [33]. A broad range of substrates, including electron-rich as well as electron-deficient alkenes, are smoothly transformed by this procedure, furnishing diacetoxylation products **18** in good to excellent yields with high diastereoselectivity (up to >19 : 1 dr). Iodobenzene or 4-iodotoluene can be used as catalysts in this reaction.

Braddock and coworkers have demonstrated that suitably *ortho*-substituted iodobenzenes act as organocatalysts for the transfer of electrophilic bromine from *N*-bromosuccinimide to alkenes via the intermediacy of bromoiodinanes [34]. Particularly active catalyst is the *ortho*-substituted iodoarene **19,** as illustrated by a bromolactonization reaction shown in Scheme 4.9. An alternative procedure for the bromolactonization of

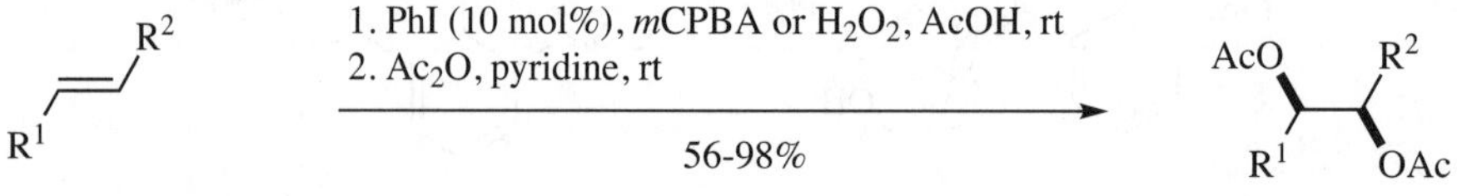

$R^1CH{=}CHR^2$ → 1. PhI (10 mol%), mCPBA or H_2O_2, AcOH, rt; 2. Ac_2O, pyridine, rt; 56-98% → $R^1CH(OAc){-}CH(OAc)R^2$ (18)

R^1 and R^2 = alkyl, aryl

Scheme 4.8

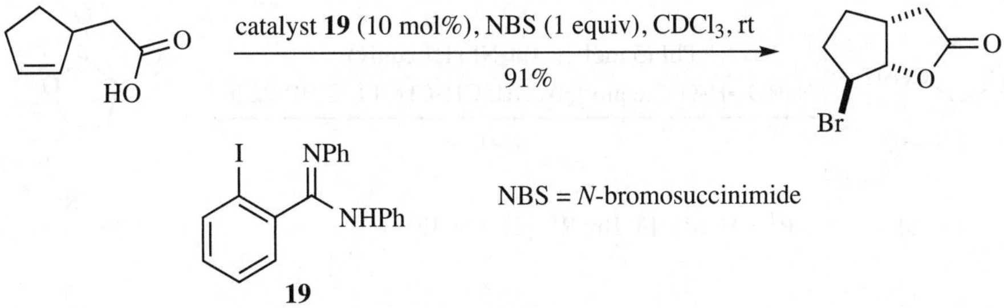

catalyst **19** (10 mol%), NBS (1 equiv), $CDCl_3$, rt
91%

HO O Br O O

I NPh NHPh **19**

NBS = *N*-bromosuccinimide

Scheme 4.9

alkenoic acids employs iodobenzene as a catalyst, sodium bromide as the source of bromine and Oxone as the terminal oxidant in CF_3CH_2OH at room temperature [35].

Based on the hypervalent iodine-catalyzed bromocarbocyclization of appropriate alkenoic precursors **21**, Gulder and coworkers have developed an efficient synthetic approach to 3,3-disubstituted oxoindoles **22** (Scheme 4.10) [36]. These cyclizations are catalyzed by 2-iodobenzamide **20** at room temperature using NBS as the source of electrophilic bromine. Alternatively, KBr can be used as the source of bromine in the presence of Oxone as a terminal oxidant. The synthetic utility of this cyclization has been demonstrated by the

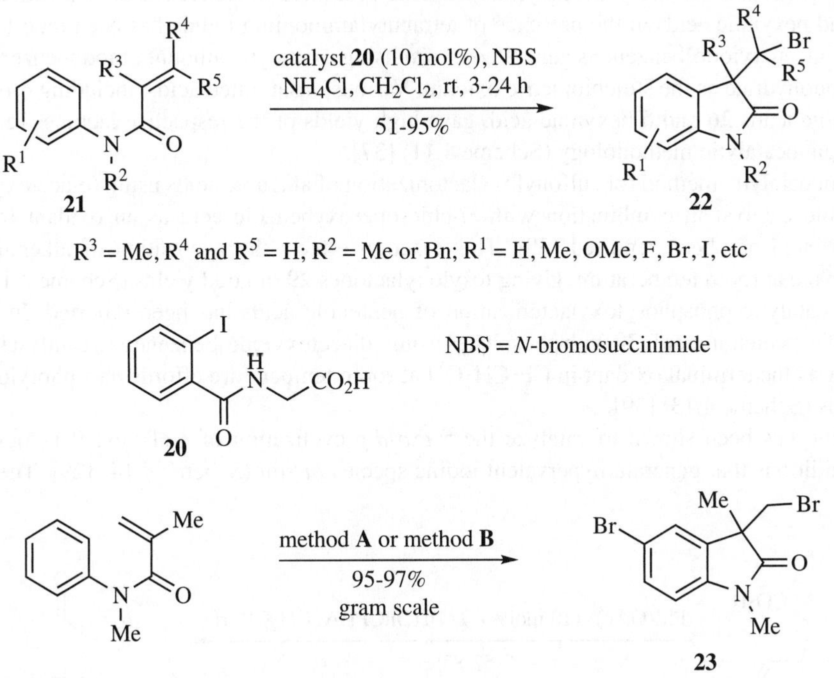

catalyst **20** (10 mol%), NBS
NH_4Cl, CH_2Cl_2, rt, 3-24 h
51-95%

21 **22**

R^3 = Me, R^4 and R^5 = H; R^2 = Me or Bn; R^1 = H, Me, OMe, F, Br, I, etc

I H N CO_2H O **20**

NBS = *N*-bromosuccinimide

method **A** or method **B**
95-97%
gram scale

Me Br Br O N Me **23**

A: catalyst **20** (10 mol%), NBS (2.4 equiv), NH_4Cl (10 mol%), CH_2Cl_2, rt, 12 h
B: catalyst **20** (10 mol%), KBr (2.4 equiv), Oxone (2.4 equiv), CH_2Cl_2, rt, 12 h

Scheme 4.10

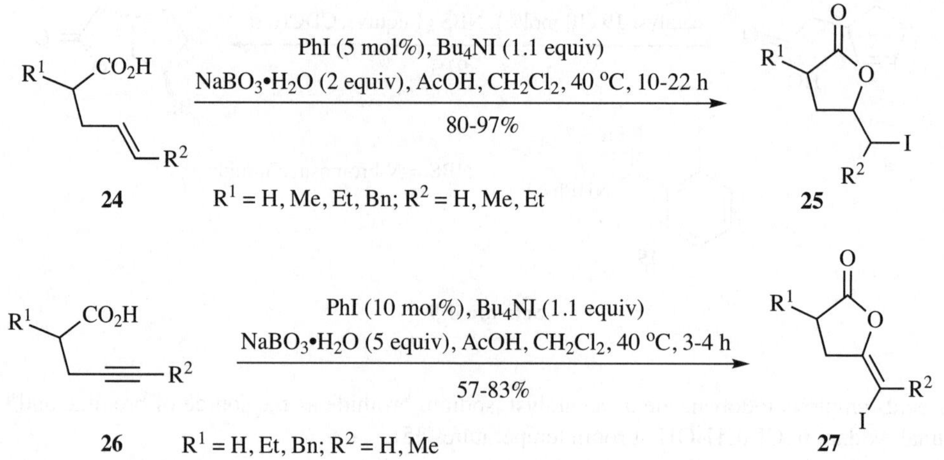

PhI (5 mol%), Bu_4NI (1.1 equiv)
$NaBO_3 \cdot H_2O$ (2 equiv), AcOH, CH_2Cl_2, 40 °C, 10-22 h
80-97%

24 → **25**

R^1 = H, Me, Et, Bn; R^2 = H, Me, Et

PhI (10 mol%), Bu_4NI (1.1 equiv)
$NaBO_3 \cdot H_2O$ (5 equiv), AcOH, CH_2Cl_2, 40 °C, 3-4 h
57-83%

26 → **27**

R^1 = H, Et, Bn; R^2 = H, Me

Scheme 4.11

preparation of product **23,** which is the key intermediate in the formal synthesis of the acetylcholinesterase inhibitor physostigmine [36].

A catalytic procedure for the (diacetoxyiodo)benzene-promoted oxidative iodolactonization of pentenoic, pentynoic and hexynoic acids in the presence of tetrabutylammonium iodide has been reported [37]. In this procedure, (diacetoxyiodo)benzene is generated *in situ* using a catalytic amount of iodobenzene with sodium perborate monohydrate as the stoichiometric oxidant. Various unsaturated acids, including δ-pentenoic acids **24,** δ-pentynoic acids **26** and δ-hexynoic acid, gave high yields of the respective lactones (e.g., **25** and **27**) using this organocatalytic methodology (Scheme 4.11) [37].

An efficient catalytic method for sulfonyloxylactonization of alkenoic acids using (diacetoxyiodo)benzene as a recyclable catalyst in combination with *m*-chloroperoxybenzoic acid as an oxidant in the presence of a sulfonic acid has been reported [38]. This reaction effects the cyclization of alkenoic acids **28** in dichloromethane at room temperature, giving tosyloxylactones **29** in good yields (Scheme 4.12).

A similar catalytic phosphoryloxylactonization of pentenoic acids has been reported. In particular, the cyclization of 4-pentenoic acids **30** with phosphates using (diacetoxyiodo)benzene as a catalyst in combination with *m*CPBA as the terminal oxidant in CF_3CH_2OH at room temperature affords phosphoryloxylactones **31** in good yields (Scheme 4.13) [39].

Iodobenzene has been shown to catalyze the 5-*exo-dig* cyclization of δ-alkynyl β-ketoesters **32** under oxidative conditions that generate hypervalent iodine species *in situ* (Scheme 4.14) [29]. The cyclopentane

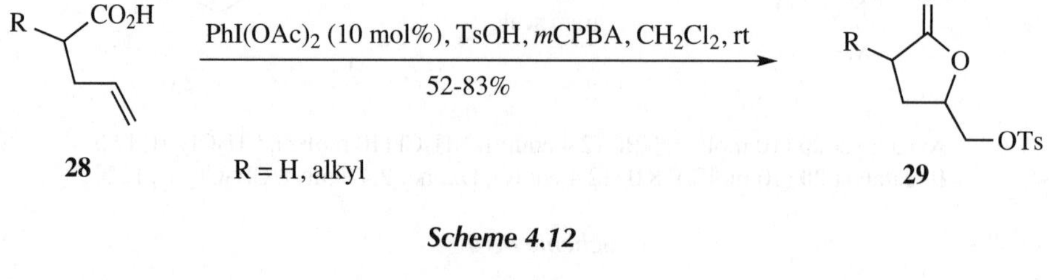

$PhI(OAc)_2$ (10 mol%), TsOH, *m*CPBA, CH_2Cl_2, rt
52-83%

28 → **29**

R = H, alkyl

Scheme 4.12

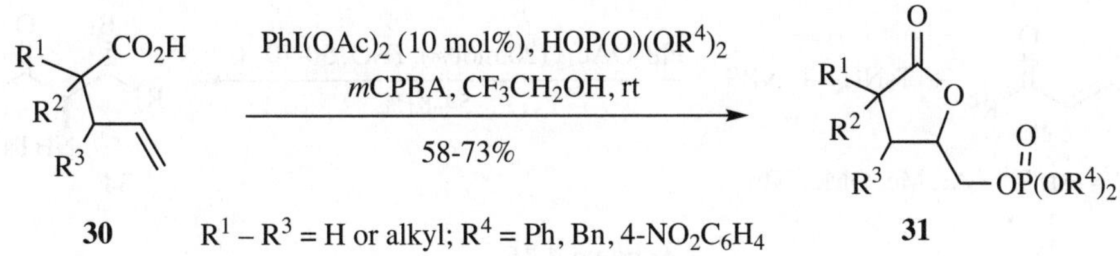

R^1–R^3 = H or alkyl; R^4 = Ph, Bn, 4-$NO_2C_6H_4$

PhI(OAc)$_2$ (10 mol%), HOP(O)(OR^4)$_2$, *m*CPBA, CF_3CH_2OH, rt, 58-73%; **30** → **31**

Scheme 4.13

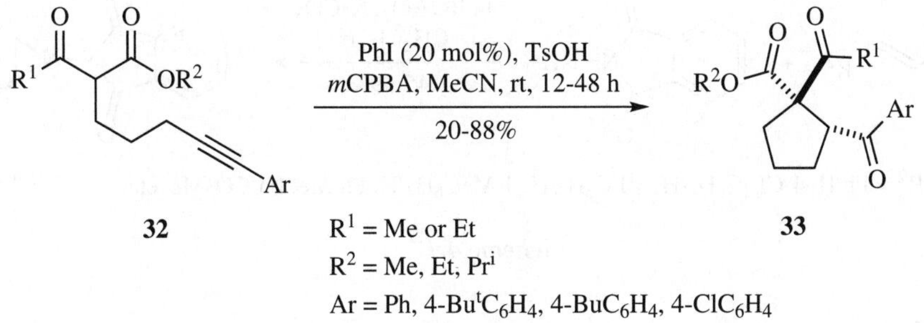

PhI (20 mol%), TsOH, *m*CPBA, MeCN, rt, 12-48 h, 20-88%; **32** → **33**

R^1 = Me or Et
R^2 = Me, Et, Pr^i
Ar = Ph, 4-Bu^tC_6H_4, 4-BuC_6H_4, 4-ClC_6H_4

Scheme 4.14

products **33** contain adjacent quaternary and tertiary stereocenters, which are generated with excellent diastereoselectivity (up to over 20 : 1 dr).

Ochiai and coworkers have developed an efficient iodoarene-catalyzed oxidative cleavage of alkenes and alkynes using *m*CPBA as a terminal oxidant [40]. Various cyclic and acyclic alkenes as well as aliphatic and aromatic alkynes are smoothly cleaved to carboxylic acids under these organocatalytic conditions (Scheme 4.15) [40].

A convenient procedure for the aminobromination of electron-deficient olefins using *N*-bromosuccinimide/tosylamide (Scheme 4.16) or Bromamine-T promoted by (diacetoxyiodo)benzene has been reported [41, 42]. This efficient metal-free protocol affords the vicinal bromamines **34** with excellent stereoselectivities. A similar (diacetoxyiodo)benzene-catalyzed aminochlorination can be performed by using Chloramine-T as nitrogen and chlorine source [43].

Wirth and coworkers published in 2007 a detailed study of the aziridination of alkenes with the PhI(OAc)$_2$/N-substituted hydrazine system (Scheme 4.17) and, in particular, reported tentative evidence that this reaction proceeds through the formation of an aminoiodane that reacts directly with the alkene [44].

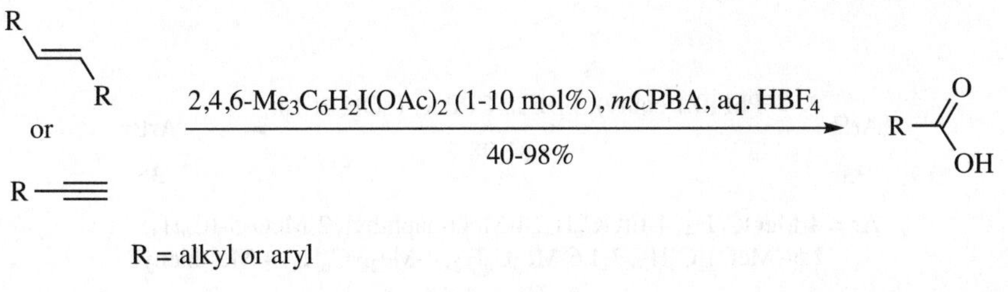

R–CH=CH–R or R–C≡CH → 2,4,6-$Me_3C_6H_2I(OAc)_2$ (1-10 mol%), *m*CPBA, aq. HBF_4, 40-98% → RCO$_2$H

R = alkyl or aryl

Scheme 4.15

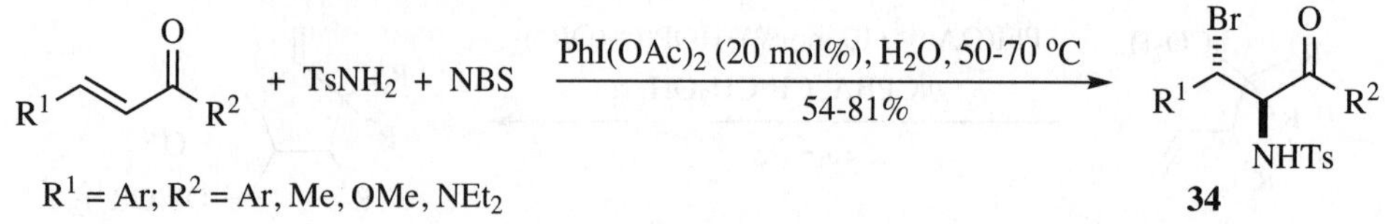

R^1-CH=CH-C(O)-R^2 + $TsNH_2$ + NBS $\xrightarrow[54\text{-}81\%]{PhI(OAc)_2\ (20\ mol\%),\ H_2O,\ 50\text{-}70\ ^{\circ}C}$ **34**

R^1 = Ar; R^2 = Ar, Me, OMe, NEt_2

Scheme 4.16

R^1-CH=CH-R^2 + N-aminophthalimide $\xrightarrow[45\text{-}99\%]{PhI(OAc)_2,\ K_2CO_3,\ CH_2Cl_2,\ rt}$ N-phthalimidoaziridine

R^1/R^2 = Ph/H, 4-$CF_3C_6H_4$/H, 4-FC_6H_4/H, 4-MeC_6H_4/H, Ph/Me, Ph/CO_2Me, etc.

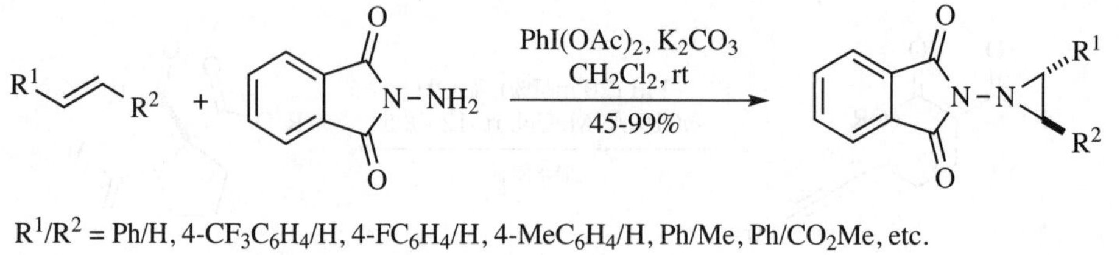

Scheme 4.17

Furthermore, the authors of this publication have analyzed the requirements to make this reaction catalytic in iodoarene. This reaction requires an oxidant that will oxidize iodoarenes but that does not oxidize alkenes and it is possible that no such oxidant actually exists [44]. However, the catalytic variant of this aziridination has been developed more recently employing an iodide–hypoiodite catalytic cycle (Section 4.4) [45].

4.1.3 Oxidative Bromination of Aromatic Compounds

Oxidative bromination of arenes can be achieved by the reaction with a source of bromide anion and an appropriate oxidant, possibly via intermediate formation of electrophilic bromoiodanes. In particular, an efficient and regioselective monobromination of electron-rich aromatic compounds has been developed, in which iodobenzene is used as the catalyst in combination with *m*CPBA as the terminal oxidant. The bromination of arenes **35** with lithium bromide is fast in THF at room temperature, providing regioselective monobrominated products **36** in good yields (Scheme 4.18) [46].

The proposed catalytic cycle for this reaction includes initial formation of [hydroxyl(tosyloxy)iodo]benzene **37** by oxidation of iodobenzene in the presence of toluenesulfonic acid followed by its conversion into the bromoiodane **38** via ligand exchange and then the bromination of arene to form the aryl bromide (Scheme 4.19). The reduced by-product, iodobenzene, is again transformed into the hypervalent iodine reagent by oxidation with *m*CPBA [46].

ArH (**35**) $\xrightarrow[81\text{-}98\%]{PhI\ (0.1\ equiv),\ LiBr,\ mCPBA,\ TsOH,\ rt\ 1\ h}$ ArBr (**36**)

Ar = 4-$MeOC_6H_4$, 4-$EtOC_6H_4$, 4-MeO-naphthyl, 2-MeO-5-IC_6H_3, 2,5-$(MeO)_2C_6H_3$, 2,4,6-$Me_3C_6H_2$, 4-$Me_2NC_6H_4$, 4-Me-thienyl

Scheme 4.18

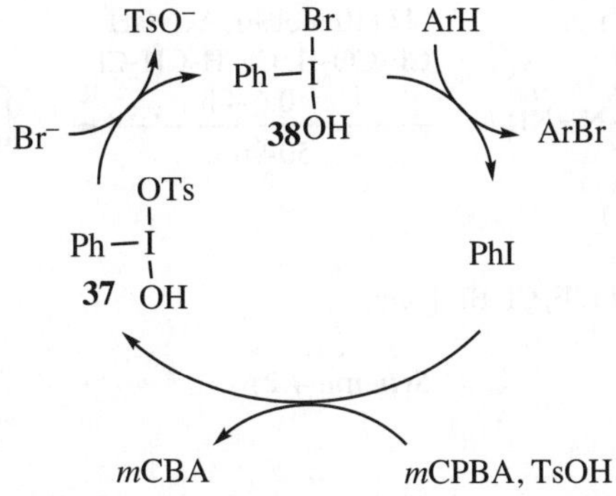

TsO⁻ Br ArH
Ph—I
Br⁻ 38 OH ArBr
OTs
Ph—I
37 OH PhI
mCBA mCPBA, TsOH

Scheme 4.19

4.1.4 Oxidative Amination of Aromatic Compounds

Aromatic C–H bonds can be aminated in intermolecular or intramolecular mode using amides as the nitrogen source, catalytic iodoarene and an appropriate oxidant, such as peroxycarboxylic acid. An atom-economical, environmentally friendly, direct oxidative intermolecular procedure for the amination and hydrazination of non-functionalized arenes has been developed by Antonchick and coworkers [47]. A wide range of arenes (e.g., **39**), including simple benzene, can be aminated using *N*-methoxybenzamides **40** as amination reagent in the presence of peracetic and trifluoroacetic acids to give products **41** (Scheme 4.20). Even electron-poor arenes like chlorobenzene and fluorobenzene can be selectively functionalized in the *para*-position using this mild method. The reactions of electron-rich arenes afford products of amination in up to 93% yield. Out of several iodoarenes tested, 2,2′-diiodo-4,4′,6,6′-tetramethylbiphenyl (**42**) has shown the highest catalytic activity in this reaction.

This procedure was further extended to organocatalytic hydrazination of non-functionalized arenes to products **44** using *N*-(1,3-dioxoisoindolin-2-yl)acetamide (**43**) as the nitrogen source (Scheme 4.21) [47].

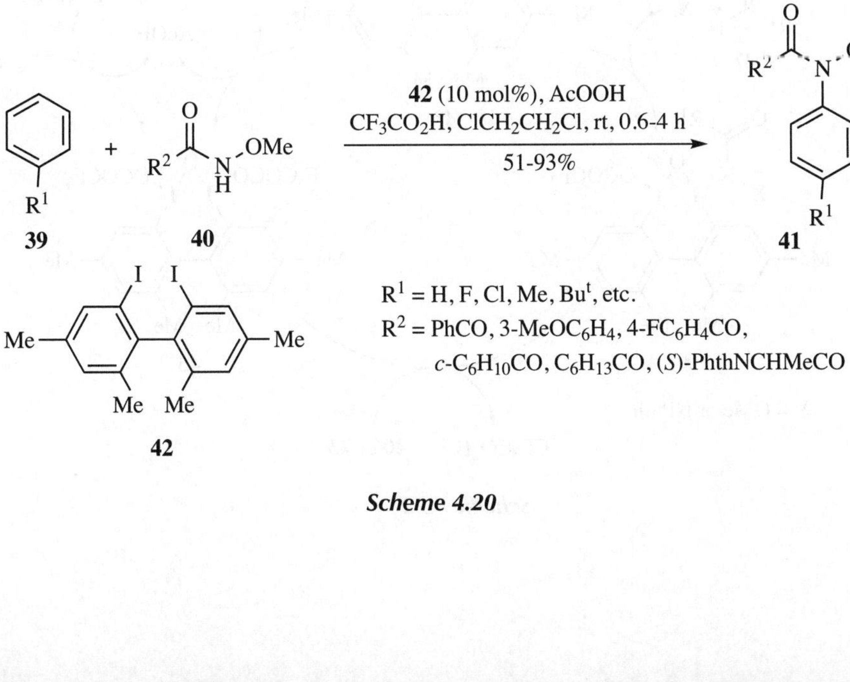

39 + 40 → 41

42 (10 mol%), AcOOH
CF_3CO_2H, $ClCH_2CH_2Cl$, rt, 0.6-4 h
51-93%

R^1 = H, F, Cl, Me, Bu^t, etc.
R^2 = PhCO, 3-MeOC$_6$H$_4$, 4-FC$_6$H$_4$CO,
c-C_6H_{10}CO, C_6H_{13}CO, (*S*)-PhthNCHMeCO

42

Scheme 4.20

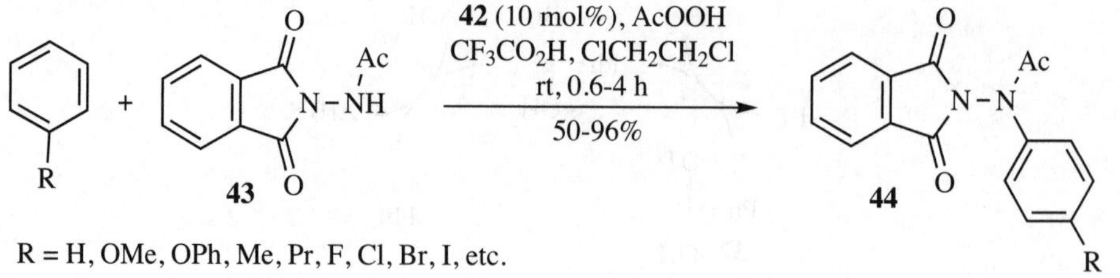

Scheme 4.21

The mechanism of this reaction involves initial oxidation of aryl iodide **42** by peracetic acid to the active hypervalent iodine species **45,** followed by ligand substitution at iodine(III) by **40** or **43** to generate species **46,** which undergo oxidative fragmentation to form nitrenium ions **47.** Reaction of an arene with the electron-deficient nitrenium ion **47** affords the final products of amination and the reduced intermediate **48,** which is reoxidized to the species **45** (Scheme 4.22) [47].

Several examples of cyclizations through intramolecular C–N bond formation catalyzed by hypervalent iodine species have been reported. Antonchick and coworkers developed an efficient organocatalytic method for the preparation of carbazoles through catalytic oxidative C–N bond formation [48]. The best yields of products were obtained in hexafluoro-2-propanol using 2,2′-diiodo-4,4′,6,6′-tetramethylbiphenyl (**42**) as the catalyst and peracetic acid as the oxidant, as illustrated by a representative example shown in Scheme 4.23.

Togo and Moroda have reported a (diacetoxyiodo)benzene-mediated cyclization reaction of 2-aryl-*N*-methoxyethanesulfonamides **49** using iodobenzene as a pre-catalyst (5–10 mol%) and *m*-chloroperoxybenzoic

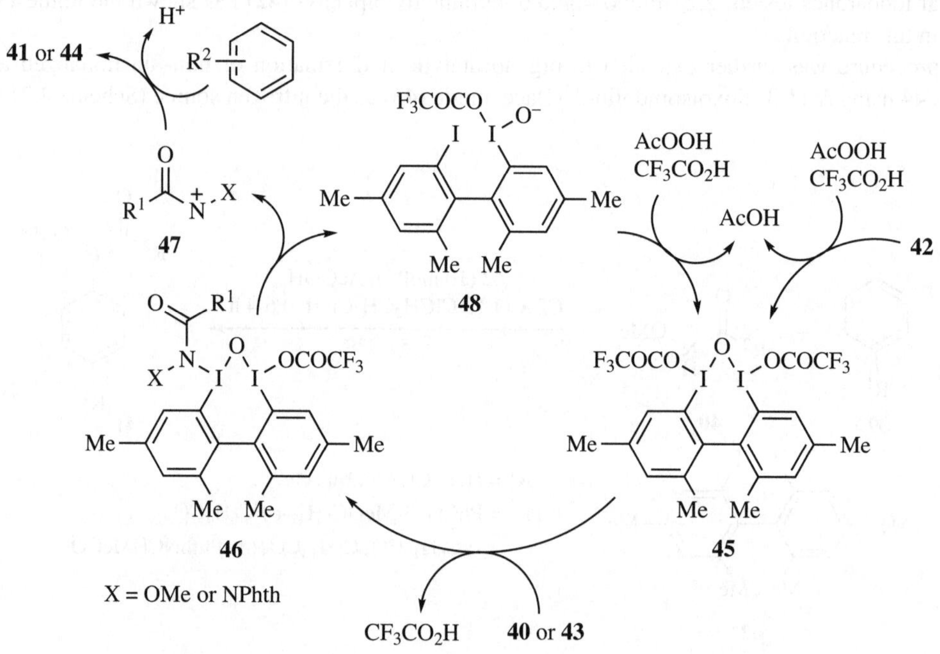

Scheme 4.22

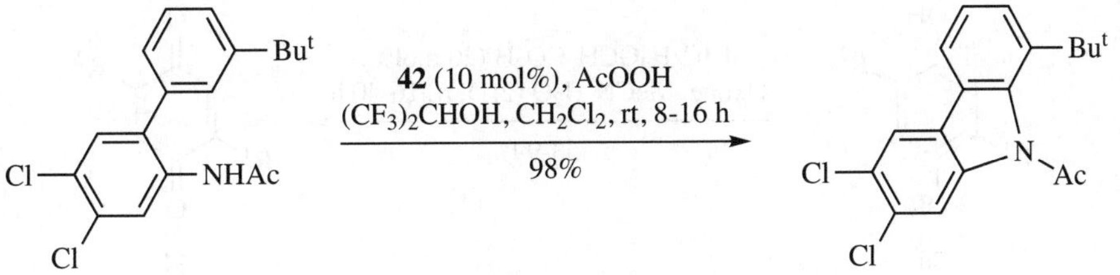

42 (10 mol%), AcOOH
$(CF_3)_2CHOH$, CH_2Cl_2, rt, 8-16 h
98%

Scheme 4.23

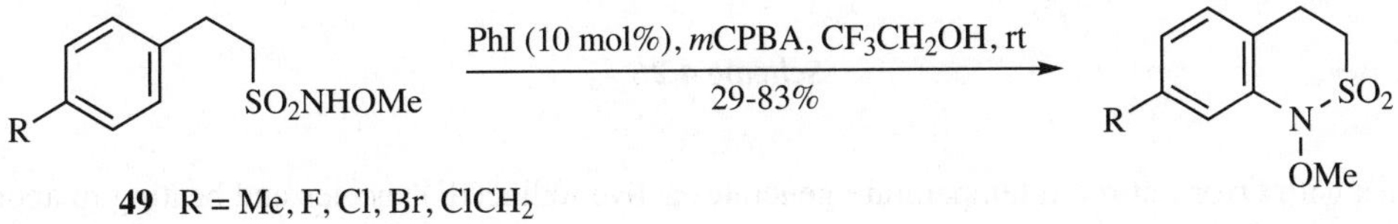

PhI (10 mol%), *m*CPBA, CF_3CH_2OH, rt
29-83%

49 R = Me, F, Cl, Br, $ClCH_2$

Scheme 4.24

acid (*m*CPBA) as the stoichiometric oxidant (Scheme 4.24) [49]. A similar catalytic cyclization has also been achieved using Oxone in acetonitrile [50].

Likewise, the oxidative C–H amination of *N*″-aryl-*N*′-tosyl/*N*′-methylsulfonylamidines and *N,N*′-bis(aryl)amidines has been accomplished using iodobenzene as a catalyst to furnish 1,2-disubstituted benzimidazoles in the presence of *m*CPBA as a terminal oxidant at room temperature (Scheme 4.25) [51]. The reaction is general and the target benzimidazoles can be obtained in moderate to high yields.

4.1.5 Oxidation of Phenolic Substrates to Quinones and Quinols

The oxidation of phenols or *o*- and *p*-hydroquinones with stoichiometric [bis(acyloxy)iodo]arenes to the corresponding benzoquinones is one of the most typical synthetic applications of hypervalent iodine reagents (Section 3.1.11). The catalytic version of this reaction was first reported by Yakura and Konishi in 2007 [52]. The reaction of *p*-alkoxyphenols **50** with a catalytic amount of 4-iodophenoxyacetic acid in the presence of Oxone as the terminal oxidant in aqueous acetonitrile at room temperature affords *p*-quinones **51** in high yields (Scheme 4.26) [52]. 4-Iodophenoxyacetic acid is a readily available and water-soluble aromatic iodide that has a particularly high catalytic activity in this reaction.

In the initial publication [52], a catalytic cycle based on the iodine(V) species was proposed for this reaction (Scheme 4.26); however, more recent studies (Section 4.2) have demonstrated that the oxidation of

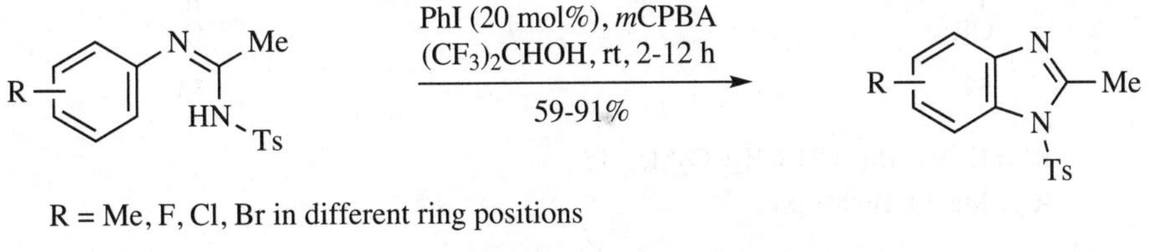

PhI (20 mol%), *m*CPBA
$(CF_3)_2CHOH$, rt, 2-12 h
59-91%

R = Me, F, Cl, Br in different ring positions

Scheme 4.25

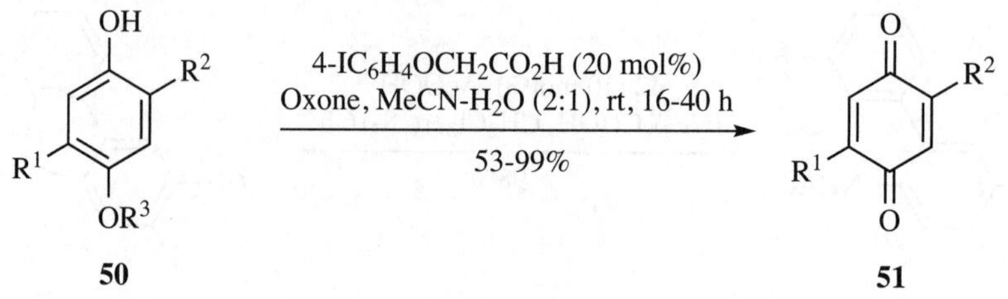

OH, R^2, R^1, OR^3 — **50** → $4\text{-}IC_6H_4OCH_2CO_2H$ (20 mol%), Oxone, $MeCN\text{-}H_2O$ (2:1), rt, 16-40 h, 53-99% → **51**

R^1 = H or $MeCH_2CH_2CMe_2$
R^2 = H, Bu^t, $MeCH_2CH_2CMe_2$, $Bu^tPh_2SiOCH_2$, N_3CH_2, phthalimide
R^3 = Me or Et

Scheme 4.26

iodoarenes with Oxone at room temperature generates active iodine(III) species and heating to about 70 °C or the presence of a ruthenium catalyst is required to oxidize ArI to $ArIO_2$ [53].

Several modifications of the original procedure shown in Scheme 4.26 have been reported [54–56]. In particular, the reaction of *p*-dialkoxybenzenes **52** with a catalytic amount of 4-iodophenoxyacetic acid in the presence of Oxone as a co-oxidant in 2,2,2-trifluoroethanol–water (1 : 2) gives the corresponding *p*-quinones **53** in excellent yields (Scheme 4.27) [54, 56]. The same solvent system, 2,2,2-trifluoroethanol–water (1 : 2), can also be used for the efficient oxidation of *p*-alkoxyphenols to *p*-quinones [55].

A similar catalytic oxidation of *p*-substituted phenols bearing an alkyl or aryl group in the *para* position affords the corresponding *p*-quinols. In particular, the reaction of *p*-substituted phenols **54** with a catalytic amount of 4-iodophenoxyacetic acid and Oxone (4 equiv) at room temperature in an aqueous tetrahydrofuran or 1,4-dioxane solution gave *p*-quinols **55** in generally high yields (Scheme 4.28) [55, 57].

4.1.6 Oxidative Spirocyclization of Aromatic Substrates

The oxidative dearomatization of appropriately substituted phenolic substrates resulting in intramolecular cyclization with the formation of spirocyclic products represents one of the most powerful synthetic tools in modern organic synthesis (Section 3.1.11). Kita and coworkers were the first to report a catalytic variant of the oxidative spirocyclization reaction based on the *in situ* regeneration of a [bis(trifluoroacetoxy)iodo]arene from iodoarene using *m*CPBA as a terminal oxidant [2]. In a representative example, the oxidation of

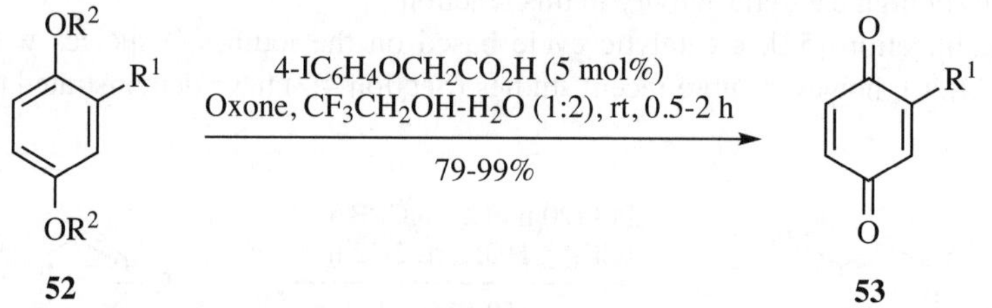

OR^2, R^1, OR^2 — **52** → $4\text{-}IC_6H_4OCH_2CO_2H$ (5 mol%), Oxone, $CF_3CH_2OH\text{-}H_2O$ (1:2), rt, 0.5-2 h, 79-99% → **53**

R^1 = H, Me, Bu^t, $CH_2CH_2CO_2Me$, etc.
R^2 = Me, Et, Bu^tMe_2Si

Scheme 4.27

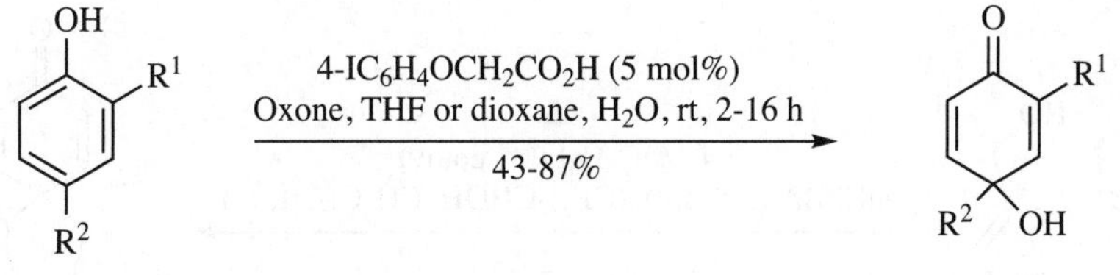

4-$IC_6H_4OCH_2CO_2H$ (5 mol%)
Oxone, THF or dioxane, H_2O, rt, 2-16 h
43-87%

54 → **55**

R^1 = H, Me, Pr, Ph, Br, etc.
R^2 = Me, Et, Bu^tCOOCH_2, Br, CN

Scheme 4.28

the phenolic substrate **56** with *m*CPBA in dichloromethane in the presence of a catalytic amount of *p*-[bis(trifluoroacetoxy)iodo]toluene (1 mol%) and trifluoroacetic acid at room temperature affords the respective spirolactone **57** in good yield (Scheme 4.29). Various other [bis(trifluoroacetoxy)iodo]arenes [e.g., $PhI(OCOCF_3)_2$, 4-$MeOC_6H_4I(OCOCF_3)_2$ and 2,4-$F_2C_6H_3I(OCOCF_3)_2$] and different acidic additives (acetic acid, $BF_3{\cdot}OEt_2$, TMSOTf, molecular sieves) can be used as catalysts in this reaction; however, the 4-$MeC_6H_4I(OCOCF_3)_2$/trifluoroacetic acid system generally provides the best catalytic efficiency. Under optimized conditions, various phenolic substrates **58** were oxidized to spirolactones **59** in the presence of catalytic amounts of *p*-iodotoluene (Scheme 4.29) [2]. Further modification of this catalytic procedure involved the use of peracetic acid as the terminal oxidant in fluoroalcohol solvents [58].

More recently, the Kita's catalytic spirocyclization has been tested in the synthesis of bioactive polyspirocyclohexa-2,5-dienones, as illustrated by Scheme 4.30 [59]. The target product **61** was isolated

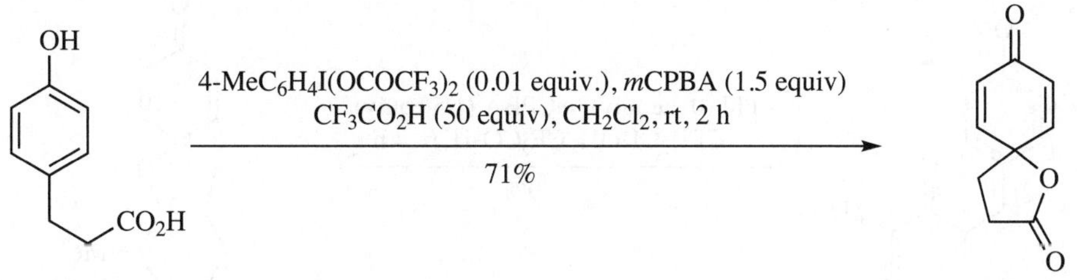

4-$MeC_6H_4I(OCOCF_3)_2$ (0.01 equiv.), *m*CPBA (1.5 equiv)
CF_3CO_2H (50 equiv), CH_2Cl_2, rt, 2 h
71%

56 → **57**

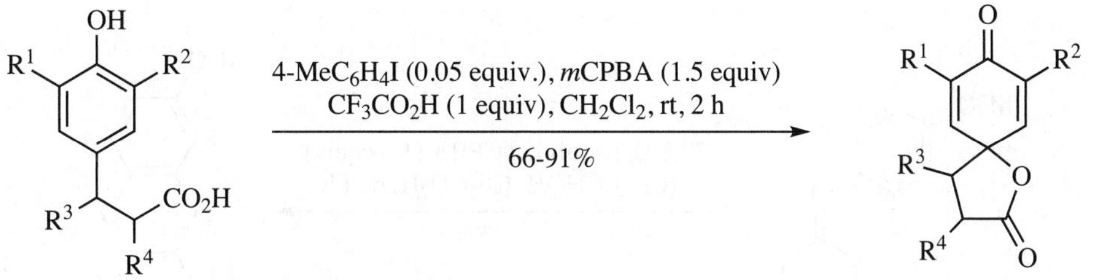

4-MeC_6H_4I (0.05 equiv.), *m*CPBA (1.5 equiv)
CF_3CO_2H (1 equiv), CH_2Cl_2, rt, 2 h
66-91%

58 → **59**

R^1, R^2, R^3, R^4 = H, Me, Br
or R^1, R^2 = H, R^3 + R^4 = (=CH-CH=CH-CH=)

Scheme 4.29

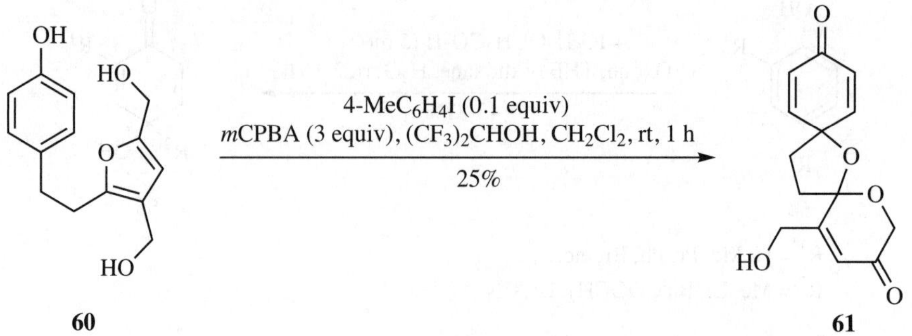

60 → 61: 4-MeC_6H_4I (0.1 equiv), mCPBA (3 equiv), $(CF_3)_2CHOH$, CH_2Cl_2, rt, 1 h; 25%

Scheme 4.30

in a low yield, probably due to the competitive oxidation of substrate **60** directly by *m*CPBA to give unwanted side products.

Hutt, Lupton and coworkers have further investigated the synthetic utility of the catalytic spirocyclization reaction and developed a procedure for the iodobenzene-catalyzed synthesis of spirofurans and benzopyrans by oxidative cyclization of vinylogous esters [60]. In particular, vinylogous esters bearing *para* or *meta* methoxy benzyl groups undergo oxidative cyclization with 5–20 mol% iodobenzene and *m*CPBA to give spirofuran- or benzopyran-containing heterocycles as illustrated by two specific examples shown in Scheme 4.31. Both of these reactions proceed via radical–cation intermediates (Section 3.1.12). These cyclizations allow rapid generation of skeletally complex products in good to excellent yields [60].

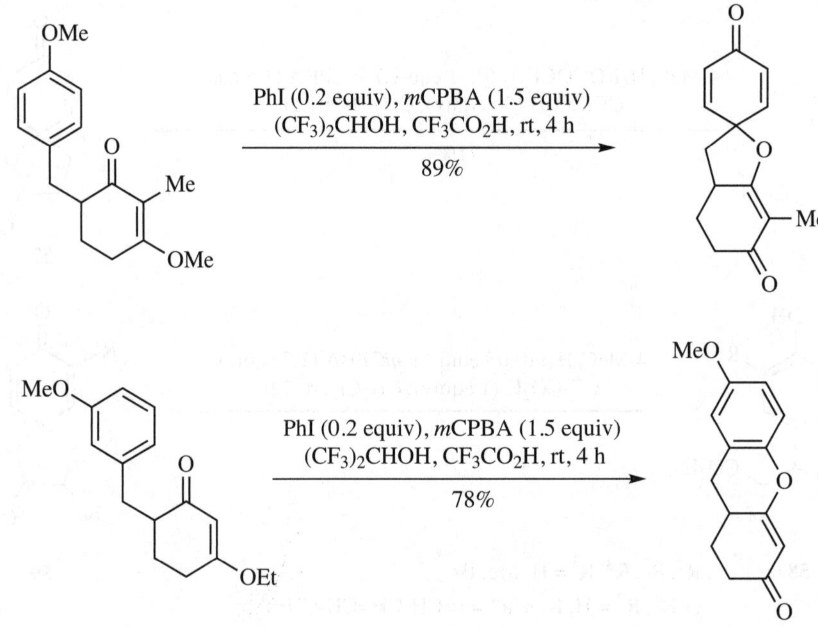

PhI (0.2 equiv), mCPBA (1.5 equiv), $(CF_3)_2CHOH$, CF_3CO_2H, rt, 4 h; 89%

PhI (0.2 equiv), mCPBA (1.5 equiv), $(CF_3)_2CHOH$, CF_3CO_2H, rt, 4 h; 78%

Scheme 4.31

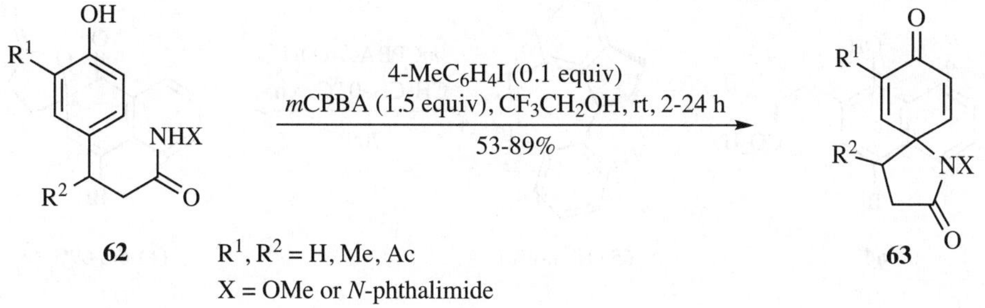

$$\mathbf{62} \xrightarrow[\text{53-89\%}]{\text{4-MeC}_6\text{H}_4\text{I (0.1 equiv)},\ m\text{CPBA (1.5 equiv)},\ \text{CF}_3\text{CH}_2\text{OH, rt, 2-24 h}} \mathbf{63}$$

R^1, R^2 = H, Me, Ac
X = OMe or *N*-phthalimide

Scheme 4.32

In a similar procedure, the amide derivatives of phenolic substrates **62** can be catalytically oxidized to the respective N-fused spirolactams **63** using catalytic amounts of *p*-iodotoluene and *m*CPBA as the terminal oxidant (Scheme 4.32) [61].

The catalytic spirocyclization procedure has been further improved by using 2,2′-diiodo-4,4′,6,6′-tetramethylbiphenyl (**42,** Section 4.1.4) as the catalyst instead of *p*-iodotoluene. Diiodide **42** has shown high catalytic activity in this reaction in the presence of peracetic acid as the terminal oxidant (Scheme 4.33) [62]. It is assumed that *in situ* generated μ-oxo-bridged hypervalent iodine(III) species (similar to species **45,** see Scheme 4.22) are the actual active species in this catalytic cycle [62].

The catalytic oxidative spirocyclization of phenolic substrates can be performed as an enantioselective reaction using chiral organic iodides as the catalysts [63–65]. In particular, the chiral iodoarene **65** with a rigid spirobiindane backbone has been found to enantioselectively dearomatize naphtholic substrates **64** in a highly selective manner to give optically active products **66** with up to 69% ee (Scheme 4.34) [63]. Ishihara and coworkers have designed a conformationally flexible C_2-symmetric iodoarene catalyst (**68**) for a similar enantioselective oxidative spirolactonization [64, 66, 67]. In a specific example, hydroxynaphthalenyl propanoic acid derivatives **67** undergo dearomatization and spirocyclization in the presence of catalyst **68** to afford the corresponding spirolactones **69** in yields of up to 94% with enantioselectivity up to 90% (Scheme 4.34) [64].

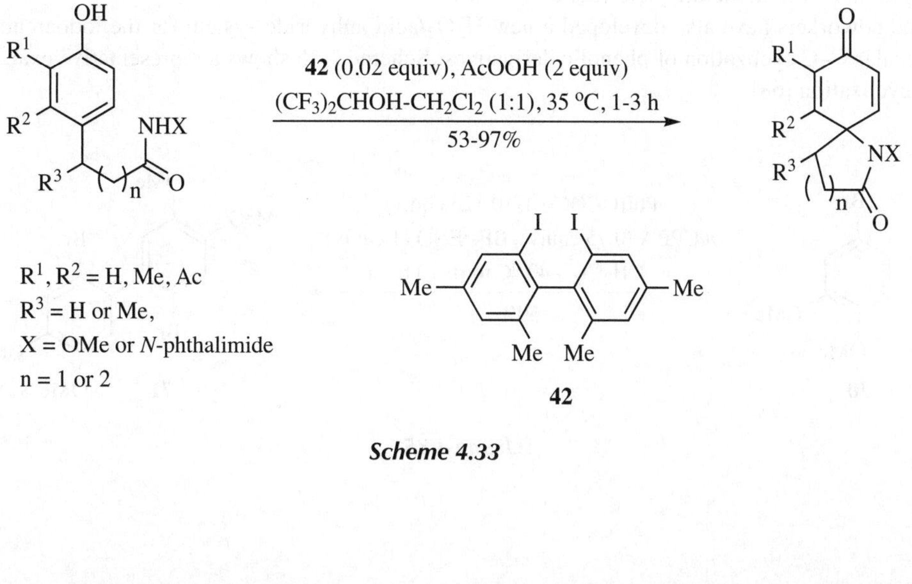

$$\xrightarrow[\text{53-97\%}]{\mathbf{42}\text{ (0.02 equiv), AcOOH (2 equiv)},\ (\text{CF}_3)_2\text{CHOH-CH}_2\text{Cl}_2\text{ (1:1), 35 °C, 1-3 h}}$$

R^1, R^2 = H, Me, Ac
R^3 = H or Me,
X = OMe or *N*-phthalimide
n = 1 or 2

42

Scheme 4.33

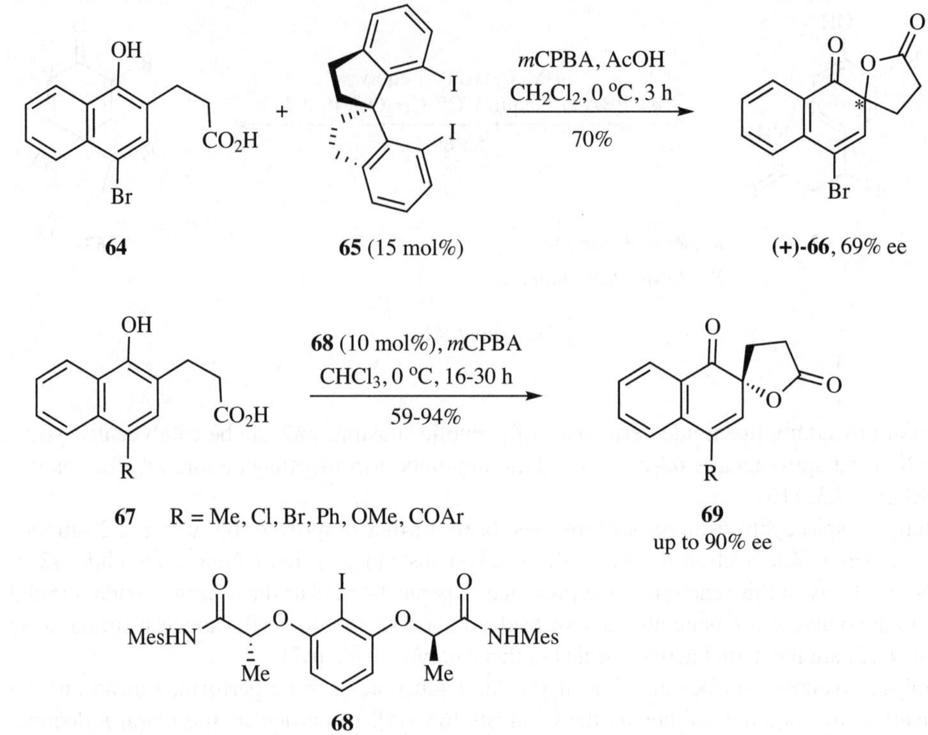

OH
CO_2H
Br
64
+
I
I
65 (15 mol%)
*m*CPBA, AcOH
CH_2Cl_2, 0 °C, 3 h
70%
O
O
O
*
Br
(+)-66, 69% ee

OH
CO_2H
R
67 R = Me, Cl, Br, Ph, OMe, COAr
68 (10 mol%), *m*CPBA
$CHCl_3$, 0 °C, 16-30 h
59-94%
O
O
O
R
69
up to 90% ee

MesHN
O
O
Me
I
O
O
Me
NHMes
68

Scheme 4.34

4.1.7 Carbon–Carbon Bond-Forming Reactions

Only a few examples of hypervalent iodine-catalyzed reactions leading to the formation of new C–C bonds have been reported. In seminal work, Kita and coworkers reported in the 2005 a single example of an intermolecular C–C bond formation reaction catalyzed by an iodoarene [2]. Specifically, the oxidative coupling of phenolic ether **70** using [bis(trifluoroacetoxy)iodo]benzene as a catalyst and *m*CPBA as a terminal oxidant afforded product **71** in moderate yield (Scheme 4.35).

Kita and coworkers have also developed a new H_2O_2/acid anhydride system for the iodoarene-catalyzed intramolecular C–C cyclization of phenolic derivatives; Scheme 4.36 shows a representative example of this catalytic cyclization [68].

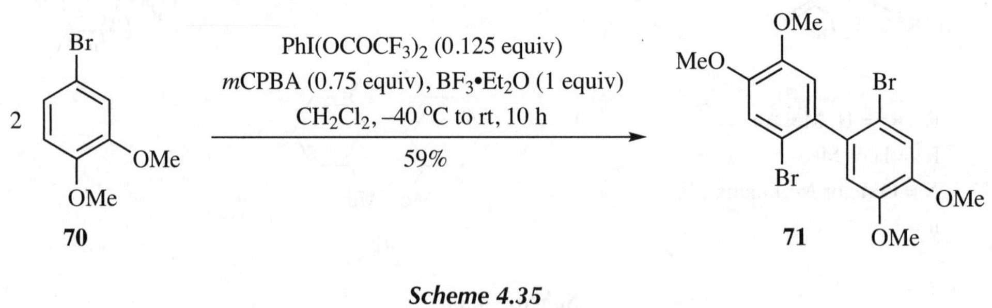

2
Br
OMe
OMe
70
$PhI(OCOCF_3)_2$ (0.125 equiv)
*m*CPBA (0.75 equiv), BF_3•Et_2O (1 equiv)
CH_2Cl_2, –40 °C to rt, 10 h
59%
OMe
MeO
Br
Br
OMe
OMe
71

Scheme 4.35

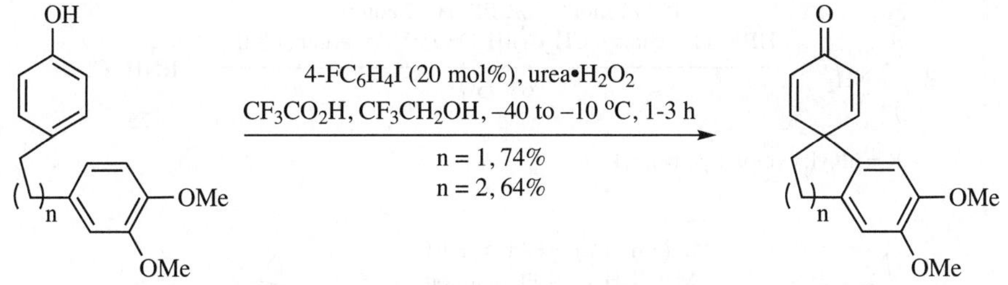

4-FC_6H_4I (20 mol%), urea•H_2O_2
CF_3CO_2H, CF_3CH_2OH, –40 to –10 °C, 1-3 h
n = 1, 74%
n = 2, 64%

Scheme 4.36

Likewise, the intramolecular oxidative coupling of substituted 4-hydroxyphenyl-*N*-phenylbenzamides **72** has been realized in a catalytic manner by using iodobenzene as catalyst and *m*CPBA or urea–H_2O_2 as terminal oxidant (Scheme 4.37). This reaction constitutes an efficient method for the synthesis of spirooxindoles **73** [69].

4.1.8 Hofmann Rearrangement of Carboxamides

Hypervalent iodine reagents have been employed in numerous synthetic works as oxidants for Hofmann-type rearrangements (Section 3.1.13.3). Synthetic procedures for Hofmann rearrangement using stoichiometric organohypervalent iodine species generated *in situ* from iodoarenes and appropriate oxidants have also been reported [70, 71]. In particular, alkylcarboxamides can be converted into the respective amines by Hofmann rearrangement using hypervalent iodine species generated *in situ* from stoichiometric amounts of iodobenzene and Oxone in aqueous acetonitrile [70]. Likewise, the Hofmann-type rearrangement of aromatic and aliphatic imides using a hypervalent iodine(III) reagent generated *in situ* from PhI, *m*CPBA and TsOH·H_2O has been reported [71].

The first catalytic version of the Hofmann rearrangement using aryl iodides as catalysts and *m*CPBA as terminal oxidant was reported by Ochiai and coworkers in 2012 [72]. A study of the catalytic efficiency of substituted iodobenzenes and some aliphatic alkyl iodides in the iodane(III) catalyzed Hofmann rearrangement of benzylic carboxamides has demonstrated that iodobenzene is the best catalyst. The introduction of both electron-donating (4-methyl, 3,5-dimethyl and 2,4,6-trimethyl) and electron-withdrawing groups (4-Cl and 4-CF_3) into iodobenzene decreased the yield of rearranged products. Aliphatic iodides such as methyl, trifluoroethyl and 1-adamantyl iodides showed no catalytic efficiency. Under optimized reaction conditions,

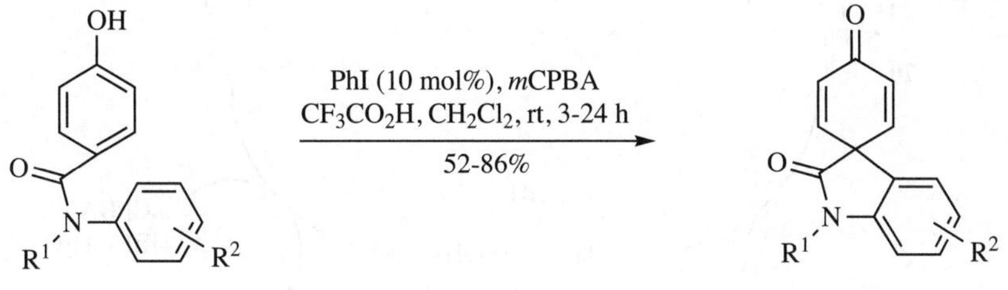

PhI (10 mol%), *m*CPBA
CF_3CO_2H, CH_2Cl_2, rt, 3-24 h
52-86%

72 **73**

R^1 = Me, $PhCH_2$, $CH_2{=}CHCH_2$
R^2 = H, Me, OMe, OEt, Cl, F in *o*- or *p*-position to N

Scheme 4.37

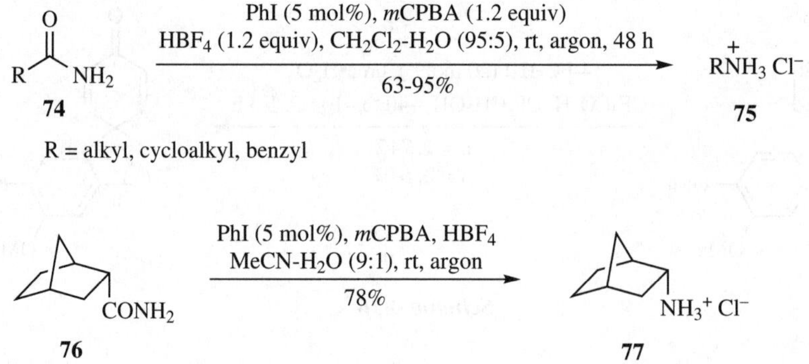

PhI (5 mol%), *m*CPBA (1.2 equiv)
HBF_4 (1.2 equiv), CH_2Cl_2-H_2O (95:5), rt, argon, 48 h
63-95%

74 → $\overset{+}{R}NH_3\ Cl^-$ **75**

R = alkyl, cycloalkyl, benzyl

PhI (5 mol%), *m*CPBA, HBF_4
MeCN-H_2O (9:1), rt, argon
78%

76 ($CONH_2$) → **77** ($NH_3^+\ Cl^-$)

Scheme 4.38

various linear, branched and cyclic aliphatic carboxamides **74** afford alkylammonium chlorides **75** in high yields (Scheme 4.38). The catalytic conditions are compatible with the presence of various functionalities such as halogens (F, Cl, Br), sulfonamides, amines and methoxy and nitro groups. Bicyclic amide **76** affords *endo*-ammonium chloride **77** stereoselectively in a good yield, which suggests retention of the stereochemistry of the migrating groups in the catalytic λ^3-iodane-promoted rearrangement of carboxamides – as observed in the classical Hofmann rearrangement.

The catalytic cycle for this reaction probably involves the *in situ* generation of a tetracoordinated square-planar bis(aqua)(hydroxy)phenyl-λ^3-iodane complex **78** as an active oxidant from a catalytic amount of iodobenzene by the reaction with *m*CPBA in the presence of aqueous HBF_4 (Scheme 4.39).

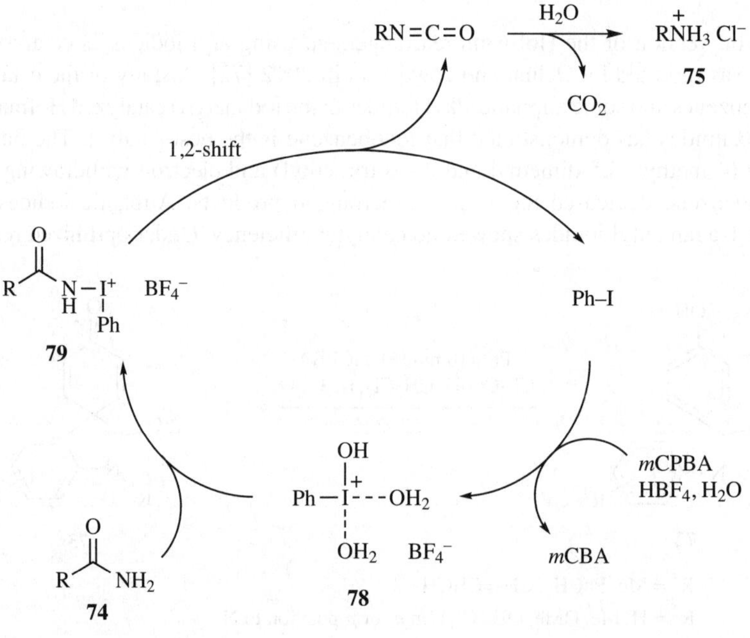

RN=C=O → (H_2O, −CO_2) → $\overset{+}{R}NH_3\ Cl^-$ **75**

1,2-shift

Ph–I

*m*CPBA, HBF_4, H_2O; *m*CBA

78: $Ph-\overset{+}{I}(OH)(OH_2)_2\ BF_4^-$

79: $RC(O)NH-I^+(Ph)\ BF_4^-$

74: $RC(O)NH_2$

Scheme 4.39

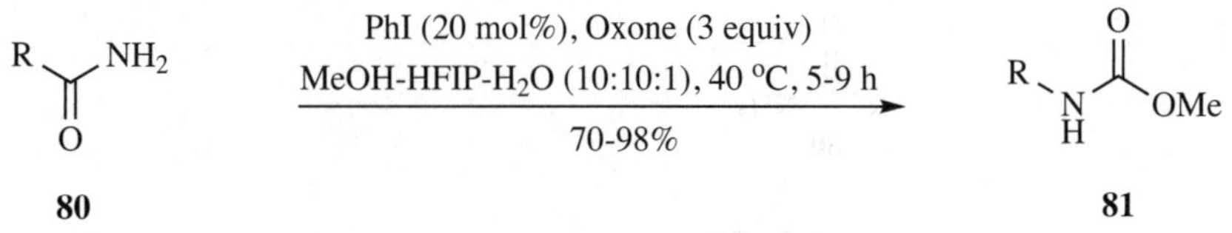

$$\underset{\mathbf{80}}{\mathrm{RC(O)NH_2}} \xrightarrow[\text{70-98\%}]{\text{PhI (20 mol\%), Oxone (3 equiv)} \atop \text{MeOH-HFIP-H}_2\text{O (10:10:1), 40 °C, 5-9 h}} \underset{\mathbf{81}}{\mathrm{RNHC(O)OMe}}$$

R = $PhCH_2$, $4\text{-}MeC_6H_4CH_2$, $4\text{-}FC_6H_4CH_2$, $4\text{-}ClC_6H_4CH_2$, $3\text{-}ClC_6H_4CH_2$, $2\text{-}ClC_6H_4CH_2$, $4\text{-}BrC_6H_4CH_2$, $4\text{-}CF_3C_6H_4CH_2$, Ph(Et)CH, C_5H_{11}, $PhCH_2CH_2$, C_6H_{13}, $BuMe_2C$, cyclopentyl, cyclohexyl, 1-adamantyl, etc.

Scheme 4.40

(Hydroxy)phenyl-λ^3-iodane complex **78** promotes the Hofmann rearrangement of carboxamides **74,** probably via intermediate formation of *N*-(phenyliodanyl)carboxamides **79** [72].

Hypervalent-iodine-catalyzed Hofmann rearrangement of carboxamides to carbamates using Oxone as the oxidant has been reported [73]. This reaction involves hypervalent iodine species generated *in situ* from catalytic amounts of PhI and Oxone in the presence of 1,1,1,3,3,3-hexafluoroisopropanol (HFIP) in aqueous methanol solutions. Under these conditions, Hofmann rearrangement of various carboxamides **80** affords the corresponding carbamates **81** in high yields (Scheme 4.40). The addition of small amount of water is required to dissolve Oxone in the reaction mixture and the presence of HFIP in the mixture dramatically improves the yield of carbamates **81.** Iodobenzene has the most pronounced catalytic effect; the use of other iodine-containing pre-catalysts (2,4,6-$Me_3C_6H_2I$, 4-MeC_6H_4I, 4-$CF_3C_6H_4I$, 3-$HO_2CC_6H_4I$, Bu_4NI) instead of PhI gives poor results. In general, all benzylcarboxamides with either electron-donating or electron-withdrawing substituents afford corresponding carbamates **81** in good yields. Various aliphatic amides, including primary, secondly, tertiary and cyclic alkylcarboxamides, also smoothly react under the same conditions. Compared to the previous method of Hofmann rearrangement with the stoichiometric hypervalent iodine species generated *in situ* [70], the catalytic method affords carbamates **81** in similar yields.

The proposed reaction catalytic cycle involves the reactive species **82** [hydroxy(phenyl)iodonium ion possibly activated by HFIP] generated from PhI and Oxone in aqueous HFIP, which further react with carboxamide **80** to give the hypervalent amidoiodane **83** via ligand exchange (Scheme 4.41). Amidoiodane **83** then undergoes the reductive elimination of iodobenzene and a 1,2-shift at the electron-deficient nitrenium nitrogen atom to give isocyanate **84.** Subsequently, the addition of methanol to isocyanate **84** gives the final carbamate **81.** The regenerated PhI continues the catalytic cycle. The presence of HFIP may help to generate more electron-deficient active species **82** and **83** via ligand exchange with hydroxy(phenyl)iodonium ion or hypervalent aminoiodine, which help to accelerate further steps of the catalytic cycle, such as ligand exchange and 1,2-shift.

4.1.9 Oxidation of Anilines

Anilines can be oxidized to azobenzenes under hypervalent iodine catalysis using peracetic acid as the terminal oxidant (Scheme 4.42) [74]. 2-Iodobenzoic acid has shown the most pronounced catalytic effect compared to other aryl iodides (PhI, 4-$MeOC_6H_4I$, 4-$NO_2C_6H_4I$, 3-$IC_6H_4CO_2H$, 4-$IC_6H_4CO_2H$). This metal-free oxidation system demonstrates wide substituent tolerance: alkyls, halogens and several versatile functional groups, such as amino, ethynyl and carboxyl substituents, are readily compatible and the corresponding products are formed with good to excellent yields. The large-scale preparation of azo compounds could also be carried out successfully by this method.

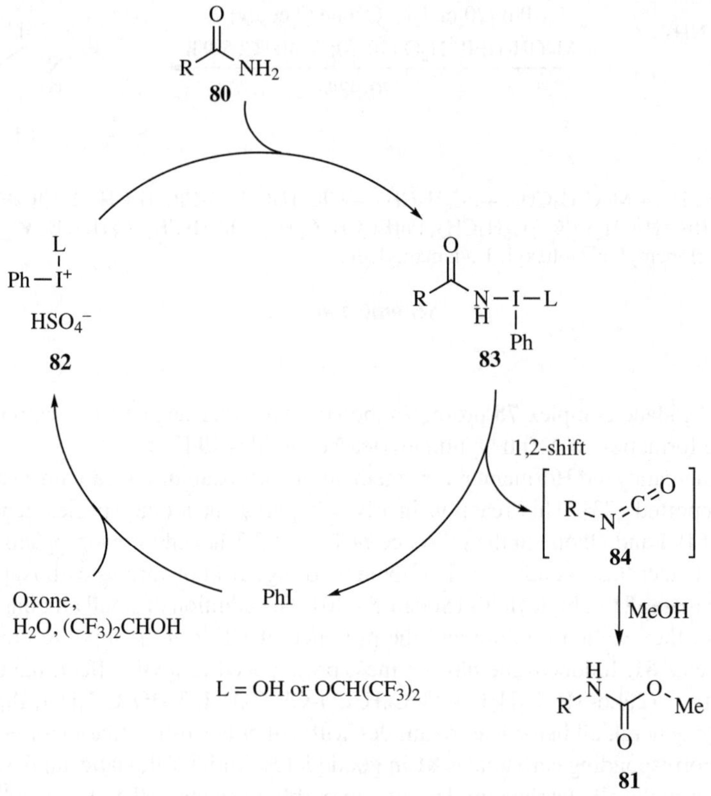

Scheme 4.41

The asymmetrical azo compounds **88** have also been prepared under these conditions in reasonable yields by coupling 3-ethynylaniline (**87**) with a twofold excess of less reactive aniline (Scheme 4.43) [74].

4.2 Catalytic Cycles Based on Iodine(V) Species

Hypervalent iodine(V) reagents, such as IBX (2-iodoxybenzoic acid) and DMP, have found widespread synthetic application as stoichiometric oxidants for the facile and selective oxidation of primary alcohols and secondary alcohols to the respective carbonyl compounds and for other important oxidative transformations

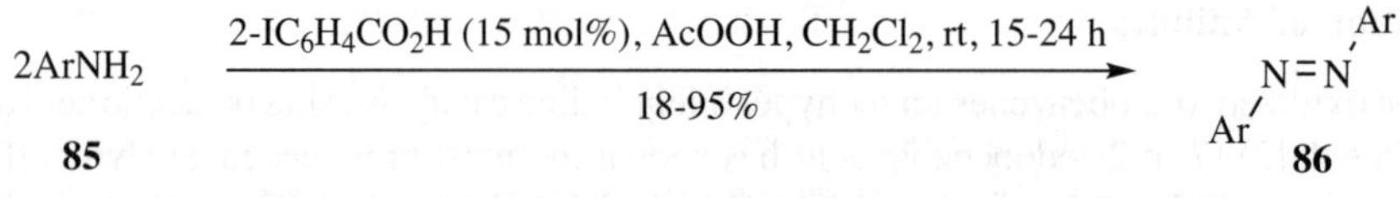

$2ArNH_2$ (**85**) → [2-$IC_6H_4CO_2H$ (15 mol%), AcOOH, CH_2Cl_2, rt, 15-24 h; 18-95%] → Ar–N=N–Ar (**86**)

Ar = H, 2-ClC_6H_4, 3-ClC_6H_4, 4-ClC_6H_4, 2-BrC_6H_4, 3-BrC_6H_4, 4-BrC_6H_4, 4-FC_6H_4, 4-MeC_6H_4, 3,5-$Me_2C_6H_4$, 4-$MeOC_6H_4$, 4-$NH_2C_6H_4$, 2-$HO_2CC_6H_4$, 4-(4'-NH_2-3'-MeC_6H_3)-2-MeC_6H_3, 3-$HC{\equiv}CC_6H_4$

Scheme 4.42

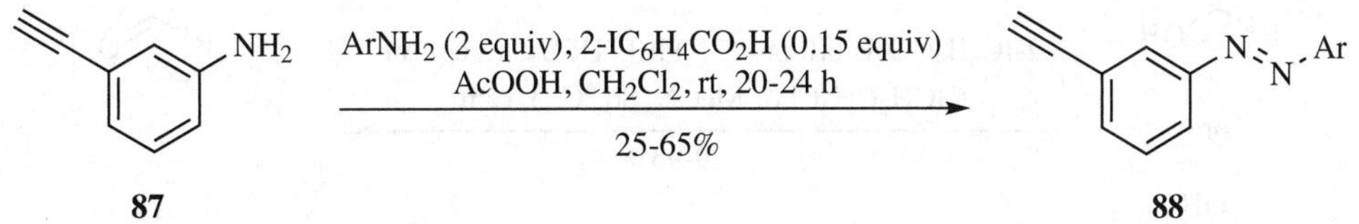
NH2
ArNH2 (2 equiv), 2-IC6H4CO2H (0.15 equiv)
AcOOH, CH2Cl2, rt, 20-24 h
25-65%
N
N
Ar

87 **88**

Ar = 3-ClC$_6$H$_4$, 4-ClC$_6$H$_4$, 3-BrC$_6$H$_4$, 4-BrC$_6$H$_4$,
4-FC$_6$H$_4$, 4-MeOC$_6$H$_4$

Scheme 4.43

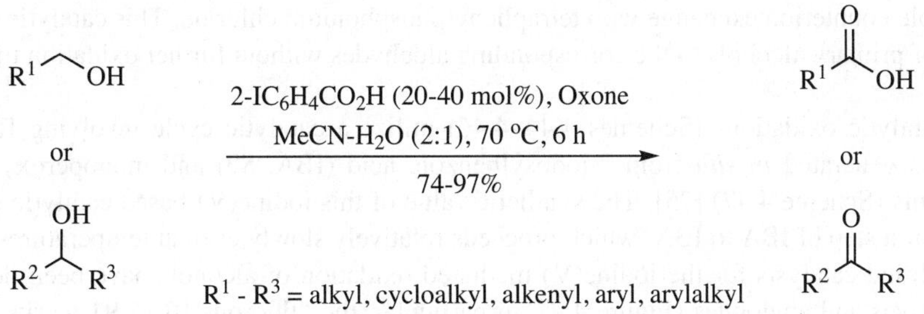
R1
OH
or
OH
R2
R3
2-IC6H4CO2H (20-40 mol%), Oxone
MeCN-H2O (2:1), 70 °C, 6 h
74-97%
R1 - R3 = alkyl, cycloalkyl, alkenyl, aryl, arylalkyl
O
R1
OH
or
O
R2
R3

Scheme 4.44

(Section 3.2). Catalytic application of the iodine(V) species in the oxidation of alcohols has been reviewed by Uyanik and Ishihara [10, 75]. The first examples of a catalytic application of an iodine(V) species (i.e., IBX) in the oxidation of alcohols using Oxone as a stoichiometric oxidant were independently reported by the groups of Vinod [4] in 2005 and Giannis [5] in 2006. Vinod's group employed 20–40 mol% of 2-iodobenzoic acid in a water–acetonitrile biphasic solvent system, in which primary and secondary alcohols were oxidized to carboxylic acids and ketones, respectively (Scheme 4.44) [4].

For a similar catalytic oxidation, Giannis' group utilized a water–ethyl acetate biphasic solvent system in the presence of 10 mol% of 2-iodobenzoic acid and tetrabutylammonium hydrogen sulfate as a phase-transfer catalyst. Under these conditions, primary benzylic alcohols were oxidized to the corresponding aldehydes, which, in contrast to the Vinod's procedure, did not undergo further oxidation (Scheme 4.45) [5].

Page and coworkers have demonstrated that primary and secondary alcohols can be oxidized to the respective aldehydes and ketones under reflux conditions in acetonitrile or dichloroethane in the presence of a catalytic amount of 2-iodobenzoic acid and using tetraphenylphosphonium monoperoxysulfate, $Ph_4P^+HSO_5^-$,

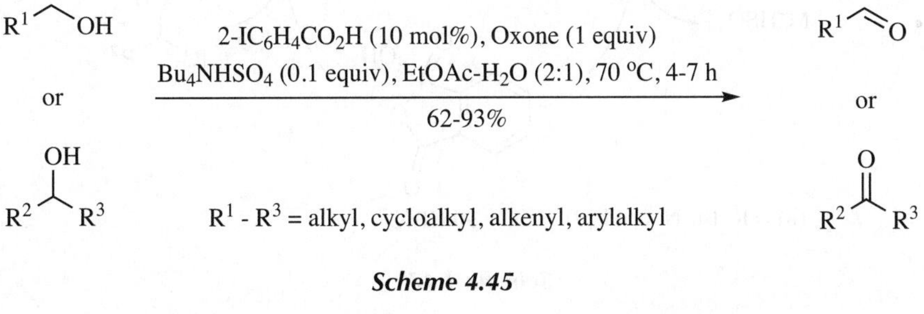
R1
OH
or
OH
R2
R3
2-IC6H4CO2H (10 mol%), Oxone (1 equiv)
Bu4NHSO4 (0.1 equiv), EtOAc-H2O (2:1), 70 °C, 4-7 h
62-93%
R1 - R3 = alkyl, cycloalkyl, alkenyl, arylalkyl
R1
O
or
O
R2
R3

Scheme 4.45

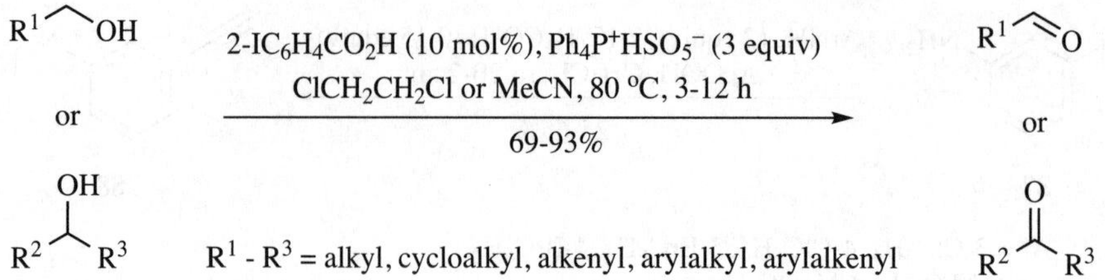

R^1–CH_2OH or R^2CH(OH)R^3 → [$2\text{-}IC_6H_4CO_2H$ (10 mol%), $Ph_4P^+HSO_5^-$ (3 equiv), $ClCH_2CH_2Cl$ or MeCN, 80 °C, 3-12 h; 69-93%] → R^1CHO or R^2C(O)R^3

R^1 - R^3 = alkyl, cycloalkyl, alkenyl, arylalkyl, arylalkenyl

Scheme 4.46

as the terminal oxidant (Scheme 4.46) [76]. Tetraphenylphosphonium monoperoxysulfate is prepared from Oxone by simple counterion exchange with tetraphenylphosphonium chloride. This catalytic system enables the oxidation of primary alcohols to the corresponding aldehydes without further oxidation to the carboxylic acids.

All these catalytic oxidations (Schemes 4.44–4.46) utilize a catalytic cycle involving IBX (**90**) as the reactive species generated *in situ* from 2-iodosylbenzoic acid (IBA, **89**) and monoperoxysulfate salts as terminal oxidants (Scheme 4.47) [75]. The synthetic value of this iodine(V)-based catalytic cycle is limited by the reoxidation step of IBA to IBX, which proceeds relatively slowly even at temperatures above 70 °C.

Several modified catalysts for the iodine(V)-mediated oxidation of alcohols have been developed based on IBX derivatives and analogues (Figure 4.2). In particular, the "fluorous IBX" **91** works efficiently as a catalyst for the oxidation of various alcohols to the corresponding carbonyl compounds in good to high yields

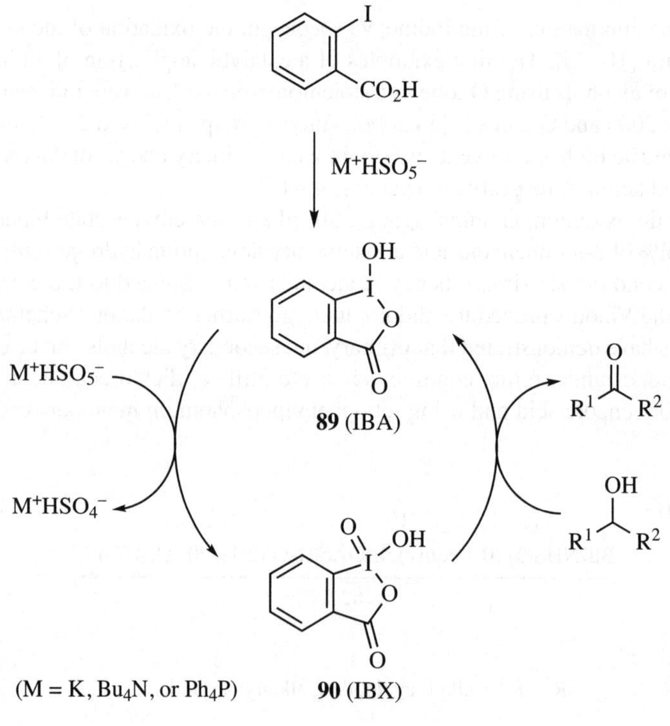

2-Iodobenzoic acid (CO_2H, I) → [$M^+HSO_5^-$] → **89** (IBA)

89 (IBA) + $M^+HSO_5^-$ → **90** (IBX) + $M^+HSO_4^-$

90 (IBX) + R^1CH(OH)R^2 → **89** (IBA) + R^1C(O)R^2

(M = K, Bu_4N, or Ph_4P)

Scheme 4.47

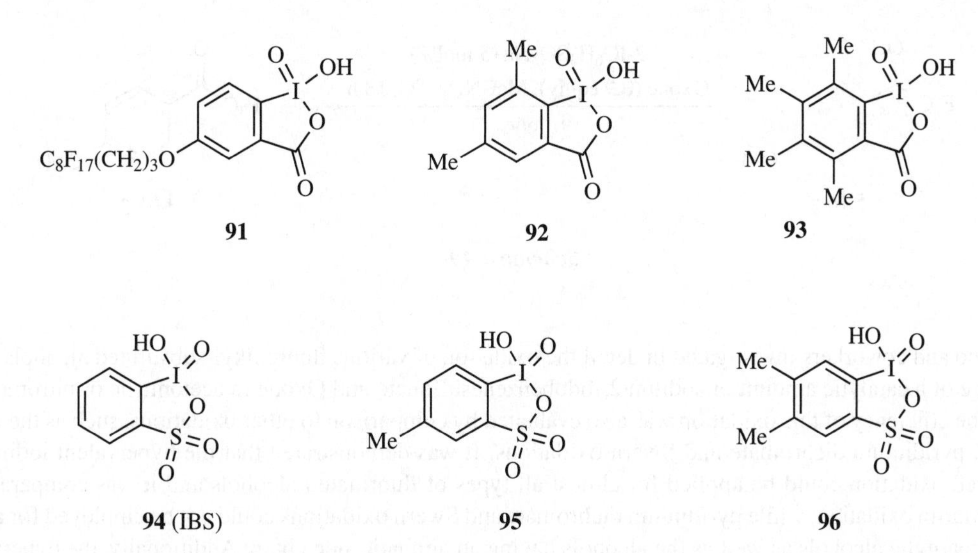

Figure 4.2 *IBX derivatives and analogues used as catalysts for the iodine(V)-mediated oxidation of alcohols.*

[77]. Fluorous IBX **91** can be regenerated from "fluorous IBA," which is readily recovered from the reaction mixture by simple filtration. The recovered reagent **91,** without further purification, retains its catalytic activity for at least five cycles.

The "twisted" dimethyl-IBX **92** and tetramethyl-IBX **93** have especially high catalytic activity and can catalyze the oxidations of alcohols and sulfides with Oxone at room temperature in common organic solvents [78]. The *ortho*-methyl groups in reagents **92** and **93** lower the activation energy corresponding to the rate-determining hypervalent twisting (Section 1.4.2) and also the steric relay between successive methyl groups twists the structure, which manifests itself in significant solubility in common organic solvents.

Ishihara and coworkers have found that 2-iodylbenzenesulfonic acid (IBS **94** [79], see Section 2.2.3) is an extremely active catalyst in the oxidations of alcohols with Oxone as the terminal oxidant [80]. Methyl substituted IBS derivatives **95** and **96** have even slightly higher catalytic activity. Catalysts **94–96** can be conveniently generated *in situ* by using sodium or potassium salts of the respective 2-iodylbenzenesulfonic acids and Oxone in such solvents as acetonitrile, ethyl acetate, or nitromethane. Scheme 4.48 illustrates the catalytic oxidation of alcohols using IBS and Oxone. Primary alcohols are oxidized to aldehydes, when a smaller quantity of Oxone (0.6–0.8 equiv) is used, or to carboxylic acids in the presence of excess Oxone (1.2 equiv) [80]. This reaction (Scheme 4.48) has been used for a selective larger scale oxidation of 4-bromobenzyl alcohol (6 g) to the corresponding aldehyde or carboxylic acid in excellent yields by controlling the amount of Oxone added in the presence of 1 mol% of the pre-catalyst, potassium 2-iodo-5-methylbenzenesulfonate [75].

$R^1CH(OH)R^2 \xrightarrow[\text{92-99\%}]{\text{2-IC}_6\text{H}_4\text{SO}_3\text{Na (1-5 mol\%), Oxone (0.6 equiv), MeCN, 70 °C}} R^1C(O)R^2$

R^1 = alkyl, cycloalkyl, alkenyl, aryl
R^2 = H or alkyl

Scheme 4.48

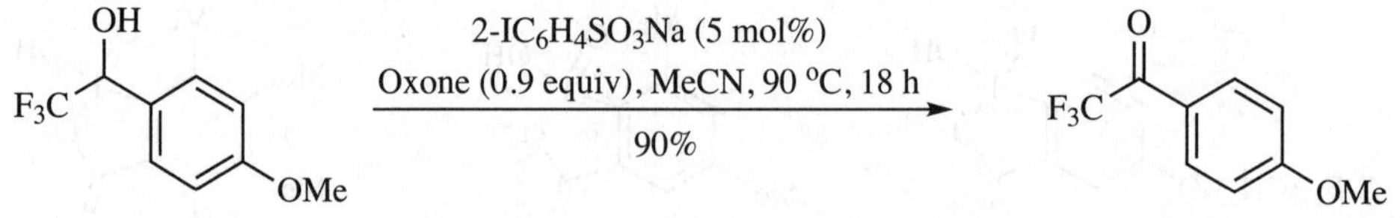

2-$IC_6H_4SO_3Na$ (5 mol%)
Oxone (0.9 equiv), MeCN, 90 °C, 18 h
90%

5.16 g → 4.61 g

Scheme 4.49

Konno and coworkers investigated in detail the oxidation of various fluoroalkyl-substituted alcohols in the presence of a catalytic amount of sodium 2-iodobenzenesulfonate and Oxone in acetonitrile or nitromethane [81]. The efficiency of this oxidation was also evaluated by comparison to other oxidations, such as the Dess–Martin, pyridinium dichromate and Swern oxidations. It was demonstrated that the hypervalent iodine(V)-catalyzed oxidation could be applied for almost all types of fluorinated alcohols and it was comparable to Dess–Martin oxidation, while pyridinium dichromate and Swern oxidations could not be employed for allylic and propargylic alcohols as well as the alcohols having an aliphatic side chain. Additionally, the hypervalent iodine-catalyzed oxidation could be applied for a larger scale reaction (Scheme 4.49) without any decrease in reaction efficiency [81].

Ishihara and coworkers have developed an oxidative rearrangement of tertiary allylic alcohols **97** to enones **98** with Oxone promoted by catalytic quantities of sodium 2-iodobenzenesulfonate (Scheme 4.50) [82]. Under these conditions, 5-methyl-IBS **95** is generated *in situ* and serves as the actual catalyst for the oxidation. Cyclic and acyclic substrates afford the corresponding enones in moderate to high yields. Notably, sterically demanding steroidal alcohol **99** has been converted into enone **100** in high yield [82].

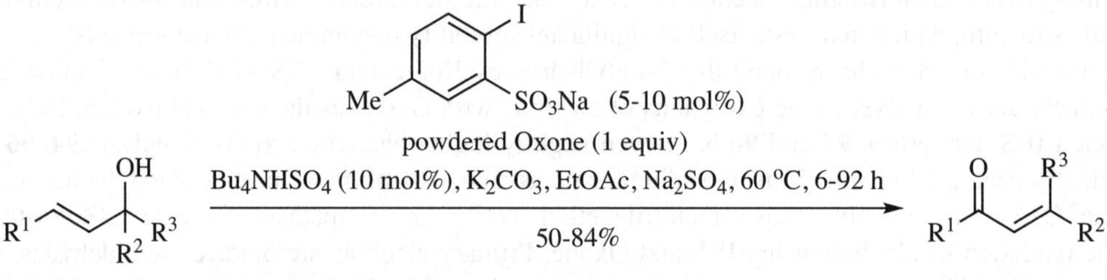

(5-10 mol%)
powdered Oxone (1 equiv)
Bu_4NHSO_4 (10 mol%), K_2CO_3, EtOAc, Na_2SO_4, 60 °C, 6-92 h
50-84%

97 → **98**

R^1 - R^3 = alkyl, phenyl, vinyl, ethynyl

conditions as above
69%

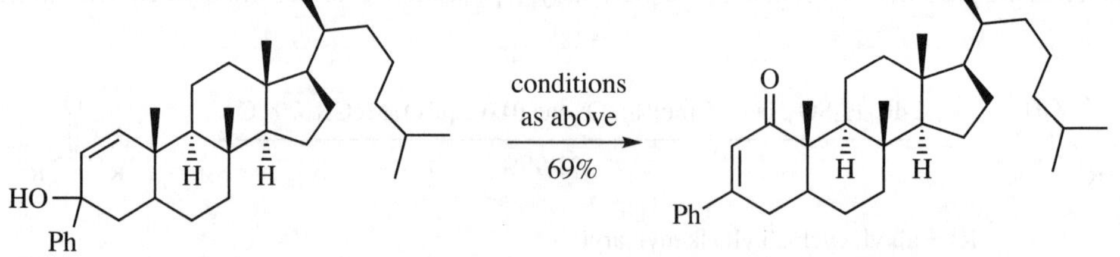

99 → **100**

Scheme 4.50

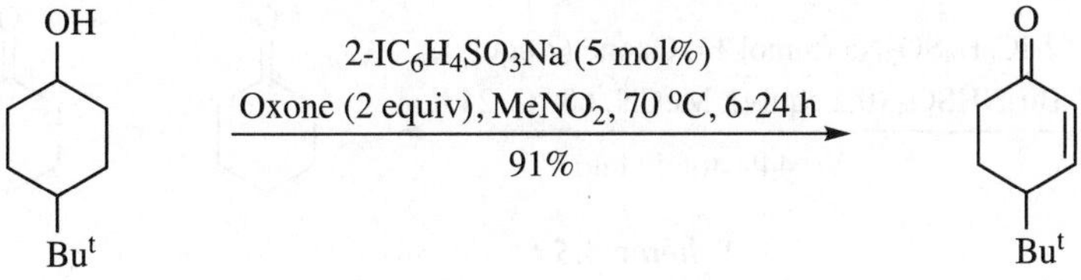

OH

Bu^t

101

$2\text{-}IC_6H_4SO_3Na$ (5 mol%)
Oxone (2 equiv), $MeNO_2$, 70 °C, 6-24 h
91%

Bu^t

102

$OSiPh_2Bu^t$

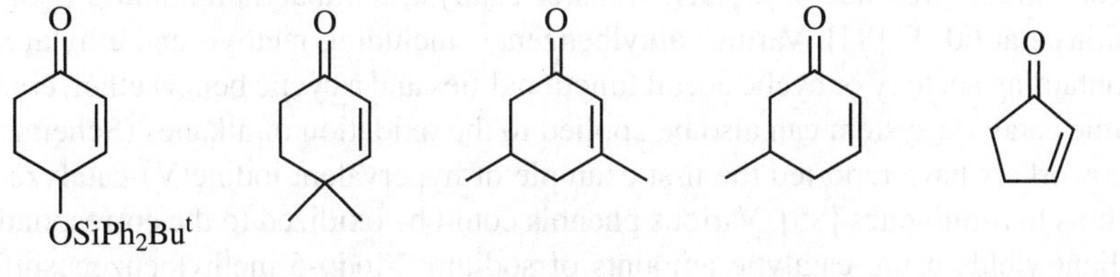

103 (88%) **104** (61%) **105** (74%) **106** (86%) **107** (70%)

Scheme 4.51

During research into the oxidation of cycloalkanols, Ishihara and coworkers also found that the selective oxidation of 4-*tert*-butylcyclohexanol (**101**) to 4-*tert*-butyl-2-cyclohexenone (**102**) can be achieved in excellent yield in the presence of a catalytic amount of sodium 2-iodobenzenesulfonate and two equivalents of Oxone (Scheme 4.51) [80]. Using this procedure, five- and six-membered cycloalkanols can be transformed into the corresponding enones **103–107** in good yields [80].

Vinod and coworkers were the first to develop a selective procedure for the oxidation of benzylic C–H bonds to the corresponding carbonyl functionalities using a catalytic amount of 2-iodobenzoic acid and Oxone as a stoichiometric oxidant in aqueous acetonitrile under reflux conditions (Scheme 4.52) [83]. The authors hypothesized that the active hypervalent iodine oxidant generated *in situ* might not be IBX (**90**) (Scheme 4.47) but, instead, a soluble derivative of IBX (**108**) that incorporates a peroxysulfate ligand. This intermediate is believed to oxidize a benzylic C–H bond via a single-electron transfer (SET) mechanism [83].

Zhang and coworkers have further improved the procedure for catalytic oxidation of benzylic C–H bonds using IBS as a catalyst, which is generated *in situ* by the oxidation of sodium 2-iodobenzenesulfonate

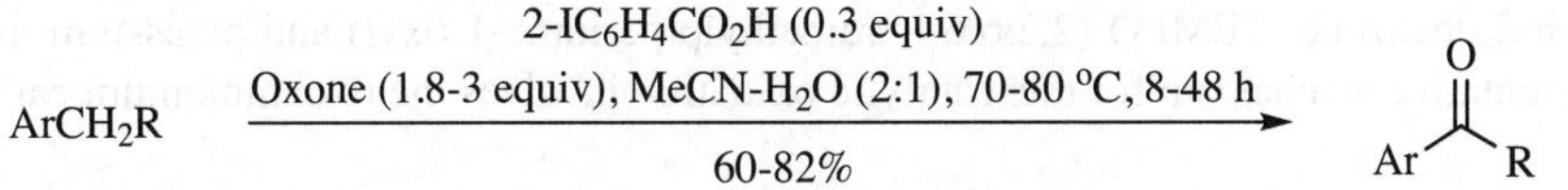

$ArCH_2R$

R = alkyl, aryl

$2\text{-}IC_6H_4CO_2H$ (0.3 equiv)
Oxone (1.8-3 equiv), $MeCN\text{-}H_2O$ (2:1), 70-80 °C, 8-48 h
60-82%

Ar–CO–R

SO_3K

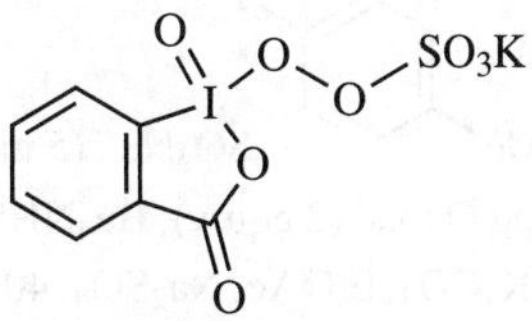

108

Scheme 4.52

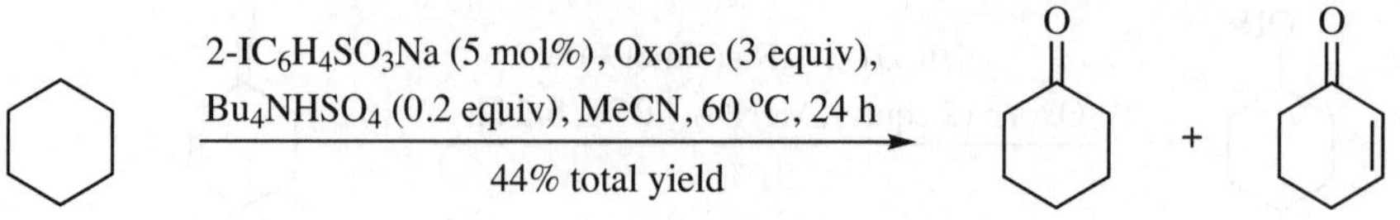

2-$IC_6H_4SO_3Na$ (5 mol%), Oxone (3 equiv), Bu_4NHSO_4 (0.2 equiv), MeCN, 60 °C, 24 h; 44% total yield

Scheme 4.53

(5 mol%) by Oxone in the presence of a phase-transfer catalyst, tetrabutylammonium hydrogen sulfate, in anhydrous acetonitrile at 60 °C [84]. Various alkylbenzenes, including methyl- and ethylarenes, substituted alkylbenzenes containing acetoxy or cyclic acetal functionalities and a cyclic benzyl ether, could be efficiently oxidized. The same catalytic system can also be applied to the oxidation of alkanes (Scheme 4.53) [84].

Ishihara and coworkers have reported the first example of hypervalent iodine(V)-catalyzed regioselective oxidation of phenols to *o*-quinones [85]. Various phenols could be oxidized to the corresponding *o*-quinones in good to excellent yields using catalytic amounts of sodium 2-iodo-5-methylbenzenesulfonate and stoichiometric amounts of Oxone as a co-oxidant under mild conditions; Scheme 4.54 shows a representative example [85].

4.3 Tandem Catalytic Systems Involving Hypervalent Iodine and other Co-catalysts

Several catalytic systems involving two or more sequential catalytic cycles and utilizing hypervalent iodine species as catalysts have been developed. An efficient, catalytic aerobic oxidation of primary and secondary alcohols to the corresponding aldehydes and ketones by using catalytic amounts of iodylbenzene, bromine and sodium nitrite in water (Scheme 4.55) has been reported by Liu and coworkers [86].

The proposed reaction mechanism includes three redox cycles (Scheme 4.55). In the first redox cycle, $PhIO_2$ is the active species that oxidizes the alcohol to the corresponding carbonyl compound and is reduced to (dihydroxyiodo)benzene, $PhI(OH)_2$. In the second cycle, $PhI(OH)_2$ is reoxidized to $PhIO_2$ with Br_2, which is reduced to HBr. In the third and final cycle, the oxidation of NO with O_2 produces NO_2, which reoxidizes HBr to Br_2 [86]. Notably, however, Ishihara's group was unable to reproduce the oxidation of alcohols under Liu's conditions and it was suggested, on the basis of several control experiments, that the actual oxidant of the alcohols in this case is Br_2 rather than $PhIO_2$ [87].

Li and coworkers developed an effective system for the oxidation of alcohols under an atmosphere of oxygen, without the need for any additional solvent or transition metal catalyst, by using catalytic amounts of (diacetoxyiodo)benzene, TEMPO (2,2,6,6-tetramethylpiperidine-1-oxyl) and potassium nitrate (Scheme 4.56) [88]. A tentative mechanism for this catalytic oxidation involves the oxoammonium cation **109,** which

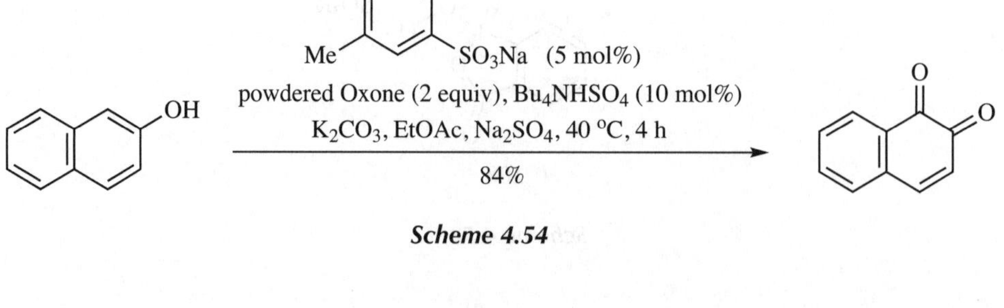

I; Me; SO_3Na (5 mol%); OH; powdered Oxone (2 equiv), Bu_4NHSO_4 (10 mol%); K_2CO_3, EtOAc, Na_2SO_4, 40 °C, 4 h; 84%; O; O

Scheme 4.54

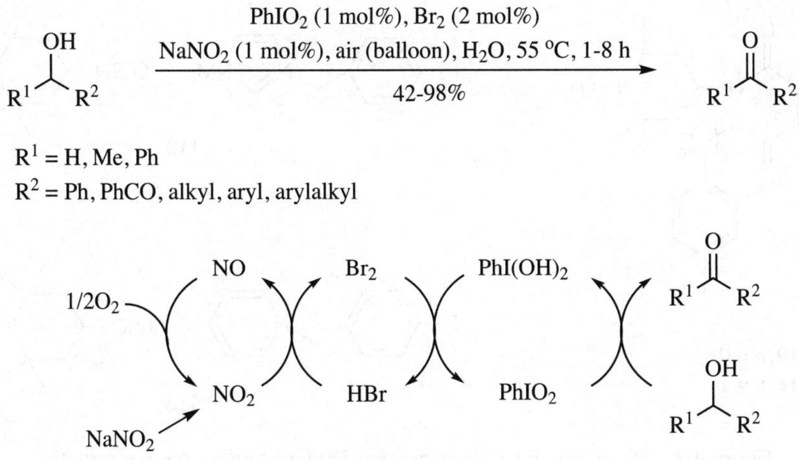

$PhIO_2$ (1 mol%), Br_2 (2 mol%)
$NaNO_2$ (1 mol%), air (balloon), H_2O, 55 °C, 1-8 h
42-98%

R^1 = H, Me, Ph
R^2 = Ph, PhCO, alkyl, aryl, arylalkyl

1/2O_2, NO, NO_2, $NaNO_2$, Br_2, HBr, $PhI(OH)_2$, $PhIO_2$

Scheme 4.55

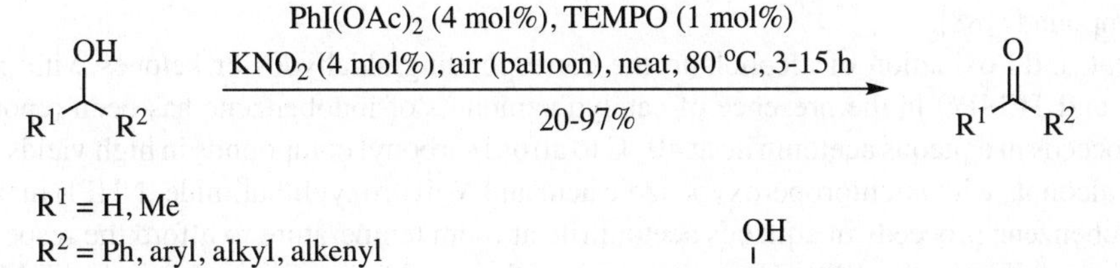

$PhI(OAc)_2$ (4 mol%), TEMPO (1 mol%)
KNO_2 (4 mol%), air (balloon), neat, 80 °C, 3-15 h
20-97%

R^1 = H, Me
R^2 = Ph, aryl, alkyl, alkenyl

1/2O_2, NO, NO_2, KNO_2, $PhI(OAc)_2$, PhI + AcOH, **109**, TEMPO

Scheme 4.56

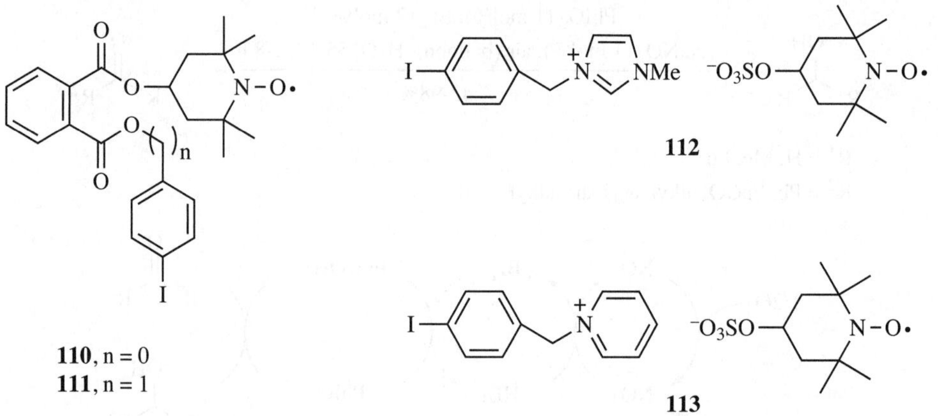

Figure 4.3 *Bifunctional catalysts bearing TEMPO and iodoarene moieties.*

serves as the active oxidant in this reaction, responsible for the oxidation of alcohols to the corresponding carbonyl compounds [88].

An efficient, mild oxidation of alcohols to the corresponding aldehydes or ketones with potassium peroxodisulfate and TEMPO in the presence of catalytic amounts of iodobenzene has been reported [89]. The oxidation proceeds in aqueous acetonitrile at 40 °C to afford carbonyl compounds in high yields. Likewise, the oxidation of alcohols with *m*-chloroperoxybenzoic acid and *N*-hydroxyphthalimide (NHPI) in the presence of catalytic iodobenzene proceeds in aqueous acetonitrile at room temperature to afford the respective carbonyl compounds in excellent yields [90]. The mechanism of these oxidations is similar to the TEMPO-promoted reaction shown in Scheme 4.56 and involves the oxidation of iodobenzene by *m*CPBA or $K_2S_2O_8$ *in situ* to form a λ^3-iodane species, which act as reoxidant of NHPI or TEMPOH.

Yakura and coworkers have developed bifunctional catalysts bearing TEMPO and iodobenzene moieties (structures **110** and **111**, Figure 4.3), which are useful for the environmentally benign oxidation of primary alcohols to carboxylic acids. Reaction of primary alcohols with a catalytic amount of catalyst **111** in the presence of peracetic acid as the terminal oxidant under mild conditions affords the corresponding carboxylic acids in excellent yields [91].

Likewise, efficient recyclable bifunctional catalysts **112** and **113** (Figure 4.3) bearing ionic liquid-supported TEMPO and iodoarene moieties have been developed and used for the environmentally benign catalytic oxidation of alcohols [92]. Compounds **112** and **113** have been tested as catalysts for the oxidation of alcohols to the corresponding carbonyl compounds using peracetic acid as a green and practical oxidant. Ion-supported catalysts **112** and **113** can be conveniently recovered from the reaction mixture and reused without any loss of catalytic activity (Section 5.5).

Hypervalent iodine species were demonstrated to have a pronounced catalytic effect on the metalloporphyrin-mediated oxygenations of aromatic hydrocarbons [93]. In particular, the oxidation of anthracene (**114**) to anthraquinone (**115**) with Oxone readily occurs at room temperature in aqueous acetonitrile in the presence of 5–20 mol% of iodobenzene and 5 mol% of a water-soluble iron(III)-porphyrin complex (**116**) (Scheme 4.57) [93]. 2-*tert*-Butylanthracene and phenanthrene also can be oxygenated under similar conditions in the presence of 50 mol% of iodobenzene. The oxidation of styrene in the presence of 20 mol% of iodobenzene leads to a mixture of products of epoxidation and cleavage of the double bond. Partially hydrogenated aromatic hydrocarbons (e.g., 9,10-dihydroanthracene, 1,2,3,4-tetrahydronaphthalene

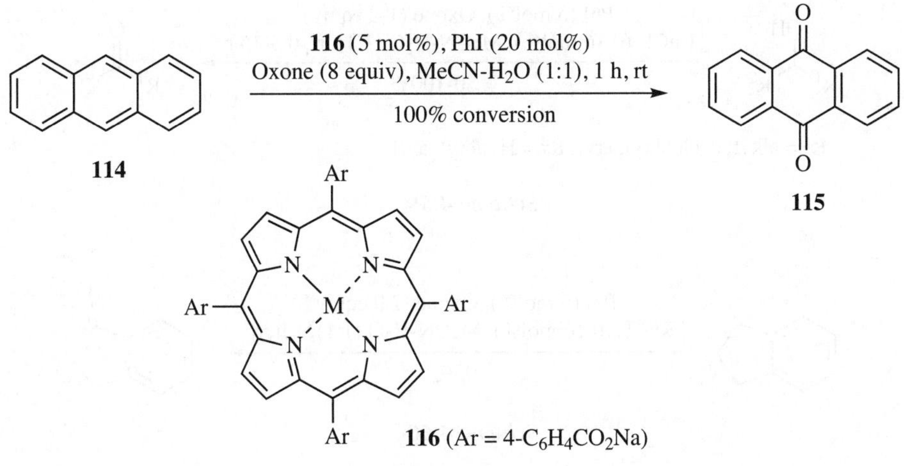

116 (5 mol%), PhI (20 mol%)
Oxone (8 equiv), MeCN-H_2O (1:1), 1 h, rt
100% conversion

114 → **115**

116 (Ar = 4-$C_6H_4CO_2Na$)

Scheme 4.57

and 2,3-dihydro-1*H*-indene) afford under these conditions the products of oxidation at the benzylic position in moderate yields.

The proposed mechanism for these catalytic oxidations includes two catalytic redox cycles: (i) initial oxidation of iodobenzene with Oxone, producing hydroxy(phenyl)iodonium ion and hydrated iodosylbenzene and (ii) the oxidation of iron(III)-porphyrin to the oxoiron(IV)-porphyrin cation-radical complex by the intermediate iodine(III) species (Scheme 4.58) [93]. The oxoiron(IV)-porphyrin cation-radical complex acts as the actual oxygenating agent toward aromatic hydrocarbons. The presence of the $[PhI(OH)]^+$ and $PhI(OH)_2$ species in solutions containing PhI and Oxone has been confirmed by ESI mass spectrometry [93].

Based on studies of the $RuCl_3$-catalyzed disproportionation of iodine(III) species to iodobenzene and iodylbenzene [53, 94–96], a mild and efficient tandem catalytic system for the oxidation of alcohols and hydrocarbons via a Ru(III)-catalyzed reoxidation of ArIO to $ArIO_2$ using Oxone as a stoichiometric oxidant has been developed [53, 96, 97]. In particular, various alcohols are smoothly oxidized in the presence of catalytic PhI and $RuCl_3$ in aqueous acetonitrile at room temperature to afford the respective ketones from secondary alcohols, or carboxylic acids from primary alcohols, in excellent isolated yields (Scheme 4.59) [97].

Likewise, various alkylbenzenes are selectively oxidized under these mild catalytic conditions to respective aromatic ketones in high yield; Scheme 4.60 shows a representative example [97]. Compared to the

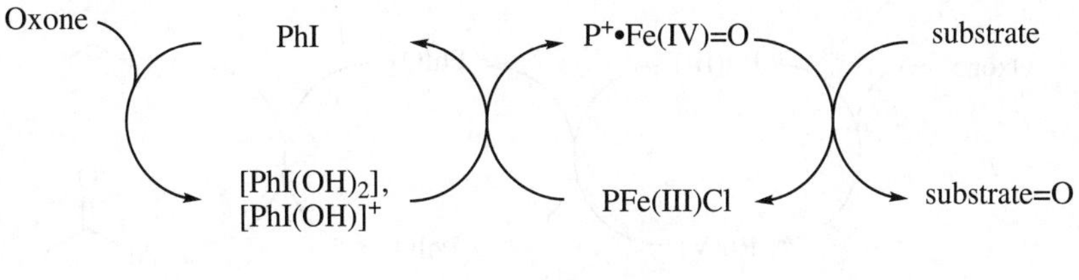

Oxone, PhI, $[PhI(OH)_2]$, $[PhI(OH)]^+$, $P^{+\bullet}Fe(IV)=O$, PFe(III)Cl, substrate, substrate=O

P = porphyrin

Scheme 4.58

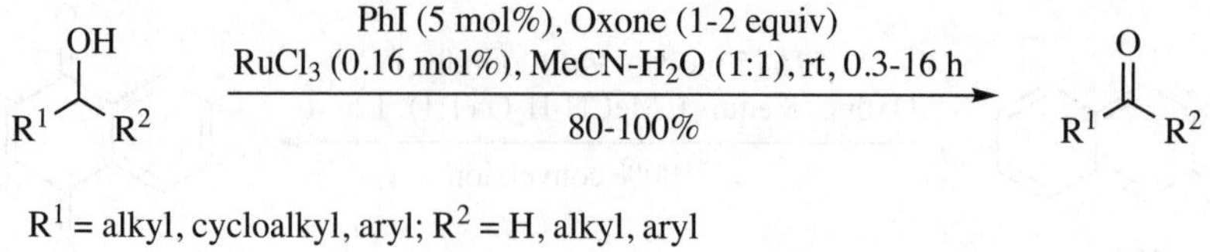

PhI (5 mol%), Oxone (1-2 equiv)
$RuCl_3$ (0.16 mol%), MeCN-H_2O (1:1), rt, 0.3-16 h
80-100%

R^1 = alkyl, cycloalkyl, aryl; R^2 = H, alkyl, aryl

Scheme 4.59

PhI (5 mol%), Oxone (2.0 equiv)
$RuCl_3$ (0.16 mol%), MeCN-H_2O (1:1), 1 h, rt
95%

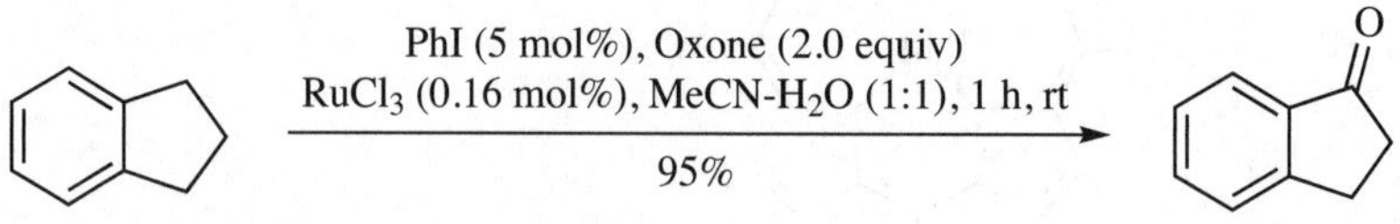

Scheme 4.60

high-temperature IBX/Oxone procedure (Section 4.2) [83], this protocol is much more selective and generally does not afford products of C–C bond cleavage and carboxylic acids.

A plausible, simplified mechanism for these catalytic oxidations includes two catalytic redox cycles (Scheme 4.61). The reaction starts with the initial oxidation of PhI to PhIO and then to $PhIO_2$ by the Oxone/Ru(III,V) system. The *in situ* generated highly active monomeric $PhIO_2$ species is responsible for the actual oxidation of organic substrates by known mechanisms [98, 99].

Efficient and recyclable bifunctional catalysts bearing silica-supported $RuCl_3$ and iodoarene moieties have been developed and used for the environmentally benign oxidation of alcohols or alkylarenes at the benzylic position. In the presence of these catalysts, the oxidation of alcohols or alkylbenzenes by Oxone as the stoichiometric oxidant affords the corresponding carbonyl compounds in high yields under mild conditions and convenient work-up. Furthermore, these SiO_2-supported bifunctional catalysts can be recovered by simple filtration and directly reused (Section 5.5) [100].

4.4 Catalytic Cycles Involving Iodide Anion or Elemental Iodine as Pre-catalysts

Catalytic reactions, utilizing iodide anion or elemental iodine as the pre-catalysts, may involve the iodine cation, hypoiodic acid, or inorganic hypervalent iodine(III) species as active oxidants. Numerous examples of such reactions have been reported since 2010; however, no systematic mechanistic studies have been published. The oxidative catalytic reactions utilizing iodide anion or elemental iodine as a catalyst or pre-catalyst have been summarized in several reviews [15, 101, 102].

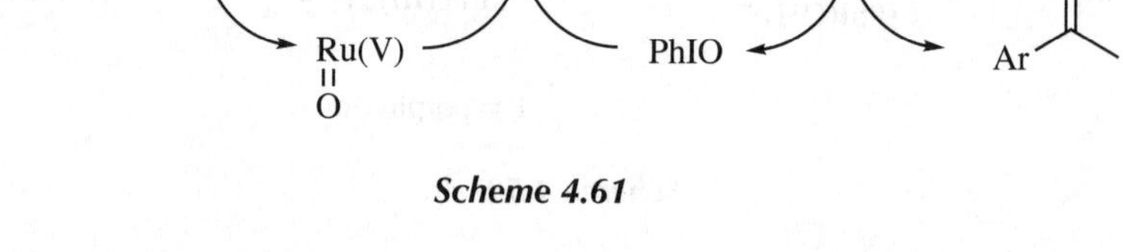

Oxone — Ru(III) — $PhIO_2$ — Ar
Ru(V)=O — PhIO — Ar(C=O)

Scheme 4.61

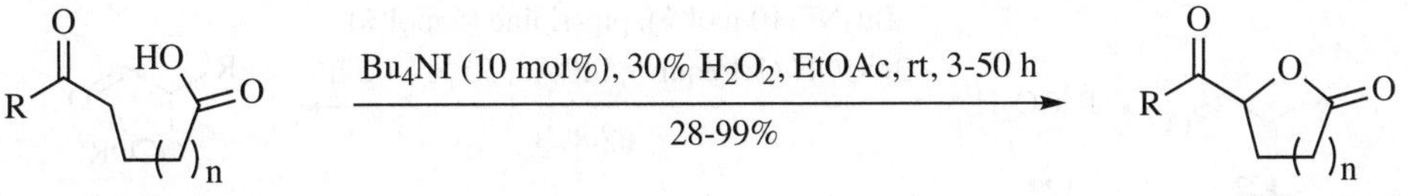

Bu_4NI (10 mol%), 30% H_2O_2, EtOAc, rt, 3-50 h; 28-99%

117 → **118**

R = aryl, hetaryl, alkyl
n = 1 or 2

Scheme 4.62

Ishihara and coworkers first found that tetrabutylammonium iodide can be used as a highly effective pre-catalyst for the oxylactonization of oxocarboxylic acids **117** with aqueous hydrogen peroxide even at room temperature (Scheme 4.62) [103]. These results are comparable with the procedure using iodobenzene with *m*CPBA (Section 4.1.1). Importantly, no Baeyer–Villiger products were obtained under these reaction conditions. Both γ-aryl- and γ-heteroarylcarbonyl-γ-butyrolactones **118** (R = aryl or heteroaryl, $n = 1$) were obtained in excellent yields and γ-alkylcarbonyl-γ-butyrolactones **118** (R = alkyl, $n = 1$) and δ-valerolactones **118** ($n = 2$) in moderate yields.

This catalytic procedure has been further extended to the oxidative coupling of ketones as well as 1,3-dicarbonyl compounds with carboxylic acids using a catalytic amount of Bu_4NI and *tert*-butyl hydroperoxide (TBHP) as the terminal oxidant [103]. Hydrogen peroxide is not an effective oxidant for this reaction; however, the use of a commercially available solution of TBHP in anhydrous nonane or decane at moderate heating gives excellent results. Various ketones (**119**) as well as 1,3-dicarbonyl compounds as substrates react with carboxylic acids **120** under these conditions to give the corresponding α-acyloxy ketones **121** in good to excellent yields (Scheme 4.63) [103]. A similar TBAI-catalyzed oxidative coupling of β-ketoesters with carboxylic acid has been reported in more recent work of Li, Zhou and Xu [104].

The α-oxyacylation of aldehydes can also be achieved under similar conditions in the presence of a catalytic amount of piperidine [103]. Thus, equimolar amounts of aldehydes **122** and acids **123** react upon mild heating in the presence of catalytic amounts of Bu_4NI and piperidine and TBHP as the terminal oxidant, in ethyl acetate to give α-acyloxy aldehydes **124** in high yields (Scheme 4.64). Several functional groups such as terminal or internal alkenyl, benzyloxy, silyloxy, acetal, halogen and ester are tolerated under these conditions.

The same Bu_4NI/TBHP catalytic system has also been applied towards the α-oxyacylation of ethers with carboxylic acids in ethyl acetate at 80 °C [105], which possibly occurs via a radical mechanism.

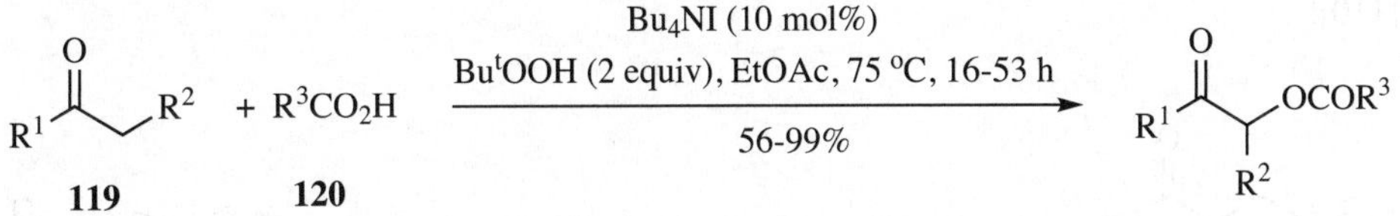

119 + R^3CO_2H (**120**) → **121** ($R^1COCH(R^2)OCOR^3$)

Bu_4NI (10 mol%), Bu^tOOH (2 equiv), EtOAc, 75 °C, 16-53 h; 56-99%

R^1 = aryl, hetaryl, alkyl
R^2 = H, alkyl, COMe, COPh, etc.
R^3 = aryl, Me, CH_2=CH, CH_2=C(Me)

Scheme 4.63

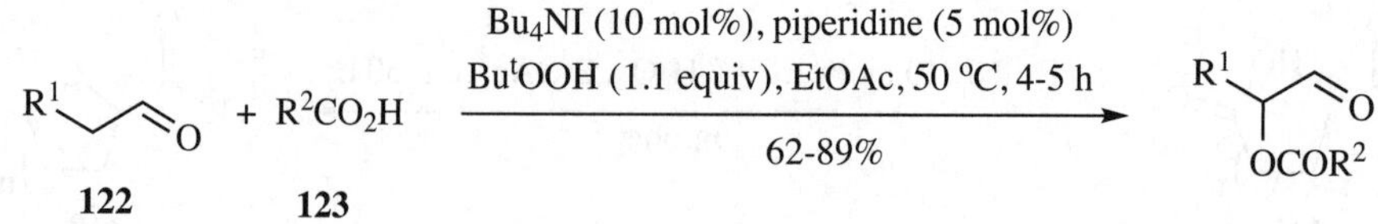

Bu_4NI (10 mol%), piperidine (5 mol%)
Bu^tOOH (1.1 equiv), EtOAc, 50 °C, 4-5 h
62-89%

122 + R^2CO_2H **123** → **124** ($OCOR^2$)

R^1 = benzyl, alkyl, alkenyl
R^2 = aryl, Me, CH_2=CH, CH_2=C(Me)

Scheme 4.64

During studies on the oxylactonization of oxocarboxylic acids **117** (Scheme 4.62), Ishihara and coworkers found that the catalytic oxidative system Bu_4NI/H_2O_2 could be applied for the oxidative cycloetherification of oxo-substituted phenols [101]. Thus, the oxidation of phenolic substrate **125** with two equivalents of 30% aqueous H_2O_2 in the presence of a catalytic amount of Bu_4NI in THF or diethyl ether at room temperature gave the corresponding 2-acyldihydrobenzofuran **126** in quantitative yield (Scheme 4.65). Notably, excellent chemoselectivity has been observed in this reaction (Scheme 4.65) with no phenol oxidation products detected.

Based on this reaction (Scheme 4.65), Ishihara and coworkers have developed a highly enantioselective oxidative cycloetherification of substrates **127** using hydrogen peroxide as the stoichiometric oxidant in the presence of the chiral quaternary ammonium iodide catalyst **128** (Scheme 4.66) [106]. The optically active 2-acyl-2,3-dihydrobenzofuran skeleton **129,** created in this cycloetherification, is a key structural unit in several biologically active compounds, such as entremirol and entremiridol.

In the later works by Li, Xue and coworkers, the Ishihara's hypoiodite-catalytic cycloetherification has been employed for the synthesis of bis-benzannelated spiro[5.5]ketals [107] and as a key step in a new synthesis of γ-rubromycin [108].

The results of several control experiments suggested that either tetrabutylammonium hypoiodite ($Bu_4N^+IO^-$) or iodite ($Bu_4N^+IO_2^-$), which should be generated *in situ* from tetrabutylammonium iodide and a co-oxidant, might be the active species in all these oxidations (Schemes 4.62–4.66) [103]. The presence of highly unstable iodite anions (IO_2^-) in reaction mixtures containing Bu_4NI/H_2O_2 was confirmed by means of negative ion ESI-MS analysis [109]. Although hypoiodite (IO^-) species were also detected, this species disappeared too quickly to obtain a reliable measurement.

A one-pot procedure for the α-tosyloxylation of ketones by the reaction of ketones with *m*CPBA and $TsOH{\cdot}H_2O$ in the presence of catalytic amounts of NH_4I and benzene in a mixture of MeCN and trifluoroethanol (8 : 2) has been reported (Scheme 4.67) [110]. This method has some advantages, such as mild reaction conditions with a simple procedure and it is suitable for preparing not only α-tosyloxy ketones but also other α-sulfonyloxy ketones. It has been suggested that [hydroxyl(tosyloxy)iodo]benzene, generated by the reaction of iodide anion with *m*CPBA, benzene and TsOH, serves as the active species in this reaction [110].

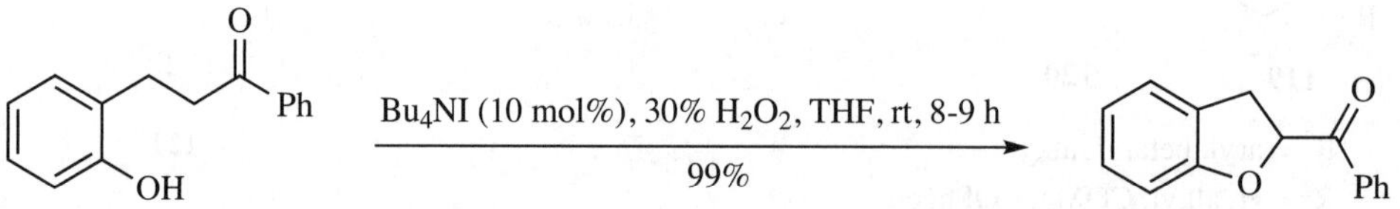

125 → **126**
Bu_4NI (10 mol%), 30% H_2O_2, THF, rt, 8-9 h
99%

Scheme 4.65

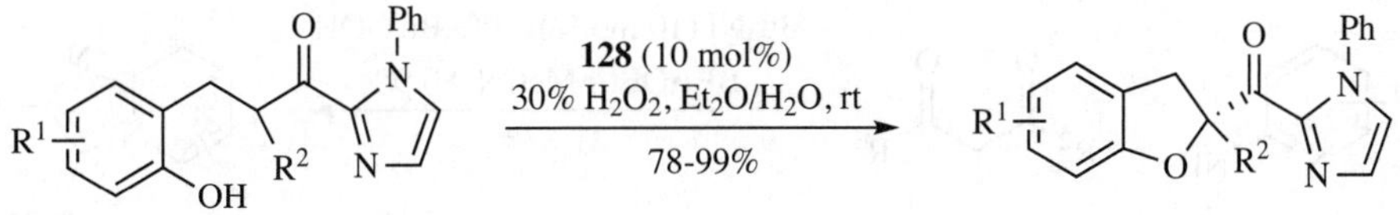

128 (10 mol%)
30% H_2O_2, Et_2O/H_2O, rt
78-99%

127

129
up to 96% ee

R^1 = 4-Cl, 4-F, 4-OR, 3,5-$(OMe)_2$, etc.
R^2 = H or Me

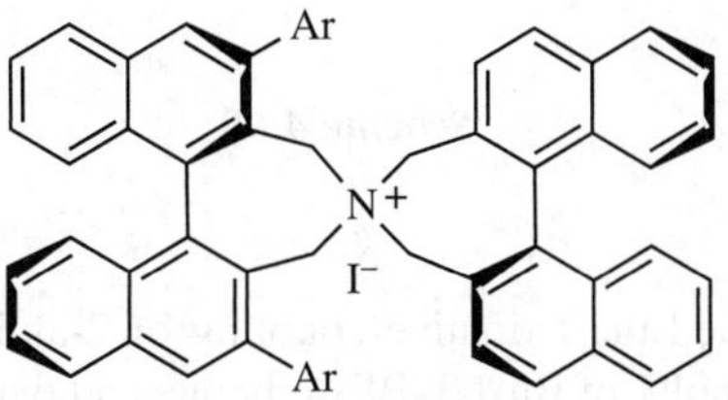

128, Ar = 3, 5-[3, 5-$(CF_3)_2C_6H_3]C_6H_3$

Scheme 4.66

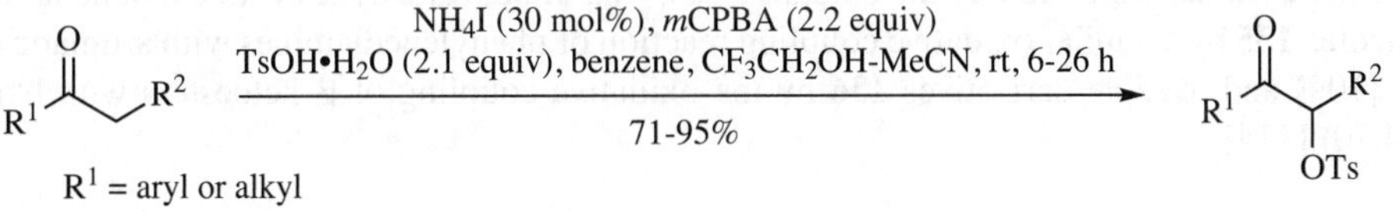

NH_4I (30 mol%), *m*CPBA (2.2 equiv)
$TsOH{\bullet}H_2O$ (2.1 equiv), benzene, CF_3CH_2OH-MeCN, rt, 6-26 h
71-95%

R^1 = aryl or alkyl
R^2 = H or alkyl

Scheme 4.67

The first examples of a hypoiodite-catalyzed oxidative C–N coupling reaction were independently reported by Nachtsheim's group [111] and Yu and Han's group [112] in 2011. Nachtsheim and coworkers found that the reaction of benzoxazoles **130** with various amines in the presence of a catalytic amount of Bu_4NI and 30% aqueous H_2O_2 or 70% aqueous TBHP as a terminal oxidant gave the corresponding C–N coupling products **131** in moderate to high yields (Scheme 4.68) [111]. The reaction was generally faster and the chemical yield of the products was higher if TBHP was used as an oxidant. The authors suggested that the *in situ* generated acetyl hypoiodite is the actual oxidant in this reaction [111].

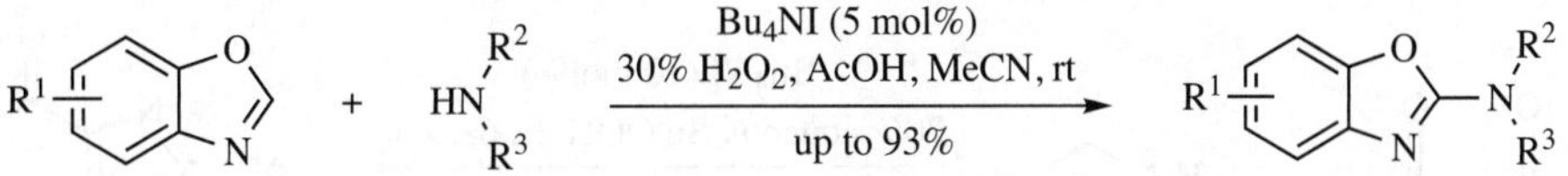

Bu_4NI (5 mol%)
30% H_2O_2, AcOH, MeCN, rt
up to 93%

130

131

R^1 = H, alkyl, Cl, NO_2
R^2 = H, alkyl
R^3 = alkyl

Scheme 4.68

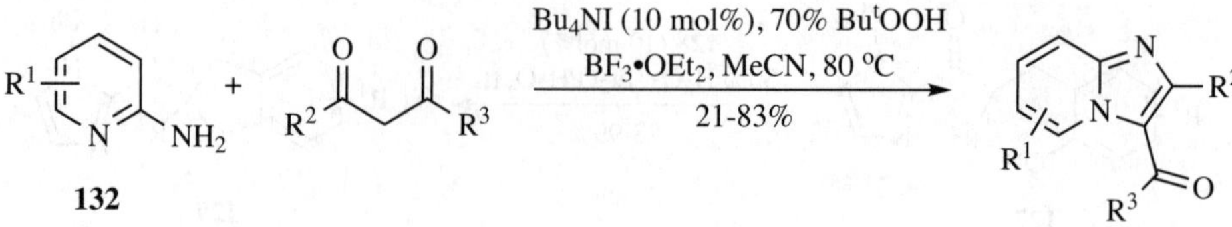

Bu_4NI (10 mol%), 70% Bu^tOOH
$BF_3•OEt_2$, MeCN, 80 °C
21-83%

132

R^1 = H, Me, Cl
R^2 = alkyl, aryl
R^3 = Me, OMe, OEt

133

Scheme 4.69

Yu, Han and coworkers reported the oxidative coupling of 2-aminopyridines **132** with β-ketoesters or 1,3-diones in the presence of 10 mol% of Bu_4NI, BF_3-etherate and two equivalents of 70% aqueous TBHP in acetonitrile (Scheme 4.69) [112]. The corresponding imidazo[1,2-*a*]pyridines **133** were obtained in moderate to high yields. The *in situ* generated hypoiodite or iodite species were suggested to be the actual oxidants under these conditions.

The catalytic oxidative system I^-/oxidant has also been applied for the synthesis of the following heterocyclic systems: 2-imidazolines **134** by the oxidative coupling of benzaldehydes with ethylenediamines [113], benzimidazoles **135** by a similar oxidative coupling reaction of phenylenediamines with aromatic or aliphatic aldehydes [109] and oxazole derivatives **136** by the oxidative coupling of β-ketoesters with benzylamines (Scheme 4.70) [114].

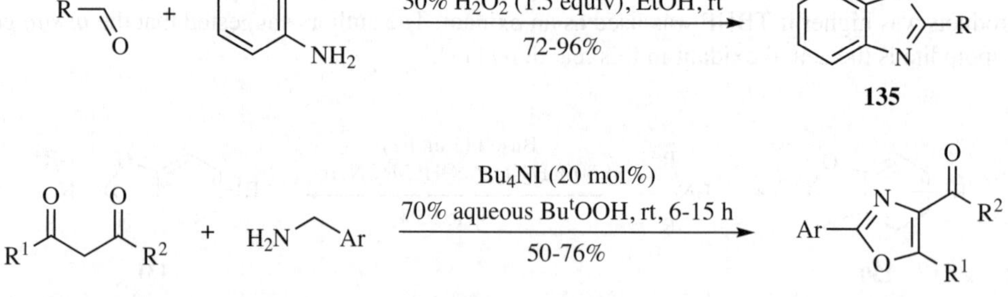

NaI (40 mol%), 30% H_2O_2 (excess)
Bu^tOH, anhydrous $MgSO_4$, 80 °C
62-99%

134

Bu_4NI (10 mol%)
30% H_2O_2 (1.5 equiv), EtOH, rt
72-96%

135

Bu_4NI (20 mol%)
70% aqueous Bu^tOOH, rt, 6-15 h
50-76%

136

R^1 = alkyl or Ph
R^2 = OMe, OEt, OBu^t, Me, Ph
Ar = Ph, 4-ClC_6H_4, 2-ClC_6H_4, 4-FC_6H_4, 4-MeC_6H_4, 2-furyl

Scheme 4.70

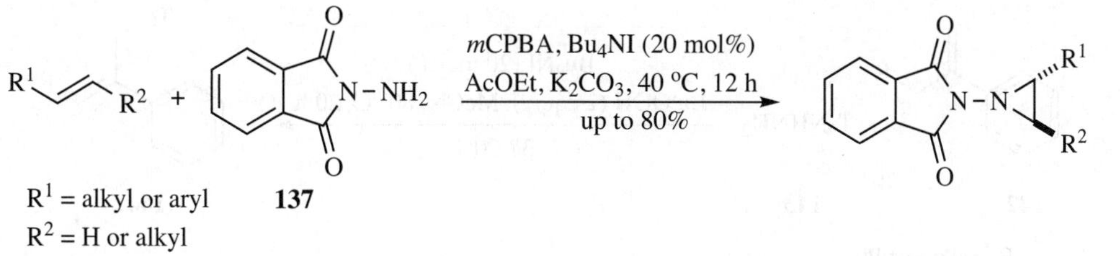

Scheme 4.71

An efficient metal-free catalytic procedure for aziridination of alkenes using tetrabutylammonium iodide as a catalyst, *m*-chloroperoxybenzoic acid as the terminal oxidant and *N*-aminophthalimide (**137**) as a nitrenium precursor has been developed (Scheme 4.71) [45].

The proposed mechanism of this catalytic aziridination is outlined in Scheme 4.72 [45]. The active species, hypoiodous acid (**138**) (or iodine 3-chlorobenzoate, IOCOAr), generated from Bu_4NI and *m*CPBA, further reacts with alkene to give the iodonium ion **139,** which is then opened at the benzylic position by *N*-aminophthalimide (**137**) (or the corresponding potassium salt, PhthNHK, formed from **137** in the presence of K_2CO_3). This sequence of reactions gives β-iodo-*N*-aminophthalimide **140,** cyclization of which affords the aziridine product **141** and iodide anion. The regenerated iodide anion continues the catalytic cycle [45].

Li and coworkers have reported Bu_4NI-catalyzed allylic sulfonylation of α-methyl styrene derivatives with sulfonyl hydrazides **143** using TBHP (Bu^tOOH) as the terminal oxidant (Scheme 4.73) [115]. The mechanism of this reaction involves the generation of sulfonyl radicals, Ts•, from sulfonyl hydrazides **143** by the Bu_4NI/TBHP catalytic system, followed by the addition of Ts• to α-methyl styrene derivatives **142** to give the corresponding allylic sulfones **144.**

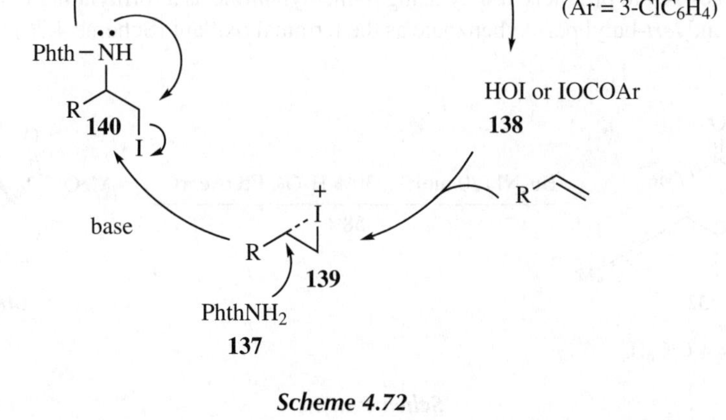

Scheme 4.72

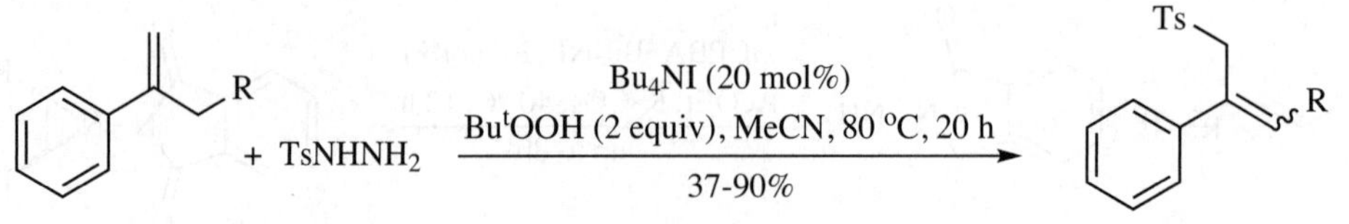

Bu_4NI (20 mol%), Bu^tOOH (2 equiv), MeCN, 80 °C, 20 h, 37-90%

142 + $TsNHNH_2$ **143** → **144**

R = alkyl or Ph

Scheme 4.73

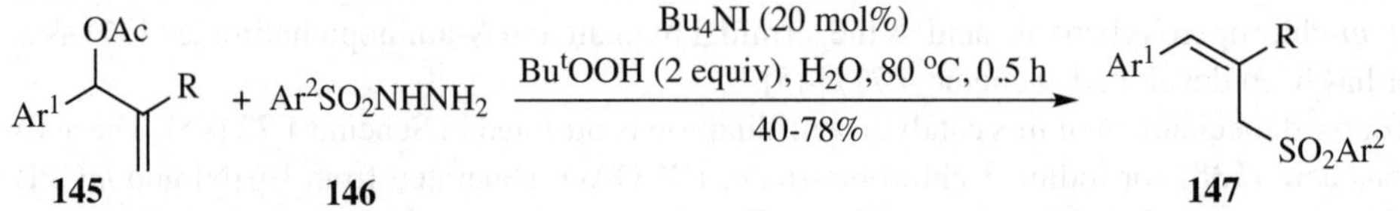

Bu_4NI (20 mol%), Bu^tOOH (2 equiv), H_2O, 80 °C, 0.5 h, 40-78%

145 + $Ar^2SO_2NHNH_2$ **146** → **147**

R = CO_2Me, CO_2Et, CO_2Bu, CN
Ar^1 = Ph, 4-MeC_6H_4, 4-ClC_6H_4, 4-$Pr^iC_6H_4$, 2-$MeOC_6H_4$, 3-$NO_2C_6H_4$, 2-thienyl, etc.
Ar^2 = Ph, 4-MeC_6H_4, 4-BrC_6H_4

Scheme 4.74

This reaction has been further extended to a catalytic procedure for the synthesis of allyl aryl sulfone derivatives**147** from Baylis–Hillman acetates **145** and sulfonyl hydrazides **146** using Bu_4NI as the catalyst and TBHP as an oxidation agent in water (Scheme 4.74) [116].

Only two examples of the C–C bond-forming Bu_4NI-catalyzed reactions have been reported. In 2011, during their studies on the hypervalent iodine-catalyzed oxidative cyclization of δ-alkynyl β-ketoesters with *m*CPBA (see Scheme 4.14 in Section 4.1.2), Moran and coworkers found that treatment of starting material **32** with a catalytic amount of Bu_4NI and 30% aqueous H_2O_2 gave product **148** in moderate yield (Scheme 4.75) [117]. Although the mechanism is not clear, this is the first example of a hypoiodite-catalyzed C–C coupling reaction.

The second example of the hypoiodite-catalyzed C–C coupling reaction is represented by the C3-selective formylation of indoles **149** to products **150** by using *N*-methylaniline as a formylating reagent in the presence of catalytic Bu_4NI and *tert*-butyl peroxybenzoate as the terminal oxidant (Scheme 4.76) [118]. Pivalic acid is

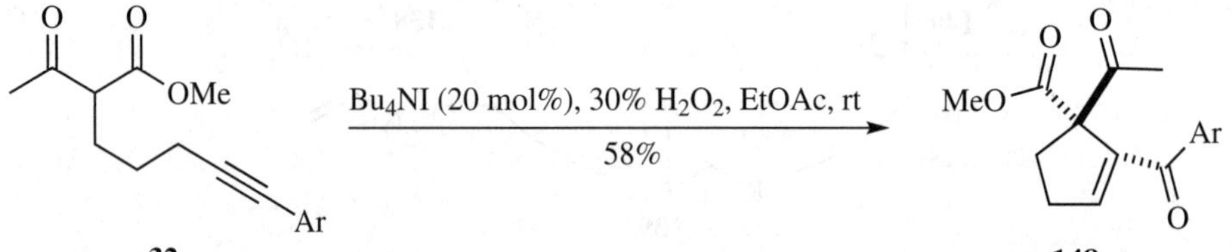

Bu_4NI (20 mol%), 30% H_2O_2, EtOAc, rt, 58%

32 → **148**

Ar = 4-ClC_6H_4

Scheme 4.75

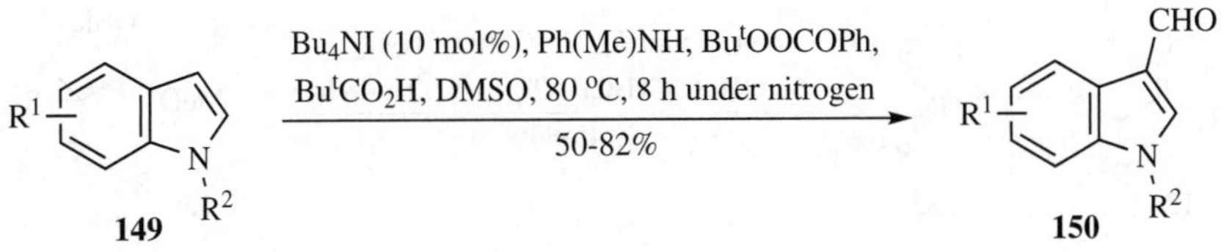

Bu_4NI (10 mol%), Ph(Me)NH, $Bu^tOOCOPh$, Bu^tCO_2H, DMSO, 80 °C, 8 h under nitrogen; 50-82%

149 → **150**

R^1 = H, F, Cl, Et, Br, I, OCH_2Ph, OMe, CN, Me, CO_2Me in different ring positions
R^2 = H, Me, CH_2Ph, Ph

Scheme 4.76

used as an additive since it has been shown to suppress decomposition of indoles under oxidative conditions. This reaction probably proceeds via a free radical process [118].

Several oxidative catalytic systems utilizing elemental iodine as the catalyst have been developed. Wang and colleagues have reported several tandem oxidative cyclization reactions using I_2 as a catalyst and 70% aqueous TBHP (Bu^tOOH) as a stoichiometric oxidant (Scheme 4.77) [119–122]. Heteroaromatic compounds such as oxazoles **151,** quinazolines **152** and pyridine derivatives **153** were synthesized in moderate to high yields under these catalytic conditions. The authors suggested that the I_2/I^- catalytic cycle might play an important role in the radical mechanism under these conditions [122].

A highly efficient α-amination of sterically divergent aldehydes **154** using secondary amines **155** as nitrogen source, iodine as the pre-catalyst and sodium percarbonate as an environmentally benign co-oxidant has been

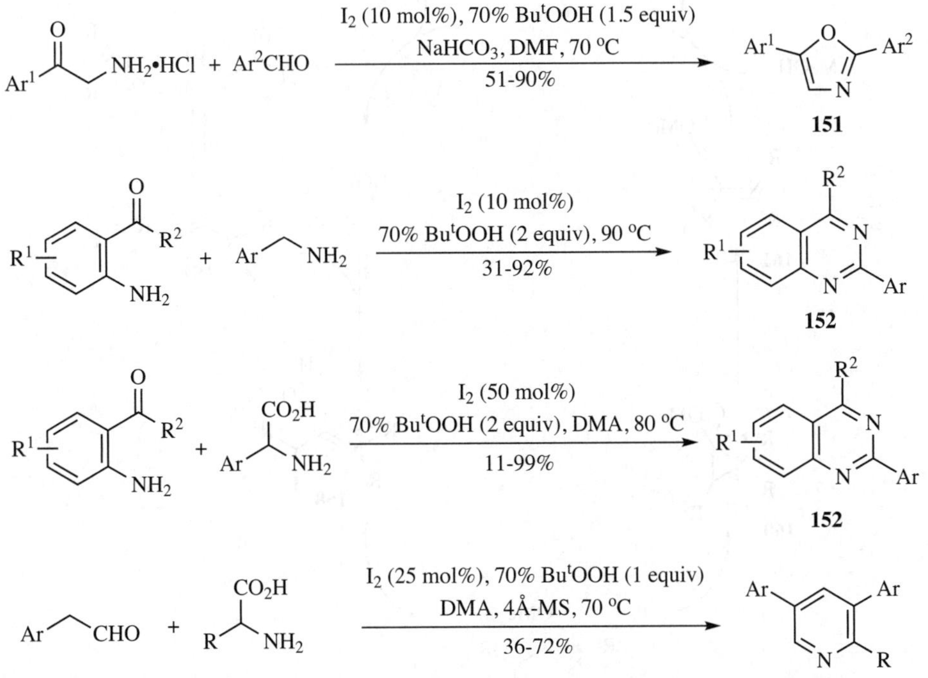

$Ar^1C(O)CH_2NH_2$•HCl + Ar^2CHO → **151**; I_2 (10 mol%), 70% Bu^tOOH (1.5 equiv), $NaHCO_3$, DMF, 70 °C; 51-90%

+ $ArCH_2NH_2$ → **152**; I_2 (10 mol%), 70% Bu^tOOH (2 equiv), 90 °C; 31-92%

+ $ArCH(CO_2H)NH_2$ → **152**; I_2 (50 mol%), 70% Bu^tOOH (2 equiv), DMA, 80 °C; 11-99%

$ArCH_2CHO$ + $RCH(CO_2H)NH_2$ → **153**; I_2 (25 mol%), 70% Bu^tOOH (1 equiv), DMA, 4Å-MS, 70 °C; 36-72%

Scheme 4.77

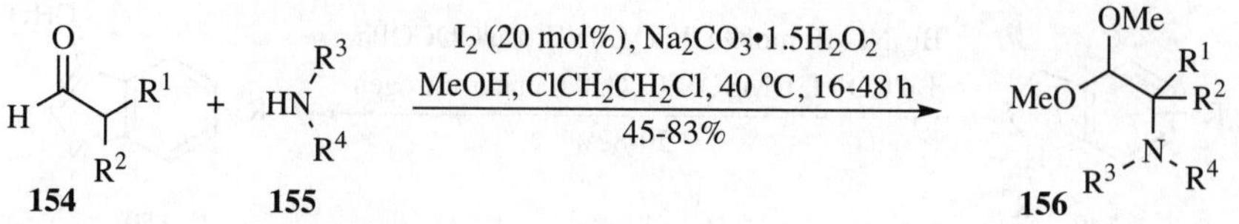

R^1 = aryl, alkyl and R^2 = H or $R^1 + R^2 = (CH_2)_5$
R^3 and R^4 = CH_2Ph, allyl, CH_2CO_2Me, Me, CH(Me)Ph

Scheme 4.78

reported (Scheme 4.78) [123]. The reaction affords synthetically useful α-amino acetals **156** in good yields and tolerates a wide range of functionalities, such as benzyl, allyl, or ester groups, as well as bulky aldehydes and secondary amine derivatives.

On the basis of control experiments a mechanism for the α-amination of aldehydes catalyzed by *in situ* generated hypoiodite has been proposed (Scheme 4.79) [123]. In the first step, the active cationic iodine species, hypoiodite acid, which is thought to function as a one-electron oxidizing reagent or electrophilic

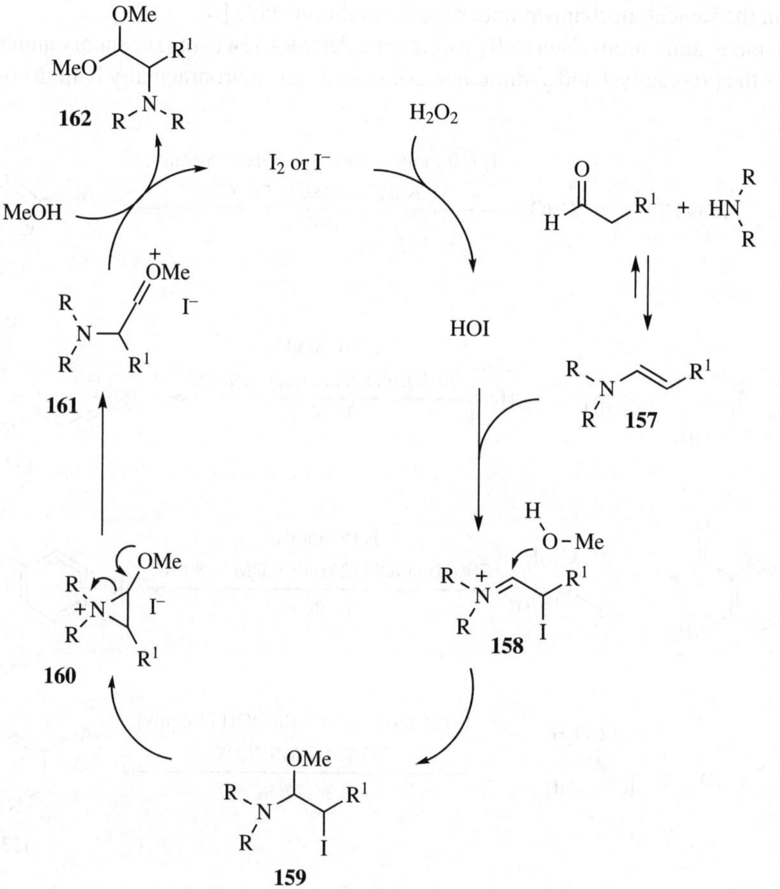

Scheme 4.79

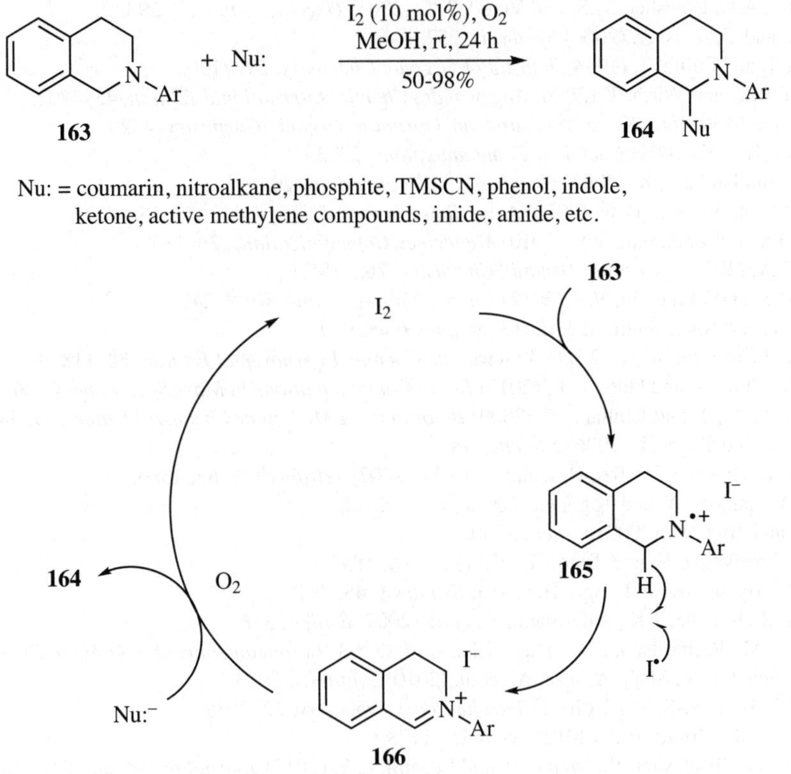

Scheme 4.80

reagent, is formed by oxidation of iodine (I_2) or iodide (I^-) with hydrogen peroxide. In the second step, hypoiodite reacts with enamine **157** to provide iminium ion **158,** the existence of which was confirmed by a control experiment; hydrolysis of the intermediate **158** led to the corresponding α-iodoaldehyde. In the third step, methanol attacks iminium ion **158** to give the iodo-substituted intermediate **159,** which undergoes intramolecular cyclization to afford aziridinium ion **160.** Finally, an additional methanol molecule captures the ring-opened intermediate **161** to afford the final product **162.**

A versatile iodine-based aerobic catalytic system (I_2 and O_2) for C–H functionalization of tetrahydroisoquinolines **163** using various nucleophiles (Nu:) has been developed by Prabhu and coworkers (Scheme 4.80) [124]. This cross-dehydrogenative coupling reaction is compatible with a large number of nucleophiles and is performed under ambient reaction conditions. A tentative reaction mechanism includes the generation of iminium iodide **166** by the reaction of tetrahydroisoquinoline **163** with molecular I_2 through a radical–cation intermediate **165,** followed by the reaction of **166** with the nucleophile and O_2 furnishing the coupled product **164** and regenerating I_2 (Scheme 4.80) [124].

References

1. Akiba, K.y. (1999) *Chemistry of Hypervalent Compounds*, Wiley-VCH Verlag GmbH, Weinheim.
2. Dohi, T., Maruyama, A., Yoshimura, M., *et al.* (2005) *Angewandte Chemie, International Edition*, **44**, 6193.
3. Ochiai, M., Takeuchi, Y., Katayama, T., *et al.* (2005) *Journal of the American Chemical Society*, **127**, 12244.

4. Thottumkara, A.P., Bowsher, M.S. and Vinod, T.K. (2005) *Organic Letters*, **7**, 2933.
5. Schulze, A. and Giannis, A. (2006) *Synthesis*, 257.
6. Fuchigami, T. and Fujita, T. (1994) *Journal of Organic Chemistry*, **59**, 7190.
7. Richardson, R.D. and Wirth, T. (2006) *Angewandte Chemie, International Edition*, **45**, 4402.
8. Ochiai, M. and Miyamoto, K. (2008) *European Journal of Organic Chemistry*, 4229.
9. Dohi, T. and Kita, Y. (2009) *Chemical Communications*, 2073.
10. Uyanik, M. and Ishihara, K. (2009) *Chemical Communications*, 2086.
11. Ngatimin, M. and Lupton, D.W. (2010) *Australian Journal of Chemistry*, **63**, 653.
12. Yusubov, M.S. and Zhdankin, V.V. (2010) *Mendeleev Communications*, **20**, 185.
13. Zhdankin, V.V. (2011) *Journal of Organic Chemistry*, **76**, 1185.
14. Yusubov, M.S. and Zhdankin, V.V. (2012) *Current Organic Synthesis*, **9**, 247.
15. Finkbeiner, P. and Nachtsheim, B.J. (2013) *Synthesis*, **45**, 979.
16. Liang, H. and Ciufolini, M.A. (2011) *Angewandte Chemie, International Edition*, **50**, 11849.
17. Wang, H.-Y., Zhou, J. and Guo, Y.-L. (2012) *Rapid Communications in Mass Spectrometry*, **26**, 616.
18. Uyanik, M., Yasui, T. and Ishihara, K. (2009) *Bioorganic & Medicinal Chemistry Letters*, **19**, 3848.
19. Yamamoto, Y. and Togo, H. (2006) *Synlett*, 798.
20. Yamamoto, Y., Kawano, Y., Toy, P.H. and Togo, H. (2007) *Tetrahedron*, **63**, 4680.
21. Akiike, J., Yamamoto, Y. and Togo, H. (2007) *Synlett*, 2168.
22. Tanaka, A. and Togo, H. (2009) *Synlett*, 3360.
23. Tanaka, A., Moriyama, K. and Togo, H. (2011) *Synlett*, 1853.
24. Kikui, H., Moriyama, K. and Togo, H. (2013) *Synthesis*, **45**, 791.
25. Richardson, R.D., Page, T.K., Altermann, S., *et al.* (2007) *Synlett*, 538.
26. Altermann, S.M., Richardson, R.D., Page, T.K., *et al.* (2008) *European Journal of Organic Chemistry*, 5315.
27. Farooq, U., Schafer, S., Shah, A.-u.-H.A., *et al.* (2010) *Synthesis*, 1023.
28. Yu, J., Cui, J., Hou, X.-S., *et al.* (2011) *Tetrahedron: Asymmetry*, **22**, 2039.
29. Rodriguez, A. and Moran, W.J. (2012) *Synthesis*, 1178.
30. Guilbault, A.-A., Basdevant, B., Wanie, V. and Legault, C.Y. (2012) *Journal of Organic Chemistry*, **77**, 11283.
31. Guilbault, A.-A. and Legault, C.Y. (2012) *ACS Catalysis*, **2**, 219.
32. Pu, Y., Gao, L., Liu, H. and Yan, J. (2012) *Synthesis*, 99.
33. Zhong, W., Liu, S., Yang, J., *et al.* (2012) *Organic Letters*, **14**, 3336.
34. Braddock, D.C., Cansell, G. and Hermitage, S.A. (2006) *Chemical Communications*, 2483.
35. He, Y., Pu, Y., Shao, B. and Yan, J. (2011) *Journal of Heterocyclic Chemistry*, **48**, 695.
36. Fabry, D.C., Stodulski, M., Hoerner, S. and Gulder, T. (2012) *Chemistry–A European Journal*, **18**, 10834.
37. Liu, H. and Tan, C.-H. (2007) *Tetrahedron Letters*, **48**, 8220.
38. Yan, J., Wang, H., Yang, Z. and He, Y. (2009) *Synlett*, 2669.
39. Zhou, Z.-S. and He, X.-H. (2010) *Tetrahedron Letters*, **51**, 2480.
40. Miyamoto, K., Sei, Y., Yamaguchi, K. and Ochiai, M. (2009) *Journal of the American Chemical Society*, **131**, 1382.
41. Wu, X.-L. and Wang, G.-W. (2009) *Tetrahedron*, **65**, 8802.
42. Xia, J.-J., Wu, X.-L. and Wang, G.-W. (2008) *ARKIVOC*, (xvi), 22.
43. Wu, X.-L. and Wang, G.-W. (2008) *European Journal of Organic Chemistry*, 6239.
44. Richardson, R.D., Desaize, M. and Wirth, T. (2007) *Chemistry–A European Journal*, **13**, 6745.
45. Yoshimura, A., Middleton, K.R., Zhu, C., *et al.* (2012) *Angewandte Chemie, International Edition*, **51**, 8059.
46. Zhou, Z. and He, X. (2011) *Synthesis*, 207.
47. Samanta, R., Bauer, J.O., Strohmann, C. and Antonchick, A.P. (2012) *Organic Letters*, **14**, 5518.
48. Antonchick, A.P., Samanta, R., Kulikov, K. and Lategahn, J. (2011) *Angewandte Chemie, International Edition*, **50**, 8605.
49. Moroda, A. and Togo, H. (2008) *Synthesis*, 1257.
50. Ishiwata, Y., Suzuki, Y. and Togo, H. (2010) *Heterocycles*, **82**, 339.
51. Alla, S.K., Kumar, R.K., Sadhu, P. and Punniyamurthy, T. (2013) *Organic Letters*, **15**, 1334.
52. Yakura, T. and Konishi, T. (2007) *Synlett*, 765.
53. Yusubov, M.S., Nemykin, V.N. and Zhdankin, V.V. (2010) *Tetrahedron*, **66**, 5745.

54. Yakura, T., Yamauchi, Y., Tian, Y. and Omoto, M. (2008) *Chemical & Pharmaceutical Bulletin*, **56**, 1632.
55. Yakura, T., Tian, Y., Yamauchi, Y., *et al.* (2009) *Chemical & Pharmaceutical Bulletin*, **57**, 252.
56. Yakura, T., Omoto, M., Yamauchi, Y., *et al.* (2010) *Tetrahedron*, **66**, 5833.
57. Yakura, T. and Omoto, M. (2009) *Chemical & Pharmaceutical Bulletin*, **57**, 643.
58. Minamitsuji, Y., Kato, D., Fujioka, H., *et al.* (2009) *Australian Journal of Chemistry*, **62**, 648.
59. Traore, M., Ahmed-Ali, S., Peuchmaur, M. and Wong, Y.-S. (2010) *Tetrahedron*, **66**, 5863.
60. Ngatimin, M., Frey, R., Andrews, C., *et al.* (2011) *Chemical Communications*, 11778.
61. Dohi, T., Maruyama, A., Minamitsuji, Y., *et al.* (2007) *Chemical Communications*, 1224.
62. Dohi, T., Takenaga, N., Fukushima, K.-i., *et al.* (2010) *Chemical Communications*, 7697.
63. Dohi, T., Maruyama, A., Takenage, N., *et al.* (2008) *Angewandte Chemie, International Edition*, **47**, 3787.
64. Uyanik, M., Yasui, T. and Ishihara, K. (2010) *Angewandte Chemie, International Edition*, **49**, 2175.
65. Dohi, T., Takenaga, N., Nakae, T., *et al.* (2013) *Journal of the American Chemical Society*, **135**, 4558.
66. Uyanik, M., Yasui, T. and Ishihara, K. (2010) *Tetrahedron*, **66**, 5841.
67. Uyanik, M. and Ishihara, K. (2012) *Yuki Gosei Kagaku Kyokaishi*, **70**, 1116.
68. Dohi, T., Minamitsuji, Y., Maruyama, A., *et al.* (2008) *Organic Letters*, **10**, 3559.
69. Yu, Z., Ju, X., Wang, J. and Yu, W. (2011) *Synthesis*, 860.
70. Zagulyaeva, A.A., Banek, C.T., Yusubov, M.S. and Zhdankin, V.V. (2010) *Organic Letters*, **12**, 4644.
71. Moriyama, K., Ishida, K. and Togo, H. (2012) *Organic Letters*, **14**, 946.
72. Miyamoto, K., Sakai, Y., Goda, S. and Ochiai, M. (2012) *Chemical Communications*, 982.
73. Yoshimura, A., Middleton, K.R., Luedtke, M.W., *et al.* (2012) *Journal of Organic Chemistry*, **77**, 11399.
74. Ma, H., Li, W., Wang, J., Xiao, G., *et al.* (2012) *Tetrahedron*, **68**, 8358.
75. Uyanik, M. and Ishihara, K. (2010) *Aldrichimica Acta*, **43**, 83.
76. Page, P.C.B., Appleby, L.F., Buckley, B.R., *et al.* (2007) *Synlett*, 1565.
77. Miura, T., Nakashima, K., Tada, N. and Itoh, A. (2011) *Chemical Communications*, 1875.
78. Moorthy, J.N., Senapati, K., Parida, K.N., *et al.* (2011) *Journal of Organic Chemistry*, **76**, 9593.
79. Koposov, A.Y., Litvinov, D.N., Zhdankin, V.V., *et al.* (2006) *European Journal of Organic Chemistry*, 4791.
80. Uyanik, M., Akakura, M. and Ishihara, K. (2009) *Journal of the American Chemical Society*, **131**, 251.
81. Tanaka, Y., Ishihara, T. and Konno, T. (2012) *Journal of Fluorine Chemistry*, **137**, 99.
82. Uyanik, M., Fukatsu, R. and Ishihara, K. (2009) *Organic Letters*, **11**, 3470.
83. Ojha, L.R., Kudugunti, S., Maddukuri, P.P., *et al.* (2009) *Synlett*, 117.
84. Cui, L.-Q., Liu, K. and Zhang, C. (2011) *Organic & Biomolecular Chemistry*, **9**, 2258.
85. Uyanik, M., Mutsuga, T. and Ishihara, K. (2012) *Molecules*, **17**, 8604.
86. Mu, R., Liu, Z., Yang, Z., *et al.* (2005) *Advanced Synthesis & Catalysis*, **347**, 1333.
87. Uyanik, M., Fukatsu, R. and Ishihara, K. (2010) *Chemistry–An Asian Journal*, **5**, 456.
88. Herrerias, C.I., Zhang, T.Y. and Li, C.-J. (2006) *Tetrahedron Letters*, **47**, 13.
89. Zhu, C., Ji, L. and Wei, Y. (2010) *Monatshefte für Chemie*, **141**, 327.
90. Zhu, C., Ji, L., Zhang, Q. and Wei, Y. (2010) *Canadian Journal of Chemistry*, **88**, 362.
91. Yakura, T. and Ozono, A. (2011) *Advanced Synthesis & Catalysis*, **353**, 855.
92. Zhu, C., Yoshimura, A., Wei, Y., *et al.* (2012) *Tetrahedron Letters*, **53**, 1438.
93. Yoshimura, A., Neu, H.M., Nemykin, V.N. and Zhdankin, V.V. (2010) *Advanced Synthesis & Catalysis*, **352**, 1455.
94. Yusubov, M.S., Chi, K.-W., Park, J.Y., *et al.* (2006) *Tetrahedron Letters*, **47**, 6305.
95. Koposov, A.Y., Karimov, R.R., Pronin, A.A., *et al.* (2006) *Journal of Organic Chemistry*, **71**, 9912.
96. Yusubov, M.S. and Zhdankin, V.V. (2010) *Mendeleev Communications*, **20**, 185.
97. Yusubov, M.S., Zagulyaeva, A.A. and Zhdankin, V.V. (2009) *Chemistry–A European Journal*, **15**, 11091.
98. Zhdankin, V.V. and Stang, P.J. (2008) *Chemical Reviews*, **108**, 5299.
99. Ladziata, U. and Zhdankin, V.V. (2006) *ARKIVOC*, (ix), 26.
100. Zeng, X.-M., Chen, J.-M., Middleton, K. and Zhdankin, V.V. (2011) *Tetrahedron Letters*, **52**, 5652.
101. Uyanik, M. and Ishihara, K. (2011) *Chimica Oggi*, **29**, 18.
102. Uyanik, M. and Ishihara, K. (2012) *ChemCatChem*, **4**, 177.
103. Uyanik, M., Suzuki, D., Yasui, T. and Ishihara, K. (2011) *Angewandte Chemie, International Edition*, **50**, 5331.
104. Li, X., Zhou, C. and Xu, X. (2012) *ARKIVOC*, (ix), 150.

105. Chen, L., Shi, E., Liu, Z., *et al.* (2011) *Chemistry–A European Journal*, **17**, 4085.
106. Uyanik, M., Okamoto, H., Yasui, T. and Ishihara, K. (2010) *Science*, **328**, 1376.
107. Wei, W., Wang, Y., Yin, J., *et al.* (2012) *Organic Letters*, **14**, 1158.
108. Wei, L., Xue, J., Liu, H., *et al.* (2012) *Organic Letters*, **14**, 5302.
109. Zhu, C. and Wei, Y. (2011) *ChemSusChem*, **4**, 1082.
110. Hu, J., Zhu, M., Xu, Y. and Yan, J. (2012) *Synthesis*, **44**, 1226.
111. Froehr, T., Sindlinger, C.P., Kloeckner, U., *et al.* (2011) *Organic Letters*, **13**, 3754.
112. Ma, L., Wang, X., Yu, W. and Han, B. (2011) *Chemical Communications*, 11333.
113. Bai, G.-y., Xu, K., Chen, G.-f., *et al.* (2011) *Synthesis*, 1599.
114. Xie, J., Jiang, H., Cheng, Y. and Zhu, C. (2012) *Chemical Communications*, 979.
115. Li, X., Xu, X. and Zhou, C. (2012) *Chemical Communications*, 12240.
116. Li, X., Xu, X. and Tang, Y. (2013) *Organic & Biomolecular Chemistry*, **11**, 1739.
117. Rodriguez, A. and Moran, W.J. (2011) *Organic Letters*, **13**, 2220.
118. Li, L.-T., Huang, J., Li, H.-Y., *et al.* (2012) *Chemical Communications*, 5187.
119. Wan, C., Gao, L., Wang, Q., *et al.* (2010) *Organic Letters*, **12**, 3902.
120. Zhang, J., Zhu, D., Yu, C., *et al.* (2010) *Organic Letters*, **12**, 2841.
121. Wang, Q., Wan, C., Gu, Y., *et al.* (2011) *Green Chemistry*, **13**, 578.
122. Yan, Y. and Wang, Z. (2011) *Chemical Communications*, 9513.
123. Tian, J.-S., Ng, K.W.J., Wong, J.-R. and Loh, T.-P. (2012) *Angewandte Chemie, International Edition*, **51**, 9105.
124. Dhineshkumar, J., Lamani, M., Alagiri, K. and Prabhu, K.R. (2013) *Organic Letters*, **15**, 1092.

5

Recyclable Hypervalent Iodine Reagents

Despite their useful reactivity and environmentally benign nature, common hypervalent iodine reagents derived from iodobenzene or simple aryl iodides are not perfect with respect to the principles of green chemistry. According to one of the basic principles of green chemistry, recyclable reagents are superior to stoichiometric reagents [1]. The most widely used hypervalent iodine(III) reagents, such as (dichloroiodo)benzene, (diacetoxyiodo)benzene, iodosylbenzene and [hydroxy(tosyloxy)iodo]benzene, are based on the oxidized form of iodobenzene and the reactions of these reagents with organic substrates produce stoichiometric amount of iodobenzene as a waste product, which results in low atom economy and complicated isolation and purification of the products. Likewise, the most common hypervalent iodine(V) reagents, IBX and DMP (Section 3.2), produce 2-iodobenzoic acid as a waste by-product. Polymer-supported hypervalent iodine reagents, as well as recyclable nonpolymeric reagents, overcome these drawbacks and in general have the same reactivity as their monomeric analogues. Moreover, polymer-supported reagents have found broad application in solid-phase and combinatorial high-throughput synthesis techniques. Attachment of a chemical reagent to an insoluble polymer matrix enables easy reaction work up by simple filtration, automated parallel synthesis and fast reaction optimization [2–4]. Numerous polymer-supported hypervalent iodine(III) and iodine(V) reagents have been developed since the beginning of the twenty-first century. Several reviews on polymer-supported, as well as nonpolymeric recyclable hypervalent iodine reagents, have been published [5–9].

5.1 Polymer-Supported Iodine(III) Reagents

The first polymer-supported hypervalent iodine reagent, poly[(dichloroiodo)styrene], was prepared by chlorination of iodinated polystyrene in the early 1980s [8]. This method, however, involves the initial preparation of iodinated polystyrene under harsh conditions (160 h, 110 °C), requires the use of hazardous chlorine gas and affords poly[(dichloroiodo)styrene] with a relatively low loading of active chlorine. An optimized one-pot preparation of polystyrene-supported (dichloroiodo)benzene **2** (loading of $-ICl_2$ up to 1.35 mmol g^{-1}) from polystyrene **1**, iodine and aqueous sodium hypochlorite (bleach) was reported in 2011 (Scheme 5.1) [10].

Recyclable reagent **2** is a convenient chlorinating reagent with a reactivity pattern similar to that of (dichloroiodo)benzene. Scheme 5.2 shows several representative chlorinations of organic substrates [10]. The final products are conveniently separated from the polymeric by-product by simple filtration and isolated in

Hypervalent Iodine Chemistry: Preparation, Structure and Synthetic Applications of Polyvalent Iodine Compounds, First Edition. Viktor V. Zhdankin.

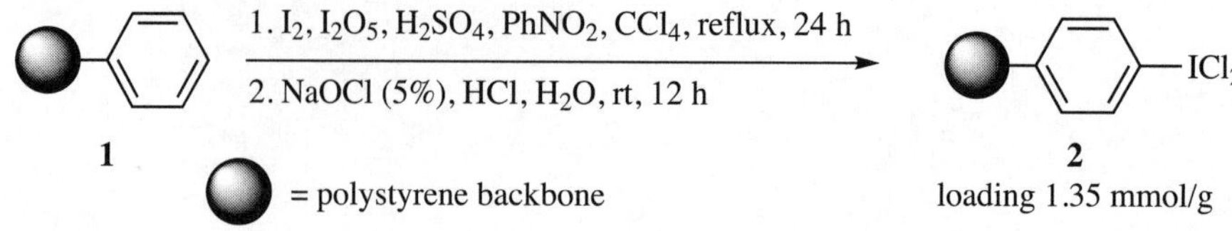

Scheme 5.1

good purity after evaporation of solvent. The polymeric by-product, poly(iodostyrene), can be converted back into reagent **2** by treatment with bleach and aqueous HCl in about 90% overall yield.

The preparation and applications of polystyrene-supported hypervalent iodine(III) reagents bearing (diacetoxy)iodo, (dihalo)iodo, hydroxy(tosyloxy)iodo, hydroxy(phosphoryloxy)iodo, aryliodonium and 1,2-benziodoxol-3-one groups were reviewed by Togo and Sakuratani in 2002 [5].

Several variants of polystyrene-supported [bis(acyloxy)iodo]arenes have been developed [11–21]. Poly[(diacetoxyiodo)styrene] (**4**) can be prepared in two steps from commercial polystyrene **1** with an average molecular weight ranging from 45 000 to 250 000 [11–13, 19–21]. In the first step, polystyrene **1** is iodinated with iodine and iodine pentoxide in sulfuric acid to give poly(iodostyrene) **3,** which is subsequently converted into the diacetate **4** by treatment with peracetic acid (Scheme 5.3) [11, 13]. The loading capacity of the polymeric reagent **4** obtained by this procedure (Scheme 5.3) varies from 2.96 to 3.5 mmol g^{-1} as measured by iodometry and elemental analysis [11–13].

The original version of reagent **4,** reported by Yokoyama, Togo and coworkers in 1998, is insoluble in organic solvents [12]. An insoluble crosslinked poly[(diacetoxyiodo)styrene] has been prepared by a similar

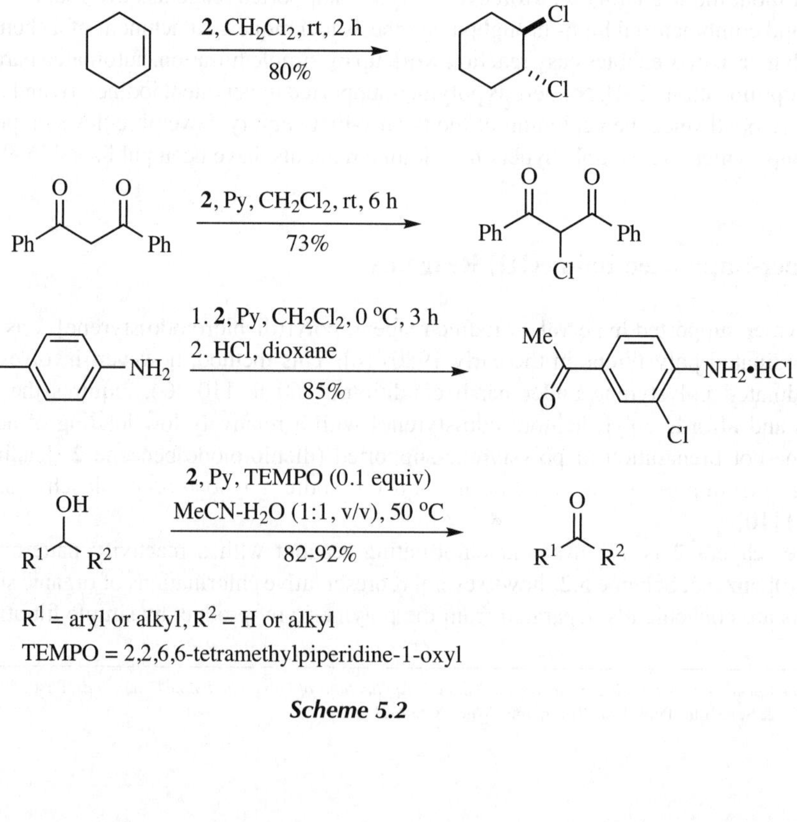

Scheme 5.2

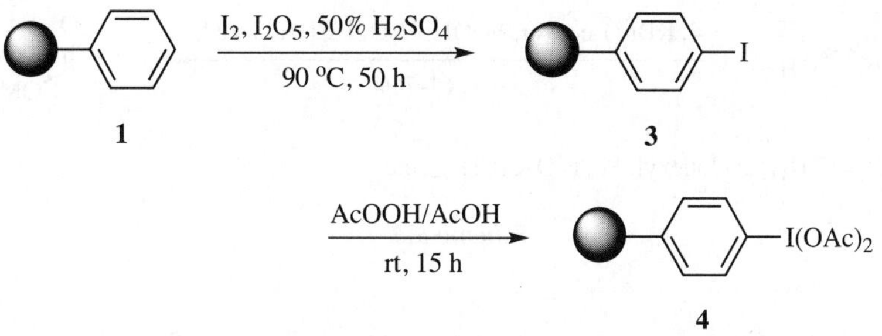

I_2, I_2O_5, 50% H_2SO_4
90 °C, 50 h

1 3

AcOOH/AcOH
rt, 15 h

I(OAc)$_2$

4

Scheme 5.3

procedure from commercially available 2% crosslinked polystyrene [15, 16]. A soluble version of diacetate **4** can be prepared from linear polystyrene (MW = 250 000) [13].

Poly[(diacetoxyiodo)styrene] (**4**) is a convenient, practical application reagent. After reaction with an organic substrate, the iodinated resin can be easily isolated from the reaction mixture by filtration and reoxidized with peracetic acid. Polymer-supported reagent **4** can be reused many times with no loss of activity [14]. Moreover, it can be used as a starting material for the preparation of several other polymer-supported hypervalent iodine reagents, such as polymeric analogs of [bis(trifluoroacetoxy)iodo]benzene [14, 17], iodosylbenzene [22], iodosylbenzene sulfate [23] and [hydroxy(tosyloxy)iodo]benzene [24, 25]. In particular, the polymer-supported [bis(trifluoroacetoxy)iodo]benzene can be prepared by the reaction of diacetate **4** with trifluoroacetic acid [17, 26], or by the oxidation of poly(iodostyrene) (**3**) with 30% $H_2O_2/(CF_3CO)_2O$ [17]. Compared to the diacetate **4,** the polymer-supported [bis(trifluoroacetoxy)iodo]benzene has better solubility in dichloromethane and is a more effective oxidant in the oxidative biaryl coupling reaction of phenol ether derivatives [17].

Polymer-supported (diacetoxyiodo)arenes **8** with a different linker have been prepared from the commercially available aminomethylated polystyrene **5** (Scheme 5.4) [18]. In the first step, the reaction of polymer **5** with 4-iodobenzoic acid (**6,** $n = 0$) or 4-iodophenylacetic acid (**6,** $n = 1$) affords polymer-supported iodides **7** in excellent yield. In the second step, iodides **7** are converted into diacetates **8** by treatment with peracetic acid at 40 °C overnight [18].

Polymer-supported reagents **8** have a reactivity pattern similar to that of polystyrene derivative **4.** Diacetates **8** are stable for storage and can be regenerated after their use without a measurable loss of activity [18].

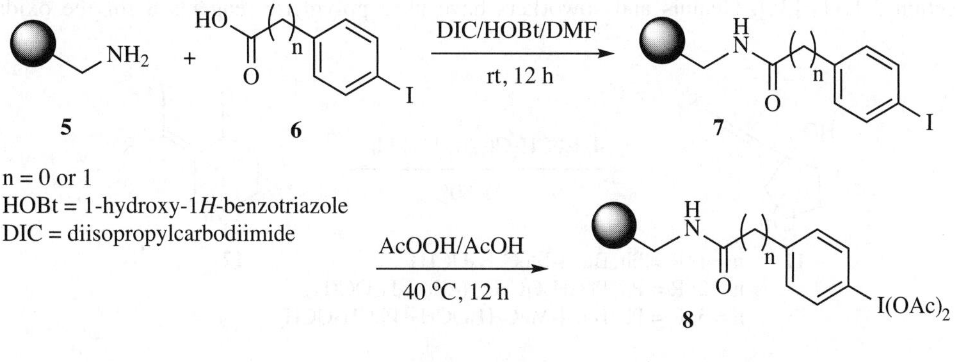

NH_2 + HO (O) n I

DIC/HOBt/DMF
rt, 12 h

5 6 7

n = 0 or 1
HOBt = 1-hydroxy-1*H*-benzotriazole
DIC = diisopropylcarbodiimide

AcOOH/AcOH
40 °C, 12 h

I(OAc)$_2$

8

Scheme 5.4

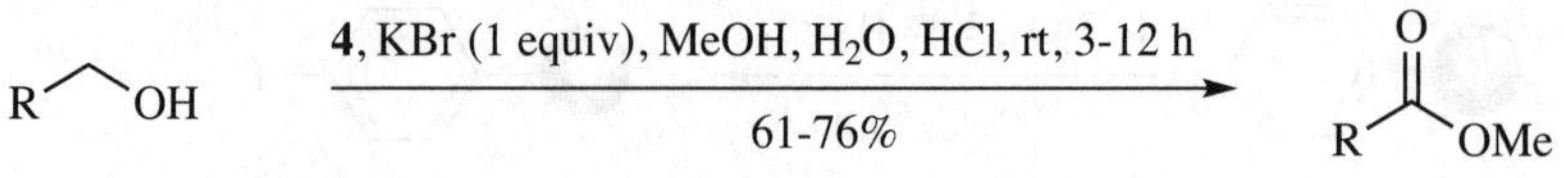

R⌒OH —[4, KBr (1 equiv), MeOH, H_2O, HCl, rt, 3-12 h; 61-76%]→ RC(O)OMe

R = C_5H_{11}, cyclohexyl, Ph, $EtO_2C(CH_2)_4$, etc.

Scheme 5.5

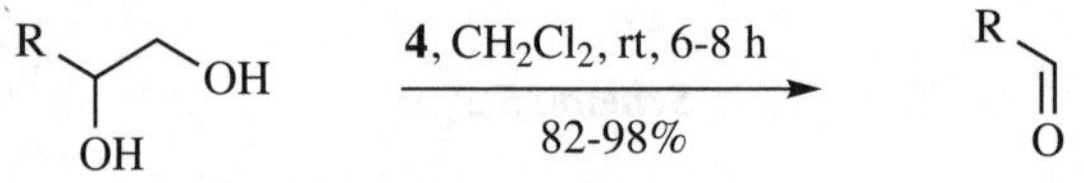

9 —[4, CH_2Cl_2, rt, 6-8 h; 82-98%]→ 10 (RCHO)

R = alkyl, aryl, etc.

Scheme 5.6

Polymer-supported (diacetoxyiodo)arenes have found broad synthetic application. Poly[(diacetoxyiodo)styrene] (**4**) has been used for the oxidation of alcohols catalyzed by TEMPO (2,2,6,6-tetramethylpiperidine-1-oxyl) [27–30], or in the presence of KBr [31]. In a specific example, primary alcohols are readily oxidized to methyl esters upon treatment with reagent **4** in the presence of KBr in the acidic aqueous methanol solution (Scheme 5.5) [31]. Likewise, organic sulfides are selectively oxidized to the respective sulfoxides by reagent **4** in water in the presence of KBr [32].

Various glycols **9** can be oxidized to aldehydes **10** using reagent **4** at room temperature (Scheme 5.6) [20]. Under these reaction conditions, *cis*-1,2-cyclohexandiol is converted into 1,6-hexandial. Protecting groups, such as OAc, OR, OBn, OBz and isopropylidene, in substrates **9** are compatible with this oxidation [20].

Treatment of 1-alkynylcycloalkanols **11** with reagent **4** and iodine affords (*Z*)-2-(1-iodo-1-organyl)methylenecycloalkanones **12** resulting from the alkoxyl radical-promoted ring-expansion reaction (Scheme 5.7) [33].

Oxidative iodination of aromatic compounds by the combination of a hypervalent iodine reagent with iodine is a synthetically important reaction (Section 3.1.4) [34]. Polymer-supported diacetate **4** is a particularly convenient reagent for oxidative iodination since it can be regenerated and reused many times. Reagent **4** gives the best results for the iodination of electron-rich arenes **13,** with predominant formation of the *para*-substituted products **14** (Scheme 5.8) [12, 21].

The reaction of [bis(acyloxy)iodo]arenes with phenols can lead to various synthetically useful products (Section 3.1.11) [34]. Giannis and coworkers have used polymeric reagents **8** for the oxidation of

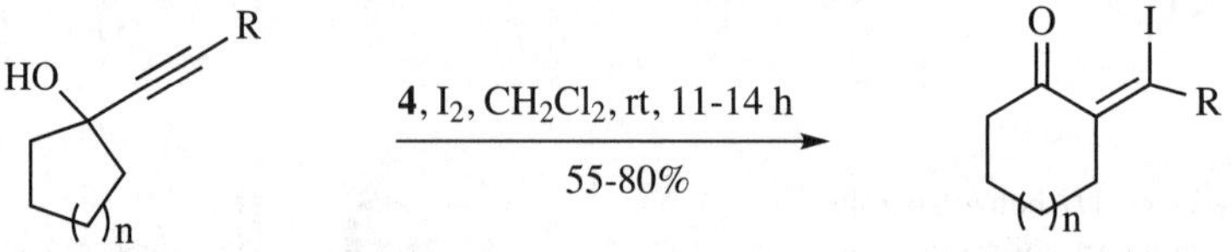

11 —[4, I_2, CH_2Cl_2, rt, 11-14 h; 55-80%]→ **12**

n = 1; R = Ph, Bu, 4-$MeC_6H_4OCH_2$
n = 2; R = Ph, $PhCH_2OCH_2$, 4-$MeC_6H_4OCH_2$
n = 3; R = Ph, Bu, 4-$MeC_6H_4OCH_2$, $PhCH_2OCH_2$

Scheme 5.7

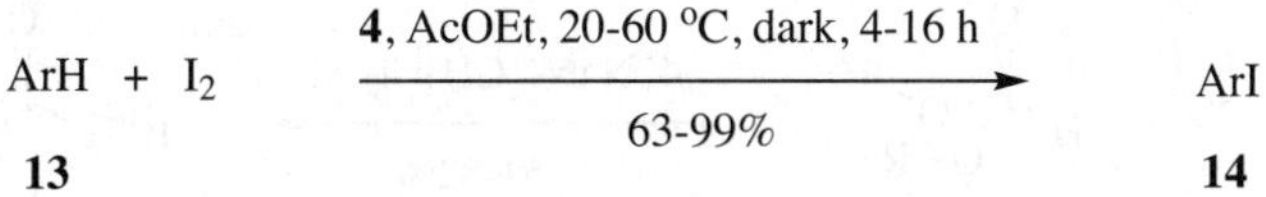

$$\underset{\mathbf{13}}{\mathrm{ArH}} + \mathrm{I_2} \xrightarrow[\text{63-99\%}]{\mathbf{4}\text{, AcOEt, 20-60 °C, dark, 4-16 h}} \underset{\mathbf{14}}{\mathrm{ArI}}$$

ArH = $MeOC_6H_5$, $HO(CH_2)_2OC_6H_5$, 4-BrC_6H_4OMe, $Bu^tC_6H_5$, 1,3,5-$Me_3C_6H_3$, 1,3,5-$Pr^i{}_3C_6H_3$, PhC_6H_5, $PhOC_6H_5$, naphthalene

Scheme 5.8

polysubstituted *p*-hydroquinones **15** and phenols **17** to obtain the respective benzoquinones **16** and **18** (Scheme 5.9) [18]. These reactions proceed under mild conditions and resins **8** can be regenerated without a measurable loss of activity. The polystyrene-supported reagent **4** oxidizes various substituted *p*-hydroquinones to benzoquinones in quantitative yields under similar conditions [14]. The oxidation of appropriate phenolic precursors with reagent **4** has been used in the synthesis of the alkaloids (±)-oxomaritidine and (±)-epimaritidine using a multistep sequence of polymer-supported reagents [35].

Polymer-supported [bis(acyloxy)iodo]arenes have also been used in the synthesis of numerous heterocyclic systems. Representative examples include the preparation of thiazolo[2, 3-*c*]-*s*-triazoles by the reaction of arenecarbaldehyde-4-arylthiazol-2-ylhydrazones with poly[(diacetoxyiodo)styrene] **4** [36], synthesis of 3,4-dihydro-2,1-benzothiazine 2,2-dioxides by oxidative cyclization of 2-arylethanesulfonamides with reagent **4** [37], synthesis of 3,5-disubstituted 1,2,4-thiadiazoles by oxidative dimerization of thioamides using reagent **4** [38], synthesis of isoflavones from 2′-hydroxychalcones using reagent **4** [39, 40], preparation of various 2-substituted 1,3-dioxolanes, 1,3-dioxanes and tetrahydrofurans from alcohols and reagent **4** in the presence of iodine under irradiation conditions [41] and the synthesis of 1-benzoyltetrahydroisoquinoline derivatives using polymer-supported [bis(trifluoroacetoxy)iodo]benzene [42].

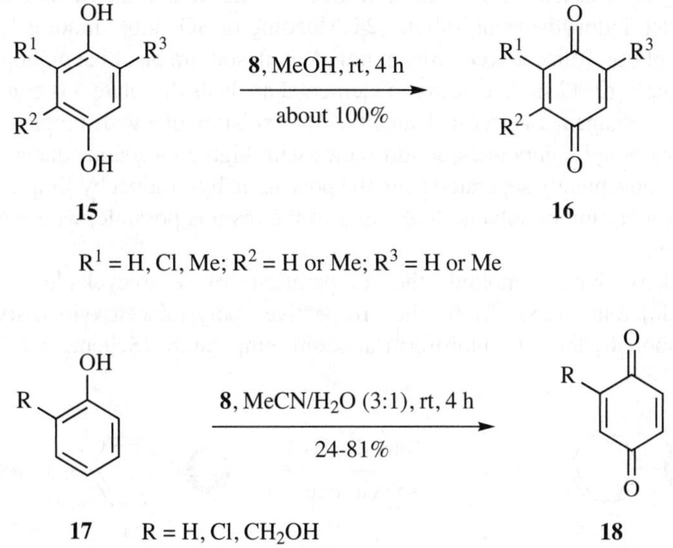

15 → **16**: **8**, MeOH, rt, 4 h; about 100%

R^1 = H, Cl, Me; R^2 = H or Me; R^3 = H or Me

17 → **18**: **8**, MeCN/H_2O (3:1), rt, 4 h; 24-81%

17 R = H, Cl, CH_2OH **18**

Scheme 5.9

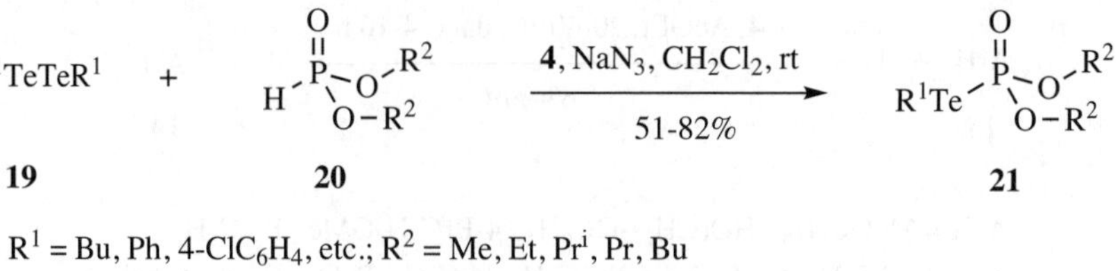

$R^1 = Bu, Ph, 4\text{-}ClC_6H_4$, etc.; $R^2 = Me, Et, Pr^i, Pr, Bu$

Scheme 5.10

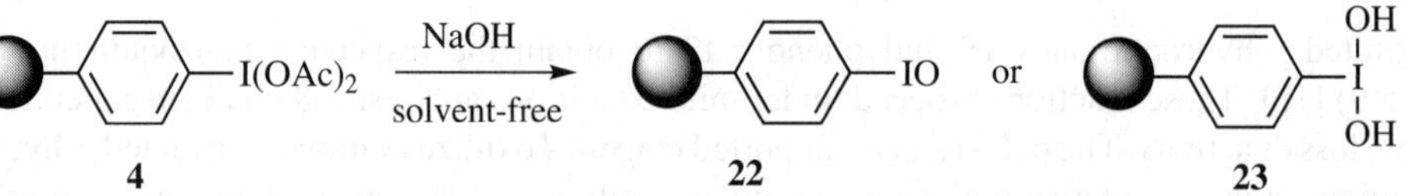

Scheme 5.11

Organyltellurophosphates **21** can be synthesized smoothly in moderate to good yields by the free radical reaction of diorganyl phosphates **20** with diorganyl ditellurides **19** using reagent **4** and sodium azide (Scheme 5.10) [43]; the polymer reagent can be regenerated and recycled with no loss of reactivity.

Polystyrene-supported iodosylbenzene (**22**) (loading of –IO groups up to 1.50 mmol g^{-1}) has been prepared by a solvent-free reaction of poly[(diacetoxyiodo)styrene] (**4**) with sodium hydroxide (Scheme 5.11) [22]. Elemental analysis of polymer **22** indicates that the –IO groups are partially hydrated as shown in structure **23.** This resin has been successfully used for efficient oxidation of a diverse collection of alcohols to aldehydes and ketones in the presence of $BF_3{\cdot}Et_2O$. Reagent **22** can also be employed as efficient co-catalyst in combination with $RuCl_3$ in the catalytic oxidation of alcohols and aromatic hydrocarbons, respectively, to the corresponding carboxylic acids and ketones using Oxone® as the stoichiometric oxidant [22].

Polystyrene-supported iodosylbenzene sulfate (**24**) (loading of SO_3 0.68 mmol g^{-1}) can be conveniently prepared by treatment of poly[(diacetoxyiodo)styrene] (**4**) with sodium bisulfate monohydrate under solvent-free conditions (Scheme 5.12) [23]. According to elemental analysis diacetate **4** is converted into the sulfate **24** almost quantitatively. Reagent **24** effects clean, efficient oxidation of a wide range of alcohols and sulfides to the corresponding carbonyl compounds or sulfoxides with high conversions under mild conditions [23]. The final products are conveniently separated from the polymeric by-product by simple filtration and isolated in good purity after evaporation of solvent. Recycling of the resin is possible, with minimal loss of activity after several reoxidations.

Togo and coworkers have reported the preparation of a recyclable, polystyrene-supported [hydroxy(tosyloxy)iodo]arene (**25**) from the respective poly[(diacetoxyiodo)styrene] (**4**) and *p*-toluenesulfonic acid monohydrate in chloroform at room temperature (Scheme 5.13) [24, 25]. According

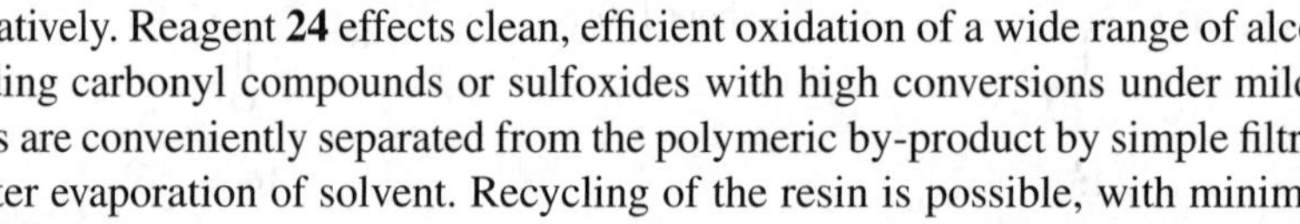

Scheme 5.12

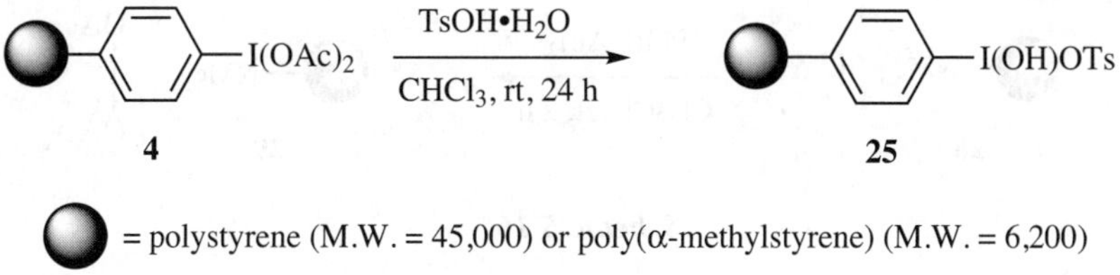

$$\text{(polymer)}-C_6H_4-I(OAc)_2 \xrightarrow[\text{CHCl}_3\text{, rt, 24 h}]{\text{TsOH}\bullet\text{H}_2\text{O}} \text{(polymer)}-C_6H_4-I(OH)OTs$$

4 → **25**

= polystyrene (M.W. = 45,000) or poly(α-methylstyrene) (M.W. = 6,200)

Scheme 5.13

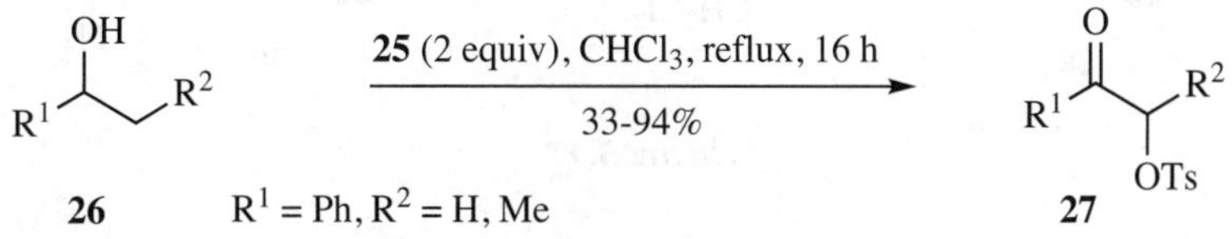

$$R^1CH(OH)CH_2R^2 \xrightarrow[\text{33-94\%}]{\textbf{25}\text{ (2 equiv), CHCl}_3\text{, reflux, 16 h}} R^1C(O)CH(OTs)R^2$$

26 → **27**

R^1 = Ph, R^2 = H, Me
or $R^1 + R^2 = (CH_2)_4$

Scheme 5.14

to analytical data, 100% of the (diacetoxy)iodo groups of polymer **4** are converted into the corresponding hydroxy(tosyloxy)iodo groups in product **25** under these reaction conditions.

Poly[hydroxy(tosyloxy)iodo]styrene (**25**) shows similar reactivity to the monomeric [hydroxy (tosyloxy)iodo]benzene (HTIB) [44] in the α-tosyloxylation of ketones. Moreover, polymeric reagent **25** can be used for the direct conversion of alcohols **26** into α-tosyloxyketones **27** (Scheme 5.14), while HTIB has a low reactivity toward alcohols [24,25].

In general, poly[hydroxy(tosyloxy)iodo]styrene (**25**) can be used in the same oxidative transformations as the monomeric HTIB [39, 45–48]. In particular, it can serve as an oxidant in the Hofmann-type degradation of carboxamides to the respective amines. In a specific example, primary alkyl- and benzylcarboxamides are converted into the corresponding alkylammonium tosylates by treatment with reagent **25** in acetonitrile at reflux in yields ranging from 60% to 90% [45]. Reagent **25** has been used for direct α-hydroxylation of ketones [46,47] and in the synthesis of isoflavones from 2′-benzoyloxychalcones [39]. Poly[hydroxy(tosyloxy)iodo]styrene (**25**) has also been used in the halotosyloxylation reaction of alkynes with iodine or *N*-bromosuccinimide (NBS) or *N*-chlorosuccinimide (NCS) (Scheme 5.15) [48]. The polymeric reagent can be regenerated and reused numerous times.

Kirschning and coworkers have developed several polymer-supported halogen-ate(I) complexes [49–53]. Specifically, the polymer-supported diacetoxyiodine-ate(I) complex **29** was prepared by the oxidation of polystyrene-bound iodide **28** using (diacetoxyiodo)benzene (Scheme 5.16) [49].

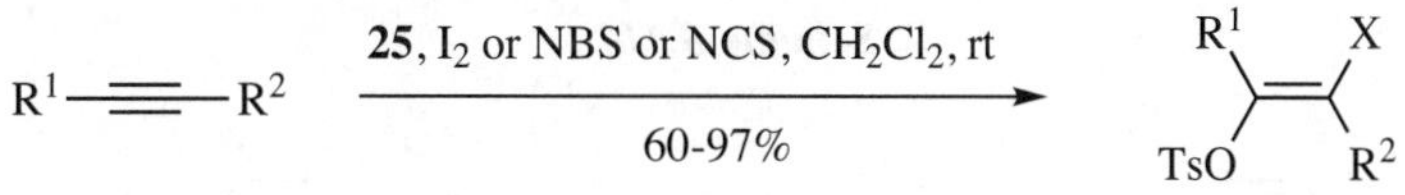

$$R^1-C\equiv C-R^2 \xrightarrow[\text{60-97\%}]{\textbf{25}\text{, I}_2\text{ or NBS or NCS, CH}_2\text{Cl}_2\text{, rt}} (TsO)R^1C{=}C(X)R^2$$

R^1 = Ph, Bu, Bu^t, CH_3OCH_2, H
R^2 = H, Ph, 4-$MeC_6H_4C(O)$, 4-$ClC_6H_4C(O)$, Ts, $P(O)Ph_2$, CO_2Me, TMS
X = I, Br, Cl

Scheme 5.15

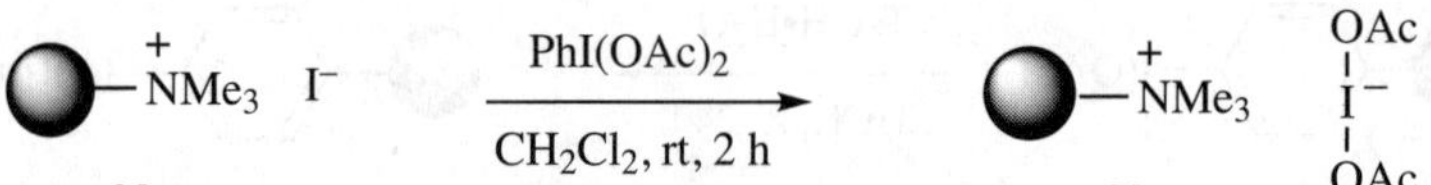

$\overset{+}{N}Me_3\ I^-$ (28) → $PhI(OAc)_2$, CH_2Cl_2, rt, 2 h → $\overset{+}{N}Me_3\ [I(OAc)_2]^-$ (29)

Scheme 5.16

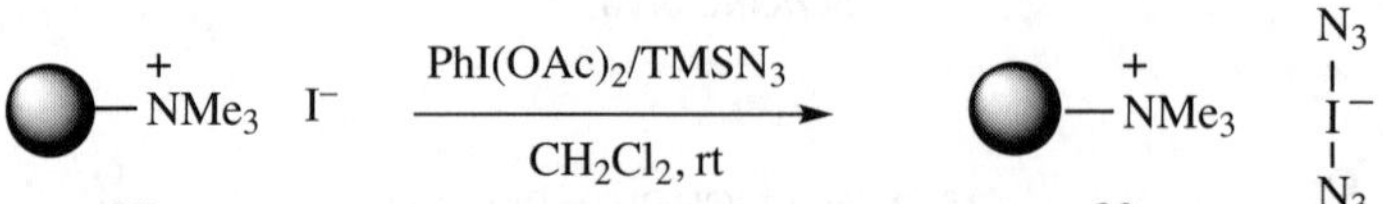

$\overset{+}{N}Me_3\ I^-$ (28) → $PhI(OAc)_2/TMSN_3$, CH_2Cl_2, rt → $\overset{+}{N}Me_3\ [I(N_3)_2]^-$ (30)

Scheme 5.17

Complex **29** and similar polymer-supported reagents are useful for the electrophilic functionalization of alkenes and for the oxidation of alcohols to carbonyl compounds [49–54].

Kirschning and coworkers have also reported the preparation of reagent systems generated *in situ* from (diazidoiodo)benzene and tetraalkylammonium halides [51]. Particularly useful is a stable polymer-bound bis(azido)iodate (**30**), which can be readily prepared by the reaction of polystyrene-supported iodide **28** with (diazidoiodo)benzene generated *in situ* from (diacetoxyiodo)benzene and azidotrimethylsilane (Scheme 5.17) [51].

Reagent **30** reacts with alkenes **31** to give the corresponding products of β-azido-iodination (**32**) (Scheme 5.18). These reactions predominantly afford products of *anti*-addition with Markovnikov regioselectivity [51].

Polystyrene-supported diaryliodonium salts **33** can be prepared by the reaction of diacetate **4** with arenes in the presence of sulfuric acid (Scheme 5.19) [55, 56]. Polymer-supported aryliodonium salts are useful aryl transfer reagents in Pd(II)-catalyzed cross-coupling reactions with salicylaldehydes [56] and aromatic hydrocarbons [55].

Polymer-supported alkynyliodonium tosylates **34** can be prepared by treatment of reagent **4** with terminal alkynes in the presence of *p*-toluenesulfonic acid (Scheme 5.20) [57]. Polymers **34** are effective alkynylating reagents toward sodium sulfinates and benzotriazole [57].

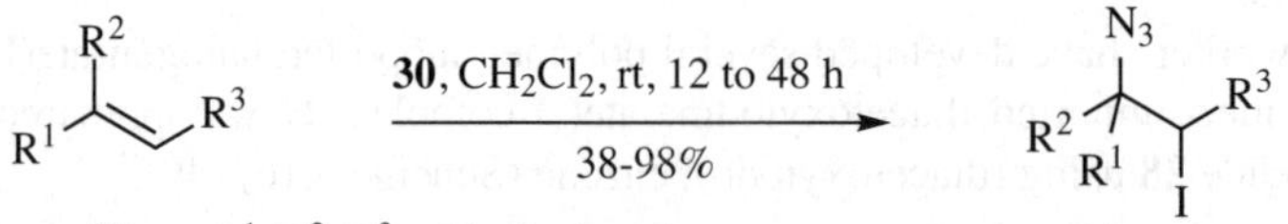

31 → **30**, CH_2Cl_2, rt, 12 to 48 h; 38-98% → **32**

R^1, R^2, R^3 = H, alkyl, aryl, etc.

Scheme 5.18

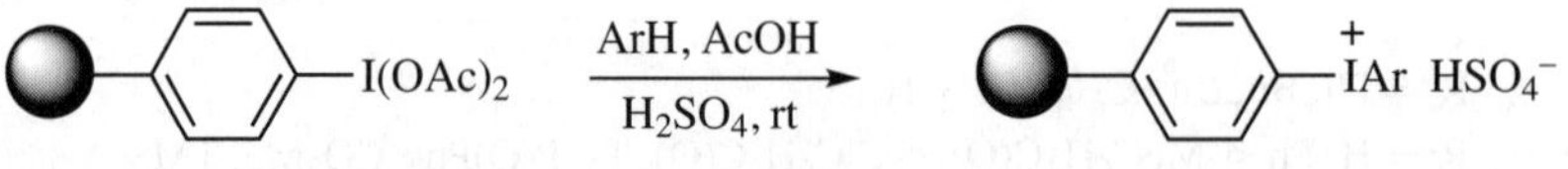

4 ($-I(OAc)_2$) → ArH, AcOH, H_2SO_4, rt → **33** ($-\overset{+}{I}Ar\ HSO_4^-$)

Ar = Ph, 4-MeC_6H_4, 4-$MeOC_6H_4$

Scheme 5.19

4 $I(OAc)_2$ $\xrightarrow[\text{2. R—≡—H, } CHCl_3\text{, reflux}]{\text{1. TsOH, } CH_2Cl_2\text{, rt}}$ **34** I^+—≡—R TsO^-

R = Ph or Bu^t

Scheme 5.20

5.2 Polymer-Supported Iodine(V) Reagents

The first preparation of polymer-supported hypervalent iodine(V) reagents derived from IBX was reported in 2001 [58,59]. Giannis and Mülbaier have developed the aminopropylsilica gel based reagent **35** (Figure 5.1), which can oxidize various primary and secondary alcohols to the respective carbonyl compounds in excellent yields at room temperature in THF under heterogeneous conditions and can be regenerated by oxidation with Oxone without any loss of activity [58]. Rademann and coworkers prepared the polystyrene-based polymeric analog of IBX **36,** which was characterized by IR spectroscopy, elemental analysis and MAS-NMR spectroscopy [59]. Reagent **36** oxidizes various primary, secondary, benzylic, allylic, terpene alcohols

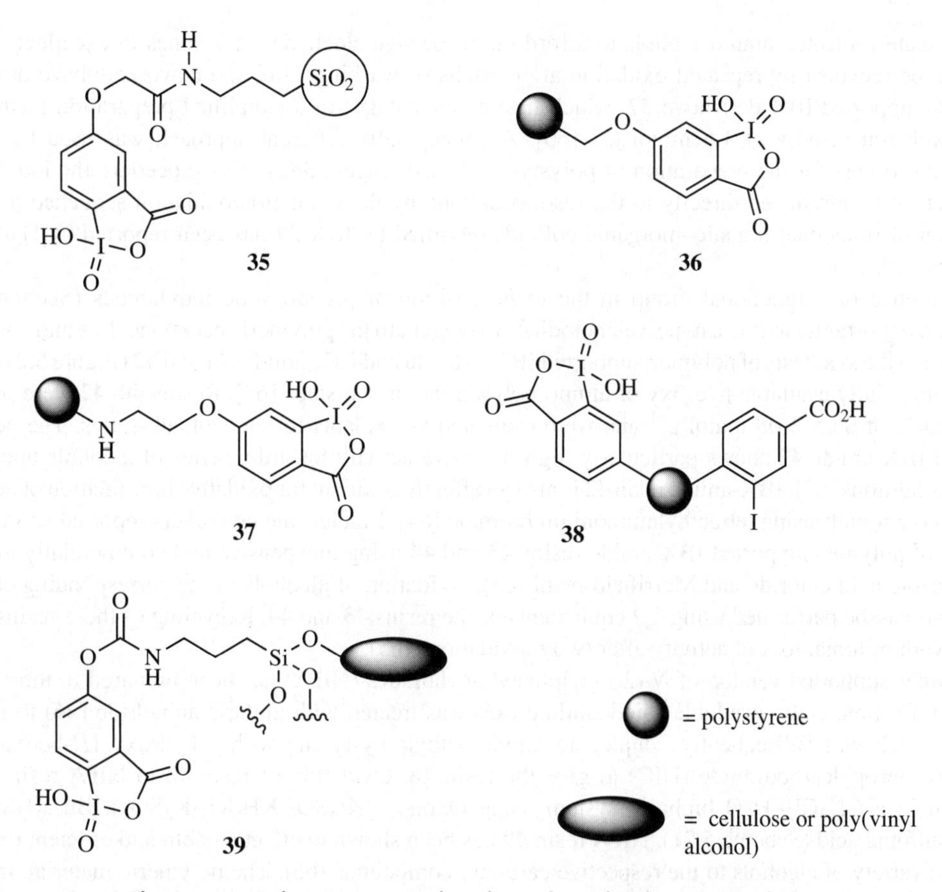

Figure 5.1 *Polymer-supported analogs of 2-iodoxybenzoic acid (IBX).*

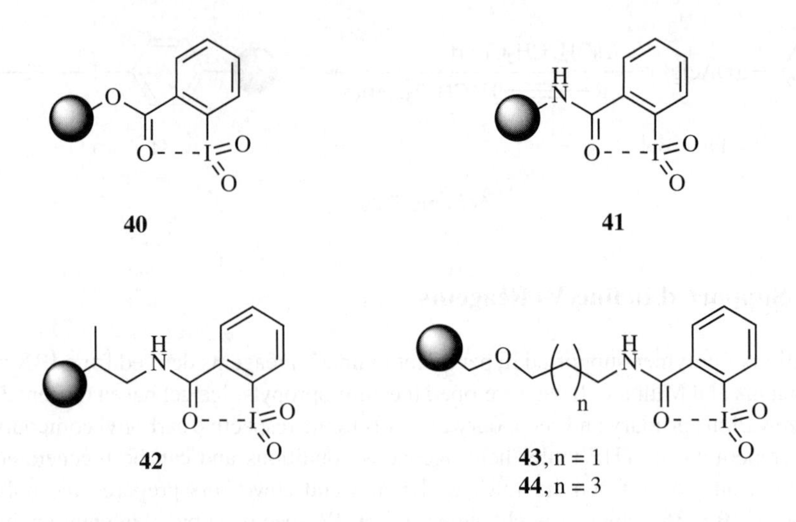

Figure 5.2 *Polymer-supported IBX-ester **40** and IBX-amides **41–44**.*

and carbamate-protected aminoalcohols to afford the respective aldehydes or ketones in excellent yields – and it can be recycled by repeated oxidation after extensive washings. Lei and coworkers have developed a polymer-supported IBX derivative **37,** which has the advantages of a simplified preparation method and a high oxidation activity of 1.5 mmol g^{-1} [60]. A conceptually different approach was used by Sutherland and coworkers for the preparation of polystyrene-based reagent **38;** in this procedure the iodobenzoic acid moiety was introduced directly to the resin backbone by the iodination/oxidation sequence [61]. The preparation of functional organic–inorganic colloids modified by IBX **39** has been reported by Hatton and coworkers [62].

The presence of a functional group in the *ortho*-position of pseudocyclic iodylarenes (Section 2.2.2) provides an opportunity to link a hypervalent iodine(V) reagent to the polymeric backbone. Lee and coworkers have reported the synthesis of polymer-supported IBX-ester **40** and IBX-amides **41** and **42** (Figure 5.2) starting from commercially available hydroxy or amino polystyrene in two steps [63]. Resins **40–42** were prepared with loadings of 0.65–1.08 mmol g^{-1} and were evaluated as oxidants for alcohol substrates. The polymer-supported IBX-amide **42** shows particularly high oxidative activity toward a series of alcohols under mild reaction conditions [63]. IBX-amide resin **42** is also an effective oxidant for oxidative bromination of activated aromatic compounds using tetraethylammonium bromide [64]. Linclau and coworkers reported an improved synthesis of polymer-supported IBX-amide resins **43** and **44** using inexpensive and commercially available 2-iodobenzoic acid chloride and Merrifield resin [65]. Oxidation of alcohols to the corresponding carbonyl compounds can be performed using 1.2 equivalents of the resins **43** and **44.** Recycling of these resins is also possible with minimal loss of activity after two reoxidations [65].

A polymer-supported version of *N*-(2-iodylphenyl)acylamides (NIPA) has been prepared in three simple steps [66]. Commercially available 2-iodoaniline (**45**) was treated with glutaric anhydride (**46**) to give the acid **47,** which was subsequently coupled to aminomethylpolystyrene with 1-hydroxy-1*H*-benzotriazole (HOBt)/diisopropylcarbodiimide (DIC) to give the resin **48.** Oxidation of resin **48** to NIPA resin **49** was performed in a CH_2Cl_2–H_2O biphasic system using Oxone ($2KHSO_5{\cdot}KHSO_4{\cdot}K_2SO_4$), Bu_4NHSO_4 and methanesulfonic acid (Scheme 5.21). NIPA resin **49** has been shown to effect smooth and efficient oxidation of a broad variety of alcohols to the respective carbonyl compounds [66]. The polymeric material, resulting

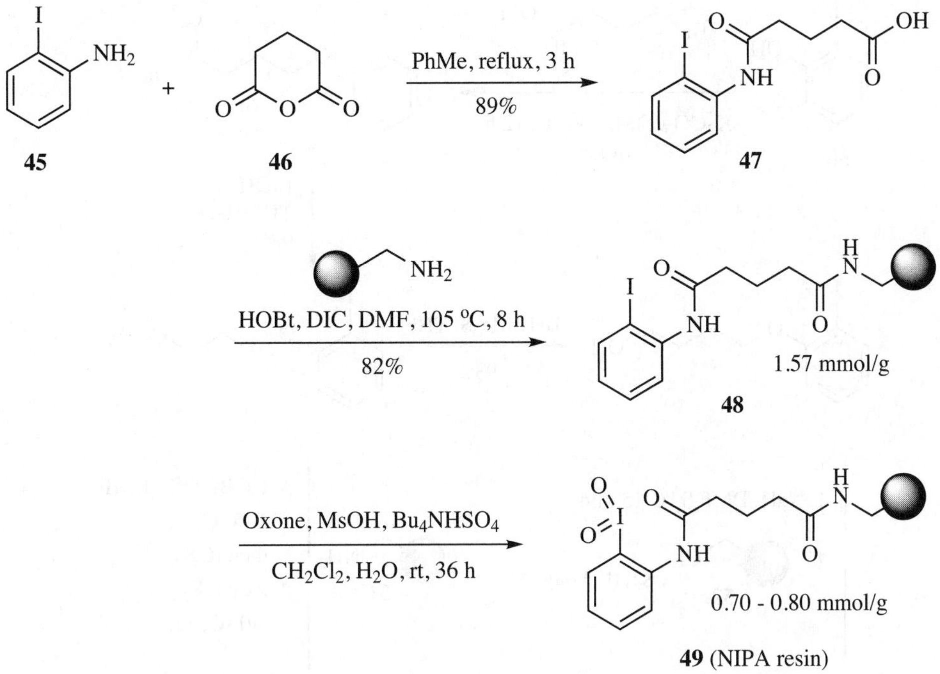

Scheme 5.21

from the reduction of NIPA resin, can be collected and reoxidized. A moderate decline in oxidative activity is observed after multiple recovery steps.

Polymer-supported analogs of 2-iodylphenol ethers, resins **53** and **54** (Scheme 5.22), have been prepared starting from commercially available 2-iodophenol (**50**) and aminomethylated polystyrene **51** or Merrifield resin **52** using 4-hydroxybutanoic acid and 1,4-butanediol, respectively, as linkers. These polymer-supported reagents effect clean, efficient conversion of a wide range of alcohols, including heteroatomic and unsaturated structures, into the corresponding carbonyl compounds. Recycling of the resins is possible with minimal loss of activity after several reoxidations [67].

5.3 Recyclable Nonpolymeric Hypervalent Iodine(III) Reagents

Polymer-supported hypervalent reagents are useful recyclable oxidants, but they have some serious disadvantages compared to nonpolymeric reagents. In particular, polymer-bound reagents often have lower reactivities than the monomeric analogues, the loading of the hypervalent iodine sites is difficult to control and, moreover, repeated use of these resins leads to significant degradation due to benzylic oxidation of the polystyrene chain. The simple, nonpolymeric, recyclable iodanes have similar reactivity to the original hypervalent iodine(III) reagents and are free of the disadvantages of polymeric reagents. The reduced form of a recyclable nonpolymeric reagent can be effectively recovered from a reaction mixture using one of the biphasic separation techniques discussed in the following subsections.

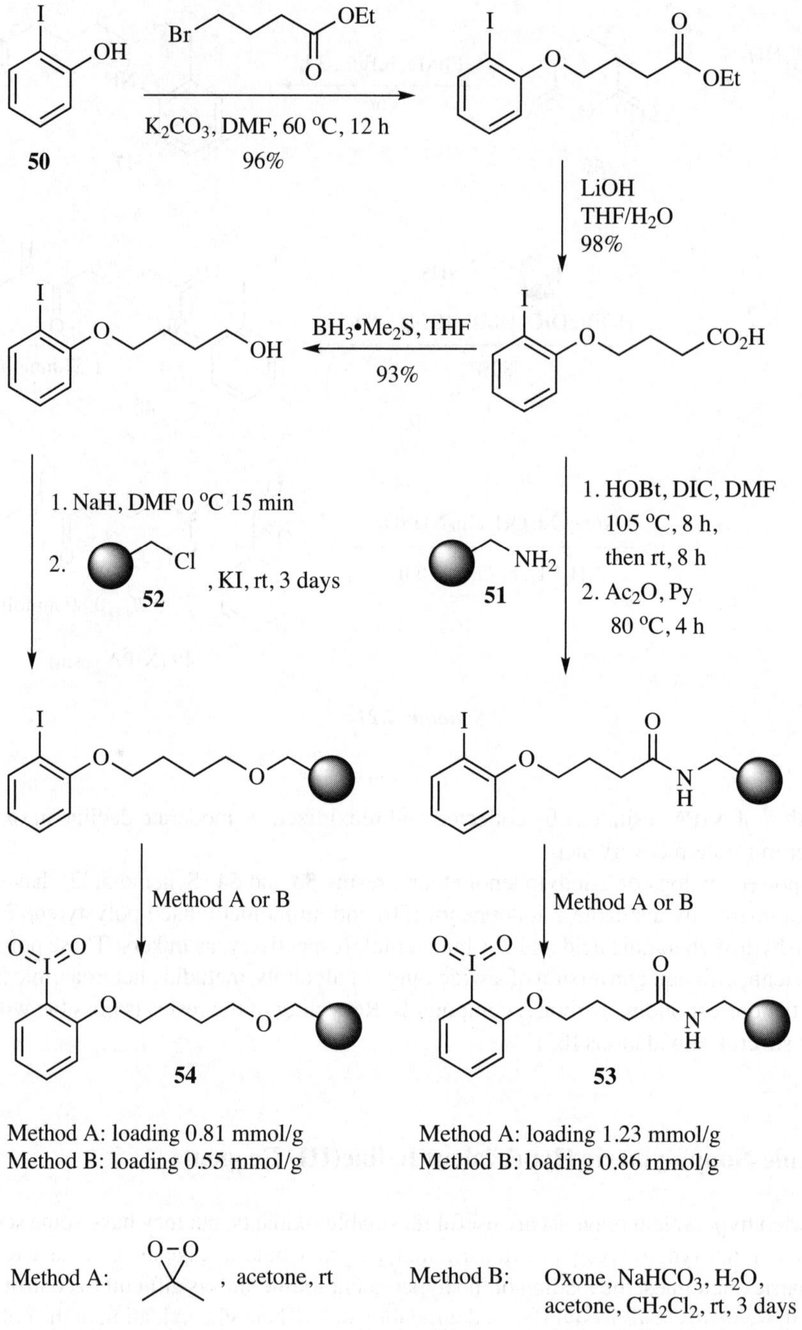

I
OH
50
Br OEt O
K_2CO_3, DMF, 60 °C, 12 h
96%
I O O OEt
LiOH
THF/H_2O
98%
I O CO_2H
$BH_3{\bullet}Me_2S$, THF
93%
I O OH
1. NaH, DMF 0 °C 15 min
2. Cl 52 , KI, rt, 3 days
NH_2 51
1. HOBt, DIC, DMF
105 °C, 8 h,
then rt, 8 h
2. Ac_2O, Py
80 °C, 4 h
I O O
I O O N H
Method A or B
Method A or B
O I O O O
54
O I O O O N H
53

Method A: loading 0.81 mmol/g
Method B: loading 0.55 mmol/g

Method A: loading 1.23 mmol/g
Method B: loading 0.86 mmol/g

Method A: O–O, acetone, rt

Method B: Oxone, $NaHCO_3$, H_2O, acetone, CH_2Cl_2, rt, 3 days

Scheme 5.22

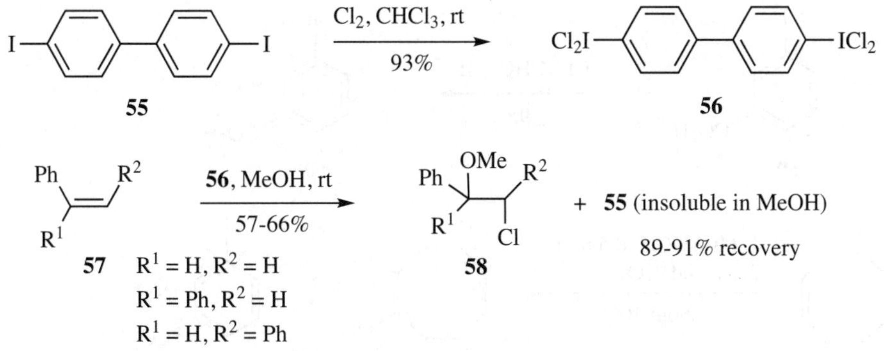

Cl_2, $CHCl_3$, rt
93%
55
56

56, MeOH, rt
57-66%
57 $R^1 = H, R^2 = H$
$R^1 = Ph, R^2 = H$
$R^1 = H, R^2 = Ph$
58
+ **55** (insoluble in MeOH)
89-91% recovery

Scheme 5.23

5.3.1 Recyclable Iodine(III) Reagents with Insoluble Reduced Form

Several recoverable hypervalent iodine reagents of this type have been developed since 2004. A recyclable chlorinating reagent, 4,4′-bis(dichloroiodo)biphenyl (**56**), can be prepared in high yield by saturating a solution of 4,4′-diiodobiphenyl (**55**) in chloroform with gaseous chlorine for one hour at room temperature under stirring (Scheme 5.23). Compound **56** is a convenient chlorinating reagent; for example, its reaction with styrene derivatives **57** in methanol affords products of electrophilic chloromethoxylation (**58**) in good yield. The reduced form of reagent **56**, 4,4′-diiodobiphenyl (**55**), is practically insoluble in methanol and can be recovered by simple filtration and used for the regeneration of reagent **56** [68,69]. 4,4′-Bis(dichloroiodo)biphenyl (**56**) has also been used as a recyclable oxidizing reagent for the oxidation of primary and secondary alcohols to the corresponding carbonyl compounds [70], as well as for the preparation of α-azido ketones and esters by direct oxidative azidation of carbonyl compounds using sodium azide as the source of azide anions [71].

Another recyclable chlorinating reagent, 3-(dichloroiodo)benzoic acid (**60**), can be conveniently prepared by the chlorination of commercially available 3-iodobenzoic acid (**59**). A reduced form of reagent **60,** 3-iodobenzoic acid (**59**), can also be easily separated as a solid from the products of chlorination by basic aqueous work-up followed by acidification with HCl [68,69]. Scheme 5.24 shows an example of a chlorination cycle using reagent **60** [69]. Alternatively, **59** can be separated from the reaction products by treatment with anionic exchange resin Amberlite™ IRA 900 (Section 5.3.2).

A green, recyclable and efficient catalytic system for the oxidation of alcohols and sulfides in water has been developed based on 3-(dichloroiodo)benzoic acid (**60**) as a stoichiometric oxidant and SiO_2-supported $RuCl_3$ **61** as the catalyst (Scheme 5.25) [72]. This catalytic oxidative system effects clean and efficient oxidation of a wide range of alcohols to the corresponding aldehydes and ketones, or sulfides to sulfoxides, in high conversions with excellent chemoselectivity, under mild conditions and with an easy work-up procedure. Furthermore, SiO_2–$RuCl_3$ catalyst **61** can be recovered by simple filtration and recycled up to six consecutive runs without significant loss of activity. The reduced form of 3-(dichloroiodo)benzoic acid (**60**), 3-iodobenzoic acid (**59**), can be easily separated from reaction mixtures and converted back into reagent **60** by treatment with bleach and aqueous HCl in about 90% overall yield [72].

Zhang and Li have reported a convenient preparation of 1-chloro-1,2-benziodoxol-3(1*H*)-one **63** and its application as a recyclable reagent for an efficient TEMPO-catalyzed oxidation of alcohols (Scheme 5.26) [73]. Reagent **63** can be prepared by a convenient, large-scale and high-yielding procedure from 2-iodobenzoic acid (**62**) using sodium chlorite as the stoichiometric oxidant in dilute hydrochloric acid at room temperature. Various alcohols can be oxidized to their corresponding carbonyl compounds in excellent yields using reagent

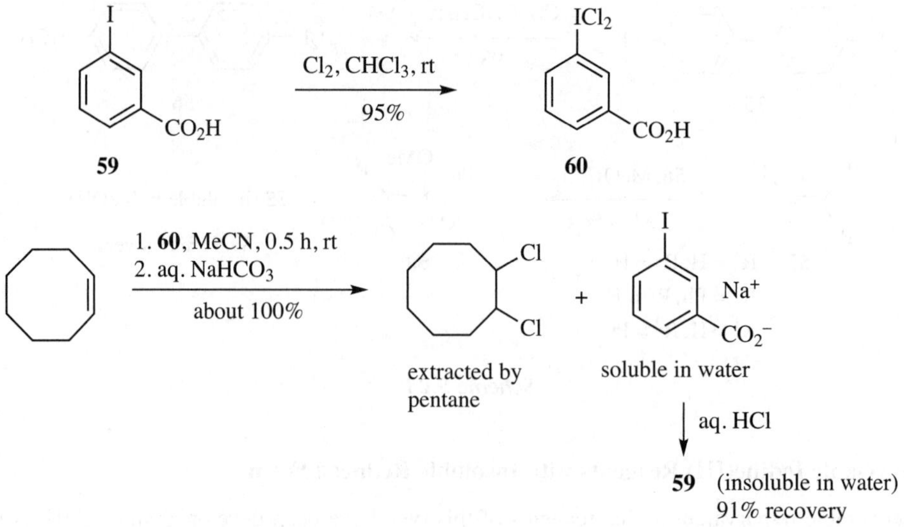

I
Cl_2, $CHCl_3$, rt
95%
ICl_2
CO_2H
59
CO_2H
60

1. **60**, MeCN, 0.5 h, rt
2. aq. $NaHCO_3$
about 100%
Cl
Cl
extracted by pentane
+
I
Na^+
CO_2^-
soluble in water
aq. HCl
59 (insoluble in water)
91% recovery

Scheme 5.24

63 as a recyclable reagent in the presence of TEMPO; a primary hydroxy group is preferentially oxidized over a secondary hydroxyl as exemplified in Scheme 5.26. 1-Chloro-1,2-benziodoxol-3(1*H*)-one **63** can be easily recycled after simple solid/liquid-phase separation and the subsequent regeneration sequence [73].

Various recyclable hypervalent iodine reagents based on the frameworks of 1,3,5,7-tetraphenyladamantane (**65–67**) [74], tetraphenylmethane (**69–71**) [75], biphenyl (**73** and **74**) [76] and terphenyl (**76**) [76] have been developed (Figure 5.3). Similarly to previously discussed reagent **56** (Scheme 5.23), aryl iodides **64, 68, 72**

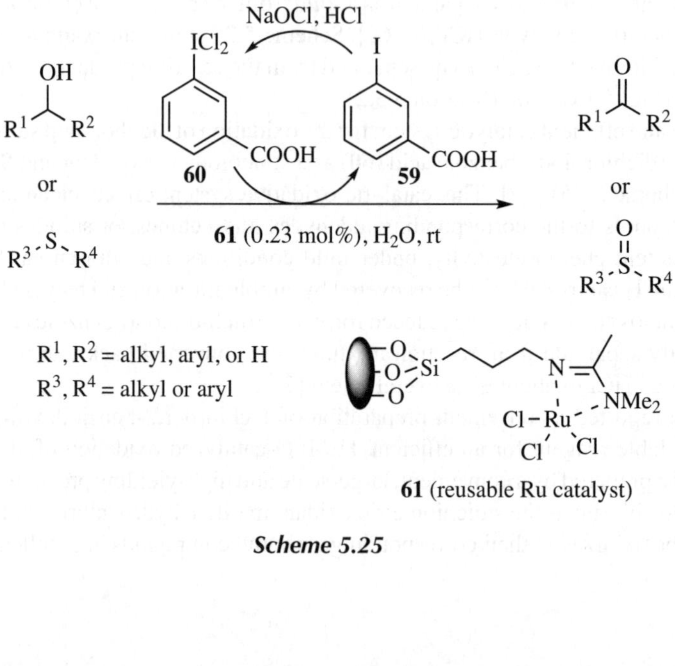

NaOCl, HCl
ICl_2
I
OH
R^1 R^2
COOH
60
COOH
59
O
R^1 R^2
or
or
61 (0.23 mol%), H_2O, rt
R^3 S R^4
O
R^3 S R^4

R^1, R^2 = alkyl, aryl, or H
R^3, R^4 = alkyl or aryl

O O Si N NMe₂ Cl–Ru Cl Cl

61 (reusable Ru catalyst)

Scheme 5.25

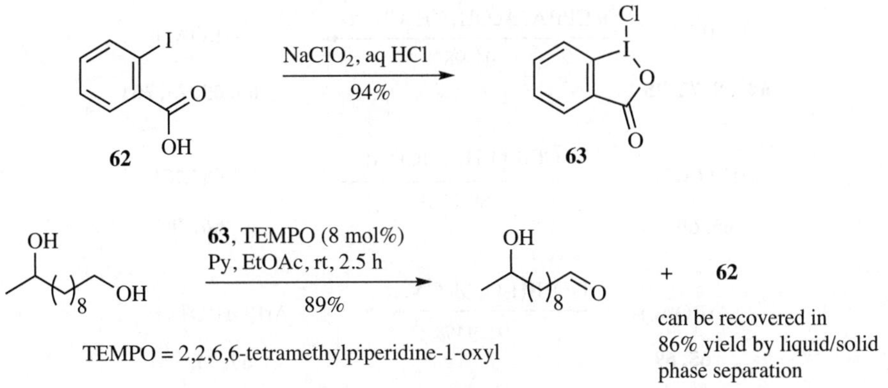

62 → 63: $NaClO_2$, aq HCl, 94%

63, TEMPO (8 mol%), Py, EtOAc, rt, 2.5 h, 89% + 62 (can be recovered in 86% yield by liquid/solid phase separation)

TEMPO = 2,2,6,6-tetramethylpiperidine-1-oxyl

Scheme 5.26

and **75,** formed in the reactions of these compounds with organic substrates, are insoluble in methanol and can be recovered by simple filtration.

Recyclable (diacetoxyiodo)arenes **65, 69, 73** and **76** (Figure 5.3) are prepared in excellent yields by the oxidation of appropriate iodides **64, 68, 72** and **75** using *m*-chloroperoxybenzoic acid (*m*CPBA) in acetic acid [74–76]. The diacetates **65** and **69** can be subsequently converted into the respective

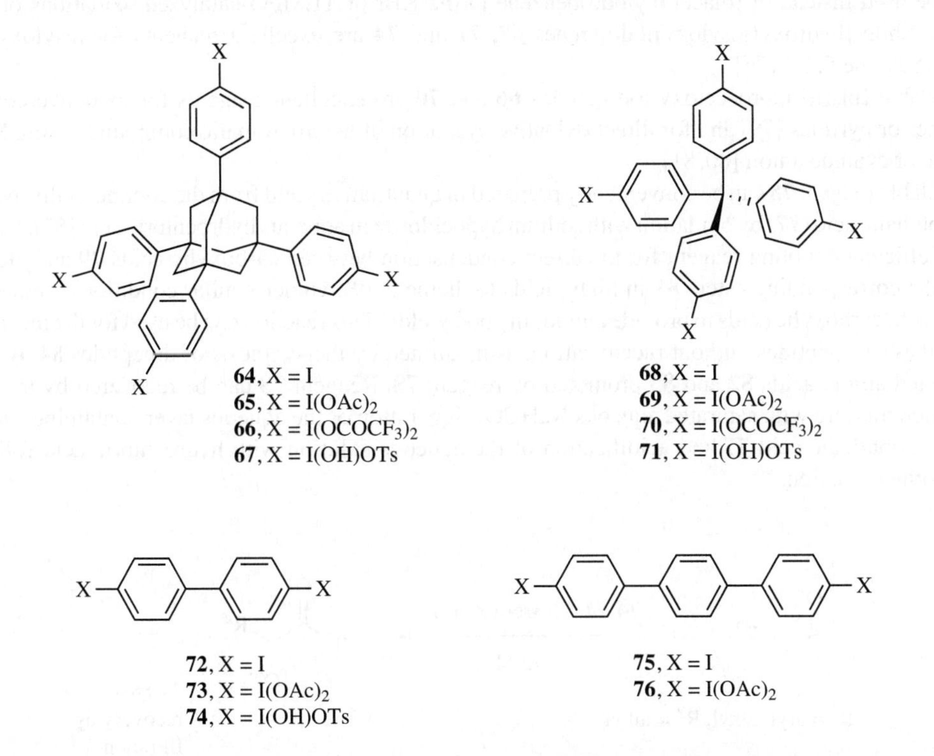

Figure 5.3 *Recyclable hypervalent iodine reagents based on the frameworks of 1,3,5,7-tetraphenyladamantane (**65–67**), tetraphenylmethane (**69–71**), biphenyl (structures **73** and **74**) and terphenyl (structure **76**).*

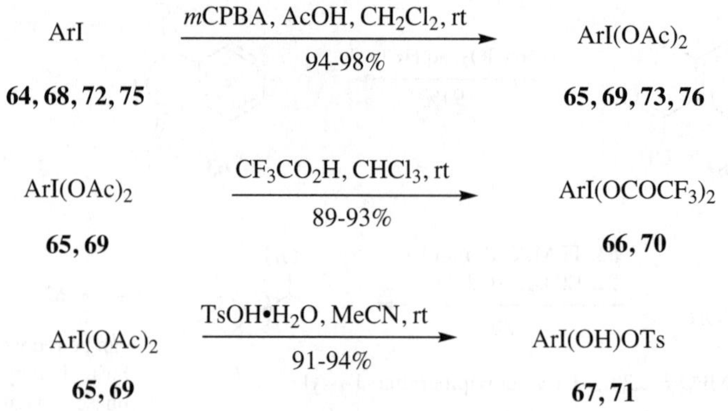

ArI $\xrightarrow[\text{94-98\%}]{m\text{CPBA, AcOH, CH}_2\text{Cl}_2\text{, rt}}$ ArI(OAc)$_2$

64, 68, 72, 75 → **65, 69, 73, 76**

ArI(OAc)$_2$ $\xrightarrow[\text{89-93\%}]{\text{CF}_3\text{CO}_2\text{H, CHCl}_3\text{, rt}}$ ArI(OCOCF$_3$)$_2$

65, 69 → **66, 70**

ArI(OAc)$_2$ $\xrightarrow[\text{91-94\%}]{\text{TsOH•H}_2\text{O, MeCN, rt}}$ ArI(OH)OTs

65, 69 → **67, 71**

Scheme 5.27

[bis(trifluoroacetoxy)iodo]arenes or [hydroxy(tosyloxy)iodo]arenes by treatment with trifluoroacetic or toluenesulfonic acids (Scheme 5.27) [74, 75]. The biphenyl-based tosylate **74** is prepared directly from the iodide **72** by oxidation with *m*CPBA in the presence of toluenesulfonic acid [76].

The reactivity pattern of recyclable hypervalent iodine reagents **65–67, 69–71, 73, 74** and **76** is similar to the common iodobenzene-based reagents [34]. For example, the recyclable (diacetoxyiodo)arenes **65, 69, 73** and **76** can be used instead of (diacetoxyiodo)benzene in the KBr or TEMPO-catalyzed oxidations of alcohols [74–78], while [hydroxy(tosyloxy)iodo]arenes **67, 71** and **74** are excellent reagents for tosyloxylation of ketones (Scheme 5.28) [76].

Recyclable [bis(trifluoroacetoxy)iodo]arenes **66** and **70** are excellent reagents for oxidative coupling of thiophenes or pyrroles [79] and for direct oxidative cyanation of heteroaromatic compounds using Me$_3$SiCN as source of cyanide anion [80, 81].

Recyclable reagent **78** can be conveniently prepared in quantitative yield from the commercially available 2-iodoisophthalic acid (**77**) by oxidation with sodium hypochlorite in aqueous hydrochloric acid [82]. Compound **78** is an efficient coupling reagent for the direct condensation between carboxylic acids **79** and alcohols **80** to give the corresponding esters **81** in high yields (Scheme 5.29). Under similar conditions, amines can be coupled with carboxylic acids to provide amides in good yields. This reaction can be used for the racemization-free synthesis of peptides without racemization, as illustrated by the synthesis of dipeptides **84** by coupling of protected amino acids **82** and **83** promoted by reagent **78.** Reagent **78** can be recovered by treatment of the reaction mixture with saturated aqueous NaHCO$_3$, separation of the aqueous layer containing sodium salt of 2-iodoisophthalic acid **77** and acidification of the aqueous solution with hydrochloric acid followed by hypochlorite oxidation.

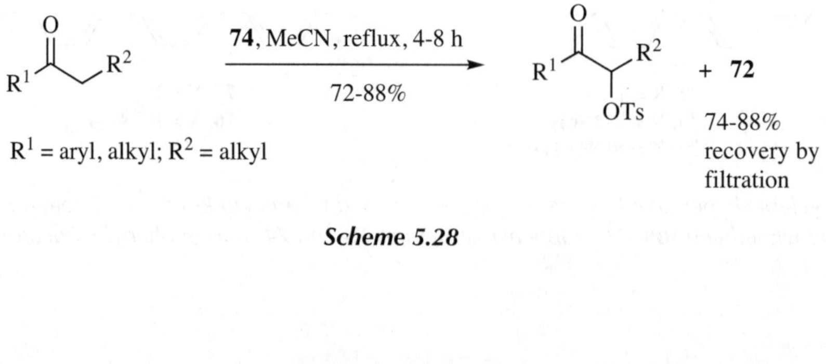

$R^1C(O)CH_2R^2$ $\xrightarrow[\text{72-88\%}]{\textbf{74}\text{, MeCN, reflux, 4-8 h}}$ $R^1C(O)CH(OTs)R^2$ + **72**

R^1 = aryl, alkyl; R^2 = alkyl

74-88%
recovery by
filtration

Scheme 5.28

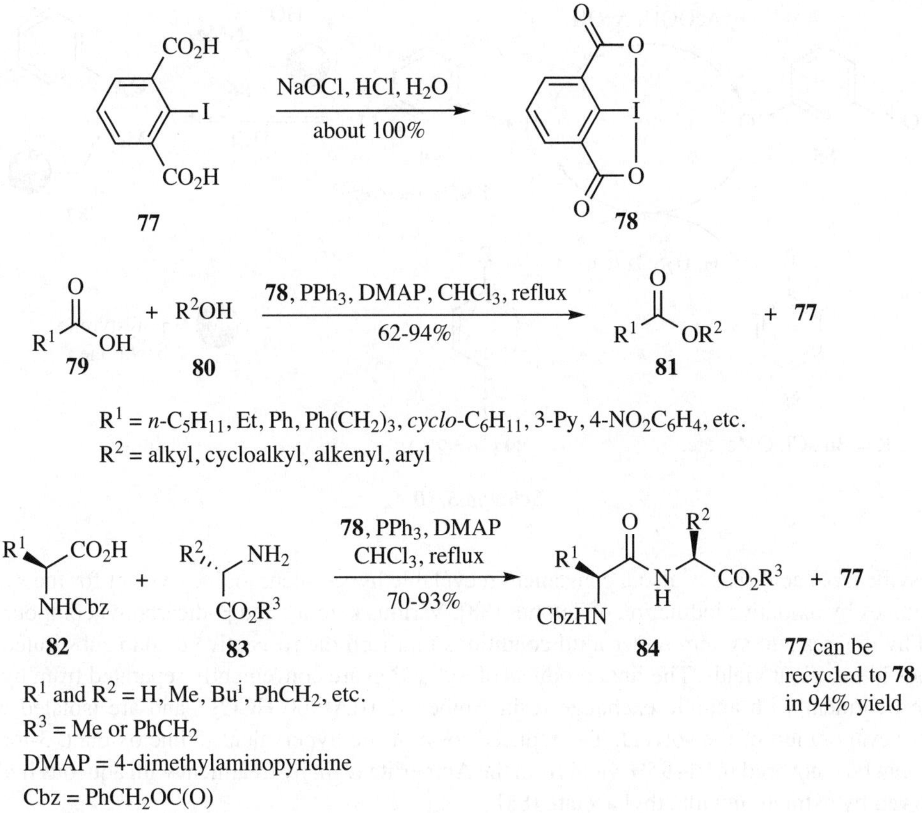

NaOCl, HCl, H_2O
about 100%

77 **78**

78, PPh_3, DMAP, $CHCl_3$, reflux
62-94%

79 **80** **81** + **77**

R^1 = n-C_5H_{11}, Et, Ph, $Ph(CH_2)_3$, cyclo-C_6H_{11}, 3-Py, 4-$NO_2C_6H_4$, etc.
R^2 = alkyl, cycloalkyl, alkenyl, aryl

78, PPh_3, DMAP
$CHCl_3$, reflux
70-93%

82 **83** **84** + **77**

77 can be recycled to **78** in 94% yield

R^1 and R^2 = H, Me, Bu^i, $PhCH_2$, etc.
R^3 = Me or $PhCH_2$
DMAP = 4-dimethylaminopyridine
Cbz = $PhCH_2OC(O)$

Scheme 5.29

5.3.2 Recovery of the Reduced Form of a Hypervalent Iodine Reagent Using Ion-Exchange Resins

Ion-exchange recycling methodology employs the so-called "tagged" hypervalent iodine reagents [83], the reduced form of which can be removed from the reaction mixture as an anion using anionic ion-exchange resins. Important recyclable hypervalent iodine reagents of this type are represented by 3-iodosylbenzoic acid (**85**) and its derivatives. The tagging approach can be illustrated by the oxidative iodination of aromatic substrates (Scheme 5.30). The iodination of aromatic compounds **88** is performed by treatment with iodine and 3-iodosylbenzoic acid (**85**) in acetonitrile in the presence of catalytic sulfuric acid [84]. The products of iodination **89** are separated from the acidic by-products by treatment with anionic exchange resin Amberlite IRA 900 (hydroxide form) **86** and isolated as pure compounds after evaporation of solvent. The reduced form of the hypervalent iodine oxidant, 3-iodobenzoic acid (**59**), can be easily recovered in 91–95% yield from the Amberlite resin **87** by treatment with aqueous HCl followed by extraction with ethyl acetate. A similar purification and recovery procedure using Amberlite IRA 900 (hydroxide or bicarbonate forms) has been used for the oxidative iodination of alkenes with 3-iodosylbenzoic acid/iodine [84,85], for the α-tosyloxylation of ketones with 3-iodosylbenzoic acid/toluenesulfonic acid recyclable reagent system [86] and for the selective oxidation of primary and secondary alcohols to the respective carbonyl compounds with 3-iodosylbenzoic acid in the presence of $RuCl_3$ (0.5 mol%) at room temperature in aqueous acetonitrile [87].

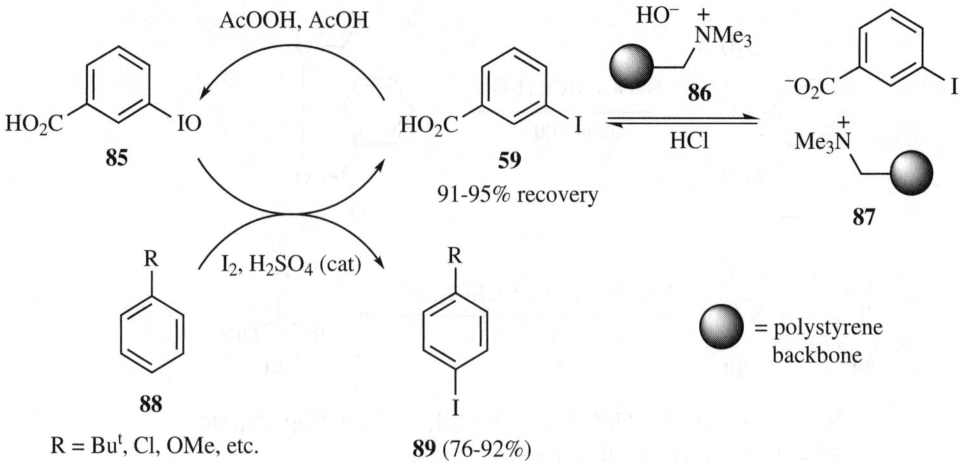

Scheme 5.30

3-Iodosylbenzoic acid (**85**) is also a convenient recyclable hypervalent iodine oxidant for the synthesis of α-iodo ketones by oxidative iodination of ketones [88]. Various ketones and β-dicarbonyl compounds can be iodinated by this reagent system under mild conditions to afford the respective α-iodo substituted carbonyl compounds in excellent yields. The final products of iodination are conveniently separated from by-products by simple treatment with anionic exchange resin Amberlite IRA 900 HCO_3^- and are isolated with good purity after evaporation of the solvent. The reduced form of the hypervalent iodine oxidant, 3-iodobenzoic acid (**59**), can be recovered in 91–95% yield from the Amberlite resin by treatment with aqueous hydrochloric acid followed by extraction with ethyl acetate [88].

Two recyclable nonpolymeric hypervalent iodine reagents, 3-[bis(trifluoroacetoxy)iodo]benzoic acid (**90**) and 3-[hydroxy(tosyloxy)iodo]benzoic acid (**91**) can be conveniently prepared in high yield by treatment of 3-iodosylbenzoic acid (**85**) with trifluoroacetic acid or *p*-toluenesulfonic acid, respectively (Scheme 5.31) [89,90].

The reactivity pattern of compounds **90** and **91** is similar to common hypervalent iodine reagents, as illustrated by their use for the oxidation of sulfides, oxidative iodination of arenes and α-tosyloxylation of ketones (Scheme 5.32) [89]. The products of all these reactions can be conveniently separated from the by-product, 3-iodobenzoic acid, by simple treatment with ion-exchange resin IRA-900 according to

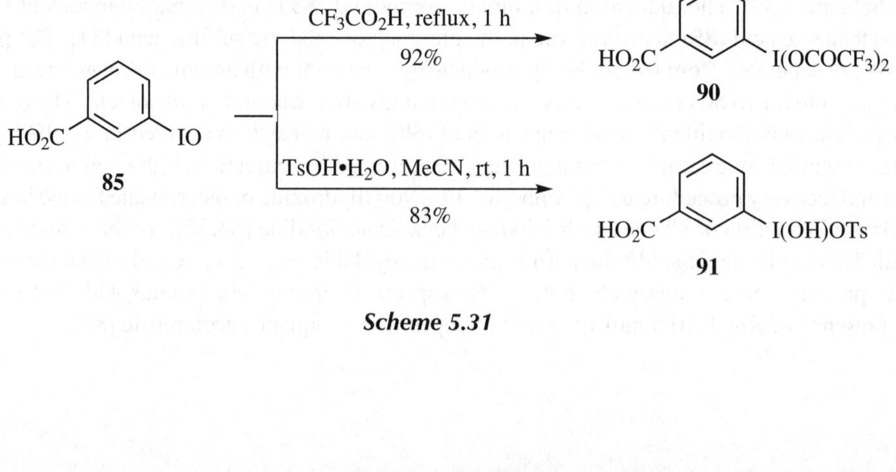

Scheme 5.31

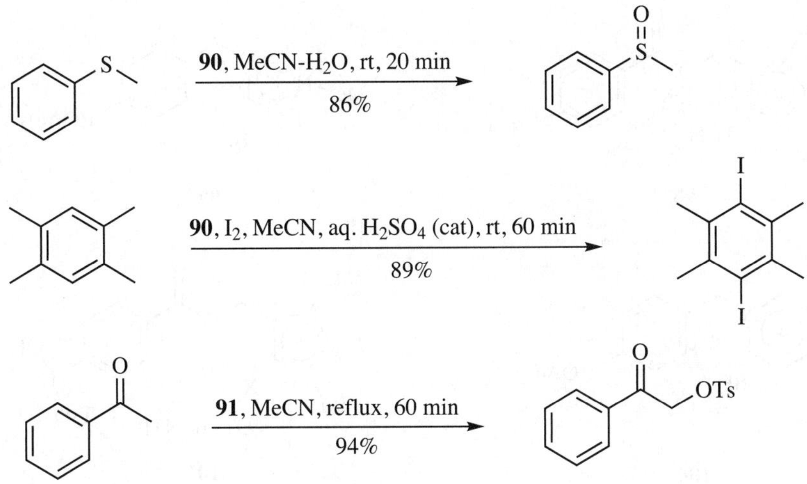

90, MeCN-H_2O, rt, 20 min
86%

90, I_2, MeCN, aq. H_2SO_4 (cat), rt, 60 min
89%

91, MeCN, reflux, 60 min
94%

Scheme 5.32

Scheme 5.30. 3-Iodobenzoic acid can be regenerated from IRA 900 resin by treatment with aqueous HCl and can be reoxidized to reagent **77** without additional purification. An alternative procedure for product separation involves basic aqueous work-up followed by acidification with HCl (Section 5.3.1).

A different tagging approach using a phenylsulfonate-tagged (diacetoxyiodo)benzene **92** has been developed by Kirschning and coworkers [91]. This approach is based on ion exchange and is initiated by an azide-promoted S_N2-reaction of isobutyl sulfonate **93** with the N_3^--resin **94** followed by trapping of the resulting aryl sulfonate anion with the ion-exchange resin. The trapped iodine reagent **93** can be recovered from the resin **95.** This concept has been successfully employed for the $RuCl_3$-catalyzed oxidation of alcohols **96** to the respective carbonyl compounds **97** (Scheme 5.33), as well as for the glycosidation of thioglycosides and the Suarez reaction of pyranoses [91,92].

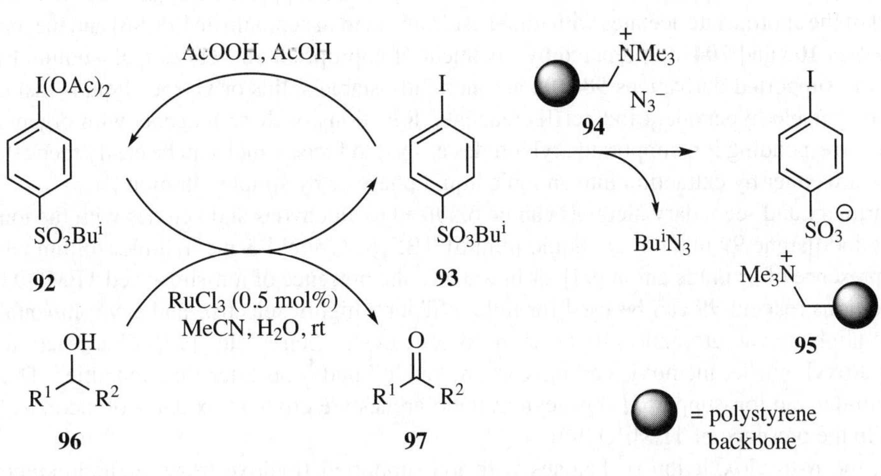

AcOOH, AcOH
$I(OAc)_2$
SO_3Bu^i
92
I
SO_3Bu^i
93
$RuCl_3$ (0.5 mol%)
MeCN, H_2O, rt
OH
R^1 R^2
96
O
R^1 R^2
97
$\overset{+}{N}Me_3$
N_3^-
94
Bu^iN_3
I
$SO_3^{\ominus}$
$Me_3\overset{+}{N}$
95
= polystyrene backbone

Scheme 5.33

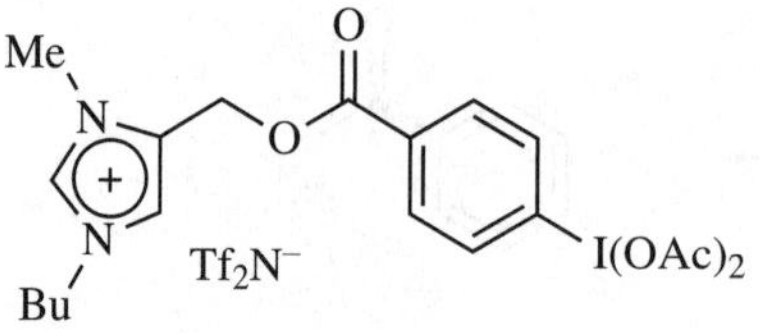

Figure 5.4 Recyclable ion-supported hypervalent iodine(III) reagents.

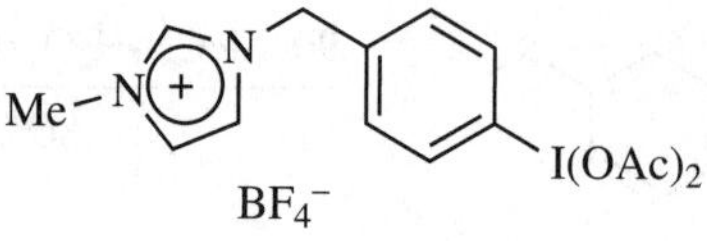

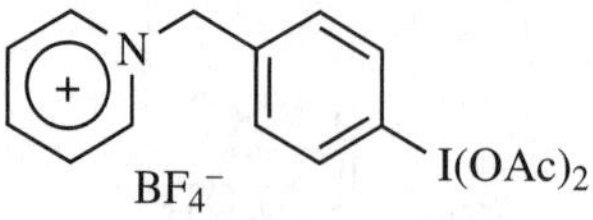

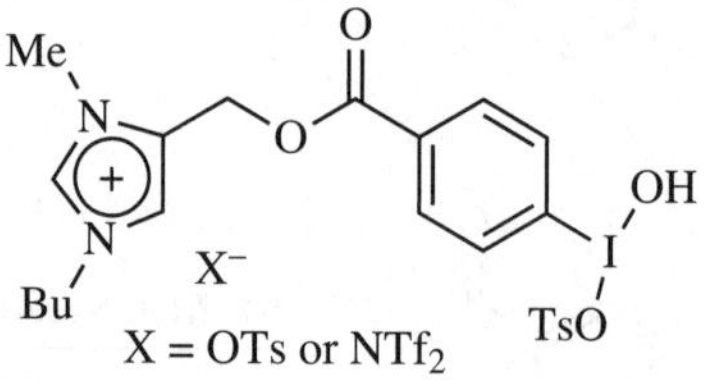

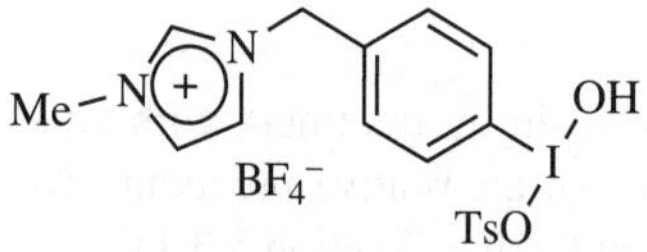

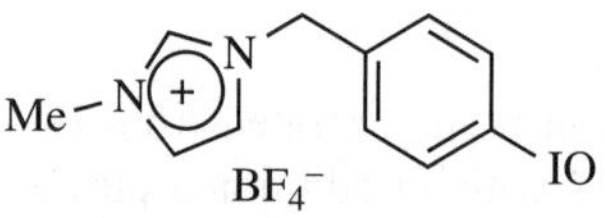

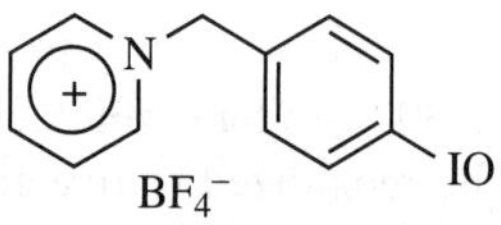

5.3.3 Ionic-Liquid-Supported Recyclable Hypervalent Iodine(III) Reagents

Figure 5.4 shows several examples of ionic-liquid-supported (also called ion-supported) hypervalent iodine(III) reagents [93]. Ion-supported (diacetoxyiodo)arenes **98–100** are prepared by the peracetic oxidation of appropriate ion-supported aryl iodides [93–95]. The ion-supported tosylates **101** and **102** are made by treatment of the appropriate acetates with toluenesulfonic acid in acetonitrile [95, 96] and the ion-supported iodosylbenzenes **103** and **104** are prepared by treatment of appropriate diacetates with sodium hydroxide in water [93]. Ion-supported derivatives **98–104** are thermally stable solids or viscous liquids that can be used as efficient recyclable hypervalent iodine(III) reagents. Reactions of these reagents with organic substrates produce the corresponding ion-supported aryl iodides as by-products, which can be easily recovered from the reaction mixture either by extraction into an ionic liquid phase or by simple filtration.

Various primary and secondary alcohols can be oxidized to aldehydes and ketones with the ion-supported [bis(acyloxy)iodo]arene **99** in the ionic liquid $[emim]^+[BF_4]^-$ (1-ethyl-3-methylimidazolium tetrafluoroborate) in the presence of bromide anion [94], or in water in the presence of ion-supported TEMPO [97]. Under similar conditions reagent **99** can be used for mild, efficient, highly selective and environmentally friendly oxidation of aliphatic and aromatic sulfides to sulfoxides in excellent yields [98]. This reaction is compatible with hydroxyl, nitrile, methoxy, carbon–carbon double bonds and ester functionalities. The analogous pyrrolidinium-derived ion-supported [bis(acyloxy)iodo]arenes are efficient oxidants of alcohols to carbonyl compounds in the presence of TEMPO [99].

Likewise, the α-tosyloxylation of ketones with ion-supported [hydroxy(tosyloxy)iodo]arenes **101** proceeds smoothly in acetonitrile even at 0 °C (Scheme 5.34) [95]. These reactions are faster than those with

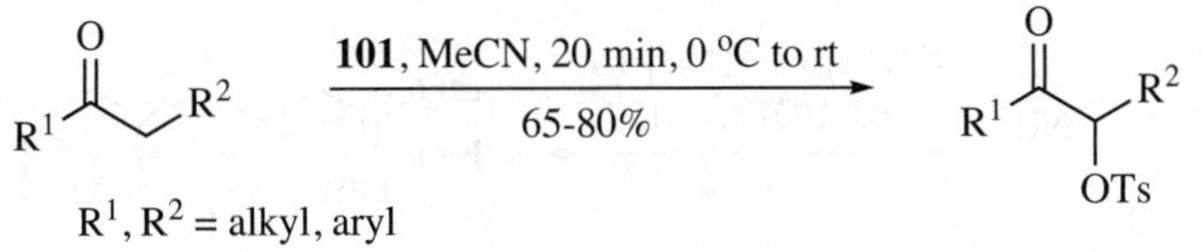

101, MeCN, 20 min, 0 °C to rt
65-80%
R^1, R^2 = alkyl, aryl
OTs

Scheme 5.34

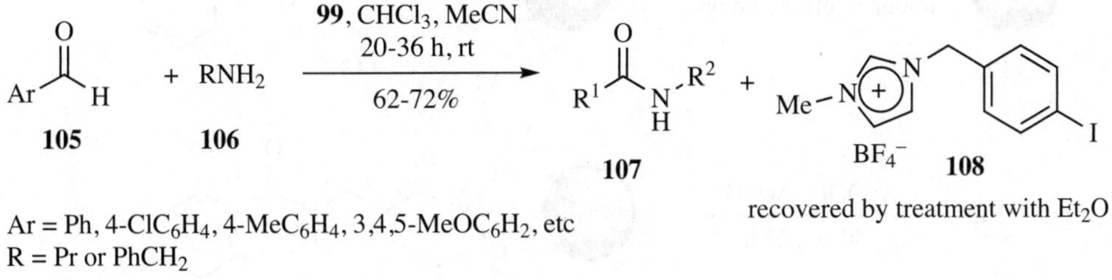

99, $CHCl_3$, MeCN
20-36 h, rt
62-72%
105 + RNH_2 **106** → **107** + **108** (BF_4^-)
recovered by treatment with Et_2O
Ar = Ph, 4-ClC_6H_4, 4-MeC_6H_4, 3,4,5-$MeOC_6H_2$, etc
R = Pr or $PhCH_2$

Scheme 5.35

PhI(OH)OTs, probably due to the presence of the electron-withdrawing ester group on the benzene ring of reagent **101.**

A metal-free direct oxidative preparation of amides **107** from aldehydes **105** and amines **106** using the ion-supported hypervalent iodine(III) reagent **99** as a recyclable oxidant under mild conditions has been reported (Scheme 5.35) [100]. The oxidant and its reduced form **108** are completely insoluble in diethyl ether; consequently, products **107** can be extracted directly from the reaction mixture and the reagent **99** can be easily recycled.

Charette and coworkers have developed tetraarylphosphonium (TAP)-supported (diacetoxyiodo)benzene **109** (Figure 5.5), which can be used as a recyclable reagent or a catalyst for the α-acetoxylation of ketones [101]. Similarly to the imidazolium-supported [bis(acyloxy)iodo]arene **99,** the reduced form of the TAP-supported reagent **109** can be recovered from the reaction mixture by simple filtration after treatment with ether.

5.3.4 Magnetic Nanoparticle-Supported Recyclable Hypervalent Iodine(III) Reagent

The magnetic nanoparticle-supported (diacetoxyiodo)benzene **114** has been prepared in three steps starting from 4-iodobenzoic acid (Scheme 5.36) [102]. The organic silane groups were used for immobilization of

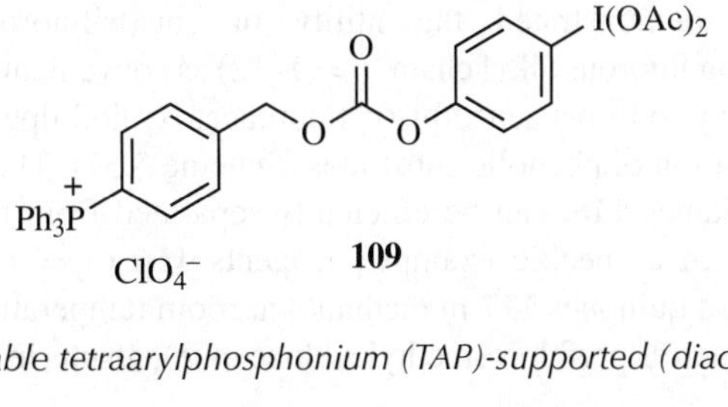

$I(OAc)_2$
Ph_3P^+
ClO_4^-
109

Figure 5.5 *Recyclable tetraarylphosphonium (TAP)-supported (diacetoxyiodo)benzene.*

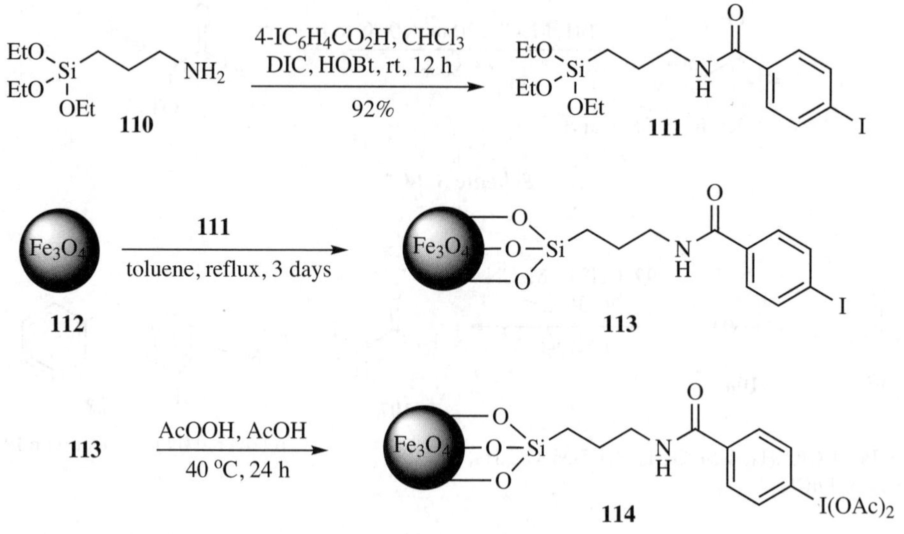

Scheme 5.36

iodobenzene groups to the surface of magnetite nanoparticles, because silanes have a large affinity for undercoordinated surface sites of metal oxide particles. The reaction of 3-aminopropyltriethoxysilane (**110**) with 4-iodobenzoic acid using *N,N*-diisopropylcarbodiimide (DIC) in combination with 1-hydroxybenzotriazole (HOBt) afforded product **111** in excellent yields. Polyvinylpyrrolidone-protected magnetic nanoparticles **112** of 8–20 nm size were prepared by coprecipitation of iron(III) and iron(II) ions in basic solution. The obtained nanoparticles **112** were sonicated for one hour before being treated with excess of compound **111** in anhydrous toluene under reflux to give the magnetic nanoparticle-supported iodobenzene **113**. Subsequent oxidation of product **113** using freshly prepared peracetic acid at 40 °C overnight afforded the diacetate derivative **114** with a loading of 0.68 mmol g^{-1} as determined by elemental analysis.

The magnetic nanoparticle reagent **114** can be used for efficient oxidation of a wide range of benzylic, allylic, heterocyclic, alicyclic and aliphatic alcohols to the corresponding aldehydes or ketones in excellent yield. Product separation can be achieved by simple decantation of the organic layer with the assistance of an external magnet. The separated magnetic material **113** is then washed and regenerated by treatment with peracetic acid. The reagent can be recycled many times with no apparent loss of activity [102].

5.3.5 Fluorous Recyclable Hypervalent Iodine(III) Reagents

Tesevic and Gladysz have demonstrated the utility of [bis(trifluoroacetoxy)iodo]perfluoroalkanes $C_nF_{2n+1}I(OCOCF_3)_2$ with a long fluorous alkyl chain ($n = 7$–12) as convenient recyclable oxidants [103,104]. Similarly to [bis(trifluoroacetoxy)iodo]benzene, [bis(trifluoroacetoxy)iodo]perfluoroalkanes **116** can serve as excellent reagents for the oxidation of phenolic substrates (Scheme 5.37). The reduced form of the reagents, the respective iodoperfluoroalkanes **118,** can be efficiently separated from the reaction mixture using fluorous techniques and reused. In a specific example, reagents **116** (n = 8, 10, 12) can rapidly oxidize 1,4-hydroquinones **115** to afford quinones **117** in methanol at room temperature (Scheme 5.37). Subsequent addition of a fluorous solvent, such as perfluoro(methylcyclohexane), results in a liquid/liquid biphasic system.

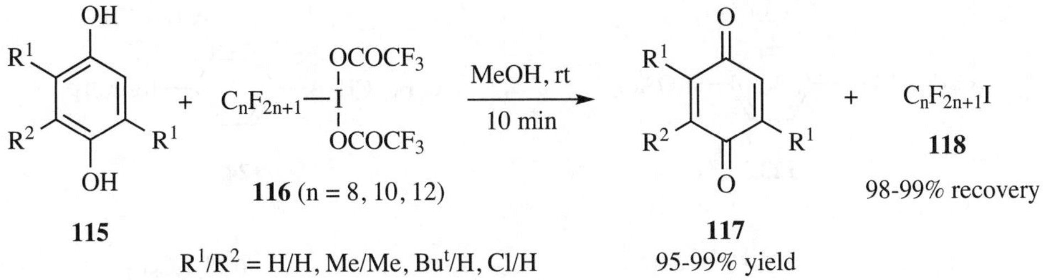

$C_nF_{2n+1}I(OCOCF_3)_2$ **116** (n = 8, 10, 12); MeOH, rt, 10 min; **115** → **117** (95-99% yield) + $C_nF_{2n+1}I$ **118** (98-99% recovery)

R^1/R^2 = H/H, Me/Me, Bu^t/H, Cl/H

Scheme 5.37

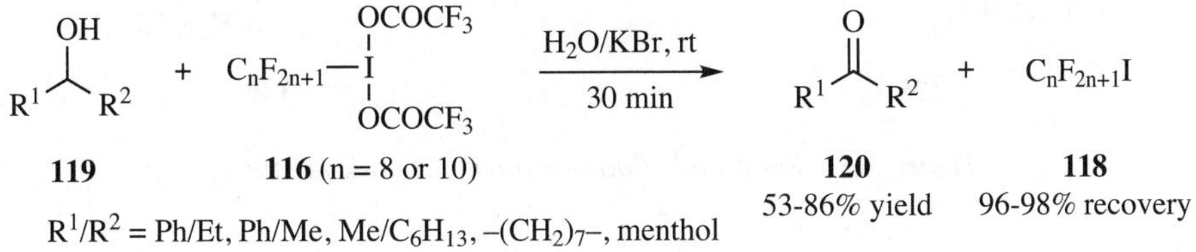

$C_nF_{2n+1}I(OCOCF_3)_2$ **116** (n = 8 or 10); H_2O/KBr, rt, 30 min; **119** → **120** (53-86% yield) + $C_nF_{2n+1}I$ **118** (96-98% recovery)

R^1/R^2 = Ph/Et, Ph/Me, Me/C_6H_{13}, –$(CH_2)_7$–, menthol

Scheme 5.38

The product quinones **117** are generally isolated in about 95% yields from the methanol phase and iodoperfluoroalkanes **118** are isolated in 98–99% yields from the fluorous phase. The recovered iodoperfluoroalkanes **118** may be reoxidized by peroxytrifluoroacetic acid to give initial reagents **116** in high yield [104]. A convenient new procedure for the oxidation of iodoperfluoroalkanes **118** to [bis(trifluoroacetoxy)iodo]perfluoroalkanes **116** using Oxone ($2KHSO_5{\cdot}KHSO_4{\cdot}K_2SO_4$) in the presence of trifluoroacetic acid has been reported [105].

[Bis(trifluoroacetoxy)iodo]perfluoroalkanes are effective recyclable reagents for the oxidation of aliphatic and benzylic secondary alcohols **119** to ketones **120** in the presence of KBr in aqueous solution (Scheme 5.38) [103]. The reduced form of the reagent, the respective iodoperfluoroalkanes **118,** can be efficiently isolated from the reaction mixture in 96–98% yield by adding three to five volumes of methanol and separating the resulting fluorous/methanolic liquid/liquid biphasic system. The recovered iodoperfluoroalkanes **118** can be reoxidized to reagents **116** and reused [103].

The fluorous bis(trifluoroacetates) **116** (Scheme 5.38) oxidize secondary alcohols in the presence of bromide ions much more rapidly than the common hypervalent iodine(III) reagents, such as iodosylbenzene or (diacetoxyiodo)benzene. The higher reactivity may be due to the directly bound electron-withdrawing perfluoroalkyl substituents in compounds **116,** which enhance their oxidizing strength [103].

Westwell and coworkers investigated the oxidation of hydroxylated stilbenes **121** using [bis(trifluoroacetoxy)iodo]perfluorohexane (Scheme 5.39) [106]. Instead of the expected products of the

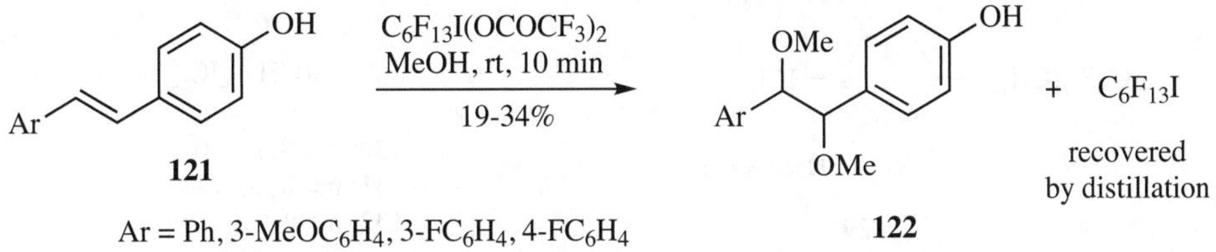

$C_6F_{13}I(OCOCF_3)_2$, MeOH, rt, 10 min, 19-34%; **121** → **122** + $C_6F_{13}I$ (recovered by distillation)

Ar = Ph, 3-MeOC_6H_4, 3-FC_6H_4, 4-FC_6H_4

Scheme 5.39

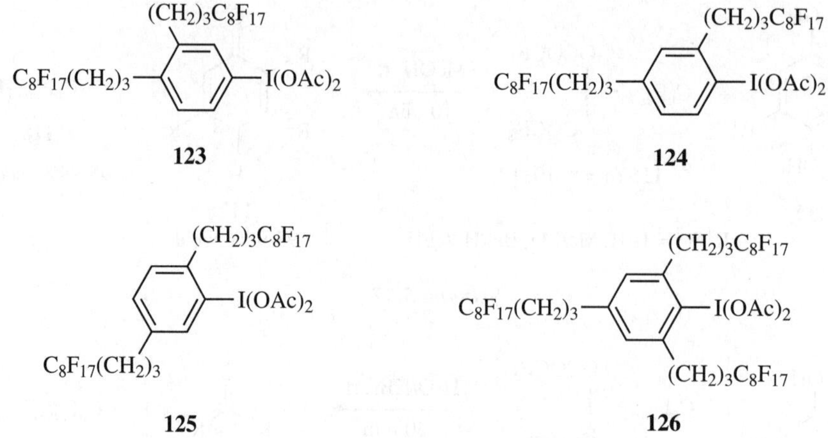

$(CH_2)_3C_8F_{17}$ / $C_8F_{17}(CH_2)_3$ — $I(OAc)_2$

123 **124** **125** **126**

Figure 5.6 Recyclable fluorous (diacetoxyiodo)arenes.

phenolic oxidation, diaryl-1,2-dimethoxyethanes **122** as mixtures of diastereoisomers were isolated from this reaction in moderate yields. The perfluorohexyl iodide by-product (bp 140 °C) could be recovered simply by distillation of the reaction mixture under reduced pressure [106].

Gladysz and Rocaboy have reported the application of fluorous (diacetoxyiodo)arenes **123–126** (Figure 5.6) in oxidations of hydroquinones to quinones; in this procedure the fluorous reagents can be conveniently recovered by simple liquid/liquid biphasic workups [107].

Likewise, the series of fluorous (dichloroiodo)arenes **127–129** and alkyl iodine(III) dichlorides **130–132** (Figure 5.7) have been prepared in 71–98% yields by reactions of the corresponding fluorous iodides with chlorine [69]. These compounds are effective reagents for the chlorination of alkenes (e.g., cyclooctene) and aromatic compounds (e.g., anisole, 4-*tert*-butylphenol and acetophenone). The organic chlorinated products and fluorous iodide co-products are easily separated by organic/fluorous liquid/liquid biphasic workups. The fluorous iodides can be recovered in 90–97% yields and reoxidized with chlorine [69].

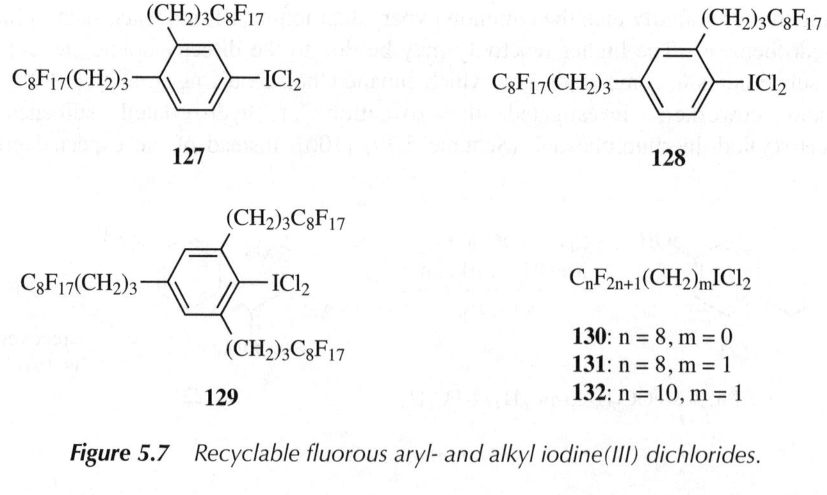

$(CH_2)_3C_8F_{17}$ / $C_8F_{17}(CH_2)_3$ — ICl_2

127 **128** **129**

$C_nF_{2n+1}(CH_2)_mICl_2$

130: n = 8, m = 0
131: n = 8, m = 1
132: n = 10, m = 1

Figure 5.7 Recyclable fluorous aryl- and alkyl iodine(III) dichlorides.

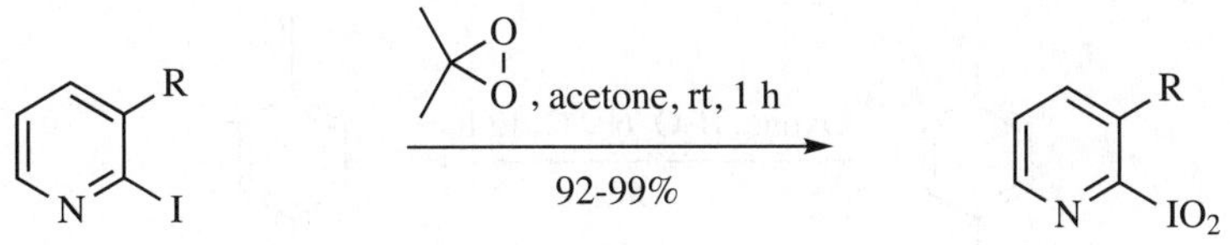

133 R = H, OMe, OPr^i, OPr, OBu 134

acetone, rt, 1 h

92-99%

Scheme 5.40

5.4 Recyclable Nonpolymeric Hypervalent Iodine(V) Reagents

The most important and practically useful hypervalent iodine(V) reagents, IBX and DMP, have a serious disadvantage with respect to the principles of green chemistry since they are normally used as the non-recyclable, stoichiometric reagents. In the only available report, it has been shown that IBX can be recovered in a low yield (about 50%) after reaction with alcohols in ethyl acetate [108].

The first nonpolymeric recyclable hypervalent iodine(V) reagents based on the readily available derivatives of 2-iodopyridine were reported in 2011 [109]. 2-Iodylpyridines **134** were prepared by oxidation of the respective 2-iodopyridines **133** with 3,3-dimethyldioxirane (Scheme 5.40). Structures of 2-iodylpyridines **134** (R = H, OPr^i and OBu) were established by single crystal X-ray crystallography [109].

2-Iodyl-3-propoxypyridine (**136**) has moderate solubility in organic solvents (e.g., 1.1 mg ml^{-1} in acetonitrile) and it can be used as a recyclable reagent for the oxidation of sulfides to sulfoxides and alcohols to the respective carbonyl compounds (Scheme 5.41) [109]. The reduced form of this reagent, 2-iodo-3-propoxypyridine (**135**), can be effectively separated from the reaction mixture by treatment with diluted sulfuric acid and recovered from the acidic aqueous solution by adding aqueous sodium hydroxide.

Another recyclable hypervalent iodine(V) reagent, potassium 4-iodylbenzenesulfonate (**138**), was prepared by the oxidation of 4-iodobenzensulfonic acid (**137**) with Oxone ($2KHSO_5{\cdot}KHSO_4{\cdot}K_2SO_4$) in water (Scheme 5.42) [110]. A single-crystal X-ray structure analysis of product **138** revealed the presence of polymeric chains in the solid state due to a combination of numerous intra- and intermolecular interactions.

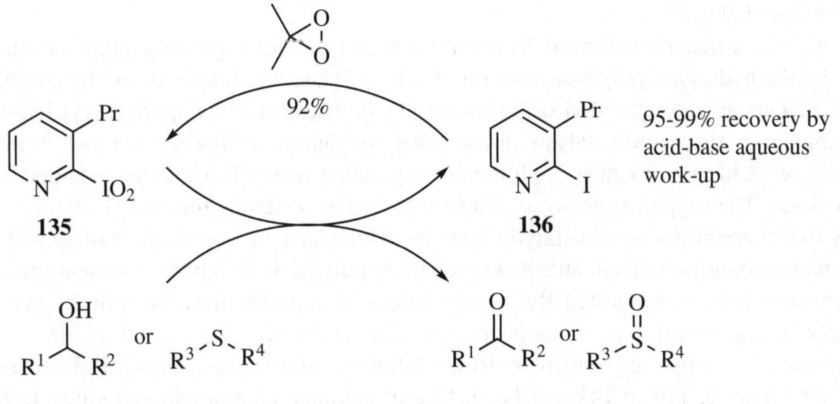

92%

135 136

95-99% recovery by acid-base aqueous work-up

or

R^1, R^3, R^4 = alkyl or aryl; R^2 = H or alkyl

Scheme 5.41

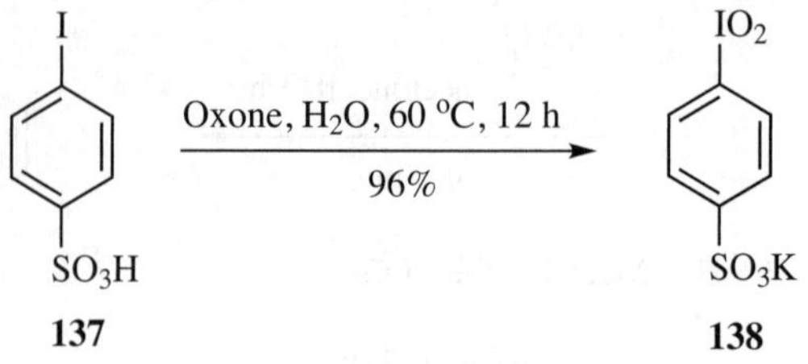

I
SO$_3$H
137

Oxone, H_2O, 60 °C, 12 h
96%

IO$_2$
SO$_3$K
138

Scheme 5.42

Potassium 4-iodylbenzenesulfonate (**138**) is a thermally stable and water-soluble hypervalent iodine(V) oxidant that is particularly useful as a reagent for oxidative iodination of aromatic substrates [110]. Reagent **138** can be effectively recovered from the reaction mixture (92% recovery) after removal of organic products by treatment of the aqueous layer with Oxone at 60 °C for 2 h, followed by filtration of the precipitate of compound **138.**

5.5 Recyclable Iodine Catalytic Systems

Polymer-supported iodanes and iodoarenes can in principle be used as recyclable catalysts. For example, polystyrene-supported iodosylbenzene (**22**) (Section 5.1) has been employed as an efficient co-catalyst in combination with $RuCl_3$ in the catalytic oxidation of alcohols and aromatic hydrocarbons, respectively, to the corresponding carboxylic acids and ketones using Oxone as the stoichiometric oxidant [22].

Polystyrene-supported iodobenzene has been used instead of iodobenzene as a catalyst for the α-tosyloxylation of ketones and alcohols with *m*CPBA and *p*-toluenesulfonic acid [111]. In particular, two kinds of polymer-supported PhI, the standard linear poly(4-iodostyrene) and the macroporous crosslinked poly(4-iodostyrene), have been used as recyclable catalysts, which could be recovered from the reaction mixture by simple filtration in 90–100% yield. Recovered poly(4-iodostyrene) could be reused for the same reaction to provide the corresponding α-tosyloxyketone in good yields [111]. Recyclable ionic-liquid-supported iodoarenes have also been used as catalysts in the α-tosyloxylation of ketones with *m*CPBA and *p*-toluenesulfonic acid [112].

The polymeric silica materials derived from silylated aryl iodides have been prepared via sol–gel processes, either by the hydrolytic polycondensation of a bis-silylated monomer or by the co-gelification of a monosilylated precursor with tetraethyl orthosilicate [113]. These silica-supported aryl iodides have been successfully applied as supported catalysts in the α-tosyloxylation of aliphatic ketones in the presence of *m*-chloroperbenzoic acid as an oxidant, with the corresponding α-tosyloxyketones obtained in moderate to good isolated yields. The supported catalysts can be recycled by a simple filtration [113].

Particularly useful are bifunctional catalytic systems, which have the iodoarene moiety and a co-catalyst, such as a complex of transition metal, attached to the same polymeric backbone. Efficient recyclable bifunctional catalysts bearing SiO_2-supported $RuCl_3$ and iodoarene moieties have been developed and used for environmentally benign oxidation of alcohols or alkylarenes at the benzylic position [114].

The synthesis of SiO_2-supported iodoarene–$RuCl_3$ bifunctional catalysts **142** and **143** consists of building a suitable ligand structure, **140** or **141,** on the surface of commercial aminopropyl silica **139,** followed by complexation with $RuCl_3$ (Scheme 5.43) [114]. The amounts of $RuCl_3$ and iodine loaded on the surface of silica gel were determined by elemental analyses of chlorine and iodine. The loadings of iodine and ruthenium for catalyst **142** are 0.60 and 0.06 mmol g^{-1} and for catalyst **143** 0.62 and 0.05 mmol g^{-1}, respectively. For

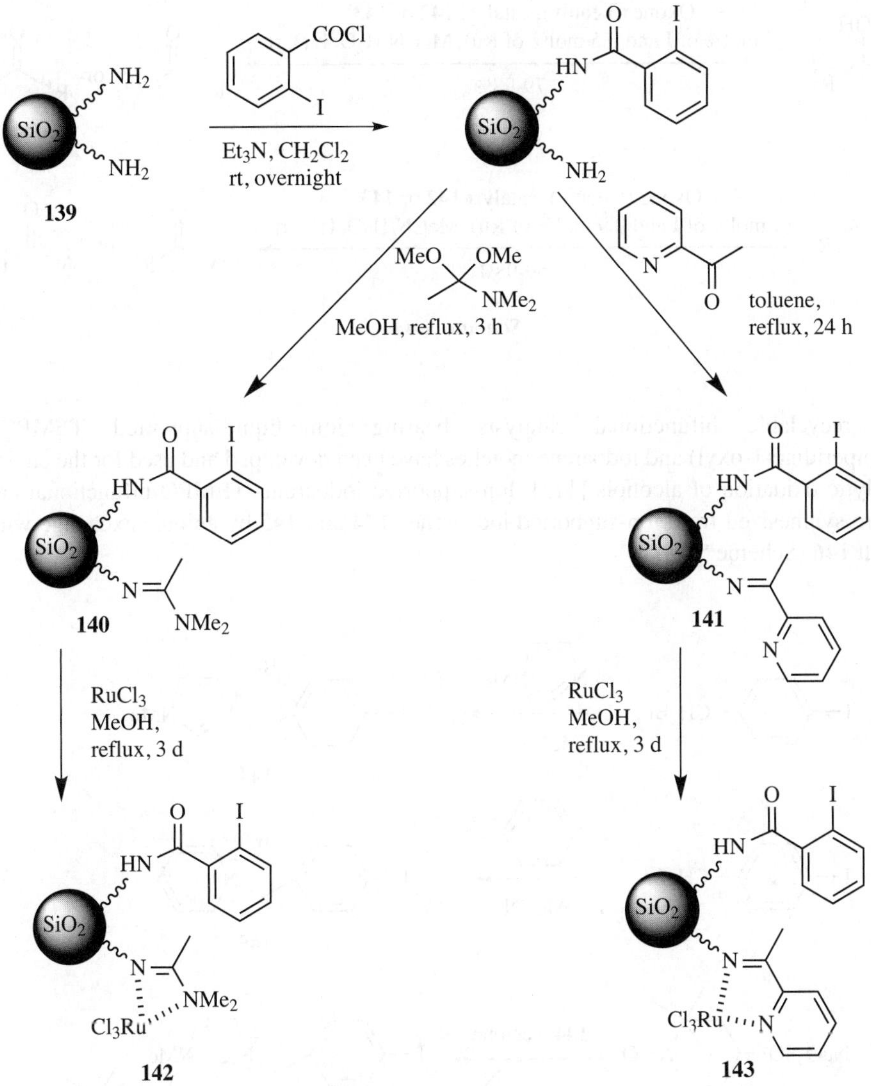

Scheme 5.43

comparison, the ratio of catalytic iodine to ruthenium in the non-recyclable tandem catalytic system $PhI/RuCl_3$ was 31 : 1 (Section 4.3).

Bifunctional catalysts **142** and **143** can be used for environmentally benign oxidation of alcohols or alkylarenes at the benzylic position (Scheme 5.44) [114]. These reactions, using Oxone as stoichiometric oxidant, afford the corresponding carbonyl compounds (or carboxylic acids in the oxidations of benzyl alcohol or toluene) in high yields under mild conditions and convenient work-up. Both catalysts **142** and **143** show similar catalytic activity in the oxidation of alcohols, while catalyst **143** has noticeably higher catalytic reactivity in oxidation at the benzylic position. Furthermore, these SiO_2-supported bifunctional catalysts can be recovered by simple filtration and directly reused [114].

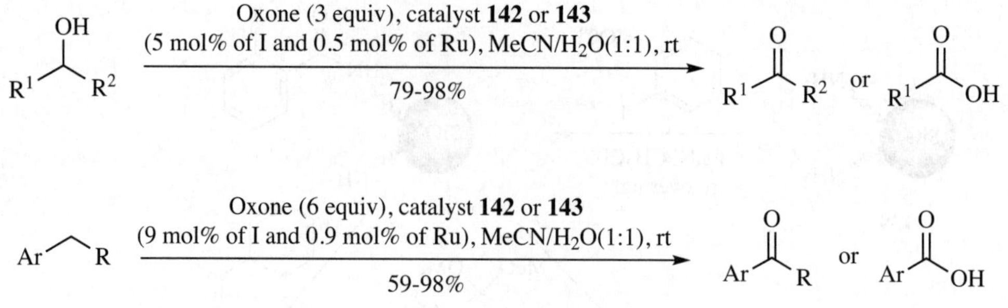

OH, R^1, R^2 → Oxone (3 equiv), catalyst **142** or **143** (5 mol% of I and 0.5 mol% of Ru), MeCN/H_2O(1:1), rt; 79-98% → O, R^1, R^2 or O, R^1, OH

Ar, R → Oxone (6 equiv), catalyst **142** or **143** (9 mol% of I and 0.9 mol% of Ru), MeCN/H_2O(1:1), rt; 59-98% → O, Ar, R or O, Ar, OH

Scheme 5.44

Efficient recyclable bifunctional catalysts bearing ionic-liquid-supported TEMPO (2,2,6,6-tetramethylpiperidine-1-oxyl) and iodoarene moieties have been developed and used for the environmentally benign catalytic oxidation of alcohols [115]. Ion-supported iodoarene–TEMPO bifunctional catalysts **147** and **148** were synthesized from ion-supported iodoarenes **144** and **145** by anionic exchange with TEMPO-sulfonate salt **146** (Scheme 5.45).

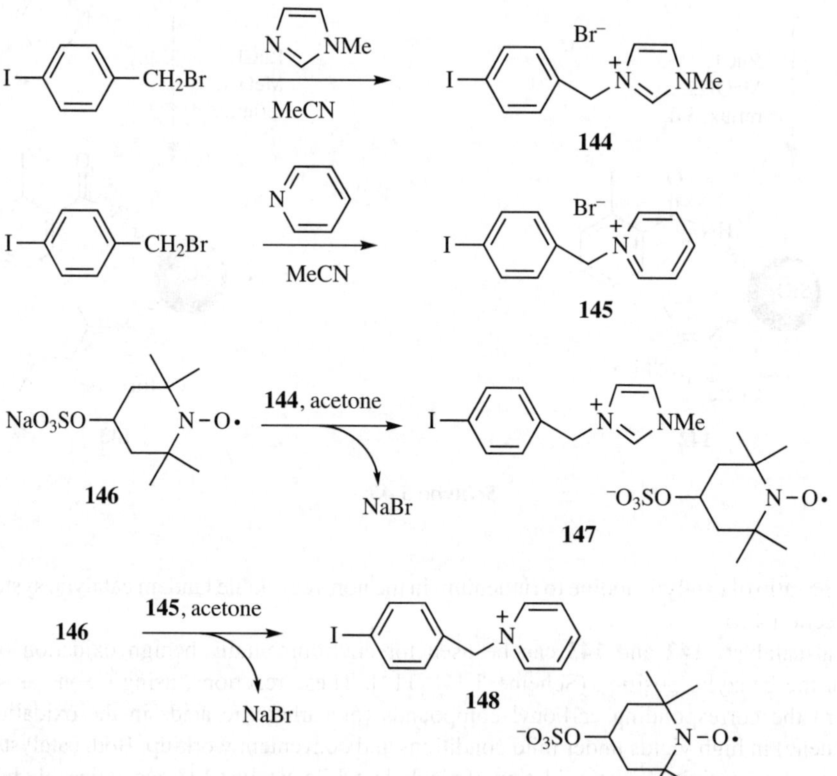

I, CH_2Br → N, NMe; MeCN → I, Br^-, +N, NMe; **144**

I, CH_2Br → N; MeCN → I, Br^-, +N; **145**

NaO_3SO, N–O•; **146** → **144**, acetone; NaBr → I, +N, NMe; ^-O_3SO, N–O•; **147**

146 → **145**, acetone; NaBr → I, +N; **148**; ^-O_3SO, N–O•

Scheme 5.45

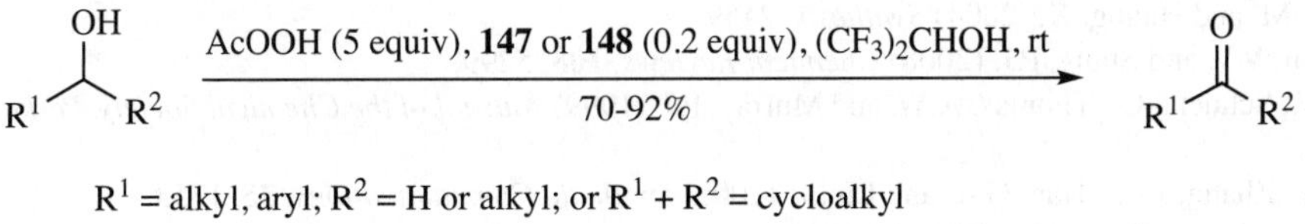

$$R^1CH(OH)R^2 \xrightarrow[\text{70-92\%}]{\text{AcOOH (5 equiv), } \mathbf{147} \text{ or } \mathbf{148} \text{ (0.2 equiv), } (CF_3)_2CHOH\text{, rt}} R^1C(O)R^2$$

R^1 = alkyl, aryl; R^2 = H or alkyl; or $R^1 + R^2$ = cycloalkyl

Scheme 5.46

Ion-supported iodoarene–TEMPO bifunctional catalysts **147** and **148** have been employed as catalysts for the oxidation of alcohols to the corresponding carbonyl compounds using peracetic acid as a green and practical oxidant (Scheme 5.46). Furthermore, catalysts **147** and **148** could be conveniently recovered and reused without any loss of catalytic activity [115].

References

1. Anastas, P.T., Levy, I.J. and Parent, K.E. (eds) (2009) *Green Chemistry Education: Changing the Course of Chemistry*, ACS Symposium Series, vol. **1011**.
2. McNamara, C.A., Dixon, M.J. and Bradley, M. (2002) *Chemical Reviews*, **102**, 3275.
3. Hodge, P. (2003) *Current Opinion in Chemical Biology*, **7**, 362.
4. Nam, N.-H., Sardari, S. and Parang, K. (2003) *Journal of Combinatorial Chemistry*, **5**, 479.
5. Togo, H and Sakuratani, K. (2002) *Synlett*, 1966.
6. Togo, H. (2002) *Eco Industry*, **7**, 5.
7. Yusubov, M.S and Zhdankin, V.V. (2012) *Current Organic Synthesis*, **9**, 247.
8. Yusubov, M.S. and Zhdankin, V.V. (2010) *Mendeleev Communications*, **20**, 185.
9. Ochiai, M. and Miyamoto, K. (2008) *European Journal of Organic Chemistry*, 4229.
10. Chen, J.-M., Zeng, X.-M., Middleton, K. and Zhdankin, V.V. (2011) *Tetrahedron Letters*, **52**, 1952.
11. Chen, D.-J. and Chen, Z.-C. (2001) *Synthetic Communications*, **31**, 421.
12. Togo, H., Nogami, G. and Yokoyama, M. (1998) *Synlett*, 534.
13. Wang, G.-P. and Chen, Z.-C. (1999) *Synthetic Communications*, **29**, 2859.
14. Ley, S.V., Thomas, A.W. and Finch, H. (1999) *Journal of the Chemical Society, Perkin Transactions 1*, 669.
15. Huang, X. and Zhu, Q. (2001) *Synthetic Communications*, **31**, 111.
16. Xian, H., Zhu, Q. and Zhang, J. (2001) *Synthetic Communications*, **31**, 2413.
17. Tohma, H., Morioka, H., Takizawa, S., *et al.* (2001) *Tetrahedron*, **57**, 345.
18. Ficht, S., Mulbaier, M. and Giannis, A. (2001) *Tetrahedron*, **57**, 4863.
19. Shang, Y., But, T.Y.S., Togo, H. and Toy, P.H. (2007) *Synlett*, 67.
20. Chen, F.-E., Xie, B., Zhang, P., *et al.* (2007) *Synlett*, 619.
21. Togo, H., Abe, S., Nogami, G. and Yokoyama, M. (1999) *Bulletin of the Chemical Society of Japan*, **72**, 2351.
22. Chen, J.-M., Zeng, X.-M., Middleton, K., *et al.* (2011) *Synlett*, 1613.
23. Chen, J.-M., Zeng, X.-M. and Zhdankin, V.V. (2010) *Synlett*, 2771.
24. Abe, S., Sakuratani, K. and Togo, H. (2001) *Synlett*, 22.
25. Abe, S., Sakuratani, K. and Togo, H. (2001) *Journal of Organic Chemistry*, **66**, 6174.
26. Chen, D.-J., Cheng, D.-P. and Chen, Z.-C. (2001) *Synthetic Communications*, **31**, 3847.
27. Herrerias, C.I., Zhang, T.Y. and Li, C.-J. (2006) *Tetrahedron Letters*, **47**, 13.
28. But, T.Y.S., Tashino, Y., Togo, H. and Toy, P.H. (2005) *Organic & Biomolecular Chemistry*, **3**, 970.
29. Sakuratani, K. and Togo, H. (2003) *Synthesis*, 21.
30. Tashino, Y. and Togo, H. (2004) *Synlett*, 2010.
31. Tohma, H., Maegawa, T. and Kita, Y. (2003) *Synlett*, 723.
32. Tohma, H., Maegawa, T. and Kita, Y. (2003) *ARKIVOC*, (vi), 62.

33. Chen, J.-M. and Huang, X. (2004) *Synthesis*, 2459.
34. Zhdankin, V.V. and Stang, P.J. (2008) *Chemical Reviews*, **108**, 5299.
35. Ley, S.V., Schucht, O., Thomas, A.W. and Murray, P.J. (1999) *Journal of the Chemical Society, Perkin Transactions 1*, 1251.
36. Liu, S.-j., Zhang, J.-z., Tian, G.-r. and Liu, P. (2005) *Synthetic Communications*, **35**, 1753.
37. Sakuratani, K. and Togo, H. (2003) *ARKIVOC*, (vi), 11.
38. Cheng, D.-P. and Chen, Z.-C. (2002) *Synthetic Communications*, **32**, 2155.
39. Kawamura, Y., Maruyama, M., Tokuoka, T. and Tsukayama, M. (2002) *Synthesis*, 2490.
40. Kawamura, Y., Maruyama, M., Yamashita, K. and Tsukayama, M. (2003) *International Journal of Modern Physics B*, **17**, 1482.
41. Teduka, T. and Togo, H. (2005) *Synlett*, 923.
42. Huang, H.-Y., Hou, R.-S., Wang, H.-M. and Chen, L.-C. (2005) *Heterocycles*, **65**, 1881.
43. Chen, J.-M., Lin, X.-J. and Huang, X. (2004) *Journal of Chemical Research*, 43.
44. Koser, G.F. (2001) *Aldrichimica Acta*, **34**, 89.
45. Liu, S.J., Zhang, J.Z., Tian, G.R. and Liu, P. (2005) *Synthetic Communications*, **36**, 823.
46. Xie, Y.-Y. and Chen, Z.-C. (2002) *Synthetic Communications*, **32**, 1875.
47. Wan, D.-B. and Chen, J.-M. (2006) *Journal of Chemical Research*, 32.
48. Chen, J.-M. and Huang, X. (2004) *Synthesis*, 1577.
49. Kirschning, A., Jesberger, M. and Monenschein, H. (1999) *Tetrahedron Letters*, **40**, 8999.
50. Monenschein, H., Sourkouni-Argirusi, G., Schubothe, K.M., *et al.* (1999) *Organic Letters*, **1**, 2101.
51. Kirschning, A., Monenschein, H. and Schmeck, C. (1999) *Angewandte Chemie, International Edition*, **38**, 2594.
52. Sourkouni-Argirusi, G. and Kirschning, A. (2000) *Organic Letters*, **2**, 3781.
53. Kirschning, A., Kunst, E., Ries, M., *et al.* (2003) *ARKIVOC*, (vi), 145.
54. Kirschning, A., Monenschein, H. and Wittenberg, R. (2001) *Angewandte Chemie, International Edition*, **40**, 650.
55. Kalberer, E.W., Whitfield, S.R. and Sanford, M.S. (2006) *Journal of Molecular Catalysis A: Chemical*, **251**, 108.
56. Chen, D.-J. and Chen, Z.-C. (2000) *Synlett*, 1175.
57. Dohi, T., Maruyama, A., Yoshimura, M., *et al.* (2005) *Angewandte Chemie, International Edition*, **44**, 6193.
58. Mülbaier, M. and Giannis, A. (2001) *Angewandte Chemie, International Edition*, **40**, 4393.
59. Sorg, G., Mengei, A., Jung, G. and Rademann, J. (2001) *Angewandte Chemie, International Edition*, **40**, 4395.
60. Lei, Z.Q., Ma, H.C., Zhang, Z. and Yang, Y.X. (2006) *Reactive & Functional Polymers*, **66**, 840.
61. Lei, Z., Denecker, C., Jegasothy, S., *et al.* (2003) *Tetrahedron Letters*, **44**, 1635.
62. Bromberg, L., Zhang, H. and Hatton, T.A. (2008) *Chemistry of Materials*, **20**, 2001.
63. Chung, W.-J., Kim, D.-K. and Lee, Y.-S. (2003) *Tetrahedron Letters*, **44**, 9251.
64. Kim, D.-K., Chung, W.-J. and Lee, Y.-S. (2005) *Synlett*, 279.
65. Lecarpentier, P., Crosignani, S. and Linclau, B. (2005) *Molecular Diversity*, **9**, 341.
66. Ladziata, U., Willging, J. and Zhdankin, V.V. (2006) *Organic Letters*, **8**, 167.
67. Karimov, R.R., Kazhkenov, Z.-G.M., Modjewski, M.J., *et al.* (2007) *Journal of Organic Chemistry*, **72**, 8149.
68. Yusubov, M.S., Drygunova, L.A. and Zhdankin, V.V. (2004) *Synthesis*, 2289.
69. Podgorsek, A., Jurisch, M., Stavber, S., *et al.* (2009) *Journal of Organic Chemistry*, **74**, 3133.
70. Telvekar, V.N. and Herlekar, O.P. (2007) *Synthetic Communications*, **37**, 859.
71. Telvekar, V.N. and Patile, H.V. (2011) *Synthetic Communications*, **41**, 131.
72. Zeng, X.-M., Chen, J.-M., Yoshimura, A., *et al.* (2011) *RSC Advances*, **1**, 973.
73. Li, X.-Q. and Zhang, C. (2009) *Synthesis*, 1163.
74. Tohma, H., Maruyama, A., Maeda, A., *et al.* (2004) *Angewandte Chemie, International Edition*, **43**, 3595.
75. Dohi, T., Maruyama, A., Yoshimura, M., *et al.* (2005) *Chemical Communications*, 2205.
76. Moroda, A. and Togo, H. (2006) *Tetrahedron*, **62**, 12408.
77. Takenaga, N., Goto, A., Yoshimura, M., *et al.* (2009) *Tetrahedron Letters*, **50**, 3227.
78. Dohi, T., Fukushima, K.-i., Kamitanaka, T., *et al.* (2012) *Green Chemistry*, **14**, 1493.
79. Dohi, T., Morimoto, K., Ogawa, C., *et al.* (2009) *Chemical & Pharmaceutical Bulletin*, **57**, 710.
80. Dohi, T., Morimoto, K., Takenaga, N., *et al.* (2006) *Chemical & Pharmaceutical Bulletin*, **54**, 1608.
81. Dohi, T., Morimoto, K., Takenaga, N., *et al.* (2007) *Journal of Organic Chemistry*, **72**, 109.

82. Tian, J., Gao, W.-C., Zhou, D.-M. and Zhang, C. (2012) *Organic Letters*, **14**, 3020.
83. Yoshida, J. and Itami, K. (2002) *Chemical Reviews*, **102**, 3693.
84. Kirschning, A., Yusubov, M.S., Yusubova, R.Y., *et al.* (2007) *Beilstein Journal of Organic Chemistry*, **3**, 19.
85. Yusubov, M.S., Yusubova, R.Y., Kirschning, A., *et al.* (2008) *Tetrahedron Letters*, **49**, 1506.
86. Yusubov, M.S., Funk, T.V., Yusubova, R.Y., et al. (2009) *Synthetic Communications*, **39**, 3772.
87. Yusubov, M.S., Gilmkhanova, M.P., Zhdankin, V.V. and Kirschning, A. (2007) *Synlett*, 563.
88. Yusubov, M.S., Yusubova, R.Y., Funk, T.V., et al. (2010) *Synthesis*, 3681.
89. Yusubov, M.S., Funk, T.V., Chi, K.-W., *et al.* (2008) *Journal of Organic Chemistry*, **73**, 295.
90. Yusubov, M.S. and Zhdankin, V.V. (2010) *Mendeleev Communications*, **20**, 185.
91. Kunst, E., Gallier, F., Dujardin, G., *et al.* (2007) *Organic Letters*, **9**, 5199.
92. Kunst, E., Gallier, F., Dujardin, G. and Kirschning, A. (2008) *Organic & Biomolecular Chemistry*, **6**, 893.
93. Zhang, J., Zhao, D., Wang, Y., *et al.* (2011) *Journal of Chemical Research*, **35**, 333.
94. Qian, W., Jin, E., Bao, W. and Zhang, Y. (2005) *Angewandte Chemie, International Edition*, **44**, 952.
95. Handy, S.T. and Okello, M. (2005) *Journal of Organic Chemistry*, **70**, 2874.
96. Su, F., Zhang, J., Jin, G., *et al.* (2009) *Journal of Chemical Research*, 741.
97. Qian, W., Jin, E., Bao, W. and Zhang, Y. (2006) *Tetrahedron*, **62**, 556.
98. Qian, W. and Pei, L. (2006) *Synlett*, 709.
99. Iinuma, M., Moriyama, K. and Togo, H. (2013) *Tetrahedron*, **69**, 2961.
100. Fang, C., Qian, W. and Bao, W. (2008) *Synlett*, 2529.
101. Roy, M.-N., Poupon, J.-C. and Charette, A.B. (2009) *Journal of Organic Chemistry*, **74**, 8510.
102. Zhu, C. and Wei, Y. (2012) *Advanced Synthesis & Catalysis*, **354**, 313.
103. Tesevic, V. and Gladysz, J.A. (2006) *Journal of Organic Chemistry*, **71**, 7433.
104. Tesevic, V. and Gladysz, J.A. (2005) *Green Chemistry*, **7**, 833.
105. Zagulyaeva, A.A., Yusubov, M.S. and Zhdankin, V.V. (2010) *Journal of Organic Chemistry*, **75**, 2119.
106. Lion, C.J., Vasselin, D.A., Schwalbe, C.H., *et al.* (2005) *Organic & Biomolecular Chemistry*, **3**, 3996.
107. Rocaboy, C. and Gladysz, J.A. (2003) *Chemistry–A European Journal*, **9**, 88.
108. More, J.D. and Finney, N.S. (2002) *Organic Letters*, **4**, 3001.
109. Yoshimura, A., Banek, C.T., Yusubov, M.S., *et al.* (2011) *Journal of Organic Chemistry*, **76**, 3812.
110. Yusubov, M.S., Yusubova, R.Y., Nemykin, V.N., *et al.* (2012) *European Journal of Organic Chemistry*, 5935.
111. Yamamoto, Y., Kawano, Y., Toy, P.H. and Togo, H. (2007) *Tetrahedron*, **63**, 4680.
112. Akiike, J., Yamamoto, Y. and Togo, H. (2007) *Synlett*, 2168.
113. Guo, W., Monge-Marcet, A., Cattoen, X., *et al.* (2013) *Reactive & Functional Polymers*, **73**, 192.
114. Zeng, X.-M., Chen, J.-M., Middleton, K. and Zhdankin, V.V. (2011) *Tetrahedron Letters*, **52**, 5652.
115. Zhu, C., Yoshimura, A., Wei, Y., *et al.* (2012) *Tetrahedron Letters*, **53**, 1438.

6

Reactions of Hypervalent Iodine Reagents in Green Solvents and under Solvent-Free Conditions

Typical reactions of common hypervalent iodine reagents are performed in non-recyclable organic solvents (dichloromethane, dimethyl sulfoxide, acetonitrile, etc.), which have potentially damaging environmental properties. The development of reactions using environmentally friendly solvents or solvent-free procedures is one of the active fields in organic synthesis due to a recent demand for realization of green chemical processes. The "greener" chemical syntheses in water or recyclable solvents, or under solvent-free procedures using mechanochemical mixing, microwave and ultrasound irradiation, have previously been summarized in two reviews [1, 2].

6.1 Reactions of Hypervalent Iodine Reagents in Water

The development of aqueous-phase reactions using environmentally friendly reagents is one of the active fields in organic synthesis [3–5]. Several protocols for reactions of hypervalent iodine reagents in water and environmentally-friendly aqueous solvent systems have been developed.

Iodosylbenzene, $(PhIO)_n$, in the presence of a hydroxylic solvent (water or alcohols) or an appropriate catalyst (Lewis acid, bromide or iodide anions, transition metal complex, etc.) can be effectively depolymerized, generating highly reactive monomeric species. Numerous examples of oxidations using iodosylbenzene in aqueous solutions have been reported. Moriarty and coworkers reported a series of oxidations of organic substrates with iodosylbenzene in hydroxylic solvents [6, 7]. For example, the oxidation of dihydropyran, cyclohexene and styrene with iodosylbenzene in aqueous solution leads predominantly to rearranged products [7]. Thus, the oxidation of dihydropyran (**1**) with iodosylbenzene in water affords tetrahydro-2-furaldehyde (**2**) via carbocationic ring contraction (Scheme 6.1) [7].

Kita, Tohma and coworkers have found that iodosylbenzene can be activated in aqueous solutions by addition of bromide salts to the reaction mixture [8, 9]. In particular, the oxidation of sulfides with iodosylbenzene in the presence of catalytic amounts of quaternary ammonium bromides selectively affords respective sulfoxides in high yields [8, 10]. The iodosylbenzene/KBr system in aqueous solution can be used for the

Hypervalent Iodine Chemistry: Preparation, Structure and Synthetic Applications of Polyvalent Iodine Compounds, First Edition. Viktor V. Zhdankin.

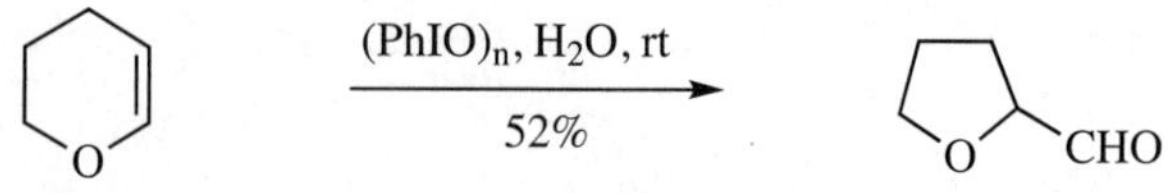

$(PhIO)_n$, H_2O, rt

52%

1 → 2 (CHO)

Scheme 6.1

$RCH_2OH \xrightarrow[76\text{-}92\%]{(PhIO)_n \text{ (2.2 equiv)}, KBr \text{ (0.2-1 equiv)}, H_2O, \text{rt}, 2 \text{ h}} RCO_2H$

R = $Ph(CH_2)_2$, $BnO(CH_2)_3$, $EtO_2C(CH_2)_4$, $N_3(CH_2)_4$, etc.

Scheme 6.2

oxidation of various primary and secondary alcohols, even in the presence of sensitive functional groups such as ether, ester, sulfonamide and azido groups. Primary alcohols under these conditions afford carboxylic acids (Scheme 6.2), while the oxidation of secondary alcohols under similar conditions furnishes ketones in almost quantitative yield [9, 11].

Activation of iodosylbenzene in the presence of KBr has been explained by the initial depolymerization of $(PhIO)_n$ with the formation of a highly reactive intermediate $PhI(Br)O^- K^+$, which reacts with alcohols to yield the corresponding carbonyl compound with regeneration of KBr [11].

[Bis(acyloxy)iodo]arenes in the presence of bromide anion in water also oxidize primary and secondary alcohols similarly to the $(PhIO)_n$/KBr system [11,12]. The oxidation of primary alcohols using $ArI(OAc)_2$/KBr in water or aqueous methanol affords carboxylic acids or esters [9, 13], while the oxidation of secondary alcohols under similar conditions results in the formation of the respective ketones in excellent yields [14]. Aldehydes can be converted into methyl esters by a similar procedure using $PhI(OAc)_2$/NaBr in an acidic aqueous methanol solution [15]. Likewise, acetals **3** can be converted into the corresponding hydroxyalkyl carboxylic esters **4** by oxidation with $PhI(OAc)_2$/LiBr in water (Scheme 6.3) [16].

The oxidation of alcohols with (diacetoxyiodo)benzene in the presence of catalytic amounts of TEMPO (2,2,6,6-tetramethylpiperidin-1-oxyl) in aqueous solutions is a common synthetic procedure (Section 3.1.5) [17–23]. An optimized protocol, published in *Organic Synthesis* for the oxidation of nerol **5** to nepal **6** (Scheme 6.4), consists of the treatment of the alcohol **5** solution in buffered (pH 7) aqueous acetonitrile with (diacetoxyiodo)benzene and TEMPO (0.1 equivalent) at 0 °C for 20 min [17].

Various organic substrates **7** with a benzylic C–H bond can be effectively oxidized to aryl ketones **8** by iodosylbenzene in the presence of KBr and montmorillonite-K10 (M-K10) clay in water (Scheme 6.5) [24].

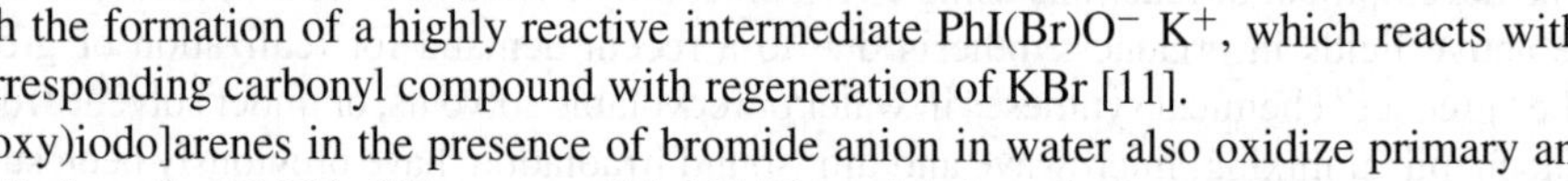

$PhI(OAc)_2$, LiBr, H_2O, rt, 15 min to 3 h

46-94%

3 → **4**

n = 1 or 2

R = $4\text{-}NO_2C_6H_4$, $2\text{-}NO_2C_6H_4$, $4\text{-}BrC_6H_4$, $3\text{-}ClC_6H_4$, $4\text{-}FC_6H_4$, $4\text{-}MeC_6H_4$, $PhCH_2$, C_5H_{11}, C_6H_{13}, C_8H_{17}, $C_{13}H_{27}$, *cyclo*-C_6H_{11}, $Ph(CH_3)CH$

Scheme 6.3

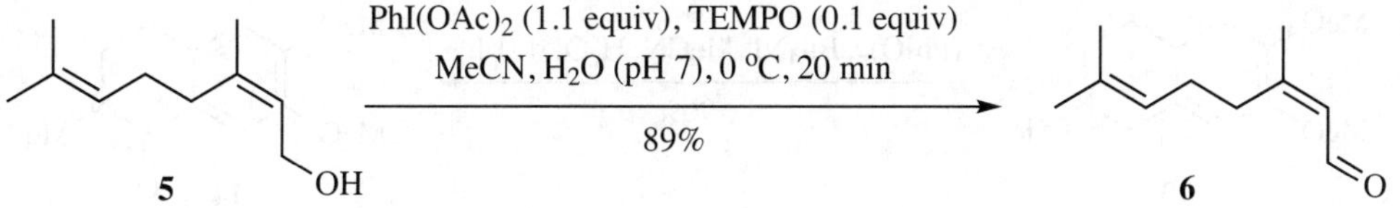

$PhI(OAc)_2$ (1.1 equiv), TEMPO (0.1 equiv)
MeCN, H_2O (pH 7), 0 °C, 20 min
89%

5 OH **6** O

Scheme 6.4

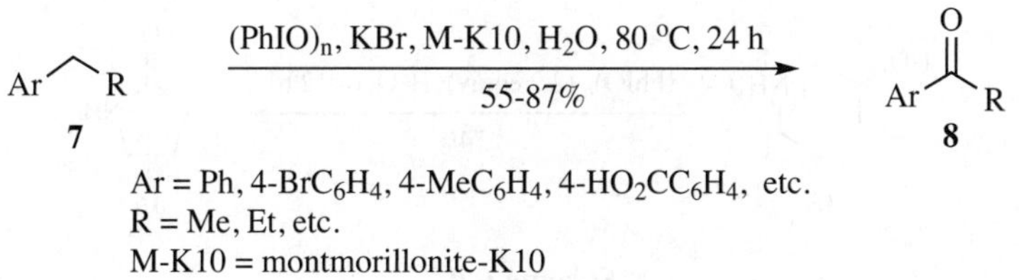

$(PhIO)_n$, KBr, M-K10, H_2O, 80 °C, 24 h
55-87%

Ar R **7** → O Ar R **8**

Ar = Ph, 4-BrC_6H_4, 4-MeC_6H_4, 4-$HO_2CC_6H_4$, etc.
R = Me, Et, etc.
M-K10 = montmorillonite-K10

Scheme 6.5

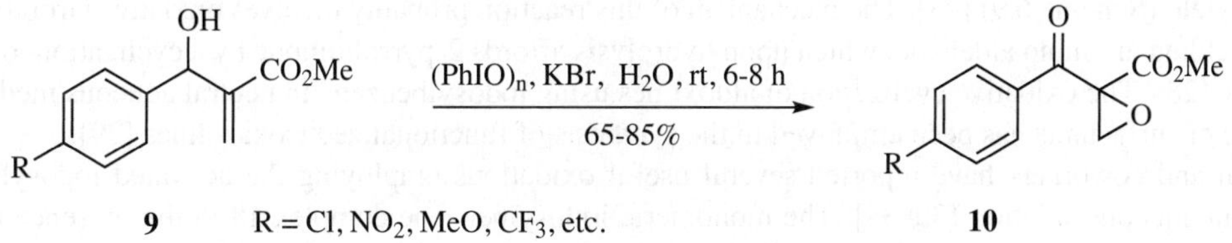

OH CO_2Me R **9** → $(PhIO)_n$, KBr, H_2O, rt, 6-8 h, 65-85% → O CO_2Me O R **10**

R = Cl, NO_2, MeO, CF_3, etc.

Scheme 6.6

Treatment of Baylis–Hillman adducts **9** with iodosylbenzene in the presence of catalytic KBr in water at room temperature affords the corresponding acyloxiranes **10** in good yields (Scheme 6.6) [25].

An efficient aqueous oxidative cyclization of Michael adducts (e.g., compound **11,** Scheme 6.7) promoted by the combination of iodosylbenzene with tetrabutylammonium iodide provides a convenient route to functionalized fused dihydrofuran derivatives (e.g., product **12**) in moderate to excellent yields with high diastereoselectivities [26].

The oxidation of *N*-methyltetrahydroisoquinolines (e.g., **13**) with iodosylbenzene in the presence of a catalytic amount of tetrabutylammonium iodide in aqueous solvents affords the respective lactams (e.g., **14**) in almost quantitative yields, as exemplified in Scheme 6.8 [27].

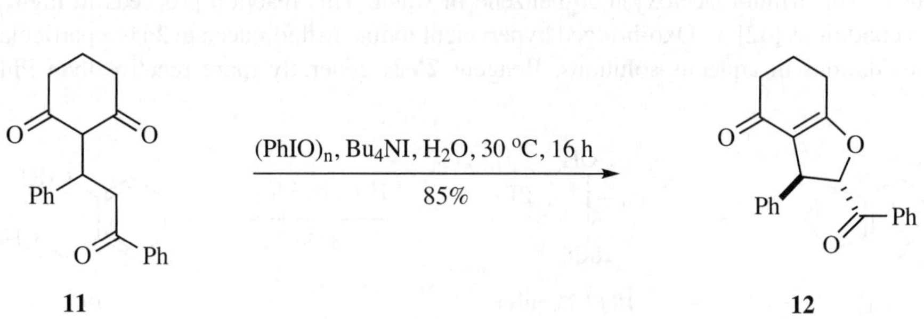

O O Ph O Ph **11** → $(PhIO)_n$, Bu_4NI, H_2O, 30 °C, 16 h, 85% → O O Ph O Ph **12**

Scheme 6.7

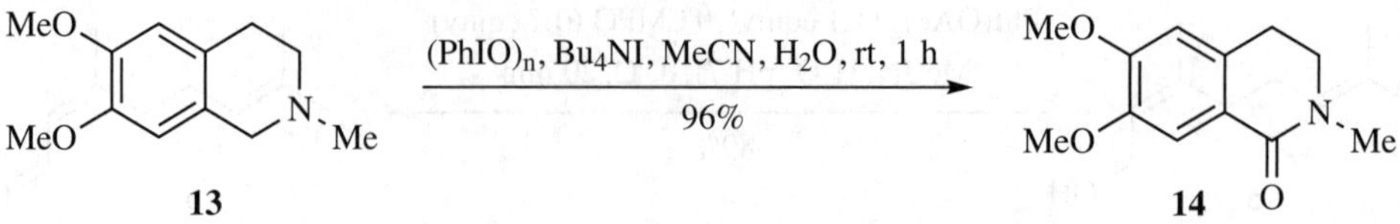

MeO, MeO, N–Me **13** → $(PhIO)_n$, Bu_4NI, MeCN, H_2O, rt, 1 h; 96% → MeO, MeO, N–Me, O **14**

Scheme 6.8

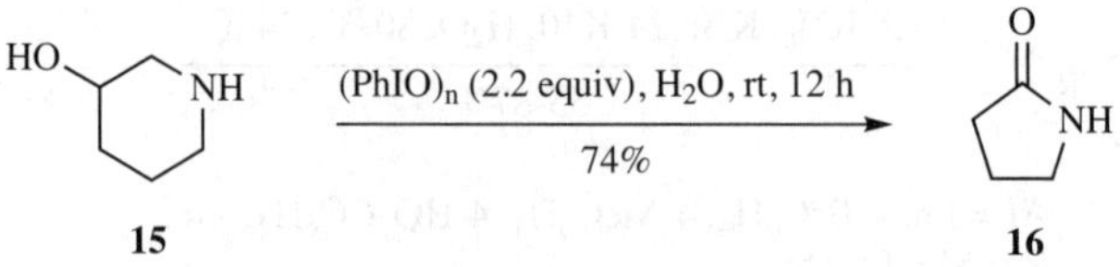

HO, NH **15** → $(PhIO)_n$ (2.2 equiv), H_2O, rt, 12 h; 74% → O, NH **16**

Scheme 6.9

The oxidation of 3-hydroxypiperidine (**15**) with iodosylbenzene in water affords 2-pyrrolidinone **16** directly in good yield (Scheme 6.9) [28]. The mechanism of this reaction probably involves oxidative Grob fragmentation yielding an imino aldehyde, which upon hydrolysis affords 2-pyrrolidinone by a cyclization–oxidation sequence [28]. The oxidative cyclization of aldoximes using iodosylbenzene in neutral aqueous media in the presence of surfactants has been employed in the synthesis of functionalized isoxazolines [29].

Ochiai and coworkers have reported several useful oxidations employing the activated iodosylbenzene species in aqueous solution [30–34]. The monomeric iodosylbenzene complex **18** in the presence of water can cleave the carbon–carbon double bond of indene **17** to form dialdehyde **19** (Scheme 6.10) [30]. Similar oxidative cleavage of various alkenes can be performed by using iodosylbenzene in water in the presence of HBF_4. This convenient procedure provides a safe alternative to the ozonolysis of alkenes [30].

The oligomeric iodosylbenzene sulfate **20** is a readily available, stable and water-soluble reagent with a reactivity pattern similar to that of activated iodosylbenzene [35–40]. It reacts with alkenes, alcohols and aryl alkyl sulfides in aqueous acetonitrile at room temperature to afford the respective products of oxidation **21**–**24** in good yields (Scheme 6.11) [35, 36].

Zhu and Wei have reported a mild and highly efficient procedure for the oxidation of alcohols to the corresponding aldehydes or ketones using oligomeric iodosylbenzene sulfate **20** in water in the presence of β-cyclodextrin, which serves as a biomimetic catalyst. The oxidation proceeded in water to afford aldehydes or ketones in excellent yields and high selectivity without overoxidation to carboxylic acids [41].

Kita and coworkers have developed a method for preparing *p*-quinone derivatives from phenol ether derivatives using [bis(trifluoroacetoxy)iodo]benzene in water. This reaction proceeds in high yields under mild reaction conditions [42]. μ-Oxo-bridged hypervalent iodine trifluoroacetate **26** is a particularly effective reagent for oxidations in aqueous solutions. Reagent **26** is generally more reactive than $PhI(O_2CCF_3)_2$,

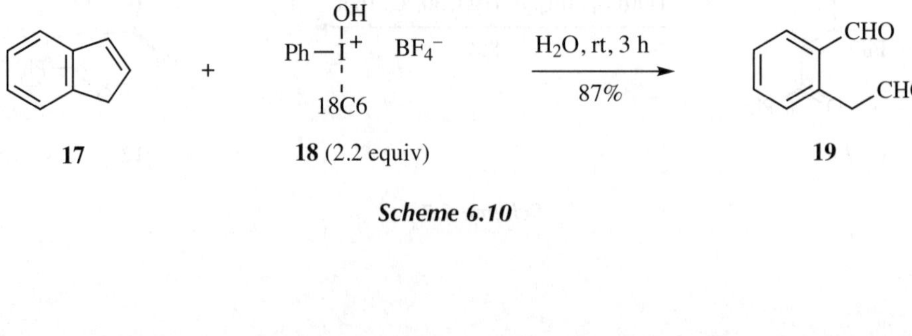

17 + Ph–I^+(OH)(18C6) BF_4^- **18** (2.2 equiv) → H_2O, rt, 3 h; 87% → CHO, CHO **19**

Scheme 6.10

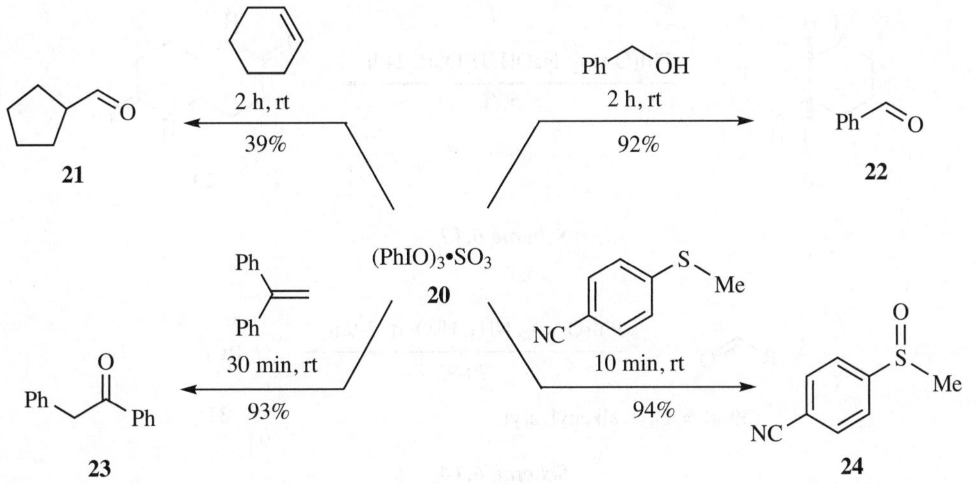

Scheme 6.11

especially toward phenolic substrates. In a specific example, μ-oxo-bridged trifluoroacetate **26** in aqueous media converts phenols **25** into the dearomatized quinones **27** in excellent yields (Scheme 6.12) [43].

Kozlowski and coworkers have reported an oxidative rearrangement of *cis*- and *trans*-1,5-diazadecalins promoted by (diacetoxyiodo)benzene in aqueous solution. In a specific example, upon treatment with (diacetoxyiodo)benzene in aqueous NaOH, 1,5-diaza-*cis*-decalin (**28**) undergoes oxidation along with fragmentation to yield the ring-expanded bislactam **29** (Scheme 6.13) [44].

A mild and high yielding protocol has been reported for the transformation of aldehydes **30** into nitriles **31** using (diacetoxyiodo)benzene in aqueous ammonia under mild reaction conditions at room temperature (Scheme 6.14) [45].

Several oxidations with iodine(V) reagents in aqueous solutions have been reported in the literature. Kuhakarn and coworkers have found that IBX can be used for the oxidation of alcohols in a water/dichloromethane (1 : 10) mixture in the presence of tetrabutylammonium bromide [46]. Vinod and coworkers have developed water-soluble analogs of IBX, *m*-iodoxyphthalic acid (mIBX) **32** [47] and a similar derivative of terephthalic acid **33** (Figure 6.1) [48], which can oxidize benzylic and allylic alcohols to

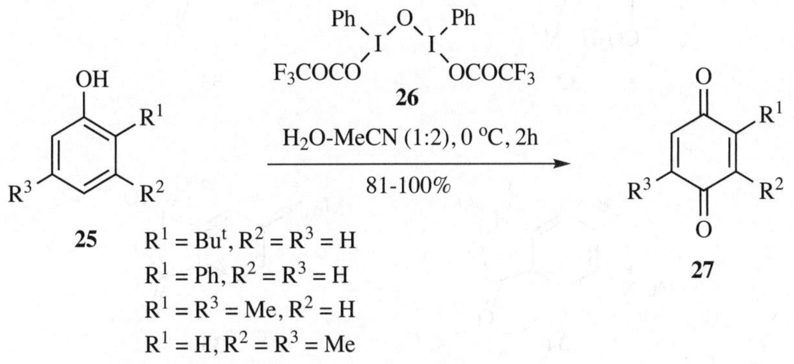

Scheme 6.12

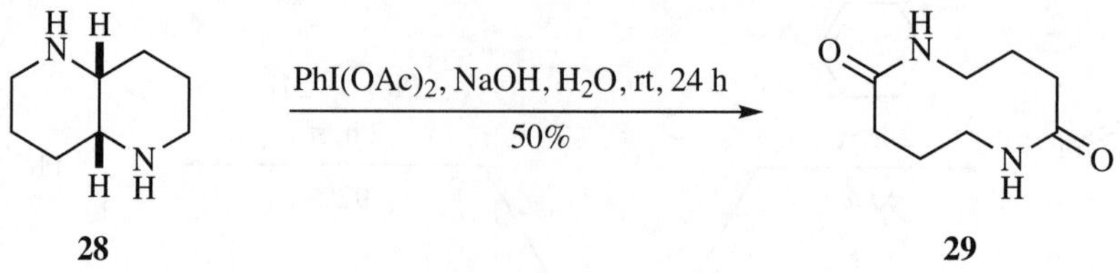

$PhI(OAc)_2$, NaOH, H_2O, rt, 24 h; 50%

28 → **29**

Scheme 6.13

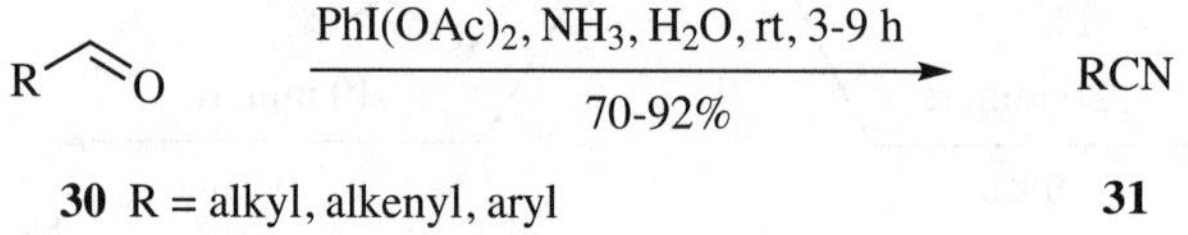

$PhI(OAc)_2$, NH_3, H_2O, rt, 3-9 h; 70-92%

30 R = alkyl, alkenyl, aryl → RCN **31**

Scheme 6.14

carbonyl compounds in water. Zhang and coworkers have prepared the IBX derivatives **34** and **35** bearing a trimethylammonium group, which possess excellent solubility in water [49]. The structure of compound **34** was established by single-crystal X-ray diffraction analysis. Compound **34** is a useful reagent for the oxidation of various β-keto esters to the corresponding dehydrogenated products in aqueous solution [49].

6.2 Reactions of Hypervalent Iodine Reagents in Recyclable Organic Solvents

Ionic liquids have gained recognition as environmentally benign alternatives to the more volatile organic solvents [50, 51]. Ionic liquids possess many attractive properties, such as wide liquid range, negligible vapor pressure, high thermal stability, good solvating ability for a wide range of substrates, easy recoverability and

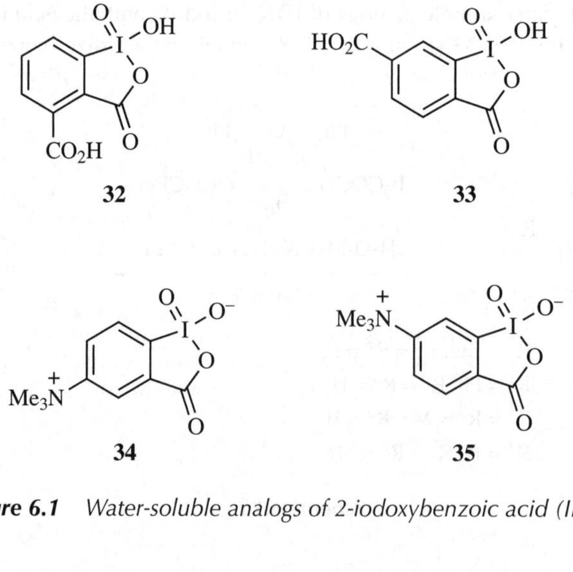

32 **33** **34** **35**

Figure 6.1 Water-soluble analogs of 2-iodoxybenzoic acid (IBX).

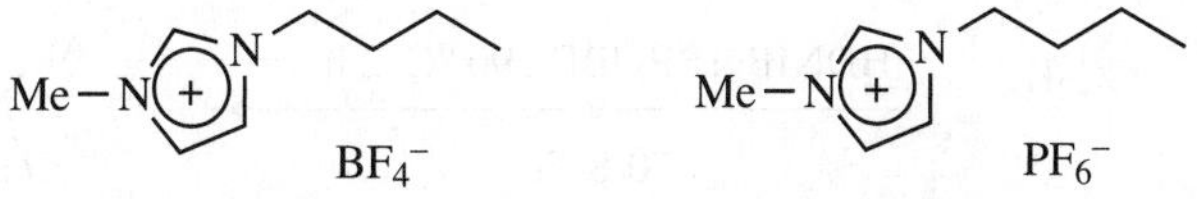

[bmim]BF_4

36

[bmim]PF_6

37

Figure 6.2 *Structures of hydrophilic [bmim]BF4 (**36**) and hydrophobic [bmim]PF6 (**37**) ionic liquids.*

reusability. Their nonvolatile nature can reduce the emission of toxic organic compounds and facilitate the separation of products and/or catalysts from the reaction solvents.

The use of ion-supported [bis(acyloxy)iodo]arene in the ionic liquid $[emim]^+[BF_4]^-$ (1-ethyl-3-methylimidazolium tetrafluoroborate) in the presence of bromide anion or ion-supported TEMPO for the oxidation of primary and secondary alcohols is discussed in Section 5.3.3 [52,53].

Several research groups have used ionic liquids for the oxidation of alcohols with *o*-iodoxybenzoic acid (IBX) or Dess–Martin periodinane (DMP). Alcohols undergo smooth oxidation with IBX or with DMP in hydrophilic [bmim]BF_4 (structure **36,** Figure 6.2) and hydrophobic [bmim]PF_6 (structure **37**) ionic liquids at room temperature under mild conditions to afford the corresponding carbonyl compounds in excellent yields with high selectivity [54]. Similar results were obtained for the oxidation of alcohols with IBX and DMP using ionic liquid [bmim]Cl (1-butyl-3-methylimidazolium chloride) [55,56]. IBX- and DMP-promoted oxidations are faster in ionic liquids compared to conventional solvents such as DMSO, DMF, ethyl acetate and water. Recovery of the by-product iodosobenzoic acid is especially simple in ionic liquids. The recovered ionic liquids can be recycled in subsequent reactions with consistent activity.

IBX in the ionic liquid [bmim]Br was found to be an efficient and eco-friendly reagent for the oxidation of 17α-methylandrostan-3β,17β-diol (**38**) to mestanolone (**39**) in good yield (Scheme 6.15) [57]. The product is easily separated from the reaction mixture by extraction with diethyl ether.

An ionic liquid of a different type (*n*-butylpyridinium tetrafluoroborate, [BPy]BF_4) has been used as a recyclable solvent for the reaction of 4-benzoylbutyric acids **40** with [hydroxy(2,4-dinitrobenzenesulfonyloxy)iodo]benzene (HDNIB) leading to 5-benzoyldihydro-2(3*H*)-furanones **41** (Scheme 6.16) [58].

A different recyclable solvent, PEG-400 [poly(ethylene glycol-400)], has been used in the one-pot synthesis of 2-arylimidazo[1,2-*a*]pyrimidines by the reaction of ketones with HDNIB and 2-aminopyrimidine. Significant rate enhancements and improved yields have been observed when compared with regular organic solvents [59].

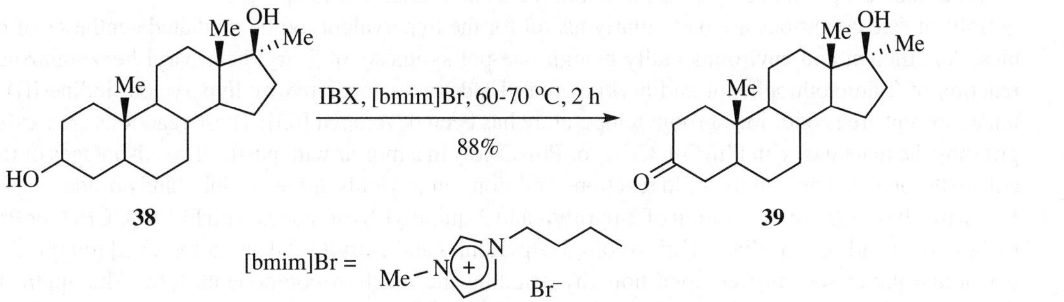

IBX, [bmim]Br, 60-70 °C, 2 h

88%

38 → **39**

[bmim]Br =

Scheme 6.15

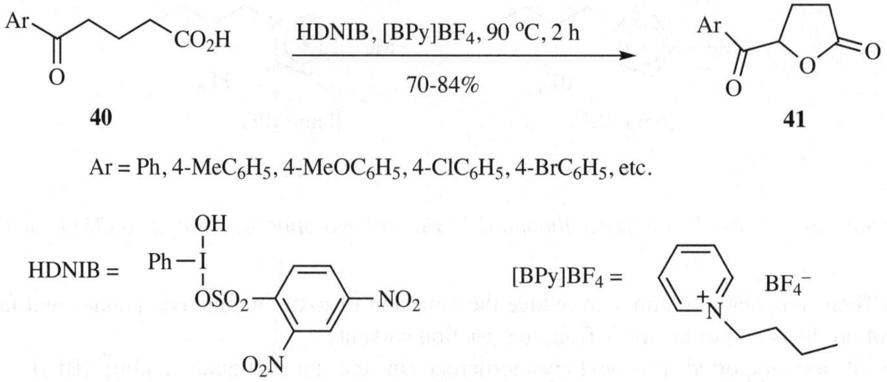

HDNIB, [BPy]BF_4, 90 °C, 2 h

70-84%

40 → 41

Ar = Ph, 4-MeC_6H_5, 4-$MeOC_6H_5$, 4-ClC_6H_5, 4-BrC_6H_5, etc.

HDNIB = PhI(OH)OSO_2-2,4-$(NO_2)_2C_6H_3$ [BPy]BF_4 = N-butylpyridinium BF_4^-

Scheme 6.16

6.3 Reactions of Hypervalent Iodine Reagents under Solvent-Free Conditions

Solvent-free reactions have many advantages such as reduced pollution, lower costs and the simplicity of the processes involved [60]. The solvent-free preparation of several important hypervalent iodine reagents has been reported [61,62]. [Hydroxy(sulfonyloxy)iodo]arenes (**42**) have been prepared in excellent yields by the solid-state reaction simply by grinding (diacetoxyiodo)arenes and appropriate sulfonic acids for several minutes in an agate mortar [61]. Tosyloxy- and mesyloxy benziodoxoles **44** can be prepared by a similar solvent-free procedure starting from 2-iodosylbenzoic acid (**43**) [61]. Likewise, [hydroxy(phosphoryloxy)iodo]arenes **46** have been conveniently prepared by a solvent-free method from (diacetoxyiodo)arenes and phosphate esters **45** (Scheme 6.17) [62].

Several examples of solvent-free reactions of hypervalent iodine reagents have been reported in the literature [61, 63–67]. Tosyloxylation of ketones or 1,3-diketones using [hydroxy(tosyloxy)iodo]benzene (HTIB) under solvent-free conditions takes place in a few minutes to give the respective α-tosyloxycarbonyl compounds in reasonable yields [61]. Likewise, α-tosyloxy β-keto sulfones **48** can be prepared under solvent-less conditions at room temperature (Scheme 6.18) [63]. This simple procedure consists of grinding together a neat mixture of β-keto sulfone **47** and HTIB at room temperature using a pestle and mortar; the reaction is complete in 4–10 min, giving products **48** in high yields [63].

A series of unsaturated compounds **49,** including α,β-unsaturated ketones, cinnamates, cinnamides and styrenes, have been aminobrominated to give products **50** with good yields and excellent diastereoselectivities under mechanical milling conditions, using $TsNH_2$ and *N*-bromosuccinimide (NBS) as the nitrogen and bromine sources, promoted by (diacetoxyiodo)benzene (Scheme 6.19) [64].

Solvent-free conditions are particularly useful for the hypervalent iodine mediated syntheses of heterocycles. An efficient and environmentally benign one-pot synthesis of 2-aryl/heteroaryl-benzothiazoles by the reaction of 2-aminothiophenol and aryl/heteroaryl aldehydes mediated by hypervalent iodine(III) reagents under solvent-free condition at room temperature has been developed [65]. These reactions are carried out by grinding the reactants with $PhI(O_2CCF_3)_2$ or $PhI(OAc)_2$ in a mortar with pestle. The advantages of this protocol are the one-step procedure, mild reaction conditions, high yields of the products and no side reactions [65]. Likewise, the oxidative cyclization of 2-pyridyl- and 2-quinolyl-hydrazones with $PhI(O_2CCF_3)_2$ or $PhI(OAc)_2$ to the corresponding 3-aryl[1,2,4]triazolo[4,3-*a*]pyridines and 1-aryl[1,2,4]triazolo[4,3-*a*]quinolines has been performed under solvent-free conditions by grinding the reaction components [66]. The application of a solvent-free procedure for the synthesis of various 2-arylbenzimidazoles **53** from *o*-phenylenediamine (**51**) and aldehydes **52** in the presence of hypervalent iodine as the oxidant is illustrated by Scheme 6.20 [67].

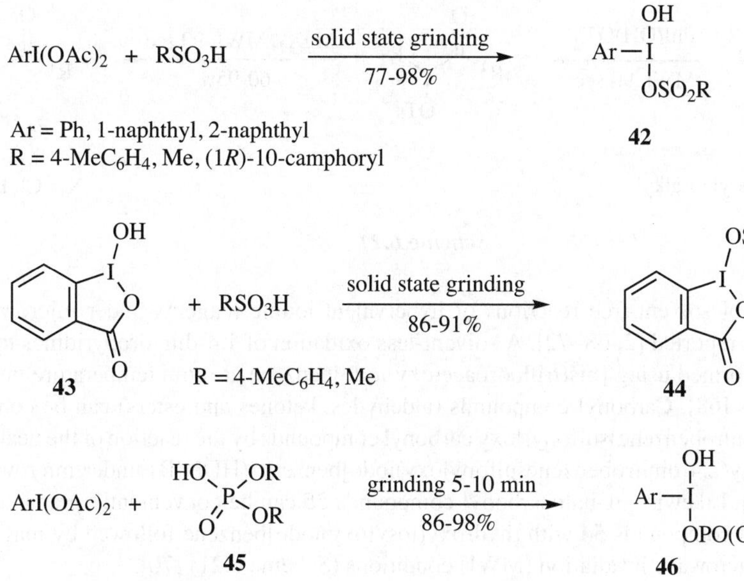

$ArI(OAc)_2$ + RSO_3H —solid state grinding, 77-98%→ $Ar{-}I(OH)OSO_2R$ **42**

Ar = Ph, 1-naphthyl, 2-naphthyl
R = 4-MeC_6H_4, Me, (1*R*)-10-camphoryl

43 + RSO_3H —solid state grinding, 86-91%→ **44**

R = 4-MeC_6H_4, Me

$ArI(OAc)_2$ + **45** —grinding 5-10 min, 86-98%→ $Ar{-}I(OH)OPO(OR)_2$ **46**

Ar = Ph, 4-ClC_6H_4
R = Ph, $PhCH_2$, (*R*)-(–)-1,1'-binaphthyl-2,2'-diyl

Scheme 6.17

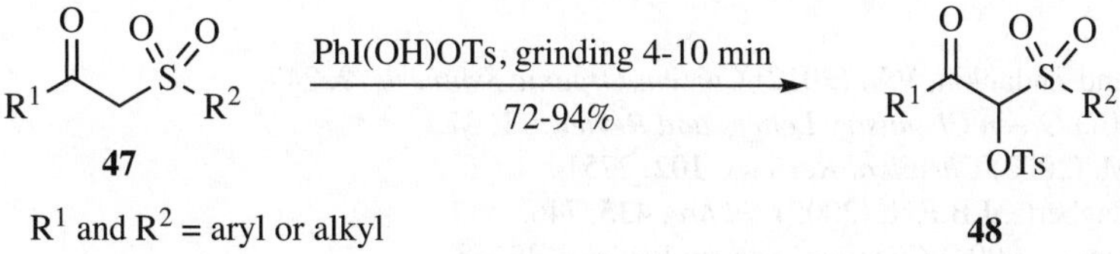

47 —PhI(OH)OTs, grinding 4-10 min, 72-94%→ **48**

R^1 and R^2 = aryl or alkyl

Scheme 6.18

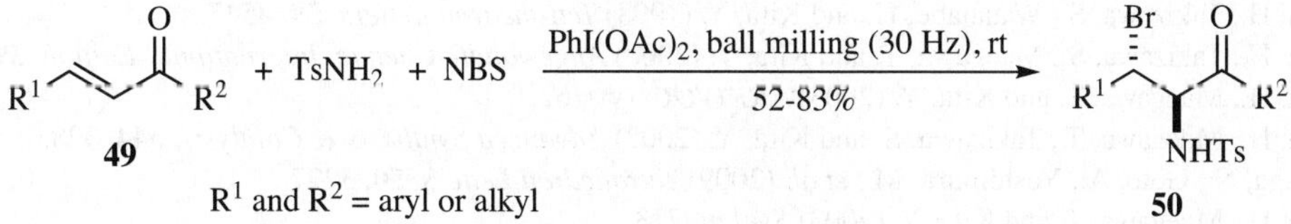

49 + $TsNH_2$ + NBS —$PhI(OAc)_2$, ball milling (30 Hz), rt, 52-83%→ **50**

R^1 and R^2 = aryl or alkyl

Scheme 6.19

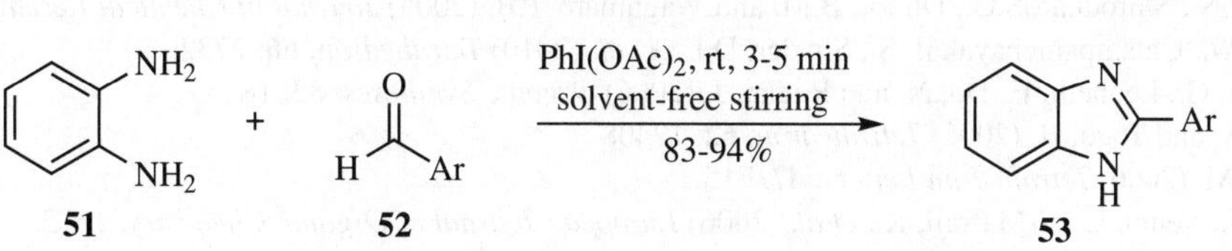

51 + **52** —$PhI(OAc)_2$, rt, 3-5 min, solvent-free stirring, 83-94%→ **53**

Ar = Ph, 4-ClC_6H_4, 2-ClC_6H_4, 4-EtC_6H_4, 4-$MeOC_6H_4$, 2-$MeOC_6H_4$, 4-BrC_6H_4, 4-FC_6H_4, 4-$Bu^tC_6H_4$, 2-MeC_6H_4, 2-furyl, 2-thienyl, 2-pyrrolyl, 2-pyridinyl

Scheme 6.20

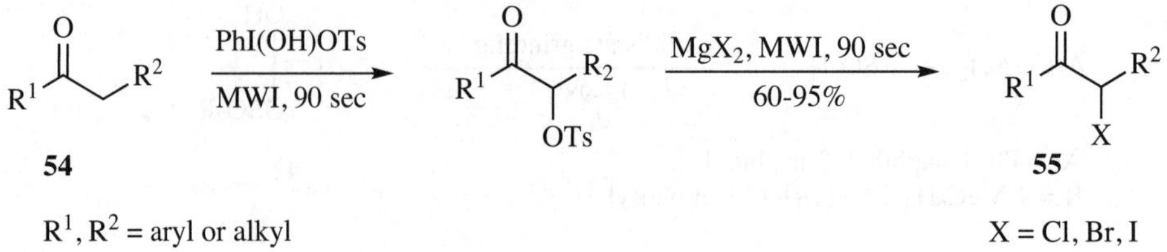

R^1, R^2 = aryl or alkyl — **54** → PhI(OH)OTs, MWI, 90 sec → α-OTs ketone → MgX_2, MWI, 90 sec, 60-95% → **55**, X = Cl, Br, I

Scheme 6.21

Several examples of solvent-free reactions of hypervalent iodine reagents under microwave irradiation conditions have been reported [2, 68–72]. A solvent-less oxidation of 1,4-dihydropyridines to the respective pyridines can be performed using [bis(trifluoroacetoxy)iodo]benzene at room temperature under microwave irradiation conditions [68]. Carbonyl compounds (aldehydes, ketones and esters) can be converted into the respective α-[(2,4-dinitrobenzene)sulfonyl]oxy carbonyl compounds by the reaction of the neat carbonyl compounds with [hydroxy(2,4-dinitrobenzenesulfonyloxy)iodo]benzene (HDNIB) under microwave irradiation in less than 40 s [69]. Likewise, α-halocarbonyl compounds **55** can be conveniently prepared by sequential treatment of carbonyl compounds **54** with [hydroxy(tosyloxy)iodo]benzene followed by magnesium halides under solvent-free microwave irradiation (MWI) conditions (Scheme 6.21) [70].

Several examples of microwave-promoted reactions of hypervalent iodine reagents in organic solvents, such as methanol [73], chloroform [73], DMF [74] and THF [75], have also been reported.

References

1. Yusubov, M.S. and Zhdankin, V.V. (2012) *Current Organic Synthesis*, **9**, 247.
2. Varma, R.S. (2007) *Green Chemistry Letters and Reviews*, **1**, 37.
3. Lindstroem, U.M. (2002) *Chemical Reviews*, **102**, 2751.
4. Klijn, J.E. and Engberts, J.B.F.N. (2005) *Nature*, **435**, 746.
5. Li, C.-J. and Chen, L. (2006) *Chemical Society Reviews*, **35**, 68.
6. Moriarty, R.M., Rani, N., Condeiu, C., *et al.* (1997) *Synthetic Communications*, **27**, 3273.
7. Moriarty, R.M., Prakash, O., Duncan, M.P., *et al.* (1996) *Journal of Chemical Research (S)*, 432.
8. Tohma, H., Takizawa, S., Watanabe, H. and Kita, Y. (1998) *Tetrahedron Letters*, **39**, 4547.
9. Tohma, H., Takizawa, S., Maegawa, T. and Kita, Y. (2000) *Angewandte Chemie, International Edition*, **39**, 1306.
10. Tohma, H., Maegawa, T. and Kita, Y. (2003) *ARKIVOC*, (vi), 62.
11. Tohma, H., Maegawa, T., Takizawa, S. and Kita, Y. (2002) *Advanced Synthesis & Catalysis*, **344**, 328.
12. Takenaga, N., Goto, A., Yoshimura, M., *et al.* (2009) *Tetrahedron Letters*, **50**, 3227.
13. Tohma, H., Maegawa, T. and Kita, Y. (2003) *Synlett*, 723.
14. Dohi, T., Maruyama, A., Yoshimura, M., *et al.* (2005) *Chemical Communications*, 2205.
15. Karade, N.N., Shirodkar, S.G., Dhoot, B.M. and Waghmare, P.B. (2005) *Journal of Chemical Research*, 274.
16. Panchan, W., Chiampanichayakul, S., Snyder, D.L., *et al.* (2010) *Tetrahedron*, **66**, 2732.
17. Piancatelli, G., Leonelli, F., Do, N. and Ragan, J. (2006) *Organic Syntheses*, **83**, 18.
18. Moroda, A. and Togo, H. (2006) *Tetrahedron*, **62**, 12408.
19. Vatele, J.-M. (2006) *Tetrahedron Letters*, **47**, 715.
20. Vugts, D.J., Veum, L., al-Mafraji, K., *et al.* (2006) *European Journal of Organic Chemistry*, 1672.
21. Zhao, X.-F. and Zhang, C. (2007) *Synthesis*, 551.
22. Hansen, T.M., Florence, G.J., Lugo-Mas, P., *et al.* (2002) *Tetrahedron Letters*, **44**, 57.
23. Li, Y. and Hale, K.J. (2007) *Organic Letters*, **9**, 1267.
24. Dohi, T., Takenaga, N., Goto, A., *et al.* (2008) *Journal of Organic Chemistry*, **73**, 7365.
25. Das, B., Holla, H., Venkateswarlu, K. and Majhi, A. (2005) *Tetrahedron Letters*, **46**, 8895.

26. Ye, Y., Wang, L. and Fan, R. (2010) *Journal of Organic Chemistry*, **75**, 1760.
27. Huang, W.-J., Singh, O.V., Chen, C.-H., *et al.* (2002) *Helvetica Chimica Acta*, **85**, 1069.
28. Tada, N., Miyamoto, K. and Ochiai, M. (2004) *Chemical & Pharmaceutical Bulletin*, **52**, 1143.
29. Chatterjee, N., Pandit, P., Halder, S., *et al.* (2008) *Journal of Organic Chemistry*, **73**, 7775.
30. Miyamoto, K., Tada, N. and Ochiai, M. (2007) *Journal of the American Chemical Society*, **129**, 2772.
31. Ochiai, M., Miyamoto, K., Shiro, M., *et al.* (2003) *Journal of the American Chemical Society*, **125**, 13006.
32. Ochiai, M., Miyamoto, K., Yokota, Y., *et al.* (2005) *Angewandte Chemie, International Edition*, **44**, 75.
33. Miyamoto, K., Hirobe, M., Saito, M., *et al.* (2007) *Organic Letters*, **9**, 1995.
34. Ochiai, M. (2006) *Coordination Chemistry Reviews*, **250**, 2771.
35. Koposov, A.Y., Netzel, B.C., Yusubov, M.S., *et al.* (2007) *European Journal of Organic Chemistry*, 4475.
36. Yusubov, M.S., Yusubova, R.Y., Funk, T.V., *et al.* (2009) *Synthesis*, 2505.
37. Nemykin, V.N., Koposov, A.Y., Netzel, B.C., *et al.* (2009) *Inorganic Chemistry*, **48**, 4908.
38. Geraskin, I.M., Pavlova, O., Neu, H.M., *et al.* (2009) *Advanced Synthesis & Catalysis*, **351**, 733.
39. Neu, H.M., Yusubov, M.S., Zhdankin, V.V. and Nemykin, V.N. (2009) *Advanced Synthesis & Catalysis*, **351**, 3168.
40. Yusubov, M.S., Nemykin, V.N. and Zhdankin, V.V. (2010) *Tetrahedron*, **66**, 5745.
41. Zhu, C. and Wei, Y. (2011) *Catalysis Letters*, **141**, 582.
42. Tohma, H., Morioka, H., Harayama, Y., *et al.* (2001) *Tetrahedron Letters*, **42**, 6899.
43. Dohi, T., Nakae, T., Takenaga, N., *et al.* (2012) *Synthesis*, 1183.
44. Li, X., Xu, Z., DiMauro, E.F. and Kozlowski, M.C. (2002) *Tetrahedron Letters*, **43**, 3747.
45. Bag, S., Tawari, N.R. and Degani, M.S. (2009) *ARKIVOC*, (xiv), 118.
46. Kuhakarn, C., Kittigowittana, K., Ghabkham, P., *et al.* (2006) *Synthetic Communications*, **36**, 2887.
47. Thottumkara, A.P. and Vinod, T.K. (2002) *Tetrahedron Letters*, **43**, 569.
48. Kommreddy, A., Bowsher, M.S., Gunna, M.R., *et al.* (2008) *Tetrahedron Letters*, **49**, 4378.
49. Cui, L.-Q., Dong, Z.-L., Liu, K. and Zhang, C. (2011) *Organic Letters*, **13**, 6488.
50. Welton, T. (1999) *Chemical Reviews*, **99**, 2071.
51. Wasserscheid, P. and Keim, W. (2000) *Angewandte Chemie, International Edition*, **39**, 3772.
52. Qian, W., Jin, E., Bao, W. and Zhang, Y. (2005) *Angewandte Chemie, International Edition*, **44**, 952.
53. Qian, W., Jin, E., Bao, W. and Zhang, Y. (2006) *Tetrahedron*, **62**, 556.
54. Yadav, J.S., Reddy, B.V.S., Basak, A.K. and Narsaiah, A.V. (2004) *Tetrahedron*, **60**, 2131.
55. Liu, Z., Chen, Z.-C. and Zheng, Q.-G. (2003) *Organic Letters*, **5**, 3321.
56. Karthikeyan, G. and Perumal, P.T. (2003) *Synlett*, 2249.
57. Chhikara, B.S., Chandra, R. and Tandon, V. (2004) *Tetrahedron Letters*, **45**, 7585.
58. Hou, R.-S., Wang, H.-M., Lin, Y.-C. and Chen, L.-C. (2005) *Heterocycles*, **65**, 649.
59. Cheng, H.-T., Hou, R.-S., Wang, H.-M., *et al.* (2009) *Journal of the Chinese Chemical Society*, **56**, 632.
60. Tanaka, K. and Toda, F. (2000) *Chemical Reviews*, **100**, 1025.
61. Yusubov, M.S. and Wirth, T. (2005) *Organic Letters*, **7**, 519.
62. Zhu, M., Cai, C.G., Ke, W. and Shao, J. (2010) *Synthetic Communications*, **40**, 1371.
63. Kumar, D., Sundaree, M.S., Patel, G., *et al.* (2006) *Tetrahedron Letters*, **47**, 8239.
64. Wu, X.-L., Xia, J.-J. and Wang, G.-W. (2008) *Organic & Biomolecular Chemistry*, **6**, 548.
65. Kumar, P., Meenakshi, Kumar, S., *et al.* (2012) *Journal of Heterocyclic Chemistry*, **49**, 1243.
66. Kumar, P. (2012) *Chemistry of Heterocyclic Compounds*, **47**, 1237.
67. Du, L.-H. and Luo, X.-P. (2010) *Synthetic Communications*, **40**, 2880.
68. Varma, R.S. and Kumar, D. (1999) *Journal of the Chemical Society, Perkin Transactions 1*, 1755.
69. Lee, J.C., Kim, S., Ku, C.H., *et al.* (2002) *Bulletin of Korean Chemical Society*, **23**, 1503.
70. Lee, J.C., Park, J.Y., Yoon, S.Y., *et al.* (2004) *Tetrahedron Letters*, **45**, 191.
71. Polshettiwar, V. and Varma, R.S. (2008) *Pure and Applied Chemistry*, **80**, 777.
72. Wurz, R.P. and Charette, A.B. (2003) *Organic Letters*, **5**, 2327.
73. Hossain, M.M., Kawamura, Y., Yamashita, K. and Tsukayama, M. (2006) *Tetrahedron*, **62**, 8625.
74. Yu, X. and Huang, X. (2002) *Synlett*, 1895.
75. French, A.N., Cole, J. and Wirth, T. (2004) *Synlett*, 2291.

7

Practical Applications of Polyvalent Iodine Compounds

Practical applications of polyvalent iodine derivatives are restricted by the low stability of many of these compounds. Among inorganic derivatives of polyvalent iodine, only iodine pentafluoride, iodine pentoxide and iodic and periodic acids and their salts are commercially available, stable products. These stable inorganic compounds have found some industrial use, mainly as powerful oxidants.

The vast majority of organic λ^3- and λ^5-iodanes lack thermal stability and some of them are explosive. Only several representatives of hypervalent iodanes with one carbon ligand are commercially available products, namely, (diacetoxyiodo)benzene (DIB or PIDA), [bis(trifluoroacetoxy)iodo]benzene (BTI or PIFA), [hydroxy(tosyloxy)iodo]benzene (HTIB), 2-iodosylbenzoic acid (IBA), 2-iodoxybenzoic acid (IBX) and Dess–Martin periodinane (DMP). These commercially available compounds are widely used in chemical laboratories as versatile and environmentally benign reagents for organic synthesis (Chapters 3–6). Organic hypervalent iodine(III) derivatives with two aryl substituents on iodine, or diaryliodonium salts, possess a higher thermal stability. Several representatives of diaryliodonium salts are commercially available products. Stable iodonium salts have found numerous practical applications, such as cationic photoinitiators in polymer chemistry, biologically active compounds and ^{18}F-fluoridating agents in positron emission tomography (PET). The biological properties of iodonium salts were summarized in 1996 in a comprehensive review on polyvalent iodine compounds [1] and a brief overview of the application of iodonium salts in PET was given in 2011 in a review on iodonium salts [2].

7.1 Applications of Inorganic Polyvalent Iodine Derivatives

Potassium iodate, KIO_3, the most thermodynamically stable and naturally occurring compound of polyvalent iodine, has found some application as a dietary supplement and a food additive. It can be used as a source of iodine in iodized salt and also as a dough conditioner [3]. In fact, because potassium iodate is more stable than iodide in the presence of air, most health authorities preferentially recommend iodate as an additive to salt for correcting iodine deficiency. Iodate is rapidly reduced to iodide in the body; iodide is essential for thyroid function. However, high levels of iodate (0.600 mg per day) have been shown to cause damage to the retina, resulting in ocular toxicity [4]. The recommended level of iodine in iodized salt is between 20 and

Hypervalent Iodine Chemistry: Preparation, Structure and Synthetic Applications of Polyvalent Iodine Compounds, First Edition. Viktor V. Zhdankin.

80 mg per kilogram of salt, which is equal to 28 to 110 mg of potassium iodate per kilogram of salt. For a maximal daily salt intake of 15 g, this results in a human exposure of 440–1700 μg of iodate per day, which is well below the acute or chronic toxicity levels [3]. Available toxicology data indicate a negligible risk of the small oral long-term doses achieved with iodate-fortified salt.

Metal iodates are of particular interest as second-order nonlinear optical (NLO) materials [5]. NLO materials are of current interest and great importance due to their applications in photonic technologies, such as laser frequency conversion, optical parameter oscillators (OPOs) and signal communication. Lithium iodate is one of the most widely used and commercially manufactured representatives of such materials [5]. Many inorganic iodates, such as α-$LiIO_3$, NaI_3O_8 [6] and α-$Cs_2I_4O_{11}$ [7], have been reported to be promising new second-harmonic generation (SHG) materials with wide transparent wavelength regions, large SHG coefficients and high optical-damage thresholds as well as high thermal stabilities.

Iodine pentoxide, I_2O_5, is the most stable oxide of the halogens [8]. It has found some practical application as a mild oxidant, especially useful in analytical chemistry. Iodine pentoxide is one of the few chemicals that can oxidize carbon monoxide rapidly and completely at room temperature. The reaction forms the basis of a useful analytical method for determining the concentration of CO in the atmosphere or in other gaseous mixtures [8].

Periodic acid and periodates are thermodynamically potent and kinetically facile oxidants. In acid solution periodic acid is one of the few reagents that can rapidly and quantitatively convert Mn(II) into Mn(VII) [8]. In organic chemistry it specifically cleaves 1,2-diols and related compounds such as α-diketones, α-ketols, α-aminoalcohols and α-diamines; all these reactions have been widely used in carbohydrate and nucleic acid chemistry. Periodate oxidation is widely used in chemical and instrumental methods of analysis of polysaccharides, oligosaccharides, glycosides, glycoproteins and other organic products of biological origin [9–12]. Applications of periodate oxidations in instrumental methods of microanalysis were summarized in a review in 2009 [12].

Periodate salts have seen some application in propellant formulations and their use as environmentally friendly oxidizers in pyrotechnic formulations has been reported [13]. In particular, Moretti, Sabatini and Chen have developed incendiary mixtures featuring periodate salts KIO_4 and $NaIO_4$ that perform similarly to perchlorates but are less toxic [13]. Combined with the alloy magnalium (magnesium–aluminium, 1 : 1) as the fuel, these mixtures produce a bright flash of light.

Iodine pentafluoride, IF_5, is one the most thermodynamically stable fluorides among the fluorides of Cl, Br and I. It is an important industrial fluorinating reagent, manufactured in the USA on a scale of several hundred tons per year. Iodine pentafluoride is available as a liquid in steel cylinders up to 1350 kg capacity; the price in 1992 was circa $50 kg [8]. Iodine pentafluoride is a relatively mild fluorinating agent and can be handled in glass apparatus. Nuclear power engineering, in particular, production process for volatile uranium hexafluoride, is one of the areas of practical uses of IF_5 [14]. Iodine pentafluoride is also used for the preparation of graphite fluorides, which are known for their application as lubricants or as cathode materials in primary lithium batteries [15, 16].

7.2 Applications of Hypervalent Iodine(III) Compounds as Polymerization Initiators

Diaryliodonium salts are widely used as photoinitiators for cationic photopolymerizations [17–20]. Photoinitiated cationic polymerization is of great practical interest due to its applicability for the curing of coatings and printing inks and for photoresist technology used in lithography [19,20]. General synthetic methods, properties and photochemistry of diaryliodonium salts as photoinitiators were reviewed by Crivello in 1984 [18].

Crivello and coworkers were the first to discover that iodonium salts having BF_4^-, PF_6^-, AsF_6^-, or SbF_6^- counterions are efficient photoinitiators for the polymerization of various cationically polymerizable

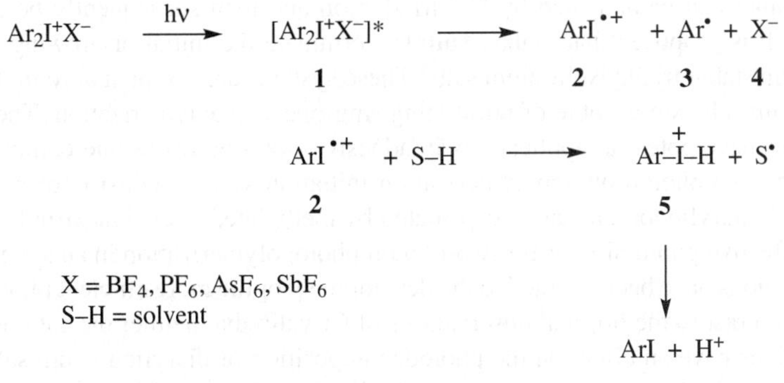

$$Ar_2I^+X^- \xrightarrow{h\nu} [Ar_2I^+X^-]^* \longrightarrow ArI^{\bullet +} + Ar^{\bullet} + X^-$$

1 → 2, 3, 4

$$ArI^{\bullet +} + S\text{–}H \longrightarrow Ar\text{–}\overset{+}{I}\text{–}H + S^{\bullet}$$

2 → 5

5 → $ArI + H^+$

X = BF_4, PF_6, AsF_6, SbF_6
S–H = solvent

Scheme 7.1

monomers [21]. The study of the mechanism of photolysis of iodonium salts confirmed a pathway involving radical–cations and aryl radicals as key intermediates. The major pathway involves the facile decomposition of the excited iodonium compound **1** to aryliodo radical–cation **2**, aryl radical **3** and anion **4** (Scheme 7.1) [21]. This process should be highly efficient due to the very low bond energy of the C–I bond (26–27 kcal mol^{-1}). Interaction of the aryliodo radical–cation with the solvent (S–H) generates a protonated iodoaromatic compound **5,** which rapidly deprotonates and a radical $S^{\bullet}$ derived from the solvent. During photolysis, the BF_4^-, PF_6^-, AsF_6^-, or SbF_6^- counterions associated with the diaryliodonium salts remain unchanged and appear in the products as the corresponding Brønsted acids. These acids (HX) are credited as the true initiators of cationic polymerization when diaryliodonium salts are employed in the cationic photopolymerization of various monomers [21]. According to this mechanism of action, diaryliodonium salts belong to an important class of photoacid generators (PAGs), which find broad applications in the manufacturing of protective coatings, smart cards, 3D rapid prototyping, UV adhesives, semiconductor devices, antireflective coatings, holograms and so on.

In the same groundbreaking paper [21], Crivello and coworkers also demonstrated that the anion plays no role in determining the photosensitivity of the iodonium salt and the photolysis rates of diaryliodonium salts having the same cations but different non-nucleophilic counterions (BF_4^-, PF_6^-, AsF_6^-, or SbF_6^-) are identical. Likewise, the cation structure has little effect on the photodecomposition of diaryliodonium salts. The utility of iodonium salts as photoinitiators has been demonstrated in several cationic polymerizations using olefins, epoxides, cyclic ethers, lactones and cyclic sulfides as the monomers [21].

More recently, iodonium salts have been widely used as photoinitiators in the polymerization studies of various monomeric precursors, such as copolymerization of butyl vinyl ether and methyl methacrylate by combination of radical and radical promoted cationic mechanisms [22], thermal and photopolymerization of divinyl ethers [23], photopolymerization of vinyl ether networks using an iodonium initiator [24, 25], dual photo- and thermally-initiated cationic polymerization of epoxy monomers [26], preparation and properties of elastomers based on a cycloaliphatic diepoxide and poly(tetrahydrofuran) [27], photoinduced crosslinking of divinyl ethers [28], cationic photopolymerization of 1,2-epoxy-6-(9-carbazolyl)-4-oxahexane [29], preparation of interpenetrating polymer network hydrogels based on 2-hydroxyethyl methacrylate and *N*-vinyl-2-pyrrolidone [30], photopolymerization of unsaturated cyclic ethers [31] and many other works.

Different initiation techniques have been investigated in polymerizations induced by iodonium salts, such as visible laser irradiation [32], dual photo- and thermally initiated cationic polymerization [23, 26] and a two-photon photopolymerization initiation system [33, 34]. For example, dual photo- and thermal-initiation systems based on selective inhibition of the photoinitiated cationic ring-opening polymerization of epoxides by dialkyl sulfides have been developed [26]. Such a dual system, iodonium salt/dialkyl sulfide, in the presence of a monomer can be activated by UV irradiation and then subsequently be polymerized by the application of heat. It is proposed that dialkyl sulfides terminate the initial or growing polyether chains at an early stage to form stable trialkylsulfonium salts. These systems are dormant at room temperature but, on heating, the sulfonium salts are capable of reinitiating ring-opening polymerization. These dual photo- and thermal-cure systems have potential applications in adhesives, potting resins and composites [26]. Another initiation system, the two-photon photopolymerization initiation system, consists of a photosensitizer dye and the photoinitiator diaryliodonium salt encapsulated by methylated-β-cyclodextrin [34]. Such a complex can be used as an effective photoinitiator for two-photon photopolymerization in an aqueous system.

Numerous publications have been devoted to the development of the more efficient photoinitiators based on iodonium salts. In contrast to the original observations of Crivello that neither the cationic part of iodonium salt nor the counterions have an effect on the photodecomposition of diaryliodonium salts [21], it has been demonstrated that both anionic and cationic portions of iodonium salts may play an important role in the overall effectiveness of the photoinitiator. Park and coworkers have published a detailed study on the participation of anion and alkyl substituents of diaryliodonium salts in photoinitiated cationic polymerization reactions of epoxides [35]. It was found that the alkyl-substituted diphenyliodonium cations, such as bis(4-*tert*-butylphenyl)iodonium and 4-cumenyl-4′-tolyliodonium salts, have higher photoacid generation efficiency than the unsubstituted diphenyliodonium salts. The lower nucleophilicity and large volume size of the anions play a decisive role in enhancing the rate of polymerization, with the general order of reactivity found to be $PF_6^- < AsF_6^- < B(C_6F_5)_4^-$ [35]. In general, the larger the anion is, the more loosely it is bound to the end of the growing cationic chain and the more active the propagating cationic species is in the polymerization. For this series, $B(C_6F_5)_4^-$ is the largest anion and the most loosely bound, while PF_6^- is the smallest and, therefore, the most tightly bound anion. For comparison, in the case of the most nucleophilic trifluoromethanesulfonate anion, cationic polymerization of epoxides was not observed [35].

Crivello and Lee have described the synthesis and characterization of a series of (4-alkoxyphenyl)phenyliodonium salts **7**, which are excellent photo- and thermal-initiators for the cationic polymerization of vinyl and heterocyclic monomers [17]. Iodonium salts **7** are conveniently prepared by the reaction of alkoxyphenols **6** with [hydroxy(tosyloxy)iodo]benzene followed by anion exchange with sodium hexafluoroantimonate (Scheme 7.2). Products **7** have very good solubility and photoresponse characteristics, which make them especially attractive for use in UV curing applications. Compounds **7** with alkoxy chains of eight carbons and longer are essentially nontoxic, compared to diphenyliodonium hexafluoroantimonate, which has an oral LD_{50} of 40 mg kg^{-1} (rats) [17].

Shirai, Kubo and Takahashi have prepared and examined a series of new substituted diaryliodonium hexafluorophosphates aiming at improved solubility and lower toxicity of the photoinitiators [36]. The alkyl substituted iodonium salts **8–13** (Figure 7.1) in combination with 2-ethyl-9,10-dimethoxyanthracene as the photosensitizer showed especially high photocuring ability [36].

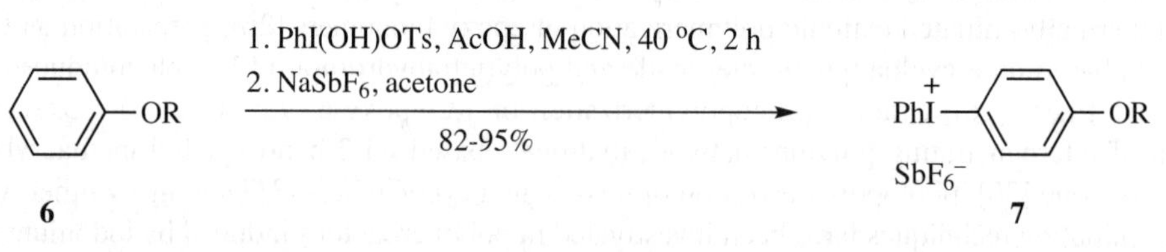

OR (**6**) → 1. PhI(OH)OTs, AcOH, MeCN, 40 °C, 2 h; 2. $NaSbF_6$, acetone; 82-95% → $Ph\overset{+}{I}$–C_6H_4–OR SbF_6^- (**7**)

R = Me, C_8H_{17}, $C_{10}H_{21}$, $C_{11}H_{23}$, $C_{14}H_{29}$, $C_{15}H_{31}$, $C_{16}H_{33}$, $C_{18}H_{37}$

Scheme 7.2

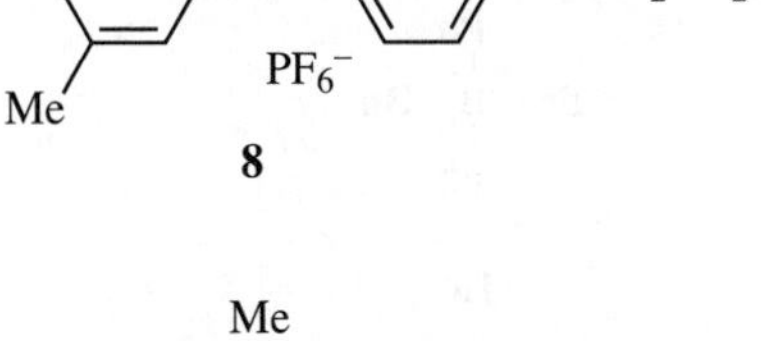

Figure 7.1 *Efficient photoinitiators with high solubility and low toxicity.*

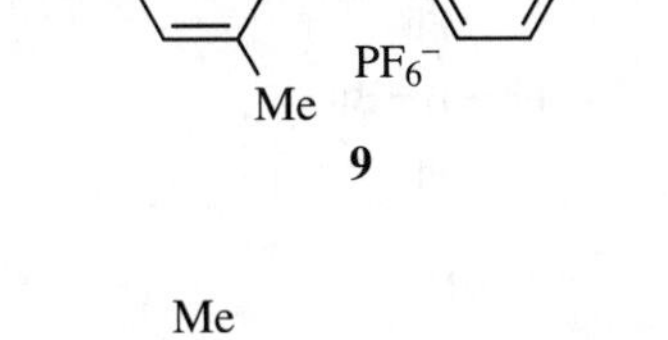

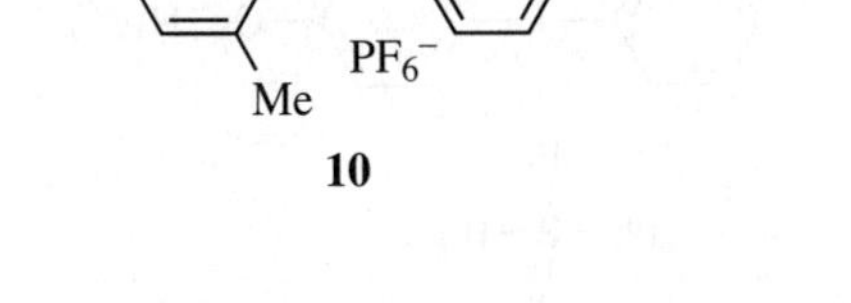

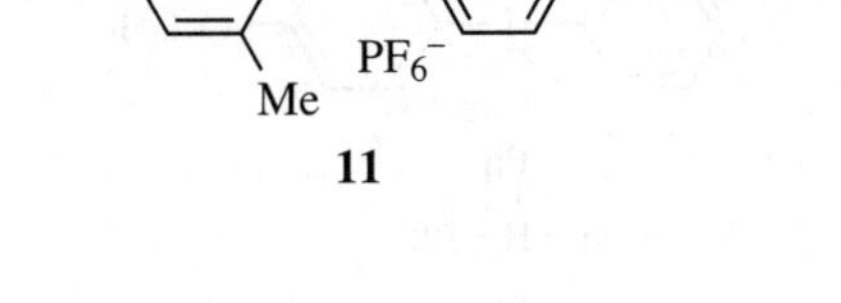

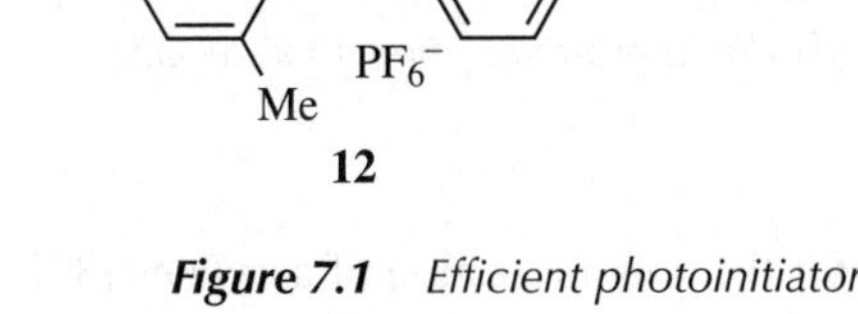

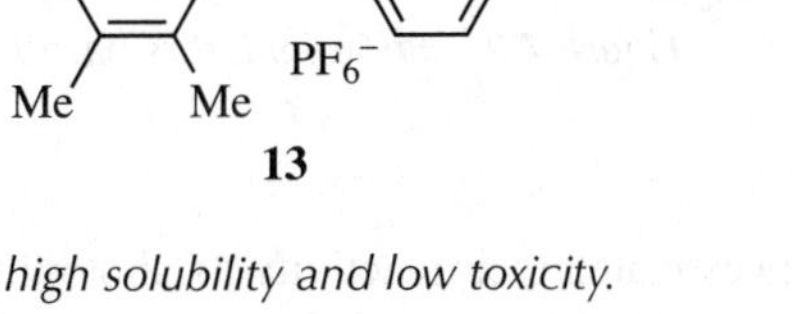

The preparation and properties of (9-oxo-9*H*-fluoren-2-yl)phenyliodonium hexafluoroantimonate (**14**) as a new photoinitiator for the cationic polymerization of epoxides have been reported [37]. Compound **14** was prepared by the reaction of (diacetoxyiodo)benzene with fluorenone followed by treatment with sodium hexafluoroantimonate (Scheme 7.3). Photoinitiator **14** has the advantage of intramolecular photosensitization and it is a more effective initiator than the conventional iodonium salts [37].

An alkynyliodonium salt, namely, phenyl(phenylethynyl)iodonium hexafluorophosphate, has been tested for application as cationic photoinitiator [38]. The high activity of phenyl(phenylethynyl)iodonium salt as a photoinitiator was verified by photo differential scanning calorimetry (photo-DSC) experiments in direct irradiation and in photosensitized initiation using 9,10-dibutylanthracene, 2-isopropylthioxanthone and benzophenone as sensitizers [38].

Neckers and coworkers have prepared diaryliodonium butyltriphenylborate salts **15–18** (Figure 7.2) by anion exchange of the respective diaryliodonium halides with tetramethylammonium butyltriphenylborate

1. $PhI(OAc)_2$, Ac_2O, AcOH, H_2SO_4, 10-35 °C
2. NH_4Cl
3. $NaSbF_6$

54%

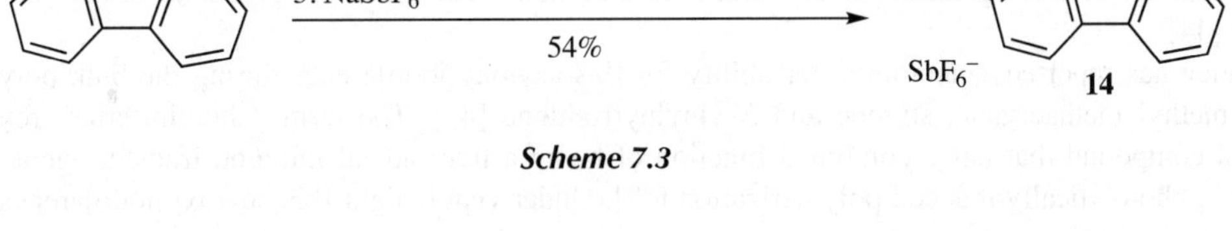

Scheme 7.3

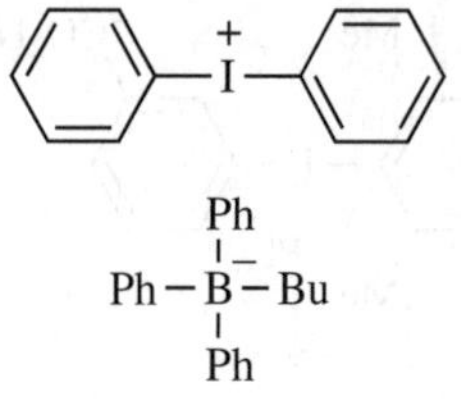

15

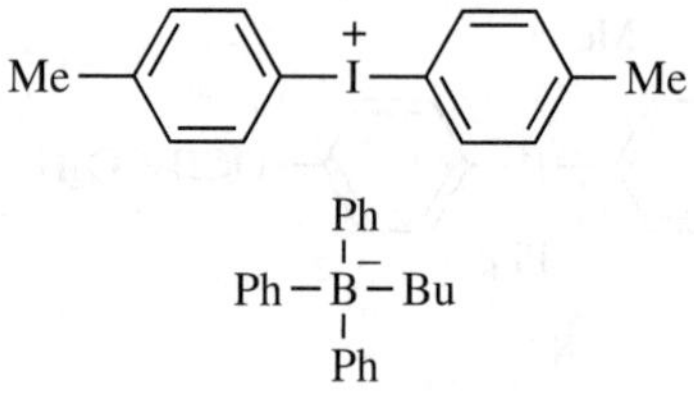

16

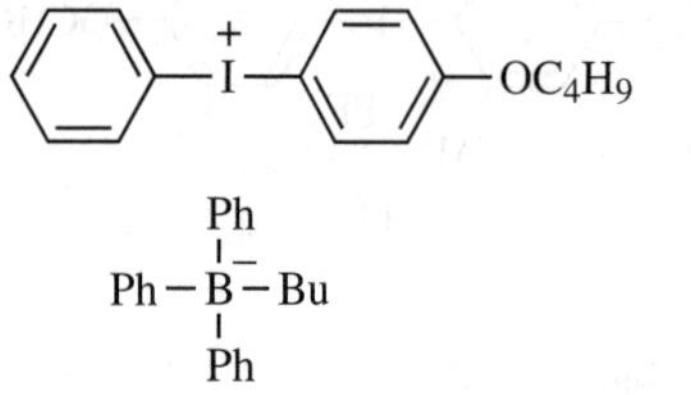

17

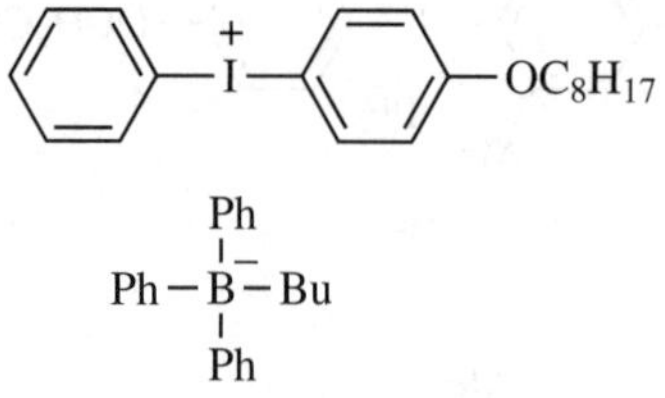

18

Figure 7.2 *Photoinitiators based on diaryliodonium butyltriphenylborate salts.*

and have investigated their reactivity as photoinitiators for the polymerization of acrylates [39]. Butyltriphenylborate salts **15–18** were found to be more efficient photoinitiators than iodonium tetraphenylborate salts, Ar_2IBPh_4. It was found from a study of the photoreaction of iodonium borate salts with a model monomer, methyl methacrylate, that iodonium butyltriphenylborate salts **15–18** simultaneously produce a butyl radical from the borate anion and an aryl radical from the iodonium cation upon irradiation. Both radicals initiate polymerization. Iodonium tetraphenylborate salts, Ar_2IBPh_4, were found to release an aryl radical, but only from the iodonium cation. Iodonium borate salts exhibit strong absorption below 300 nm, with a tail absorption above 400 nm. Thus, iodonium butyltriphenyl borate salts **15–18** are efficient photoinitiators even when used with visible light [39].

In 1985, Georgiev, Spyroudis and Varvoglis first reported that [bis(acyloxy)iodo]arenes, such as $PhI(OAc)_2$ and $PhI(OCOCF_3)_2$, are effective photoinitiators of cationic and radical polymerization [40]. In particular, under photochemical conditions $PhI(OAc)_2$ and $PhI(OCOCF_3)_2$ are efficient initiators for the homopolymerizations of 2-(dimethylaminoethyl) methacrylate (DMAEM) and methyl methacrylate (MMA) and also for the copolymerizations of DMAEM with MMA or styrene [40,41]. The proposed mechanism for the photoinitiation involves initial homolytic decomposition of the λ^3-iodane, for example, $PhI(OAc)_2$ (**19**), producing acyl **21** and iodanyl **20** radicals (Scheme 7.4) [41–45]. The actual initiators of radical polymerization are methyl radicals **22** generated by the decarboxylation of acyl radical **21,** as has been proved by the radical scavenger method [46]. It was proposed that the iodanyl radical **20** can further undergo both homolytic and heterolytic decomposition; homolytic fragmentation produces additional acyl and methyl radicals, while heterolytic fragmentation gives the phenyl iodide cation–radical **23,** which is a precursor of the true cationic initiator [44].

Georgiev has reported a photoiniferter ability for [bis(acyloxy)iodo]arenes during the bulk polymerization of methyl methacrylate, styrene and *N*-vinylpyrrolidone [44]. The term "photoiniferter" refers to a chemical compound that has a combined function of being a free radical initiator, transfer agent and terminator in photolytically induced polymerization [47]. Under visible light [bis(acyloxy)iodo]arenes initiate

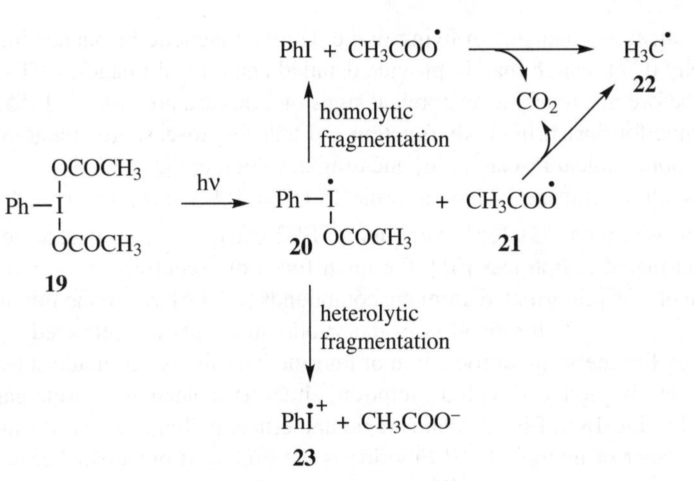

PhI + $CH_3COO^\bullet$ → $H_3C^\bullet$ **22** + CO_2

homolytic fragmentation

Ph–I($OCOCH_3$)$_2$ **19** —hν→ Ph–$\dot{I}$–$OCOCH_3$ **20** + $CH_3COO^\bullet$ **21**

heterolytic fragmentation

$PhI^{\bullet+}$ + CH_3COO^- **23**

Scheme 7.4

the “pseudo-living” radical polymerization, while a conventional radical or cationic polymerization are the consequences of the iodane decomposition under UV irradiation. It is suggested that the spectral selectivity of the iodanyl radical **20** (Scheme 7.4) decomposition and the relative instability of the ends of the iodane macromolecule are the reasons of this unusual iodane ability [44].

The acetoxy groups in (diacetoxyiodo)benzene can exchange with methacrylic acid in various solvents, yielding [acetoxy(methacryloyloxy)iodo]benzene or (dimethacryloyloxyiodo)benzene. These two λ^3-iodanes can serve as inimers due to the presence of polymerizable moiety and the easy generation of radicals upon thermal or light-induced homolysis of the I–O bonds [48]. When $PhI(OAc)_2$ is added to mixtures of methacrylic acid and methyl methacrylate and upon heating to 80 °C, branched or transiently crosslinked polymers are formed. In contrast, when homopolymerization of methyl methacrylate is initiated by $PhI(OAc)_2$ in the absence of the monomer with carboxylic acid group, no branching or gelation is observed [48].

Tsarevsky has found that hypervalent iodine compounds can be used for the direct azidation of polystyrene and consecutive click-type functionalization [49]. In particular, polystyrene can be directly azidated in 1,2-dichloroethane or chlorobenzene using a combination of trimethylsilyl azide and (diacetoxyiodo)benzene. 2D NMR HMBC spectra indicate that the azido groups are attached to the polymer backbone and also possibly to the aryl pendant groups. Approximately one in every 11 styrene units can be modified by using a ratio of $PhI(OAc)_2$ to trimethylsilyl azide to styrene units of 1 : 2.1 : 1 at 0 °C for 4 h followed by heating to 50 °C for 2 h in chlorobenzene. The azidated polymers have been further used as backbone precursors in the synthesis of polymeric brushes with hydrophilic side chains via a copper-catalyzed click reaction with poly(ethylene oxide) monomethyl ether 4-pentynoate [49].

7.3 Application of Iodonium Salts for Fluoridation in Positron Emission Tomography (PET)

Nuclear medicine – in particular, positron emission tomography (PET) – is one of the emerging, important fields of practical application of iodonium salts. PET is a powerful and rapidly developing area of molecular imaging that is used to study and visualize human physiology by the detection of positron-emitting radio-pharmaceuticals [50–55]. PET experiments provide direct information about metabolism, receptor/enzyme function and biochemical mechanisms in living tissue. Unlike magnetic resonance imaging (MRI) or computerized tomography (CT), which mainly provide detailed anatomical images, PET can measure chemical changes that occur before macroscopic anatomical signs of a disease are observed [55]. PET is emerging as a revolutionary method for measuring body function and tailoring disease treatment in living subjects and it is widely applied in both clinical research [56] and drug development [51, 57–60].

The short-lived positron-emitting isotope fluorine-18 ($t_{\frac{1}{2}} = 109.7$ min) has gained great importance as a radiolabel for probes used with PET [55]. Although [^{18}F]-2-deoxy-2-fluoro-D-glucose is currently the most widely used ^{18}F-fluorinated radiotracer [61], the main focus of recent efforts in radiotracer synthesis has been the preparation of [^{18}F]-fluorinated aromatic compounds [62–66]. Since the initial report from Pike and Aigbirhio in 1995 [67], the radiofluorination of diaryliodonium salts has attracted significant interest as a valuable methodology for late-stage introduction of fluorine into diverse aromatic substrates.

Fluorine-18 is generally produced with a cyclotron, either as molecular fluorine gas or as [^{18}F]-fluoride. Any application of fluorine-18 in PET demands rapid and efficient chemical transformation to introduce the fluorine-18 into the tracer of interest. [^{18}F]-Fluoride is the preferred precursor because it can be produced in higher radioactivity than molecular [^{18}F]-fluorine gas. There are two common pathways for the ^{18}F-labeling of the aromatic ring. The electrophilic ^{18}F-fluorination leads only to carrier-added products because of the unavoidable addition of elemental fluorine to the target gas. The other pathway, via nucleophilic displacement of adequate leaving groups (e.g., NO_2 or $^{+}NMe_3$), which are activated by electron-withdrawing substituents (e.g., CHO, COMe, COOMe, NO_2, CN, etc.), by no-carrier-added (NCA) [^{18}F]-fluoride, is generally used for the fluorination of electron-deficient arenes. The methodology introduced by Pike and Aigbirhio complements typical approaches based on nucleophilic aromatic substitution by providing a means to fluorinate electron-rich, as well as problematic electron-poor aromatic rings not easily accessed by direct substitution [67].

Diaryliodonium salts are becoming highly popular reagents for the efficient introduction of [^{18}F]-fluoride (fluoridation) via aromatic nucleophilic substitution. ^{18}F-Labeled imaging agents for PET have a short lifetime (due to the short radioactive half-life of ^{18}F) and require fast and convenient methods for introduction of ^{18}F into organic substrate molecules. Reactions of diaryliodonium salts provide a fast and convenient method of [^{18}F]-fluoridation as outlined in Scheme 7.5 [67–69].

Several problems, mainly due to the low selectivity of reactions, are associated with the use of diaryliodonium salts for [^{18}F]-fluoridation. In the case of the reactions of symmetrical diaryliodonium salts, $Ar_2I^+X^-$, there is no problem with the regioselectivity of fluoridation; however, only half of a molecule of substrate is converted into the target product and the second half gives aryl iodide as a by-product (Scheme 7.5), which results in a low atom economy. In addition, in this case the separation of aryl iodide and aryl fluoride may be complicated due to their similar structure and a chromatographic purification procedure (usually HPLC) is required for separation and purification of the target fluorinated product.

The regioselectivity of fluoridation plays an especially important role in the reactions of nonsymmetrical iodonium salts (Scheme 7.6). The distribution of the fluorine-18 containing products depends on the nature of substituents in the benzene ring; in general, the presence of electron-withdrawing substituents in aromatic

$$Ar_2I^+\,X^- \xrightarrow[80\text{-}120\ ^{\circ}C,\ 35\text{-}40\ min]{^{18}F^-\ M^+,\ MeCN\ or\ DMSO} Ar^{18}F + ArI$$

$M = K^+$/crown ether or Bu_4N^+

Scheme 7.5

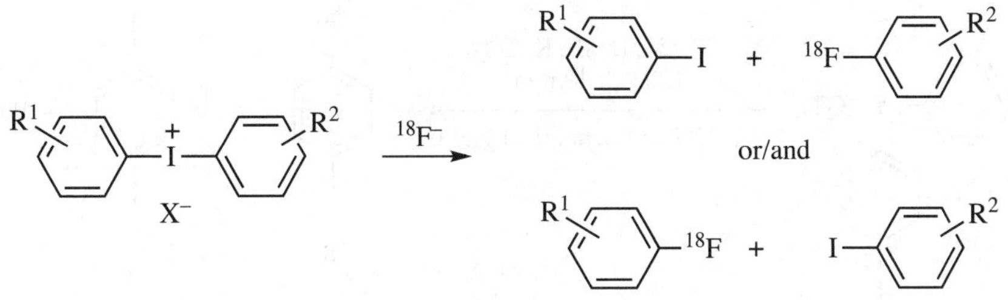

Scheme 7.6

ring is favorable for introduction of the fluoride nucleophile (see Section 3.1.22.1 for mechanistic discussion). The problem of low selectivity of the [^{18}F]-fluoridations can in principle be solved by modification of the electronic and steric properties of substituents R^1 and R^2 and by optimizing the reaction conditions.

The possibility of the synthesis of aryl fluorides by thermal decomposition of diaryliodonium tetrafluoroborates was first demonstrated by Van der Puy in 1982 [70]. It was found that the reactions of diphenyliodonium salts with different anions (BF_4^-, CF_3COO^-, TsO^-, Cl^-) in DMF in the presence of KF upon heating afford fluorobenzene in 11–85% yield. The lowest yield (11%) was observed in the reaction of diphenyliodonium chloride with KF in DMF at 115 °C, while the thermolysis of $Ph_2I^+BF_4^-$ in the presence of KF at 160–170 °C without solvent gave PhF in 85% yield. The formation of benzene (2–9%) due to a parallel radical decomposition process was also observed in all these reactions [70].

Later, in 1995, Pike and Aigbirhio applied for the first time diaryliodonium salts for the preparation of ^{18}F-labeled aryl fluorides using potassium [^{18}F]-fluoride in the presence of the diaza-crown ether Kryptofix (K2.2.2; structure **24** in Scheme 7.7) in acetonitrile at 85 °C or 110 °C [67]. Under these conditions, the reaction of diphenyliodonium chloride provided [^{18}F]-fluorobenzene in 31–78% radiochemical yield. The use of Kryptofix is required for phase transfer of the [^{18}F]-fluoride ion obtained by the nuclear reaction in the cyclotron as a solution in water enriched with oxygen-18.

Further studies have demonstrated that the regioselectivity of [^{18}F]-fluoridation is controlled by electronic factors and by the bulk of the *ortho*-substituents on the rings, with the latter being the dominant factor. Pike and coworkers have performed a detailed investigation of the reactions of a wide range of iodonium salts, which contain *ortho*-substituents in aromatic rings, with [^{18}F]-fluoride in acetonitrile at 85 °C [69]. It was found that the electronic effects of substituents on aromatic rings in radiochemical fluoridation processing

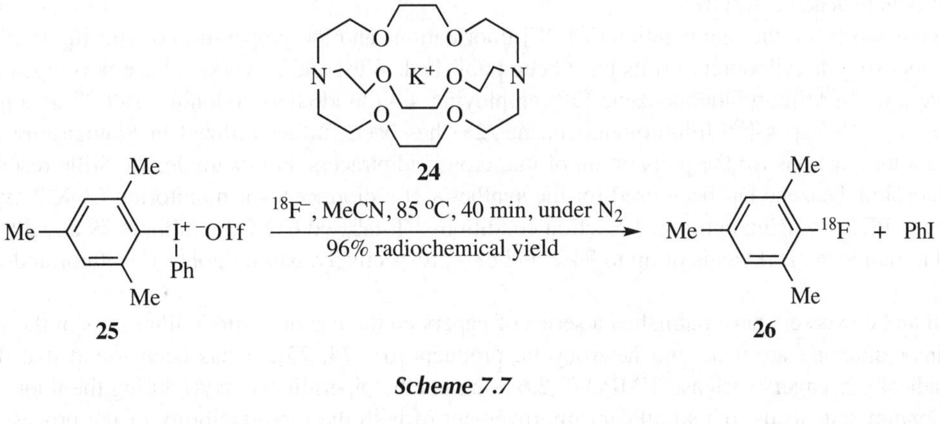

Scheme 7.7

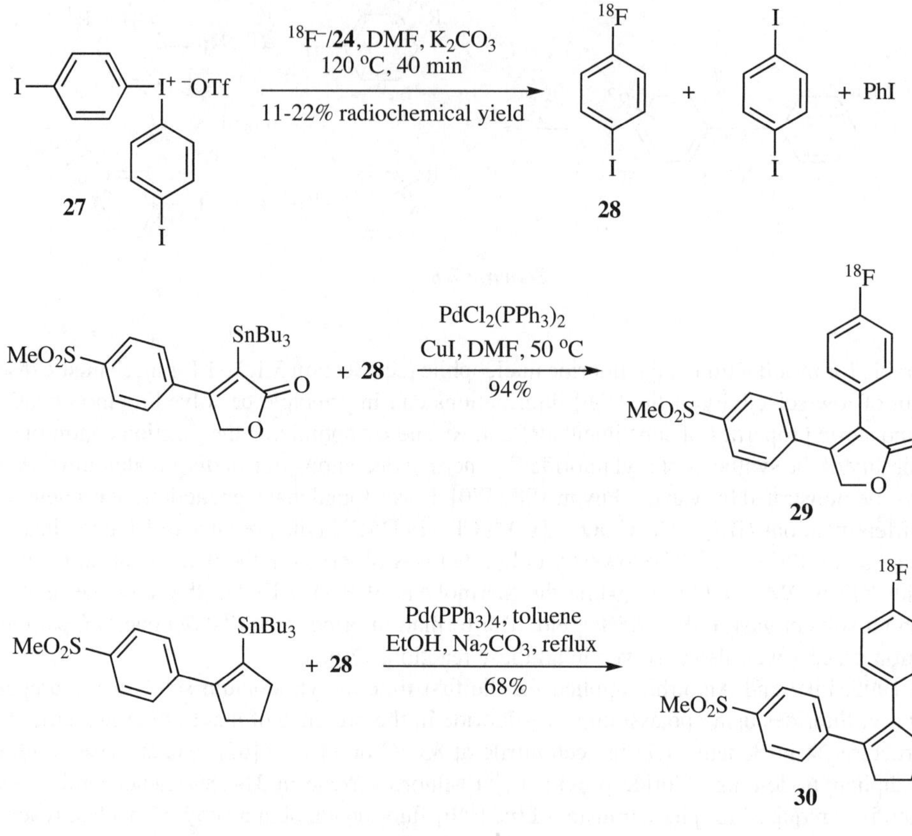

Scheme 7.8

are similar to those in the reactions of iodonium salts with other nucleophiles and fluorine-18 is introduced into the aromatic ring containing electron-withdrawing substituents. However, the presence of a bulky *ortho*-substituent changes the regioselectivity, allowing fluoridation of the electron-rich *ortho*-substituted ring. For example, the reaction of 2,4,6-trimethylphenyl(phenyl)iodonium triflate (**25**) with potassium [^{18}F]-fluoride in the presence of Kryptofix **24** exclusively affords 1-fluoro-2,4,6-trimethylbenzene **26** along with iodobenzene as a by-product (Scheme 7.7) [69].

Numerous works on the optimization of [^{18}F]-fluoridations and the preparation of specific [^{18}F]-labeled radiotracers using diaryliodonium salts have been published. Wüst and coworkers have developed a convenient access to 4-[^{18}F]fluoroiodobenzene (**28**) employing 4,4′-diiododiaryliodonium salt **27** as a precursor (Scheme 7.8) [71–73]. 4-[^{18}F]Fluoroiodobenzene (**28**) has been further utilized in Sonogashira or Stille cross-coupling reactions for the preparation of numerous radiotracers. For example, the Stille reaction with 4-[^{18}F]fluoroiodobenzene has been used for the synthesis of radiotracers for monitoring COX-2 expression by means of PET. By using optimized reaction conditions ^{18}F-labeled COX-2 inhibitors **29** and **30** could be obtained in radiochemical yields of up to 94% and 68%, respectively, based upon 4-[^{18}F]fluoroiodobenzene (**28**) [72].

Carroll and coworkers have published a series of papers on the use of aryliodonium salts in the synthesis of fluorine-containing aromatic and heterocyclic products [63, 74, 75]. It has been found that the addition of radical scavengers such as TEMPO (2,2,6,6-tetramethylpiperidine-1-oxyl) during the fluoridation of diaryliodonium salts leads to a significant improvement of both the reproducibility of the process and the

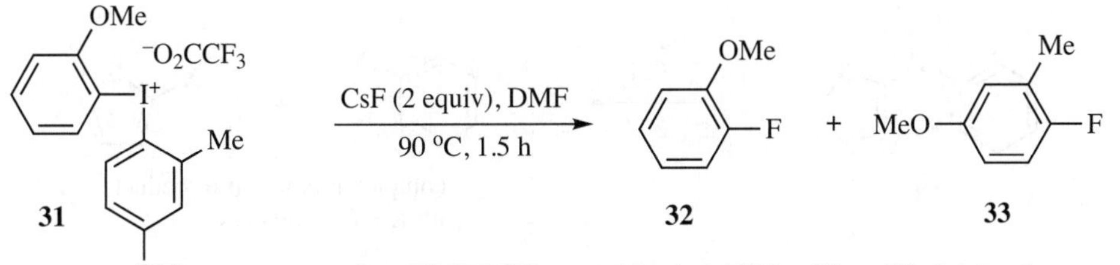

OMe $^{-}O_2CCF_3$ I^+ Me OMe **31**

CsF (2 equiv), DMF
90 °C, 1.5 h

OMe F **32** + MeO Me F **33**

0 mol% TEMPO: combined yield **32** + **33** = <5%, 1:1.1 ratio
20 mol% TEMPO: combined yield **32** + **33** = 27-35%, 1:1.3 ratio

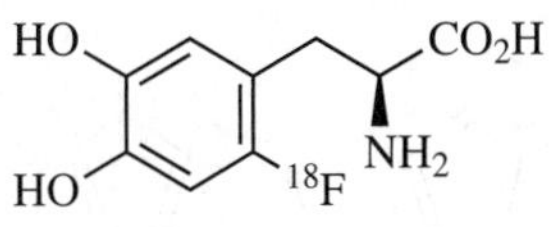

HO HO CO_2H NH_2 ^{18}F **34**

Scheme 7.9

material yield of the desired fluoroarene products without affecting the regioselectivity of the process [74]. In a specific example, the reaction of diaryliodonium trifluoroacetate **31** with cesium fluoride in different solvents (DMF, DMSO, acetonitrile, dimethylacetamide) in the absence of a radical trap affords mixture of fluoroarenes **32** and **33** in the ratio 1 : 1 with a combined yield below 5%. Carrying out this reaction in the presence of 20 mol% TEMPO leads to increased yields of **32** and **33** of up to 35% with almost unchanged regioselectivity (Scheme 7.9) [74]. This methodology is potentially applicable in the production of fluorine-18 labeled radiopharmaceuticals including L-6-[^{18}F]fluoroDOPA **34,** which is an important radioligand for the study of brain dopaminergic neuron density in movement disorders, such as Parkinson's disease.

The fluoridation reactions of several classes of heteroaromatic iodonium salts have been studied extensively by different research groups [63, 75–78]. In general, a theoretical prediction that the nucleophilic substitution of aryl(heteroaryl)iodonium salts by fluoride ion is regioselective for the aryl ring has been confirmed by experimental observation [76]. Coenen and coworkers have reported a highly efficient fluoridation method using aryl-(2-thienyl)iodonium salts **35** (Scheme 7.10) [77]. The 2-thienyl group is a very electron-rich group that allows ^{18}F to be introduced directly into even electron-rich arenes like anisoles. It has been also found that the selectivity of the fluoridation of iodonium salts **35** depends on the nature of the counteranion X^- with the highest yields of Ar^{18}F (up to 60% radiochemical yield) achieved in the reactions of iodonium bromides [77].

In contrast to Coenen's results [77], a detailed study of the reaction of aryl(thienyl)iodonium salts with CsF by Carroll and coworkers has demonstrated a very low selectivity of this reaction, producing a mixture of six products as illustrated by Scheme 7.11 [75]. The authors suggested that the previous reports of the absence of

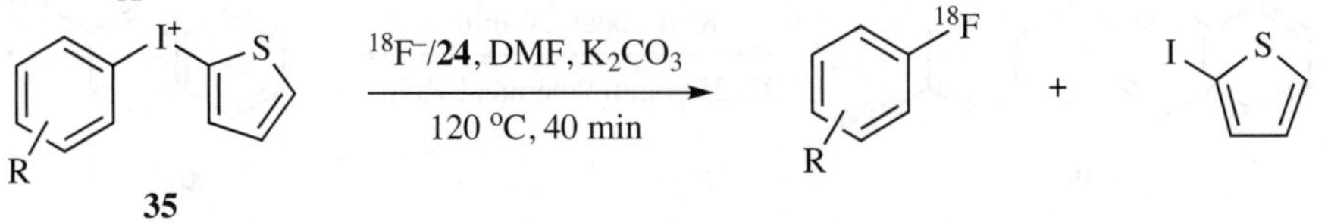

X^- I^+ S R **35**

$^{18}F^-$/**24**, DMF, K_2CO_3
120 °C, 40 min

^{18}F R + I S

X = Br, I, OTs, OTf
R = 2-OMe, 3-OMe, 4-OMe, 4-OBn, H, 4-I, 4-Br, 4-Cl

Scheme 7.10

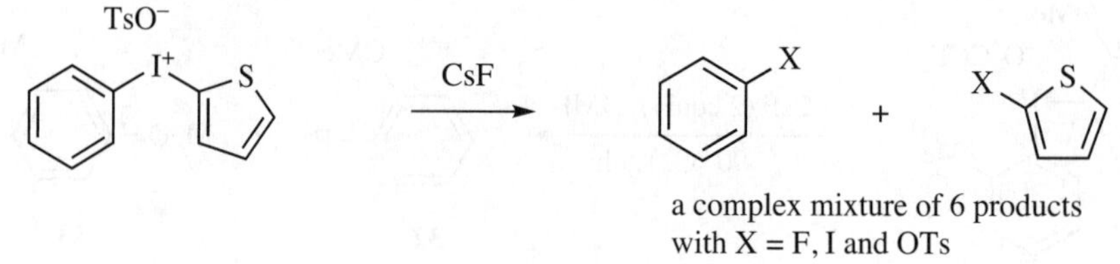

TsO⁻ I⁺ S — CsF → X + X S

a complex mixture of 6 products
with X = F, I and OTs

Scheme 7.11

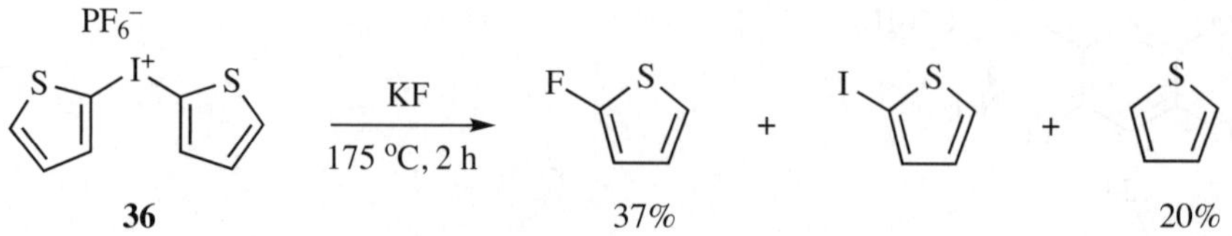

PF_6^- I⁺ S S **36** — KF, 175 °C, 2 h → F S (37%) + I S + S (20%)

Scheme 7.12

2-fluorothiophene among the reaction products of aryl-(2-thienyl)iodonium salts were misleading. This lack of detection may be due the highly volatile nature of 2-fluorothiophene (boiling point 82 °C), which may be lost under the reaction conditions or on workup/analysis [75].

Onys'ko, Gakh and coworkers have shown that 2-fluorothiophene can be synthesized selectively by heating bis(2-thienyl)iodonium salts with potassium fluoride [78]. Specifically, treatment of bis(2-thienyl)iodonium hexafluorophosphate (**36**) with potassium fluoride (as a mechanical mixture) at 172–175 °C for 2 h yields 2-fluorothiophene, 2-iodothiophene and thiophene (Scheme 7.12). Bis(2-thienyl)iodonium salts with more nucleophilic anions, such as trifluoroacetate, yielded only trace amounts of the desired 2-fluorothiophene [78].

Carroll and coworkers have developed a practical and selective route to fluorine-18 labeled 3-fluoropyridine (**38**) and 3-fluoroquinoline (**40**) by [^{18}F]-fluoridation of iodonium salts **37** and **39** (Scheme 7.13) [63]. The use of 4-methoxyphenyl as the other aromatic ring in iodonium salts **37** and **39** provides the necessary degree of

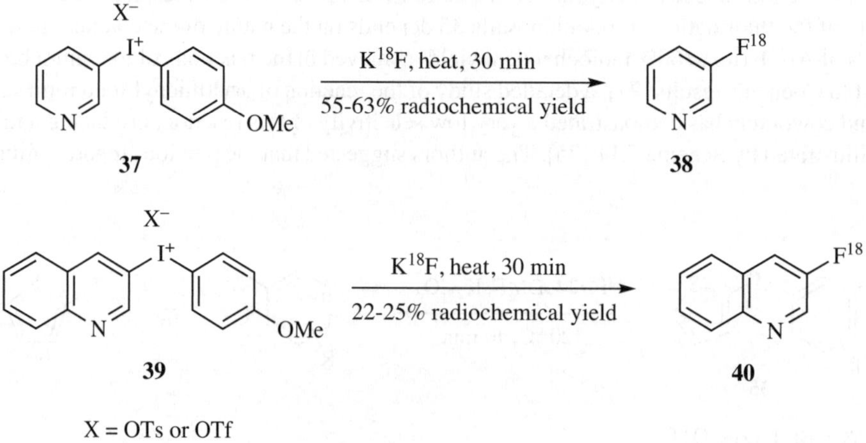

X⁻ I⁺ N OMe **37** — $K^{18}F$, heat, 30 min; 55-63% radiochemical yield → F^{18} N **38**

X⁻ I⁺ N OMe **39** — $K^{18}F$, heat, 30 min; 22-25% radiochemical yield → F^{18} N **40**

X = OTs or OTf

Scheme 7.13

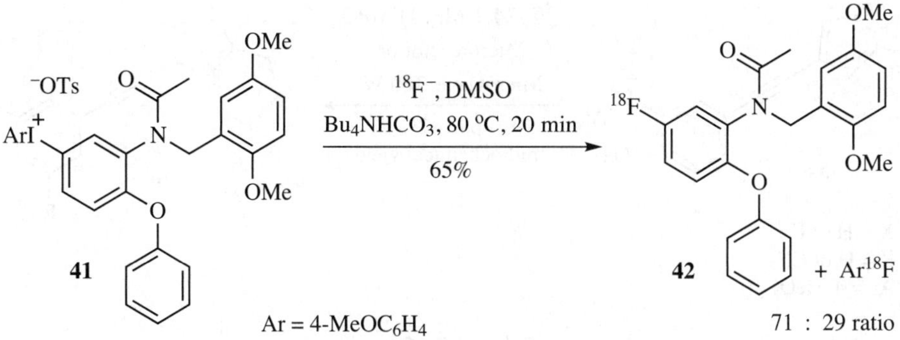

$^{-}$OTs

ArI^{+}

OMe

OMe

41

$^{18}F^-$, DMSO
Bu_4NHCO_3, 80 °C, 20 min
65%

Ar = 4-$MeOC_6H_4$

^{18}F

OMe

OMe

42 + $Ar^{18}F$

71 : 29 ratio

Scheme 7.14

selectivity in the fluoridation process. Fluorine-18 labeled fluoropyridines have found increasing applications in medical imaging technique by positron emission tomography (PET) [63].

Zhang and coworkers reported a successful synthesis of the PET ligand [^{18}F]DAA1106 (**42**) from diaryliodonium salt **41** with the radioactive ^{18}F anion (Scheme 7.14) [79]. The electron-rich 4-methoxyphenyl is essential as the second aromatic substituent in iodonium salt **41;** the analogous phenyliodonium salt gave the desired product **42** in only 3% yield. Compound **42** is used as a PET ligand for imaging a peripheral-type benzodiazepine receptor [79].

Katzenellenbogen and coworkers have reported the radiochemical synthesis and evaluation of two ^{18}F-labeled analogues of the potent and selective PPARγ agonist farglitazar [80]. In particular, the radioligand **44** has been prepared by the fluoridation of phenyliodonium salt **43** in good radiochemical yield (Scheme 7.15) [80]. Interestingly, the reactions of iodonium salts **43** bearing of 3-methoxyphenyl or 2-thienyl substituents instead of the phenyl did not afford any fluorinated product **44.**

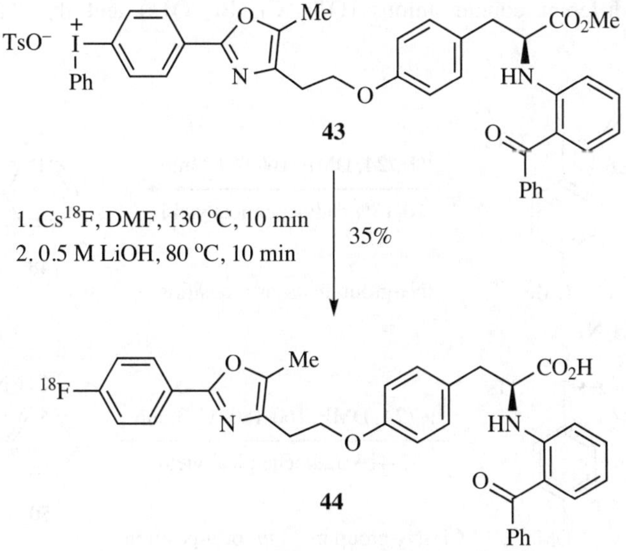

TsO$^-$ I$^+$ Ph

Me

CO_2Me

HN

Ph

43

1. $Cs^{18}F$, DMF, 130 °C, 10 min
2. 0.5 M LiOH, 80 °C, 10 min

35%

^{18}F

Me

CO_2H

HN

Ph

44

Scheme 7.15

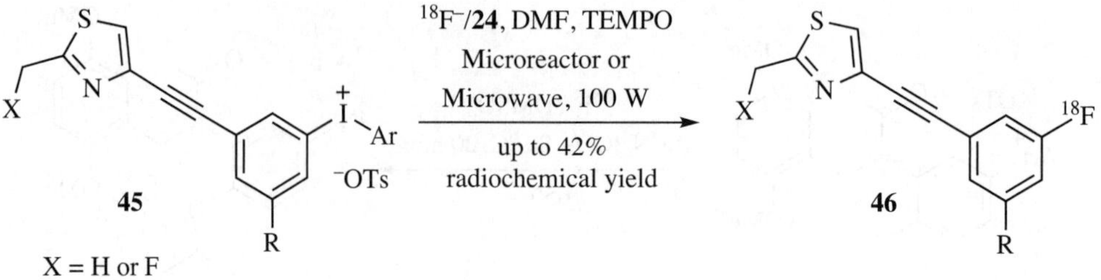

$^{18}F^-$/**24**, DMF, TEMPO
Microreactor or
Microwave, 100 W
up to 42%
radiochemical yield

45 → **46**

X = H or F
R = H or CN
Ar = 4-$MeOC_6H_4$

Scheme 7.16

Pike and coworkers have explored the scope of the radiofluoridation of diaryliodonium salts with [^{18}F]fluoride ion for the preparation of otherwise difficult to access *meta*-substituted [^{18}F]fluoroarenes [81, 82]. These studies have led to the development of a synthetic approach to 3-fluoro-1-[(thiazol-4-yl)ethynyl]benzenes **46** through the radiofluorination of diaryliodonium tosylates **45** (Scheme 7.16) [82]. 3-Fluoro-1-[(thiazol-4-yl)ethynyl]benzenes constitute an important class of high-affinity metabotropic glutamate subtype 5 receptor (mGluR5) ligands; fluorine-18 labeled compounds **46** are used as radioligands for molecular imaging by PET of brain mGluR5 in living animal and human subjects [82].

Chun and Pike described the rapid, single-step radiosynthesis of azido- or azidomethyl-bearing [^{18}F]fluoroarenes **48** and **50** from the reactions of diaryliodonium salts **47** or **49,** respectively, with no-carrier-added [^{18}F]fluoride ion within a microfluidic apparatus to obtain these previously poorly accessible ^{18}F-labeled click synthons in good radiochemical yields (Scheme 7.17) [83]. The radiosynthesis of synthons **50** was also possible with "wet" cyclotron-produced NCA [^{18}F]fluoride ion, in the presence of about 70 vol.% water and thus obviating the need to dry the cyclotron-produced [^{18}F]fluoride ion and greatly enhancing the practicality of the method [83].

Griffiths and coworkers reported the syntheses and characterization of phenyl(3-formylphenyl)iodonium salts containing four different counter-anions (OTf, Cl, Br, OTs) and the ^{18}F-fluoridation of these

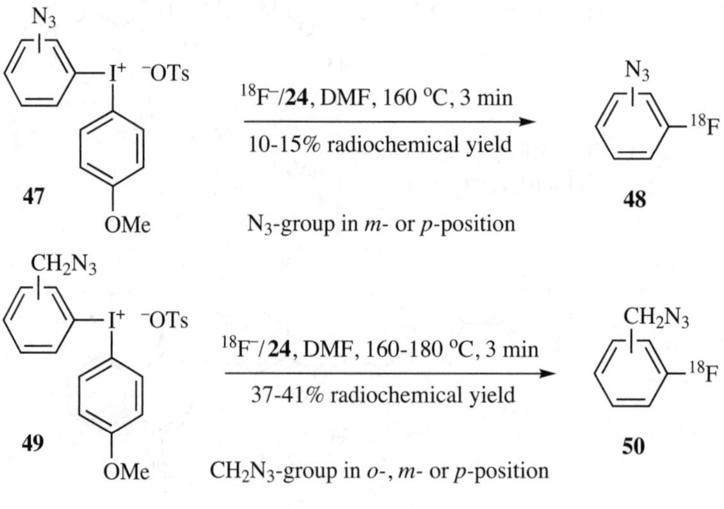

47 → **48**: $^{18}F^-$/**24**, DMF, 160 °C, 3 min; 10-15% radiochemical yield

N_3-group in *m*- or *p*-position

49 → **50**: $^{18}F^-$/**24**, DMF, 160-180 °C, 3 min; 37-41% radiochemical yield

CH_2N_3-group in *o*-, *m*- or *p*-position

Scheme 7.17

iodonium salts leading to *m*-[^{18}F]fluorobenzaldehyde and *m*-[^{18}F]fluorobenzyl bromide [84]. In particular, *m*-[^{18}F]fluorobenzaldehyde was prepared by the reaction of phenyl(3-formylphenyl)iodonium bromide with $Cs^{18}F/Cs_2CO_3$ in dimethylformamide at 100 °C for 5 min in a microwave in the presence of 1 equivalent of TEMPO [84]. The obtained 3-[^{18}F] fluorobenzaldehyde was further reduced to benzyl alcohol and converted into 3-[^{18}F] fluorobenzyl bromide. 3-[^{18}F]Fluorobenzyl bromide was subsequently used in the synthesis of ^{18}F-radiolabeled lapatinib, a potential tracer for PET imaging of ErbB1/ErbB2 tyrosine kinase activity [85].

Several mechanistic and structural studies of fluoridation reactions of iodonium salts have been published [86–88]. DiMagno and coworkers have found that diaryliodonium salts undergo rapid, fluoride-promoted aryl exchange reactions at room temperature in acetonitrile [86]. This exchange is highly sensitive to the concentration of fluoride ion in solution; the fastest exchange is observed as the fluoride concentration approaches a stoichiometric amount at 50 mM substrate concentration. It was demonstrated that free fluoride ion or a four-coordinate anionic I(III) species may be responsible for the exchange [86]. The fluoride-promoted aryl exchange reaction is general and allows direct measurement of the relative stabilities of diaryliodonium salts featuring different aryl substituents.

It has also been found that the selectivity of fluoridation and yields of products can be improved by changing the reaction conditions [87]. In particular, the use of low polarity aromatic solvents (benzene or toluene) and/or the removal of inorganic salts result in dramatically increased yields of fluorinated arenes from diaryliodonium salts [87].

Lee, Pike and coworkers have investigated the conformational structure and energetics of 2-methylphenyl(2-methoxyphenyl)iodonium chloride [88]. In particular, X-ray structural analysis revealed that 2-methylphenyl(2-methoxyphenyl)iodonium chloride has a conformational dimeric structure with hypervalent iodine as a stereogenic center in each conformer. In addition, LC-MS of this iodonium chloride showed the presence of dimeric and tetrameric anion-bridged clusters in organic solution. This evidence of the existence of dimeric and higher order clusters of iodonium salts in solution is important for a general understanding of the mechanism and outcome of reactions of diaryliodonium salts in organic media with nucleophiles, such as the [^{18}F]fluoride ion [88].

DiMagno and coworkers have shown that exceptionally electron-rich arene rings can be fluoridated with high regioselectivity by the reductive elimination reactions of 5-methoxy[2.2]paracyclophan-4-yl diaryliodonium salt **51** (Scheme 7.18) [89, 90]. Application of the sterically hindered cyclophane directing group permits a high degree of control in fluorination reactions of diaryliodonium salts. Despite excellent selectivity, this approach has obvious disadvantages, such as the use of inaccessible starting compounds and complex synthetic procedures.

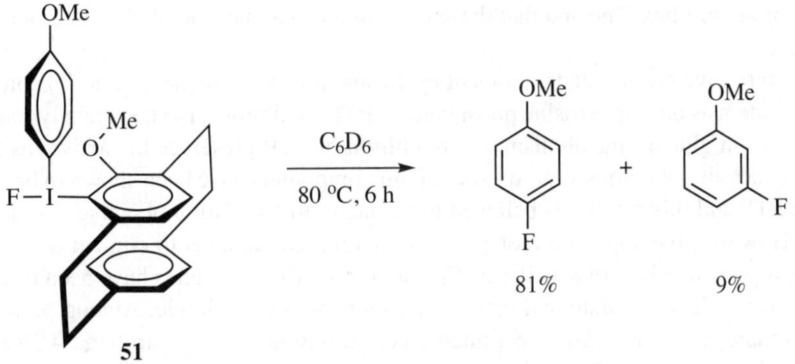

OMe
Me
O
F–I–
51

C_6D_6
80 °C, 6 h

OMe
F
81%
+
OMe
F
9%

Scheme 7.18

Pike and coauthors have developed a microreactor for [^{18}F]fluoridations using diaryliodonium salts. This apparatus allows the reaction between the [^{18}F]fluoride anion and iodonium salt to be carried out rapidly and efficiently [81, 91].

7.4 Biological Activity of Polyvalent Iodine Compounds

A summary of the biological properties of iodonium salts has been provided in a 1996 review [1]. Among the numerous known structural types of polyvalent organic iodine compounds, only aryl- and heteroaryliodonium salts and iodonium ylides have considerable, practically useful biological activity. Table 7.1 provides a brief description of the specific biological activity of several patented iodonium derivatives.

The data published in patents have established the biocidal and antimicrobial activity of numerous diaryl- and heteroaryliodonium salts [92–103]. They describe the activities of iodonium salts against a wide variety of both Gram positive and Gram negative bacterial species, such as *Staphylococcus aureus, Mycobacterium phlei, Trichophyton mentagrophytes, Bacillus subtilis, Cephalvaucus fragons* and *Escherichia coli* (Table 7.1). Another patent claims antimicrobial activity for a wide range of iodonium ylides [104]. While exhibiting significant activity against microorganisms most iodonium salts have relatively low toxicity towards mammals: the LD_{50} in mice for Ph_2ICl is 56 mg kg^{-1} of body weight and for (4-chlorophenyl)(thienyl)iodonium chloride it is over 4000 mg kg^{-1} [99]. Potential applications of iodonium salts due to their potent antifungal and antimicrobial activity include disinfectants as well as preservatives for diverse materials such as paints, adhesives, inks, paper, textiles, lubricants, cosmetics and so on [92–103, 105].

Menkissoglu-Spiroudi and coworkers have found that hypervalent iodine compounds are potent antibacterial agents against ice nucleation active (INA) *Pseudomonas syringae* [106]. *Pseudomonas syringae* is a potent bacterial plant pathogen, which can reduce the productivity of many plant species, including tree fruits such as citrus, pear and almonds. Several hypervalent iodine compounds belonging to aryliodonium salts, aryliodonium ylides and (diacyloxyiodo)arenes were tested for their antibacterial activities against INA *Pseudomonas syringae* and the MIC and EC_{50} values were determined [106]. All of the compounds examined caused a dose-dependent decrease in bacterial growth rates. Aryliodonium salts, especially those with electron-withdrawing groups, exhibit higher antibacterial activities with MIC = 8–16 ppm, whereas the nature of the anion does not seem to affect the activities of the diaryliodonium salts [106].

Goldstein and coworkers have investigated the *in vitro* activities of several iodonium salts against oral and dental anaerobes [107]. In particular, the comparative *in vitro* activities of 11 iodonium salts, chlorhexidine and four other antimicrobial agents against 322 anaerobic and fastidious potential dental and periodontal bacterial pathogens have been studied. It has been demonstrated that the activities of iodonium salts are comparable to that of chlorhexidine and that these compounds may be suitable for incorporation into an oral mouthwash [107].

Just and coworkers have synthesized a series of cyclic and noncyclic organoiodine(V) compounds (periodinanes) and tested them as protein tyrosine phosphatase (PTP) inhibitors [108]. Protein tyrosine phosphatases play important roles in glucose metabolism and inhibition of PTP preserves the active insulin receptor and mimics the insulin activity. Previously, peroxovanadium compounds have been shown to be very active in the inhibition of the PTP and this activity is believed to be due to the oxidation of active cysteine residue by the metal complex. However, the application of peroxovanadium compounds to clinical use is limited by their toxicity, poor absorption and lack of specificity. Organoiodine(V) compounds have been found to have better PTP inhibition activity than vanadate and their synthesis is relatively simple. Among periodinanes studied, the noncyclic iodylarenes with a *para*-substituted electron-withdrawing group (e.g., $4\text{-}NO_2C_6H_4IO_2$) have been shown to be the strongest PTP inhibitors [108].

Table 7.1 *Biological activity of some iodonium salts and ylides.*

Compound	Biological activity and application	Reference
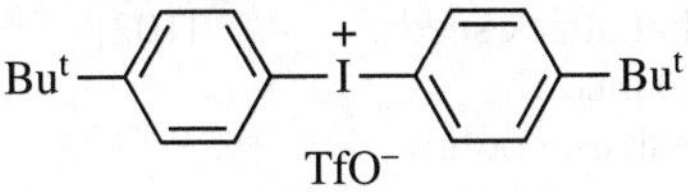 Bu^t, Bu^t, TfO^-	Antimicrobial activity against *Staphylococcus aureus, Bacillus subtilis, Escherichia coli* and *Pseudomonas aeruginosa*. Can be used as components of effective industrial antimicrobial agents with low mutagenicity	[92]
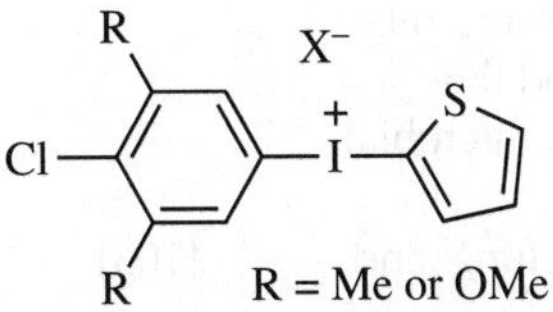 R = Me or OMe X = Cl or Br	Active against *Escherichia coli* and *Streptococcus aureus* and nontoxic toward human epithelial cells. Can be used as highly active antimicrobial agents nontoxic toward plants and mammals	[93]
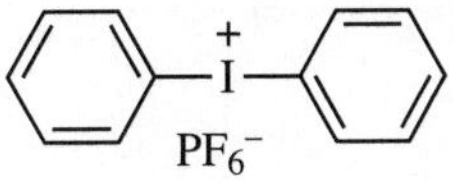 PF_6^-	Antimicrobial component that is added to hardenable antimicrobial dental compositions	[94]
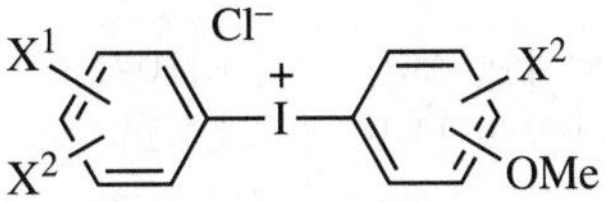 X^1, X^2, X^3 = F, Cl, Br	Antibacterial and anthelmintic agents	[95]
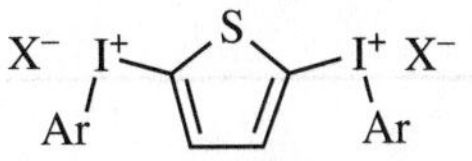 X = Cl, Br, I, HSO_4, NO_3, BF_4, CF_3CO_2, lactate	Antimicrobials for inhibition of the growth of many bacterial and fungal organisms that attack seeds, roots and above-ground portions of plants	[96]
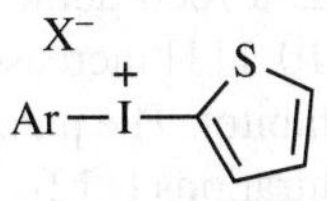 X = Cl, CF_3CO_2, CCl_3CO_2	Controls several bacterial organisms as well as molds, mildews, fungi and slimes	[97–99]
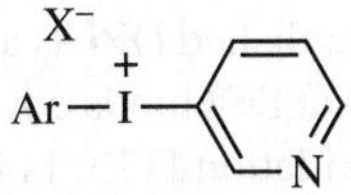 X = F, Cl, Br, I, CF_3CO_2, CCl_3CO_2, HSO_4, NO_3, BF_4	Inhibition against growth of bacteria, fungi and organisms that attack seeds, roots and above-ground portions of plants and microorganisms responsible for mold, mildew, rot, decay	[100]
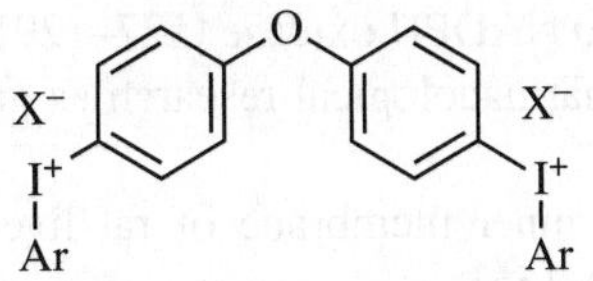 X = F, Cl, Br, I, CF_3CO_2, CCl_3CO_2, HSO_4, NO_3, BF_4, lactate	Antimicrobial agents against bacteria, fungi and organisms that attack seeds, roots and above-ground portions of plants; viricidal against small RNA viruses	[101]

(continued)

Table 7.1 (Continued)

Compound	Biological activity and application	Reference
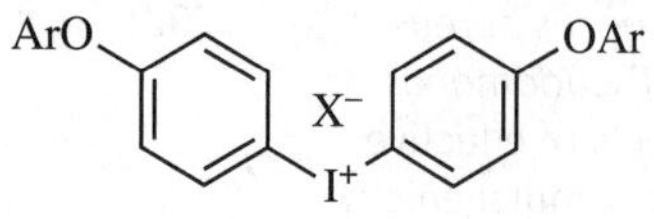 ArO–C6H4–I⁺–C6H4–OAr, X^-; X = F, Cl, Br, HSO_4, etc.	Antimicrobial agents against bacteria that attack seeds, roots and above-ground portions of plants. The compounds may be included in adhesives, cooling waters, inks, plasticizers, latices, polymers, resins, fuels, greases, soaps, detergents, cutting oils and oil or latex paints to prevent mold and mildew and the degradation of such products resulting from microbial attack	[102]
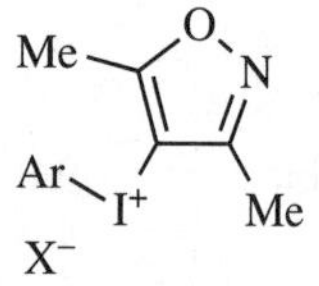 Ar–I⁺–(3,5-dimethylisoxazol-4-yl) X^-; X = Cl, Br, CF_3CO_2, CCl_3CO_2	Antimicrobial agents against various bacteria, fungi and yeasts	[103]
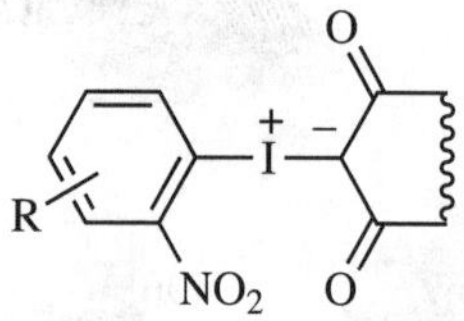 R-substituted 2-nitrophenyl iodonium ylide of cyclic 1,3-dione (NO_2, $\overset{+}{I}$–$\overset{-}{C}$, O, O)	Active against *Bacillus subtilis, Pseudomonas* species, *Escherichia coli, Staphylococcus* species, fungi, algae and yeasts	[104]

Bis(*p*-tolyl)iodonium chloride, Tol_2ICl, is a potent inhibitor of microbes that deaminate amino acids in ruminating animals, thereby preventing maximum food utilization. The use of Tol_2ICl as a food additive (about 25 mg kg^{-1}), along with a methane inhibitor, in lambs [109] and growing cattle [110, 111] increased feed utilization efficiency by about 10%. Presumably, iodonium salts act as a deaminase inhibitor. The parent bis(phenyl)iodonium chloride, Ph_2ICl, inhibits casein degradation at extremely low concentrations [112].

More recently, the physiological effects of several diaryliodonium salts and in particular bis(phenyl)iodonium chloride (BPI), Ph_2ICl and diphenyleneiodonium chloride (DPI, structure **52** in Figure 7.3), have been extensively explored. The initial work by Lardy and coworkers established DPI as a potent hypoglycaemic agent at a dose as low as 4 mg kg^{-1} body weight [113]. It is assumed that DPI binds covalently to a 23.5 kDa protein within Complex I causing irreversible inhibition of NADH oxidation [113, 114]. DPI inhibits gluconeogenesis in isolated rat hepatocytes [115], causes swelling of rat liver mitochondria [116] and induces cardiomyopathy [117]. Diaryliodonium salts have been shown to be effective in the treatment of hypertension [118]. Both BPI and DPI induce mitochondrial myopathy [119, 120], inhibit the superoxide production of neutrophils [121–124], as well as nitric oxide synthase [125, 126] and NADPH oxidase [127–129]. In particular, DPI has found broad application in modern biochemical and pharmacological research as an NADPH oxidase inhibitor [130–153].

The 1,3-dichloro substituted DPI catalyzes chloride exchange across the inner membrane of rat liver mitochondria and also inhibits succinate oxidation and stage 3 respiration [154, 155].

BPI has also been shown to inhibit cytochrome P450 reductase [156]. Cohen, Gallop and coworkers demonstrated that DPI as well as the bis(2-thienyl)iodonium triflate inhibit superoxide anion formation *in vivo* in rabbit aorta [157]. It was proposed that the physiological activity of iodonium species is due to their

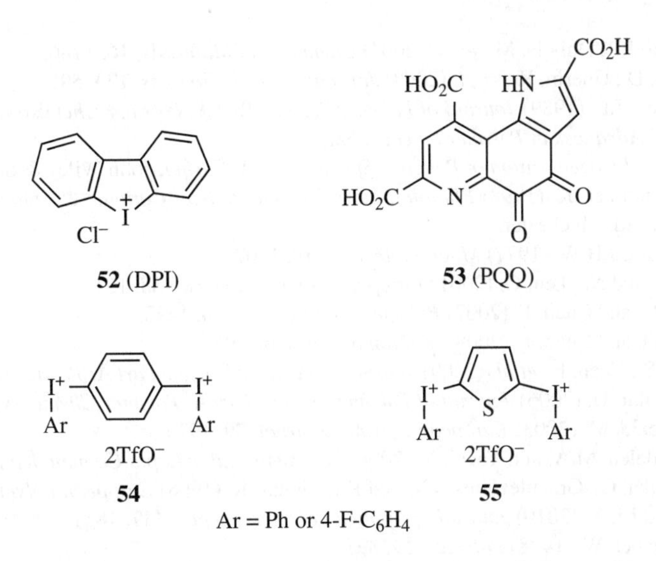

Figure 7.3 *PQQ (**53**) and iodonium salts active as PQQ inhibitors.*

interaction with PQQ (**53,** Figure 7.3) [158, 159]. Methoxatin or PQQ (**53**), a pyrroloquinoline quinone, is an organic cofactor, analogous to flavins, found in an increasing number of biological redox processes. It is widely distributed in microorganisms, animal cells, tissues and fluids and was first isolated from cultures of a methylotrophic soil bacterium *Pseudomonas* sp. in 1979 [160].

Various iodonium salts are potent sequestering agents for PQQ [161]. In particular, bis-iodonium salts **54** and **55** inhibit PQQ at a nanomolar level (7–13 nM) and are nearly a 1000 times better inhibitors of PQQ than Ph_2ICl and DPI (**52**). The potential implication of this activity is in the possible uses of bis-iodonium salts as biocides and in particular as a new class of antimicrobial agents [161].

References

1. Stang, P.J. and Zhdankin, V.V. (1996) *Chemical Reviews*, **96**, 1123.
2. Yusubov, M.S., Maskaev, A.V. and Zhdankin, V.V. (2011) *ARKIVOC*, (i), 370.
3. Burgi, H., Schaffner, T. and Seiler, J.P. (2001) *Thyroid*, **11**, 449.
4. Snyder, S.A., Vanderford, B.J. and Rosario-Ortiz, F.L. (2009) in *Comprehensive Handbook of Iodine* (eds V. Preedy, G. Burrow and R. Watson), Elsevier, p. 287.
5. Kong, F., Sun, C.-F., Yang, B.-P. and Mao, J.-G. (2012) in *Structure–Property Relationships in Non-Linear Optical Crystals 1* (eds X.T. Wu and L. Chen), Structure and Bonding, vol. **144**, Springer, Berlin, p. 43.
6. Phanon, D. and Gautier-Luneau, I. (2007) *Angewandte Chemie, International Edition*, **46**, 8488.
7. Ok, K.M. and Halasyamani, P.S. (2004) *Angewandte Chemie, International Edition*, **43**, 5489.
8. Greenwood, N.N. and Earnshaw, A. (1997) *Chemistry of the Elements*, Butterworth-Heinemann, Oxford.
9. Fukuda, N., Shan, S., Tanaka, H. and Shoyama, Y. (2006) *Journal of Natural Medicines*, **60**, 21.
10. Hounsell, E.F., Davies, M.J. and Smith, K.D. (2002) in *Protein Protocols Handbook* (ed. J.M. Walker), 2nd edn, Humana Press, Totowa, p. 803.
11. Nilsson, B. (1994) *Molecular Biotechnology*, **2**, 243.
12. Vlessidis, A.G. and Evmiridis, N.P. (2009) *Analytica Chimica Acta*, **652**, 85.
13. Moretti, J.D., Sabatini, J.J. and Chen, G. (2012) *Angewandte Chemie, International Edition*, **51**, 6981.
14. Ostvald, R.V., Shagalov, V.V., Zherin, I.I., *et al.* (2009) *Journal of Fluorine Chemistry*, **130**, 108.

15. Guerin, K., Pinheiro, J.P., Dubois, M., *et al.* (2004) *Chemistry of Materials*, **16**, 1786.
16. Giraudet, J., Claves, D., Guerin, K., *et al.* (2007) *Journal of Power Sources*, **173**, 592.
17. Crivello, J.V. and Lee, J.L. (1989) *Journal of Polymer Science Part A: Polymer Chemistry*, **27**, 3951.
18. Crivello, J.V. (1984) *Advances in Polymer Science*, **62**, 1.
19. Roffey, C.G. (1997) *Photogeneration of Reactive Species for UV Curing*, John Wiley & Sons, Ltd, Chichester.
20. Crivello, J.V. and Dietliker, K. (1998) *Photoinitiators for Free Radical Cationic & Anionic Photopolymerisation*, John Wiley & Sons, Ltd, Chichester.
21. Crivello, J.V. and Lam, J.H.W. (1977) *Macromolecules*, **10**, 1307.
22. Braun, H., Yagci, Y. and Nuyken, O. (2001) *European Polymer Journal*, **38**, 151.
23. Chen, S., Cook, W.D. and Chen, F. (2007) *Polymer International*, **56**, 1423.
24. Chen, S., Cook, W.D. and Chen, F. (2009) *Macromolecules*, **42**, 5965.
25. Cook, W.D., Chen, S., Chen, F., *et al.* (2009) *Journal of Polymer Science Part A: Polymer Chemistry*, **47**, 5474.
26. Crivello, J.V. and Bulut, U. (2006) *Journal of Polymer Science Part A: Polymer Chemistry*, **44**, 6750.
27. Hartwig, A. and Sebald, M. (2003) *European Polymer Journal*, **39**, 1975.
28. Kahveci, M.U., Tasdelen, M.A. and Yagci, Y. (2008) *Macromolecular Rapid Communications*, **29**, 202.
29. Lazauskaite, R., Buika, G., Grazulevicius, J.V. and Kavaliunas, R. (1998) *European Polymer Journal*, **34**, 1171.
30. Wang, J., Sun, F. and Li, X. (2010) *Journal of Applied Polymer Science*, **117**, 1851.
31. Zhu, Q.Q. and Schnabel, W. (1998) *Polymer*, **39**, 897.
32. Fouassier, J.P. and Chesneau, E. (1991) *Makromolekulare Chemie*, **192**, 1307.
33. Li, C., Luo, L., Wang, S., *et al.* (2001) *Chemical Physics Letters*, **340**, 444.
34. Li, S., Li, L., Wu, F. and Wang, E. (2009) *Journal of Photochemistry and Photobiology, A: Chemistry*, **203**, 211.
35. Park, C.H., Takahara, S. and Yamaoka, T. (2006) *Polymers for Advanced Technologies*, **17**, 156.
36. Shirai, A., Kubo, H. and Takahashi, E. (2002) *Journal of Photopolymer Science and Technology*, **15**, 29.
37. Hartwig, A., Harder, A., Luhring, A. and Schroder, H. (2001) *European Polymer Journal*, **37**, 1449.
38. Hoefer, M. and Liska, R. (2009) *Journal of Polymer Science Part A: Polymer Chemistry*, **47**, 3419.
39. Feng, K., Zang, H., Martin, D., *et al.* (1998) *Journal of Polymer Science Part A: Polymer Chemistry*, **36**, 1667.
40. Georgiev, G., Spyroudis, S. and Varvoglis, A. (1985) *Polymer Bulletin*, **14**, 523.
41. Georgiev, G., Kamenska, E., Khristov, L., *et al.* (1992) *European Polymer Journal*, **28**, 207.
42. Georgiev, G., Kamenska, E., Karayannidis, G., *et al.* (1987) *Polymer Bulletin*, **17**, 169.
43. Khristov, L., Georgiev, G., Sideridou-Karayannidou, I., *et al.* (1991) *Polymer Bulletin*, **26**, 617.
44. Georgiev, G.S. (1999) *Polymer Bulletin*, **43**, 223.
45. Georgiev, G.S., Kamenska, E.B., Tsarevsky, N.V. and Christov, L.K. (2001) *Polymer International*, **50**, 313.
46. Alberti, A., Benaglia, M. and Vismara, E. (1989) *Research on Chemical Intermediates*, **11**, 117.
47. Otsu, T., Yoshida, M. and Kuriyama, A. (1982) *Polymer Bulletin*, **7**, 45.
48. Han, H. and Tsarevsky, N.V. (2012) *Polymer Chemistry*, **3**, 1910.
49. Tsarevsky, N.V. (2010) *Journal of Polymer Science Part A: Polymer Chemistry*, **48**, 966.
50. Kim, D.W., Jeong, H.-J., Lim, S.T. and Sohn, M.-H. (2010) *Nuclear Medicine and Molecular Imaging*, **44**, 25.
51. Ametamey, S.M., Honer, M. and Schubiger, P.A. (2008) *Chemical Reviews*, **108**, 1501.
52. Cai, L., Lu, S. and Pike, V.W. (2008) *European Journal of Organic Chemistry*, 2853.
53. Hollingworth, C. and Gouverneur, V. (2012) *Chemical Communications*, 2929.
54. Lee, S., Xie, J. and Chen, X. (2010) *Chemical Reviews*, **110**, 3087.
55. Miller, P.W., Long, N.J., Vilar, R. and Gee, A.D. (2008) *Angewandte Chemie, International Edition*, **47**, 8998.
56. Wong Dean, F., Grunder, G. and Brasic James, R. (2007) *International Review of Psychiatry*, **19**, 541.
57. Lee, C.-M. and Farde, L. (2006) *Trends in Pharmacological Sciences*, **27**, 310.
58. Burns, H.D., Hamill, T.G., Eng, W.-S., *et al.* (1999) *Current Opinion in Chemical Biology*, **3**, 388.
59. Miller, J.M., Kumar, D., Mann, J.J. and Parsey, R.V. (2008) *Current Radiopharmaceuticals*, **1**, 12.
60. Halldin, C., Gulyas, B., Langer, O. and Farde, L. (2001) *The Quarterly Journal of Nuclear Medicine*, **45**, 139.
61. Dolle, F., Roeda, D., Kuhnast, B. and Lasne, M.-C. (2008) *Fluorine and Health*, 3.
62. Lemaire, C., Gillet, S., Guillouet, S., *et al.* (2004) *European Journal of Organic Chemistry*, 2899.
63. Carroll, M.A., Nairne, J. and Woodcraft, J.L. (2007) *Journal of Labelled Compounds and Radiopharmaceuticals*, **50**, 452.

64. Sun, H. and DiMagno, S.G. (2007) *Chemical Communications*, 528.
65. Sun, H. and DiMagno, S.G. (2007) *Journal of Fluorine Chemistry*, **128**, 806.
66. Pike, V.W. (2009) *Trends in Pharmacological Sciences*, **30**, 431.
67. Pike, V.W. and Aigbirhio, F.I. (1995) *Journal of the Chemical Society, Chemical Communications*, 2215.
68. Shah, A., Pike, V.W. and Widdowson, D.A. (1997) *Journal of the Chemical Society, Perkin Transactions 1*, 2463.
69. Shah, A., Pike, V.W. and Widdowson, D.A. (1998) *Journal of the Chemical Society, Perkin Transactions 1*, 2043.
70. Van der Puy, M. (1982) *Journal of Fluorine Chemistry*, **21**, 385.
71. Wüst, F.R. and Kniess, T. (2003) *Journal of Labelled Compounds and Radiopharmaceuticals*, **46**, 699.
72. Wüst, F.R., Hoehne, A. and Metz, P. (2005) *Organic & Biomolecular Chemistry*, **3**, 503.
73. Wüst, F.R. and Kniess, T. (2004) *Journal of Labelled Compounds and Radiopharmaceuticals*, **47**, 457.
74. Carroll, M.A., Nairne, J., Smith, G. and Widdowson, D.A. (2007) *Journal of Fluorine Chemistry*, **128**, 127.
75. Carroll, M.A., Jones, C. and Tang, S.-L. (2007) *Journal of Labelled Compounds and Radiopharmaceuticals*, **50**, 450.
76. Martin-Santamaria, S., Carroll, M.A., Carroll, C.M., *et al.* (2000) *Chemical Communications*, 649.
77. Ross, T.L., Ermert, J., Hocke, C. and Coenen, H.H. (2007) *Journal of the American Chemical Society*, **129**, 8018.
78. Onys'ko, P.P., Kim, T.V., Kiseleva, O.I., *et al.* (2009) *Journal of Fluorine Chemistry*, **130**, 501.
79. Zhang, M.-R., Kumata, K. and Suzuki, K. (2007) *Tetrahedron Letters*, **48**, 8632.
80. Lee, B.C., Dence, C.S., Zhou, H., *et al.* (2009) *Nuclear Medicine and Biology*, **36**, 147.
81. Chun, J.-H., Lu, S. and Pike, V.W. (2011) *European Journal of Organic Chemistry*, 4439.
82. Telu, S., Chun, J.-H., Simeon, F.G., *et al.* (2011) *Organic & Biomolecular Chemistry*, **9**, 6629.
83. Chun, J.-H. and Pike, V.W. (2012) *European Journal of Organic Chemistry*, 4541.
84. Basuli, F., Wu, H. and Griffiths, G.L. (2011) *Journal of Labelled Compounds and Radiopharmaceuticals*, **54**, 224.
85. Basuli, F., Wu, H., Li, C., Shi, Z.-D., *et al.* (2011) *Journal of Labelled Compounds and Radiopharmaceuticals*, **54**, 633.
86. Wang, B., Cerny, R.L., Uppaluri, S., *et al.* (2010) *Journal of Fluorine Chemistry*, **131**, 1113.
87. Wang, B., Qin, L., Neumann, K.D., *et al.* (2010) *Organic Letters*, **12**, 3352.
88. Lee, Y.-S., Hodoscek, M., Chun, J.-H. and Pike, V.W. (2010) *Chemistry–A European Journal*, **16**, 10418.
89. Wang, B., Graskemper, J.W., Qin, L. and DiMagno, S.G. (2010) *Angewandte Chemie, International Edition*, **49**, 4079.
90. Graskemper, J.W., Wang, B., Qin, L., *et al.* (2011) *Organic Letters*, **13**, 3158.
91. Chun, J.-H., Lu, S., Lee, Y.-S. and Pike, V.W. (2010) *Journal of Organic Chemistry*, **75**, 3332.
92. Hirayama, M. and Terashima, K. (2006) JP Patent 2006232801.
93. Wojcik, L.H. Jr. and Cornell, D.D. (2011) U.S. Patent 8053478.
94. Velamakanni, B.V., Mitra, S.B., Wang, D., *et al.* (2006) U.S. Patent 20060205838.
95. Doub, L. (1969) U.S. Patent 3422152.
96. Jezic, Z. (1973) U.S. Patent 3712920.
97. Moyle, C.L. (1973) U.S. Patent 3763187.
98. Moyle, C.L. (1975) U.S. Patent 3885036.
99. Moyle, C.L. (1976) U.S. Patent 3944498.
100. Jezic, Z. (1971) U.S. Patent 3622586.
101. Jezic, Z. (1973) U.S. Patent 3734928.
102. Jezic, Z. (1973) U.S. Patent 3759989.
103. Plepys, R.A. and Jezic, Z. (1975) U.S. Patent 3896140.
104. Relenyi, A.G., Koser, G.F., Walter, R.W. Jr., *et al.* (1992) U.S. Patent 5106407.
105. Cornell, S.W., Cornell, D.D. and Cornell, P.W. (2002) WO Patent 2002013612.
106. Menkissoglu-Spiroudi, U., Karamanoli, K., Spyroudis, S. and Constantinidou, H.-I.A. (2001) *Journal of Agricultural and Food Chemistry*, **49**, 3746.
107. Goldstein, E.J.C., Citron, D.M., Warren, Y., *et al.* (2004) *Antimicrobial Agents and Chemotherapy*, **48**, 2766.
108. Leung, K.W.K., Posner, B.I. and Just, G. (1999) *Bioorganic & Medicinal Chemistry Letters*, **9**, 353.
109. Horton, G.M.J. (1980) *Canadian Journal of Animal Science*, **60**, 169.

110. Chalupa, W., Patterson, J.A., Parish, R.C. and Chow, A.W. (1983) *Journal of Animal Science (Savoy, IL, United States)*, **57**, 201.
111. Horton, G.M. (1980) *Journal of Animal Science (Savoy, IL, United States)*, **50**, 1160.
112. Broderick, G.A. and Balthrop, J.E. Jr. (1979) *Journal of Animal Science (Savoy, IL, United States)*, **49**, 1101.
113. Holland, P.C., Clark, M.C., Bloxham, D.P. and Lardy, H.A. (1973) *Journal of Biological Chemistry*, **248**, 6050.
114. Gatley, S.J. and Sherratt, H.S.A. (1976) *Biochemical Journal*, **158**, 307.
115. Gatley, S.J., Al-Bassam, S.S., Taylor, J.R. and Sherratt, H.S.A. (1975) *Biochemical Society Transactions*, **3**, 333.
116. Holland, P.C. and Sherratt, H.S.A. (1972) *Biochemical Journal*, **129**, 39.
117. Brosnan, M.J., Hayes, D.J., Challiss, R.A.J. and Radda, G.K. (1986) *Biochemical Society Transactions*, **14**, 1209.
118. Pettibone, D.J., Sweet, C.S., Risley, E.A. and Kennedy, T. (1985) *European Journal of Pharmacology*, **116**, 307.
119. Cooper, J.M., Petty, R.K.H., Hayes, D.J., *et al.* (1988) *Journal of the Neurological Sciences*, **83**, 335.
120. Cooper, J.M., Petty, R.K.H., Hayes, D.J., *et al.* (1988) *Biochemical Pharmacology*, **37**, 687.
121. Doussiere, J. and Vignais, P.V. (1991) *Biochemical and Biophysical Research Communications*, **175**, 143.
122. Ellis, J.A., Mayer, S.J. and Jones, O.T.G. (1988) *Biochemical Journal*, **251**, 887.
123. Cross, A.R. and Jones, O.T.G. (1986) *Biochemical Journal*, **237**, 111.
124. Cross, A.R. (1987) *Biochemical Pharmacology*, **36**, 489.
125. Stuehr, D.J., Fasehun, O.A., Kwon, N.S., *et al.* (1991) *FASEB Journal*, **5**, 98.
126. Lee, H.-R., Do, H., Lee, S.-R., *et al.* (2007) *Journal of Food Science and Nutrition*, **12**, 74.
127. O'Donnell, V.B., Tew, D.G., Jones, O.T.G. and England, P.J. (1993) *Biochemical Journal*, **290**, 41.
128. Doussiere, J. and Vignais, P.V. (1992) *European Journal of Biochemistry*, **208**, 61.
129. Hancock, J.T., White, J.I., Jones, O.T.G. and Silver, I.A. (1991) *Free Radical Biology & Medicine*, **11**, 25.
130. Moody, T.W., Osefo, N., Nuche-Berenguer, B., *et al.* (2012) *Journal of Pharmacology and Experimental Therapeutics*, **341**, 873.
131. Hino, S., Kito, A., Yokoshima, R., *et al.* (2012) *Biochemical and Biophysical Research Communications*, **421**, 329.
132. Gong, H., Chen, G., Li, F., *et al.* (2012) *Biologia Plantarum*, **56**, 422.
133. Liu, J., Zhou, J. and Xing, D. (2012) *PLoS One*, **7**, e33817.
134. Tsai, K.-H., Wang, W.-J., Lin, C.-W., *et al.* (2012) *Journal of Cellular Physiology*, **227**, 1347.
135. Pakizeh, M., Kuliev, A.A., Mammadov, Z.M., *et al.* (2011) *World Applied Sciences Journal*, **14**, 67.
136. Lu, L., Gu, X., Li, D., Liang, L., *et al.* (2011) *Molecular Genetics and Metabolism*, **104**, 241.
137. Ishibashi, Y., Matsui, T., Takeuchi, M. and Yamagishi, S. (2011) *Hormone and Metabolic Research*, **43**, 619.
138. Miraglia, E., Lussiana, C., Viarisio, D., *et al.* (2010) *Fertility and Sterility*, **93**, 2437.
139. Van De Veerdonk, F.L., Smeekens, S.P., Joosten, L.A.B., *et al.* (2010) *Proceedings of the National Academy of Sciences of the United States of America*, **107**, 3030.
140. Sairam, R.K., Kumutha, D., Ezhilmathi, K., *et al.* (2009) *Biologia Plantarum*, **53**, 493.
141. Hu, Z.-H., Shen, Y.-B., Shen, F.-Y. and Su, X.-H. (2009) *Acta Physiologiae Plantarum*, **31**, 995.
142. Souza, V., Escobar, M. d. C., Bucio, L., *et al.* (2009) *Toxicology Letters*, **187**, 180.
143. Beffagna, N. and Lutzu, I. (2007) *Journal of Experimental Botany*, **58**, 4183.
144. Masamune, A., Watanabe, T., Kikuta, K., *et al.* (2008) *American Journal of Physiology*, **294**, G99.
145. Chen, J.-R., Shankar, K., Nagarajan, S., *et al.* (2008) *Journal of Pharmacology and Experimental Therapeutics*, **324**, 50.
146. Choi, H.W., Kim, Y.J., Lee, S.C., *et al.* (2007) *Plant Physiology*, **145**, 890.
147. Miller, F.J., Filali, M., Huss, G.J., *et al.* (2007) *Circulation Research*, **101**, 663.
148. Ho, C., Lee, P.-H., Huang, W.-J., *et al.* (2007) *Nephrology*, **12**, 348.
149. Serrander, L., Jaquet, V., Bedard, K., *et al.* (2007) *Biochimie*, **89**, 1159.
150. Barger, S.W., Goodwin, M.E., Porter, M.M. and Beggs, M.L. (2007) *Journal of Neurochemistry*, **101**, 1205.
151. Hutchinson, D.S., Csikasz, R.I., Yamamoto, D.L., *et al.* (2007) *Cellular Signalling*, **19**, 1610.
152. Sauer, H., Klimm, B., Hescheler, J. and Wartenberg, M. (2001) *FASEB Journal*, **15**, 2539.
153. Wyatt, C.N., Weir, E.K. and Peers, C. (1994) *Neuroscience Letters*, **172**, 63.
154. Gatley, S.J. and Sherratt, H.S.A. (1974) *Biochemical Society Transactions*, **2**, 517.

155. Gennimata, D., Kotali, E., Varvoglis, A., *et al.* (1989) *Journal of Chemotherapy*, **1**, 229.
156. Tew, D.G. (1993) *Biochemistry*, **32**, 10209.
157. Pagano, P.J., Ito, Y., Tornheim, K., *et al.* (1995) *American Journal of Physiology*, **268**, H2274.
158. Bishop, A., Paz, M.A., Gallop, P.M. and Karnovsky, M.L. (1995) *Free Radical Biology & Medicine*, **18**, 617.
159. Bishop, A., Paz, M.A., Gallop, P.M. and Karnovsky, M.L. (1994) *Free Radical Biology & Medicine*, **17**, 311.
160. Salisbury, S.A., Forrest, H.S., Cruse, W.B.T. and Kennard, O. (1979) *Nature (London)*, **280**, 843.
161. Gallop, P.M., Paz, M.A., Fluckiger, R., *et al.* (1993) *Journal of the American Chemical Society*, **115**, 11702.

Index

Hypervalent Iodine Chemistry: Preparation, Structure and Synthetic Applications of Polyvalent Iodine Compounds, First Edition. Viktor V. Zhdankin.